Shallow Lakes '95

Developments in Hydrobiology 119

Series editor
H. J. Dumont

Shallow Lakes '95

Trophic Cascades in Shallow Freshwater and Brackish Lakes

Edited by

Lech Kufel, Andrzej Prejs & Jan Igor Rybak

Reprinted from Hydrobiologia, vols 342/343 (1997)

Kluwer Academic Publishers
Dordrecht / Boston / London

Library of Congress Cataloging-in-Publication Data

Shallow lakes '95 : trophic cascades in shallow freshwater and
brackish lakes / edited by Lech Kufel and Andrzej Prejs and Jan Igor
Rybak.
p. cm. -- (Developments in hydrobiology ; 119)
Papers presented at an international conference held in Mikolajki,
Poland, August 21-26, 1995.
ISBN 0-7923-4421-9 (hb : alk. paper)
1. Eutrophication--Congresses. 2. Lake ecology--Congresses.
3. Eutrophication--Control--Congresses. I. Kufel, Lech.
II. Prejs, Andrzej. III. Rybak, Jan Igor. IV. Series.
QH96.8.E9S48 1997
577.63'158--dc21 96-54608

ISBN 0-7923-4421-9

Published by Kluwer Academic Publishers,
P.O. Box 17, 3300 AA Dordrecht, The Netherlands

Kluwer Academic Publishers incorporates
the publishing programmes of
D. Reidel, Martinus Nijhoff, Dr W. Junk and MTP Press.

Sold and distributed in the U.S.A. and Canada
by Kluwer Academic Publishers,
101 Philip Drive, Norwell, MA 02061, U.S.A.

In all other countries, sold and distributed
by Kluwer Academic Publishers Group,
P.O. Box 322, 3300 AH Dordrecht, The Netherlands

Printed on acid-free paper

Printed in Belgium

Contents

Preface

For many years hydrobiological studies were directed more to deep, stratified lakes than to shallow water bodies. However, in recent years, the attention of aquatic scientists has been paid to shallow, non-stratified lakes. The interest has been growing in parallel to the development and maturation of the biomanipulation approach to the lake eutrophication control. This tendency will probably increase, since recent studies show that changes in the biotic structure of shallow lakes are more likely to result in improved water quality than are changes in deep lakes.

This volume of *Hydrobiologia* contains the papers presented at an international conference on "Trophic cascades in shallow freshwater and brackish lakes" held in Mikołajki, Poland, 20–26 August 1995. The conference was organized by the Institute of Ecology, Polish Academy of Sciences, University of Warsaw and the Polish Hydrobiological Society. One hundred scientists from 14 countries participated in the conference, delivering 40 oral and 42 poster presentations. Oral contributions were divided into 5 main sessions, each preceded by a keynote lecture:

- **Nutrient Fluxes** – keynote lecture: "Are shallow lakes really more susceptible to nutrient loadings?", by Lech Kufel
- **Biotic Structure** – keynote lecture: "Differences between fishpond and shallow lakes littorals in Central Europe", by Jan Kvet
- **Trophic Interactions** – keynote lecture: "Top-down regulation in shallow lakes with special emphasis on the importance of fish and submerged macrophytes", by Eric Jeppesen
- **Whole Lake Studies** – keynote lecture: "Vertically-challenged limnology – contrasts between deep and shallow lakes", by Brian Moss, Meryem Beklioglu, Laurence Carvalho, Sabri Kilinc, Suzanne McGowan & Deborah Stephen
- **Biomanipulation, Restoration and Management** – keynote lecture: "Food-web manipulation in shallow eutrophic lakes: bridging the gap between whole lake approach and behavioural and demographic studies", by Joanna Pijanowska & Andrzej Prejs

In addition there were four workshops on benthos in shallow water bodies, land/water ecotones, picoplankton ecology and biomanipulation.

We are greatly indebted to a number of people who helped us in various ways. Anna Hillbricht-Ilkowska, Ewa Pieczyńska, Kazimierz Dobrowolski and Stanisław Radwan pre-reviewed all delivered abstracts and made a lot of suggestions as to the structure of the sessions. Our special thanks go to the staff of the Hydrobiological Station in Mikołajki, in particular to Jolanta Ejsmont-Karabin, Andrzej Karabin and Irena Kufel for helpful assistance before and during the conference, and to Irena Sawicka and Jarosław Sawicki for all practical arrangements. And, last but not least, we are grateful to all the Referees and to Teresa Węgleńska, whose hard work significantly improved final versions of the published papers; and to Prof. Henri Dumont, editor-in-chief of *Hydrobiologia* and Kluwer Academic Publishers for their interest in the production of this volume.

The conference was supported financially or materially by the Ministry of Education, State Committee for Scientific Research, Stefan Batory Foundation, Merck Poland, Wielkopolska Brewery and the Mayor of Mikołajki.

LECH KUFEL
ANDRZEJ PREJS
JAN IGOR RYBAK
Chairmen and Guest Editors

List of Participants

Brigitte Auer, Jutta Mayer, Margit Schönberger (Austria), Liubov L. Nagorskaja, Tatjana V. Zhukova (Belarus), Eduardo von Sperling (Brazil), Vladimir Korinek, Jan Kvet (Czech Republic), Frede Ø. Andersen, Lars A. Angantyr, Søren Berg, Klaus P. Brodersen, Peter Wiberg-Larsen, Lene Jacobsen, Eric Jeppesen, Claus Lindegaard, Tom Rugaard, Martin Søndergaard, Julie A. Wolin (Denmark), Tiina Nõges, Peeter Nõges (Estonia), Wolfgang Arp, Horst Behrendt, Frank Gervais, Maria Kapfer, Andreas Kleeberg, Bernd Koppelmeyer, Armin E. Liepelt, Alexandra Makulla, Jens Meisel, Brigitte Nixdorf, Wolfgang Ostendorp, Kerstin Ramm, Andree Riemer, Jacqueline Rücker, Volker Scheps, Mechthild Schmitt, Gisela Stolpe, Claudia Wiedner (Germany), Katalin V.-Balogh, Gyula Lakatos, Ilona Musko, Imre Oldal, Matyas Presing, Lajos Vörös (Hungary), Joost Backx, Marcel S. van den Berg, Carolien P.M. Breukers, Hugo Coops, Edith M. van Dam, Jan H. Janse, Maria-Luisa Meijer (The Netherlands), Ryszard J. Chróst, Piotr Dawidowicz, Wojciech Donderski, Kazimierz Dobrowolski, Jolanta Ejsmont-Karabin, Andrzej Giziński, Ryszard Gołdyn, Andrzej Górniak, Jan Gromadzki, Iwona Jasser, Zdzisław Kajak, Krystyna Kalinowska, Andrzej Karabin, Rozwita Kornatowska, Janusz Kornatowski, Paweł Koperski, Ryszard Kornijów, Anna Kozak, Joanna Królikowska, Irena Kufel, Lech Kufel, Krzysztof Lewandowski, Tomasz Mieszczankin, Zbigniew Mudryk, Ewa Pieczyńska, Joanna Pijanowska, Ryszard Piotrowicz, Andrzej Prejs, Stanisław Radwan, Jan Igor Rybak, Marek Rzepecki, Anna Stańczykowska-Piotrowska, Elżbieta Szeląg-Wasielewska, Mirosław Ślusarczyk, Andrzej Tatur, Jan Trojanowski, Teresa Węgleńska, Ryszard Wiśniewski, Maciej Zalewski, Ludwik Żmudziński (Poland), Victor P. Beliakov, Galina M. Lavrentyeva, Vladimir V. Skvortsov (Russia), John A. Strand (Sweden), Brian Moss, Martin R. Perrow, Deborah C. Stephen (United Kingdom), Cathleen Wigand (USA)

Hydrobiologia **342/343:** 1–8, 1997.
L. Kufel, A. Prejs & J. I. Rybak (eds), Shallow Lakes '95.

A model of nutrient dynamics in shallow lakes in relation to multiple stable states

Jan H. Janse
RIVM, Laboratory of Water and Drinking Water Research, P.O. Box 1, NL-3720 BA Bilthoven, The Netherlands

Key words: model, phosphorus, eutrophication, hysteresis, lake, restoration

Abstract

There is increasing evidence that, within a range of nutrient loadings, shallow lakes may have two alternative stable states. One is dominated by phytoplankton and the other one by submerged macrophytes as the main primary producer. The question arises at what level of nutrient loading a transition may occur between the two states. This question was addressed by means of the integrated lake model *PCLake*. The model describes the competition between phytoplankton and macrophytes, within the framework of closed nutrient cycles in the lake system, including the upper sediment. Top-down effects via the food web were regarded as well. The model was run for a hypothetical shallow lake, representative for the situation in The Netherlands. Long-term simulations were carried out for a realistic range of nutrient loadings and starting from different initial conditions. The results showed a highly non-linear response, which also showed hysteresis: the loading level at which a transition occurs turned out to be dependent on the initial conditions. The results were compared with empirically derived chlorophyll *a* to phosphorus relations. Factors influencing the 'critical nutrient level' were the lake dimensions and the net sedimentation rate. The model was also used to evaluate the role of food web management in lake restoration. The results suggest that a long-term effect of additional management is possible only if combined with a decrease in nutrient loading.

Introduction

As a result of high nutrient loadings during the past decades, many shallow lakes have become highly eutrophic. They are now characterized by dense algal blooms of cyanobacteria, high turbidity, absence of vegetation and a fish community dominated by bream. Although these effects were caused by high nutrient loadings, restoration of the former macrophyte-dominated clear-water state often could not be achieved by external load reduction alone: eutrophic lakes often show resistance to recovery. Apparently, once the system has switched from a clear to a turbid state, this switch cannot simply be reversed (e.g. Jeppesen et al., 1991; Gulati et al., 1990b; De Haan et al., 1993; Boers et al., in press). Several, often interacting, mechanisms for this resistance have been proposed. Firstly, a prolonged internal loading from nutrient-rich sediments may delay the response (Ryding & Forsberg, 1977; Sas, 1989). Secondly, an increase of the nutrient utilization efficiency of the phytoplankton makes them produce the same biomass with less nutrient (Riegman, 1985; Van Liere & Janse, 1992). Thirdly, the grazing pressure on the phytoplankton is low, both because of the poor edibility of cyanobacteria and the strong predation by bream (Gulati et al., 1990a). Finally, the large amount of detritus accumulated in the system keeps the water turbid and impedes return of the vegetation (Van Dijk & Van Donk, 1991). Clearly, both direct effects of nutrients and indirect effects through the food web may contribute to the often observed resistance to recovery. Therefore, additional measures are sometimes considered apart from, or combined with, nutrient load reduction (Gulati et al., 1990a).

On the other hand, also the clear-water state of shallow lakes, dominated by submerged macrophytes, shows a certain resistance to external forcings, like a moderate increase in nutrient loading (Moss, 1990).

Several stabilising mechanisms may play a role. Nutrient uptake by macrophytes may suppress algal growth due to nutrient limitation (Van Donk et al., 1993), they may provide favourable conditions for predatory fish and they may reduce wind-induced resuspension by stabilising the sediment.

The question addressed in this paper is how the probability of a transition from the clear-water state to the turbid state, or vice versa, is related to the external nutrient loading. This topic is approached by means of a mathematical model, in order to facilitate a systematic analysis. The model used in this study, called *PCLake*, combines a description of the dominant biological components with a description of the nutrient cycle in shallow lake ecosystems. The model differs both from many eutrophication models, which confine themselves mainly to the nutrient cycling, as well as from more detailed biological models. It also differs from so-called minimodels (e.g. Scheffer, 1990) in that it is based on closed nutrient cycles, allowing a more quantified analysis. The aim of this study is to analyse the system's long-term response, in terms of phosphorus, algal and macrophytes biomass, as a function of the external nutrient loading and the initial conditions. Some implications for lake management are discussed.

Model structure

The *PCLake* model calculates the water quality parameters chlorophyll *a*, transparency, phytoplankton types and the density of submerged macrophytes. It also calculates the distribution and fluxes of the nutrients N and P. Inputs to the model are: lake hydrology, nutrient loading, dimensions (mean depth and size) and sediment characteristics. An extensive description of the model may be found in Janse & Aldenberg (1996).

The model describes a completely mixed water body and comprises both the water column and the upper sediment layer (Figure 1). A default sediment depth of 0.1 m has been used. Any further horizontal or vertical distinction is not taken into account. At the base of the model are the water and nutrient budgets (in- and outflow). The physico-chemical module describes the exchange of detritus, inorganic matter and nutrients between sediment and water. Processes involved are sedimentation, resuspension, diffusion, burial and chemical adsorption. Mineralization processes are described in both layers. Two modules, for phytoplankton and macrophytes, respectively, describe the primary production. The food web module is kept as simple as possible and comprises zooplankton, macrozoobenthos, whitefish and predatory fish. All biota are modelled on the basis of functional groups. In this way, effects of the food-web structure on the nutrient cycles are accounted for, and biomanipulation measures may be simulated. The user may choose between different model configurations according to his questions and the available data. In this study, only the phytoplankton has been split into three functional groups, *viz.* cyanobacteria, diatoms and other small edible algae, because of their different characteristics and because of management's interests. It is assumed that the zooplankton has a lower food preference for cyanobacteria and detritus than for the other groups. It is further assumed that the macrophytes may extract nutrients from both the water and the sediment pore water. Apart from mass fluxes (food relations etc.), some 'empirical relations' are included in the model: the resuspension rate is positively affected by the amount of whitefish and negatively by the vegetation density, while the vegetation is assumed to have a positive influence on the growth of predatory fish.

The overall nutrient cycles for N, P and Si are described as completely closed (except for in- and outflow and denitrification). This was done by modelling all nutrient- to-dry-weight ratios dynamically (as indicated by the doubled blocks in Figure 1). Mechanisms were included to cope with the often observed increase of the weight-specific nutrient contents of the organisms at higher trophic levels (e.g. Gulati et al., 1991) and with variations in algal nutrient contents. It should be stressed that all organisms are considered as dependent, directly or indirectly, on the nutrients that are available in the lake or the lake sediment.

The model has been implemented in the simulation package ACSL, version 10. Parameter values were derived from literature and from calibration on several case studies (Janse et al., 1992, 1993; Van Dijk & Janse, 1993). A partial calibration study on a multi-lake data set using Bayesian statistics has also been carried out (Aldenberg et al., 1995). The model has been used for several scenario analyses. These include studies on nutrient load reduction (Janse et al., 1992), biomanipulation (Janse et al., 1995; Janse et al., in press) and combinations of these with dredging (Janse et al., 1993; Zamurović-Nenad, 1993; Aysever, 1994).

In this study, simulations have been performed for a hypothetical lake system which may be considered as representative for many shallow lakes occurring in The Netherlands. Its main characteristics are: mean depth = 2 m, areal hydraulic loading = 20 mm

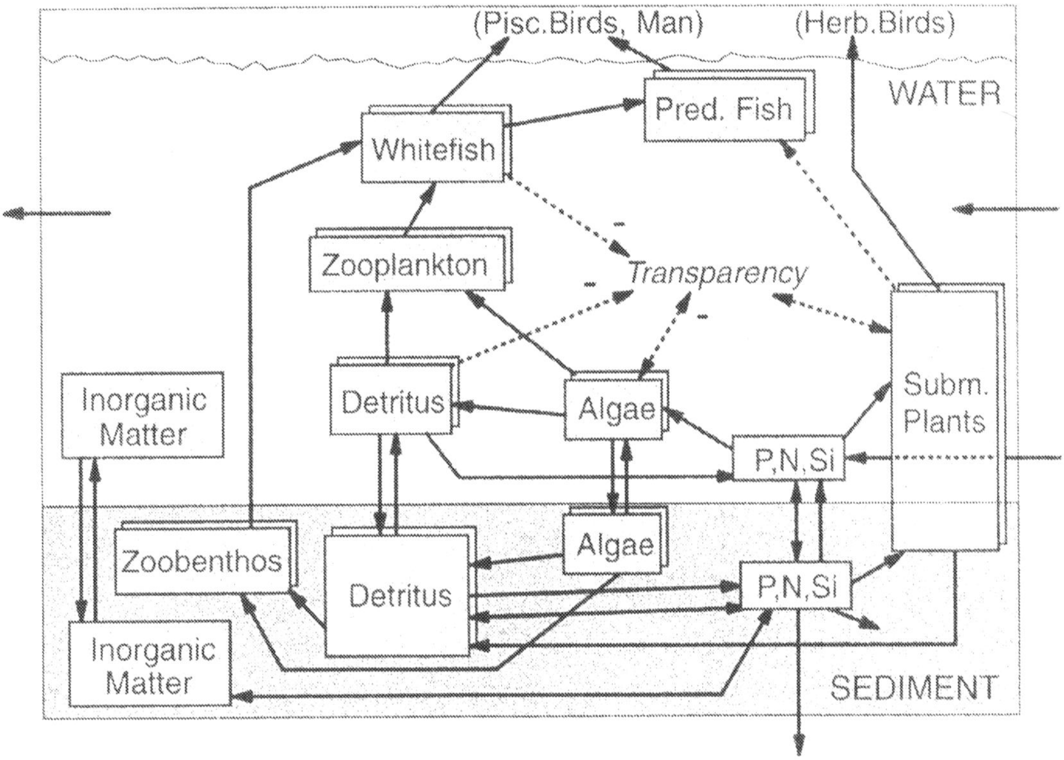

Figure 1. PCLake model structure. Doubled blocks denote compartments modelled in both dry weight and nutrient units. Three functional groups of phytoplankton are distinguished: cyanobacteria, diatoms and other small edible algae. Arrows with solid lines denote mass fluxes (e.g. food relations), arrows with dotted lines denote 'empirical' relations (minus sign denotes negative influence, otherwise positive influence). Egestion and mortality fluxes of animal groups and respiration fluxes are not shown.

d^{-1} (=7.2 m y^{-1}), no seepage, porosity of upper sediment = 0.9, sediment Fe content = 10 mg g^{-1}. Some of the simulations have also been carried out for a mean depth of 5 m and a mean depth of 10 m, with the same hydraulic loading. The (theoretical) total P concentration in the inflow to Dutch shallow lakes ranges from 0.1 to 1.0 mg P l^{-1}. Typical in-lake total phosphorus concentrations are between 0.05 and 0.5 mg P l^{-1}, while summer averaged chlorophyll-*a* concentrations are around 100–200 μg l^{-1} (Lijklema et al., 1989).

Results

Variation of nutrient loading and initial conditions

The long-term impact of different nutrient loadings on the above-mentioned 'average shallow lake', with a depth of 2 m, was simulated. Simulations have been performed for ten years, with the nutrient loading set to a fixed value, ranging from 0 to 2.5 g Pm $^{-2}$ d^{-1}. The nitrogen loading has been set to 10 times the phosphorus loading. For every loading value, the simulation has been done twice, starting either from a macrophyte-dominated state, or from a phytoplankton-dominated state with high chlorophyll *a*, respectively. The resulting summer-averaged total-P and chlorophyll *a* concentrations in the tenth year are shown (Figure 2a, b), as well as the relation between them (Figure 2c). The relations between nutrient input and both variables were highly non-linear, with a rather sudden switch between a macrophyte-dominated state with low chlorophyll *a* values and a phytoplankton-dominated state with high ones. Moreover, the response showed hysteresis, so that the 'critical' nutrient loading was considerably different whether one started from the 'clear' or from the 'turbid' side. The calculations were also repeated for some different parameter values. The values of

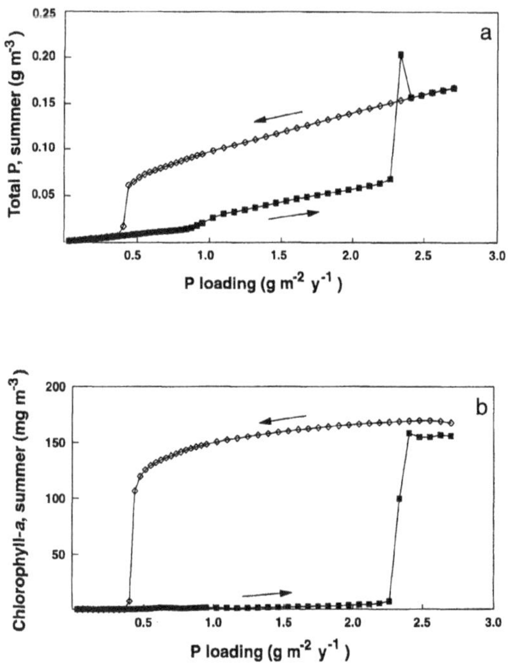

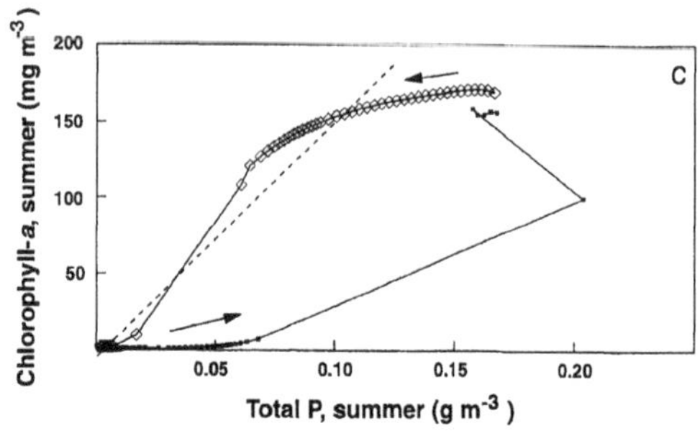

Figure 2. Simulations for an average shallow lake (see text) with a depth of 2 m, for a range of P loadings. The N loading is assumed to be always 10 times the P loading. All simulations have been done for two initial states, a clear-water state dominated by macrophytes (solid symbols) and a turbid state with a blue-green algal dominance (open symbols). Output values are summer-averages after 10 years with the same loading conditions. a. in-lake total P as a function of the P loading; b. chlorophyll *a* as a function of the P loading; c. relation between chlorophyll *a* and in-lake total P concentration. The broken line in Figure 2c is the empirically derived maximum for a data base of Dutch lakes (Lijklema et al., 1989).

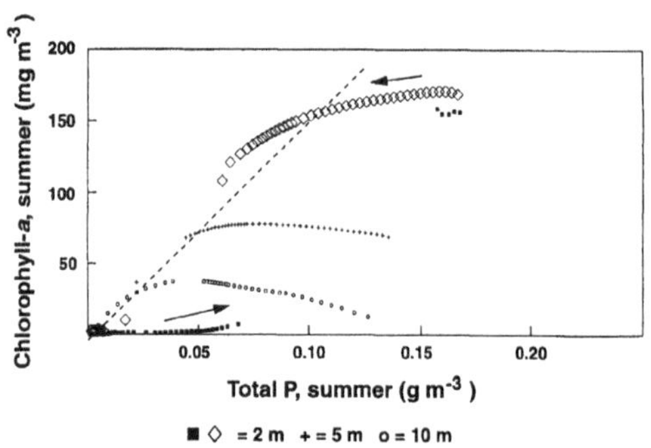

Figure 3. Chlorophyll *a* to in-lake total P relation for three lakes with a depth of 2 m, 5 m and 10 m, respectively, with equal water inflow, for a range of P loadings. Simulations as in Figure 2. Solid squares: depth 2 m, starting from clear state; diamonds: depth 2 m, starting from turbid state; plus signs: depth 5 m; circles: depth 10 m. The broken line is the empirically derived maximum for a data base of Dutch lakes (Lijklema et al., 1989).

the switch points were notably sensitive for the parameters determining resuspension and sedimentation, which are dependent on lake dimensions and sediment characteristics. The lower the net sedimentation rate, the lower were also the critical loading levels.

In comparable simulations of deeper lakes, where the mean depth had been set to 5 m and 10 m, respectively, this hysteresis phenomenon did not occur (Figure 3). These lakes were apparently too deep for macrophyte development so that a phytoplankton dominance appeared in all simulations, but with much lower chlorophyll *a* concentrations than in the simulations of the shallow lakes. The relation between phosphorus and chlorophyll *a* was more or less continuous in these cases. In the left part of the graphs, representing a nutrient-limited situation, all curves overlap. To the right, where light limitation occurs, the chlorophyll *a* concentrations decrease with lake depth, as expected.

The points in the left part of the graphs (Figures 2c and 3) coincide with the line of the maximum summer-averaged chlorophyll-*a* concentration at a given P concentration, empirically derived from a data set of 120 lakes in The Netherlands (Lijklema et al., 1989). All of these lakes were dominated by phytoplankton, the lakes with the higher chlorophyll *a* to total P ratios, up till 1.4 mg mg^{-1}, being dominated by cyanobacteria. These data are in agreement with the chlorophyll *a* and total-P model simulations.

Eutrophication, de-eutrophication and management

In order to study the dynamical behaviour of the model, the same hypothetical shallow lake with a depth of 2 m was simulated during the course of eutrophication and restoration (Figure 4). The simulation started under mesotrophic conditions with the inflowing water

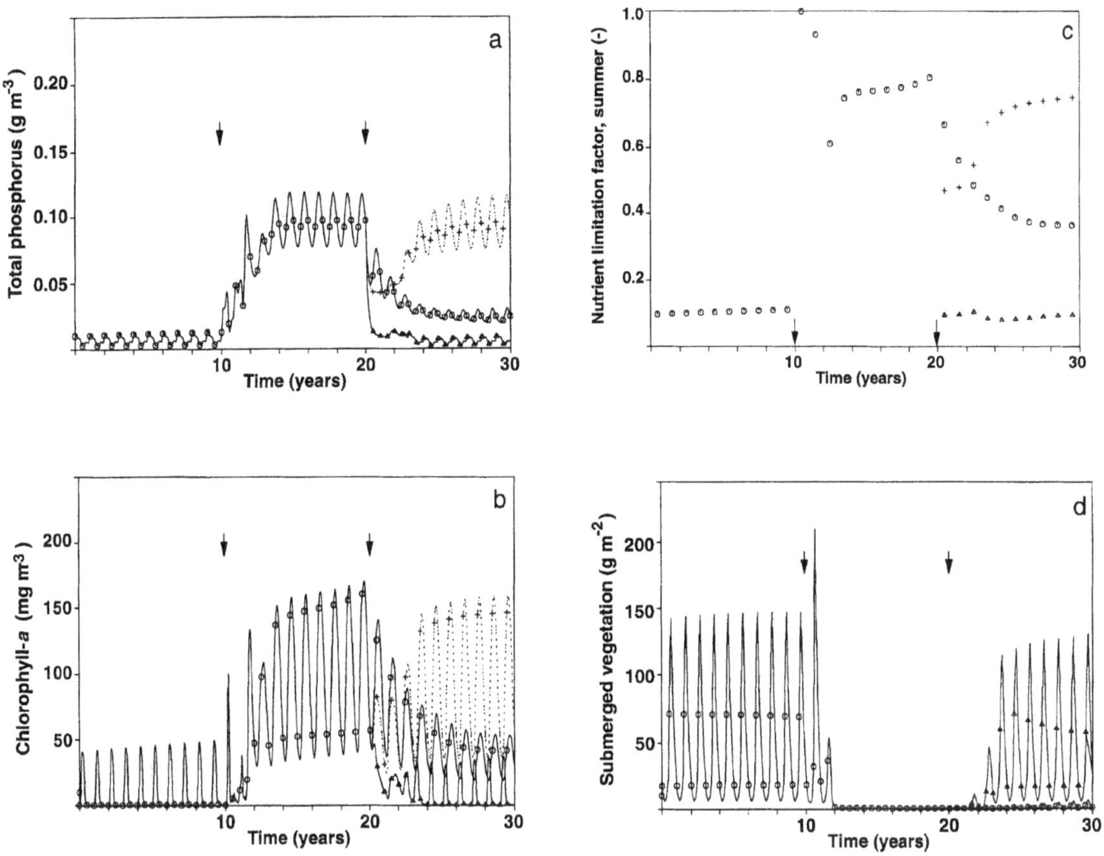

Figure 4. Simulations for an average shallow lake (see text) with a depth of 2 m. After 10 years, nutrient loadings have been increased sevenfold. After another 10 years, three options were simulated: (1) decrease in nutrient loading down to the starting value (circles); (2) idem combined with dredging and food web management (triangles); (3) dredging and fish management only (dashed line with 'plus' signs). a. total phosphorus; b. chlorophyll *a*; c. nutrient limitation function of phytoplankton; d. submerged vegetation.

containing 0.03 mg l^{-1} total phosphorus and 0.3 mg l^{-1} total nitrogen. After 10 years, these concentrations were increased sevenfold, thus bringing the lake into a hypertrophic state. After another 10 years, three options were simulated:

(a) cutting down the nutrient loadings to the former, low, values.

(b) idem, combined with dredging and fish management, each carried out once. This was done in the model by removing 80% of the sediment detritus, half of the zoobenthos and 80% of the whitefish, and introducing 0.1 g d.w. m^{-2} predatory fish.

(c) dredging and fish management without nutrient load reduction.

The effects of the increase and subsequent decrease in nutrient loading are discussed first. The simulated in-lake total phosphorus concentration (Figure 4a) generally followed the changes in inflow concentration with a lag time of about 4 years. This delay may be caused by a slower reaction of the sediment with respect to the water. After the load reduction at year 20, however, the concentration did not quite drop down to the starting value. The chlorophyll *a* concentration (Figure 4b) remained low during the 'mesotrophic period', with small edible algae dominating (data not shown). Their growth rate in mid-summer is severely limited by the availability of nutrients (Figure 4c). Within two years after the increase in loading, the algae were replaced by blue-greens, in high density. Their growth rate was hardly limited by nutrient supply. Algal biomass decreased again following the load reduction at year 20, but the phytoplankton remained to be dominated by blue-greens. As a consequence of the reduced loading, the algal growth rate became nutrient limited again to some extent. The zooplankton almost completely vanished when the blue-greens dominance established, while the whitefish density gradually increased and the predatory fish disappeared. The development of the submerged vegetation (Figure 4d) was more or less inverse to that of the blue-greens. The lake started in a stable, macrophyte-dominated situation. By uptake of the scarce nutrients, they clearly contributed to the nutrient limitation of the algae. The macrophyte density first increased after the load increase, followed by an almost complete disappearance. This was caused by a deteriorating light climate (decreasing Secchi disk depth) due to increasing seston concentrations. The load reduction 10 years later failed to restore the macrophyte dominance. Summarizing, eutrophication made the system change from a clear, macrophyte-dominated state to a turbid state, dominated by blue-greens. This process could not be reverted by nutrient load reduction alone, in agreement with field observations in comparable lakes.

Load reduction combined with dredging and fish stock management, however, proved to be able to make the system 'switch back': the blue-greens were replaced by small edible algae again, in low density, and the vegetation reestablished, as well as the piscivorous fish and the zooplankton. The phytoplankton growth rate became nutrient limited again. Apparently, the sudden improvement in light climate created favourable conditions for return of the vegetation, which was able to stabilize the new situation. To compare, the direct system manipulation as stand-alone measure, while the nutrient loading remained high, was only temporarily effective and failed to reestablish a stable clear state of the system.

Discussion

The model results confirm the existing evidence that shallow lakes may have two alternative states, a clear-water state dominated by macrophytes and a turbid state dominated by phytoplankton (Moss, 1990; Scheffer, 1990; Jeppesen et al., 1990). Several factors determine which state prevails in a certain case. A general constraint is set by the external nutrient loading. At a very high loading, only the turbid state is stable, whereas the opposite is true for very low loadings. In the intermediate range, both states may exist and switches between the two states are possible. Because both states possess a number of self-stabilizing buffering mechanisms, the critical loading level at which a shift occurs is dependent on the initial state of the system: the shift from turbid to clear occurs at a much lower loading level than the opposite one (hysteresis). Either state has a certain attraction field, which are separated by a so-called 'separation line'. A shift may be invoked by a (natural or anthropogenic) disturbance of the system, moving it across the separation line (cf. Scheffer, 1990). Such a shift seems to be always correlated with a dramatic change in water transparency, a key factor for survival of submerged macrophytes. An example of a disturbance might be food web management, possibly combined with physical measures as dredging. The model suggests that additional management is, on long term, more (or only) effective if combined with load reduction, as was also concluded from biomanipulation experiments (Gulati et al., 1990a). Analyses like this one may contribute to improved predictions of the chance of success of additional management at different nutrient levels.

The model analyses also stress the close interrelationship between the nutrient cycle on the one hand and the biological structure on the other. The competition between the different primary producers, phytoplankton and macrophytes, is related to both light climate, nutrient availability and food web interactions. Changes in trophic state may cause changes in food web structure, while the latter may influence the system's (resistance to) response to nutrient loading. Cascading effects in the food web may be analyzed also in terms of changes in nutrient cycles (cf. Carpenter et al., 1992), while the overall trophic state of the system sets a constraint to its total productivity and may be regarded as a general boundary condition. It may be concluded that it is a useful approach to integrate these aspects in one mathematical model in order to study the combined effects of bottom-up and top-down control. It also allows evaluation of different ecological hypotheses and mechanisms. This approach may be complementary to the existing modelling tools for the analysis of subsystems (e.g. Scheffer, 1990) and to the more detailed eutrophication models.

The results of the long-term simulations for total phosphorus and chlorophyll *a* are, in general, within realistic ranges, if compared with empirically derived relations. Also the general shape of the response is in agreement with observations from various shallow lake studies (*op. cit.*). The calculation of the 'critical loading value', the switch point between the turbid and the clear state, might be useful for the derivation of ecological standards for water quality. The calculated values should, however, only be regarded as indicative values at this stage, because they are dependent on model parameters which have only partly been calibrated. This is due to the rather limited range in the present water quality data in Dutch lakes, where, more specifically, clear lakes are lacking. More extensive calibration on a wider data set is needed. Secondly, many model parameters are more or less uncertain due to natural variability or other reasons. As (part of) this uncertainty is inavoidable, the results should be interpreted in a probabilistic rather than a deterministic way. The uncertainty in the results as a function of the combined uncertainty in the parameters should be established by means of appropriate statistical tools, such as Bayesian uncertainty analysis. The model outcome may then be expressed as, for instance, the chance for recovery of a lake or lake type under different circumstances or with different management options, or in terms of the minimum load reduction to be achieved. This approach resembles the ecological risk assessment adopted for toxic substances and other environmental issues.

Conclusions

The model results are in agreement with the concept of two possible stable states, dominated by macrophytes and phytoplankton, respectively. The simulated long-term response of a hypothetical 'average shallow lake system' as a function of the nutrient input was highly non-linear and showed hysteresis. The modelling approach presented here, taking into account both the biological structure and the nutrient cycle, may be a promising tool for the derivation of nutrient regulation and lake management strategies.

References

Aldenberg, T., I. H. Janse & P. R. G. Kramer, 1995. Fitting the dynamic lake model *PCLake* to a multi-lake survey through Bayesian statistics. Ecol. Mod. 78: 83–99.

Aysever, S. S., 1994. Integrated water management; A case study for Lake Kortenhoef, The Netherlands. I.H.E., Delft, M.Sc. Thesis E. E. 162, 67 pp.

Boers, P., D. Van der Molen, E. De Deckere, J. Van der Does & W. Rip, 1995. Experiences with lake restoration projects in The Netherlands.

Carpenter, S. E., C. E. Kraft, R. Wright, X. He, P. A. Soranno & J. R. Hodgson, 1992. Resilience and resistance of a lake phosphorus cycle before and after food web manipulation. Am. Nat. 140: 781–798.

De Haan, H., L. Van Liere, S. P. Klapwijk & E. Van Donk, 1993. The structure and function of fen lakes in relation to water table management in The Netherlands. Hydrobiologia 265: 155–177.

Gulati, R. D., E. H. R. R. Lammens, M.-L. Meijer & E. Van Donk (eds), 1990a. Biomanipulatlon – Tool for Water Management. Developments in Hydrobiology 61. Kluwer Academic Publishers, Dordrecht, 628 pp. Reprinted from Hydrobiologia 200–201.

Gulati, R. D., K. Siewertsen & L. Van Liere, 1991. Carbon and phosphorus relationships of zooplankton and its seston food in Loosdrecht lakes. In Giussani, G., L. Van Liere & B. Moss (eds), Ecosystem research in freshwater environment recovery. Mem. Ist. ital. Idrobiol. 48: 279–298.

Gulati, R. D., L. Van Liere & K. Siewertsen, 1990b. The Loosdrecht lake system: Man's role in its creation, perturbation and rehabilitation. In Ravera, O. (ed.), Terrestrial and Aquatic Ecosystems, Perturbation and Recovery. Ellis Harwood Ltd., Chichester, UK: 593–606.

Janse, J. H. & T. Aldenberg, 1996. The eutrophication model *PCLake*. RIVM, Bilthoven, report no. 732404005.

Janse, J. H., T. Aldenberg & P. R. G. Kramer, 1992. A mathematical model of the phosphorus cycle in Lake Loosdrecht and simulation of additional measures. Hydrobiologia 233: 119–136.

Janse, J. H., J. Van der Does & J. C. Van der Vlugt, 1993. *PCLake*; Modelling eutrophication and its control measures in Reeuwijk Lakes. In G. Giussani & C. Callieri (eds), Proc. 5th Int. conf. on conserv. and managem. of lakes, Stresa, Italy: 117–120.

Janse, J. H., E. Van Donk & T. Aldenberg, 1996. A model study on the stability of the macrophyte-dominated clear-water as affected by biological factors. Wat. Res. (in press)

Janse, J. H., E. Van Donk & R. D. Gulati, 1995. Modelling nutrient cycles in relation to food web structure in a biomanipulated shallow lake. Neth. J. aquat.. Ecol. 29: 67–79.

Jeppesen, E., J. P. Jensen, P. Kristensen, M. Søndergaard, E. Mortensen, O. Sortkjaer & K. Olrik, 1990. Fish manipulation as a lake restoration tool in shallow, eutrophic, temperate lakes 2: threshold levels, long-term stability and conclusions. Hydrobiologia 200/201: 219–227.

Jeppesen, E., P. Kristensen, J. P. Jensen, M. Søndergaard, E. Mortensen & T. Lauridsen, 1991. Recovery resilience following a reduction in external phosphorus loading of shallow, eutrophic Danish lakes: duration, regulating factors and methods for overcoming resilience. In G. Giussani, L. Van Liere & B. Moss (eds), Ecosystem research in freshwater environment recovery. Mem. Ist. ital. Idrobiol. 48: 127–148.

Lijklema, L., J. H. Janse & R. I. Roijackers, 1989. Eutrophication in The Netherlands. Wat. Sci. Tech. 21: 1899–1902.

Moss, B., 1990. Engineering and biological approaches to the restoration from eutrophication of shallow lakes in which aquatic plant communities are important components. Hydrobiologia 200/201: 367–377.

Riegman, R., 1985. Phosphate-phytoplankton interactions. Ph. D. Thesis, University of Amsterdam, 135 pp.

Ryding, S. O. & C. Forsberg, 1977, Sediments as a nutrient source in shallow polluted lakes. In H. Golterman (ed.), Interaction between sediments and fresh water. Dr W. Junk Publishers, The Hague: 227–235.

Sas, H. (ed.), 1989. Lake restoration by reduction of nutrient loading: expectations, experiences, extrapolations. Academia-Verlag Richarz, St. Augustin, 497 pp.

Scheffer, M., 1990. Multiplicity of stable states in freshwater systems. Hydrobiologia 200/201: 475–486.

Van Dijk, G. M. & J. H. Janse, 1993. Modelling resource allocation in *Potamogeton pectinatus* L. J. aquat. Plant Mgmt, 31: 128–134.

Van Dijk, G. M. & E. Van Donk, 1991. Perspectives for submerged macrophytes in shallow lake restoration projects in The Netherlands. Hydrobiol. Bull. 24: 125–131.

Van Donk, E., R. D. Gulati, A. Iedema & J. Meulemans, 1993. Macrophyte-related shifts in the nitrogen and phosphorus contents of the different trophic levels in a biomanipulated shallow lake. Hydrobiologia 251: 19–26.

Van Liere, L. & J. H. Janse, 1992. Restoration and resilience to recovery of the Lake Loosdrecht ecosystem in relation to its phosphorus flow. Hydrobiologia 233: 95–104.

Zamurović-Nenad, Ž., 1993. Eutrophication modelling as a management tool for the Vecht lakes. M.Sc. Thesis E. E.118, I.H.E., Delft, The Netherlands, 99 pp.

Hydrobiologia **342/343:** 9–26, 1997.
L. Kufel, A. Prejs & J. I. Rybak (eds), Shallow Lakes '95.

Phosphorus release in Lake Großer Müggelsee and its implications for lake restoration

Andreas Kleeberg[1] & Hans-Peter Kozerski[2]
[1] *Technical University Cottbus, Dept. Environmental Sciences, Research Station Bad Saarow, Seestr. 45, D-15526 Bad Saarow, Germany*
[2] *Institute of Freshwater Ecology and Fisheries, Dept. Limnology of Lowland Rivers and Shallow Lakes, Müggelseedamm 260, D-12562 Berlin, Germany*

Key words: shallow lake, phosphorus release, lake restoration, dissolved oxygen, nitrate, dredging

Abstract

The shallow, hypertrophic L. Großer Müggelsee, Germany, needs to be converted into a meso- to eutrophic level by means of restoration of the catchment area. Despite efforts in phosphorus (P) loading reduction in the inflow, this polymictic lake still demonstrates an year-to-year fluctuating and strong internal P loading with increasing pore-water phosphate concentrations during summer due to formation of anoxic microlayers at the sediment surface. This P release is indirectly governed by external factors including the supply of dissolved oxygen and nitrate via inflow (high runoff), changes in land use (e.g., decreasing nitrogen fertilization) and climatic factors, and is facilitated by wind-induced rapid changes of stratification and mixing events. Important management implications are: first, declines in nitrate concentrations in the inflow may result in depletion of electron acceptors with respect to phosphate desorption and increase the P release. Second, the P release will continue for a long period and thus delay the recovery of the lake. In dredging the uppermost 1 m sediment layer, the P content of the sediment would decrease from 3.5 to between 0.5 and 1.0 mg P g dry weight^{-1} and result in reduced capability of internal P loading. Because of the high costs, a decision should be made about dredging or whether the limnological improvement of the lake can be achieved via dephosphorization between 11 and 16 years.

Introduction

Phosphorus (P) has long been recognized as the most critical nutrient limiting phytoplankton growth in lakes (e.g., Golterman, 1984) and the biological structure of shallow lakes was found to be extremely dependent on in-lake P concentration (Jeppesen et al., 1991). Recently a restoration programme has been started to reclaim the culturally eutrophicated L. Großer Müggelsee (Klein et al., 1993). This hypertrophic lake (OECD, 1982; Dudel & Kohl, 1992) needs to be converted into a meso- to eutrophic level. Thus, efforts are primarily focused on reduction of P concentration in the inflow from currently about 200 mg P m^3 to reach first in-lake P concentrations between 40 and 60 mg P m^3 (reduction between 66 and 75 %) and finally between 10 and 20 mg m^3 (Klein et al., 1993). However, following external nutrient reduction, shallow lakes are more resistant to recovery than deep lakes (Sas, 1989; Jeppesen et al., 1991; Phillips et al., 1994). As in other lakes, this resilience to change is mainly due to the benthic P release (Marsden, 1989; Lijklema, 1994), so in L. Müggelsee too. To overcome this problem in several lakes the P-rich top layers were carefully removed by dredging, leading to successful restoration in all cases (Björk, 1985; Cooke et al., 1986). Consequently, following a very high summer P release in 1992 – threefold the external P load – the Senate for Metropolitan Development and Environmental Protection of Berlin considered a partial sediment removal as a lake restoration technique.

In order to optimize the lake management, knowledge of the factors influencing its internal P cycling is needed. The most important questions are: (i) what is the potential mobility of P in the sediments, (ii) what are the main environmental factors which control immobilization/mobilization of P in the sediment, (iii) what is the magnitude of the internal P loading in the lake, (iv) under which circumstances dredging is to recommend.

Although, the P content of the uppermost layer of sediment may give some indication to the extent of internal loading during recovery (Sas, 1989; Jensen et al., 1992), the actual P release from sediment is influenced by a number of mechanisms being redox (oxidation-reduction) reactions, adsorption/desorption, mineral phase solubility, mineralization of organic matter (e.g., Marsden, 1989; Boström et al., 1988). Related factors that have been identified as controlling the benthic P release are: temperature (e.g., Holdren & Armstrong, 1986), dissolved oxygen (Einsele (1936)/Mortimer (1941; 1942) model), nitrate (e.g., Andersen, 1982; Jansson, 1986), sulfate (e.g., Caraco et al., 1993), pH (e.g., Holdren & Armstrong, 1986; Kleeberg & Schlungbaum, 1993), wind (De Groot, 1981; Osgood, 1988), and sediment Fe content (e.g., Jensen et al., 1992). Moreover, several authors have shown that a release of P and Fe from 'aerobic' sediment surface in shallow polymictic lakes occurs (e.g., Marsden, 1989; Drake & Heaney, 1987) following the formation of anoxic microlayers at the sediment–water interface (Löfgren & Boström, 1989; Boström et al., 1988) due to the enhanced microbial activity in the sediment during summer (Sinke & Cappenberg, 1988; Gächter et al., 1988).

In addition, the release of sediment P is a function of its bindings forms as well as its mobility (e.g., Istvanovics, 1994; Boström et al., 1985). Chemical extractions of different forms of P allow to determine which portion is potentially mobil or not involved in exchange processes (Psenner et al., 1984).

Based upon investigations of sediment quality including a P speciation and interstitial water components, as well as pelagic water quality in relation to various environmental parameters of three years with very different discharge characteristics, the aim of this paper is to evaluate the factors which are regulating the extent of internal P loading in L. Müggelsee, with regard to management alternatives currently being considered for remediation of the lake.

Material and methods

Study site

The shallow, polymictic L. Großer Müggelsee (Figure 1; lat. 58°13′, long. 34°08′) is one of the intercalated lakes of the River Spree. Therefore, the amount and quality of lake water are mainly influenced by the inflowing river water (Driescher et al., 1993). Since early 1970s L. Müggelsee changed from a macrophyte rich to a phytoplankton-dominated state (Dudel & Kohl, 1992). Despite efforts to restore the lake by reducing external nutrient loading, large algal biomasses in low diversity, i.e. *Limnothrix redekeii* (Van Goor) Meffert, *Planktothrix agardhii* (GOM.) and *Aphanizomenon flos-aquae* (L.) Ralfs. ex Bornet et Flah., reappear every summer, and chlorophyll a in the epilimnion commonly reaches 56 μg Chl *a* l^{-1} (Nixdorf, 1994).

Due to its periglacial genesis the lake is bowl-shaped and has flat margins with a relatively regular shoreline without bays (Figure 1), giving intensive wind impacts to the lakes surface. Hence, the lake stratifies seldom for longer periods, and earlier studies (Behrendt et al., 1993) even found that stratifications last between one and two weeks and that the lake in general is well supplied with oxygen. The lake and its environment is described in detail in Driescher et al. (1993).

Water sampling and analysis

Samples from 6 stations has been taken biweekly in 1992, 1993 and weekly in 1994 and pooled to get a representative sample for the whole lake (Dudel & Kohl, 1992). Additionally, depth profiles were surveyed at the deepest point (Figure 1). Dissolved oxygen (DO) and temperature were measured in the field using a microprocessor oximeter (OXI 196, WTW). After filtration (0.45 μm) soluble reactive P (SRP) and nitrate (NO_3) were measured spectrophotometrically with the stannous chloride method (Freeman et al., 1990) and the diazotization method with sulfanilamide (Henriksen & Selmer-Olsen, 1970) using flow injection analysis (FIA; EVA-Analyzer, EPPENDORF), respectively. From unfiltered sample total P (TP) was determined after digestion (1 N HCl, 16 h, 200 °C; Vogler, 1976) as SRP. All other routine measurements are described in detail elsewhere (Dudel & Kohl, 1992).

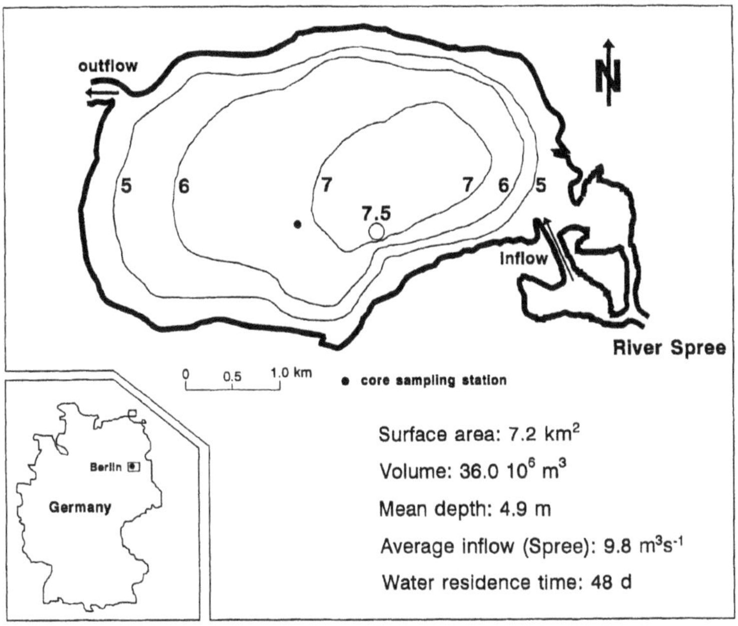

Figure 1. Location, bathymetric map and general characteristics of L. Großer Müggelsee, Germany showing the inflow and outflow of River Spree and the sediment core sampling station.

Sediment sampling and analysis

For interstitial water parameter intact sediment cores up to 25 cm long were taken biweekly from March to November in 1992 and 1993 with an UWITEC-corer (flutter valve, acrylic glass tubes, i.d. = 6.0 cm) at a water depth of about 6 m (Figure 1). This site represents an accumulation bottom area. In the laboratory, cores were kept no longer than 2 h at 4 °C until analysis. From the core the overlying lake water was siphoned off in two steps (20–10 cm and 10–0 cm) downcore. The sediment was sliced into 5 mm segments between 0–5 cm, and into 2 cm intervals from 5–20 cm depth and the pore-water was pressed out. Interstitial NO_3, sulphate (SO_4) and ortho phosphate (PO_4) concentrations were determined using ion chromatography with a Dionex model DX-100 (DIONEX, Idstein) equipped with an anion self-regenerating supressor ASRS-I. Ammonium (NH_4) concentration was measured by FIA and gas diffusion according to Schulze et al. (1988). Concentration differents between the 0 and 2 cm sediment layer and the deepest water layer (10–0 cm) were used to calculate diffusive flux (f) of PO_4, given as P release rates PRR_f, towards the sediment surface according to Fick's first law of diffusion (Berner, 1980; Sinke et al., 1990).

For sediment quality parameter estimation parallel sampling was performed in October 1993 using a LIVINGSTONE-sampler (Livingstone, 1955) for cores up to 1.10 m length. 14 cores were selected from a 500 * 500 m sampling grid on the lakes surface. Each core was sliced at 2 cm intervals. After determining dry weight (DW; 48 h, 60 °C) samples were homogenized and analyzed for total carbon (TC) and total nitrogen (TN) in a C/N-Analyzer (HERAEUS, CHNO-Rapid). Ca, Al, K, Na, Mn and Mg were determined by ICP (AES UNICAM 701) after wet acid digestion (2:1:1 $HClO_4$:HNO_3:HF, 4 h, 180 °C). The standard solutions have to be prepared in the same reagent ($HClO_4$) as the samples to be measured. After determination of loss on ignition (LOI; 3 h, 550 °C) the ash was digested (0.5 h, 110 °C, HCl 1 mol l^{-1}) for the determination of TP and total iron (TFe) (Andersen, 1976; Nausch, 1981). After filtration (0.45 μm) the P was measured as SRP (see above) and Fe according to Legler et al. (1986).

Microcosm experiments

Experimental aerobic (bubbled with air) and anaerobic (non-aerated) PRR were determined fourfold in intact sediment cores, incubated 30 days in the dark at in-situ temperature (± 3 °C) in a continuous flow system as described in Boers et al. (1984) and Boers & Van Hese (1988), respectively. In another experiment, running over 19 days, the addition of freshly precipitated iron(oxi)hydroxides ($FeCl_3$ 0.2 N solution) amounted to 3.0 g Fe m^{-2}. The addition of a KNO_3 solution gave initial NO_3 concentrations of 2.5 mg l^{-1}. Freeze-dried phytoplankton detritus (10 mg seston DW l^{-1}) containing about 80% *Aphanizomenon flos-aque* (L.) Ralfs. ex Bornet et Flah. and 20% *Microcystis aeruginosa* Kützing (7.4 mg P g DW^{-1}), respectively, was added. Aerated cores as a control remained untreated. The flow through rate (10 ml d^{-1}) was controlled by a peristaltic pump and corresponded to water residence times between 96 and 138 days. Water samples for analysis (see above) were daily taken in the supernatant and replaced by filtered (0.45 μm) lake surface water. The PRR were calculated from the mass balance given in Boers et al. (1984).

Phosphorus speciation

Wet sediment samples were successively extracted in triplicate with (a) deionized water, (b) $NaHCO_3/Na_2S_2O_4$ 0.11 mol l^{-1} at 40 °C (Bicarbonate-Dithionite, BD), (c) NaOH 1 mol l^{-1}, (d) HCl 0.5 mol l^{-1} and (e) NaOH 1 mol l^{-1} at 85 °C according to the stepwise fractionating scheme of Psenner et al. (1984). In this way fractionation of sedimentary P into water soluble P (H_2O-P), reductant-soluble P (BD-P), Fe- and Al-bound P (NaOH-P), Ca-bound P (HCl-P) and refractory P (NaOH(85)-P) was attempted. Non-reactive (organic) P (NRP) was deduced from the difference between TP – determined by digestion (H_2SO_4, 16 h, 180 °C) – and the sum of SRP in the extraction steps (a)–(d). Residual P is the difference between TP and the sum of the fractions (a)–(e) and amounted on average 5.0 ± 0.8% TP.

Data analysis

P release rates budget (PRR_b) were calculated (Mothes et al., 1986; Dudel & Kohl, 1992) on the basis of biweekly TP, SRP measurements in L. Großer Müggelsee and River Spree and known terms for external inputs and outputs, being the daily water volumes of inflow, outflow and waterworks withdrawal via bank filtration. Wind and precipitation data used are from daily averages at Berlin-Tempelhof Airport, a few kilometers away, measured by the airport authority, released in Wetterkarte (Amtsblatt des Deutschen Wetterdienstes, ISSN 09365818). All univariate and multiple regressions were performed using the Statgraphics (1985) statistical computer program.

Results

Sediment quality and phosphorus speciation

The maximum depth of the sediments in L. Großer Müggelsee amounts 11.4 m, the total volume 35,179,400 m^3 (Eyrich, 1980). At a density of 1.05 g cm^{-3} it corresponds to 36,938,300 m^3 fresh sediment. The organogenic mud layer is situated within the area at 5 m water depth (4,050,000 m^2) which comes to 56.5% of the total lake surface area. The uppermost 1 m sediment layer considered for removal consists of a soft, blackish, fine-grained material (mean grain size 0.060 ± 0.022 mm) rich in organic substances as well as different nutrients (Table 1).

Organic bound P (NRP) dominates the P pool. This fraction constitutes between 58.6 and 63.6% TP (61.4 ± 1.9% TP), while between 3.2 and 5.8% are dissolved in interstitial water (H_2O-P; 4.3 ± 1.0% TP), 27.3 ± 2.9% is reductant-soluble P (BD-P, ranging from 22.9 to 29.9% TP). Between 32.5 and 39.4% TP (35.6 ± 2.5%) has been found in the cold NaOH extract. The SRP portion of this fraction between 22.0 and 25% is mainly metal bound P (Al- and adsorbed P and $FePO_4$, if the latter is present (Psenner et al., 1984)), while the NRP portion of this fraction (75 to 78% TP) is attributed to organically bound P. The apatite P constitutes 13.4 ± 2.0% of TP (between 11.1 and 15.5%) and the refractory P, extracted with hot NaOH, 12.9 ± 0.6% of TP (12.3 to 14.2% TP), respectively (Figure 2).

Pore-water dynamics

The PO_4 concentration in interstitial water increases on a regular seasonal basis (Figure 3). This process is facilitated by readily degradable organic material fed to the sediment in spring and early summer. Up to March, the SO_4 concentration at the sediment–water interface is similar to that of the overlying water of 140 to 150 mg l^{-1}. The NO_3 concentration in the pelagical water is subject to a steady decline (see below), being as low as 0.1 mg NO_3 l^{-1} in the interstitial water in the uppermost 5 mm layer from May onward. With the DO concentration declining and water temperature rising, a distinct summer SO_4 gradient up to a

Table 1. Composition of L. Großer Müggelsee sediment: mean values and standard deviations of three variants of a partial sediment removal.

Parameter (unit)	Sediment layer to dredge 0.3 m ($n = 15$)	0.5 m ($n = 25$)	1.0 m ($n = 50$)
DW (% fresh sediment)	13.5 ± 1.8	13.7 ± 1.7	13.7 ± 1.2
LOI (% DW)	30.6 ± 1.4	31.8 ± 1.9	33.4 ± 2.2
C (mg mg DW^{-1})	0.147 ± 0.007	0.149 ± 0.006	0.172 ± 0.027
N (mg mg DW^{-1})	0.020 ± 0.004	0.019 ± 0.004	0.020 ± 0.003
C:N (atomic)	9.14 ± 1.91	9.66 ± 1.71	10.29 ± 1.41
P (mg g DW^{-1})	2.10 ± 0.80	1.98 ± 0.66	1.66 ± 0.62
Fe (mg g DW^{-1})	100.39 ± 7.07	108.23 + 6.57	102.62 ± 8.27
Fe:P (atomic)	33.06 ± 12.60	32.98 ± 9.81	37.73 ± 11.03
Al (mg g DW^{-1})	15.37 ± 4.96	13.2 ± 4.67	9.63 ± 5.96
K (mg g DW^{-1})	5.29 ± 0.58	4.45 ± 1.08	3.19 ± 1.89
Na (mg g DW^{-1})	2.45 ± 0.55	2.34 ± 0.51	1.80 ± 0.83
Mn (mg g DW^{-1})	1.42 ± 0.34	1.52 ± 0.33	1.58 ± 0.34
Mg (mg g DW^{-1})	2.15 ± 0.54	1.87 ± 0.54	1.53 ± 0.62
Ca (mg g DW^{-1})	44.06 ± 8.65	30.24 ± 4.25	51.09 ± 4.45

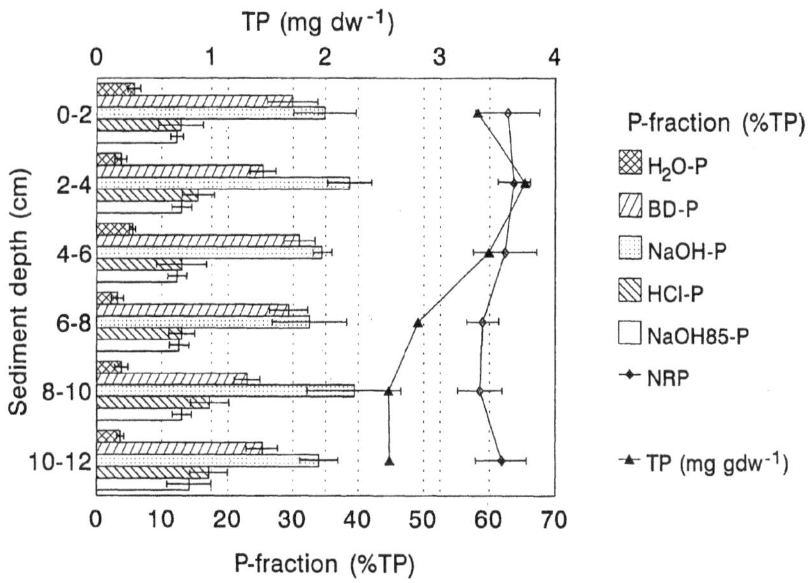

Figure 2. Representative example of total P (TP) and of percentages of different P fractions according to Psenner et al. (1984) – water soluble P (H_2O-P), Fe- and Al-bound P (NaOH-P), Ca-bound P (HCl-P) and non-reactice (organic) P (NRP) – in the top 12 cm of the sediment in October 1991 (means of triplicate samples with standard deviations).

sediment depth of 10 cm becomes obvious (Figure 3). SO_4 reduction activity was not measured; however, the correlation between PO_4 and SO_4 indicates that SO_4 reduction might be very important for PO_4 mobilization ($\ln PO_4$ (mg l^{-1}) = $-2.013 \times \ln SO_4$ (mg l^{-1}) + ln 8.008; $n = 118$; $R^2 = 0.901$; $p < 0.01$). Correlation coefficients between water temperature and interstitial water PO_4 and SO_4 concentration, respectively, of the different sediment layers are given in Table 2. In addition, a linear relationship has been established between the PO_4 and NH_4 concentration of the interstitial water (0–15 cm) PO_4 (mg l^{-1}) = $0.558 \times NH_4$ (mg l^{-1})

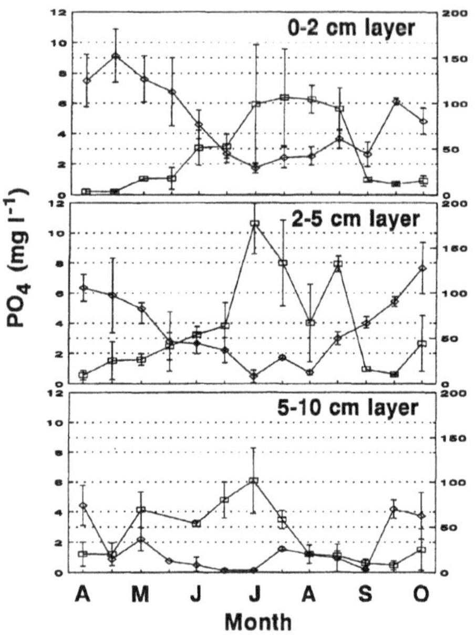

Figure 3. Annual course (April to October) of interstitial water orthophosphate (PO_4) and sulphate (SO_4) concentration in the uppermost 0–2, 2–5 and 5–10 cm sediment layer in 1992 (means of dublicate cores with standard deviations).

-0.051; ($n = 21$; $R^2 = 0.952$; $p < 0.01$) and for interstitial PO_4 and TDFe concentration: PO_4 (mg l^{-1}) $= -2.172 \times$ TDFe (mg l^{-1}) + 15.177; $R^2 = 0.556$; $p < 0.01$).

Climatic and hydrological factors

Following a P input into the sediment in spring (negative PRR_b), the P concentration increases with annual regularity but in different extent (Figure 4). Water temperature in summer months of 1992 was on average about 4 °C above the temperatures recorded in 1993 and 1994 (Figure 5). The long-term yearly mean of total precipitation (1951–1980) is 580 mm and the catchment area of the lake (7,000 km^2) equals that of the inflow of the River Spree, respectively (Driescher et al., 1993). Hence, the different amounts of precipitation (1992: 592 mm; 1993: 618 mm; 1994: 713 mm) led to different mean hydraulic inflow (1992: 6.3 m^3 s^{-1}; 1993: 7.7 m^3 s^{-1}; 1994: 8.5 m^3 s^{-1}), which, however, had different effects on the outflow (Figure 5). With an annual sum of outflow of 56.3 m^3 s^{-1} in 1992, 75.7 m^3 s^{-1} in 1993 and 102.3 m^3 s^{-1} in 1994, respectively, 31.6% of the yearly total was discharged from May to September 1993, whereas it was only 25.7% in 1992 and 26.7% in 1994, thus approximately 40% less in these both years. Obviously, L. Großer Müggelsee is changing its system condition through the modified discharge. To demonstrate this, the theoretical water residence times computed with respect to the daily means were summarized in terms of monthly means (Figure 5). This way L. Großer Müggelsee slowly run through at a mean water residence time of 630 days between May and Semptember 1992 (in several incidents Q_{out} was zero) became a river lake with a water residence time of 155 and 141 days in 1993 and 1994, respectively.

As all three years did not differ in their average wind velocity and their seasonal distribution (Figure 5), a possible influence of the wind on the P release in L. Müggelsee was investigated for each year separately using a model by De Groot (1981). Because of its characteristics and especially due to the fact, that 70% of the lake water are concentrated in the upper 0–4 m (Driescher et al., 1993), the wind impact is highly probable. In general, the wind velocity is highest between November and April (mean 4.1 m s^{-1}), a period with cyclonic weather situations. Between May and September the range of wind velocities (0–7 m s^{-1}) was considerable, but no periodicity was recognisable. Whereas, the lowest velocities are observed in August and September (mean 3 m s^{-1}) due to prevailing anticyclonic weather situations. In regard to the values of the daily and biweekly changes of wind velocity and the pertaining timely differences of TP concentration (Figure 4), as for PRR_b (1992–1994) and PRR_f (1992 and 1993) no distinct relation could be proven. Intervals with low PRR steadily changed with those of high PRR.

Dissolved oxygen

For 1992 as well as for 1993 and 1994 no anaerobic conditions could be ascertained throughout the entire water column and the mean DO concentration was equally high (Figure 6). The temperature rapidly increases over a short time interval (Figure 5) and because of an increased supply of organic material from the phytoplankton collapse of spring diatom bloom and its microbial degradation DO is heavily declined from April onwards. The DO concentration never reached values lower than 2.4 mg l^{-1}, except at the deepest point from early July to early August when values between 0.04 and 0.1 mg l^{-1} were reached. However, in 1992 and 1993 the DO deficiency between

Table 2. Correlation coefficients between water temperature and interstitial water PO_4 and SO_4 concentration, respectively, of different sediment layers (* indicates $p < 0.01$; 1992 $n = 13$, 1993 $n = 7$).

	1992			1993		
	0–2 cm	2–5 cm	5–10 cm	0–2 cm	2–5 cm	5–10 cm
PO_4	0.850*	0.716*	0.472*	0.633*	0.109	0.189
SO_4	−0.694*	−0.964*	−0.748*	−0.740*	−0.458*	−0.590*

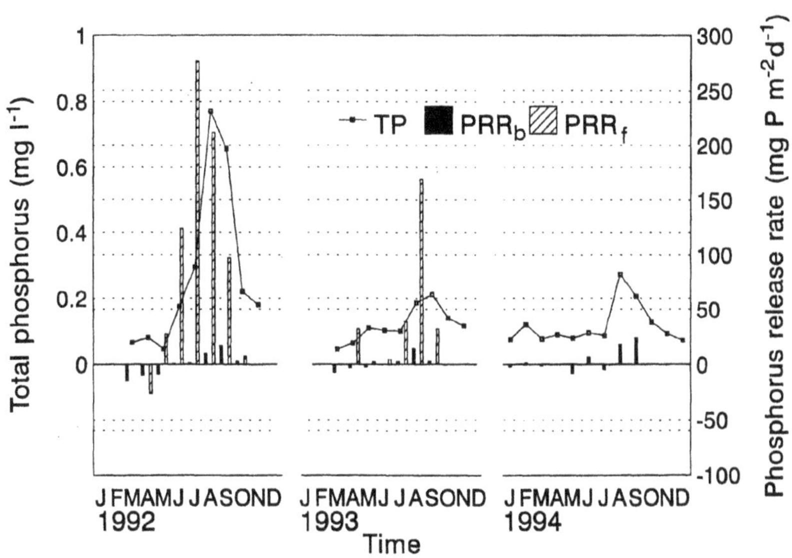

Figure 4. Annual course of lake total P (TP) concentration and P release rates (PRR) calculated from a mass balance approach (PRR_b; 1992–1994) and based on flux calculations of interstitial water orthophosphate gradients (PRR_f; 1992–1993), respectively.

April and May (in 1994 later), which was due to the small inflow Q_{in} is maintained for a longer time period, while with increasing Q_{in} in 1992 a larger amount of DO is fed into the lake (Figure 6).

Important binding partners of P in relation to DO are the particulate and the dissolved iron compounds. According to the relationship found by Jensen et al. (1992) between P release and Fe:P ratio by weight in the uppermost 5 cm of sediment layer, P release in L. Großer Müggelsee with a Fe:P ratio of 35 with 2.72 mg P m^{-2} d^{-1} is supposed to be minimal (Figure 7). Indeed significant higher anaerobic and aerobic PRR were calculated from P balance and microcosm experiments (Figures 4 and 7).

Nitrate

In general, the seasonal course of inflow NO_3 concentration and load were determined by changes in hydraulic load (Q_{in}; Figure 8), respectively.

- 1992 ($n = 18$; $p = 0.01$) $NO_3\text{-}N_{in}$ [mg l^{-1}] = 0.176 * Q_{in} −0.457; $R^2 = 0.771$; NO_3-N [mg m^{-2} h^{-1}] = 0.922 * Q_{in} −3.004; $R^2 = 0.800$
- 1993 ($n = 19$; $p = 0.01$) $NO_3\text{-}N_{in}$ [mg l^{-1}] = 0.210 * Q_{in} −0.667; $R^2 = 0.500$; NO_3-N [mg m^{-2} d^{-1}] = 1.030 * Q_{in} −4.074; $R^2 = 0.625$
- 1994 ($n = 39$; $p = 0.01$) $NO_3\text{-}N_{in}$ [mg l^{-1}] = 0.207 * Q_{in} −0.572; $R^2 = 0.801$; NO_3-N [mg m^{-2} h^{-1}] = 1.896 * Q_{in} −9.322; $R^2 = 0.888$

According to Boström et al. (1988) no P release seems to occur at a NO_3-N supply < 0.5 mg m^{-2} h^{-1}. The time period during which the NO_3-N supply < 0.5 mg m^{-2} h^{-1} is about two weeks shorter in summer 1993 than in 1992 and 1994 (Figure 8), respectively,

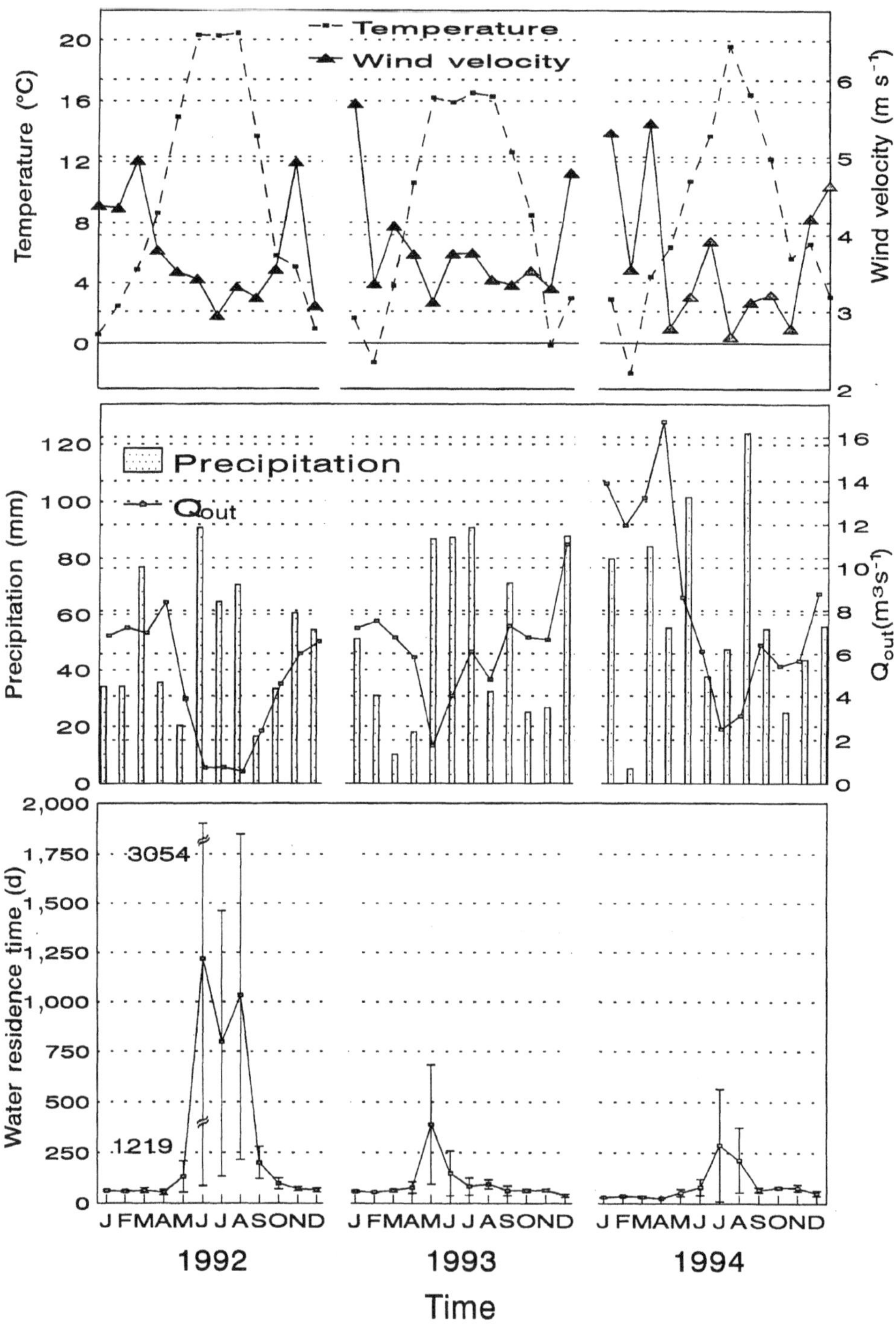

Figure 5. Annual course of monthly means of water temperature, wind velocity, amounts of precipitation (rain) and water volumes of outflow (Q_{out}) as well as the theoretical water residence time of L. Großer Müggelsee from 1992 to 1994. (The monthly means were calculated on basis of daily means of runoff Q_{out}, which was several times zero in 1992.)

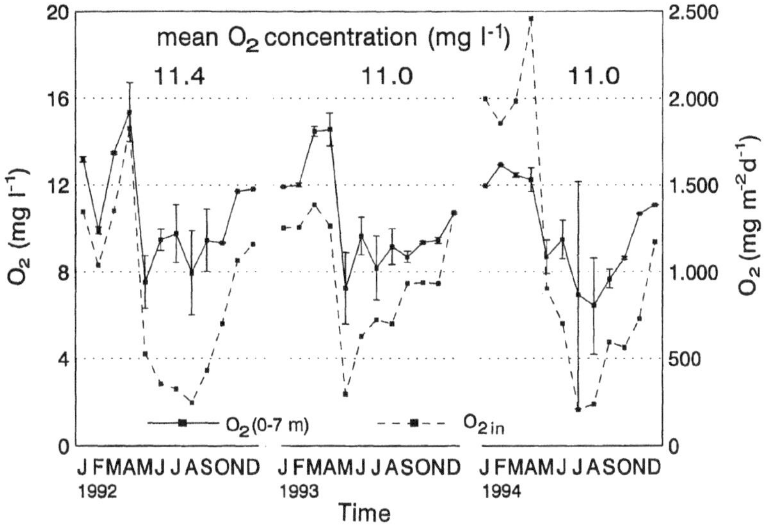

Figure 6. Annual course of in-lake dissolved oxygen concentration (means for water depth 0–7 m with standard deviations) and oxygen supply via inflow (O_{2in}) from 1992 to 1994.

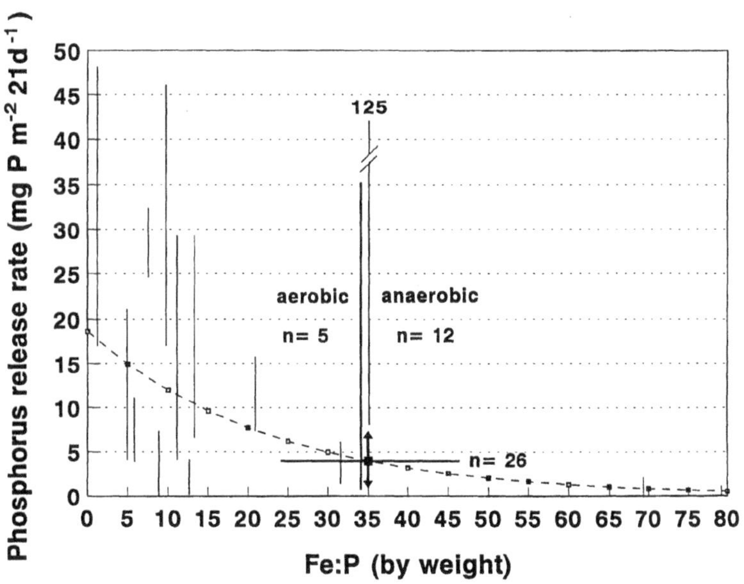

Figure 7. Cumulative P release rates (PRR) given as standard deviations during 21 days from intact sediment cores of 15 shallow Danish lakes (vertikal thin lines) and of L. Großer Müggelsee from aerobic and anaerobic incubations (vertical thick lines) plotted against surface sediment (0–5 cm) Fe:P ratio. The broken line represents the best fit for this relationship (PRR = 12.7 × $e^{-0.044 \text{ Fe:P}}$; R^2 = 0.58, $p < 0.001$) found by Jensen et al. (1992). The vertical arrows show the PRR of L. Großer Müggelsee according to this model at the corresponding mean Fe:P ratio (horizontal line = standard deviation).

and a significant correlation between the lake TP and NO_3 concentration has been established.

– 1992: (n= 18; p = 0.01) TP [mg l^{-1}] = e exp(−0.025 * NO_3-N[mg l^{-1}] + 2.616); R^2 = 0.688

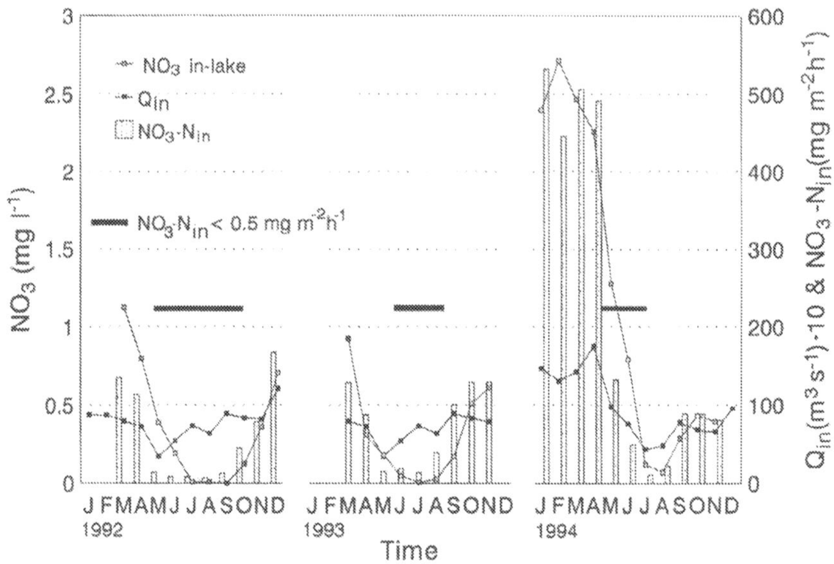

Figure 8. Annual course of in-lake nitrate (NO_3) concentration, inflowing water volumes (Q_{in}) and nitrate (NO_3-N_{in}) supply from 1992 to 1994. The vertical bars show the time period where the NO_3 supply according to Boström et al. (1988) is to low to suppress a P release effectively.

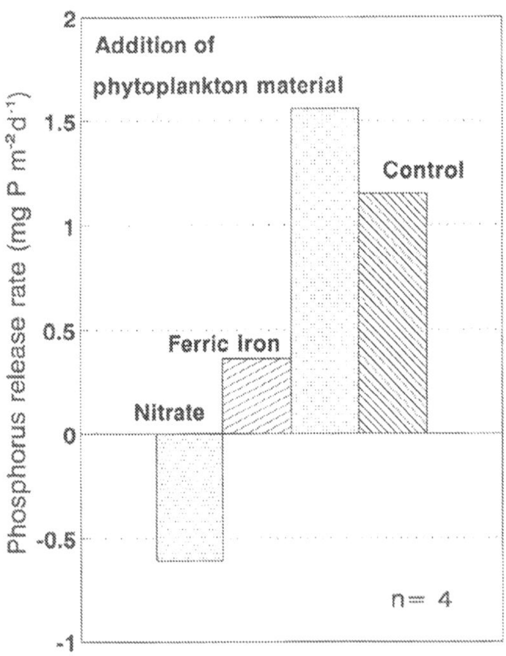

Figure 9. Mean P release rates (PRR) of microcosmos experiments with additions of phytoplankton detritus (10 mg DW l^{-1}), ferric iron (3.0 g m^{-2}) and nitrate (2.5 mg NO_3 l^{-1}), respectively, running from 7/23/ to 8/10/1992 over 19 days.

– 1993: (n = 19; p = 0.01) TP [mg l^{-1}] = e exp(−0.016 * NO_3-N[mg l^{-1}] + 1.543); R^2 = 0.450

– 1994: (n = 39; p = 0.01) TP [mg l^{-1}] = e exp(−1.926 * NO_3-N[mg l^{-1}] + 0.252); R^2 = 0.543

Although it is difficult to assess the effect of inflowing NO_3 concentration, it becomes obvious that P release is considerably affected by NO_3 reduction in the lake's inflow (Figure 8). The TP load varied between 3.6 (1980) and 7.2 g P m^{-2} yr^{-1} (1981 and 1982) (Kozerski et al., 1993), 1992 it accounted for 2.7, 1993 3.3 and 1994 3.1 g P m^{-2} yr^{-1}, with the TP concentration in the inflow being constant.

Microcosm experiments

Two different series of sediment core incubation were run to confirm the lake studies. In the first series over 30 days, the water temperature increases slightly from 20 to 24.5 °C, while the DO saturation was maintained constantly at 85–8% for the aerobic cores. In the anaerobic cores the DO saturation of 92% decreased down to anaerobic conditions after 11 days. Thus, oxygen consumption rates between 15 and 85 mg O_2 m^{-2} d^{-1} were found.

In the second set of experiments at the same temperature, PRR were determined after treatment with different additions (Figure 9). These cores were gently destratified by cooling the top of the water (temperature difference of 0.5–1 °C) resulting in lower DO consumption rates between 2 and 8 mg O_2 m^{-2} d^{-1}. At 30% DO saturation – similar to the corresponding in-lake DO deficiency – chemicals were added.

Estimation of volume, mass, costs and time of a partial sediment removal

Based on previous experiences with dredging measures in waters of Berlin (Kleeberg, 1994; Kleeberg, 1995), U.S. $ 40.0 m^{-3} fresh sediment are to assume for the three different variants of dredging depth (Table 3).

Discussion

Seasonal course of phosphorus

The seasonal course of TP in L. Großer Müggelsee is primarily a result of a high fluctuating internal P load (Figure 4). For example, calculations conducted by Mothes et al. (1986) and Kozerski et al. (1993) yielded maximal PRR of 100 mg P m^{-2} d^{-1} (weekly means during summer). The P release from the sediment varies seasonally and is mainly restricted to the period from May to September, when the PRR_b over about two weeks is as high as 25 mg P m^{-2} d^{-1}. The course of PRR_f is unaffected by sedimentation and co-precipitation, and thus can be regarded as potential, possible maximum PRR. On the assumption that the deep sediments are inhomogenous, Mothes et al. (1986) believe, however, that 'random maximum rates within days' of a magnitude of 200 mg P m^{-2} d^{-1} may occur. While 1992 in spring the PRR_b as well as the PRR_f were mainly negative and positive in summer, in 1993 they show both irregular positive and negative distributions (Figure 4). Thus, a distinctly higher P release holds true for 1992, which is reflected in distinct differences in P concentrations (Figure 10).

Pore-water

In contrast to most other investigations (e.g., Tessenow, 1972), the sediment layers subject to changes in P concentration in L. Großer Müggelsee are deep. This is probably due to the loose nature of sediment and is also substantiated by the homogeneous NRP distribution within the uppermost 12 cm (Figure 2), and the increase in PO_4 concentration is provable up to 10 cm down-core (Figure 3). As for the course of PO_4 as well as of SO_4 concentrations, a correlation with temperature up to a sediment depth of 10 cm has been evidenced (Table 2) indicating an enhanced intensity of biological processes with increasing temperature. As a result, the P mobilization was stimulated and thus also P release. From this it follows, that P release is significantly dependent upon temperature (e.g., Marsden, 1989; Sondergaard et al., 1990). The PO_4 increase in deeper sediment layers is important. Because Sondergaard (1989) and Löfgren & Boström (1989) found that P was mainly released from 2–6 cm depth – the upper 1–2 cm of the sediment seem to be in a state of equilibrium and did not release P – and mobilized at sediment depth of 3–8 cm, respectively. Moreover, the relationship between PO_4 and NH_4 shows, that a large proportion of the mobilized P supposedly stems from the degradation by bacteria of the sedimenting and intrinsic organic sediment material (Boers & De Bles, 1991; Sinke et al., 1990). These findings imply, that the P release from L. Großer Müggelsee sediment is strongly associated with microbial activity, on the one hand primarily through mineralization including direct release of cell-bound P, and on the other hand related to the enhanced hierarchical consumption of electron acceptors (O_2, NO_3, SO_4, CO_2). Thus, binding capacity of iron in relation to the reduced sulfur is of importance.

Phosphorus speciation

H_2O-P was mainly PO_4, and it represents a seasonally variable pool of P compounds dissolved in the interstitial water. BD-P is, in general, considered a potentially mobile pool of P. It may be released from anaerobic sediments and becomes an internal P source to water bodies which suffer from occasional DO depletion (e.g., Einsele, 1936; Mortimer 1941, 1942; Nürnberg, 1988). The present fractionation shows that only about one third of TP is BD-P, although the Fe content of the sediment is relatively high (Table 1), indicating that specific humic substances and/or SO_4 reduction (Figure 3) may inhibit binding of P by Fe, possibly by competition for binding sites. Therefore, only part of the sedimentary P in L. Großer Müggelsee is ascribe to the classical release mechanism. Compared to it P is probably released to some extent also from the NRP, which has been found mainly in the cold NaOH extract. While, the SRP portion of this fraction between 22.0

Table 3. Volume-, mass-, costs- and time-estimation of three variants of a partial sediment removal in L. Großer Müggelsee.

Dredging up to (m)	Mud volume (10^6 m^3)	Mud mass (t)	Portion on total mud volume (%)	Duration of dredging measure (yr)	Costs (10^6 U.S.$)
0.3	1.215	125,659	3.4	1.11	48.6
0.5	2.025	209,431	5.8	1.85	81.0
1.0	4.050	418,862	11.5	3.70	162.0

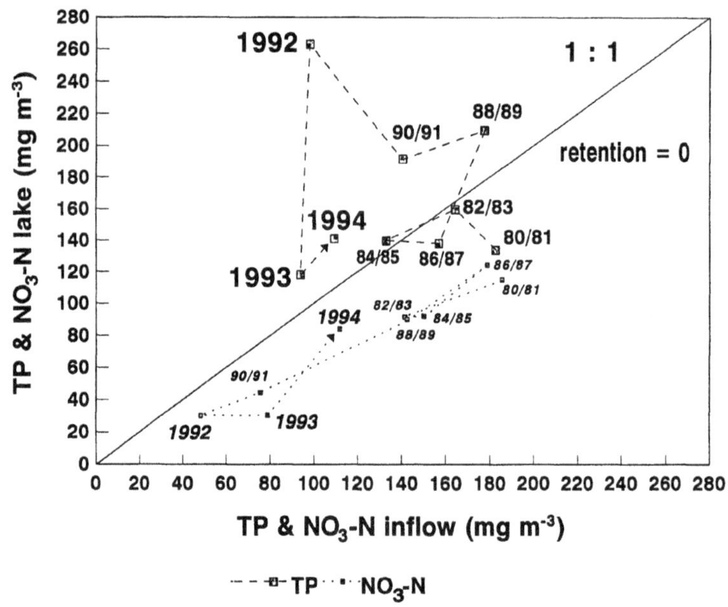

Figure 10. Annual means of in-lake total P (TP) and nitrate-nitrogen (NO_3-N) concentration in dependence on the inflow TP and NO_3-N concentration of River Spree between 1980 and 1994.

and 25% is mainly Al- and adsorbed P and $FePO_4$, if the latter is present (Psenner et al., 1984) and contributes to the capacity of the surface sediment to immobilize P.

The different P forms in the sediment, however, seem to be very mobile and interacting. Therefore, no significant trend of in- or decrease of one or the other fraction could be shown (Figure 2). Moreover, the non-apatite P is usually considered as the algal available fraction and among the most easily releasable P components in the sediment (Boström et al., 1988). The HCl-P portion, representing the apatite P is relatively small. This means that a major part of the P in the sediment of L. Großer Müggelsee potentially can be released to the water phase. This assumption seems to be confirmed by the small portion of the NaOH85-P

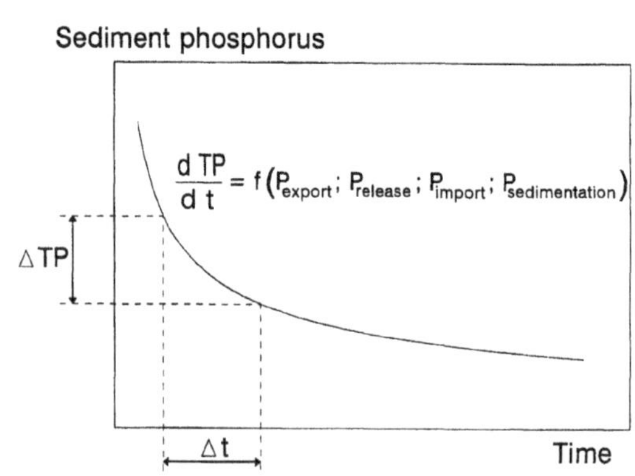

Figure 11. Simplified model of decrease of sedimentary P – dephosphorization over time as a function of P export, P release, P import and P sedimentation.

(between 12.3 and 14.2% TP), which is regarded as a refractory residual P fraction (Psenner et al., 1984), hence, is not subject to exchange procresses with the overlying or interstitial water. The NRP in the sediment must be an important factor of the overall P dynamics in the sediment of L. Großer Müggelsee and constitutes an important pathway for the flux of P. Increasing temperature, as during early summer probably leads to increased decomposition and release of P from this P pool.

Impact of climatic and hydrological factors on phosphorus release

A complex of climatic, hydrological as well as hydrochemical factors is regarded as the cause of the vast difference between the P release in 1992 and in 1993/94 (Figure 4). The trend for the wind distribution found leads to the assumption, that it is just the frequent changing between stratification and mixing (polymixis), which is based upon wind-induced circulation of the water body, as a result of which a positive influence is exerted on P release. Bottom profiles of DO and temperature measured at the lake deepest have shown that wind velocities > 4 m s^{-1} to be capable of completely mixing the water body up to a depth of 6 m. With respect to the wind velocities > 5 m s^{-1} a positive relationship between the wind and the DO could be established for all water depths for 1993 (data not shown).

By creating waves and circulation currents, the wind may have two major effects. First, it may increase the quantity of particles in suspension leading to a short increase of P concentration (resuspension; Sondergaard et al., 1992) or to an uptake of P (adsorption/precipitation; Peters & Cattaneo, 1984), mainly depending on the sediment's character (Hieltjes, 1980; Rossi & Premazzi, 1991). Secondly, the wind might, by increasing turbulence (mixing), increase the random convective exchange of water between the lake and its sediment, thus decreasing the gradient of redox potential (oxygen supply, Behrendt et al., 1993) and pH (adsorption; De Groot, 1981), or maintaining the phosphate gradient across the sediment-water interface due to increased water flow velocity (Fowler et al., 1987).

Dissolved oxygen

In shallow lakes such as L. Großer Müggelsee, wind mixing of water columns can maintain aerobic conditions, which in turn, maintains an oxidized layer at

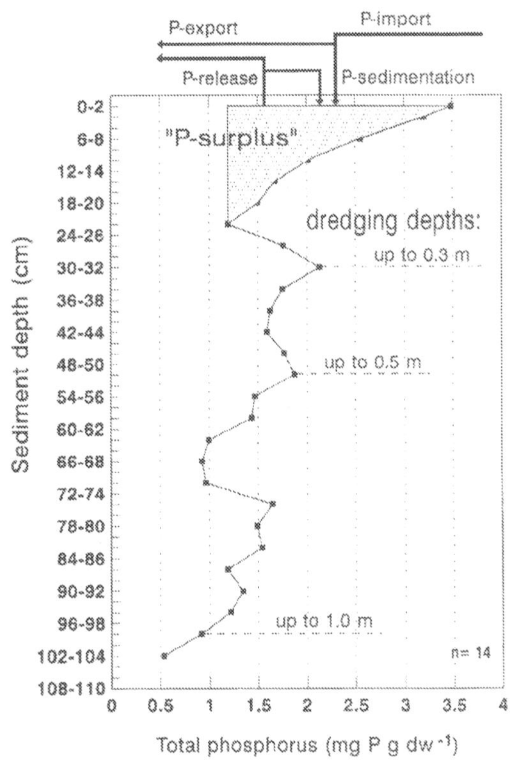

Figure 12. Conceptual model of main pathways of P turnover at the sediment-water interface including the 'P-surplus' to dilute. The depth profile of total P (TP, means of 14 cores) shows three variants of dredging depths currently considered for sediment removal.

the sediment-water interface, causing a reduction in the rate of P release. However, during summer months when the winds are minimal in their impact and DO demand is high (Figure 6), surface sediments may become anaerobic. Like in the years described here (Figure 6) in previous years, too, greater deficiencies, above all in the summer months, and/or several depletions of DO at the lake's deepest occured occasionally, which in most cases coincided with a P release (Mothes et al., 1986; Behrendt et al., 1993). However, up to date no unambiguous relationship could be deduced from the differences of SRP concentration betwen 7 and 0 m and the content of DO in the deeper layers of the lake.

Although, obviously only part of sedimentary P is reductant-soluble the presence of effective binding partners (e.g., ferric iron compounds) is decisive for the P release taking place from May to September. Evidence of Fe controlling P behaviour is the fact that anaerobic fluxes from intact sediment cores are orders of magnitude higher than that under aerobic conditions (Figure 7). After Holdren & Armstrong (1986) the PO_4 can only be immobilized by the oxidation of the ferrous iron, if the atomic Fe:P ratio in the pore-water is > 1.8. At PO_4 concentrations ranging from 3.5 (0–2 cm layer) to 5.0 mg l^{-1} (6–8 cm layer) and TDFe concentrations ranging from 5.5 to 5.0 mg l^{-1} regarding these layers, the Fe:P ratio is found to lie between 0.9 and 0.6. Due to the steep DO gradient the oxidation of ferrous iron cannot take place at the sediment-water interface, or, if DO is present, ferric iron is formed too slowly to immobilize PO_4 which moves upward from deeper layers. This way a release of PO_4 both into anaerobic and aerobic water can be occur (Figures 4 and 6). The mechanism controlling P concentrations under reducing conditions in L. Großer Müggelsee sediments seems to be the the SO_4 reduction. Because of the presence of DO in the water column, SO_4 reduction (Figure 3) and subsequent H_2S production is probably restricted to certain sediment layers (Figure 3) and can lead to inactivation of Fe through FeS and FeS_2 formation; both of which are known to enhance P release (Boström et al., 1988; Caraco et al., 1993). This is supported by the relationship found for interstitial PO_4 and TDFe concentration. Moreover, it is important to note that if large Fe amounts are bound as insoluble FeS, the capacity of surface sediments to re-precipitate P under the re-appearance of aerobic conditions will be reduced (Carignan & Tessier, 1988).

Nitrate

The presence of NO_3 can both enhance and reduce benthic P release due to either increased decomposition through denitrification (Jansson, 1987; Kleeberg & Schlungbaum, 1993) or improved redox conditions in the absence of DO (Ripl, 1976; Andersen, 1982). This was attributed to the ability of NO_3 to buffer the redox potential of the surface sediments at a concentration > 1 mg N l^{-1}, which is high enough to prevent a release of iron-bound P according to the Einsele-Mortimer model. The effect of NO_3 on P release in shallow polymictic lakes was less clear. So, in most cases, no P release seems to occur if NO_3 supply remains high, i.e. > 0.5 mg N m^{-2} h^{-1} throughout the summer (Boström et al., 1988).

It was shown by Dudel & Kohl (1992) that especially during the summer NO_3 minimum in L. Großer Müggelsee, the fluvial NO_3 pulses originating from heavy rainfall (e.g., as shown above) were levelled within the lake (Figure 8) and this NO_3 elimination after heavy rain input is more pronounced in years with lower discharge. The most important loss process is the denitrification which exceeds the effect of sedimentation (Dudel & Kohl, 1992). The seasonality of NO_3 import and denitrification are similar in the years studied. However, in comparison with 1992 in 1993 the NO_3 load was relatively high during late summer (Figure 8) due to increasing Q_{in}. While elimination was probably comparably low and in 1994 large NO_3 amounts are supplied in spring with high inflowing water volumes, which first maintains a high in-lake NO_3 level (Figure 8), respectively. Hence, P release in L. Großer Müggelsee, which often bears the character of a 'sudden event', obviously can be much decreased because of the accumulated amount and the batch of DO and NO_3 contained in the inflow or included in exchange processes. So, the P release in 1993 and 1994 was effectively suppressed by a supply of NO_3 (shorter duration where NO_3-N was < 0.5 mg m^{-2} h^{-1}; Figures 8 and 10).

Microcosm experiments

The most direct approach to quantify internal P loading is by mass balance. This method, however, does not give information concerning the nature of loading and factors controlling the P release (Sondergaard, 1989). Therefore, microcosm experiments were conducted to verify the findings from the entire lake approach. Basis for the high PRR (Figures 4 and 7) is the large pool of P accumulated in the sediment, creating a steep P gradient between different sediment layers (Figure 2) and between sediment and water, thereby promoting a release of P. The P released between day 0 and 2 may have partially originated from sinking matter suspended in the supernatant water. However, predicting the pattern and orders of magnitude of P release from the sediment of L. Großer Müggelsee, several points can be made. NO_3 additions in presence of low DO concentration decreases PRR substantially obviously due to improved redox conditions at a level of DO deficiency (Andersen, 1982; Jansson, 1986). Additions of ferric iron reduces the PRR partly. Either the dose was too small – e.g., Boers et al. (1994) used 100 g ferric iron m^{-2} to decrease the PRR substantially – or the durability was only temporary at a low DO level. Additions of organic matter (cyanobacteria detritus) resulted in higher PRR indicating both, DO consumption and consequently an increased P release probably from the reductant-soluble P pool and a P flux from the NRP. In spite of the limited applicability of microcosms experiments due to difficulties in relating in-vitro to in-situ conditions, the magnitudes of PRR recorded (Figure 7) are in good agreement with anaerobic PRR (Mothes et al., 1986; Kozerski et al., 1993) and completing the orders of magnitudes for aerobic PRR, which are not to differentiate by mass balance studies.

Consequences for lake restoration

While L. Großer Müggelsee retained P in the early 1980s (Kozerski et al., 1993) and it probably would have yet responded to a reduction of P load, this is obviously currently not the case with regard to a drastic decrease of nutrient load from 1988/89 onward (Figure 10). Along with stoppage of area cultivation and the decreased use of mineral fertilizers in agriculture there is less leaching of NO_3 from the soils and thus NO_3 concentration in inflow is decreasing. The consequence of the reduced supply of electron acceptors to the lake is an enhanced P release and the loss of P retention. This internal P loading counteracts the decreasing external P load as a prime aim of lake restoration practice.

Sediment dredging

Dredging is only useful after the main sources of pollution have been removed and its success largely depends on the effectiveness of the removal of the unconsolidated top-layer loaded with phosphates (Table 1) and reduced compounds. Dredging is usually an effective, but costly method (Table 3). Since this material is fine-grained and has a high organic content (Table 1), dewatering is difficult or time consuming. Thus, it appears that we cannot as yet predict the effect of this single restoration measure sufficiently accurate for a convincing cost-benefit analysis. In L. Großer Müggelsee the accumulation of toxic chlorinated hydrocarbons (Terytze & Goschin, 1991) and the surface enrichment of different heavy metals are regarded as not hazardous (Terytze, 1993). Hence, the sediment needs no additional and expensive special treatment and in its elemental composition (Table 1) it is suitable as raw material for agricultural purposes (Kleeberg, 1994).

Alternatives to dredging are in-situ treatments with ferric chloride or alum compounds to chemically immobilize phosphates or wait and see. However, chemical treatments inherent ecological risks and must therefore be used with caution (Jeppesen et al., 1991).

Wait and see: This method only gives acceptable results when the lake can be refreshed permanently with inflowing river water low in P. The aim is to reach a sediment with a TP content similar or smaller as 1.0 mg g DW^{-1} (Sas, 1989; Kleeberg, 1995). The question arises how long will it take till the P storaged ('P-surplus'; Figures 11 and 12) at the sediment surface is diluted up to 1 mg g DW^{-1}?

Following the reduction of P loading in the inflow the P content of the sediment sediment will theoretically decrease gradually via dephosphorization (Figure 11). Although our knowledge of factors controlling internal P loading has increased rapidly in recent years it is still inadequate for precise predictions for P dilution in water and sediment (e.g., Sas, 1989; Van der Molen & Boers, 1994; Jeppesen et al., 1991;). In general, the interaction between sediments and overlying waters is reciprocal: the loading of the sediments with P – generally in particulate form – from the overlying water is controlled by its cycling and incorporation in different trophic levels in the water column, but this is in turn affected by the rate of release from the sediments which in a complex way is related to the loading of the sediment and the subsequent accumulation, redistribution and speciation (Figure 12). Especially in a transitional phase, when reduced external inputs start to restrict the productivity we have to know what the role of the sediment.

It is to assume, that initially the P is released stronger, than decreasing and it is transported away, which is in general favoured by a relatively short water residence time. However, the dilution process is retarded by that a part of the P released reaches the sediment surface again via sedimentation (Figures 11 and 12). Since the change of this differential quotient (Figure 11) for the time period of P loading reduction (no proportionality of magnitudes among each other) is not exactly to predict, the decrease of sedimentary P over time is calculated based on annual means of the status quo of the lake in 1992 (March to November; Dudel et al., 1992) and on the aim of the restoration programm (state 1993; see above). In the most unfavourable case, on the assumption that, the P release first remains constant at a rate of 3.9 mg P m^{-2} d^{-1} (weighted annual mean 1992) and, the P sedimentation is not substantial lowered, i.e. the P export is minimal, it would theoretically last ca 18 yr, till the sediment P content of the uppermost 15 cm layer would decrease from currently 3.5 to 1.0 mg P g DW^{-1} (Figure 12). At simultanous reduction of P sedimentation following a continous reduction in P loading it would last ca 16 yr and ca 11 yr, assuming the P sedimentation is zero, i.e. the P export is maximal, respectively.

As shown by Keizer & Sinke (1992) the dephosphorization process of the uppermost sediment layers can last considerably longer as estimated. In the Dutch Loosdrecht Lakes in 1984 the external P loading was decreased from 3.3 to 1.0 mg P m^{-2} d^{-1} (69.7% reduction). As a result the internal P loading decreased from 1984 to 1990 from 1.0 to 0.3 mg P m^{-2} d^{-1} (70% reduction) and the sedimentary P of the uppermost 2 cm decreased between 1982 and 1991 from 0.94 to 0.60 mg P g DW^{-1}. This dephosphorization by 36% of the original P content over 9 years corresponds to 0.038 mg P g DW^{-1} yr^{-1}. At such rate the decrease of sedimentary P in L. Großer Müggelsee from 3.5 to 1.0 mg P g DW^{-1} would theoretically last 66 years!

Jeppesen et al. (1991) reviewed the recovery resilience of 27 shallow, eutrophic Danish lakes following a reduction in external P loading. Here, the P loading reduction ranged from 34 to 94%. In most cases a new steady state concerning P was not reached within 4–16 yr following loading reduction. There were no differences for low or high hydraulic retention time (tw). The delay, expressed in terms of tw, was more than 10–300 times longer than expected from the simple dilution theory. The relatively slow response in lakes with a low tw was attributed to a higher P pool in the sediment reflecting a higher P loading. In addition, the slow process of diffusion from the deeper layers of the sediment and the seasonal fluctuation in tw, with low flushing rates during summer, when the internal loading is high, may have contributed to the relative long delay period in lakes with a low annual mean tw.

Conclusions and management implications

The redox potential at the sediment–water interface, which is governed by biological activity and the supply of electron acceptors, such as O_2 and NO_3, can vary from oxidized to reduced conditions where SO_4 reduction and Fe immobilization takes place, making it the most important variable with respect to internal P cycling in L. Großer Müggelsee. These findings have important management implications for the lake. First, declines in NO_3 concentration due to reductions in agricultural inputs may result in electron acceptor depletion with respect to redox reactions and increases in the release of P from sediments to the water column. Second, since the demand and the supply of electron acceptors is governed by 'external' factors the future internal P loading is difficult to predict. Third, although the P dilution process will continue for quite a few years, dredging should be rejected at present time and measures of lake restoration should, be directed to a further reduction of loading of the River Spree to decrease the productivity and the ability of the sediment to release large amounts of P.

Acknowledgements

These investigations were funded in part by the Senate for Metropolitan Development and Environmental Protection of Berlin (SenStadtUm). The logistic support of the staff of the Institute of Biology (Ecological Group), Humboldt-University is gratefully acknowledged.

References

Andersen, J. M., 1976. An ignition method for determination of total phosphorus in lake sediments. Wat. Res. 10: 329–331.

Andersen, J. M., 1982. Effect of nitrate concentration in lake water on phosphate release from the sediment. Wat. Res. 16: 1119–1126.

Behrendt, H., B. Nixdorf & W.-G. Pagenkopf, 1993. Phenomenological description of polymixis and influence on oxygen budget and phosphorus release in Lake Müggelsee. Int. Revue ges. Hydrobiol. 78: 411–421.

Berner, R. A., 1980. Early diagenesis. A theoretical approach. Princeton Univ. Press, New York, 241 pp.

Björk, S., 1985. Scandinavian lake restoration activities. International congress on lakes pollution and recovery. Rome 1985, Proceedings: 239–301.

Boers, P. C. M, J. W. T. Bongers, A. G. Wisselo & Th. E. Cappenberg, 1984. Loosdrecht lakes restoration project: sediment phosphorus distribution and release from the sediments. Verh. int. Ver. Limnol. 22: 842–847.

Boers, P. C. M. & F. De Bles, 1991. Ion concentrations in interstitial water as indicators for phosphorus release processes and reactions. Wat. Res. 25: 591–598.

Boers, P. C. M, J. Van der Does, M. Quaak & J. Van der Vlugt, 1994. Phosphorus fixation with iron(III)chloride: a new method to combat internal phosphorus loading in shallow lakes? Arch. Hydrobiol. 129: 339–351.

Boers, P. C. M. & O. Van Hese, 1988. Phosphorus release from the peaty sediments of the Loosdrecht Lakes (The Netherlands). Wat. Res. 22: 355–363.

Boström, B., I. Ahlgren & R. Bell, 1985. Internal nutrient loading in a eutrophic lake, reflected in seasonal variations of some sediment parameters. Verh. int. Ver. Limnol. 22: 3335–3339.

Boström, B., J. M. Andersen, S. Fleischer & M. Jansson, 1988. Exchange of phosphorus across the sediment–water interface. Hydrobiologia 170: 229–244.

Caraco, N. F., J. J. Cole & G. E. Likens, 1993. Sulfate control of phosphorus availability in lakes – A test and re-evaluation of Hasler and Einsele's model. Hydrobiologia 253: 275–280.

Carignan, R. & A. Tessier, 1988. The co-diagenesis of sulfur and iron in acid lake sediments of southwestern Quebec. Geochim. Cosmochim. Acta 52: 1179–1188.

Cooke, G. D., E. B. Welch, S. A. Peterson & P. R. Newroth, 1986. Lake and reservoir restoration Butterworth, Stoneham, 506 pp.

De Groot, W. T., 1981. Phosphate and wind in a shallow lake. Arch. Hydrobiol. 91: 475–489.

Drake, J. C. & S. I. Heaney, 1987. Occurrence of phosphorus and its potential remobilization in the littoral sediments of a productive English lake. Freshwat. Biol. 17: 513–523.

Driescher, E., H. Behrendt, G. Schellenberger & R. Stellmacher, 1993. Lake Müggelsee and its environment – Natural conditions and anthropogenic impacts. Int. Revue ges. Hydrobiol. 78: 327–343.

Dudel, G. E. & J.-G. Kohl, 1992. The nitrogen budget of a shallow lake (Großer Müggelsee, Berlin). Int. Revue ges. Hydrobiol. 77: 43–72.

Dudel, G. E., A. Kleeberg, K. Hilbig & B. Rechenberg, 1992. Phosphorfreisetzung aus dem Sediment des Großen Müggelsees unter besonderer Berücksichtigung des Stickstoff- und Eisenhaushaltes. Limnological Report, Senate for Metropolitan Development and Environmental Protection of Berlin, Berlin, 46 pp.

Einsele, W., 1936. Über die Beziehung des Eisenkreislaufs zum Phosphatkreislauf im eutrophen See. Arch. Hydrobiol. 29: 644–686.

Eyrich, A., 1980. Limnogeologische Untersuchungen in Berliner Gewässern zur Ermittlung der Verbreitung rezenter und subrezenter Sedimente und deren Bedeutung für die Uferfiltration. PhD Thesis, Humboldt-Univ., Berlin, 99 pp.

Fowler, B., J. Drake, D. Hemenway & S. I. Heaney, 1987. An inexpensive water circulation system for studies of chemical exchange using intact sediment cores. Freshwat. Biol. 17: 509–511.

Freeman, P. R., I. D. McKelvie, B. T. Hart & T. J. Cardwell, 1990. A flow injection analysis method for determination of low levels of phosphorus in natural waters. Anal. chim. Acta 234: 409–416.

Gächter, R., J. S. Meyer & A. Mares, 1988. Contribution of bacteria to release and fixation of phosphorus in lake sediments. Limnol. Oceanogr. 33: 1542–1558.

Golterman, H. L., 1984. Sediments, modifying and equilibrating factors in the chemistry of freshwater. Verh. int. Ver. Limnol. 22: 23–59.

Henriksen, A. & A. R. Selmer-Olsen, 1970. Automatic methods for determining nitrate and nitrite in water and soil extracts. Analyst 95: 514–518.

Hieltjes, A. H. M., 1980. Properties and behaviour of phosphate in sediments. PhD Thesis, Technical Univ. Twente, 301 pp.

Holdren, G. C. & D. E. Armstrong, 1986. Interstitial ion concentrations as an indicator of phosphorus release and mineral formation in lake sediments. In Sly, P. G. (ed.), Sediments and Water Interactions, Springer, New York: 133–147.

Istvanovics, V., 1994. Fractional composition, adsorption and release of sediment phosphorus in the Kis-Balaton reservoir. Wat. Res. 28: 717–726.

Jansson, M., 1986. Nitrate as a catalyst for phosphorus mobilization in sediments. In Sly, P. G. (ed.), Sediments and Water Interactions. Springer-Verlag, New York: 382–387.

Jansson, M., 1987. Anaerobic dissolution of iron-phosphorus complexes in sediment due to the activity of nitrate reducing bacteria. Microb. Ecol. 14: 81–89.

Jensen, H. S., P. Kristensen, E. Jeppesen & A. Skytthe, 1992. Iron-phosphorus ratio in surface sediment as an indicator of phosphate release from aerobic sediments in shallow lakes. Hydrobiologia 235/236: 731–743.

Jeppesen, E., P. Kristensen, J. P. Jensen, M. Sondergaard, E. Mortensen & T. Lauridsen, 1991. Recovery resilience following a reduction in external phosphorus loading of shallow, eutrophic Danish lakes: Duration, regulating factors and methods for overcoming resilience. Mem. Ist. ital. Idrobiol. 48: 127–148.

Keizer, P. & A. J. S. Sinke, 1992. Phosphorus in the sediment of the Loosdrecht lakes and its implications for lake restoration perspectives. Hydrobiologia 233: 39–50.

Kleeberg, A., 1994. Auswirkungen einer Sedimentbaggerung im Großen Müggelsee auf die Phosphorfreisetzung - Experimentelle Simulation und Versuch einer Prognose. Limnological report, Senate for Metropolitan Development and Environmental Protection of Berlin, Berlin, 31 pp.

Kleeberg, A., 1995. Die Sanierung/Restaurierung des Großen Müggelsees - mit oder ohne eine Baggerung der phosphorreichen Sedimentschichten. In Jaeger, D. & R. Koschel (eds), Verfahren zur Sanierung und Restaurierung stehender Gewässer. Gustav Fischer Verlag, Stuttgart; Limnologie Aktuell 8: 281–293.

Kleeberg, A. & G. Schlungbaum, 1993. In situ phosphorus release experiments in the Warnow River (Mecklenburg, northern Germany). Hydrobiologia 253: 263–274.

Klein, M., D. Jahn, H. Witt & D. Glase, 1993. Der Große Müggelsee und sein Einzugsgebiet Nutzungen, Belastungen und Sanierungskonzeption. Annual report, Senate for Metropolitan Development and Environmental Protection of Berlin, Berlin, 112 pp.

Kozerski, H.-P., J. Gelbrecht & R. Stellmacher, 1993. Seasonal and long-term variability of nutrients in lake Müggelsee. Int. Rev. ges. Hydrobiol. 78: 423–437.

Legler, Ch., G. Breitig, G. Steppuhn & V. Vobach, 1986. Ausgewählte Methoden der Wasseruntersuchung. Gustav Fischer Verlag, Jena, 517 pp.

Lijklema, L., 1994. Nutrient dynamics in shallow lakes: effects of changes in loading and role of sediment–water interactions. Hydrobiologia 275/276: 335–348.

Livingstone, D. A., 1955. A lightweight piston sampler for lake deposits. J. Ecol. 36: 137–139.

Löfgren, S. & B. Boström, 1989. Interstitial water concentrations of phosphorus, iron and manganese in a shallow, eutrophic swedish lake - implications for phosphorus cycling. Wat. Res. 23: 1115–1125.

Marsden, M. W., 1989. Lake restoration by reducing external phosphorus loading: the influence of sediment phosphorus release. Freshwat. Biol. 21: 139–162.

Mortimer, C. H., 1941. The exchange of dissolved substances between mud and water in lakes. I., J. Ecol. 29: 280–329.

Mortimer, C. H., 1942. The exchange of dissolved substances between mud and water in lakes. II., J. Ecol. 30: 147–201.

Mothes, G., L. Dollan, I. Hoffmann, H.-P. Kozerski & J. Steinmann, 1986. Der Phosphoraustausch zwischen Sediment und Pelagial im Großen Müggelsee. Acta hydrophys. 30: 91–137.

Nausch, G., 1981. Die Sedimente der Darß-Zingster-Boddengewässer - Zustandsanalyse und Stellung im Phosphorkreislauf. PhD Thesis, Rostock Univ., Rostock, 112 pp.

Nixdorf, B., 1994. Polymixis of a shallow lake (Großer Müggelsee, Berlin) and its influence on seasonal phytoplankton dynamics. Hydrobiologia 275/276: 173–186.

Nürnberg, G. K., 1988. Predicting of phosphorus release rates from total and reductant soluble phosphorus of anoxic lake sediments. Can. J. Fish. aquat. Sci. 45: 453–462.

OECD, 1982. Eutrophications of waters. Monitoring, Assessment and Control. - OECD report, OECD Paris, 154 pp.

Osgood, R. A., 1988. Lake mixis and internal phosphorus dynamics. Arch. Hydrobiol. 113: 629–638.

Peters, R. H. & A. Cattaneo, 1984. The effects of turbulence on phosphorus supply in a shallow bay of Lake Memphremagog. Verh. int. Ver. Limnol. 22: 185–189.

Phillips, G., R. Jackson, C. Bennett & A. Chilvers, 1994. The importance of sediment phosphorus release in the restoration of very shallow lakes (The Norfolk Broads, England) and implications for biomanipulation. Hydrobiologia 275/276: 445–456.

Psenner, R., R. Pucsko & M. Sager, 1984. Fractionation of organic and inorganic phosphorus compounds in lake sediments. Arch. Hydrobiol. Suppl. 70: 111–155.

Ripl, W., 1976. Biochemical oxidation of polluted lake sediments with nitrate. A new lake restoration method. Ambio 5: 132–135.

Rossi, G. & G. Premazzi, 1991. Delay in lake recovery caused by internal loading. Wat. Res. 25: 567–575.

Sas, H., 1989. Lake restoration by reduction of nutrient loading. Academia-Verlag Richarz GmbH, Sankt Augustin, 497 pp.

Schulze, G., C. Y. Liu, M. Brodowski, O. Elsholz, W. Frenzel & J. Möller, 1988. Different approaches to the determination of ammonium ions at low levels by flow injection analyses. Anal. chim. Acta 214: 121–136.

Sinke, A. J. C. & Th. E. Cappenberg, 1988. Influence of bacterial processes on the phosphorus release in the eutrophic Loosdrecht lakes, The Netherlands. Arch. Hydrobiol. Beih. 30: 5–13.

Sinke, A. J. C., A. A. Cornelese, P. Keizer, O. F. R. Van Tongeren & Th. E. Cappenberg, 1990. Mineralization, pore water chemistry and phosphorus release from peaty sediments in the eutrophic Loosdrecht lakes, The Netherlands. Freshwat. Biol. 23: 587–599.

Søndergaard, M., 1989. Phosphorus release from hypertrophic lake sediment: experiments with intact sediment cores in a continous flow system. Arch. Hydrobiol. 116: 45–59.

Søndergaard, M., E. Jeppesen, P. Kristensen & O. Sortkjaer, 1990. Interaction between sediment and water in a shallow and hypertrophic lake: a study on phytoplankton collapses in Lake Sobygard, Denmark. Hydrobiologia 191: 139–148.

Søndergaard, M., P. Kristensen & E. Jeppesen, 1992. Phosphorus release from suspended sediment in the shallow and wind-exposed Lake Arreso, Denmark. Hydrobiologia 228: 91–99.

Statgraphics, 1985. STSC Users' Guide. Software Publishing Group, Rockville, 635 pp.

Terytze, K., 1993. Anreicherung und Verteilung von Schwermetallen und anderen Elementen in Oberflächensedimenten der Berliner Gewässer und ihres Umlandes. Acta hydrochim. hydrobiol. 21: 8–20.

Terytze, K. & M. Goschin, 1991. Chlorkohlenwasserstoffe in aquatischen Sedimenten in Berlin und Umland. Wasserwirtschaft 81: 3–8.

Tessenow, U., 1972. Lösungs-, Diffusions- und Sorptionsprozesse in der Oberschicht von Sedimenten. I. Ein Langzeitexperiment unter aeroben und anaeroben Bedingungen im Fließgleichgewicht. Arch. Hydrobiol. Suppl. 38: 353–398.

Van der Molen, D. T. & P. C. M. Boers, 1994. Influence of internal loading on phosphorus concentration in shallow lakes before and after reduction of the external loading. Hydrobiologia 275/276: 379–389.

Vogler, P., 1976. Analysenautomation in Wasserlaboratorien mit flow-stream Automaten. Acta hydrochim. hydrobiol. 4: 115–127.

Hydrobiologia **342/343**: 27–34, 1997.
L. Kufel, A. Prejs & J. I. Rybak (eds), Shallow Lakes '95.

Do rooted macrophytes increase sediment phosphorus release?

Debbie Stephen[1], Brian Moss[1] & Geoff Phillips[2]
[1]*Department of Environmental and Evolutionary Biology, University of Liverpool, Liverpool, L69 3BX, England*
[2]*NRA Anglian region, Cobham Road, Ipswich, Suffolk, IP3 9JE, England*

Key words: sediment phosphorus release, submerged macrophytes, Norfolk Broads, Cheshire Meres

Abstract

Sediment cores were collected from two of the Norfolk Broads and two of the West Midland Meres, to test the hypothesis that rooted macrophytes increase the sediment release rate of phosphorus during the growing season.

These were taken from inside and outside plant beds (mixed *Myriophyllum spicatum* and *Chara* sp. in Hickling Broad and *Nuphar lutea* in Burntfen Broad, Mere Mere and Little Mere). The overlying water was replaced with a standard water designed to mimic lowland freshwater. After an overnight 'batch' incubation the water was analysed for TP and SRP. From this their release rates were calculated. Interstitial measurements of SRP were also made for some of the Broads cores at the end of the incubation period. The data were analysed using General Linear Models. TP and SRP were the responses for both data sets. Factors for the Broads were inside/outside plant beds, date, and site, whilst for the Meres the factors of date and site could not be delineated.

Release rates in this study were affected by a number of factors (plants, site and date). The picture emerging so far is that when rooted macrophytes have a significant effect on phosphorus release they increase it. However the large variability between replicates highlights the logistical difficulties in running such experiments.

Introduction

Studies investigating the mechanisms controlling sediment phosphorus (P) release have markedly increased over the last two decades. This has been in response to the resilience of many lakes which, despite reductions in external P-loadings, have maintained high P-concentrations e.g. (Jeppesen et al., 1991; Marsden, 1989; Sondergaard et al., 1993).

Mortimer's pioneering studies demonstrated that in oxygenated sediments phosphorus is immobilised by sorption with Fe(III) and that this process is reversed under reducing conditions when Fe(III) changes to Fe(II) and the P is mobilized (Mortimer, 1941, 1942). Redox and iron are still considered to be of importance, although it is now been recognised that there are many other potential controlling mechanisms. Those which have been shown to increase sediment P-release include high concentrations of H_2S in the overlying water (Driscoll, 1993; Serruya et al., 1974), increased temperature, and elevated pH (Jensen & Andersen, 1992). In contrast coprecipitation of phosphate with carbonate (Serruya et al., 1974; Otsuki & Wetzel, 1972) and high concentrations of NO_3 (Andersen, 1982) are factors which have been shown to reduce P-release.

Rooted macrophytes affect their immediate environment in various ways, some of which have the potential to influence sediment phosphorus processes. For example, increased sedimentation rates within plant stands lead to a build up of organic matter (Carpenter, 1981). The mineralization of this organic matter may be enhanced through oxygen release from roots (Carpenter et al., 1983; Wium-Andersen & Andersen, 1972) and PO_4^--P also may be released due to the very low redox conditions at the sediment surface in plant beds (Gambrell & Patrick Jr., 1978), or due to elevated pH (Rorslett et al., 1985). Bioturbation through invertebrate activity may further enhance this release (Phillips et al., 1994). Alternatively any available PO_4^--P may be taken up by the plants through both roots and

Table 1. Location and characteristics of sampling sites used.

Lake	Area (ha)	Mean depth (m)	Plant bed sampled	Date (all in 1994)
Burntfen Broad	4.6	1.0	*Nuphar lutea*	21st June & 1st August
Hickling Broad	141.1	1.3	*Myriophyllum spicatum* & *Chara* spp. (mixed)	23rd June & 3rd August
Little Mere	2.5	0.9	*Nuphar lutea*	4th July
Mere Mere	15.8	2.8	*Nuphar lutea*	19th July & 9th August

shoots (Carignan & Kalff, 1980; Denny, 1972), or in marl lakes coprecipitated with carbonate.

The aim of the study presented in this paper was to assess whether, on balance, rooted macrophytes increase or decrease the rate of sediment phosphorus release during the growing season.

Study area

Sampling took place in 1994 in two areas of England, the Norfolk Broads and the West Midland Meres (see Figure 1 for location). Two lakes were selected in each area, details of the sampling dates, lakes and of the plant species in the beds from which sediment cores were taken are given in Table 1.

The Norfolk Broads are a series of shallow lakes formed in the fourteenth century AD when peat workings in the area were flooded (Moss, 1987). As a result of agricultural intensification and a switch to mains sewerage the majority of the Broads have undergone severe eutrophication since the 1950s. As well as nutrient increases the waterways have been subjected to a massive increase in boating activity, which has lead to severe bank erosion. These changes have been mirrored by ecosystem changes with widespread loss of macrophytes (Moss, 1987). Burntfen Broad receives the majority of its water from the nutrient rich River Bure. It still retains a substantial water lily stand but very few submerged macrophytes. By contrast Hickling Broad receives water pumped from surrounding land which has relatively low nutrient concentrations. Following a near total loss of submerged plants in the 1970s, which has been attributed to guanotrophication, plants have returned (Bales et al., 1993). Hickling now has a substantial submersed macrophyte community, dominated by *Myriophyllum spicatum* L., *Potamogeton pectinatus* L. and *Chara* spp.

Lakes in the West Midlands are mostly of glacial origin formed about 10 000 years ago, and include

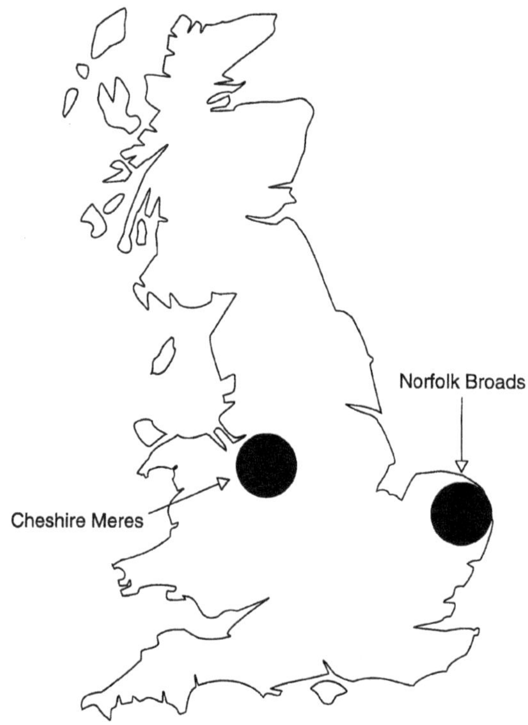

Figure 1. Map of Britain showing the location of the two study areas.

both shallow and moderately deep lakes. Mere Mere and Little Mere are two in a series of three connected lakes. The series starts with Mere Mere and water then flows from it into Little Mere. Mere Mere is a relatively pristine glacially formed lake, which is noted for its rich diversity and abundance of macrophytes. Little Mere is a relatively new basin (probably formed at the end of the last century), which received poorly treated sewage effluent over a few decades until the effluent was diverted in 1991. The plant population is still undergoing changes since diversion. *Nuphar lutea* was the only macrophyte present pre-diversion and these

Table 2. Standard Solution (all final solutions were made up as follows: 10 ml stock made up to 1 l ultra pure water, except K_2HPO_4 which was 1 ml stock made up to 1 l with ultra pure water).

Stock solution	Stock solution concentration $g\,l^{-1}$	Ions in final solution	Final conc. $mg\,l^{-1}$
$Ca(NO_3)_2.4H_2O$	1.0	Na^+	5.47
K_2HPO_4	0.05	K^+	4.01
$MgSO_4.7H_2O$	2.5	Ca^+	21.70
$NaHCO_3$	2.0	Mg^{++}	2.47
$CaCl_2$	5.54	SO_4^-	9.74
KCl	0.76	Cl^-	39.01
		HCO_3^-	14.53
		PO_4-P	0.01
		NO_3-N	1.19

beds are still substantial. In the summer of 1994 the other dominant species were *Elodea canadensis, Nitella* spp., and *Potamogeton pectinatus*.

Materials and methods

Sampling was carried out between June and August 1984, twice in each of the Broads, twice in Mere Mere and once in Little Mere. Intact sediment cores were collected in extruded acrylic tubes (length 50 cm diameter 6.9 cm) using a pole-corer. The cores had 3 mm holes drilled at 1 cm intervals to allow access to the water and sediment, these holes were sealed with waterproof tape. On each occasion six replicates were taken from within the plant bed and six from bare sediment. Due to the small core diameter, plants were not contained within the samples. The cores were sealed top and bottom with rubber bungs. Care was taken to ensure that no air was present above the water column, as this would have allowed the water to move more freely leading to sediment disturbance during transportation. On arrival at the laboratory the sediment depth was measured and, where necessary, standardised to 20 cm by releasing sediment from the base of the core. The hole nearest the sediment was pierced in order to drain the overlying water through a thin plastic tube (the hole chosen was at least 0.5 cm above the sediment surface to ensure that the sediment never came into direct contact with the air). This water was slowly replaced with a standard water, designed to mimic a 'typical' lowland freshwater (see Table 2), to a depth of 15 cm, (by standardising the overlying water we were aiming to isolate the factor of macrophytes across all the lakes). In the Broads this replacement was achieved with a peristaltic pump, whereas in the meres the water was replaced by siphoning. Black plastic was wrapped around the half of the tube containing sediment to exclude light from the core's sides. The cores were then placed overnight in a water bath at tap-water temperature (used as a surrogate for lake temperature) with a typical incubation period of 16 hours.

The next day, overlying water samples were taken using a syringe lowered from above. The water was analysed for Soluble Reactive Phosphorus (SRP) and Total Phosphorus (TP) Both methods were carried out using standard methods based on molybdenum blue formation (see (Holdway et al., 1978)). After the overlying water had been removed interstitial water samples were taken using a syringe at 1, 2, 3, 6, 9, and 12 cm depths from three of the Burntfen Broad cores and analysed for SRP (using the NRA's protocol for small volumes analysis, which was adapted from the work of John (1970), Murphy & Riley (1962) and Stephens (1963)).

Release rates were then calculated using the following equation:

$$\text{release rate (mg m}^{-2}\text{ day}^{-1}) = \frac{C * H * 1000 * 24}{i}$$

where C is concentration in mg l^{-1} (overlying water at the end of incubation-initial concentration in the standard water) H = height of overlying water in m i = incubation in hours

Results

The data for the broads and meres were analysed separately by means of an Analysis of Variance (General Linear Model) followed by Tukey LSD tests where appropriate. Due to differences in the experimental design for the two locations it was not possible to run a combined analysis for the whole data set. Summaries of the results for the broads and meres are presented separately.

Broads

The three factors in the analysis were:

1. Plants: whether the cores were taken from inside or outside plant beds (two levels: Plant and Open)
2. Site (two categories: Burntfen and Hickling)
3. Date (two occasions: June and August)

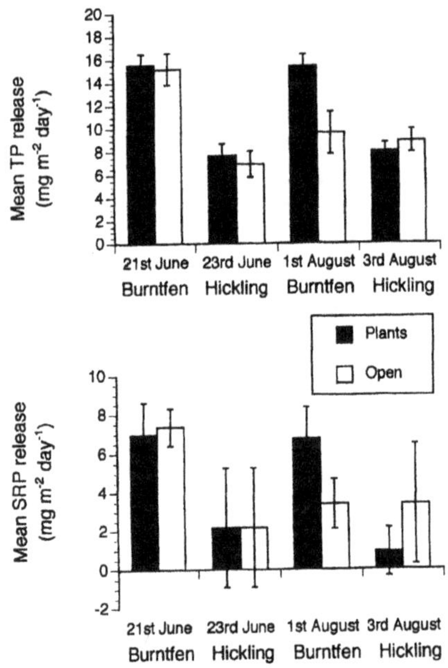

Figure 2. Sediment release of TP and SRP in Burntfen Broad and Hickling Broad (the mean and standard deviation are plotted for each data set).

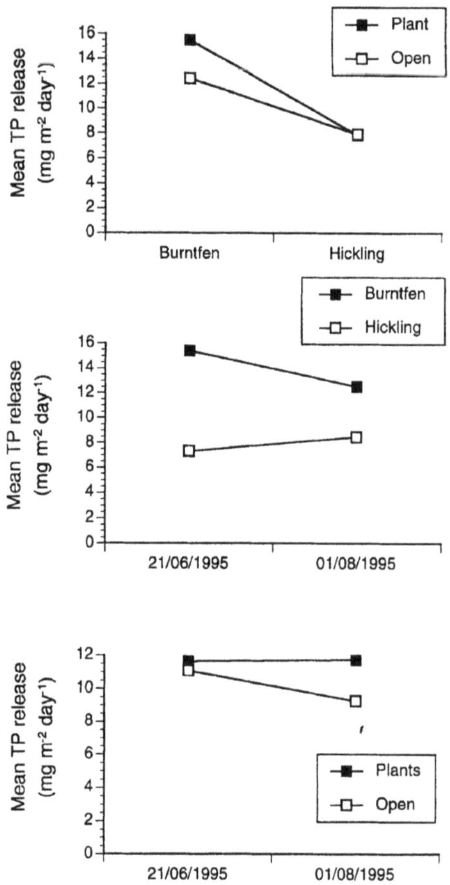

Figure 3. Interaction plots for TP sediment release in The Broads, showing the relationships between plants, site, and date.

For Total Phosphorus release (Figure 2) all the factors (plants, site, and date) were significant. All two-way and three combinations were also shown to produce significant interactions. TP release was significantly higher from those cores taken from within plant beds, higher from the cores taken from Burntfen than those from Hickling, and higher in June than in August (d.f. = 1, F = 10.98, p = 0.002; d.f. = 1, F = 215.89, p<0.0005; d.f. = 1, F = 6.56, p = 0.015).

The interactions for TP release are shown in Figure 3. Further analysis of the plant*site interaction revealed that the higher release from cores taken from plant beds only happened in Burntfen Broad (d.f. = 1, F = 13.37, p = 0.001). Analysis of the plants*date interaction showed that there was no difference in release between cores from within and outside the plant beds in June but that there was significantly higher release from those taken from within plant beds in August (d.f. = 1, F = 4.43, p = 0.043). This was due to a lower release from those taken outside the plant beds in August rather than there being an increase in release from those taken within plant beds.

A closer inspection of the date*site interaction (d.f. = 1, F = 31.16, p<0.0005) revealed that TP release from Burntfen Broad was higher than that in Hickling Broad on both occasions. There was a significant decrease in release rates from Burntfen Broad in August when compared with the release in June. Although there was greater release in Hickling Broad in August than in June, the difference was not significant.

There was also a significant three-way interaction (d.f = 1, F = 15.89, p<0.0005) between all the factors.

For SRP there was a main effect of site and a significant three-way interaction between the three factors of plants, site, and date (d.f. = 1, F = 30.61, p<0.0005; d.f. = 1, F = 4.46, p = 0.043).

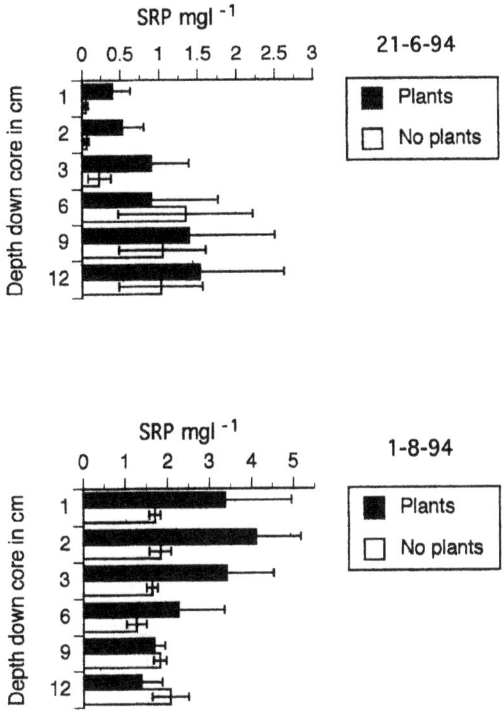

Figure 4. Interstitial SRP concentrations in Burntfen Broad cores on 21-6-94 and 1-8-94 (the mean and standard deviation of 3 replicates are plotted)

The main effect of site is demonstrated in Figure 2, which shows consistently greater release rates of SRP from the cores taken from Burntfen Broad.

The profiles of interstitial SRP concentrations, which were measured at the end of the incubation period, are plotted in Figure 4. On both sampling occasions SRP concentrations in the top three cm were higher in the cores taken from plant beds. A correlation between interstitial measurements with release rates was made (Figure 5). The interstitial measurement at 2 cm was used (to represent release potential in the top few cm of sediment), and was significantly negatively correlated with release for both cores taken from within ($r = -0.83$) and outside plant beds ($r = -0.83$). The 2 cm concentration was used as it was decided that the 1 cm sample may not be reliable due to a small amount of surface sediment disturbance during water replacement.

Meres

Two factors were available for the statistical analysis of release rates in the meres:

1. Plants: whether the cores were taken from inside or outside plant beds (two levels: Plant and Open)
2. Date (three levels: 4th July, 19th July, 9th August)

Site could not be used as a third factor as it could not be delineated from the factor of date.

For TP (Figure 6) the only significant difference was to be found in a main effect of date where release from the cores in Little Mere on the 4th July was higher than Mere Mere on the 17th of July (d.f. = 2, $F = 7.10$, $p = 0.003$). The release from Mere Mere on the 9th August did not differ significantly from either of the others.

For SRP there was a main effect of site and a significant interaction between plants and site (d.f. = 2, $F = 8.02$, $p = 0.002$; d.f. = 2, $F = 3.66$, $p = 0.038$). Release was higher from Little Mere on the 4th of July than on both of the samplings from Mere Mere, and the release rates from the two samplings of Mere Mere did not differ from each other.

Closer inspection of the interaction revealed that release from the cores taken from within the plant beds from Little Mere was higher than all the sample sets from Mere Mere but did not differ significantly from those taken outside the plant beds in Little Mere. Those taken from outside plant beds in Little Mere did not differ significantly, however, from any sample set taken from Mere Mere and the Mere Mere sample sets did not differ from each other.

Discussion

Broads

The release rates measured for Burntfen Broad were very similar to those measured on two occasions in Burntfen Broad during autumn, 1993 (Pitt, 1994). The latter procedure differed from ours in that the overlying water in the cores was filtered broad water, whereas we used a standard water. It therefore appears that the standard water did not alter the behaviour of the sediment.

Although phosphorus release from plant beds compared with that outside did not show a common pattern over the sample sets, some interesting differences were found. Analysis of the August sampling of Burntfen

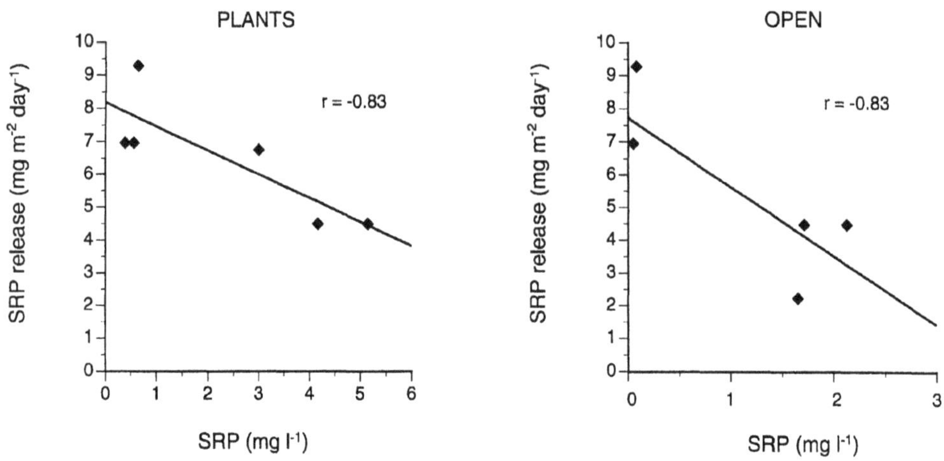

Figure 5. SRP release in Burntfen Broad plotted against final interstitial sediment SRP concentration at 2 cm.

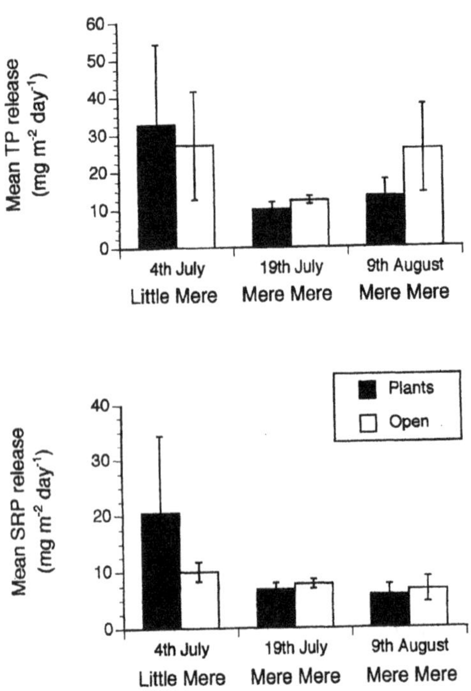

Figure 6. Sediment release of TP and SRP in Little Mere and Mere Mere (the mean and standard deviation are plotted for each data set).

Broad revealed that TP release was significantly lower from cores taken from outside plant beds than from those taken inside, but the corresponding SRP release was not significantly lower. From this we can conclude that the observed difference in TP release comes from a source other than SRP. Possible fractions are particulate P and soluble organic P. The release in the laboratory may have been linked to increased invertebrate activity in the plant cores. Whatever the mechanism the result points to a difference in the nature of the sediment within and outside plant beds on this sampling occasion, but not in the earlier (June) sampling of the broad. This difference may have arisen as a result of greater build up of easily resuspendible sediment in the plant bed (*Nuphar lutea*) compared with the open water during the growing season. This situation was found by Kairesalo & Matilainen (1994), who found a net accumulation of sediment and phosphorus in the inner areas of a bed of *Equisetum fluviatile* L.. Another possible contributing factor is that of temperature, during the three weeks preceeding the August sampling unusually high air temperatures (often exceeding 25 °C) were recorded. This may have accelerated decomposition of detritus and therefore increased the amount of easily resuspendible sediment.

In Hickling Broad there was no significant difference between release of TP or SRP from the cores taken from plant beds (mixed *Myriophyllum spicatum* and *Chara* spp.) compared with those taken from open water. This contrast with Burntfen may have been due to the different plant species involved. However it is more likely to be due to the fact that Hickling Broad being more susceptible to the effects of wind disturbance as it is a far larger water body (141 ha) than Burntfen Broad (4.6 ha). The cores were taken from the centre, and therefore most exposed part, of the lake. Such disturbance may negate any differences there may have otherwise been between release from vegetated and unvegetated areas through counteraction of any build up through sedimentation in calmer periods. This may also explain why in Hickling Broad the release of TP and SRP was significantly lower than that in Burntfen Broad, with the exception of cores taken from outside plant beds in August. Another possibility for the higher release in Burntfen Broad compared with Hickling Broad is the difference between external nutrient loadings which the two broads receive. Burntfen Broad is fed with nutrient rich river water, whereas Hickling Broad receives relatively low external nutrient inputs pumped from the surrounding land.

Interstitial measurements

The fact that SRP release and the post-release interstitial (2 cm depth) concentrations of SRP were negatively correlated in Burntfen Broad cannot be accounted for by a simple diffusion gradient operating within each core. Had this been the case a positive correlation would have been expected. From the interstitial concentrations (Figure 4) it appears that even at the end of the incubation period a diffusion gradient existed between the sediment interstitial water and the overlying water, with concentrations being higher in the interstitial water. One explanation is that some cores had released phosphate during the incubation period whilst others had not begun the process (even though a potential concentration gradient existed between the sediment and overlying water). Had the incubation period been longer (average incubation period was 16 hours) all the cores may have released an equivalent amount of SRP. The gradient between interstitial and overlying water was greater in the cores from plant beds (Figure 4), indicating a greater potential for SRP release. These results highlight the inherent variability between replicates, which is common in sediment release experiments, leading to large standard deviations.

Meres

There was no main effect of plants on TP release in the meres. Date did play a significant role in the release of TP, with release on the first date (in Little Mere) being higher than that on the second (in Mere Mere). No definite conclusion can be drawn as to whether this difference was due to the location or whether it was an effect of date (the second sampling of the meres could not be performed in Little Mere due to the 100% plant coverage by that time, thus the change of location to Mere Mere). It is, however, worth noting that Little Mere suffered from gross organic pollution for a period of approximately 30 years until 1991, whereas Mere Mere was unaffected by this.

The SRP results support the possibility of there being a site difference, with release in Little Mere being higher than either occasion in Mere Mere. Although there was no main effect of plants the plant/date interaction pointed to the higher release of SRP from the plant beds in Little Mere being responsible for the date difference found.

Release was measured in cores without plants present. Therefore any release measured does not necessarily translate to release in the lake situation where the plants may mediate this potential for release. For example, plant roots may take up available sediment phosphorus, they may also provide conditions (through root oxygen release) suitable for the formation of complexes of P with iron(III).

Although this study did not find statistical backing for a main effect of plants on sediment release, when plants were a significant factor release was higher in the presence of plants. The study also provided pointers suggesting not only that plants influence release, but that the manner in which they do so is dependent upon many factors (such as location, and date, and the large variability within sample sets). Taken together these conclusions demonstrate the need for further studies to contain greater replication and fewer variables in order to increase statistical robustness, and thus turn pointers into prizes.

Acknowledgments

This study was jointly funded by NERC and the NRA, to whom we are grateful. Thanks also to Jezz Fox for his invaluable assistance.

References

Andersen, J. M., 1982. Effect of nitrate concentration in lake water on phosphate release from the sediment. Wat. Res. 16: 1119–1126.

Bales, M., B. Moss, G. Phillips, K. Irvine & J. Stansfield, 1993. The changing ecosystem of a shallow, brackish lake, Hickling Broad, Norfolk, U.K. II. Long-term trends in water chemistry and ecology and their implications for restoration of the lake. Freshwat. Biol. 29: 141–165.

Carignan, R. & J. Kalff, 1980. Phosphorus sources for aquatic weeds: water or sediments? Science 207: 987–988.

Carpenter, S. R., 1981. Submerged vegetation: an internal factor in lake ecosystem succession. Am. Nat. 118: 372–383.

Carpenter, S. R., J. J. Elser & K. M. Olson, 1983. Effects of roots of *Myriophyllum verticillatum* L. on sediment redox conditions. Aquat. Bot. 17: 243–249.

Denny, P., 1972. Sites of nutrient absorption in aquatic macrophytes. J. Ecol. 60: 819–829.

Driscoll, C. T., 1993. Supply of phosphorus to the water column of a productive hardwater lake: controlling mechanisms and management considerations. Hydrobiologia 253: 61–72.

Gambrell, R. P. & W. H. Patrick Jr., 1978. Chemical and microbiological properties of anaerobic soils and sediments. In Hook, D. D. and R. M. M. Crawford (eds), Plant Life in Anaerobic Environments. Ann Arbor Science: 375–423.

Holdway, P. A., R. A. Watson & B. Moss, 1978. Aspects of the ecology of *Prymnesium parvum* (Haptophyta) and water chemistry of the Norfolk Broads, England. Freshwat. Biol. 8: 295–311.

Jensen, H. S. & F. O. Andersen, 1992. Importance of temperature, nitrate, and pH for phosphate release from aerobic sediments of four shallow, eutrophic lakes. Limnol. Oceanogr. 37: 577–589.

Jeppesen, E., P. Kristensen, J. P. Jensen, M. Sondergaard, E. Mortensen & T. Lauridsen, 1991. Recovery resilience following a reduction in external phosphorus loading of shallow, eutrophic Danish lakes: duration, regulating factors and methods for overcoming resilience. Mem. Ins. ital. Idrobiol. 48: 127–148.

John, M. K., 1970. Colorimetric determination of phosphorus in soil and plant materials with ascorbic acid. Soil Science 4: 214–220.

Kairesalo, T. & T. Matilainen, 1994. Phosphorus fluctuation in water and deposition into sediment within an emergent macrophyte stand. Hydrobiologia. 275/276: 285–292.

Marsden, M. W., 1989. Lake restoration by reducing external phosphorus loading: the influence of sediment phosphorus release. Freshwat. Biol. 21: 139–162.

Mortimer, C. H., 1941. The exchange of dissolved substances between mud and water in lakes. J. Ecol. 29: 280–329.

Mortimer, C. H., 1942. The exchange of dissolved substances between mud and water in lakes. J. Ecol. 30: 147–201.

Moss, B., 1987. The Broads. Biologist 34: 7–13.

Murphy, J. & J. P. Riley, 1962. A modified single solution method for the determination of phosphate in natural waters. Analyt. chim. Acta 27: 31–36.

Otsuki, A. & R. G. Wetzel, 1972. Coprecipitation of phosphate with carbonates in a marl lake. Limnol. Oceanogr. 17: 763–767.

Phillips, G., R. Jackson, C. Bennett & A. Chilvers, 1994. The importance of sediment phosphorus release in the restoration of very shallow lakes (The Norfolk Broads, England) and implications for biomanipulation. Hydrobiologia. 275/276: 445–456.

Pitt, J.-A., 1994. Phosphorus release and sediment chemistry. In Pitt, J.-A. & G. L. Phillips (eds), The Development of Biomanipulation Techniques and Control of Phosphorus Release from Sediments. National Rivers Authority Report Number 475/2/A: 19–61.

Rorslett, B., D. Berge & S. W. Johansen, 1985. Mass invasion of *Elodea canadensis* in a mesotrophic, South Norwegian lake – impact on water quality. Verh. int. Ver. Limnol. 22: 2920–2926.

Serruya, C., M. Edelstein, U. Pollingher & S. Serruya, 1974. Lake Kinneret sediments: Nutrient composition of the pore water and mud exchanges. Limnol. Oceanogr. 19: 489–508.

Sondergaard, M., P. Kristensen & E. Jeppesen, 1993. Eight years of internal phosphorus loading and changes in the sediment phosphorus profile of Lake Sobygaard, Denmark. Hydrobiologia 253: 345–356.

Stephens, K., 1963. Determination of low phosphate concentrations in lake and marine waters. Limnol. Oceanogr. 8: 361–362.

Wium-Andersen, S. & J. M. Andersen, 1972. The influence of vegetation on the redox profile of the sediment of Grane Langso, a Danish *Lobelia* lake. Limnol. Oceanogr. 17: 948–952.

Hydrobiologia **342/343:** 35–41, 1997.
L. Kufel, A. Prejs & J. I. Rybak (eds), Shallow Lakes '95.

Facilitation of phosphate assimilation by aquatic mycorrhizae of *Vallisneria americana* Michx

Cathleen Wigand[1] & J. Court Stevenson
University of Maryland System, Horn Point Environmental Laboratory, P.O. Box 775, Cambridge, MD 21613, USA
[1] *Present address: Institute of Ecosystem Studies, P.O. Box AB (Route 44A), Millbrook, NY 12545-0129, USA (E-mail: wigand@aol.com)*

Key words: mycorrhizal fungi, vascular plants, nutrients cycling

Abstract

Presence of vesicular-arbuscular mycorrhizal fungi was found to enhance phosphate uptake in the submersed plant *Vallisneria americana* compared with plants treated with a fungicidal medium (i.e., Captan). Incorporation of ^{33}P-orthophosphate into root tissue in short-term incubations was over 85% greater for plants with active mycorrhizae. In addition, we measured a fine-scale iron gradient and elevated concentrations of solid-phase phosphate in the extensive sheath surrounding the roots. The coupling of fungal symbionts with phosphorus storage in the sheath may be an important mechanism of phosphate assimilation in submersed aquatic macrophytes. Contrary to the effect on phosphate uptake, we did not find that $^{15}NH_4$ assimilation by *Vallisneria americana* roots was enhanced by the presence of the mycorrhizal association.

Introduction

Most aquatic vascular plants have specialized tissue (i.e., aerenchyma and lacunae) that channels oxygen into the roots and the surrounding sediments (e.g., Sculthorpe, 1967; Sand-Jensen et al., 1982) because of the high level of reductants in the waterlogged sediments. However, in freshwater ecosystems, oxygen release into the rhizosphere can contribute to phosphate limitation due to the formation of an iron-hydroxide sheath surrounding the roots that adsorbs phosphate, making it less available for direct uptake (Jaynes & Carpenter, 1986). A sheath composed primarily of iron and manganese surrounding the roots is characteristic of most plant species growing in more or less anoxic sediments, and has been also described for the submersed macrophyte, *Vallisneria americana* (St. Cyr et al., 1993). Vesicular-arbuscular mycorrhizal fungi which may facilitate phosphate uptake have also been identified with this macrophyte (Wigand & Stevenson, 1994).

When *V. americana* is a dominant macrophyte (128 ± 48 g dw m^{-2}) in the tidal, upper Chesapeake Bay (Maryland), the sheath surrounding the roots is 0.5 – 1.0 mm in diameter (Wigand & Stevenson, 1994). In Danish lakes sheaths with widths ten times greater (i.e., 0.5 cm) have been reported in association with isoetids (Wium-Andersen & Andersen, 1972). Sheath thickness and composition vary among plant species and sites, no doubt, because some species of submersed macrophytes oxygenate the sediments to a greater degree than others (i.e., Wium-Andersen & Andersen, 1972; Sand-Jensen et al., 1982) and redox potentials and levels of sheath forming minerals (e.g., iron and manganese) vary from site to site (Tessenow & Baynes, 1975; Jaynes & Carpenter, 1986).

Mycorrhizal fungi facilitate nutrient cycling and afford some terrestrial vascular plants a competitive advantage over non mycotrophic plants (Allen, 1991). The occurrence of mycorrhizae is widespread and most terrestrial plants are mycorrhizal (Allen, 1991). Similarly, many aquatic plants are probably able to form mycorrhizae, normally of the vesicular-arbuscular category, and endomycorrhizae of freshwater submersed macrophytes are reported from Denmark (Søndergaard & Laegaard, 1977), Great Britain (Farmer, 1985), India (Chaubal et al., 1982), New Zealand (Clayton & Bagyaraj, 1984) and most recently the United States (Wigand & Stevenson, 1994). Fungal symbionts with submersed macrophytes are speculated to facilitate nutrient cycling (Søndergaard & Laegaard, 1977; Sand-Jensen & Søndergaard, 1979; Tanner & Clayton, 1985), but no studies to our knowledge have experimentally tested this hypothesis. In this study, we hypothesized that vesicular-arbuscular mycorrhizal fungi can increase nutrient assimilation in roots of *Vallisneria americana* by enhancing the uptake of phosphate (PO_4^{-3}) and ammonium (NH_{4+}) from the sediments. We also investigated the chemical composition of the extensive sheath surrounding *V. americana* roots to determine whether the sheath may be an important phosphorus sink compared with phosphorus storage in the vegetated sediments.

Materials and methods

Nutrient uptake experiments

Field material of mycorrhizal *Vallisneria americana* was collected during June and mid-July (1992) in the freshwater, upper Chesapeake Bay, Maryland (39°32′15N, 76°05′15E) at a site described by Wigand & Stevenson (1994). In a concurrent study, we determined *V. americana* to be 80% infected with vesicular-arbuscular mycorrhizae at this site (Wigand & Stevenson, 1994). The grassbeds have low PO_4^{-3} concentrations (< 0.2 μmol l^{-1}) in the water column, and in addition, the molar N:P ratios in the sediment porewater average about 80 which suggests that phosphate is limiting at this site (Wigand, 1994). Other researchers have demonstrated that *V. americana* relies almost exclusively on the sediment for PO_4^{-3} uptake (Carignan & Kalff, 1980), and in this experiment we applied ammonium and phosphate to only the root compartment. The sediment was removed from the roots of the plants by gently shaking the plant in a tub of water, but the iron-oxide sheath was left as intact as possible.

The plants were acclimated in the dark at 25°C for 12 hours in modified Gaudet's medium (Bristow & Whitcombe, 1971), which was diluted 1:100 with deionized water and contained 0.5 mmoles l^{-1} sodium bicarbonate, no phosphorus or nitrogen compounds. Rates of phosphate and ammonium uptake were compared between fungicide-treated and non-fungicide (control) treatments. The fungicide-treated replicates were added Captan (N-trichloromethylthio-4-cyclohexene-I, 2-dicarboxyimide: ICI Americas Inc.) at a dosage rate of 50 mg l^{-1}. Captan, a comparatively mild fungicide, was used because of its low toxicity to higher plants and the ease with which it is hydrolyzed (e.g., Woodcock, 1978; Kough et al., 1987; Beare et al., 1992).

The roots in the bottom compartment (0.5 l) were effectively separated from shoots in the upper compartment (1.0 l) using bicompartmental microcosms, with a leak proof silicone plug (Hill, 1989). The shoot compartment contained only modified Gaudet's medium without phosphorus or nitrogen. Since we were limited by the number of available microcosms to conduct experiments, we ran three sets of experiments on three separate dates.

In order to simulate anoxic sediment conditions, the bottom compartments were deoxygenated by bubbling with nitrogen gas for 20 minutes prior to incubations. PO_4^{-3} and NH_4^+ levels in the bottom compartment were chosen to insure that roots did not become nutrient limited during the 4-hour incubation period, and were based on field measurements (Wigand & Stevenson, 1994) and previous nutrient uptake kinetic studies with *V. americana* (Hill, 1989; Hensel, 1992). Replicates either received 1.4 μmol l^{-1} PO_4^{-3} in bottom compartments (n = 1 control; n = 3 fungicide-treated), 2.8 μmol l^{-1} PO_4^{-3} (n = 1 control), or 3.6 μmol l^{-1} PO_4^{-3} (n = 5 control, n = 5 fungicide-treated). All bottom compartments had a concentration of 2.7 mmole l^{-1} NH_4^+, and each was spiked with 8 mg of ^{15}N ammonium nitrate ($^{15}NH_4NO_3$) (7% ambient NH_4^+) and 10 μcuries of $^{33}PO_4^{-3}$.

Bicompartmental microcosms were incubated at 25°C and received 300 μE m^{-2} s^{-1} light from a combination of cool and warm fluorescent bulbs. After four hours of incubation, shoot and root tissue were dried at 60°C to constant weight. For the $^{33}PO_4^{-3}$ analysis, dried root samples were ground, weighed, treated with sodium and magnesium sulfate, and baked at 500°C for three hours (Chambers & Fourqurean, 1991). Subsequently, root samples were treated with 0.5 N HCl for 30 minutes at 80°C, and then stored until activity was measured. Fluor (15 ml, Optifluor) was added to each vial prior to measuring the radioactive signal with a Packard 2260 Tri-carb liquid scintillation analyzer. Quenching was determined by the external standard ratio method and quenched standards (Long, 1977).

To account for physical adsorption to roots (without incorporation into tissue), root samples ($n = 3$) were placed into $^{33}PO_4^{-3}$ media of the same concentration as experimental chambers and rinsed immediately (within 30 s) to measure average adsorption as activity (i.e., disintegrations per minute, DPM) per gram dry weight of tissue (DPM g DW^{-1} control). This value was used as a correction factor (K) for the activity values measured in experimental samples. The phosphate uptake rate was determined using the following equation (Hill, 1989):

$$PO_4^{-3} \text{ uptake} = \frac{(R_o - K) \times C \times V}{R_a \times T \times \text{g DW}},$$

where:

PO_4^{-3} uptake = the rate at which phosphate is actually absorbed and incorporated into the roots per unit of time (μmol g DW^{-1} h^{-1});

R_o = activity recovered in plant tissue (DPM);

K = physical adsorption activity; will vary depending on mass of experimental sample; (g DW × DPM g DW^{-1} control);

C = concentration of non-radioactive PO_4^{-3} in the medium (μmol l^{-1});

V = volume of medium to which isotope was added;

R_a = activity of the label added (DPM); and

T = time of incubation (hours).

To measure uptake of $^{15}NH_4^+$ by roots and acropetal translocation, root and shoot samples for control and fungicide-treated plants (pooled $n = 5$ each) were ground, weighed, and placed in ampules (after waiting ten ^{33}P half lives to lower sample radioactivity). Copper and cupric oxide were added to the ampules to allow for the conversion of nitrogen species to nitrogen gas (Fiedler & Proksch, 1975). Samples were baked at 500°C for 2 hours before measuring tissue ^{15}N enrichment above background levels with mass spectrometer equipped with a nuclide 3″ 60 Sector analyzer with dual mass collection. Tank nitrogen gas was used as the atmospheric standard. Precision of analyses in the range of less than one to ten atom % enrichment was better than ±0.001 atom % (Glibert et al., 1991). The following equation was used to determine ammonium uptake rates (μmol g DW^{-1} h^{-1}), (modified from: Glibert & Capone, 1993):

$$NH_4 \text{ uptake sample} = [V \times \frac{(\% \text{ N tissue} \times \mu \text{ g DW sample})] \div \text{g DW}}{14\ \mu \text{ g}\ \mu \text{ mol}^{-1}}$$

$$V \text{ (specific uptake rate) } (h^{-1}) = \frac{\text{atom \% excess}}{\text{atom \% enrichment} \times \text{incubation time}}$$

Fungicide toxicity experiments

Toxicity tests with non-mycorrhizal laboratory grown *Vallisneria americana* were conducted to determine if significant differences in uptake of radiolabel PO_4^{-3} existed between control and Captan (50 mg l^{-1}) treated plants ($n = 5$ each). Subterranean turions (overwintering buds) collected from the field in December (1992) were grown in washed sand in 50 l aquaria that received 350 μE m^{-2} s^{-1} on a 14 h light; 10 h dark cycle. Root samples from laboratory grown plants were examined histologically (Phillips & Hayman, 1970) for the presence of arbuscules and vesicles to confirm that there was no fungal infection. The experimental methods were the same as described above for the nutrient uptake experiments. All bottom compartments received 3.6 μmol l^{-1} PO_4^{-3}.

Energy dispersive spectrometry

Transverse root sections with sheaths were examined for iron with scanning electron microscopy (JEOL model JXA840; operated at 28 KeV), interfaced with energy dispersive spectrometry (Chen et al., 1980; Mendelssohn & Postek, 1982; Vale et al., 1990) and a computerized data acquisition system (Analyzer tracor-Northern model 5502) with internal elemental standards. X-ray counts of iron were normalized with gold X-ray counts to account for rough topography of the specimen and the possibility of uneven coating. Root samples with sheaths were preserved in buffered 2.5% glutaraldehyde, subsequently rinsed in sodium cacodylate buffer, and dehydrated in a graded ethanol and ethanol/xylene series. Dehydrated samples were fractured after freezing with liquid nitrogen and sputter coated with gold-palladium (Bacha & Hossner, 1977; Chen et al., 1980). We probed for iron immediately adjacent to the root epidermis and at 125 μm intervals to the furthest extension of the sheath (500 μm), (Figure 1). Area maps were set at a 'window' (Mendelssohn & Postek, 1982) of 0.3 μm^2 along the transect. This allowed for a qualitative comparison of iron along the radially lateral axis of the sheath.

Sediment biogeochemistry

To determine if the sheath surrounding the roots of *Vallisneria americana* was acting as a sink for phosphorus, concentrations of solid phase inorganic phosphorus and iron were measured in the sheath and in the surrounding sediments ($n = 4$ cores). Sheath from

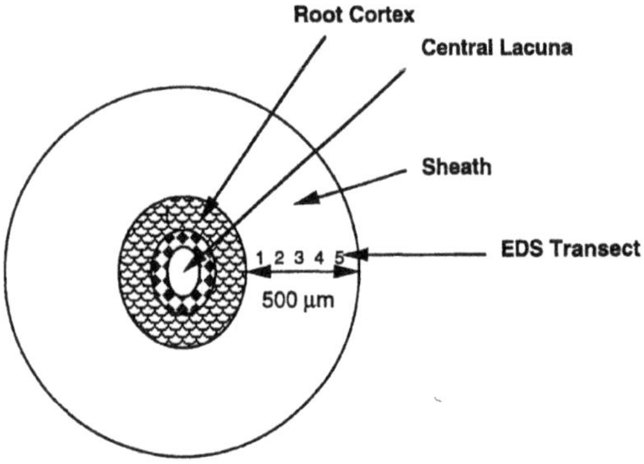

Figure 1. Location of energy dispersive spectrometry (EDS) transect (1–5) in a transverse section of *Vallisneria americana* root with sheath.

Table 1. Phosphate and ammonium root uptake (μmol g DW^{-1} h^{-1} $\pm$ se) by mycorrhizal controls and fungicide-treated *V. americana* groups

Mycorrhizal controls	(pooled n)	Fungicide-treated	(pooled n)	p
Phosphate Uptake				
0.88 $\pm$ 0.10	(6)	0.47 $\pm$ 0.09	(8)	0.008
Ammonium Uptake				
2.8 $\pm$ 0.88	(5)	4.9 $\pm$ 0.62	(5)	0.06

root samples ($n = 4$) was scraped using a scalpel, stored in separate petri dishes, and frozen until analysis. Sediment cores (7 cm diameter) from the grassbed were sliced (depths of 1, 2, 4, 6, 8, 12, 16, 20 cm) within four hours of collection, stored in plastic centrifuge tubes, and frozen. Samples for extraction of solid phase iron and inorganic phosphorus were dried at 65°C and ground with mortar and pestle. Samples were treated with 1 N HCl, set on a shaker table for 24 hours, and subsequently centrifuged (phosphorus: Aspila et al., 1976; iron: Leventhal & Taylor, 1990). Phosphate was measured using colorimetric methods (Parsons et al., 1984), and reactive iron using absorption spectrophotometry (Leventhal & Taylor, 1990).

Statistical analyses

ANOVA (F-test) and covariate models were used to measure statistical differences (significance at $p < 0.05$) between uptake rates ($x \pm$ se) of control and fungicide-treated groups. Using the covariate technique of sequential removal of terms to pool mean squares ($p > 0.25$) (Sokal & Rohlf, 1981, p. 285), we determined no significant interaction of treatment with date of experiment or effect of varying date or PO_4^{-3} concentration. Therefore our final covariate model to examine nutrient uptake between plant groups accounts only for date of experiment and treatment. For some plants, low root biomass prevented us from measuring stable and radioactive isotope uptake which contributed to unequal replication in the models.

Regression analysis and an ANOVA model ($p < 0.001$) were used to determine the relationship between inorganic phosphorus and iron in the vegetated sediments.

Results

We determined that PO_4^{-3} uptake rates of control mycorrhizal roots were nearly twice as great as fungicide-treated roots (Table 1), and the difference was statistically significant ($p < 0.05$). The $^{33}PO_4^{-3}$ signal in the upper compartment was equivalent to background levels suggesting no leakage between compartments or excretion of translocated phosphate by the shoots. In a fungicide toxicity experiment with non-mycorrhizal plants, there was no significant difference ($p > 0.05$) in phosphate uptake between control roots (1.45 $\pm$ 0.30 μmol g DW^{-1} h^{-1}; pooled $n = 5$) and fungicide-treated roots (1.40 $\pm$ 0.25 μmol g DW^{-1} h^{-1}; pooled $n = 5$).

Contrary to our hypothesis, the fungicide-treated group had 75% greater uptake of ammonium compared with the control mycorrhizal group, although the difference was not statistically significant (Table 1). No translocation of ammonium to the shoots was measured during the four-hour incubation in either the control or fungicide-treated groups.

Our observations of the sheath surrounding the root revealed that the relative amount of iron was 3.5 times greater adjacent to the root epidermis, and twice as great in the interior of the sheath, than in the outer wall of the sheath (Table 2). The concentration of solid-phase iron in the sheath (1634 $\pm$ 103 μmol g^{-1}; $n = 4$) was about six times greater than the peak iron concentration in the sediment of the root zone (287 $\pm$ 10 μmol g^{-1} at 8 cm depth; Figure 2). In the grassbed sediment, there was a highly significant ($p < 0.001$; $r = 0.99$) and positive regression ($y = 0.1 \times -1.87$) between

Table 2. Relative iron (determined from X-ray counts) along the transect from the root epidermis to the outer extent of *Vallisneria americana* sheath

Transect	[a]Normalized Iron X-rays	[b]Relative iron
(1) Adjacent to epidermis	3.14	3.5
(2) 125 μm	1.86	2.1
(3) 250 μm	2.08	2.3
(4) 375 μm	1.94	2.2
(5) 500 μm (adjacent to sediment)	0.90	1

[a] X-ray counts of iron were normalized with gold X-ray counts to account for rough topography of the specimen.
[b] Relative iron determined by dividing iron by 0.90, lowest X-ray value.

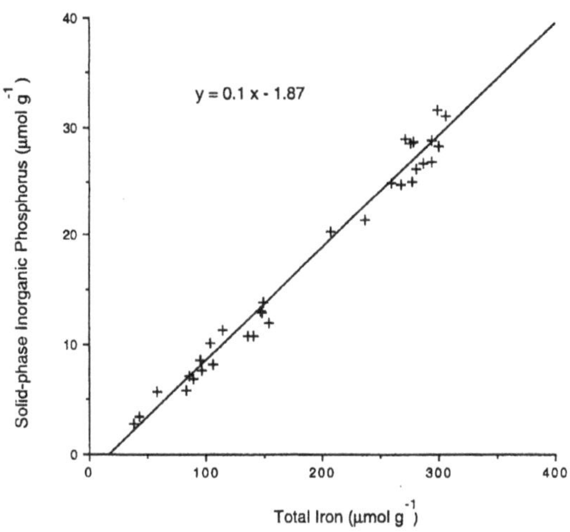

Figure 2. Relationship of solid-phase inorganic phosphorus and iron in the upper Chesapeake Bay grassbeds (+ Points represent sediment samples with depth, 1 – 20 cm in four cores; * Significant regression $p < 0.001$; $r = 0.99$; ** Inverse of the slope is equal to the Fe/P ratio of the sediments).

solid-phase inorganic phosphorus and iron (Figure 2). Although we measured twice as much inorganic phosphorus (61 ± 1 μmol g^{-1}; $n = 4$) in the sheath than the sediments of the root zone (29 ± 0.7 μmol g^{-1}; $n = 3$), the molar Fe/P ratio (i.e., 27) of the sheath was over 2.5 times greater than that ratio (i.e., 10) of the sediments. Thus, there was more solid-phase inorganic phosphorus adsorbed per mole of iron in the vegetated sediment than in the sheath surrounding the roots.

Discussion

Mycorrhizae appear to facilitate phosphate uptake by *Vallisneria americana* in the present study. Prior to this study, we demonstrated that also plant vitality, in terms of leaf elongation and chlorophyll *a* production, was significantly greater in control mycorrhizal plants than fungicide-treated plants during 5-week greenhouse experiments with *V. americana* (Wigand & Stevenson, 1994). In fungicide toxicity studies with aquatic non-mycorrhizal plants, no adverse effect of the fungicide Captan was observed neither on phosphate uptake by roots (present study) nor on plant growth (Wigand & Stevenson, 1994). These results are similar to findings of Captan toxicity with non-mycorrhizal terrestrial plants (e.g., Kough et al., 1987).

An alternative explanation for the difference in PO_4^{-3} uptake between the control mycorrhizal and fungicide-treated groups is bacterial mediation of phosphate solubilization within the sheath. In addition to inhibiting fungal activity, the fungicide could have inhibited bacterial activity in the sheath. Although Captan is reported to influence bacterial nitrogen cycling in some cases (e.g., Banerjee & Banerjee, 1987), the effect of the fungicide on bacterial phosphate cycling is not known or well-studied. However, the high fungal infection of the roots (80% mycorrhizal) suggests that fungal mediation was a major mechanism of facilitated phosphate assimilation in *V. americana*.

Short-term inactivation of nitrifying bacteria and/or stimulation of nitrogen fixers has been described for terrestrial soils due to Captan application (e.g., Anderson, 1978; Banerjee & Banerjee, 1987). These effects on nitrogen cycling might explain why the fungicide-treated group in this study had a greater $^{15}NH_4^+$ uptake rate compared with the mycorrhizal group. For example, bacterially mediated nitrification, which regularly occurs in the microoxic zone surrounding the roots of submersed macrophytes (Iizumi et al., 1980), has been reported as high as 452 nmol N cm^{-3} h^{-1} (Caffrey & Kemp, 1990) in the top 1 cm of vegetated sediments of *Potamogeton perfoliatus* L. Thus ammonium, which is the preferred nitrogen compound for uptake of submersed macrophytes (Iizumi & Hattori, 1982), would be more available to the roots of the fungicide-treated group. Therefore, it is possible that the mycorrhizal fungi also facilitate N-uptake, but this is masked by the Captan effect on nitrifying bacteria or other microorganisms involved in nitrogen cycling, and is thus not possible to demonstrate in the present experiment.

The sheath surrounding the roots (Wigand & Stevenson, 1994) was left intact during this study to promote fungal interaction with nutrient uptake. However, variations in the thickness (approximately 0.05–0.5 mm) and composition (e.g., levels of fungi and bacteria) of the sheath between plants could also explain some of the variation in uptake rates within treatment groups. In addition, facilitation of PO_4^{-3} uptake by mycorrhizae may vary depending upon degree of fungal infection. Along the longitudinal axis of roots, infection can range from 17–100% in field collected *V. americana* (Wigand & Stevenson, 1994).

We speculate that the fine-scale iron gradient in the sheath may act as a phosphate trap (via adsorption by iron-hydroxides) and a potential phosphorus storage for mycorrhizal fungi. One explanation for the higher Fe/P ratio in the sheath external to the roots compared with the surrounding sediment might be that solid-phase inorganic phosphorus is acquired by either fungal symbionts or the roots. In addition, the sheath may be composed of more hydroxyl-containing crystalline iron minerals (e.g., lepidocrocite and goethite) which have lower surface area for adsorption of PO_4^{-3} than the sediments (St.-Cyr et al., 1993). In spite of this Fe/P gradient, the level of phosphorus in the root tissue (190 μmol g^{-1}) was over three times greater than in the sheath and six times greater than the surrounding sediment. Therefore, iron and phosphate gradients in the rhizosphere of *Vallisneria americana* imply translocation and storage of PO_4^{-3} from the sediments, into the sheath, and finally into the roots. Since PO_4^{-3} is low in the overlying water at this site, and *V. americana* has been shown previously to primarily rely on the sediment as its PO_4^{-3} source (Carignan & Kalff, 1980), it was unlikely that PO_4^{-3} was translocated from the leaves to the roots. The N:P ratio in the roots was 6:1 and that of the leaves was 19:1 (100 μmol P g^{-1} leaf tissue).

Facilitation of phosphate uptake is characteristic of mycorrhizae of most terrestrial plants (Malloch et al., 1980; Allen, 1991) and as demonstrated in the present study, this is a function also of aquatic endomycorrhizae of *V. americana* in the phosphate-limited environment at the fresh headwaters of the Chesapeake Bay. Furthermore, the coupling of fungal symbionts with phosphorus storage in the sheath surrounding the roots represents a new mechanism of phosphate assimilation for submersed macrophytes. Similar to terrestrial mycorrhizal associations, other benefits of fungal endosymbionts with submersed macrophytes in aquatic ecosystems may include promotion of the growth of seedlings and resistance to pathogens. These benefits may be important in the reestablishment and persistence of submersed macrophytes in disturbed aquatic systems.

Acknowledgements

This work was partially funded by a Center for Environmental and Estuarine Studies, Horn Point Laboratories doctoral fellowship, and this is contribution #2748 for the University of Maryland Center for Estuarine and Environmental Studies. The manuscript was written at the Institute of Ecosystem Studies in winter 1995 with financial support from the Mary Flagler Cary Charitable Trust. The Strategic Environmental Programme, Denmark supported an oral presentation of this work at the Int. Conf. Shallow Lakes '95 meetings in Mikolajki, Poland. We thank Drs Stuart Findlay, David Strayer, and Charlie Canham for reviewing early drafts of this manuscript, and two anonymous reviewers for helpful comments on the final draft. We also would like to thank Dr Jeff Cornwell and Sue Hill for assistance with methods development. Dr Gene Taylor and Linda Morin with X-ray microanalysis and electron microscopy, Dr Pat Glibert and John Ludlam with mass spectrometry, and Dr Craig Carlson wtih scintillation counting.

References

Allen, M. F., 1991. The ecology of mycorrhizae. Cambridge University Press, Great Britain, 184 pp.

Anderson, J., 1978. Pesticide effects on non target soil microorganisms. In Hill, I. R. & S. J. L. Wright (eds), Pesticide Microbiology (Microbiological Aspects of Pesticide Behavior). Academic Press, New York: 313–533.

Aspila, K. I., A. Haig & A. S. Y. Chau, 1976. A semi-automated method for the determination of inorganic, organic and total phosphate in sediment. Analyst 101: 187–197.

Bacha, R. E. & L. R. Hossner, 1977. Characteristics of coatings formed on rice roots as affected by iron and manganese additions. Soil Sci. Soc. am. J. 41: 931–935.

Banerjee, A. & A. K. Banerjee, 1987. Influence of captan on some microorganisms and microbial processes related to the nitrogen cycle. Plant. Soil 102: 239–245.

Beare, M. H., R. W. Parmelee, P. F. Hendrix, W. Cheng, D. C. Coleman & D. A. Crossley, 1992. Microbial and faunal interactions and effects on litter nitrogen and decomposition in agroecosystems. Ecol. Monogr. 62: 569–591.

Bristow, J. M. & M. Whitcombe, 1971. The role of roots in the nutrition of aquatic vascular plants. Am. J. Bot. 58: 8–13.

Caffrey, J. M. & W. M. Kemp, 1990. Nitrogen cycling in sediments with estuarine populations of *Potamogeton perfoliatus* and *Zostera marina*. Mar. Ecol. Prog. Ser. 66: 147–160.

Carignan, R. & J. Kalff, Jr., 1980. Phosphorus sources for aquatic weeds: water or sediments. Science 207: 987–989.

Chambers, R. M. & J. W. Fourqurean, 1991. Alternative criteria for assessing nutrient limitation of a wetland macrophyte (*Peltandra virginica* (L. Kunth), Aquat. Bot. 40: 305–320.

Chaubal, R., G. D. Sharma & R. R. Mishra, 1982. Vesicular – arbuscular mycorrhiza in subtropical aquatic and marshy plant communities. Proc. Indian Acad. Sci. (Plant Science) 91: 69–77.

Chen, C. C., J. B. Dixon & F. T. Turner, 1980. Iron coatings on rice roots: morphology and models of development. Soil. Sci. Soc. am. J. 44: 1113–1119.

Clayton, J. S. & D. J. Bagyaraj, 1984. Vesicular-arbuscular mycorrhizas in submerged aquatic plants of New Zealand. Aquat. Bot. 19: 251–262.

Farmer, A. M., 1985. The occurrence of vesicular-arbuscular mycorrhiza in isoetid-type submerged aquatic macrophytes under naturally varying conditions. Aquat. Bot. 21: 245–249.

Fiedler, R. & G. Proksch, 1975. The determination of nitrogen – 15 by emission and mass spectrometry in biochemical analysis: a review. Analyt. Chim. Acta 78: 1–62.

Glibert, P. M. & D. G. Capone, 1993. Mineralization and assimilation in aquatic, sediment and wetland systems. In Knowles, R. & P. H. Blackburn (eds), Nitrogen Isotope Techniques, 9; Academic Press, Inc., Sand Diego: 243–272.

Glibert, P. M., C. Garside, J. A. Fuhrman & M. R. Roman, 1991. Time-dependent coupling of inorganic and organic nitrogen and NH_{4+} regeneration in the plume of the Chesapeake Bay estuary, USA, and its regulation by large heterotrophs. Limnol. Oceanogr. 36: 895–909.

Hensel, P. F., 1992. Nitrogen uptake characteristics of wild celery. M.S. Thesis, The University of Maryland, College Park, MD, 181 pp.

Hill, S. M., 1989. Phosphorus uptake and translocation in submersed aquatic *Vallisneria americana*. M.S. Thesis, The University of Maryland, College Park, MD, 55 pp.

Iizumi, H. & A. Hattori, 1982. Growth and organic production of eelgrass (*Zostera marina* L.) in temperate waters of the Pacific coast of japan. III The kinetics of nitrogen uptake. Aquat. Bot. 12: 245–256.

Iizumi, H., A. Hattori & C. P. McRoy, 1980. Nitrate and nitrite in interstitial waters of eelgrass beds in relation to the rhizosphere. J. exp. mar. Biol. Ecol. 47: 191–201.

Jaynes, M. L. & S. R. Carpenter, 1986. Effects of vascular and nonvascular macrophytes on sediment redox and solute dynamics. Ecology 67: 875–882.

Kough, J. L., V. Gianinazzi-Pearson & S. Gianinazzi, 1987. Depressed metabolic activity of vesicular-arbuscular mycorrhizal fungi after fungicide applications. New Phytologist 106: 707–715.

Leventhal, J. & C. Taylor, 1990. Comparison of methods to determine degree of pyritization. Geochim. Cosmochim. Acta 54: 2621–2625.

Long, E. C., 1977. Applications of quench monitoring by Compton edge: the #". Publication of Scientific Instruments Division, P. O. Box C – 19600, Irvine, California 92713, 17 pp.

Malloch, D. W., K. A. Pirozynski & P. H. Raven, 1980. Ecological and evolutionary significance for mycorrhizal symbioses in vascular plants (A review). Proc. natn. Acad. Sci. U.S.A. 77: 2113–2118.

Mendelssohn, I. A. & M. T. Postek, 1982. Elemental analysis of deposits on the roots of *Spartina alterniflora* Loisel. Am. J. Bot. 69: 904–912.

Parsons, T. R., Y. Maita & C. M. Lalli, 1984. A Manual of Chemical and Biological Methods for Seawater Analysis. Pergamon Press Inc., Oxford, 173 pp.

Phillips, J. M. & S. Hayman, 1970. Improved procedures for clearing roots and staining parasitic and vesicular-arbuscular Mycorrhizal fungi for rapid assessment of infection. Trans. br. Mycol. Soc. 55: 158–167.

Sand-Jensen, K., C. Prahl & H. Stokohlm, 1982. Oxygen release from roots of submerged aquatic macrophytes. Oikos 38: 349–354.

Sand-Jensen, K. & M. Søndergaard, 1979. Distribution and quantitative development of aquatic macrophytes in relation to sediment characteristics in oligotrophic Lake Kalgaard, Denmark. Freshwat. Biol. 9: 1–11.

Sculthorpe, C. D., 1967. The Biology of Aquatic Vascular Plants. Arnold, London, 610 pp.

Sokal, R. R. & F. J. Rohlf, 1981. Biometry, The Principles and Practice of Statistics in Biological Research; 2nd edn. W. H. Freeman and Co., San Francisco, 859 pp.

Søndergaard, M. & S. Laegaard, 1977. Vesicular-arbuscular mycorrhiza in some aquatic vascular plants. Nature 268: 232–233.

St.-Cyr, L., D. Fortin & P. G. Campbell, 1993. Microscopic observations of the iron plaque of submerged aquatic plant (*Vallisneria americana* Michx). Aquat. Bot. 46: 155–167.

Tanner, C. C. & J. S. Clayton, 1985. Vesicular arbuscular mycorrhiza studies with a submerged aquatic plant. Trans. br. Mycol. Soc. 85: 683–688.

Tessenow, U. & Y. Baynes, 1975. Redox-dependent accumulation of Fe and Mn in a littoral sediment supporting *Isoetes lacustris*. Naturwissenschaften 62: 342–343.

Vale, C., F. M. Catarino, C. Cortesao & M. I. Cacador, 1990. Presence of metal-rich rhizoconcretions on the roots of *Spartina maritima* from the salt marshes of the Tagus estuary, Portugal. Sci. tot. envir. 97/98: 617–626.

Wigand, C., 1994. The ecological significance of mycorrhizal fungi in the submersed macrophyte *Vallisneria americana* (Michx) in the upper Chesapeake Bay, Ph.D. dissertation, The University of Maryland, College Park, Maryland, 135 pp.

Wigand, C. & J. C. Stevenson, 1994. The presence and possible ecological significance of mycorrhizae of the submersed macrophyte, *Vallisneria americana*. Estuaries 17: 206–215.

Wium-Andersen, S. & J. M. Andersen, 1972. The influence of vegetation on the redox profile of the sediment of Grane Langso, a Danish *Lobelia* lake. Limnol. Oceanogr. 17: 948–952.

Woodcock, D., 1978. Microbial degradation of fungicides, fumigants and nematocides. In Hill, I. R. & S. J. L. Wright (eds), Pesticide Microbiology. Academic Press, New York: 731–799.

Hydrobiologia **342/343:** 43–53, 1997.
L. Kufel, A. Prejs & J. I. Rybak (eds), Shallow Lakes '95.

Phosphorus balance of a polytrophic shallow lake with the consideration of phosphorus release

Kerstin Ramm[1] & Volker Scheps[2]
[1] *Institute of Applied Freshwater Ecology, Mitschurinstr. 5, D-14469 Potsdam, Germany*
[2] *Geological Survey of Brandenburg, Stahnsdorfer Damm 77, D-14532 Kleinmachnow, Germany*

Key words: seasonal variations, mass balance, internal loading, phosphorus binding forms

Abstract

In the course of a one-year investigation (April 1994 to March 1995) the phosphorus balance of a polytrophic shallow lake with consideration of phosphorus release (P-release) was studied. The investigated Lake Blankensee covers an area of 2.9 km^2 with an average depth of 1.2 m (volume 3.43 10^6 m^3). The average summer chlorophyll-*a* concentrations range from 104 to 418 $\mu g\ l^{-1}$ (1992–1995). Secchi-depths from 0.2 to 0.3 m are typical in summer. Massive algae blooms occur frequently and submerse macrophytes can no longer exist. The trophic state is polytrophic.

In summer phosphorus concentrations in the lake were controlled by the internal processes, in winter by the inflowing River Nieplitz. The increasing total phosphorus concentrations and the soluble reactive fraction of phosphorus in the lake water from winter to summer are partly a function of net internal loading.

According to the detailed chemical studies which are representative only for this investigated period the Lake Blankensee acts as a phosphorus source, which also affects the down-stream lakes and the river system. The P-balance is determined by the river input of 16.84 10^3 kg P and by an output of 20.71 10^3 kg P. Therefore the export of about 3.87 10^3 kg P characterises the negative P-retention rate due to the strong internal loading during summer.

More than 60% of the sediment phosphorus is bound to the iron-redox system. Therefore it indicates a high potential mobility of phosphorus under anaerobic conditions in the sediment surface zone in summer during special weather conditions. This process is limited due to the high Fe:P ratio of about 21 (by weight) in the surface sediments of the Lake Blankensee (0–5 cm).

Study area

The Lake Blankensee is a shallow eutrophic and polymictic lake flowed through by the River Nieplitz. It is situated about 40 km south of Berlin (Figure 1). From the geological point of view the origin of the lake is related to a subrosion zone as a consequence of halokinetic movements.

Introduction

Due to the shallowness polymictic lakes are characterised by a high variability of physical and chemical parameters (Friedrich, 1982). Polymixis and aeration of the whole water body are the mean processes in the lake system and influence the nutrient cycling and the primary production. In the summer the phosphorus release (P-release) constituts often a major fraction of the total phosphorus load (Ryding, 1985). It mainly occurs from an aerobic sediment surface because of the polymictic nature of the shallow lakes (Jensen et al., 1992).

The variability of soluble reactive phosphorus (SRP) concentrations in the pore water is determined by several conditions and processes (Boström et al., 1982). The P-release from bottom sediments is closely related to the composition of the sedimentary organic

Figure 1. Geographic position of the lake Blankensee.

matter (Hupfer, 1993). Oxygen in the water overlying the sediments controls the redox-sensitive iron-bound phosphorus fraction. This aerobic zone is comparable with the function of the metalimnion in stratified lakes (Oláh, 1975). The mobility of phosphorus in aerobic sediments can be regulated if the Fe:P ratio is above 15 by weight (Jensen et al., 1992). Resuspension of the sediment surface contributes to the internal loading in shallow lakes too. The knowledge of the phosphorus budget, the role of the inflow and the sediments is needed to support decisions on strategies for load reduction.

Besides common limnological questions it was the special object of this study to determine and to evaluate the seasonal variability of the phosphorus retention in the lake and the relation to the process of P-release from sediments. This study reported here bases on the observations from April 1994 to March 1995 of the Lake Blankensee and its inflow River Nieplitz which was chosen as an example for the shallow lakes of the Northgerman Lowlands which are frequently polluted by anthropogenic influence in the catchment area.

It covers an area of 2.9 km^2, an average depth of 1.2 m and a volume of 3.43 10^6 m^3 (Figure 2). The location and the shape of the lake is favourable for mixing thoroughly the water body without seasonal stratification. Nearly the whole lake is surrounded by a reed-belt of several tenths to 100 m width. Submerse macrophytes are lacking because of domination of plankton. The average water retention time is about 10 days with a mean discharge of 4 m^3 s^{-1}. Up to the Lake Blankensee the River Nieplitz has a catchment area of about 738 km^2 of glacial and alluvial sediments with sandy, permeable and marshy soils. The direct lake catchment area (16 km^2) is composed of arable lands (28%), meadows (24%), forests (35%), swamps (5%), and settlements (8%).

Due to duck farming during the 1960's (Kalbe, 1973), an intensive fishery and a high nutrient loading from the catchment area the trophic state of the lake today is polytrophic. Massive algae blooms frequently occur in summer (blue green algae, Arp, 1995, pers. com.). In the past as a result of eutrophication several problems occured e.g. the dying of fishes near the outflow (Kalbe, 1973).

As measures to reduce the nutrient loading and the organic matter the duck farming was stopped at the end of the 1960's. Furthermore two holes were excavated in 1989, each of about four meters depth (Figure 2), in order to build sedimentary traps close to the aestuaric structure where the River Nieplitz enters the lake.

Materials and methods

Four different sampling surveys were carried out in order to determine chemical and physical parameters, the variability of phosphorus concentrations in the lake, in the pore water, and in the River Nieplitz. Water samples were taken from the river about 100 m upstream of the inflow and near its outflow biweekly from April 1994 to March 1995. The lake water samples derive from the centre of the lake from a depth of 0.5 m. The sediment's pore water was taken monthly by sampling sediment cores. It was gained by a dialyses method according to Enell & Löfgren (1988). More information regarding these sampling procedures are given by Vietinghoff (1994). Chemical analyses of soluble reactive phosphorus (SRP) and total phosphorus (TP) of the lake water, the river water, and the pore water were carried out as stated in 'Ausgewählte Methoden der Wasseruntersuchung, 1986'. Water samples were filtered through 640 nm 1/4 folded filters (Macherey-Nagel) for analyses of SRP. It was measured spectrophotometrically by the molybdenum blue method (Murphy & Riley, 1962). For determination of TP water samples were digested in quartz tubes in an incubator after addition of potassium peroxodisulfate ($K_2S_2O_8$) at 134 °C for half an hour. The phosphorus was measured according to the SRP-method.

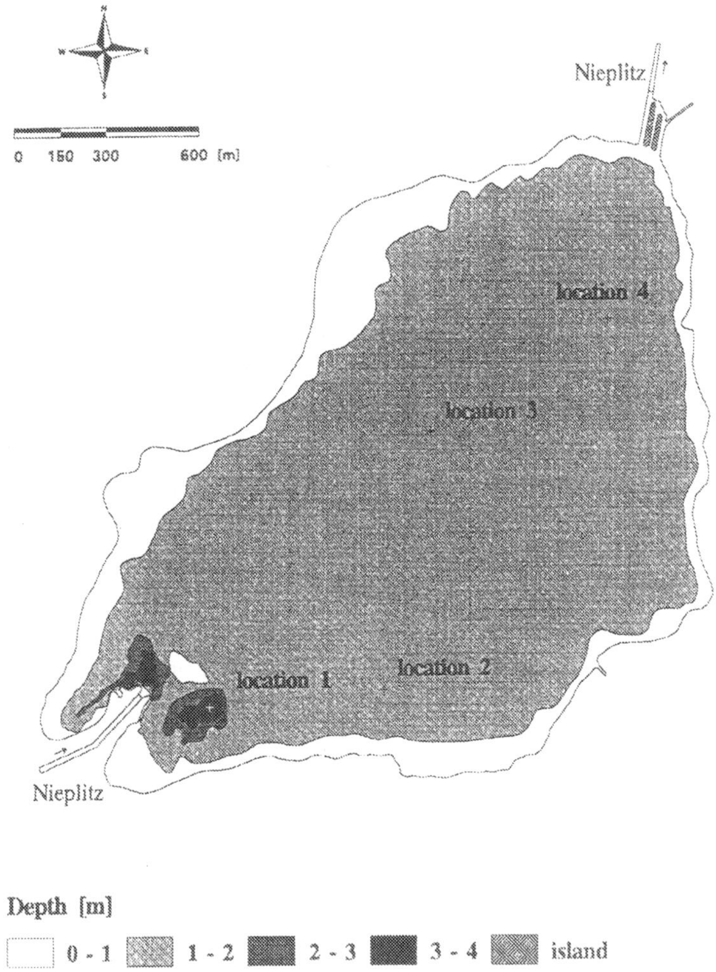

Figure 2. Bathymetric chart of the lake Blankensee.

Within the lake four mud profiles of about 20 cm depth were sampled with a sediment corer and analysed in detail in horizons of 2 cm each (Figure 2). The sediment samples were extracted by aqua regia and total phosphorus was determined by ICP/AES besides other elements. The available, iron-bound, aluminium-bound, organic-bound and calcium-bound phosphorus were analysed according to the algorithm given by Psenner et al. (1984, 1988) and Hupfer (1995). The trophic state was classified to the scheme of TGL 27885/01 (1982).

The general mass balance equation for a substance is (Kozerski et al., 1993):

$$\frac{\mathrm{d}C}{\mathrm{d}t} = \frac{1}{V} \cdot (Q_{\mathrm{in}} \cdot C_{\mathrm{in}} - Q_{\mathrm{out}} \cdot C_{\mathrm{out}}) + E + A \quad (1)$$

C = concentration in the lake;
V = volume of the lake;
Q = discharge, input and output;
C = concentration in the inflowing water, input and output;
A = exchange flux with the atmosphere;
E = exchange flux with the bottom.

In the case of the Lake Blankensee the factor A, E and the volume are negligible. Therefore the mass balance was formulated as:

$$R = (Q_{in} \cdot C_{in}) - (Q_{out} \cdot C_{out}) \qquad (2)$$

R = retention of phosphorus

The diffusion rates were calculated with the SRP-concentrations of pore water by the help of the First Fick's Law of diffusion (equation 3) due to the linear gradients of SRP-concentrations in the pore water:

$$JPO_4^{3-} = \phi D_s \cdot dCPO_4^{3-}/dX \qquad (3)$$

with $D_s = \phi^2 \cdot D$

JPO_4^{3-} = diffusion flux ($\mu g\ cm^{-2}\ s^{-1}$);
D = molecular diffusion coefficient: (for $H_2PO_4^-$: 7.15 $10^{-6}\ cm^2\ s^{-1}$, temperature: 18 °C; 8.46 $10^{-6}\ cm^2\ s^{-1}$, temperature: 25 °C; Li & Gregory, 1974);
ϕ = porosity (-);
$dCPO_4^{3-}/dX$ = linear concentration gradient ($\mu g\ cm^{-3}\ cm^{-1}$).

Results

Seasonal variability of phosphorus in the Lake Blankensee and the River Nieplitz

During the time of mean summer chlorophyll-a concentrations of 418 $\mu g\ l^{-1}$ (1994) the Secchi-depth varies from 0.2 m to 0.3 m only (Figure 3). The chlorophyll-*a* concentrations show four maxima - in April (1994) followed by a short clear-water stadium, in February (1995), the highest in the beginning of August (1994) and a smaller, autumn maximum in November (1994).

The average concentrations of SRP (0,037 mg l^{-1}) and TP (0.15 mg l^{-1}) in the lake are high (Figure 4). According to the classification scheme (TGL 27885/01, 1982) it characterises the eutrophic state of this lake.

The TP and SRP-concentrations in the Lake Blankensee during the 12-month period (1994–1995) are illustrated in Figure 4, show high seasonal variations. There is a characteristic minimum of SRP in July (1994) and February (1995) due to algae blooms.

But the level of the TP and SRP-concentrations in the lake are high. TP in the lake decreases not below the critical level of 0,04 mg TP l^{-1} for a significant phosphorus limitation (Klein, 1989). It seems that the most important limitation factor for the primary production is not the amount of nutrients but the influence of self-shadowing by explosive algae reproduction. Nitrogen-limitation could not be found. The N:P ratio (Figure 4) is always above 5 (Wetzel, 1983). It decreases to five in July and August only due to maximum P-release from sediments. To point out the influence of the River Nieplitz on the phosphorus concentrations in the lake water the pattern of the 12-month period is calculated (Figure 5). A feature of the lake and the river is the decreasing SRP-concentrations in spring due to algae blooms. Whereas SRP increases in the lake after the spring minimum strongly it increases slowly in the River Nieplitz. These differences between lake and river are caused by the internal processes. The TP-concentrations in the lake were similar to that of SRP. The seasonal variability of TP in the river is smaller than those of the lake (Figure 5).

Phosphorus balance

The phosphorus input by inflow is higher than those of the phosphorus output in spring (1994) and in winter (1994/1995). Therefore the lake becomes a P-sink in this investigated period (Figure 6, above). The situation changes in the summer due to strong P-release from sediments (Figure 6, below). The lake acts as a phosphorus source. The sum of the external TP-input is about 16.84 10^3 kg by the inflow (River Nieplitz) during the investigated 12-month period, being a load of 5.8 $g\ P\ m^{-2}\ a^{-1}$. The export from lake occurs 20.71 10^3 kg phosphorus. P-release and resuspension are quantified by the external phosphorus balance (equation 2). The lake loses in this one year about 3.87 10^3 kg phosphorus which obviously derives from the lake sediments by release and resuspension.

Phosphorus variability in pore water and P-release

To quantify the process of internal loading by diffusion the amount of SRP in the pore water and its seasonal variability during the whole vegetation period was observed. Similar to the SRP-concentrations in the lake water the maximum of SRP in the pore water of 6.6 mg l^{-1} was found in July at a depth of 8 cm. The SRP-maximum moves during the vegetation period from a depth of 10 cm in April to a depth of 2 cm in August (Figure 7).

It marks the horizons of high microbiological activity and chemical transformation. Therefore the internal loading by diffusion (calculated by means of equation

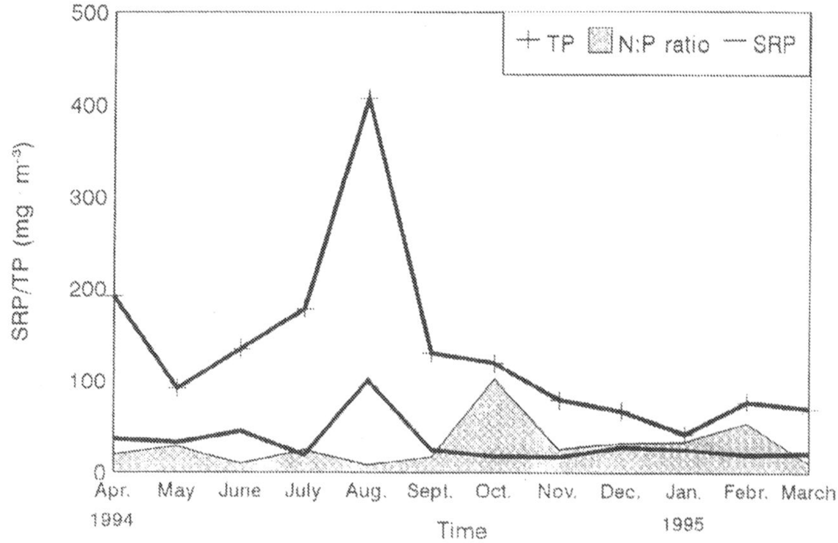

Figure 3. Secchi-Depth and chlorophyll-*a*.

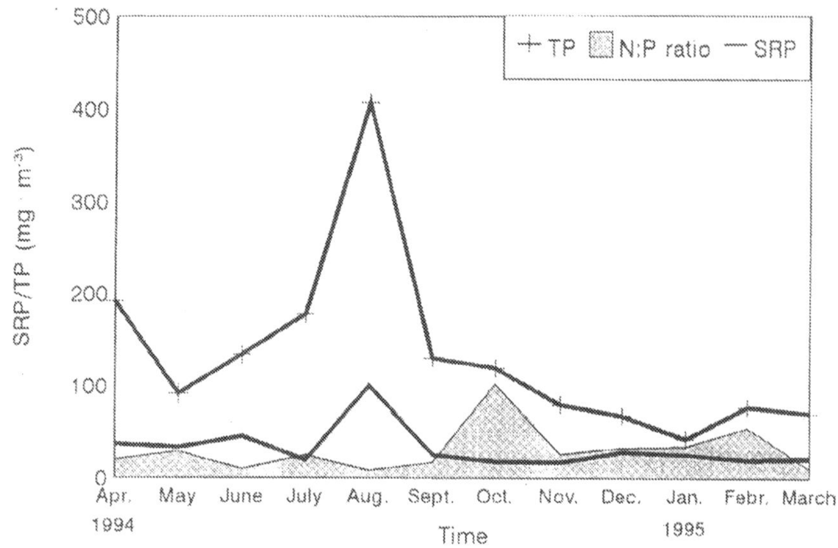

Figure 4. Variability of SRP and TP (lines) and the N:P ratio in the lake Blankensee (April 1994 to March 1995).

3) gets its maximum from the end of June to the beginning of September. In Figure 8 the external loading as the most important P-source for the lake is illustrated according to the theoretical P-release by diffusion.

The recent TP-concentrations of the sediment (Figure 9) reflect a relatively low TP-level in comparison to other data from literature (e.g. Hupfer, 1993). The highest TP-concentrations of the sediment cores were found in core no. 1 (location 1). It was sampled in a four meters deep excavation which obviously still works as a sedimentary and geochemical trap (Figure 2). Core no. 2 (location 2) was derived from a marginal relative low energy zone. Cores no. 3 and 4 (location 3 and 4) were taken within a mainstream zone. This classifica-

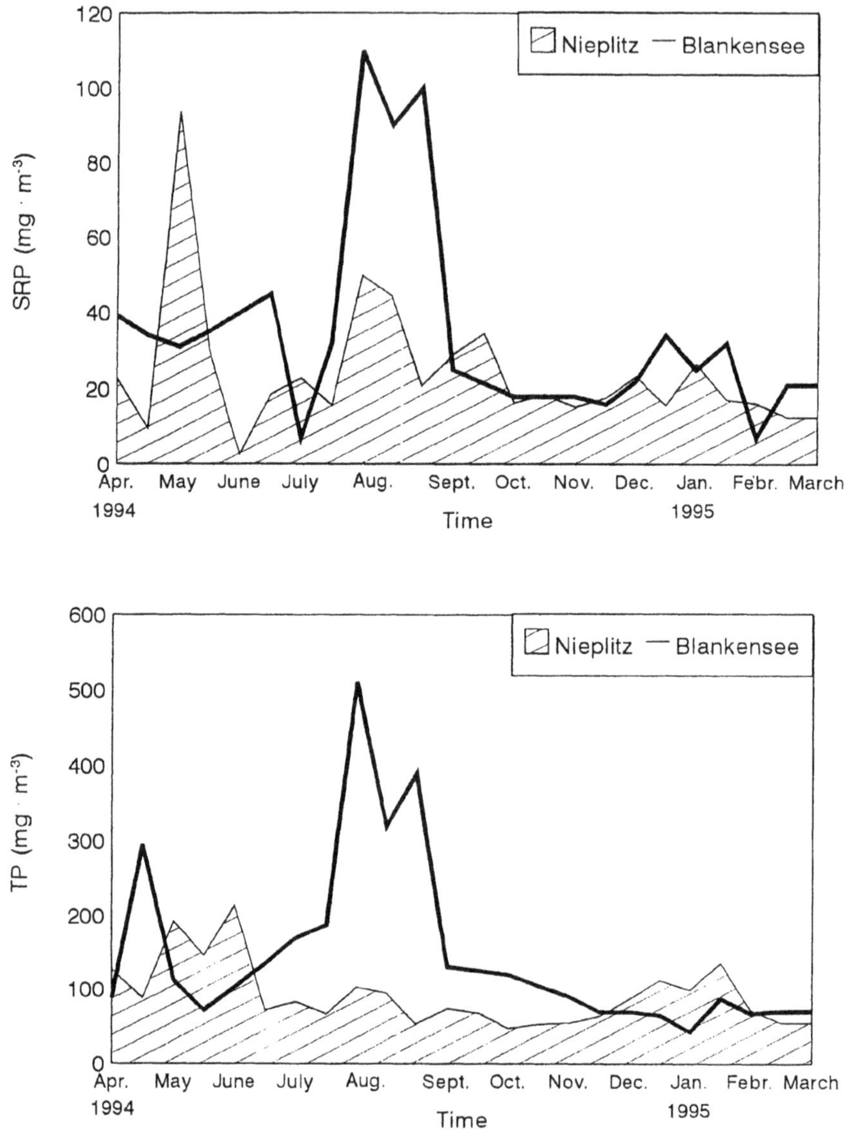

Figure 5. Seasonal pattern of SRP (above) and TP (below) in the lake and the river.

tion is supported by the results of geochemical studies carried out on the same samples by ICP/AES analyses as mentioned before.

The cores no. 2, 3, and 4 show increased phosphorus concentrations within the upper 6 to 8 cm only. According to the other authors (e.g. Hupfer, 1993) it is not possible to use the amount of the total phosphorus in the lake sediments in order to estimate the release potential from the bottom. For example, the TP-concentrations of the four sediment cores are similar to those of the 69 m deep and oligotrophic Lake Stechlin (Gonsiorczyk et al., 1995).

Therefore the phosphorus binding forms were studied additionally (Figure 10). The highest concentrations are related to the iron-bound fraction (BD-P) with 1.4 g P kg^{-1} DW in the first centimeter of the sediment.

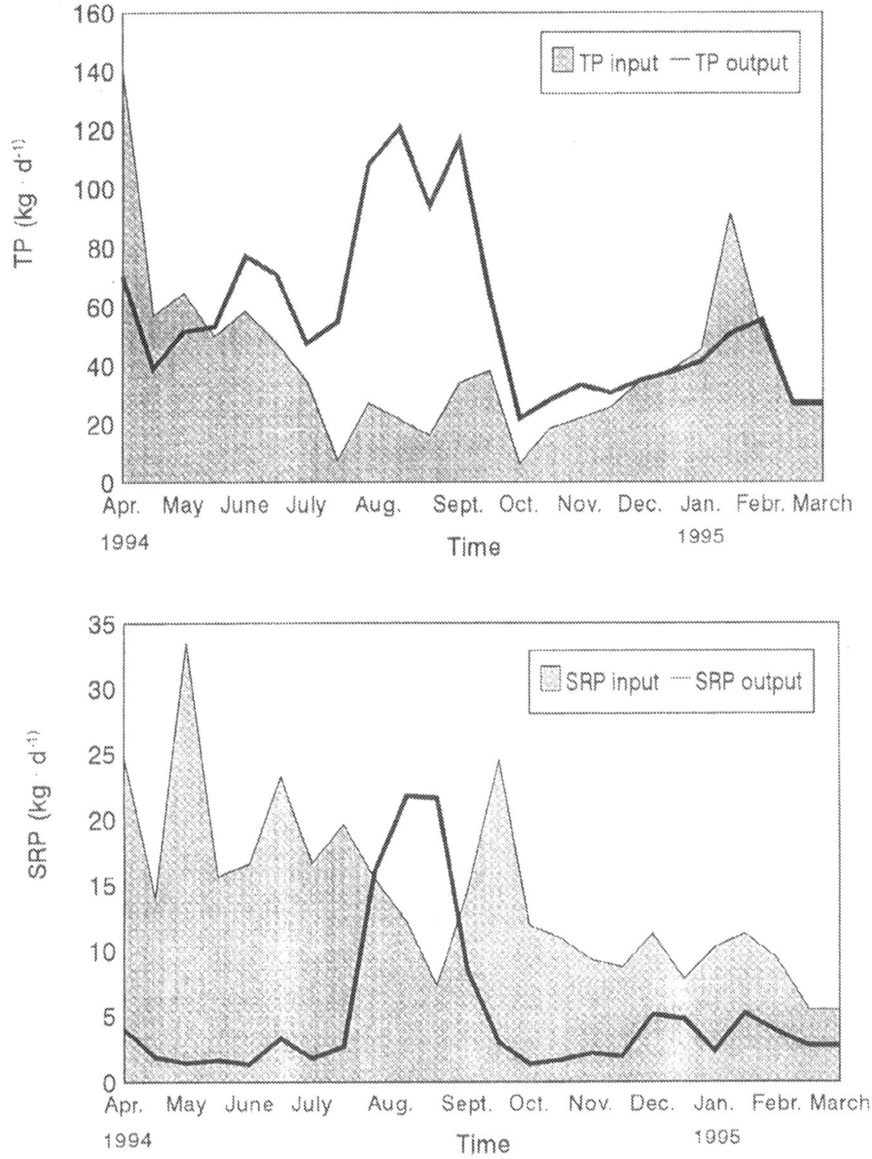

Figure 6. External loading and output of phosphorus fractions (TP and SRP) and its seasonal variability.

This is about 60% of the whole sediment phosphorus. The calcium-bound phosphorus (HCl-P) in the surface sediment is about 19% only according to the high amount of carbonate (mean 46%) and calcium (mean 229.3 g Ca kg^{-1} DW). 14% (total) of the sediment (0–1 cm) consists of the organic-bound (NaOH-NRP), the aluminium-bound, and the hydroxides-bound (NaOH-SRP) fractions. The ammonia chloride extractable phosphorus (NH_4Cl-SRP) is low similar to the residue phosphorus (Resid.-P), each of about 3%.

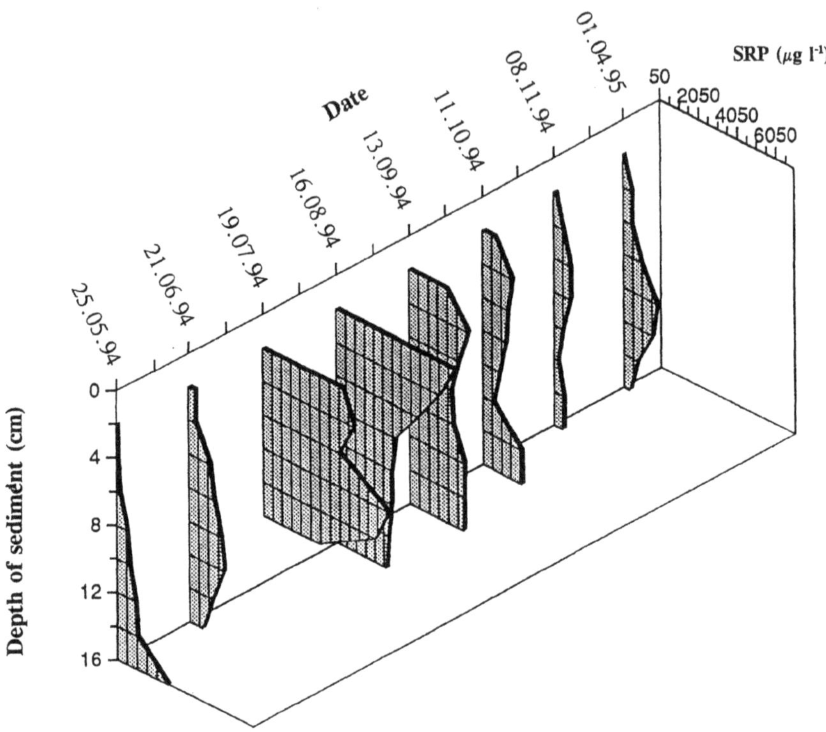

Figure 7. SRP-concentrations in the pore water (Lake Blankensee).

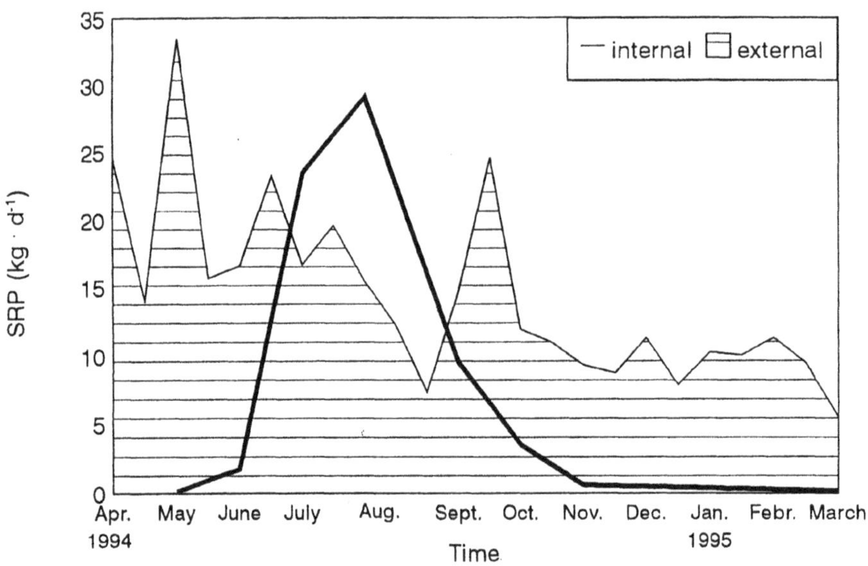

Figure 8. External loading of SRP and calculated phosphorus release from sediments.

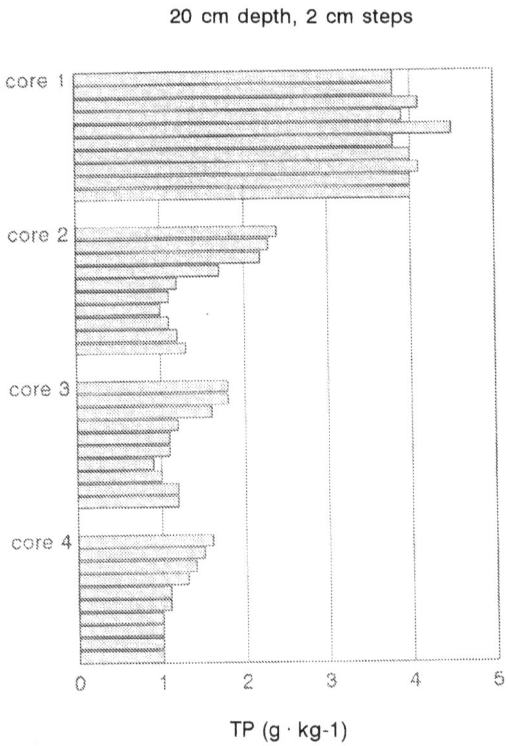

Figure 9. Total phosphorus concentrations of four sediment cores.

Discussion

The investigation of the phosphorus dynamics in the Lake Blankensee lasted one year only. It shows the high seasonal variability of phosphorus concentrations in the lake caused by nutrient loading of the River Nieplitz and by different ecological processes. The most significant characteristic of the phosphorus dynamic is the release of SRP from sediments in summer if the SRP-concentrations in the lake rised up to more than 0.11 mg P l^{-1}. It affects the down-stream lakes Grössinsee and Schiasser See because both of them are polytrophic. A load of 5.8 g P m^{-2} a^{-1}, which is representative for this one-year investigation only, characterises the lake as strongly eutrophicated.

In winter the dynamics of SRP and TP are controlled by the concentration levels of the inflowing river due to the short water retention time of 10 days (mean). When the plankton increases in spring, the influence diminishes. The whole summer-period the phosphorus variations in the lake are characterised by the internal processes. Especially in spring and in winter the lake becomes a phosphorus sink. The situation changes in the summer season with an excessive export of phosphorus. Its annual maximum was observed in August due to the maximum of the internal loading (29.2 kg P d^{-1}) and the process of resuspension which was also described as the phenomen of floating mud for lower saxonien shallow lakes (Poltz, 1980). It seems that the short water retention time of 10 days favours a low average accumulation rate due to the negative phosphorus retention rate.

The process of P-release from sediments is related to the iron-redox system. About 60% of the sediment phosphorus is iron-bound. This source can be remobilised provided that oxygen or nitrate is not available for the microbiological metabolism yet and the Fe^{3+} is reduced to Fe^{2+} subsequently. The physical conditions in the Lake Blankensee during the summer time are characterised by high water temperatures, high primary productivity, and a good aeration of the water body. This situation is normally responsible for the formation of an aerobic enviroment in the upper most millimeters of the sediment surface in shallow lakes. As a consequence of special weather conditions in summer the formation of anoxic microzones in the uppermost millimeters of the sediments is possible. In the case of the Lake Blankensee it was measured and described by Kalbe (1973). Furthermore the development of high SRP-concentration gradients between the overlying water and the sediment is an important presupposition for the P-release under aerobic conditions in the lake.

The high amount of the iron-bound fraction seems to be a sign for a good capacity of the sediments to bind phosphorus permanently due to the high Fe:P ratio in the surface sediments (0-5 cm) of about 21 (by weight). Therefore it is possible to control the P-release by keeping the surface sediment oxidised (Jensen et al., 1992).

Conclusions

The Lake Blankensee is an extremely shallow polytrophic lake which is flowed through by the River Nieplitz. In the observed time (1994–1995) it worked as a phosphorus source for the down-stream river system and the lakes (Grössinsee, Schiasser See), which are polytrophic. The phosphorus balance of the lake is partly determined by the river input, in summer by the P-release and the resuspension, respectively.

The TP-concentrations of the sediment is low in comparison to the high trophic level of the lake. The

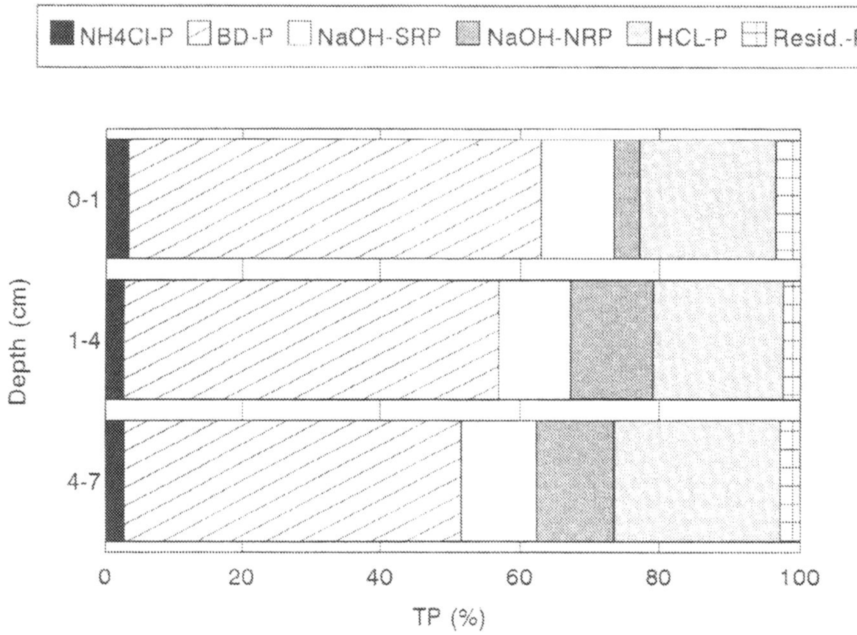

Figure 10. Share of phosphorus binding forms in sediments (repres. for cores no 2, 3).

potential mobility is obviously related to the iron-redox system. Due to the surplus of iron according to the phosphorus (Fe:P ratio of 21, by weight) in the upper most layer (0.5 cm) this process is controlable by keeping the surface sediment oxidised (Jensen et al., 1992). The most important measure to restorate this lake is the phosphorus-load reduction from the catchment area.

Acknowledgements

We are indebted to the colleagues and the technical staff of our institutes for their assistance during field work and analyses. Furthermore we thank Prof. Dr R. Koschel and Dr P. Casper from the Institute of Freshwater and Fish Ecology (Neuglobsow) for their kind help.

References

Ausgewählte Methoden der Wasseruntersuchung, 1986. Vol. I: Chemische, physikalische, elektrochemische Methoden. VEB Gustav Fischer Verlag Jena.

Boström, B., M. Jansson & C. Forsberg, 1982. Phosphorus release from lake sediments. Arch. Hydrobiol. Beih. Ergebn. Limn. 18: 5–59.

Enell, M. & S. Löfgren, 1988. Phosphorus in interstitial water: methods and dynamics. Hydrobiologia 170: 103–132.

Friedrich, G., 1982. Limnologie flacher Seen und Talsperren. Z. Wasser Abwasser Forsch. 15: 81–88.

Gonsiorczyk, T., P. Casper & R. Koschel, 1995. Vergleichende Sedimentuntersuchungen zum Kohlenstoff- und Phosphorgehalt in baltischen Seen. Limnologica 25: 353–365.

Hupfer, M., 1993. Untersuchungen zur Phosphatmobilität in Gewässersedimenten. Diss. A, TU Dresden, 144 pp.

Hupfer, M., 1995. Bindungsformen und Mobilität des Phosphors in Gewässersedimenten. In Bernhardt, Ch. H. & H. Klapper (eds), Handbuch Angewandte Limnologie. Ecomed Landsberg.

Jensen, H. S., P. Kristensen, E. Jeppesen & A. Skytthe, 1992. Iron:phosphorus ratio in surface sediment as an indicator of phosphorus release from aerobic sediments in shallow lakes. Hydrobiologia 235/236: 731–743.

Kalbe, L., 1973. Über Sauerstoff und Primärproduktion in hocheutrophen Flachseen des Havelgebietes. Habil. TU Dresden. 113 pp.

Klein, G., 1989. Anwendbarkeit des OECD-Vollenweider-Modells auf den Oligotrophierungsprozeß an eutrophierten Gewässern. Vom Wasser 73: 365–373.

Kozerski, H.-P., J. Gelbrecht & R. Stellmacher, 1993. Seasonal and long-term variability of nutrients in Lake Müggelsee. Int. Revue ges. Hydrobiol. 78: 423–437.

Li, Y.-H. & S. Gregory, 1974. Diffusion of ions in sea water and deep-sea sediments. Geochim. Cosmochim. Acta 38: 703–714.

Murphy, J. & J. P. Riley, 1962. A modified single solution method for the determination of phosphate in natural waters. Analyt. chim. Acta 27: 31–36.

Oláh, J., 1975. Metalimnion function in shallow lakes. Symp. Biol. Hung. 15: 149–155.

Poltz, J., 1980. Some studies on the problem of 'Treibmudde' in Steinhuder Meer. In: Dokulil, M., H. Metz & D. Jewson (eds), Shallow Lakes. Developments in Hydrobiology 3. Dr W. Junk Publishers, The Hague: 3–25.

Psenner, R. & R. Pucsko, 1988. Phosphorus fractionation: advantages and limits of the method for the study of sediment P origins and interactions. Arch. Hydrobiol. Beih. Ergebn. Limnol. 30: 43–59.

Psenner, R., R. Pucsko & M. Sager, 1984. Fraktionierung der Phosphorverbindungen von Sedimenten. Arch. Hydrobiol., Suppl. 70: 112–149.

Ryding, S.-O., 1985. Chemical and microbiological processes as regulators of exchange of substances between sediments and water in shallow eutrophic lakes. Int. Revue ges. Hydrobiol. 70: 657–702.

TGL 27885/01, 1982. Nutzung und Schutz der Gewässer: Stehende Gewässer, Fachbereichstandard der DDR, Ministerium für Umweltschutz und Wasserwirtschaft, Berlin, 16 pp.

Vietinghoff, H., 1994. Beiträge zur Hydrographie und Limnologie ausgewählter Seen in Ostbrandenburg sowie zum Wasserhaushalt stehender Gewässer. Diss. A. Humboldt-Universität Berlin, 125 pp.

Wetzel, R.G., 1983. Limnology. 2nd edn., Saunders College Publishing, 767 pp.

Hydrobiologia **342/343:** 55–61, 1997.
L. Kufel, A. Prejs & J. I. Rybak (eds), Shallow Lakes '95.

Relative nitrogen deficiency without occurrence of nitrogen fixing blue-green algae in a hypertrophic reservoir

Mátyás Présing, Katalin V.-Balogh, Lajos Vörös & Hesham M. Shafik
Balaton Limnological Research Institute of the Hungarian Academy of Sciences, P.O. Box 35, H-8237 Tihany, Hungary

Key words: hypertrophic reservoir, filter feeding fish, low N:P, lack of N_2-fixing cyanobacteria, enclosure experiment, ^{15}N uptake

Abstract

The N:P ratios (TN:TP, NO_3-N:TP and DIN:SRP) in the shallow, hypertrophic Marcali reservoir varied about 1, far below the physiological optimum for the planktonic algae. Moreover, N_2-fixing cyanobacteria were absent in the vegetation period. One possible explanation for this phenomenon could be the filtration of the stocked silver carp (*Hypophthalmichthys molitrix* Val.), a phytoplankton feeder, and bighead carp (*Aristichthys nobilis* Rich.), an omnivorous plankton feeder population (600 kg ha^{-1}) which strongly reduces the large sized algae.

Enclosure and uptake experiments were carried out to confirm the relative N deficiency in the reservoir and, by excluding the fish, to check whether the reservoir water is suitable for the development of N_2-fixing cyanobacteria, or the planktivorous fish are responsible for the lack of blue-greens. The enclosures were free of fish. In one of them the NO_3-N enrichment resulted in a rapid growth of diatoms and cryptophytes, confirming the relative N limitation of the phytoplankton biomass. The results of ^{15}N uptake experiments showed a high preference of algae for ammonium. Its actual uptake velocity (19.5 $\mu g\ l^{-1}\ h^{-1}$) was much higher than that for nitrate (1.0 $\mu g\ l^{-1}\ h^{-1}$). But the relatively high value of the half-saturation constant (32 $\mu g\ l^{-1}$) in the ammonium uptake experiment indicated only a moderate affinity of algae even to the preferred nitrogen source. The other enclosure was inoculated with N_2-fixing cyanobacteria. During the 29 days experiment they did not grow in the enclosure and were gradually eliminated from the water column. The results of enclosure experiments suggested that factors such as the relatively high abundance of inorganic N (30–50 $\mu g\ l^{-1}$ ammonium and 90–140 $\mu g\ l^{-1}$ nitrate, respectively) could be responsible for the absence of N_2-fixing cyanobacteria in the reservoir, rather than the size selective filtration of planktivorous fish.

Introduction

To reduce the mineral nutrient loading to Lake Balaton several reservoirs were constructed on its tributaries. One of them is the Marcali reservoir, a shallow, hypertrophic water which is heavily stocked with planktivorous fish. Though, the environmental conditions in the reservoir could favour the development of N_2-fixing cyanobacteria, it has never been observed.

There are hypotheses, summarised by Shapiro (1990) to explain the dominance of cyanobacteria in several waters. From the favourable environmental conditions listed by him the suitable water temperature, light and CO_2/pH were available for cyanobacteria in the reservoir during the vegetation period. The water temperature was above 21 °C during the summer time, exceeding 25 °C in July and August. The shallow, unstratified water was calm for long periods. The pH was between 8 and 9 during the whole year. The question of N and P concentrations and their ratios in the water is more complicated. The first N:P hypotheses formulated by Smith (1983) took the total nitrogen to total phosphorus ratio into consideration. His threshold ratio for cyanobacteria was of 29. There are results, summarised by Pick & Lean (1987) that the low inorganic nitrogen:phosphorus relationships may be a better prediction than the TN:TP for cyanobacteria dominance. McQueen & Lean (1987) found significant correlation (negative) between the percentage of cyanobacteria and NO_3-N. Below 21 °C and with NO_3-N:TP above 5, cyanobacterial blooms never occurred. The calculated TN:TP in Marcali reservoir and in the enclosures were always far below 29. The NO_3-N:TP never reached 5 from March to October in the reservoir and was extremely low (below 1) in the summer time. The DIN: SRP was below the Redfield-ratio (5.7 weight/weight) in the vegetation period and varied about 1 from May to September. By the ratios above the reservoir should be suitable for cyanobacteria and their dominance could be expected.

There are a number of works about the effect of silver and bighead carp filtration on the structure of phytoplankton. According to Vörös et al. (1997) the silver carp is able to filter only those algae larger than 10 μm and in this size category the non-mucilaginous filamentous cyanobacteria proved to be well digestible. The results of Miura (1990) show that it is possible to change an algal assemblage from a colony-forming blue-green *Microcystis*-dominated one to one dominated by green algae smaller than 10 μm by the increased filtration pressure of silver and bighead carp. In his opinion this kind of biomanipulation can be considered, if the use of the lake is for recreation or sightseeing, and if there is no condition of reducing nutrients. In enclosure experiments on a Brazilian reservoir the dominant cyanobacterium *Cylindrospermopsis raciborskii* (Wolosz.) Seenaya and Subba Raju was reduced in the presence of silver carp (Starling & Rocha, 1990). They suggested that the control of the undesirable *C. raciborskii* directly by silver carp grazing is a promising management tool. A heavy water bloom (mainly *Microcystis, Anabaena* and *Oscillatoria*) which occurred regularly in a shallow hypereutrophic Chinese lake before mid-1980s has disappeared since 1985, most probably due to the increased stocking rate of the silver and bighead carp (Shei et al., 1995). Nutrient conditions in the lake were still suitable for the growth of bloom forming algae. To clarify the possible role of silver and bighead carp in the sudden disappearance of the water bloom from the lake, enclosure experiment were conducted. In the enclosures stocked with silver and/or bighead carp, no water bloom appeared, in the no-fish enclosure a heavy water bloom emerged and it was eliminated completely within 10–20 days after silver and/or bighead carp were introduced (Shei et al., 1993).

The several N:P ratios measured in the reservoir and the literature data on the grazing effect of silver and bighead carp, have led some to conclude that the size selective filtration of stocked fish keeps back the development of the cyanobacteria biomass. Enclosure and uptake experiments were carried out to confirm the relative N deficiency in the reservoir and in the absence of fish, clarify whether the reservoir water is suitable for the development of N_2-fixing cyanobacteria or the dense silver and bighead carp population is responsible for the lack of cyanobacteria.

Material and methods

The reservoir has a surface area of 4 km^2 with average depth of 1.8 m. It was flooded and drained biennially and after the flooding three species of fish (age: 1+) were introduced regularly. The density of introduced fish was some 600 kg ha^{-1} in 1994. About 80% of the fish population was silver carp (*H. molitrix*) 15% bighead carp (*A. nobilis*) and 5% common carp (*Cyprinus carpio* L.). A detailed description of the study area is published in the papers of Vörös et al. and V.-Balogh & Vörös (1997).

Three experimental enclosures (3 m in diameter and 2 m in height) were installed on the open water close to the outflow of the reservoir on 9 July 1994. Their construction was identical with those used successfully on Lake Balaton in the previous years (Istvánovics et al., 1986). The enclosures were free of fish, while in the open water the total fish stock was approximately 1000 kg ha^{-1}. In the first experiment (from 11 to 21 July) KNO_3 was applied once (on the first day) to adjust the NO_3-N concentration in the enclosure to 3.5 mg l^{-1}. In the second experiment (from 11 July to 9 August) another enclosure was inoculated with a suspension of filamentous cyanobacteria resulting in a biomass of above 1 mg l^{-1} in the water. The suspension was prepared by carefully filtering the water from the nearby reservoir at the time of heavy N_2-fixing cyanobacteria bloom (*C. raciborskii, Aphanizomenon flos-aquae* (L.) Ralfs., *Anabaena* spp.). The treatment was repeated at 2nd August with a suspension prepared in the same way, but originating from another reservoir. The third enclosure was not treated and was used as a control. During the year the reservoir was sampled approximately monthly, with a 2 m long sampling tube, to represent the whole water column. At the time of the enclosure experiments samples were taken every 3–5 days from the centre of each enclosure and

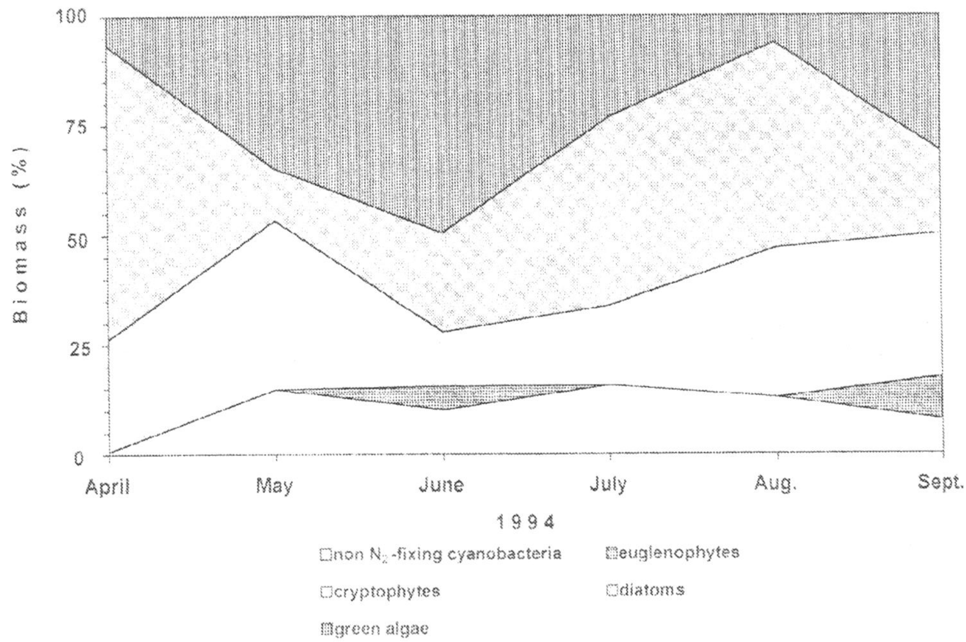

Figure 1. Contribution of various phytoplankton species to the total algal biomass in the reservoir during the vegetation period.

from the neighbouring open water as well. The samples were transported to the laboratory within one hour and kept in the dark at reservoir temperature. Samples for nutrient analyses were filtered through Whatman GF/F filters. The chemical analyses were started immediately according to the 'Water analysis' of Mackereth et al. (1978). The chlorophyll-*a* concentrations were measured according to Iwamura et al. (1970). Phytoplankton samples were preserved by Lugol's solution. Algae larger than 2 μm were identified and counted with an inverted microscope (Utermöhl, 1958). The wet weight of each species was calculated from cell volumes. The autotrophic picoplankton was studied with an epifluorescence microscope (Zeiss Axioplan) with blue excitation light (Caron, 1982). Nitrogen uptake experiments were carried out at the end of August with 99 atom% ^{15}N-enriched ammonium chloride and potassium nitrate. The relatively high nitrate concentration in the reservoir allowed the use of a tracer uptake experiment in which the ^{15}N enrichment was kept at about 10–20% of the ambient NO_3-N concentration. The uptake velocities for both the nitrate and ammonium at different concentrations were calculated by the equation of Collos (1987), using the final particulate nitrogen of the samples. In the case of ammonium the water samples were enriched in a range of initial substratum concentrations of 10–640 μg ^{15}N l^{-1} and the *in situ* uptake velocity was extrapolated from the measured rates by the Michaelis-Menten equation. With a relatively low ambient concentration of ammonium and large standing stocks of phytoplankton it is the only method of obtaining meaningful results (Paasche & Kristiansen, 1982). The samples, kept in Erlenmeyer flasks (volume 500 ml), were incubated at reservoir temperature and illuminated by white fluorescent lamps at an intensity of 110 μE m^{-2} s^{-1} which was found to be in the optimal range for the nitrogen uptake. The incubation times were 30 and 40 minutes in the NH_4 and the NO_3 experiments, respectively. The uptake was terminated by filtration through Whatman GF/F filters (pressure differential $<$50 mmHg). The filters were rinsed with unlabelled and prefiltered reservoir water and then dried at 60 °C. Small discs were punched out from the filters and wrapped in tin capsules. The total nitrogen and carbon content and the ^{14}N:^{15}N ratio of the prepared samples were determinated by an Isotope Ratio Mass Spectrometer (ANCA-MS, Europa Scientific Ltd., UK).

Table 1. Nutrient characteristic and chlorophyll-*a* concentrations of Marcali reservoir.

Concentration ($\mu g\ l^{-1}$)	Date									
	27.01.	16.02.	24.03.	20.04.	19.05.	23.06.	21.07.	11.08.	12.09.	13.10.
NH_4-N	440	30	100	40	40	30	30	50	30	170
NO_2-N	27	24	41	15	0	17	0	80	0	22
NO_3-N	2790	2500	1000	300	140	90	140	140	140	400
DIN	3257	2554	1141	355	180	407	170	270	170	592
TN	4720	5220	4140	2810	3240	2000	2040	2520	3640	4720
SRP	750	40	190	10	150	400	200	180	110	20
TP	930	160	400	200	400	440	440	440	600	600
Chl-*a*	43	86	38	143	81	74	126	122	171	205
Ratios										
TN:TP	5.1	32.6	10.4	14.1	8.1	4.6	4.6	5.7	6.1	7.8
DIN: SRP	4.3	63.8	6.0	35.5	3.6	1.0	0.9	1.5	1.5	29.6
NO_3-N:TP	3.0	15.6	2.5	1.5	0.4	0.2	0.3	0.3	0.2	0.7

DIN = Dissolved Inorganic N; TN and TP = Total N and P; SRP = Soluble reactive (PO_4) P; Chl-*a* = Chlorophyll-*a*

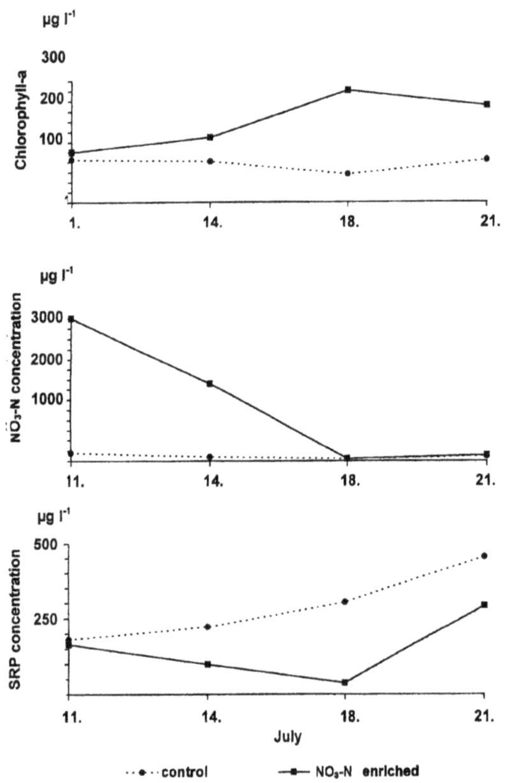

Figure 2. Nutrient and chlorophyll-*a* concentrations in the enclosure experiment with NO_3-N addition.

Results and discussion

The nutrient and chlorophyll-*a* concentrations are shown in Table 1. Both the ammonium and nitrate concentrations were high from autumn to spring and decreased in the summer months. The chlorophyll-*a* concentration ranged between 38 and 205 $\mu g\ l^{-1}$ during the year. Its annual average exceeded 100 $\mu g\ l^{-1}$. According to the OECD (1982) classification system the reservoir was hypertrophic. From April to September the phytoplankton biomass was high and varied between 9 and 26 mg l^{-1}. The most abundant organisms were green algae (*Scenedesmus* spp), small centric diatoms (*Stephanodiscus hantzschii* Grun., *Cyclotella* spp.) and *Cryptomonas* spp. (Figure 1). N_2-fixing species did not occur in the small cyanobacterial population.

In the nitrogen uptake experiments the actual uptake velocity of ammonium (19.5 $\mu g\ l^{-1}\ h^{-1}$) was much higher than the value for the nitrate (1 $\mu g\ l^{-1}\ h^{-1}$). The calculated maximum uptake velocity (V_{max}) of ammonium was very high (109.4 $\mu g\ l^{-1}\ h^{-1}$). Unfortunately, from the tracer experiment (in the case of nitrate) one could not calculate the V_{max} and K_s values as was possible from the Michaelis-Menten kinetics (in the case of ammonium). Although the actual uptake velocities were determined with different methods they should be comparable. The much higher uptake velocity (at a lower ambient concentration) for ammonium than for the nitrate shows a significant preference of this mixed algae population for ammonium. The preference for ammonium is not universal, but in the overwhelming majority of waters it is the nitrogen form preferred by algae (Dortch, 1990). However the relatively high value of the half-saturation constant (32 $\mu g\ l^{-1}$) in the ammonium uptake experiment indi-

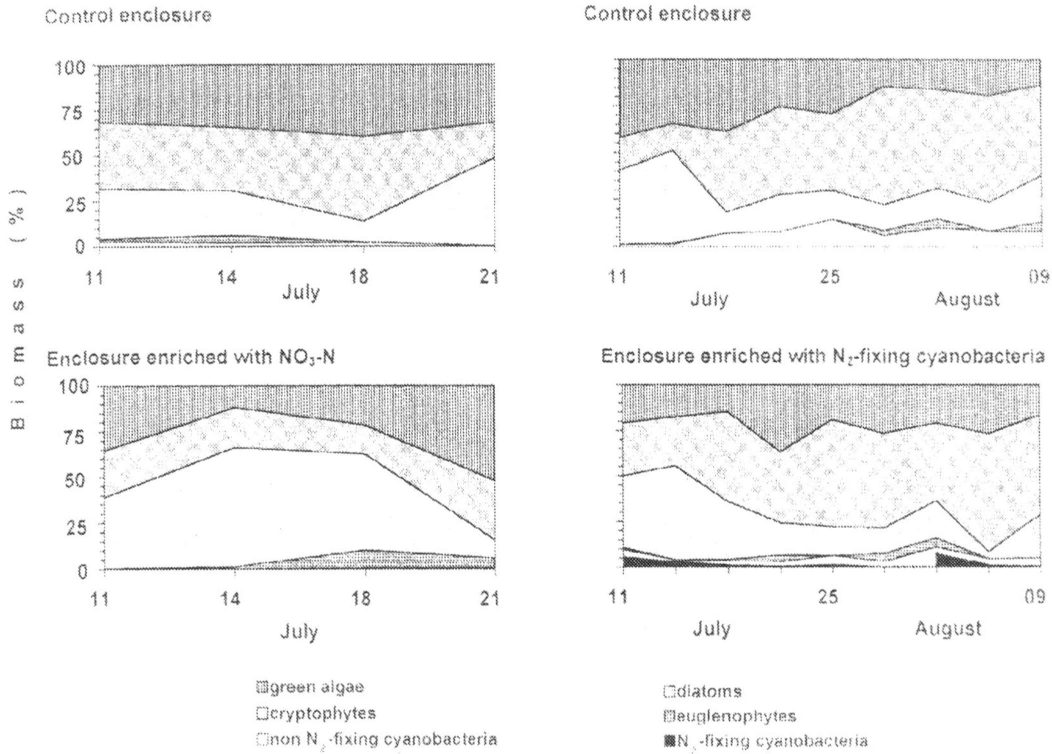

Figure 3. Contribution of phytoplankton species to the total algal biomass in the enclosures with NO_3-N and N_2-fixing cyanobacteria addition.

cated a moderate affinity of algae even to the preferred nitrogen source.

In the first enclosure experiment the NO_3-N addition was followed by a rapid growth of phytoplankton. During one week the chlorophyll-*a* concentration doubled (Figure 2), and the phytoplankton biomass increased from 24 to 59 mg l^{-1}. The biomass increase was primarily due to the rapid growth of *Cryptomonas* species and centric diatoms (Figure 3). The chlorococcales green algae did not show any response to the nitrate enrichment. After a week the increase stopped and the biomass and chlorophyll-*a* concentration decreased. The artificially high NO_3-N concentration rapidly dropped in the enclosure (Figure 2), and after a week it was at the level of the control enclosure. As the NO_3-N concentrations dropped during the first days of the experiment the SRP decreased from 164 to 37 μg l^{-1} (Figure 2). But after a week it started to increase and on the 10th day of the experiment reached its usual level. The algae reacted to the increased N supply and grew rapidly until they had depleted (after some 7 days) the extra N source. During this time they were able to utilise more phosphorus thus decreasing the SRP level in the enclosure. The results suggest that N rather than P is the limiting nutrient for the algae in the reservoir. This is a common case among hypertrophic lakes and reservoirs, especially if the N:P ratio is low in their external loading which is very often due to its sewage content.

Despite the relative nitrogen deficiency we did not recognise any species of N_2-fixing cyanobacteria in the enclosures (from which fish were absent) or in the reservoir. Although we could keep and observe the control enclosure for a further 10 days, the whole duration of the experiment (20 days) could have been short for their development without a pre-existing population. According to Reynolds (1984), a pre-existing population is one of the most important factors for the blue-green bloom formation. If the N_2-fixing cyanobacteria had been totally outfiltered by the fish in the previous years they could not enter into the population in the water from the sediment. According to the feeding experiments, the estimated filtration capacity of silver carp with 1 and 2 kg wet weight was 0.7–0.5 m^3 water d^{-1} kg $fish^{-1}$, respectively (Vörös 1990) in the reservoir. So they were able to filter the total water volume during 10–40 days, depending on the density of stocking and the age of fish.

In the second experiment we intended to eliminate the problem of pre-existing population by adding a cyanobacteria inoculum into the other enclosure. Although it was about 5% of the total algal biomass, the cyanobacteria did not grow and during 3 weeks they were slowly eliminated from the water column (Figure 3). It happened after the second treatment as well. They were not able to grow in the reservoir water.

Although every nitrogen to phosphorus ratio calculated for the reservoir should have been suitable for the development of N_2-fixing cyanobacteria, these ratios originated from relatively high ambient concentrations. According to data in the literature these nitrogen concentrations are not favourable for N_2-fixing cyanobacteria. They are capable of using nitrate as a nitrogen source, but are not able to compete with other phytoplankton for it. According to Blomqvist et al. (1994) this may be attributed to differences in the enzymatic processes by which nitrate is assimilated. Other algae make major use of nitrate, but not the N_2-fixing cyanobacteria (Toetz et al., 1977). Therefore they can be overgrown by other algae. They are not only uncompetitive for nitrate, but above a certain level (approximately 20 $\mu g\ l^{-1}$) it reduces their growth (Blomqvist et al., 1994). The general rule, that ammonium is preferred over nitrate as a nitrogen source by phytoplankton holds true especially for cyanobacteria. They take advantage of their ability to assimilate ammonium at low ambient concentrations (some $\mu g\ l^{-1}$), but a high abundance of ammonium could be detrimental for them. Branco & Senna (1994) observed heterocysts in *C. raciborskii* only in the months when low ammonium concentrations were registered in the eutrophic Paranoá reservoir. In the enclosure experiment of Barica et al.(1980) the high concentrations of inorganic nitrogen (around 1 mg l^{-1}) from ammonium or nitrate depressed the development of N_2-fixing cyanobacteria.

The results of the enclosure experiment with N_2-fixing cyanobacteria addition and the literature data suggest that factors such as the high abundance of inorganic N could be responsible for the absence of N_2-fixing cyanobacteria in this reservoir, rather than the size selective filtration of the silver and bighead carp population.

Acknowledgments

This study was supported by a grant from the South Transdanubian Environmental Inspectorate. Our special thanks go to James Arthurs for improving the English. We are grateful to Miss Terézia Horváth for her skilful preparation of the manuscript.

References

Barica, J., H. Kling & J. Gibson, 1980. Experimental Manipulation of Algal Bloom Composition by Nitrogen Addition. Can. J. Fish. aquat. Sci. 37: 1175–1183.

Blomqvist, P., A. Petterson & P. Hyenstrand, 1994. Ammonium-nitrogen: A key regulatory factor causing dominance of non-nitrogen-fixing cyanobacteria in aquatic systems. Arch. Hydrobiol. 132: 141–164.

Branco, C. W. C. & P. A. C. Senna, 1994. Factors influencing the development of *Cylindrospermopsis raciborskii* and *Microcystis aeruginosa* in the Paranoá Reservoir, Brasilia, Brazil. Arch. Hydrobiol./Suppl. 105, Algolog. Studies 75: 85–96.

Caron, D. A., 1982. Technique for Enumeration of Heterotrophic and Phototrophic Nanoplankton, Using Epifluorescence Microscopy, and Comparison with Other Procedures. Appl. envir. Microbiol. 46: 491–498.

Collos, Y., 1987. Calculations of ^{15}N Uptake Rates by Phytoplankton Assimilating One or Several Nitrogen Sources. Appl. Radiat. Isot. 38: 275–282.

Dortch, Q., 1990. The interaction between ammonium and nitrate uptake in phytoplankton. Mar. Ecol. Prog. Ser. 61: 183–201.

Istvánovics, V., L. Vörös, S. Herodek, L. G. Tóth & I. Tátrai, 1986. Changes of phosphorus and nitrogen concentration in enriched lake enclosures. Limnol. Oceanogr. 31: 798–811.

Iwamura, T., H. Nagai & S. Ishimura, 1970. Improved methods for determining contents of chlorophyll, protein, ribonucleic and desoxyribonucleic acid in planktonic populations. Int. Revue ges. Hydrobiol. 55: 131–147.

Mackereth, F. J. H., J. Heron & J. F. Talling, 1978. Water analysis: Some revised methods for limnologists. Freshwater Biological Association, Scientific Publication No. 36, Ambleside, 120 pp.

McQueen, D. J. & D. R. S. Lean, 1987. Influence of water temperature and nitrogen to phosphorus ratios on the dominance of blue-green algae in Lake St. George, Ontario. Can. J. Fish. aquat. Sci. 44: 598–604.

Miura, T., 1990. The effects of plaktivorous fishes on the plankton community in a eutrophic lake. Hydrobiologia 200/201: 567–579.

OECD, 1982. Eutrophication of waters. Monitoring, Assessment and Control. Final Report. OECD Co-operative Programme on Monitoring of Inland Waters (Eutrophication Control), Environment Directorate, OECD, Paris, 154 pp.

Paasche, E. & S. Kristiansen, 1982. Nitrogen Nutrition of the Phytoplankton in the Oslofjord. Estuar. coast. shelf Sci. 14: 237–249.

Pick, F. R. & D. R. S. Lean, 1987. The role of macronutrients (C, N, P) in controlling cyanobacterial dominance in temperate lakes. N. Zeal. J. mar. Freshwat. Res. 21: 425–434.

Reynolds, C. S., 1984. The ecology of Freshwater Phytoplankton. Cambridge University Press, 384 pp.

Shapiro, J., 1990. Current beliefs regarding dominance by blue-greens: The case for the importance of CO_2 and pH. Verh. int. Ver. Limnol. 24: 38–54.

Shei, P., J. Wang & X. Huang, 1995. Impact of the Planktivorous Silver and Bighead Carps on Plankton Community Structure in a Shallow, Hypereutrophic Chinese Lake. Proceedings of the 6th International Conference on the Conservation and management of lakes – Kasumiguara, 95: 781–784.

Shei, P., J. Wang, X. Liu & J. K. Liu, 1993. An enclosure experiment on the influence of three 'domestic' carps-silver carp, bighead carp and grass carp, on the 'water bloom' in a shallow, eutrophic Chinese lake (Lake Donghu). In Liu, J. K. (ed.), Annual Report of FEBL for 1992. International Academic Publishers, Beijing, China: 13–20.

Smith, V. H., 1983. Low nitrogen to phosphorus favour dominance by blue-green algae in lake phytoplankton. Science 221: 669–671.

Starling, F. L. R. M. & A. J. A. Rocha. 1990. Experimental study of the impacts of planktivorous fishes on plankton community and eutrophication of a tropical Brazilian reservoir. Hydrobiologia 200/201: 581–591.

Toetz, D., L. Varga & B. Huss, 1977. Observations on Uptake of Nitrate and Ammonia by Reservoir Phytoplankton. Arch. Hydrobiol. 79: 182–192.

Utermöhl, H., 1958. Zur Vervollkommnung der quantitativen Phytoplankton-Methodik. Mitt. int. Ver. Limnol. 9: 1–38.

V.-Balogh, K. & L. Vörös, 1997. High bacterial production in hypertrophic shallow reservoirs rich in humic substances. Hydrobiologia 342/343: 63–70.

Vörös, L., 1990. Hydrobiological investigation of Marcali reservoir 1990. Scientific report. (A Marcali-tározó hidrobiológiai vizsgálata, 1990. Tudományos jelentés) Balaton Limnological Research Institute, Tihany, Hungary: 75 pp.

Vörös, L., I. Oldal, M. Présing & K. V.-Balogh, 1997. Size selective filtration and taxon specific digestion of planktonic algae by silver carp (*Hypophthalmichthys molitrix* Val.). Hydrobiologia 342/343: 223–228.

Hydrobiologia **342/343**: 63–70, 1997.
L. Kufel, A. Prejs & J. I. Rybak (eds), Shallow Lakes '95.

High bacterial production in hypertrophic shallow reservoirs rich in humic substances

Katalin V.-Balogh & Lajos Vörös
Balaton Limnological Research Institute of the Hungarian Academy of Sciences, P.O.Box 35, H-8237 Tihany, Hungary

Key words: shallow reservoirs, bacterial production, primary production, dissolved humic substances, macrophytes

Abstract

Two reservoirs, constructed for pollution control in the catchment area of Lake Balaton, were studied. Both are hypertrophic due to high phosphorus loading, and both are rich in dissolved humic substances (colour in Pt units usually varies between 50–150 Pt, mg l^{-1}) as they were built on former wetland areas. We measured planktonic primary production with ^{14}C technique, and bacterioplankton production using the [*methyl*-^{3}H]thymidine incorporation method. The levels of dissolved organic carbon (DOC) and humic substances (as water colour) present were investigated. Temperature, pH and light intensity at different depths were recorded regularly. In both reservoirs studied the bacterioplankton production was as high as the phytoplankton production and in some cases significantly exceeded it. Dissolved humic substances proved to be an important substrate for planktonic bacteria in these water bodies.

Introduction

Many lakes, especially recently flooded reservoirs, are coloured brown from humic substances. Aquatic humic substances may either be produced autochthonously, or imported from terrestrial, geochemical sources (Sieburth & Jensen, 1968; Gjessing, 1976). In lakes, the pool of dissolved organic carbon (DOC) is dominated by dissolved humic substances (up to 80% of DOC) (Steinberg & Münster, 1985; Thurman, 1985).

Planktonic bacterial production in surface waters of lakes and marine systems is positively related to primary production. In most cases primary production explains much of the variation in bacterial production among ecosystems (Cole et al., 1988). Several independent studies of marine and freshwater pelagic ecosystems have estimated that the bacteria utilise a large fraction (up to 90%) of planktonic primary production (Larsson & Hagström, 1982; Lowell & Konopka, 1985; Lavandier, 1990).

Since the base of the planktonic food web is generally autochthonous primary production humic waters have often been classified as unproductive. The reason for this is that the brown colour of the water reduces light penetration and thus primary production (Kirk, 1976; Arvola, 1984). However, high bacterial biomass in humic lakes has been recorded by several investigators (Johansson, 1983; Hessen, 1985) and there have been several indications that the biota of humic lakes utilise allochthonous energy (Tranvik, 1989). The presence of an important carbon source in addition to planktonic primary producers is suggested by high ratios of respiration to primary production (Johansson, 1983; Salonen et al., 1983).

The aim of the present work was to estimate bacterial production and evaluate its relation to algal production and other variables such as dissolved organic carbon and temperature in two shallow reservoirs rich in humic substances. The presence and absence (open water) of macrophytes (reed stand, submerged macrophytes) were also taken into consideration.

Study area

The large shallow Lake Balaton is situated in western part of Hungary, and has a surface area of 596 km^2 and a mean depth of 3.2 m. The nutrient loading to the lake is discharged by tributaries – the main one being the Zala River – into the western part of the lake. Half of its watershed of 5179 km^2 is drained by the Zala River into the western basin. This part of the lake has a hypertrophic character, and the maximum chlorophyll *a* concentration usually exceeds 100 μg l^{-1}. The studied shallow reservoirs, Kis-Balaton and Marcali, were constructed for nutrient control on the west and south-west parts of the catchment area.

The Kis-Balaton water system is situated at the mouth of the Zala River. It has two reservoirs, of which the upper was flooded in 1984 and the lower in 1993. The upper Kis-Balaton is a hypertrophic reservoir with a surface area of 18 km^2, a mean depth of 1.1 m. The lower Kis-Balaton reservoir is designed to have a surface area of 51 km^2, and a mean depth of 1.2 m. In 1992 a 16 km^2 experimental area of the lower Kis-Balaton reservoir was flooded. About 95% of this area is covered by macrovegetation, primarily by reed. This wetland, the lower Kis-Balaton reservoir (LKBR) was studied by us at five different sampling sites: LKBR 1 and LKBR 2 were open waters, LKBR 3 was covered by submerged macrophytes (*Cerathophyllum demersum, Lemna trisulca*) and LKBR 4, LKBR 5 were covered by dense reed (*Phragmites australis*) vegetation.

Marcali reservoir is situated in the south-west catchment area of Lake Balaton. It was built in 1984. The reservoir has a surface area of 4 km^2 and average depth of 2 m. It is flooded and drained biennially. About 10% of Marcali reservoir is covered by submerged (*Polygonum amphibiun, Potamogeton natans*) and emerged macrophytes (*Phragmites australis, Typha latifolia*) along the shoreline. The sampling site was situated in the open water near to the outflow of the reservoir.

Materials and methods

The data presented here are based on sample collections carried out from April to October 1994. The samples were taken monthly and they were transported to the laboratory within one hour in a dark box at *in situ* temperature.The pH and water temperature were measured *in situ* by a HI (HANNA instrument) 9025 microcomputer pH meter. Water temperature ranged from 10 °C–25 °C in studied reservoirs. Water pH varied between 6.87 and 9.31 in the lower Kis-Balaton reservoir and 8.16–8.97 in Marcali reservoir. Penetration of photosynthetically active radiation (PAR, μEinstein m^{-2} s^{-2}) was measured *in situ* with a LI-COR (LI-185B) radiometer equipped with a flat (2π) sensor.

Dissolved organic carbon (DOC) and humic substances (HS)

Subsamples of 100 ml were filtered through cellulose acetate membrane filters (pore size 0.45 μm) rinsed with distilled water and sample water. The pH of each filtered sample was adjusted to 8.25 by a sodium borate buffer of 1 M (pH = 7). For DOC measurements absorbance of samples was measured at 254 nm in a quartz cell of 1 cm using a similarly buffered distilled water blank. Aquatic humic substances were measured as colour (Cuthbert & Giorgio, 1992). Absorbance of samples was measured at 440 nm in a glass cell of 10 cm. Water colour$_{440}$ units (Pt, mg l^{-1}) were determined by the method of Cuthbert & Giorgio (1992). For spectrophotometric measurements a SHIMADZU UV-160A spectrophotometer was used.

Dissolved organic carbon concentrations were calculated from a saturation equation:

$$\mathrm{DOC} = \frac{a \cdot \mathrm{ABS}}{(1 + b \cdot \mathrm{ABS})} \tag{1}$$

where: DOC = dissolved organic carbon concentration (mg l^{-1}), ABS = absorbancia value (254 nm, 1 cm), measured parameters: $a = 123.525$, $b = 4.282$, $r^2 = 0.98$. This DOC/absorbance equation was established on 18 value pairs of UV-absorbance and DOC concentration measured in water samples collected recently from three locations (Western basin of Lake Balaton, Kis-Balaton reservoir and Marcali reservoir – where the water was coloured by humic substances). The DOC concentration was measured by an ASTRO Total Organic Carbon Analyser.

Bacterioplankton production

For thymidine incorporation studies, the procedure of Fuhrman & Azam (1982) was followed. Subsamples of 20 ml (triplicates plus 2% formaldehyde killed blanks) were incubated in darkened plastic bottles in an incubator system at the actual original water temperature for 0.5 h with [*methyl*-^{3}H]thymidine [Me-^{3}HT]dR-(specific activity, 42 Ci mM^{-1}; Amersham Life Science) at a final concentration of 50 nM l^{-1}. After the 0.5 hour incubation period, formaldehyde was also added to a final concentration of 2%, and the samples were kept at 0 °C. Then 50% ice cold trichloroacetic acid (TCA) was added to each bottle to a final concentration of 5%. The incubation was terminated after 15 minutes when the trichloroacetic acid-insoluble precipitate was collected on 0.2 μm polycarbonate filters. The filters were rinsed five times with 1 ml portions of ice cold 5% TCA and finally with 1 ml of ice cold 80% ethanol, and put in scintillation vials. Radioactivity retained on the filters was assayed by a LKB Rackbeta liquid scintillation counter. Counting efficiency was determined by the internal standard addition technique.

Bacterial production was calculated from the amount of thymidine incorporation assuming a conversion factor of $2.0 * 10^{18}$ cells M^{-1} of thymidine. According to Lee & Fuhrman (1987) for a biovolume range of bacterial cells of between 0.036 and 0.073 μm^3, the average per-cell carbon content is relatively constant, 20 fg of C. As the bacterioplankton of the studied reservoirs was within this size range to calculate the bacterial carbon production from the cell production 20 fg C $cell^{-1}$ was accepted. Bacterial growth efficiency was assumed to be 30%.

Phytoplanktonic primary production

Subsamples of 50 ml with 0.4 MBq sodium bicarbonate-^{14}C were kept for 2 hours in an incubator system, at the original water temperature and at seven different light intensities between 10 and 500 μEinstein m^{-2} s^{-1}. The light intensity of the white fluorescent lamps was regulated by neutral density screens. The samples were then darkened and filtered through cellulose acetate filters (pore size 0.45 μm). The non-cellular ^{14}C was removed from the filters with fumes of hydrochloric acid. Then the filters were put into the scintillation vials and Bray-scintillation cocktail was added to them. The radioactivity was measured by a LKB Rackbeta liquid scintillation counter. The calculation of inorganic carbon fixation was based on the ^{14}C uptake and the total dissolved inorganic carbon originally present (Vollenweider, 1969).

To describe the light dependence of phytoplankton photosynthesis the exponential saturation equation was used (Webb et al., 1974):

$$P = P_{max}\{1 - \exp(-I/I_k)\} \qquad (2)$$

where: P = rate of photosynthesis, P_{max} = maximal rate of photosynthesis (μg C l^{-1} h^{-1}), I = light intensity, I_k = light intensity at which initial slope of photosynthesis reaches the P_{max} (μEinstein m^{-2} s^{-1}).

The surface related photosynthesis (mg C m^{-2} day^{-1}) was calculated from equation 2 (above), with light attenuation coefficients determined *in situ* and the current global radiation data. Forty-eight percent of the global radiation considered to be the photosynthetically active radiation (PAR) (Wetzel & Likens, 1990).

Statistical test

The statistical significance for linear correlation was determined according to Fisher (1963).

Results

In the open water part of Marcali reservoir (Figure 1a) the primary production rate varied between 2.9 and 6.1 g C m^{-2} day^{-1}. Bacterial production rates ranged from 2.4 to 12.3 g C m^{-2} day^{-1}. Comparing the monthly value pairs, the bacterial production was usually 2–4 times higher than the primary production, except for colder months (April and September), when the bacterial production was equivalent to 60–80% of primary production. The pH values varied between 8.2 and 9.0.

At various parts of lower Kis-Balaton reservoir, whether macrophytes were present or absent, the observed seasonal pattern of bacterial production was similar. Its open water area (Figure 1b) the primary production peak, 9.5 g C m^{-2} day^{-1} was observed in late summer, in August, and the minimum value, 0.62 g C m^{-2} day^{-1} was measured in April. Bacterial production rates ranged from 0.23 to 16.5 g C m^{-2} day^{-1}. The minimum value was also observed in April, and the maximum in July. Comparing the monthly value pairs, the bacterial production rate was 2–3 times higher usually, than the primary production rate, except for colder months (April and September), when the bacterial production rate was found to be equivalent to 22–35% of the primary production rate. The pH values varied between 7.2 and 8.6.

Inside dense submerged macrophyta vegetation (Figure 1c) the rate of phytoplanktonic primary production was one order of magnitude lower than in open water in summer months. The higher value 1.45 g C m^{-2} day^{-1} was observed in early autumn in September, and the lower ones, 0.15 g C m^{-2} day^{-1} were found in May and June. Bacterial production rates ranged from 0.55 to 5 g C m^{-2} day^{-1}. The minimum value was observed in April, and the maximum in July.

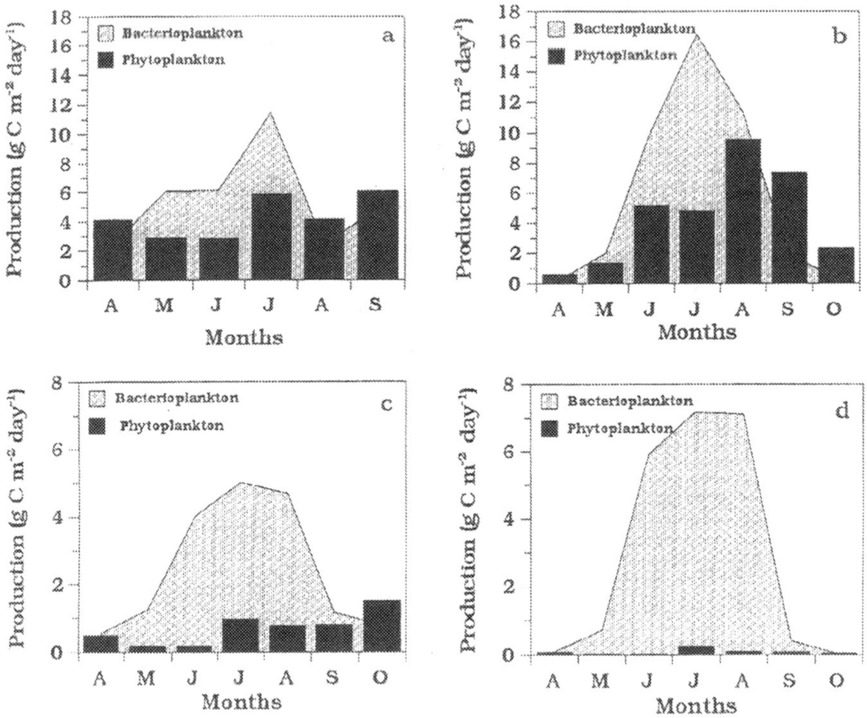

Figure 1. Seasonal changes of phytoplankton and bacterioplankton production in open water areas of Marcali reservoir (a) and lower Kis-Balaton reservoir (site LKBR 2) (b), inside submerged macrophytes of lower Kis-Balaton reservoir (site LKBR 3) (c) and inside reed stand of lower Kis-Balaton reservoir (site LKBR 5) (d) in 1994.

Comparing the monthly value pairs, the bacterioplankton production was usually 5–25 times higher than the phytoplanktonic primary production, except for colder months (April and September), when the bacterial production rate was found to be equivalent to 50–100% of the primary production rate. The pH values varied between 7.0 and 8.0.

Inside the dense reed stand (Figure 1d) primary production was negligible during the whole period, ranging from 0.02–0.24 g C m^{-2} day^{-1}. The maximum bacterioplankton production, 7.2 g C^{-2} day^{-1} was found in July, and the minimum, 0.12 g C m^{-2} day^{-1} in September. Comparing the monthly value pairs, the bacterial production in summer was 30–300 times higher, than the primary production, except for colder months (April and September), when the rate of bacterial production was close to that of primary production. The pH values varied between 6.9 and 7.8.

The concentration of humic substances estimated by colour measurement ranged between 19 Pt, mg l^{-1} and 138 Pt, mg l^{-1} in different sampling locations during the seven month period. The minimum colour values were generally found in April and maximum concentrations in July. Comparing the studied sites, lower values were found in open water areas (Marcali reservoir, site LKBR 1) and higher values in sites covered with reed stand (sites LKEBR 4 and LKBR 5) (Table 1).

The calculated DOC concentrations ranged between 15.1 mg l^{-1} (Marcali reservoir) and 21.7 mg l^{-1} (sites LKBR 4 and LKBR 5) in different sampling locations (Table 2). The seasonal patterns of DOC concentrations were matched by the seasonal courses of water temperature (Figure 2). In addition DOC concentrations were closely related to water temperature. The regression equation describing the relationship is:

$$\mathrm{DOC} = 14.09 + 0.29T \qquad (3)$$

$$r^2 = 0.656, n = 42$$

where: DOC = dissolved organic carbon concentration (mg l^{-1}), T = temperature (°C).

Table 1. Temporal changes of humic substances (as Pt-colour - Pt, mg l^{-1}) in different parts of studied reservoirs.

Sampling site	A	M	J	J	A	S	O
Marcali reservoir	34.19	19.09	51.81	48.45	42.58	66.91	–
LKBR 1	40.91	46.78	47.62	94.61	84.53	89.57	39.23
LKBR 2	41.74	55.17	55.17	117.26	84.53	89.57	49.29
LKBR 3	45.11	65.24	77.82	125.65	98.81	104.67	52.65
LKBR 4	52.65	91.24	116.42	149.98	125.65	129.01	63.56
LKBR 5	45.94	72.79	80.34	137.39	120.61	138.23	72.79

LKBR= lower Kis-Balaton reservoir

Table 2. Temporal changes of dissolved organic carbon (DOC) concentrations (mg l^{-1}) in different parts of humic substance rich reservoirs.

Sampling site	A	M	J	J	A	S	O
Marcali reservoir	15.10	17.00	18.40	19.40	20.30	20.30	18.20
LKBR 1	16.28	16.68	18.03	20.90	21.26	20.74	17.74
LKBR 2	16.00	16.98	18.26	21.16	21.23	20.76	18.75
LKBR 3	15.44	17.08	18.60	21.16	21.33	21.05	18.93
LKBR 4	17.16	19.09	20.12	21.59	21.73	21.50	19.65
LKBR 5	15.95	17.30	18.69	21.12	21.50	21.71	20.14

LKBR= lower Kis-Balaton reservoir

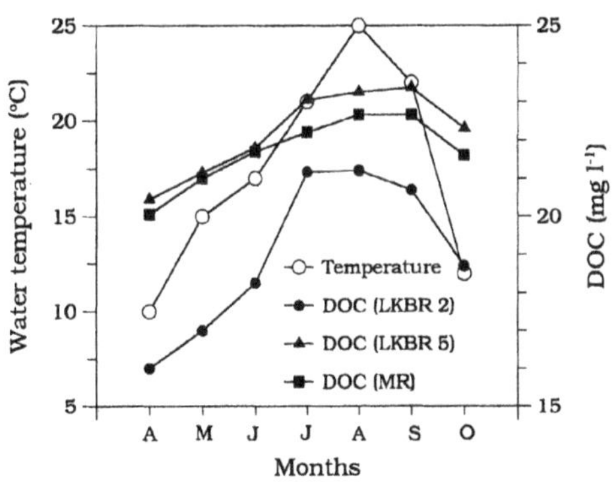

Figure 2. Seasonal changes of water temperature and dissolved organic carbon concentration in Marcali reservoir and in lower Kis-Balaton reservoir.

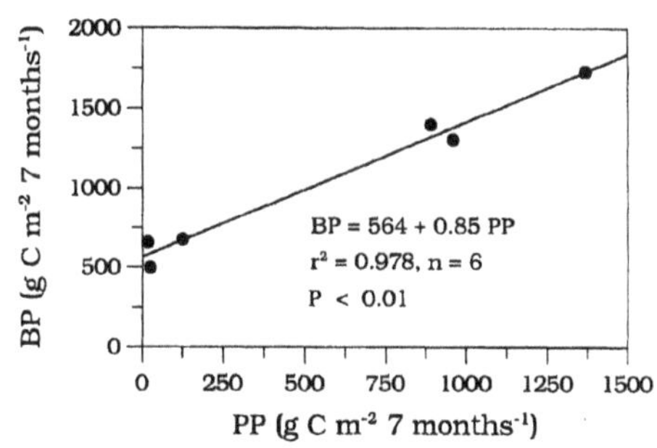

Figure 3. Relationship between phytoplanktonic primary production and bacterioplankton production in Marcali reservoir and lower Kis-Balaton reservoir (cumulative values from 6 sampling locations).

The bacterioplankton and phytoplankton production rates were integrated over the seven month period. The productivity of bacteria (Table 3) was found to be 2.5 times higher (average 1500 g C m^{-2} 7 $months^{-1}$) in the open water areas, than at the locations covered by macrophytes (average 610 g C m^{-2} 7 $months^{-1}$). A close linear relationship ($P<0.01$) was observed between phytoplankton primary production and bacterioplankton production (Figure 3). The regression equation describing the relationship is:

$$BP = 564 + 0.85PP \quad (4)$$

$$r^2 = 0.978, n = 6$$

where: BP = bacterial production (g C m^{-2} 7 $months^{-1}$), PP = phytoplankton production (g C m^{-2} 7 $months^{-1}$).

Table 3. Cumulative bacterial production in different parts of humic rich reservoirs during 7 months period (from April to October) in 1994. [* Proportion was calculated from equation 4, as percentage of intercept value (546 g C m^{-2} 7 $months^{-1}$) to total bacterial production (g C m^{-2} 7 $months^{-1}$).]

Sampling site	Characteristic	Total bacterial production (gC m^{-2})	Proportion due to humic substances* (%)
Marcali reservoir	Open water	1402	40
LKBR 1	Open water	1732	33
LKBR 2	Open water	1306	43
LKBR 3	Subm. macroph.	676	83
LKBR 4	Reed stand	497	≈100
LKBR 5	Reed stand	657	86

LKBR = lower Kis-Balaton reservoir

Discussion

In marine and freshwater pelagic systems on average about one half of the planktonic primary production is utilised by bacteria (Cole et al., 1988).The humic-free part of the shallow Lake Balaton corresponds to this general finding; on an annual basis 47% of the planktonic primary production was channelled through the bacterioplankton (Vörös et al., 1996). The investigated humic-rich reservoirs presented a very different picture. In their open waters, where the primary producers were planktonic algae, the bacterioplankton production was about two times higher than primary production. This relationship suggests that there is a carbon source for heterotrophic organisms other than phytoplankton production.

In the macrophyta covered areas the non phytoplanktonic carbon source became decisive and only a negligible portion of the bacterioplankton production was attributable to the phytoplankton production. The large scale intersystem studies suggest that the planktonic primary production explains much of the variation in bacterial production among ecosystems (Cole et al., 1988). The observed relationship between the integrated primary and bacterial production supports this statement. The observed very high intercept (equation 4) suggests to a large non phytoplanktonic carbon source. Probably the dissolved humic substances constitute this 'extra' DOC pool.

The association of aquatic humic substances with productivity was implied in Thienemann's (1925) ecological classification of lakes according to their humic concentration and level of organic production into 'clear water lakes' and 'brown water lakes'. Thienemann (1925) introduced the still commonly used term 'dystrophic' to refer to humic lakes, since the brown colour of the water reduces light penetration and thus primary production, and the base of the planktonic food web is conventionally thought to be primary production.

Humic lakes generally have higher DOC concentrations than clearwater lakes, and this means that there is a higher carrying capacity for bacterioplankton in humic lakes. In this way, the DOC of allochthonous origin in humic lakes can constitute an input to the aquatic biota, counteracting its negative effect on primary production due to light attenuation. In environments with a significant allochthonous input of bacterial substrates (like the humic lakes), the bacteria may act as a true source, passing organic matter imported from external sources up the food web (Tranvik, 1988).

Generally it is recognised that much of the colour in water comes from humic substances leached from plant and soil organic matter (Thurman, 1985). In brownwater reservoirs, such the reservoirs studied by us, a decisive fraction of the dissolved organic carbon (DOC) arises from the dissolved humic substances. In lake water, the humic substances typically account for 40–80% of the total DOC (Peuravuori & Pihlaja, 1991).

The quantitative importance of allochthonous DOC as a bacterial substrate is in general poorly understood. However, lakes receiving large amounts of humic compounds contain large standing stocks of pelagic bacteria, compared with lakes with less humic input (Hessen, 1985; Tranvik, 1988). Exceedingly high ratios of respiration to primary production were shown for humic lakes, indicating the presence of an important organic carbon source in addition to that supplied by primary production (Sarvala et al., 1981).

Our results support the hypothesis that bacterial growth in humic lakes largely occurs at the expense of allochthonous substrates (Tranvik, 1988), but in humic waters the relationship between the two different pools of bacterial substrates largely depends on their trophic state. The planktonic primary production in humic rich waters is not necessarily low as it was originally suggested by Thienemann (1925). For instance the open areas of the investigated reservoirs showed hypertrophic character. The seasonal course of bacterial production there was determined by the seasonal change in primary production and concentration of humic substances.

The production of dissolved humic substances is significantly affected by seasonal variation in water temperature. Highest concentrations typically occur during the warmer summer period and resulted in more intensive humification and microbial processes. Meyer et al. (1987) also found that highest DOC concentrations occurred in summer and autumn when C was liberated by microbial decomposition of organic debris in streams. Considerable differences occur in the seasonal patterns of DOC at ELA (Experimental Lakes Area in northwestern Ontario, Canada), as a result of climate, soils or vegetation and hydrological regimes (Schindler et al., 1992).

Moreover, in summer not only the high temperature but also the higher UV-B radiation may have been responsible for the increased availability of dissolved humic substances during the summer months (Lindell et al., 1995). We did not check this hypothesis, but the observed high bacterial production in humic-rich waters studied requires a more detailed examination of the availability of humic substance dominated DOC.

Acknowledgments

This study was partly supported by a grant from South Transdanubian Environmental Inspectorate, by a grant from West Transdanubian Water Direction and by grant No. T 016 304 of the Hungarian Research Fund (OTKA). We are greatly indebted to R. Bertoni (Italy) for dissolved organic carbon concentration measurements.

References

Arvola, L., 1984. Vertical distribution of primary production and phytoplankton in two small lakes with different humic concentration in southern Finland. Holarct. Ecol. 7: 390–398.

Cole, J. J., S. Findlay & M. L. Pace, 1988. Bacterial production in fresh- and saltwater ecosystems: a cross overview. Mar. Ecol. Prog. Ser. 43: 1–10.

Cuthbert, I. D. & P. del Giorgio, 1992. Toward a standard method of measuring colour in freshwater. Limnol. Oceanogr. 37: 1319–1326.

Fisher, R. A., 1963. Statistical Tables for Biological, Agricultural and Medical Research. 7th edn. Oliver-Boyd, London, 146 pp.

Fuhrman, J. A. & F. Azam, 1982. Thymidine incorporation as a measure of heterotrophic bacterioplankton production in marine surface waters: evaluation and field results. Mar. Biol. 66: 109–120.

Gjessing, E. T., 1976. Physical and Chemical Characteristics of Aquatic Humus. Ann. Arbor Science Publ., Ann Arbor Michigan, 120 pp.

Hessen, D. O., 1985. The relation between bacterial carbon and dissolved humic compounds in oligotrophic lakes. FEMS Microb. Ecol. 31: 215–223.

Johansson, J. Å., 1983. Seasonal development of bacterioplankton in two forest lakes in central Sweden. Hydrobiologia 91: 71–87.

Kirk, J. T. O., 1976. Yellow substance (gelbstoff) and its contribution to the attenuation of photosynthetically active radiation in some inland and coastal south-eastern Australian waters. Aust. J. mar. Freshwat. Res. 27: 61–71.

Larsson, U. & Å. Hagström, 1982. Fractionated phytoplankton primary production exudate release and bacterial production in a Baltic eutrophication gradient. Mar. Biol. 67: 57–70.

Lavandier, P., 1990. Dynamics of bacterioplankton in a mesotrophic French reservoir (Pareloup). Hydrobiologia 207: 79–86.

Lee, S. & J. A. Fuhrman, 1987. Relationships between biovolume and biomass on naturally derived marine bacterioplankton. Appl. envir. Microbiol. 53: 1298–1303.

Lindell, M. J., W. Granéli. & L. J. Tranvik, 1995. Enhanced bacterial growth in response to photochemical transformation of dissolved organic matter. Limnol. Oceanogr. 40: 195–199.

Lowell, C. R. & A. Konopka, 1985. Primary and bacterial production in two dimictic Indiana lakes. Appl. envir. Microbiol. 49: 485–491.

Meyer, J. L., C. M. Tate, R. T. Edwards & M. T. Crocker, 1987. The trophic significance of dissolved organic carbon in streams. In Swank, W. T. & D. A. Crossley (eds), Forest Hydrology and Ecology at Coweeta. Springer-Verlag, New York: 269–278.

Peuravuori, J. & K. Pihlaja, 1991. Isolation and fractionation of humic substances in lake waters. In Allard, B., H. Borén & A. Grimvall (eds), Humic Substances in the Aquatic and Terrestrial Environment. Springer: 123–134.

Salonen, K., K. Kononen & L. Arvola, 1983. Respiration of plankton in two small, polyhumic lakes. Hydrobiologia 101: 65–70.

Sarvala, J., V. Ilmavirta, L. Paasivirta & K. Salonen, 1981. The ecosystem of the oligotrophic lake Pääjärvi 3. Secondary production and en ecological energy budget of the lake. Verh. int. Ver. Limnol. 21: 454–459.

Schindler, D. W., S. E. Bayley, P. J. Curtis, B. R. Parker, M. P. Stainton & C. A. Kelly, 1992. Natural and man-caused factors affecting the abundance and cycling of dissolved organic substances in precamprian shield lakes. Hydrobiologia 229: 1–21.

Sieburth, J. McN. & A. Jensen, 1968. Studies on algal substances in the sea. I. Gelbstoff (humic material) in terrestrial and marine waters. J. exp. mar. Biol. Ecol. 2: 174–189.

Steinberg, C. & U. Münster, 1985. Geochemistry and ecological role of humic substances in lakewater. In Aiken, G. R., D. M. McKnight, R. L. Wershaw & P. Maccarthy (eds), Humic Substances in Soil, Sediment, and Water. John Wiley & Sons, New York: 105–145.

Thienemann, A., 1925. Die Binnengewässer Mitteleuropas. Binnengewässer 1, 255 pp.

Thurman, E. M., 1985. Organic Geochemistry of Natural Waters. Martinus Nijhoff/Dr W. Junk Publishers, Boston, 350 pp.

Tranvik, L. J., 1988. Availability of dissolved organic carbon for planktonic bacteria in oligotrophic lakes of differing humic content. Microb. Ecol. 16: 311–322.

Tranvik, L. J., 1989. Bacterioplankton in humic lakes – a link between allochthonous organic matter and pelagic food webs. Dissertation, Lund, 104 pp.

Vollenweider, R. A., 1969. A Manual on Methods for Measuring Primary Production in Aquatic Environments. IBP Handb. 12. Blackwell Sci. Publ., Oxford, 225 pp.

Webb, W. L., M. Newton & D. Starr, 1974. Carbon dioxide exchange of *Alnus rubra*: A mathematical model. Oecologia 17: 281–291.

Wetzel, R. A. & G. E. Likens, 1990. Limnological Analyses. 2nd edn. Springer Verlag, New York, 391 pp.

Vörös, L., K. V.-Balogh & S. Herodek, 1996. Microbial food web in a large shallow lake (Lake Balaton, Hungary) Hydrobiologia 339: 57–65.

Hydrobiologia **342/343:** 71–78, 1997.
L. Kufel, A. Prejs & J. I. Rybak (eds), Shallow Lakes '95.

The occurrence of heterotrophic bacteria decomposing some macromolecular compounds in shallow estuarine lakes

Zbigniew Mudryk[1] & Wojciech Donderski[2]
[1] *Pedagogical University, Institute of Biology, Department of Experimental Biology, Arciszewskiego 22b, 76-200 Słupsk, Poland*
[2] *Nicolaus Copernicus University, Institute of Biology, Department of Water Microbiology and Biotechnology, Gagarina 9, 87-100 Toruń, Poland*

Key words: estuary, planktonic bacteria, physiological properties, organic matter, decomposition

Abstract

Planktonic bacteria participating in decomposition processes of organic macromolecular compounds were studied in shallow estuarine lakes Łebsko, Gardno and Jamno.The abilities of heterotrophic bacteria to decompose proteins, lipids, starch, nucleic acids, pectin, chitin and cellulose were determined. Characteristic features among the bacteria in the lakes under study were to decompose a wide spectrum of organic macromolecular compounds. Most bacteria hydrolysed proteins, lipids and starch. Pectinolytic and chitinolytic bacteria were less numerous as well as microorganisms able to hydrolyse nucleic acids. The microflora hydrolysing cellulose was represented by the least abundant group of organisms.

Introduction

Heterotrophic bacteria performing the processes of organic matter decomposition are the most numerous group of microorganisms in aquatic ecosystems (Donderski et al., 1984; Kim & Hoppe, 1984; Chróst, 1991; Karner et al., 1992). These organisms are able to decompose a wide spectrum of organic compounds whose molecules differ in size from monomers to polymeres (Jacobsen & Azam, 1984; Quemeneur & Morty, 1992). Microbial hydrolysis of organic matter plays an important role in substrate turnover in aquatic environments (Kim & Hoppe, 1984). Most of the organic compounds produced in natural waters have a polymeric structure (Somville, 1984; Münster & Chróst, 1990). They cannot be directly taken by bacteria and they must be hydrolysed through the action of exo and endoenzymes and converted into monomeric low molecular substrates (Billen & Fontigny, 1987; Meyer-Reil, 1987).These are rapidly assimilated by bacteria (Chróst, 1991).

As there are differences between various types of water bodies in the composition and content of organic matter, the intensity of biodegradation by bacterial microflora differs as a function of the biochemical activity of the dominant physiological groups (Godlewska-Lipowa, 1974; Krstulović & Solić, 1988).

The present study was aimed to investigate physiological properties of bacteria in the water of estuarine lakes, their ability to carry out particular metabolic processes and to evaluate their potential possibility in degradation of certain organic macromolecular compounds.

Study area

Lakes Łebsko, Gardno and Jamno are situated close to the Polish Baltic Sea coast (Figure 1). Two lakes Łebsko and Gardno are situated in the World Biosphere Reserve – the Słowinski National Park. The investigated estuarine lakes are very shallow with an avarage depth of 1.5 m and cover large areas, ca. 20–70 km^2 each. They are 7–16 km long and 3–8 km wide. The emergent macroflora covers 3–9% of their surfaces forming a wide offshore belt of 20–200 m, which constitutes a residence for many bird species. Main species of plants are *Typha angustifolia, Phragmites australis* and *Schoenoplectus lacustris*.

These lakes are supported by rivers and, at the same time, they are influenced by Baltic Sea waters. Through narrow channel, large volumes of sea water flow into the lakes, especially in autumn and winter months. Therefore, water of these lakes often equales in salinity to that of Baltic Sea. Consistently with the venetian system, these lakes can be classified as the mixo-oligohaline type (Dethier, 1992). Lakes Łebsko and Gardno are of eutrophic type, Lake Jamno is a hypereutrophic reservoir. Some physicochemical properties of water in the lakes under study are shown in Table 1 (Trojanowski et al., 1990).

Materials and methods

Microbiological studies were carried out during the period 1986–1989. In each lake there were six stations: one near the river inflow (station 1), one in the near-sea part (station 6) and mid-lake stations (2–5) (Figure 1). All stations were situated in open water. Samples were taken four times during the year: in March, May, July and October.

Water samples were taken directly into sterile bottles from the surface (ca. 20 cm below water surface). Samples were placed in a container with ice and transported to the laboratory. The time between sample collection and performance of the analyses usually did not exceed 6–8 h.

For determining the total number of heterotrophic bacteria (CFU), the samples of water were diluted with sterile buffered water (pH 7.2) (Daubner, 1967) and inoculated by the spread plate method, in five replicates, on iron-peptone agar medium (IPA) prepared according to Ferrer et al. (1963). After 10 day incubation at 20 °C, bacterial colonies were counted and results were recalculated per 1 cm^3 of water. Then, from the whole surface of the plates or from a particular sector ca. 50 bacterial colonies, were collected at random from each station and transferred to semi-liquid IPA medium. After purity control, the bacteria were stored at 4 °C, with inoculation on fresh medium every 3 months and, subsequently they were used for further studies.

The isolated bacteria were inoculated on several test media containing various organic compounds. The following decomposition abilities of bacteria were tested:

1. The ability to decompose proteins, starch and lipids were assayed in IPA medium enriched with either gelatin (20 g per dm^3) or starch (5 g per dm^3) or tributyrin (10 ml per dm^3). The plates were point inoculated. The ability to hydrolyse proteins was tested using Frazier reagent (Weyland et al., 1970). Amylolytic bacteria were determined using Lugol's solution (Seiler et al., 1980). Lipolytic properties of the bacteria were determined on the basis of clear zones formation around bacterial colonies (Lawrence et al., 1967).
2. The ability to decompose deoxyribonucleic acid (DNA) was assayed in Oxoid medium (England) containing 2.0 g DNA per dm^3 of medium. Double-layer plates were used: DNA was added to the upper layer of medium. Hydrolysis of DNA was disclosed by pouring 1 N HCl over the plates. Clear zones around the colonies pointed to the DNA decomposition ability (Seiler et al., 1980).
3. Hydrolysis of ribonucleic acid (RNA) was determined according to Jeffries et al., (1957). The decomposition of RNA was revealed as in item 2.
4. Pectinolytic bacteria were assayed in medium prepared according to Jayasankar & Graham (1970). After incubation a 1% solution of Cetrimide (Searle Co., England) was poured over the medium. The occurrence of clear zones was accepted as a positive result.
5. Chitin hydrolysis was estimated in medium prepared according to Helmke & Weyland (1986). Colloidal chitin was prepared according to Lingappa & Lockwood (1962), using chitin from Windsor, Berkshire (England). The appearance of clear zones was accepted as a positive result.
6. The ability to decompose cellulose was investigated on a medium composed of KNO_3 – 1.0 g, K_2HPO_4 – 1.0 g, $MgSO_4 \cdot 7H_2O$ – 0.2 g, $CaCl_2$ $2H_2O$ – 0.1 g, $FeCl_3$ – 0.01 g, colloidal cellulose – 7.0 g of dry mass, agar – 15.0 g, tap water 1 dm^3. Colloidal cellulose was prepared from Cellulose Powder 11, Whatman according to Halliwell (1962). Double layer plates were used. The occurrence of clear zones was accepted as a positive result.

pH of all media were adjusted to 7.0–7.4 and were sterilized at 117 °C for 20 min. 48–72 hours cultures from slants on the IPA medium were applied as inoculum. Results were checked after 6-day incubation at 26 °C, with the exception of pectin, chitin and cellulose, where the decomposition was determined after 14-day incubation. Degree of proteolytic, lipolytic and

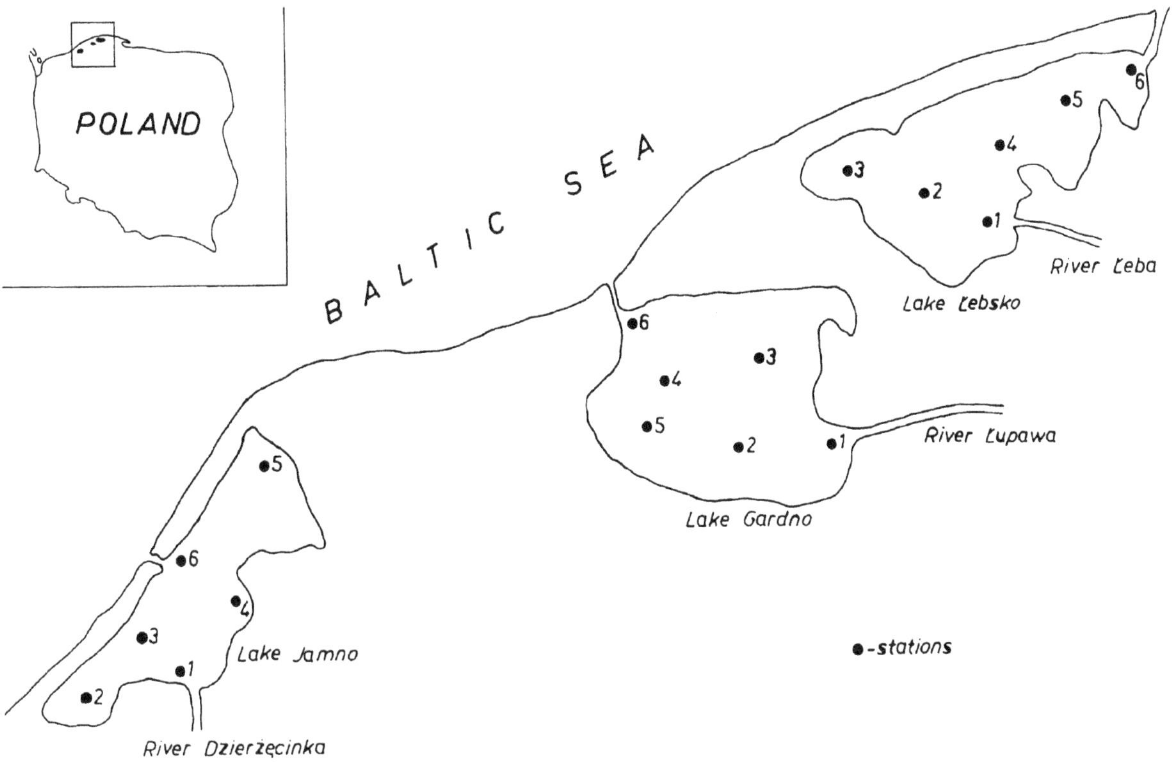

Figure 1. Location of the Łebsko, Gardno and Jamno Lakes.

Table 1. Mean values of selected physiochemical parameters of the water in lakes Łebsko, Gardno and Jamno in the period 1987–1989. (data by Trojanowski et al., 1990)

Parameter	Dimension	Lake Łebsko		Lake Gardno		Lake Jamno	
Transparency	m	0.57	(0.20–0.25)	0.35	(0.10–1.10)	0.15	(0.10–0.25)
Temperature	°C	13.3	(7.0–23.0)	14.7	(0.5–23.4)	13.8	(0.1–22.7)
pH		8.8	(7.0-10.2)	9.0	(7.7–9.7)	9.1	(7.2–10.0)
Oxygen	mg O_2 dm^{-3}	11.8	(8.3-12.5)	11.5	(8.7–12.8)	7.7	(1.6–10.9)
BOD_5	mg O_2 dm^{-3}	5.1	(2.6–6.8)	6.3	(4.3–8.4)	9.0	(6.0–10.5)
N - total	mg N dm^{-3}	3.46	(1.05–6.24)	3.62	(1.51–5.94)	5.16	(3.82–6.48)
P - total	mg P dm^{-3}	0.256	(0.171–0.393)	0.370	(0.145–0.600)	0.414	(0.235–0.691)
Salinity	mg Cl dm^{-3}	542	(21–2462)	510	(30–2090)	550	(49–2191)
Total suspended matter	mg dm^{-3}	41	(55 - 61)	52	(28–69)	75	(36–104)
Chlorophyll 'a'	mg m^{-3}	68.5		85.3		120.4	
Primary production	mg C m^{-2} h^{-1}	220	(214–229)	236	(183–241)	881	(740–961)

amylolytic activity of bacteria was determined by measured diameter clear zones around the colonies according to Lawrence et al. (1967), Strzelczyk et al. (1976).

Results

In general the results obtained suggest that abundance of heterotrophic bacteria depended on the degree of trophy of the lakes and revealed considerable seasonal

Table 2. Occurrence of bacteria able to decompose selected macromolecules in water of three estuarine lakes (average all stations)

Lake	Date of sampling	Number of strains studied	% of bacteria decomposing indicated substrates							
			protein	lipid	starch	DNA	RNA	pectin	chitin	cellulose
	08.05.86	285	51	82	53	22	14	16	12	2
	03.07.86	285	75	84	52	17	21	21	10	5
Gardno	09.10.86	287	76	85	50	33	15	25	14	8
	25.03.87	289	45	61	50	23	7	23	5	2
	20.05.87	296	52	46	46	29	21	20	32	10
	24.07.87	290	53	61	33	24	30	25	26	6
Łebsko	14.10.87	282	60	49	35	28	38	27	43	15
	16.03.88	284	38	42	29	31	27	28	23	8
	11.05.89	289	73	67	55	12	7	11	5	0
	13.07.89	291	79	61	51	14	11	13	10	2
Jamno	12.10.89	290	74	85	63	14	10	13	4	4
	09.03.89	239	54	63	48	9	6	10	5	1

variations (Figure 2). On the whole the investigations showed that planktonic bacteria occurred in the largest number in the hypereutrophic Lake Jamno and they were the least numerous in the eutrophic lakes Łebsko and Gardno. Figure 2 showed that in all three lakes the greatest number of bacteria in water appeared in summer, reaching its minimum in winter. This regularity was noticed during three years of investigation.

The occurrence of bacteria with particular physiological properties among the microorganisms isolated from the water of the lakes Łebsko, Gardno and Jamno are presented in Table 2. It is evident from these data that most strains showed the ability to decompose different macromolecular compounds. Organisms hydrolysing proteins and lipids were the most numerous group among the strains studied (Table 2). Proteolytic bacteria constituted 38–79%, lipolytic bacteria accounted for 42–85% of the total number of heterotrophic bacteria. The most rapid development of bacteria hydrolysing proteins and lipids occurred in summer and autumn. Both these physiological groups usually showed the lowest percentage in winter, in all studies lakes. The organisms capable of performing the process of starch hydrolysis also were rather numerous The largest numbers of amylolytic organisms were found in Lake Jamno, the smallest in Lake Łebsko. Their maximum development was usually noted in spring and autumn. Bacteria able to hydrolyse DNA and RNA were present at lower percentages than the above-mentioned physiological groups (Table 2) and they were more abundant in Lakes Łebsko and Gardno than in Lake Jamno. In Lakes Łebsko and Gardno throughout the investigation period, pectinolytic and chitinolytic organisms showed a higher percentage of occurrence than in the Lake Jamno were they only made up to 4–13% of the total amount of heterotrophic bacteria. The microflora hydrolysing cellulose represented the least numerous physiological group of organisms isolated from three estuarine lakes. They were somewhat more abundant in the Lake Łebsko where they constituted from 6% to 15% of the total number of bacteria.

No regular pattern in horizontal distribution of studied physiological groups were established (Figure 3). In all parts of investigated lakes the abundance of planktonic bacteria with ability to decompose organic macromolecular compounds showed a similar level with the exception for lipolytic bacteria which showed maximum density in seawater zone (st. 6).

Figure 4 illustrated the degree of proteolytic, lipolytic and amylolytic activity of bacteria isolated in the highest percentage from water of lakes Łebsko, Gardno and Jamno. About 30–40% of strains have high proteolytic and lipolytic activity. The amylolytic activity of planktonic bacteria was the lowest.

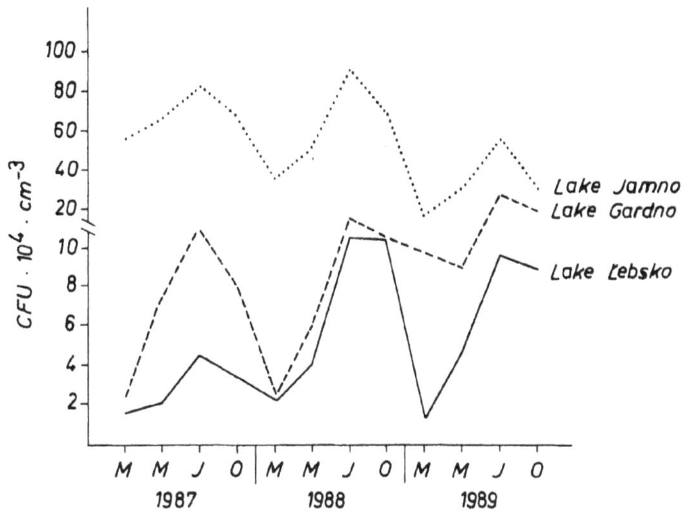

Figure 2. Seasonal fluctuations of heterotrophic bacteria

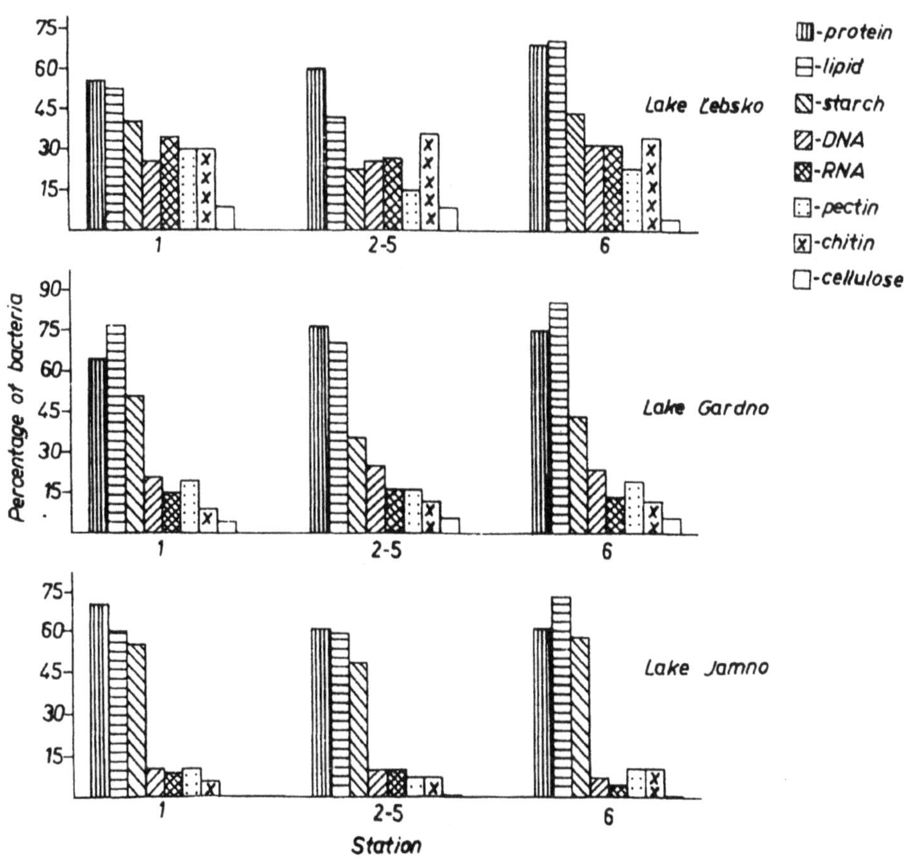

Figure 3. Spatial distribution of the studied bacterial physiological groups.

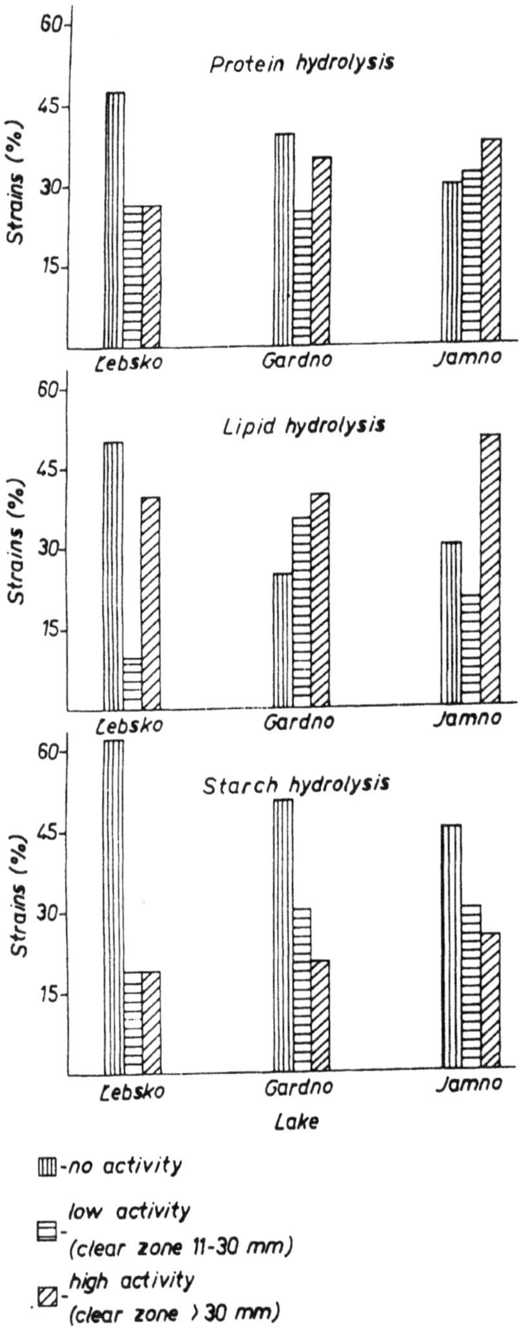

Figure 4. Degree of proteolytic, lipolytic and amylolytic activity of planktonic bacteria.

Discussion

Aquatic ecosystems are inhabited by bacteria able to carry out various metabolic processes which facilitates matter and energy transformation (Riemann, 1983; Donderski & Lalke, 1993). Due to their abilities, heterotrophic bacteria participate in the decomposition of dead bodies of plants and animals or their remains (Donderski & Stopinski, 1993).

A considerable part of heterotrophic bacteria inhabiting water of studied lakes was capable to hydrolyse different macromolecular compounds being a source of food and energy for them. In general proteolytic and lipolytic bacteria were numerous among the studied physiological groups. The occurrence of large quantities of those bacteria in other estuarine reservoirs has also been observed earlier by Sieburth (1978); Austin et al. (1977); Hoshimoto et al. (1983); Ellis-Evans (1985).

According to Little et al. (1979), the protein decomposing bacteria occur in such a great number because in water bodies proteins, peptides and amino acids are the main components of organic matter. Excretions and dead bodies of the phyto and zooplankton are their basic source (Billen & Fontigny, 1987). Hoppe et al. (1988) reported that bacteria of brackish waters display a higher metabolic activity against proteins and amino acids than against carbohydrates. In examined lakes proteolytic activity was the lowest in winter period, probably on account of low temperatures; namely according to Little et al. (1979), temperatures below 15 °C abruptly inhibit the activity and synthesis of proteases.

From our investigations it follows that lipolytic organisms also occurred in great number among bacterioplankton. Studies carried out in other brackish waters confirm the presence of a large number of lipolytic bacteria. Sieburth (1978) found in the estuarine reservoir Narragansett 13–84% of lipolytic bacteria and Austin et al. (1977) in the Cheaspeake Bay more than 50%. In the sea water lipolytic bacteria constituted 80–100% of the total number of microflora (Bölter & Rheinheimer, 1987; Prieur, 1989; Mudryk et al., 1991). Kjelleberg & Hakansson (1977) and Quemeneur & Morty (1992) explain this phenomenon by the accumulation of many lipids on water surface, like the triglycerides, phospolipides, free fatty acids, lipoproteins or wax esters, whose presence creates optimal conditions for the development of lipolytic bacteria. Particularly large amounts of lipids are accumulated in the cells of bluegreen algae, green algae as well as Copepoda in the case zooplankton (Jacobsen & Azam, 1984; Arts et al., 1992). According to Nitkowski et al. (1977) a rise of salinity of the aquatic environment leads to an increase in the lipase activity. The present results confirm this regularity. Namely, in the water of investigated lakes the percentage of lipolytic bacteria was the highest in seawater zone where the salinity was higher than that at the remaining stations.

Amylolytic bacteria represent a relatively numerous group of heterotrophic bacteria in the investigated estuarine reservoirs. They accounted for ca 40% of the total numbers of planktonic microflora. Similar results were observed earlier in other lakes by Ellis-Evans (1985); Sugita et al. (1987); Donderski & Stopiński (1993). The results of the present work point out that organisms capable to hydrolyse starch occur in the highest numbers during spring and autumn. This corresponds with the results obtained by Krstulović & Solić (1988) in the Kastela Bay.

Degradation of nucleic acids from decaying plant and animal tissues is the essential function of heterotrophic bacteria. As a result, bacteria able to decompose DNA and RNA are numerous both in inland basins and in the marine environment (Niewolak, 1980; Strzelczyk et al., 1972; Maeda & Taga, 1974; Stewart et al., 1991). On the other hand, in our investigations these organisms were present in lower percentage than other physiological groups of bacteria.

The ability of many bacteria to hydrolyse pectin and chitin and the high activity of their enzymes indicate the considerable importance of these groups of organisms for the processes of plant and animal remains decomposition in aquatic ecosystems. Occurrence of large number of chitinolytic bacteria in lakes Lebsko and Gardno can be accounted for by rapid mass growth of zooplankton (1000 ind. cm^3) mainly Cladocera and Copepoda (Żmudziński et al., 1992). Their shells may stimulate the development of chitinoclastic bacteria. Relatively small number of bacteria decomposing pectin and chitin in the water of hypereutrophic Lake Jamno confirms earlier studies by Mudryk, (1994). This is probably related to its content of very diverse food substances, many of which may be metabolized more easily and more readily than chitin and pectin.

The least numerous group among planktonic bacteria in studied lakes were microorganisms decomposing cellulose. The results agree with those published earlier by other authors (Bölter, 1977; Sugita et al., 1987; Petrycka et al., 1990).

References

Arts, D., M. S. Evans & D. Robarts, 1992. Seasonal patterns of total and energy reserve lipids of dominant crustaceans from a hypereutrophic lake. Oecologia 90: 560–571.

Austin, B., S. D. Allen, A. Milles & R. Colwell, 1977. Numerical taxonomy of heavy metal-tolerant bacteria isolated from an estuary. Can. J. Microbiol. 23: 1433–1447.

Billen, G. & A. Fontigny, 1987. Dynamics of Phaeocystis-dominant spring bloom in Belgian coastal waters. II. Bacterioplankton dynamics. Mar. Ecol. Prog. Ser. 37: 249–257.

Bölter, M., 1977. Numerical taxonomy and character analysis of saprophytic bacteria isolated from Kiel Fjord and Kiel Bight. Ecol. Stud. 25: 148–176.

Bölter, M & G. Rheinheimer, 1987. Numerical analysis of microbial and chemical characters and saprophytic bacteria from the Baltic Sea. Bot. mar. 30: 535–544.

Chróst, R. J., 1991. Environmental control of the synthesis and activity of aquatic microbial ectoenzymes. In Chróst, R. J. (ed.), Microbial Enzymes in Aquatic Environments. New York, Springer-Verlag: 29–59.

Daubner, J., 1967. Mikrobiologia Vody. Slov. Akad. Vied. Bratyslava, 354 pp.

Dethier, M. N., 1992. Classifying marine and estuarine natural communities. An alternative to the coward in system. J. nat. Areas 12: 90–100.

Donderski, W., T. Bylinska, H. Czajkowski, B. Gardocka, D. Kowalkowska, G. Myzyk & E. Poziemska, 1984. Initial studies on heterotrophic bacteria capable of decomposition of some macro-molecular compounds in waters and bottom sediments of six lakes in Iława Lake District. AUNC. Toruń Limnol. Papers. 14: 75–82.

Donderski, W. & E. Lalke, 1993. Occurrence and physiological properties of epiphytic bacteria isolated from reed (*Phragmites australis* Trin. ex Steudel). AUNC. Toruń Limnol. Papers. 18: 13–24.

Donderski, W. & M. Stopiński, 1993. Occurrence and physiological properties of heterotrophic bacteria inhabiting the tripton of lakes of different trophy. AUNC Toruri. Limnol. Papers. 18: 3–12.

Ellis-Evans, J. C., 1985. Decomposition processes in marine Antarctic lakes. Ant. Nut. Cyc. 1: 253–260.

Ferrer, E. B., E. M. Stapert & W. T. Sokalski, 1963. A medium for improved recovery of bacteria from water. Can. J. Microbiol. 9: 420–422.

Godlewska-Lipowa, W., 1974. Heterotrophic activity of bacterial microflora in Mazurian lakes of various trophy. Pol. Arch. Hydrobiol. 21: 51–58.

Halliwell, G., 1962. Cellulose. In Bergmeyer, H. V. (ed.), Methods of Enzymatic Analysis. Verlag Chemie GmbH Weinhein: 64–71.

Helmke, E. & H. Weyland, 1986. Effect of hydrostatic pressure and temperature on the activity and synthesis of chitinases of Antarctic Ocean bacteria. Mar. Biol. 91: 1–7.

Hoppe, H. G., W. Schramm & P. Bacold, 1988. Spatial and temporal distribution of pelagic microorganisms and their proteolytic activity outer a partly destroyed coral reef. Mar. Ecol. Prog. Ser. 44: 95–102.

Hoshimoto, H., N. Hayasahi, M. Sasaki, H. Kawakami, H. Nakano, M. Kiiynkia & S. Ishida, 1983. Dynamics of the bacteria flora in the estuarine region of the Ohta river. J. Fac. Appl. Biol. Sci. 22: 217–227.

Jacobsen, T. R. & F. Azam, 1984. Role of bacteria in copepod fecal pellet decomposition. Colonization, growth rates and mineralization. Bull. mar. Sci. 35: 495–502.

Jayasankar, N. P. & P. H. Graham, 1970. An agar plate method for screening and enumerating pectinolytic microorganisms. Can. J. Microbiol. 16: 10–23.

Jeffries, Ch. D. D., D. F. Holtman & D. G. Gure, 1957. Rapid method for determining the activity of microorganisms on nucleic acids. J. Bact. 73: 590–591.

Karner, M., D. Fuks & G. J. Herndel, 1992. Bacterial activity along a trophic gradient. Microb. Ecol. 24: 243–257.

Kim, S. J. & H. G. Hoppe, 1984. Microbial extracellular enzyme detection on agar by means of fluorogenic methylumbelliferyl substrates. CNRS. Brest Acta Coll. 3: 175–183.

Kjelleberg, S. & N. Hakansson, 1977. Distribution of lipolytic, proteolytic and amylolytic marine bacteria between the film and subsurface water. Mar. Biol. 39: 103–109.

Krstulović, N. & M. Solić, 1988. Distribution of proteolytic, amylolytic and lipolytic bacteria in Kastela Bay. Acta Adria. 29: 75–83.

Lawrence, R. C., T. F. Fryer & B. Reiter, 1967. Rapid method for the quantitative estimation of microbial lipases. Nature 213: 1264–1265.

Lingappa, Y. & J. L. Lockwood, 1962. Chitin media for selective isolation and culture of actinomycetes. Phytopatology 52: 318–323.

Little, A. G., R. B. Sjogern & C. R. Carson, 1979. Measurement of proteolysis in natural waters. Appl. envir. Microbiol. 37: 900–908.

Maeda, M.& N. Taga, 1974. Occurrence and distribution of deoxyribonucleic acid-hydrolyzing bacteria in sea water. J. Exp. mar. biol. Ecol. 14: 157–169.

Meyer-Reil, L. A., 1987. Seasonal and spatial distribution of extracellular enzymatic activities and microbial incorporation of dissolved organic substrates in marine sediments. Appl. envir. Microbiol. 53: 1748–1755.

Mudryk, Z., 1994. Heterotrophic bacteria in the transformation processes of organic matter in estuarine lakes. Pedagogical University Press. Słupsk, 175 pp.

Mudryk, Z., K. Korzeniewski & L. Falkowska, 1991. Bacteriological investigation of the surface microlayer of Gulf of Gdansk. Oceanologia 30: 93–103.

Münster, U. & R. J. Chróst, 1990. Origin, composition and microbial utilization of dissolved organic matter. In Overbeck, J. & R. J. Chróst (eds), Aquatic Microbial Ecology. Biochemical and Molecular Approaches. Springer-Verlag, New York: 8–46.

Niewolak, S., 1980. Occurrence of microorganisms in fertilized lakes. I. Ribonucleic acid-hydrolizing microogranisms. Pol. Arch. Hydrobiol. 27: 37–52.

Nitkowski, N. F., S. Dudley & J. T. Graikowski, 1977. Identification and characterization of lipolytic and proteolytic bacteria isolated from marine sediments. Mar. Poll. Bull. 8: 276–279.

Petrycka, H., J. Mrozowska & H. Kasza, 1990. Changes in bacterial microflora against the background of increasing eutrophication of Goczałkowice Reservoir, southern Poland. Acta Hydrobiol. 32: 55–66.

Prieur, D., 1989. Preliminary study of heterotrophic bacterial communities in water intervertebrates from deep sea hydrothermal vents. Proc. 21th EMBS Gdańsk 393–401.

Quemeneur, M. & Y. Morty, 1992. Sewage influence in a macrotidial estuary: fatty acid and sterol distribution. Estuar. Coast. mar. Sci. 34: 347–363.

Riemann, B., 1983. Biomass and production of phyto-bacterioplankton in eutrophic Lake Tystrup, Denmark. Freshwat. Biol. 13: 389–398.

Seiler, H., R. Braatz & G. Ohmayer, 1980. Numerical cluster analysis of the Coryne-form bacteria activated sludge. Zbl. Bakt. Hyg. 1: 357–375.

Sieburth, J. N., 1978. Bacterioplankton: nature, biomass and relationship to the protist plankton. J. Phycol. 14: 31–41.

Somville, M., 1984. Measurement and study of substrate specificity of exoglucosidase activity in eutrophic water. Appl. envir. Microbiol. 48: 1181–1185.

Stewart, G. J., C. D. Sinigalliano & O. Garko, 1991. Binding of exogenous DNA to marine sediments and effect of DNA sediment binding on natural transformation of *Pseudomonas stutzeri* strain Zo Bell in sediment columns. Microbiol. Ecol. 85: 1–8.

Strzelczyk, E., W. Donderski & A. Mielczarek, 1972. Studies of cultural properties of planktonic, benthic and ephytic bacteria. AUNC Toruń Limnol. Papers. 7: 3–12.

Strzelczyk, E., W. Donderski & M. Stopiński, 1976. Studies on physiological of heterotrophic bacteria isolated from water and mud three lakes. AUNC Toruń Limnol. Papers 9: 27–38.

Sugita, H., K. Oshima., T. Fushino & Y. Deguchi, 1987. Substrate specificity of heterotrophic bacteria in water and sediment of carp culture pond. Aquacult. 64: 39–46.

Trojanowski, J., Cz. Trojanowska & K. Korzeniewski, 1990. Warunki hydrochemiczne w jeziorach przymorskich (Hydrochemical condition in coastal lakes). Słupskie Prace Mat-Przyr. 8: 123–139.

Weyland, H., H. J. Rüger & H. Schwarz, 1970. Zur Isolierung und Identifizierung mariner Bakterien. Ein Beitrag zur Standardisierung und Entwicklung adäquater Methoden. Veroff. Inst. Meeresforsch. 12: 269–296.

Zmudziński, L., Z. Dobrowolski, M. Labuda, Z. Mudryk, E. Paturej & Cz. Trojanowska, 1992. Variability of the biocenoses of three Polish estuarine lakes. Proc. 12th Baltic Marine Biologist Symposium. Helsingor: 185–189.

Hydrobiologia **342/343**: 79–85, 1997.
L. Kufel, A. Prejs & J. I. Rybak (eds), Shallow Lakes '95.

Picoplankton and other size groups of phytoplankton in various shallow lakes

Elżbieta Szeląg-Wasielewska
Department of Water Protection, Adam Mickiewicz University, Drżymały 24, 60-613 Poznań, Poland

Key words: lobelian lakes, phytoplankton size groups, picoplankton, abundance, biomass

Abstract

This paper presents the results of the studies on picoplankton and other size groups of phytoplankton from twelve lakes (maximum depth from 3.5 to 7.8 m) located in the Polish part of Pomerania. Samples were taken from the pelagial zone in August 1993 and 1994. Phytoplankton composition, abundance and biomass were investigated within the three main size groups, i.e. pico-, nano- and phytoplankton >20 μm. Picoplankton numbers ranged from 3.2×10^2 to 1.16×10^6 cells ml^{-1}. Its population density, with the exception of one lake, was greater by 1–3 orders of magnitude than nano- and >20 μm populations, while its biomass values in the range from <0.001 to 0.65 mg l^{-1} did not exceed 11% of the total phytoplankton biomass. Picoplanktonic communities were composed of cyanobacteria and/or eukaryotes. A significant, negative correlation between picoplankton contribution to total biomass and total biomass was observed. Phytoplankton >20 μm was the most important constituent of the biomass in most lakes, and it reached highest values and proportions: 23.3 mg l^{-1} and 99.9%, respectively. In comparison, the nanoplankton biomass was not so high: up to 1.3 mg l^{-1} and 65.2% proportions.

Introduction

To understand the phytoplankton community as a whole, a knowledge of its smallest size group – picoplankton, is essential. The interest in picoplankton has been growing for more than ten years, which can be seen in the increasing number of articles published in literature. Some of them are of review and summary nature (Stockner & Antia, 1986; Stockner, 1988, 1991). Picoplankton is widespread in freshwater ecosystems. The predominant taxon is the cyanobacterial genus *Synechococcus* sp.. The relative importance of picoplankton is highest in the oligotrophic lakes, decreasing in importance along the eutrophic gradient, but the number of cells may be highest under eutrophic conditions. The seasonal and vertical distribution of the picoplankton community has been described from many lakes in Europe (Weisse, 1988; Happey-Wood, 1991; Hawley & Whitton, 1991; Søndergaard, 1991; Vörös et al., 1991; Amblard et al., 1995). However, there is a paucity of reports of the abundance, biomass and composition of the entire phytoplanktonic community from shallow, small lakes.

The aim of the present research was to identify photoautotrophic plankton organisms <2.0 μm in size, to define their numbers, biomass and proportion in the phytoplankton community as well as to establish the taxonomic composition and size of the remaining part of planktonic algae composition in shallow lobelian lakes located in the West Pomerania region. Results of similar research on two other lakes in this area was presented earlier (Szeląg-Wasielewska, 1994).

Study area

The studied lakes are located in the northern part of West Pomerania in the area of, and between two towns: Złocieniec and Bytów (Figure 1). Their common characteristic feature is the presence of a rare relict plant – *Lobelia dortmanna* L. The work of Kraska et al., (1994) concerning several dozen lobelian lakes in West Pomerania showed large trophic differences between

Table 1. Some characteristices of the lakes.

No	Lake	Area ha	Max depth m	Total phytoplankton biomass* mg l^{-1}
1	Cietrzewie Małe	3.5	7.5	0.2
2	Krasne	28.0	5.5	2.2
3	Lubienieckie Małe	3.3	5.8	2.7
4	Bobięcińskie Małe	33.3	4.0	4.2
5	Sękacz	15.0	4.0	5.3
6	Lubienieckie Duże	13.0	4.0	6.4
7	Borzytuchom III	9.6	4.5	11.7
8	Nowoparszczenickie	2.3	4.9	13.6
9	Nierybno	9.0	5.2	13.7
10	Ciemniak	8.0	7.8	18.1
11	Leśniówek Mały	1.6	5.0	21.4
12	Linowskie	10.5	3.5	24.2

* Calculated on the basis of microscopic measurments and expressed as a weighted mean value for the water column.

these lakes. For this research, a group of the twelve most shallow lakes out of 42 was chosen, in which in the years 1993–1994 all phytoplankton size groups were analysed. As can be seen in Table 1, the lakes are not only shallow, but for the most part, small as well. Furthermore, according to the total phytoplankton biomass values, the lakes represent various trophic states, from oligotrophic to eutrophic.

Materials and methods

The samples from 9 lakes (Borzytuchom III, Cietrzewie Małe, Krasne, Linowskie, Lubienieckie Małe, Lubienieckie Duże, Nierybno, Nowoparszczenickie, Sękacz) were collected in August 1993, and from 3 lakes (Bobięcińskie Małe, Ciemniak, Leśniówek Mały) in August 1994. Sampling stations were located in the deepest part of the lakes' pelagial. Integrated samples of epilimnetic water and in five lakes of metalimnetic water were also taken. Water for picoplankton study was fixed with glutaraldehyde to the final concentration of 1%. Subsamples of 1–10 ml volume were filtered on to black coloured, 0.2 μm pore size Nuclepore filters at a low vacuum pressure. Picoplankton was identified and calculated using the autofluorescence of its pigments under the Olympus microscope. 1000 × and 1500 × magnification and the standard optic filter set for green and blue excitation were used. Samples for study of nanoplankton (from 2 to 20 μm), and the remaining size groups (>20 μm) were preserved with Lugol solution and analysed with an inverted microscope after sedimentation in settling chambers of 9, 14 and 24 ml volume, at the 150 × and 600 × magnification. The volume of organisms was estimated on the basis of microscope measurements of their size, and then their biomass was expressed as fresh weight.

Results

Picoplankton

The presence of picoplankton was observed in all lakes and samples. It was, with the exception of Linowskie Lake, the most numerous size group. Its abundance ranged from 3.2×10^2 cells ml^{-1} in the above mentioned lake to 1.16×10^6 cells ml^{-1} in Sękacz Lake, while the proportion in total phytoplankton abundance varied from 0.9 to 99.9%. It is worth noting the very high share of picoplankton in total phytoplankton abundance from as many as 11 lakes, always above 85%, very often even above 95%. (In the lake with the lowest picoplankton proportion in total abundance, small organisms also dominated, but according to accepted criterion they had to be considered as belonging to nanoplankton.) Picoplankton biomass was not high and ranged from <0.001 to 0.65 mg l^{-1} while its proportion in the total phytoplankton biomass varied from several thousandths of one per cent in Linowskie Lake to 11% in Sękacz Lake (Figures 2 and 3).

Cyanobacteria and eukaryotes were present in picoplankton from the examined lakes. Cyanobacteria were of *Synechococcus* sp. and *Synechocystis* sp. type. They were found in 8 lakes, and their abundance ranged from 4.49×10^3 cells ml^{-1} in Ciemniak Lake to 1.16×10^6 cells ml^{-1} in Sękacz Lake. Despite a wide value range, Cyanobacteria numbers oscillated usually around 100 000 cells ml^{-1}. Cyanobacteria biomass was not high and only in Sękacz Lake did it reach a value of 0.58 mg l^{-1}, while in the remaining lakes it did not exceed a value of 0.11 mg l^{-1}. Eukaryotes were found in 6 lakes. Their greatest abundance was observed in the following lakes: Ciemniak – 2.45×10^5 cells ml^{-1} and Leśniówek Mały – 5.5×10^4 cells ml^{-1}, and their biomass was 0.65 mg l^{-1} and 0.24 mg l^{-1}, respectively (Figure 4). Eukaryotes cells were *Chlorella* sp. and *Stichococcus* sp. like, of size larger than Cyanobacteria.

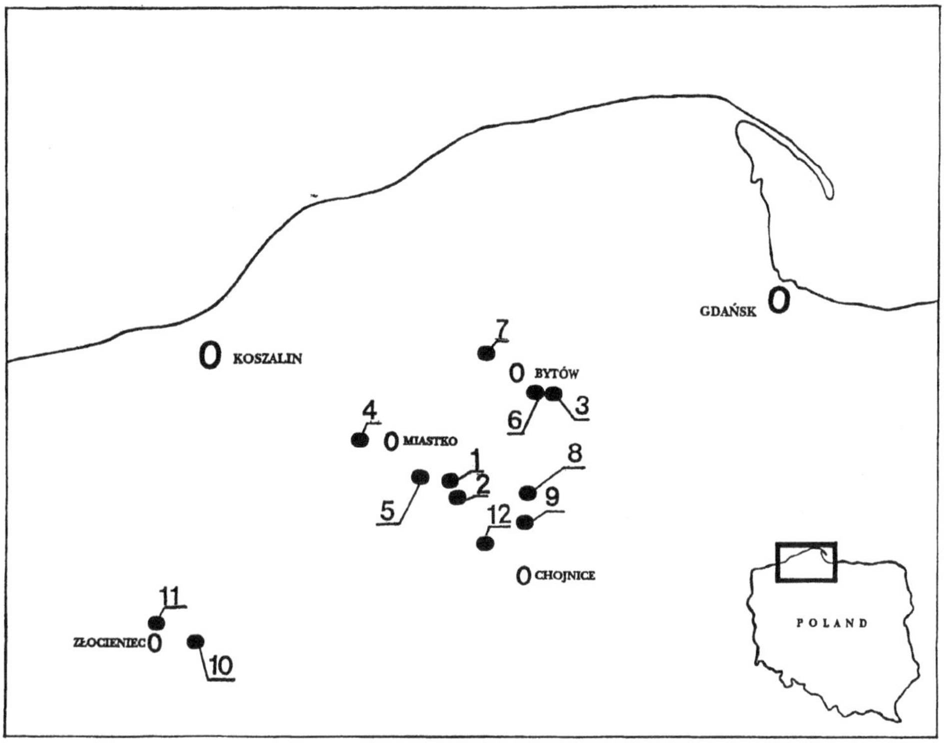

Figure 1. Location of the lakes. Numbers from 1 to 12 correspond to the names of lakes in Table 1.

Relationships between the trophic state of lakes and abundance, biomass or the percentage contribution of picoplankton biomass to total biomass of phytoplankton were analysed statistically. The total biomass of phytoplankton was used as an indicator of trophic state and non-weighted mean values for the water column were taken. It was found that the correlation between picoplankton numbers or picoplankton biomass and total biomass was not statistically significant ($r = -0.02$, $p = 0.936$) or ($r = 0.402$, $p = 0.098$, respectively). By contrast, the correlation between the contribution of picoplankton biomass to total biomass was statistically significant ($r = -0.469$, $p = 0.049$). This relationship can be expressed by linear regression. However, it is better expressed ($r = -0.549$, $p = 0.018$, $n = 18$) by the power function due to the rapid decrease and considerable variability of this percentage when total phytoplankton biomass is low (Figure 5).

Nanoplankton

The abundance and biomass of nanoplankton of studied lakes ranged from 40 ind. ml^{-1} to 21.2×10^3 ind. ml^{-1} and from 0.01 mg l^{-1} to 1.34 mg l^{-1}, respectively. Nanoplankton share in phytoplankton total abundance, with the exception of Linowskie Lake (59%), was not big, usually only a few per cent. The contribution of nanoplankton to total biomass ranged from 0.1 to 65.2% and was only high in Cietrzewie Małe Lake. In this lake phytoplankton was present in low population densities. The highest biomass concentration associated with nanoplankton was in Ciemniak Lake. This corresponded to the populations of Chlorophyceae. Very rapid decrease of nanoplankton contribution in phytoplankton biomass of Linowskie Lake (from 59% to merely 4%) in comparison to its high contribution in abundance, is interesting (Figures 2 and 3). It results from domination in the plankton of numerous nanoplankton, but of small size, just at the bottom limit of sizes for nanoplankton, Chlorophyceae-*Stichococcus* sp. taxon. General-

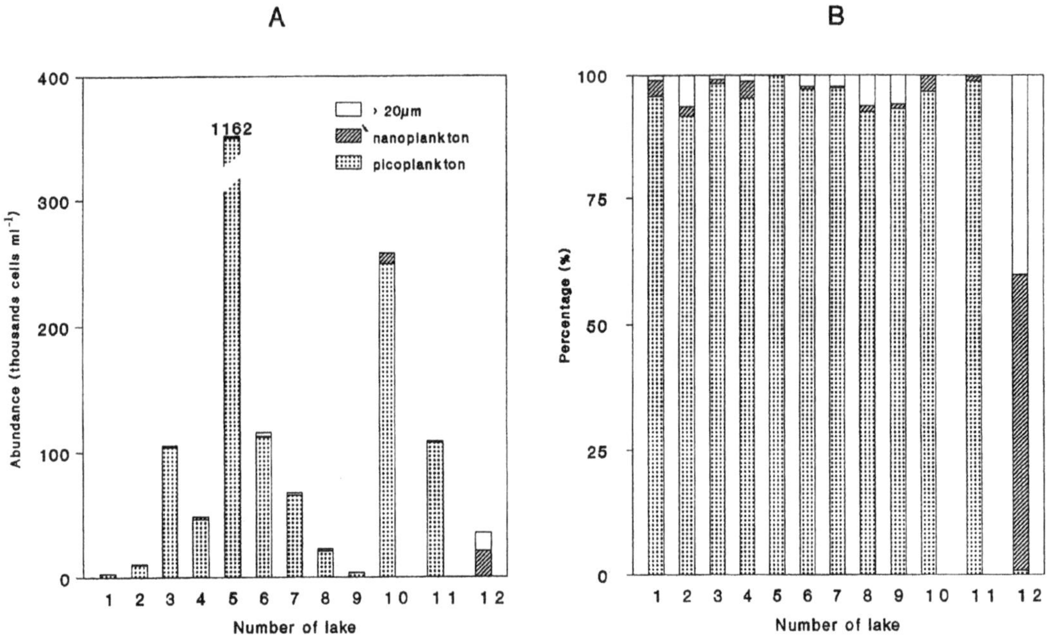

Figure 2. Abundance (A) and percentage share (B) of size groups of the phytoplankton. Numbers of lakes correspond to the names of lakes in Table 1.

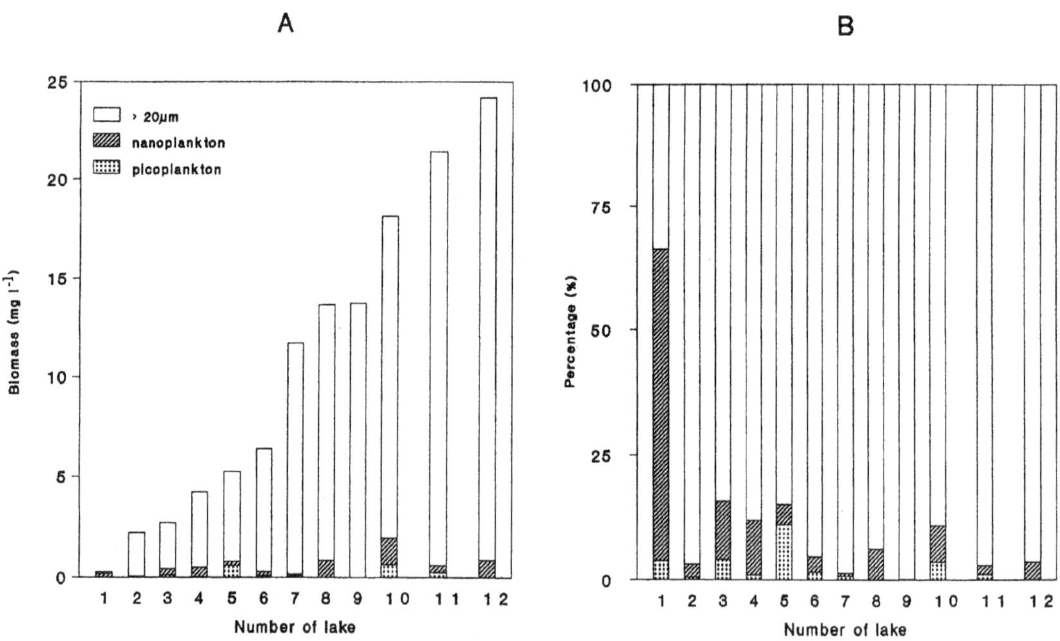

Figure 3. Biomass (A) and percentage share (B) of size groups of the phytoplankton. Explanation as Figure 2.

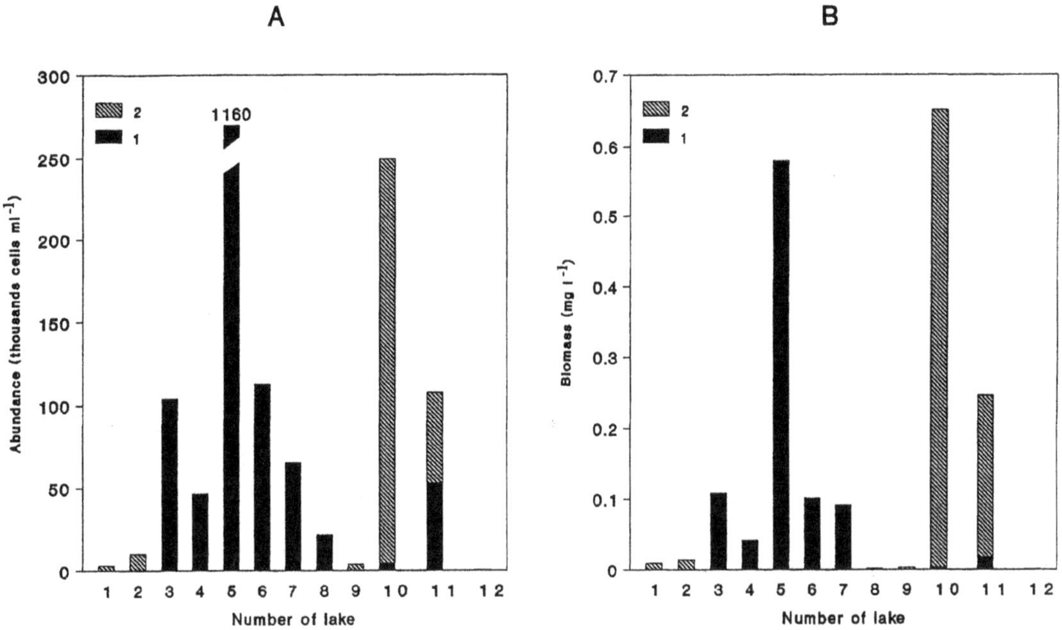

Figure 4. Abundance (A) and biomass (B) of cyanobacterial (1) and eukaryotic (2) organisms in picoplankton. Explanation as Figure 2.

ly, nanoplankton included 76 taxa, but its importance in the investigated lakes was small.

Phytoplankton >20 μm

Larger >20 μm phytoplankton contribution to total abundance, with the exception of Linowskie Lake (40%), was very small with the exception of Linowskie Lake (40%), was very small or even non-significant, while its contribution to total biomass was generally very high and ranged from 34.9 to 99.9%. In nine lakes it exceeded 85%. The greatest biomass of these organisms occured in lakes with the greatest total phytoplankton biomass, i.e. in plankton of Linowskie, Leśniówek Mały and Ciemniak: 23.3 mg l^{-1} 20.7 mg l^{-1}, 16.2 mg l^{-1}, respectively (Figures 2 and 3). Phytoplankton >20 μm included 89 taxa. Cyanobacteria had a high contribution in 3 lakes. In Borzytuchom III and Leśniówek Mały the most abundant were *Gomphosphaeria naegeliana*, while in Lubieniecke Małe Lake *Lyngbya* sp. Contrary to Dinophyceae and Raphidophyceae dominating in most lakes, Cryptophyceae were generally not very numerous and their contribution to biomass was not important. High Chrysophyceae proportion in phytoplankton biomass were observed only in Linowskie and Lubienieckie Duże lakes. They were represented by species of *Dinobryon* genus, mainly *D. divergens* and *D. bavaricum*. In general, the proportion of total biomass attributable to phytoplankton >20 μm was, except for one lake, larger than other size groups.

Discussion

In these lobelian lakes groups, picoplankton was found in all samples, and its abundance did not exceed values observed in the corresponding season in other freshwater reservoirs. In other lakes seasonal picoplankton study showed considerable variations in its abundance, both during a year as well as between successive study years in the same lake, reaching 1, 2 and even 4 orders of magnitude (Happey-Wood, 1991; Hawley & Whitton, 1991; Søndergaard, 1991; Weiss & Kenter, 1991). However, this interesting problem in lobelian lakes could not be investigated because samples were taken only once a year. It can only be supposed, taking into account trophic state, that picoplankton abundance at least in some of them can be considerably higher than that observed at the examined time. It can be seen in data from Weisse & Kenter (1991) concerning spring abundance peaks of picoplankton in some years

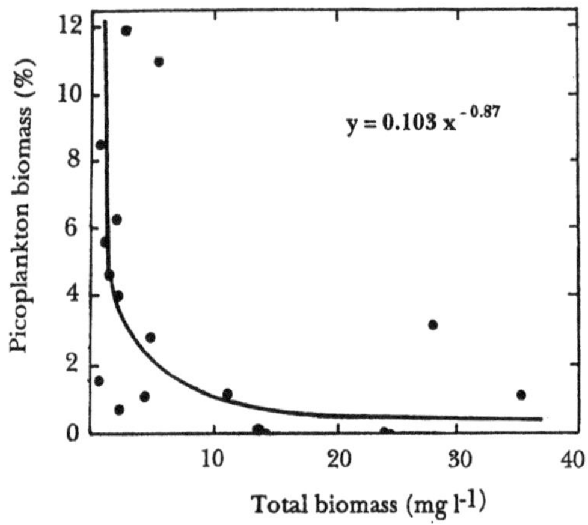

Figure 5. Relationship between percentage contribution of picoplankton to total biomass of phytoplankton and total biomass of phytoplankton.

of meso-eutrophic Lake Constance as well as in data from Stockner & Shortreed (1991) concerning spring and early-summer peaks in some oligotrophic lakes.

In studied lakes two organism groups in freshwater picoplankton i.e. prokaryotes and eukaryotes, most often quoted in literature (Stockner & Antia, 1986; Stockner, 1988, 1991), were found. Cyanobacteria were observed more often which corroborates many researchers' opinion about their common presence in oligo- and mesotrophic freshwater ecosystems. Moreover, remarks about eukaryotic picoplankton reaching larger biomass, due to their larger individual size, than prokaryotic picoplankton, have been confirmed (Burns & Stockner, 1991; Søndergaard, 1991). It should be stressed, however, that as for the taxonomic composition, still not much is known about picoplankton, and the number of determined taxa reaches a value of a dozen or so. Taxonomic problems ensue mainly from very small sizes of these organisms. Their thorough and complete identification requires using electron microscope (TEM, SEM), isolating taxa and their culture, and pigment composition analysis (Fahnenstiel et al., 1991; Weisse & Kenter, 1991; Corpe & Jansen, 1992; Andreoli et al., 1992).

Picoplankton proportion in the total phytoplankton biomass (expressed as chlorophyll or fresh weight) ranges usually from 0.2 to 43% (Stockner & Antia, 1986; Stockner, 1988; Weisse, 1988; Wehr, 1990). The latest reports give also higher maximum values: 56% (Pick & Agbeti, 1991), >60% (Happey-Wood, 1991) and 73% (Stockner & Shortreed, 1994). Thus, the results for studied lakes can be placed in the lower part of the given range. Although the data from lobelian lakes collected in August does not necessarily show maximum or mean values of their picoplanktonic biomass, correlation analysis between trophic state expressed as total phytoplanktonic biomass determined microscopically and picoplanktonic biomass, was undertaken. The results of present studies confirmed reverse dependence between picoplankton proportion in phytoplankton total biomass and the growth of trophic state of pelagial waters given, among others, by Søndergaard (1990) and Takamura & Nojiri (1994). This negative dependence, similar to waters studied by Søndergaard et al. (1991), was best described by power function. Moreover, in the analysed group of lakes, as in seven Danish lakes studied by Søndergaard (1990), correlation between picoplankton abundance and water trophy was not statistically significant. Nevertheless, Burns & Stockner (1991) in their research on six lakes of New Zealand, found that there was significant negative correlation between total picoplankton concentration and chlorophyll *a*. In lobelian lakes the dependence was proved to be non-significant even after elimination of values from some lakes which were most different from others, as was the case in lakes of British Columbia and Yukon Territory. For these lakes Stockner & Shortreed (1991) after rejecting the data coming from three most acid lakes, achieved dependence, which was both significant and positive, between mean seasonal average abundance of total picoplankton number and total phosphorus concentration. In the light of reports about lack of consistency between value of chlorophyll concentration obtained as a result of division of samples into size groups and calculated on the basis of phytoplankton volume (Søndergaard et al., 1991), especially when filters of different retention time are used (Stockner et al., 1989), it seems that there is an urgent need to study the importance of picoplankton in waters of various trophic state.

Acknowledgments

Microscopy analyses were possible due to sponsorship of The Foundation for Polish Science which bought the Olympus BX-60 microscope (programme SUBIN).

References

Amblard, C., J.-F. Carrias, G. Bourdier & N. Maurin, 1995. The microbial loop in a humic lake: seasonal and vertical variation in the structure of the different communities. Hydrobiologia 300/301: 71–84.

Andreoli, C., L. Scarabel, S. Spini & C. Grassi, 1992. The picoplankton in Antarctic lakes of northern Victoria Land during summer 1989–1990. Polar biol. 11: 575–582.

Burns, C. W. & J. G. Stockner, 1991. Picoplankton in six New Zealand lakes: abundance in relation to season and trophic state. Int. Revue ges. Hydrobiol. 76: 523–536.

Corpe, W. A. & T. E. Jensen, 1992. An electron microscopic study of picoplanktonic organisms from a small lake. Microbiol. Ecol. 24: 181–197.

Happey-Wood, C. M., 1991. Temporal and spatial patterns in the distribution and abundance of pico, nano and microphytoplankton in an upland lake. Freshwat. Biol. 26: 453–480.

Hawley, G. R. W. & B. A. Whitton, 1991. Seasonal changes in chlorophyll – containing picoplankton population of ten lakes in northern England. Int. Revue ges. Hydrobiol. 76: 545–554.

Fahnenstiel, G. L., H. J. Carrick, C. E. Rogers & I. Sicko-Goad, 1991. Red-fluorescing phototrophic picoplankton in the Laurentian Great Lakes: what are they and what are they doing? Int. Revue ges. Hydrobiol. 76: 603–616.

Kraska, M., H. Szyper & W. Romanowicz, 1994. Charakterystyka trofii wód 37 jezior lobeliowych Pojezierza Bytowskiego i Borów Tucholskich. W: Jeziora lobeliowe. Charakterystyka, funkcjonowanie i ochrona. Cz. I, Red. M Kraska. [Water trophy of 37 lobelian lakes of the Tuchola Forest and of the Bytów Lake District. In Kraska, M. (ed.), Lobelian lakes. Characterisation, performance and protection. Part 1], Idee ekologiczne 6, Ser. Szkice 4: 135–147.

Pick, F. R. & M. Agbeti, 1991. The seasonal dynamics and composition of photosynthetic picoplankton communities in temperate lakes in Ontario, Canada. Int. Revue ges. Hydrobiol. 76: 565–580.

Søndergaard, M., 1990. Picoplankton in Danish lakes. Verh. int. Ver. Limnol. 24: 609–612.

Søndergaard, M., 1991. Phototrophic picoplankton in temperate lakes: seasonal abundance and importance along a trophic gradient. Int. Revue ges. Hydrobiol. 76: 505–522.

Søndergaard, M., L. M. Jensen & G. Artebierg, 1991. Picoalge in Danish coastal waters during summer stratification. Mar. Ecol. Prog. Ser. 79: 139–149.

Stockner, J. G., 1988. Phototrophic picoplankton: an overview from marine and freshwater ecosystems. Limnol. Oceanogr. 33: 765–775.

Stockner, J. G., 1991. Autotrophic picoplankton in freshwater ecosystems: The view from the summit. Int. Revue ges. Hydrobiol. 76: 483–492.

Stockner, J. G. & N. J. Antia, 1986. Algal picoplankton from marine and freshwater ecosystems: a multidisciplinary perspective. Can. J. Fish. aquat. Sci. 43: 2472–2503.

Stockner, J. G., M. E. Klut & W. P. Cochlan, 1989. Leaky filters: a warning to aquatic ecologists. Can. J. Fish. aquat. Sci. 47: 16–23.

Stockner, J. G. & K. S. Shortreed, 1991. Autotrophic picoplankton: community composition, abundance and distribution across a gradient of oligotrophic British Columbia and Yukon Territory lakes. Int. Revue ges. Hydrobiol. 76: 581–601.

Stockner, J. G. & K. S. Shortreed, 1994. Autotrophic picoplankton community dynamics in pre-alpine lake in British Columbia, Canada. Hydrobiologia 274: 133–142.

Szeląg-Wasielewska, E., 1994. Pikoplankton na tle innych frakcji wielkościowych planktonu dwóch jezior lobeliowych na Pojezierzu Bytowskim. W: Jeziora lobeliowe. Charakterystyka, funkcjonowanie i ochrona. Cz.I. Red. M. Kraska. [Picoplankton versus other size fractions of plankton of two lobelian lakes in the Bytów Lake District. In Kraska, M. (ed.), Lobelian lakes. Characterisation, performance and protection. Part 1]. Idee ekologiczne 6, Ser. Szkice 4: 29–35.

Takamura, N. & Y. Nojiri, 1994. Picophytoplankton biomass in relation to lake trophic state and the TN:TP ratio of lake water in Japan. J. Phycol. 30: 439–444.

Vörös, L., P. Gulyás & J. Németh, 1991. Occurrence, dynamics and production of picoplankton in Hungarian shallow lakes. Int. Revue ges Hydrobiol. 75: 617–629.

Wehr, J. D., 1990. Predominance of picoplankton and nanoplankton in eutrophic Calder Lake. Hydrobiologia 203: 35–44.

Weisse, T., 1988. Dynamics of autotrophic picoplankton in Lake Constance. J. Plankton Res. 10: 1179–1188.

Weisse, T. & U. Kenter, 1991. Ecological characteristics of autotrophic picoplankton in a prealpine lake. Int. Revue ges. Hydrobiol. 76: 493–504.

Hydrobiologia **342/343:** 87–93, 1997.
L. Kufel, A. Prejs & J. I. Rybak (eds), Shallow Lakes '95.

The dynamics and importance of picoplankton in shallow, dystrophic lake in comparison with surface waters of two deep lakes with contrasting trophic status

Iwona Jasser
Department of Hydrobiology, Institute of Ecology, Polish Academy of Sciences, Dziekanów Leśny, 05-092 Łomianki, Poland

Key words: picoplankton, phytoplankton, abundance, biovolume, seasonal pattern

Abstract

The abundance, biovolume and seasonal patterns of autotrophic picoplankton and larger phytoplankton were studied in 1993 in the shallow, dystrophic Lake Flosek and in the epilimnion of up to 5 m in two non-dystrophic, deep lakes: mesotrophic Lake Majcz and eutrophic Lake Mikołajskie. In all the three lakes picoplankton exhibited well visible seasonal patterns. In non-dystrophic lakes spring/early summer abundance peaks were observed while in Lake Flosek two abundance peaks occurred: a smaller one in spring and second, bigger, in autumn.

It was noticed that maximal picoplankton abundances did not coincided with abundance peaks of larger phytoplankton. The share of picoplankton in the total phytoplankton community in two less productive lakes (Flosek and Majcz) varied between 7 and 24% during picoplankton abundance peaks and about 5% during the rest of the vegetation season. In Lake Mikołajskie picoplankton's share was noticeable only during the abundance peak (13–6%) while it was insignificant (about 1% of the total phytoplankton biovolume) during the rest of the season.

Introduction

Autotrophic picoplankton, the smallest fraction of phytoplankton, (cell size of between 0.2 to 2.0 μm) is an important component of freshwater and marine ecosystems (Stockner & Antia, 1986). Wide research of picoplankton started in 1980s after its occurrence was recorded for the first time (Johnson & Sieburth, 1979; Waterbury et al., 1979). Autotrophic picoplankton consists of several groups of algae, but is mainly represented by unicellular, coccoid cyanobacteria from genus *Synechococcus* or *Synechocystis* (Stockner & Antia, 1986). According to Stockner (1988, 1991) the abundance and biovolume of picoplankton, as it is with other groups of algae, increases with eutrophication, while its share in the total phytoplankton biovolume and production decreases. However, the overall significance of picoplankton in aquatic ecosystem is likely to be higher than if only its biovolume is taken into consideration. This seems to result from the extremely small size of these algae. This cause a high surface-to-volume ratio which brings about the efficient nutrient uptake, high primary production and high growth rate (Raven, 1986; Zevenboom, 1986). Latest results of Vörös et al. (1991) indicate that autotrophic picoplankton can account even for 50% of primary production in some eutrophic and hypertrophic, shallow lakes. We still need more research concerning seasonal patterns of picoplankton and its share in the seasonal succession of the phytoplankton community in lakes, as well as the role of picoplankton and the way of its incorporation into the lake food web.

The aim of this study was to compare the pattern of occurrence and significance of picoplankton in a shallow, dystrophic lake with water column of comparable depth in meso- and eutrophic lakes of non-dystrophic character.

Table 1. Kiejdahl nitrogen concentrations (KTN = $N\text{-}NH_4$ + N org), total phosphorus (TP), and their KTN/TP ratio in Lake Flosek and in the surface waters of two deep lakes: Majcz and Mikołajskie during spring/early summer picoplankton abundance peaks, summer and autumn.

Lake	Date	KTN $mg\,l^{-1}$	TP $\mu g\,l^{-1}$	KTN/TP
Flosek	Spring peak	1.8	55.7	32.3
	Summer	1.3.	42.1	30.8
	Autumn	1.4	27.2	51.5
Majcz	Spring peak	1.0	13.7	73.0
	Summer	2.0	45.3	44.2
	Autumn	0.9	52.0	17.3
Mikołajskie	Spring peak	1.9	44.0	43.2
	Summer	2.1	80.0	26.3
	Autumn	2.2	162.0	13.6

Material and methods

Sampling was carried out in monthly intervals in 1993, from April to October, in shallow, dystrophic Lake Flosek and in the surface waters (up to 5 m) of two deep Lakes: Majcz and Mikołajskie. All three are situated in the Mazurian Lakeland in Northeastern Poland. Lake Flosek is a small (0.4 km^2), shallow (mean depth 3 m, max. depth 7 m), low-productive (TP $\sim 40\,\mu g\,l^{-1}$) lake in the middle of a forest. Lake Majcz is a mesotrophic, lake (mean depth 6 m, max. depth 15 m), stratified in summer and with the surface area of 1.6 km^2. Summer TP concentration in surface waters is about 40 $\mu g\,l^{-1}$. Lake Mikołajskie is a eutrophic, deep lake (mean depth 11 m, max. depth 28 m) with the surface area of 4.6 km^2 and the mean summer TP concentration about 80 μg^{-1}. Kiejdahl nitrogen and total phosphorus concentrations as well as KTN/TP ratio for selected periods are summarized in Table 1.

Samples, were taken with a five-liters water sampler, subsamples for autotrophic picoplankton were fixed with glutaraldehyde, buffered with sodium cacodylate to pH 7.2, to the final concentration of 1%. They were stored in 4 °C in a dark place and investigated within three days. Nikon Optiphot-2 autofluorescence microscope with 50 W halogen lamp was used to examine picoplankton. Five or ten ml of samples were filtered on black-stained polycarbonate (Porvair) filters with a pore size of 0.2 μm. Autotrophic picoplankton cells were counted under the 1000 × magnification, using two sets of filters: B-2A (blue excitation light) for phycocyanin-containing cyanobacteria and picoplanktonic eukaryota and G-2A (green excitation light) for phycoerythrin-containing cyanobacteria (Tsuji et al., 1986). The biovolume of picoplankton was calculated from the cells numbers and diameter for cocci while from cells numbers, length and width for oval cells.

For the examination of nano- and microplanktonic algae, 100 ml-water samples were preserved with Lugol solution and by formalin to the final concentration of 1%, then stored in 4 °C in a dark place. Samples were concentrated by sedimentation to 1 ml and phytoplankton species were identified and enumerated in the Rosenthal chamber with a volume of 3.2 mm^3 and surface of 16 mm^2. The abundance of algae was converted to 1 l, and the volume of these algae were calculated on the basis of shape, size and abundance.

Results

Abundance, biovolume and seasonal patterns of picoplankton

Autotrophic picoplankton is present in all three lakes exhibiting well-visible seasonal patterns (see Figure 1A, B). In all the three lakes the spring/early summer peak of abundance was observed (4.3, 2.2 and 6.9×10^5 cells ml^{-1} in lakes Flosek, Majcz and Mikołajskie). The abundance and biovolume of picoplankton during the remaining vegetation season varied between 0.5 and 1.4×10^5 ml^{-1}, and about 10–190 mm^3 m^{-3} respectively and did not significantly differ in the three lakes. However, during the spring/early summer peak the abundance of picoplankton was related to the trophic status of each lake. The highest peak occurred in the eutrophic Lake Mikołajskie in which picoplankton abundance in May was about seven times and biovolume about five times higher than during summer months. The peak spilled over to June hut was less intensive. In Majcz, the spring abundance peak occurred in April with picoplankton abundance and biovolume of about two times higher than during summer months. The shallow, dystrophic Lake Flosek was found to he different from the two other lakes in terms of the seasonal pattern of the picoplankton community. The abundance peak occurred twice: a smaller one in spring and the second, bigger, in autumn. Spring picoplankton peak in Flosek was noticed in May with abundance about three times higher than during summer months. This peak was also almost two times higher than in Majcz. An autumn peak, which occurred only in Flosek, took place in October. Picoplankton abundance during this peak was

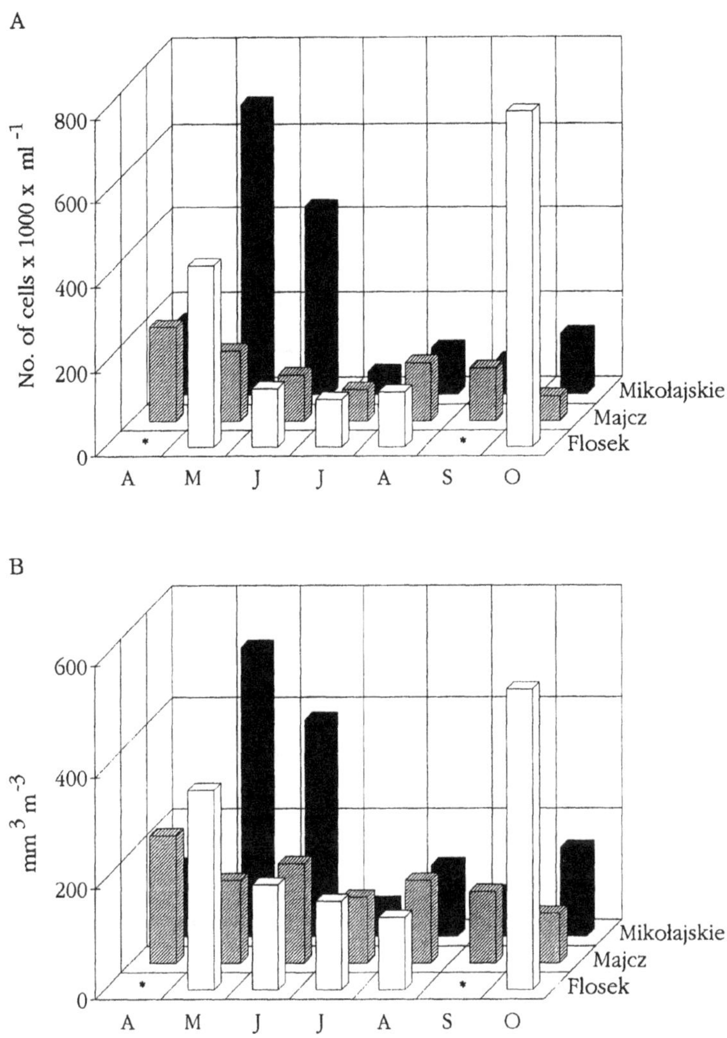

Figure 1. The abundance (A) and biovolume (B) of picoplankton in shallow Lake Flosek and in the surface waters of Lakes Majcz and Mikołajskie. * = lack of data.

almost 8.0×10^5 ml^{-1} making it even higher than the spring/early summer peak in Lake Mikołajskie.

Contribution of picoplankton to phytoplankton community

Generally, picoplankton plays a more important role (i.e. its share in total phytoplankton is higher) in lakes with lower trophic status (Flosek and Majcz) than in more productive ecosystem, such as the eutrophic Lake Mikołajskie (Figure 2). It was found that in lakes with lower trophic status (Flosek and Majcz) picoplankton accounted for about 5% of total phytoplankton biovolume during summer months and about 7% during abundance peaks. Both lakes, are characterized by spring phytoplankton peaks occurring in May, and moderate to low phytoplankton biovolume during summer (Figure 2), as in the case of mesotrophic and low-productive humic lakes (Hensen, 1962; Sommer et al., 1986). The spring picoplankton peak occurring in Lake Flosek in May and in Lake Majcz in April coincided with periods with higher abundance of larger phytoplankton (nano- and microplankton). Therefore, a higher share of picoplankton in the total phytoplankton biovolume (14 in Flosek and 9% in Majcz) occurred in both lakes not during the spring picoplankton abun-

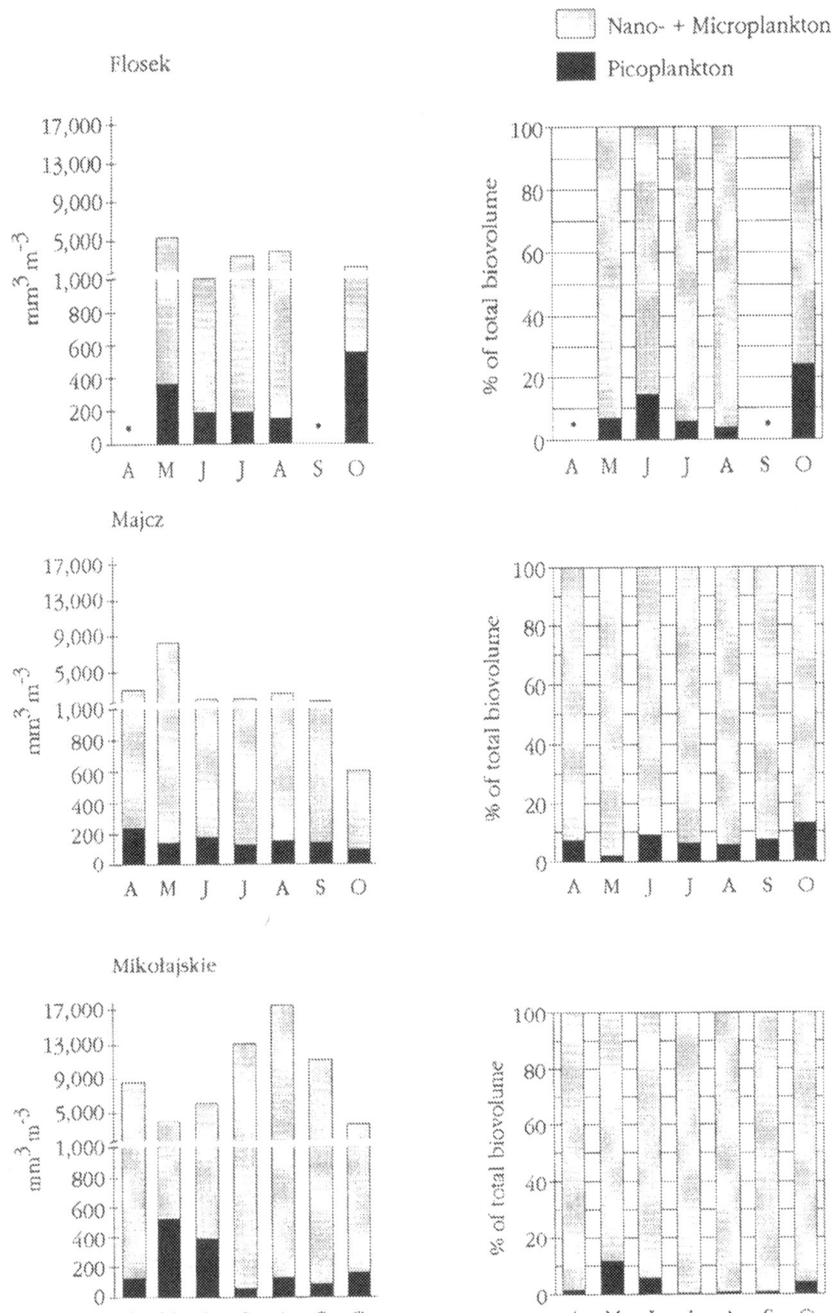

Figure 2. The biovolume of picoplankton and nano- + microplankton groups and their share in the total phytoplankton biovolume in all three lakes.

dance peak, but in June, when the biovolume of nano- and microplanktonik algae was lower.

Another interesting phenomenon in the two lakes is an increase of picoplankton's role in the phytoplankton community in autumn. In Flosek it coincided with the increase of picoplankton abundance, as it was noticed above, while in Majcz it is a result of lower abundance and biovolume of larger phytoplankton groups. At that time, autumn, the share of picoplankton in phytoplankton community was highest during vegetation season – 24% in Flosek and 13% in Majcz.

Picoplankton in Lake Mikołajskie, played less important role in the phytoplankton community (Figure 2) though its abundance was similar or even higher than in lakes with lower trophic status (Figure 1). The share of picoplankton during spring/early summer peak of abundance in May and June was 13 and 6% respectively, and similar to the two other lakes. This picoplankton peak correlated with lower abundance and biovolume of bigger algae-clear water phase (as shown in Figure 2), which is characteristic of phytoplankton succession in eutrophic lakes (Sommer et al., 1986). Before the peak, in spring, and in summer months picoplankton became insignificant in Lake Mikołajskie, with a 1.5% and less than 1% share in the total phytoplankton biovolume. This resulted from: a very high abundance of nano- and microphytoplankton, characteristic for eutrophic ecosystem in summer (Sommer et al., 1986), and from lower picoplankton abundance. As in the less productive lakes, picoplankton's share in the Lake Mikołajskie phytoplankton community slightly increased in autumn. In October it achieved 4.3% of the total phytoplankton biovolume, significantly more than in the summer months in this lake. However, it was still less than in Flosek and Majcz. As in the case of Majcz, higher share of picoplankton in the total phytoplankton biovolume in Lake Mikołajskie was rather due to a decrease of larger phytoplankton taxa, although picoplankton numbers also slightly increased.

Discussion

The following regularities concerning picoplankton's seasonal occurrence and importance in the three lakes with different trophic status were found in this study:

– The highest picoplankton numbers and biovolume in spring was noted in the eutrophic Lake Mikołajskie, but its prominence in the phytoplankton community was similar at that time to that in the two other low-productive lakes during their spring picoplankton peak.

– The highest values of abundance and biovolume of picoplankton as well as the highest share in the total algal biovolume was noted in the dystrophic, shallow Lake Flosek in autumn (up to 24%). In this period, no peak in picoplankton abundance was observed in the other lakes. Generally, the role of picoplankton in the total phytoplankton is rather similar in less productive Lakes Flosek and Majcz and is smaller in more productive Lake Mikołajskie.

Lakes Flosek and Majcz have more or less similar summer TP concentration, but different in spring and autumn (Table 1). On the other hand Flosek's nitrogen content, which is known to be a limiting nutrient for picoplankton (Wehr, 1991), was higher in periods of picoplankton peaks than during Majcz's picoplankton peaks. In this case the nutrient status of Lake Flosek and the KTN/TP ratio can partially explain the richness of picoplankton, especially in autumn.

Another hypothesis explaining picoplankton richness in Lake Flosek revolves around the shallowness of Lake Flosek and the impact of its watershed. Lake Flosek is a humic, forest lake, bordered by *Sphagnum* mats and coniferous forest. The allochthonous organic matter entering Lake Flosek seems to decomposes slowly, as it was recorded in other humic lakes (Jones, 1992; Tranvic, 1992). There is also evidence that in shallow non-humic lakes, picoplankton abundance is high and occasionally it dominates the phytoplankton community (Vörös et al., 1991). In such lakes (i.e. shallow and of small size) the input of allochthonous, humic matter is probably a more significant source of organic matter than the autochthonous pelagic production which usually dominates in deep meso- and eutrophic lakes. On the other hand in deep, oligotrophic lakes Cole & Caraco (1993) and Jones also found that the allochthonous, refractory compounds dominate in total organic matter. According to Weisse (1993), in oligotrophic lakes picoplankton is relatively numerous, and its role in phytoplankton production is higher than in eutrophic lakes. It seems that these two phenomena, i.e. the role of picoplankton and the amount of slowly decomposing organic matter may be linked. As it is known this kind and amount of organic matter can enhance the production of microbial biomass and the microbial food web. Weisse (1993) suggested that prokaryotic picophytoplankton which has lower metabolism than bigger algae and also eukaryotic picoplankton, is a natural, archaic part of microbial loop, able to utilise very low quantity of nutrients characteristic for oligotrophic conditions and slowly decomposing matter. Thus, the recycling of the slowly releasing nutrients inside the microbial loop in a lake, such as Lake Flosek, might boost the existence of autotrophic picoplankton.

As it was noticed above picoplankton in Lakes Flosek and Majcz is a significant component of the phytoplankton community in spring and autumn. Picoplankton constituted at that time up to 13 and 7% of the total phytoplankton biovolume (in Flosek and Majcz) in spring and 24 and 14% in autumn, while during summer about 5%. Similarly Amblard et al. (1995) found that in moderately productive, humic Lake Vassiviére autotrophic picoplankton accounted for about 5% of the total biomass of microorganisms, in terms of carbon, during vegetation season. In contrary, in Lake Mikołajskie, only during spring/early summer abundance peak, picoplankton's share in the total phytoplankton biovolume was noticeable, while during the rest of the vegetation season it's role was insignificant. In this case, picoplankton's role in all three studied lakes is in line with earlier findings (Stockner, 1988; Weisse, 1993) that in less productive lakes is higher (irrespective of the origin of organic matter related to depth conditions as mentioned above) while it decreases in more productive lakes.

Another phenomenon reported in this study is that picoplankton abundance peaks generally occurred when abundance of larger algae was lower, not during larger phytoplankton abundance peaks. This suggest that due to more or less similar nutrient regime there is an exploitative competition between these two components of the phytoplankton community. Picoplankton develops well when larger phytoplankton is controlled by grazing and/or resources concentrations below its threshold level. In Lake Mikołajskie the picoplankton peak occurred when larger phytoplankton was less abundant, probably because of control by zooplankton grazing, especially by cladocerans. Grazing by cladocerans, present in Mikołajskie in very high densities during clear-water phase (Kajak, 1978), could have a direct impact on picoplankton (Bogdan & Gilbert, 1987; Vörös et al., 1991). However, other results (Burns & Stockner, 1991; Stockner, 1987) indicate that cladocerans species although consume, but not digest picoplanktonic cyanobacteria. Therefore, it seems that macrozooplankton (cladocerans and copepods) might be responsible for a decrease in main picoplankton's grazers such as nanoflagellates and ciliates. In this way macrozooplankton could enhance picoplankton growth in this period bringing about a picoplankton abundance peak. The later decrease of picoplankton in summer is correlated with an increase of larger algae such as filamentows cyanobacteria i.e. *Anabaena* sp. and *Aphanizomenon* sp. These algae are able to assimilate atmospheric nitrogen, becoming competitive to picoplankton which is limited by the absence of inorganic nitrogen. There is evidence that in lake Balaton picoplankton disappear almost completely as soon as nitrogen fixing cyanobacteria occur (Vörös et al., 1991). Competitive relations between picoplankton and nitrogen fixing filamentous cyanobacteria were also described by Stockner & Shortreed (1988). In those study, an addition of nitrogen to water caused a switch from an *Anabaena*-dominated phytoplankton community to one dominated by *Synechococcus*.

In two studied low-productive lakes (Flosek and Majcz) the high abundance and significance of picoplankton is in most cases (except of the spring peak in Lake Flosek) correlated with low abundance of larger algae, as it was already pointed out. This suggests that also in these two lakes competitive relations between picoplankton and other phytoplankton groups are present. However, it seems that in humic, low-productive, shallow Lake Flosek, where allochthonous matter input co-dominates the pool of organic matter together with autotrophic production, the role of picoplankton might be higher and more stable seasonally than in eutrophic Lake Mikołajskie or in mesotrophic Majcz. It is probably due to an intensive cycling of nutrients within the microbial loop.

Conclusions

– Picoplankton is related to the trophic status of a lake only during its spring/early summer abundance peaks: the highest peak occurs in a lake with highest trophy.

– In a shallow, humic lake the microbial food web sustained by allochthonous input of organic matter is probably responsible for a generally higher and more stable role of picoplankton in relation to the mixed layers of two non-humic lakes (Majcz and Mikołajskie). Possibly, picoplankton takes advantage of intensive nutrient recycling within the microbial loop.

– Picoplankton's role in the studied lakes decreases with increasing trophy.

– The significance of picoplankton increases in periods of lower abundance of larger phytoplankton groups. This suggests that due to more or less similar nutrient regime there is an exploitative competition between these two components of phytoplankton. Picoplankton develops well when larger phytoplankton is controlled by such factors as: heavy grazing and/or less resources than its threshold level.

References

Amblard, C., J-F. Carrias, G. Bourdier & N. Maurin, 1995. The microbial loop in a humic lake: seasonal and vertical variations in the structure of different communities. Hydrobiologia 300/301: 71–84.

Bogdan, K. G. & J. J. Gilbert, 1987. Quantitative comparison of food niches in some freshwater zooplankton. A multi-tracer-cell approach. Oecologia, Berlin 72: 331–340.

Burns, C. W. & J. G. Stockner, 1991. Picoplankton in six New Zealand lakes: abundance in relation to season and trophic state. Int. Revue ges. Hydrobiol. 76: 523–536.

Cole, J. J. & N. F. Caraco, 1993. The pelagic microbial food web of oligotrophic lakes. In Ford, T. E. (ed.), Aquatic microbiology: an ecological approach. Blackwell Scientific Publications, 101–111.

Hensen, K., 1962. The dystrophic lake type. Hydrobiologia 19: 183–191.

Johnson, P. W. & J. McN. Sieburth, 1979. Chroococcoid cyanobacteria in the sea: A ubiquitous and diverse phototrophic biomass. Limnol. Oceanogr. 24: 928–935.

Jones, R. I., 1992. The influence of humic substances on lacustrin planktonic food chains. Hydrobiologia 229: 73–91.

Jones, R. I. The forms and distribution of carbon in a deep, oligotrophic lake (Loch Ness, Scotland). Verh. int. Ver. Limnol. (in press).

Kajak, Z., 1978. The characteristics of a temperate eutrophic, dimictic lake (Lake Mikołajskie, Northern Poland). Int. Revue ges. Hydrobiol. 63: 451–480.

Raven, J. A., 1986. Physiological consequences of extremely small size for autotrophic organisms in the sea. In Platt, T. D. & W. K. W. Li (eds), Photosynthetic picoplankton. Can. Bull. Fish. aquat. Sci. 214: 1–70.

Sommer, U., Z. M. Gliwicz, W. Lampert & A. Duncan, 1986. The PEG model of seasonal succession of planktonic events in fresh water. Arch. Hydrobiol. 106: 433–471.

Stockner, J. G., 1987. Lake fertilization: The enrichment cycle and lake sockeye salmon (*Oncorhyncus nerka*) production. In Smith, H. D., L. Margolis & C. C. Wood (eds), Sockeye salmon (*Oncorhyncus nerka*) population biology and future management. Can. Spec. Publ. Fish. aquat. Sci. 96: 198–215.

Stockner, J. G., 1988. Phototrophic picoplankton: An overview from marine and freshwater ecosystems. Limnol. Oceanogr. 33: 765–775.

Stockner, J. G., 1991. Autotrophic picoplankton in freshwater ecosystems: the view from the summit. Int. Revue ges. Hydrobiol. 76: 483–493.

Stockner, J. G. & N. J. Anba, 1986. Algal picoplankton from marine and freshwater ecosystems: A multidisciplinary perspective. Can. J. Fish. aquat. Sci. 43: 2472–2503.

Stockner, J. G. & K. S. Shortreed, 1988. Response of *Anabaena* and *Synechococcus* to manipulation of nitrogen: phosphorus ratios in a lake fertilization experiment. Limnol. Oceanogr. 33: 1348–1361.

Tranvic, L. J., 1992. Allochthonous dissolved organic matter as an energy source for pelagic bacteria and the concept of the microbial loop. Hydrobiologia 229: 104–114.

Tsuji, T., K. Ohki & Y. Fujita, 1986. Determination of photosynthetic pigment composition in an individual phytoplankton cell in seas and lakes using fluorescence microscopy: properties of the fluorescence emitted from picoplankton cells. Mar. Biol. 93: 343–349.

Vörös, L., P. Gulyas & J. Nemeth, 1991. Occurrence, dynamics and production of picoplankton in Hungarian shallow lakes. Int. Revue ges. Hydrobiol. 76: 607–629.

Waterbury, J. B., S. W. Watson, R. R. L. Guillard & L. E. Brand, 1979. Widespread occurrence of unicellular, marine, planktonic caynobacterium. Nature, London 277: 293–294.

Wehr, J. D., 1991. Nutrient and grazer-mediated effects on picoplankton and size structure of phytoplankton communities. Int. Revue ges. Hydrobiol. 76: 643–659.

Weisse, T., 1993. Dynamics of autotrophic picoplankton in marine and freshwater ecosystems. In Jones, J. G. (ed.), Advances in microbial ecology. Plenum Press, New York 13: 327–370.

Zevenboom, W., 1986. Ecophysiology of nutrient uptake, photosynthesis and growth. In Platt, T. D. & W. K. W. Li (eds), Photosynthetic picoplankton. Can. Bull. Fish. aquat. Sci. 214: 391–422.

Hydrobiologia **342/343:** 95–105, 1997.
L. Kufel, A. Prejs & J. I. Rybak (eds), Shallow Lakes '95.

Influence of small-scale turbulence and large-scale mixing on phytoplankton primary production

Frank Gervais, Dieter Opitz & Horst Behrendt
Department of Lowland Rivers and Shallow Lakes, Institute of Freshwater Ecology and Inland Fisheries, Müggelseedamm 260, 12587 Berlin, Germany

Key words: phytoplankton, primary production, turbulence, mixing, shallow lake, plankton tower

Abstract

To test the hypotheses that as well small scale turbulence as large scale vertical mixing cannot be neglected when quantifying primary production especially in shallow polymictic lakes, experiments were run on three different scales. (1) To achieve more natural conditions in bottles used for *in situ* incubation measurements of primary production, bottle stirrers were designed and tested *in situ*. The operation of the bottle stirrers guaranteed a homogeneous distribution of seston in the samples during incubation. Stirring increased primary production of plankton from a eutrophic lake significantly only when buoyant cyanobacteria were dominant. (2) To investigate the influence of turbulent mixing on primary production under controlled conditions, a circulating water column was maintained in large mesocosms. The comparison of static and dynamic *in situ* measurements of primary production revealed a distinct increase of production by mixing. (3) To find out the importance of mixing form, primary production was measured in the shallow, eutrophic, polymictic lake Müggelsee by moving water samples up and down with bottle lifts in two different ways. The two simulated motions (linear and circular) result in a different integral light supply of the enclosed phytoplankton. The higher light supply during circular movement in comparison to linear movement resulted in higher primary production in the circulating bottles compared to the bottles that were moved in a linear fashion.

Introduction

For the measurement of primary production in the pelagial, a wide spectrum of different methods has been developed. Besides bulk-property and optical methods, the *in vitro* measurement of primary production is established (Platt & Sathyendranath, 1993). The *in vitro* measurement tries to simulate the natural conditions of the water sample as near as possible during incubation. One possibility is to incubate the samples *in situ* under the natural light and temperature conditions of the different sampling depths. As deployment and recovery of these samples is a laborious and time consuming task especially in marine studies, however, the simulated *in situ* method is a possible alternative. Here the water samples are incubated on the deck of a ship or in a laboratory under controlled temperature and light conditions (Lohrenz, 1993).

One disadvantage of the different *in vitro* methods are the so-called bottle effects because the phytoplankton is artificially confined in a container (Venrick et al., 1977; Madden & Day, 1992; Maestrini et al., 1993). Another disadvantage is the fact that it is almost impossible to simulate the natural light climate that phytoplankton experiences in shallow lakes, estuaries and the surface mixed layer of deep lakes, and oceans under the conditions of vertical mixing (MacIntyre, 1993; Maestrini et al., 1993).

The following paper deals with the effects of small scale turbulence and large scale mixing on primary production of phytoplankton using the *in situ* method of primary production measurement in mesocosms and a shallow lake.

The lack of turbulence in incubation bottles is one of the striking differences between experimental and natural conditions when measuring primary production with an in vitro method. Inducing turbulence by rotating or shaking is easy in culture studies (e.g. Talling, 1960) or with the simulated *in situ* method that uses an incubator (Madden & Day, 1992).

When performing *in situ* measurements of primary production, however, the water samples usually remain in the incubation bottles without any water motion inside. In the present study, stirrers that make it possible to move the water samples inside the bottles during *in situ* incubation (Behrendt, 1989b) were further developed. Carbon fixation was measured in a series of experiments comparing stirred and unstirred samples incubated *in situ*.

To overcome the problem of an unnatural light climate during *in vitro* primary production measurement due to the static nature of the incubation, several authors have tried to simulate a fluctuating light regime either *in situ* (Jewson & Wood, 1975; Dokulil et al., 1978; Marra, 1978a; Nusch, 1981; Gallegos & Platt, 1982; Joiris & Bertels, 1985; Yoder & Bishop, 1985; Nixdorf et al., 1990, 1992; Nixdorf & Behrendt, 1991), in an incubator (Marra, 1978b; Savidge, 1986; Grobbelaar, 1989, 1994; Kromkamp et al., 1992; Mallin & Paerl, 1992) or by modelling (Benndorf & Baumert, 1981; Falkowski & Wirick, 1981; Farmer & Takahashi, 1982; Pahl-Wostl, 1992; Franks & Marra, 1994). Most of the practical studies have been performed in marine environments or with marine organisms. The results of these investigations show that there is no uniform reaction of primary production to turbulent mixing because this reaction is influenced by the intensity of mixing, the turbidity of the water, the degree of photoinhibition and the physiological state of the populations studied.

Besides presenting new data on the influence of dynamic incubation on *in situ* primary production measurements, the present paper focuses on the effect of the form of vertical mixing (i.e. linear or circular up and down movement) on primary production in a shallow lake. Another approach compares the measurement of primary production by static and dynamic incubation of samples in large mesocosms under controlled conditions.

Materials and methods

Site description

Experiments were carried out in the shallow, polymictic and eutrophic lake Müggelsee (area: 7.3 km^2; mean depth: 4.9 m; maximum depth 8.0 m; Driescher et al., 1993; Behrendt et al., 1993).

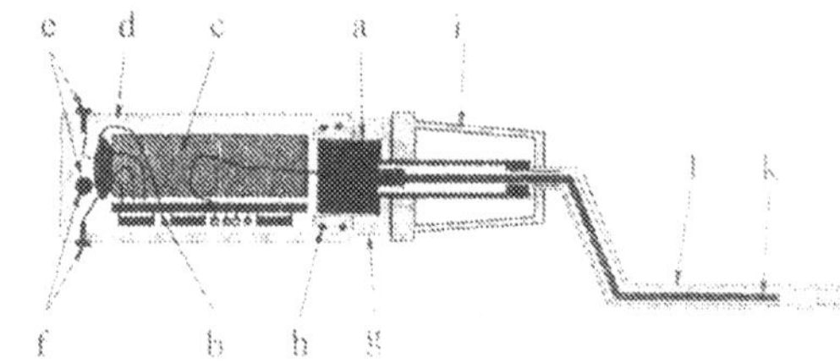

Figure 1. Schematic drawing of the bottle stirrer. a – motor (6 V, 8 mA, 5500 rotations per minute, gearbox 139:1, torque 2 Ncm); b electronic control unit; c – accumulator; d – watertight perspex casing (37 mm × 85 mm); e – contacts for motor switching; f – contacts for recharging; g – cover; h – washer; i – polypropylene stopper; k – elongated motor axis; l – silicone tube.

Small scale turbulence

To induce small scale turbulence in the incubation bottles, specially designed bottle stirrers were used (based on Behrendt, 1989a; Figure 1). The stirrer is fixed on an incubation bottle by means of a polypropylene stopper. Through the stopper and into the bottle leads the elongated axis of the motor shielded by a watertightly closed silicone tubing. An electronic control unit allows to choose between different rotation velocities and break times. On the outside of the casing are three contacts that allow recharging of the accumulator and switch on the motor when the stirrer is getting in contact with water. The following setting of the stirrers were used in all experiments: a rotation velocity of 0.6 rotations per s, a rotation time of 2.5 s, 10 s break time. This setting results in about 20 rotations of the water sample in the bottle per minute. The advantage of having a break time of 10 s is that the water motion in the bottle becomes turbulent instead of unidirectional.

Large scale mixing

Movement of the incubation samples throughout the water column (subsequently named 'dynamic incubation') was simulated with the help of bottle lifts (Behrendt, 1989b). The movement of the bottles is generated by a stepper motor that unrolls and rolls up the rope fixed to the bottle holder on a winch axis. One lift can simultaneously carry up to 4 glass bottles with a volume of 600 ml. A programmable electronic control unit enables an up and down movement of the incubation bottles simulating either a linear movement or a circular movement. The movement amplitude can be chosen between 0 and 10 meters with a period between 2 and 40 minutes. The movement velocity of the bottles in dynamic incubation experiments (see below) was in the range described for velocities in Langmuir circulations (Denman & Gargett, 1983).

Mesocosm experiments

Experiments were carried out in the Plankton Towers of the Max Planck Institute for Limnology in Plön, Germany (Lampert & Loose, 1992). They consist of two stainless steel columns of 11.2 m in height and 86 cm in diameter. The towers were filled with water from the mesotrophic lake Schöhsee filtered through 10 μm filters. The mesocosms were vertically divided by perspex plates up to a depth of 2.5 m or 4.7 m length, respectively (Figure 2). The plates were placed 10 cm below the water surface. To induce a controlled water circulation, a boat motor (Yamaha M15) was placed 50 cm above the lower edge on one side of the plate in both towers (Figure 2). The motor sent water upwards on one side of the plate, the water flowed over the top of the plate and returned on the other side. A linear correlation between propeller revolutions of the motor and mean water velocity in the towers (registered by a VEM003 Electromagnetic Current Flow Meter, Valeport Marine Scientific Limited) was recorded. The temperature of the circulating water was set at 20 °C. Below the perspex plates, water temperature was set at 10 °C. This ensured that the two different water strata did not mix and that the water masses returned to the other half of the tower about 40 cm below the lower edge of the plates. Each tower was lit by a daylight metal halide lamp (HQI-T 2000 W/D/I, Osram) with a light/dark ratio of 14/10 hours. Just below the water surface, the photosynthetically active radiation (PAR) amounted to 550 μE m^{-2} s^{-1}.

The lake water in the towers was enriched with nutrients to reach the quality of the culture medium 'WC' by Guillard & Lorenzen (1972). Instead of buffering the water, pH was readjusted to 7.2 by adding diluted HCl solution every evening. To prevent high oxygen oversaturation, the photosynthetically enlarged concentration of oxygen was reduced to 80% saturation by bubbling the tower water with gaseous nitrogen for two hours every evening.

The initially inoculated culture of *Asterionella formosa* Hass. was replaced by *Scenedesmus acutus* Meyen and autotrophic picoplankton within 10 days of the pre-experimental time. These algae must have survived the preceding cleaning of the towers or passed the filters when filling the towers with lake water.

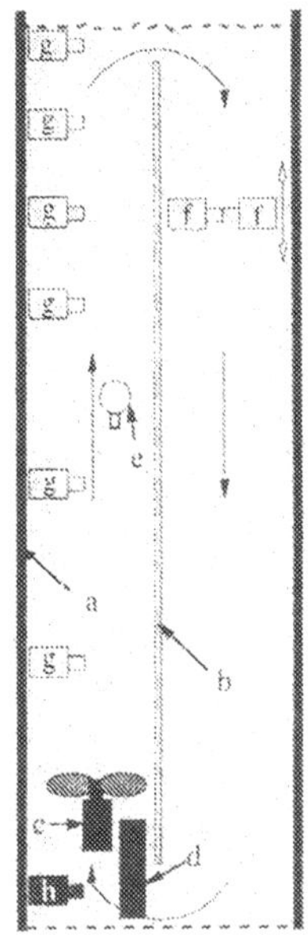

Figure 2. Schematic drawing of the experimental design in the Plankton Towers of the Max Planck Institute for Limnology in Plön. a – tower wall; b – perspex plate; c – boat motor; d – multiprobe; e – light sensor; f – dynamically incubated bottles; g – statically incubated bottles; h – dark bottle. Ropes, cords and cables are not shown.

In the towers, PAR was continuously measured by a stationary placed sensor (see below and Figure 2). Temperature, pH, oxygen concentration and specific conductance were registered with a multiparameter probe (see below and Figure 2).

Primary production measurements were carried out under two different experimental designs. The first experiment lasted 8 days, the mixing depth was regulated at 2.9 m. The second experiment lasted 4 days with a mixing depth of 5 m. During both experiments the water column circulated with a mean period of 10 min around the perspex plates.

In the course of both experiments, algal biomass increased from initially low values so that primary production measurements were carried out at a range of biomass values and underwater light attenuation coefficients on the different days. *Scenedesmus acutus* was the dominant algal species during the first experiment. Autotrophic picoplankton dominated the phytoplankton in the second experiment.

For the comparison of the static and the dynamic *in situ* method of primary production measurement, a subsample of the circulating tower plankton was incubated in the towers in bottles that were kept under static and dynamic conditions inside the towers (Figure 2). Static bottles were incubated at fixed depths (experiment 1: 0, 50, 100, 150, 250 cm; experiment 2: 0, 50, 100, 150, 250, 350 cm; two parallels per depth). The dynamic approach was achieved by moving 4 parallel bottles (2 oxygen method, 2 radiocarbon method) up and down with the bottle lift (see above) in a linear fashion for a period of 10 min at an amplitude of 2.9 and 4.9 m, respectively. Thus the algae inside the bottles faced the same light climate as the algae in the surrounding mesocosm. During incubation all samples were mixed by the bottle stirrers described above.

Lake experiments

Thirteen *in situ* experiments studying the influence of the bottle stirrers on primary production were carried out 50 m off the shore of lake Müggelsee (June–September 1994; April/May 1995). In each experiment, lake water from 1 m depth was incubated in sets of three parallel glass bottles with stirrers and three parallel bottles without stirrers in static incubation (0 cm –50 cm depth) and dynamic incubation (amplitude 2 m, period 15 min, linear movement), respectively. Three dark bottles with stirrers and three dark bottles without stirrers were also incubated.

Eight experiments comparing static and dynamic *in situ* measurement of primary production were carried out 500 m off the shore of lake Müggelsee (June–September 1994). In each experiment, lake water from 1 m depth was incubated in glass bottles with bottle stirrers in operation. Static measurement was done by incubating 1 or 2 bottles in 0, 50, 100, 200 and 500 cm water depth, respectively. Parallel to the static incubation, four different kinds of dynamic incubation were performed with two parallel bottles on each lift: lift 1: amplitude 2 m and linear movement, lift 2: amplitude 2 m and circular movement; lift 3: amplitude 5 m and linear movement, lift 4: amplitude 5 m and circular movement. In all dynamic incubations, bottles were moved with a period of 15 min for one up and down movement.

General methods

All experiments were accompanied by the following *in situ* measurements: water temperature, oxygen saturation, pH and specific conductance were registered with an H_2O Multiparameter Water Quality Data Transmitter (Hydrolab Corporation). For measuring photosynthetically active radiation (PAR), two LI 193 SB Spherical Quantum Sensors (LICor, Inc.) were used. The basis for the calculation of attenuation coefficient and euphotic depth were continuous measurements in 0.5 m and 1 m water depth (experiments on stirrer influence) or hourly measurements of PAR every 50 cm in the water column (experiments on lift influence), respectively. For analysing the influence of the linear and circular form of vertical motion on light climate, a light sensor was moved up and down with a lift.

From the original water samples, subsamples were not only taken for primary production measurement (see below) but also for the estimation of chlorophyll-*a*-concentration, dissolved inorganic carbon (DIC) and phytoplankton composition. Chlorophyll concentration was analyzed according to DIN 38412-L16 in Deutsche Einheitsverfahren (1983) in two or three replicates per sample. Alkalinity and concentration of DIC were measured and calculated according to Mackereth et al. (1989) or Gelbrecht et al. (1987) in two replicates. Independent of our investigations, phytoplankton from lake Müggelsee was analysed weekly (Driescher et al., 1993) according to Utermöhl (1958) and Rott (1981).

Primary production measurement

Productivity rates were determined using the *in situ* light and dark bottle oxygen technique (Wetzel & Likens, 1991; DIN 38412-L 13 in Deutsche Einheitsverfahren 1983) and the *in situ* radiocarbon method (Wetzel & Likens, 1991).

All measurements of primary production were carried out in horizontally orientated 600 ml clear and opaque glass bottles for an incubation period of 4 to 5 hours between 9 and 14 CET. While in the mesocosm experiments oxygen production and carbon uptake were measured in different but parallel incubated bottles, in the lake experiments oxygen production and carbon uptake were registered in the same bottle.

For the measurement of carbon uptake, bottles were filled with sample water from a carboy, injected with $NaH^{14}CO_3$ solution (1 μCi to 3 μCi per 100 ml water) and exposed *in situ* within 10 min after sampling. Following incubation, three replicates of 10 ml to 20 ml volume per sample were gently filtered through 25 mm cellulose acetate membrane filters with a pore size of

Table 1. Comparison of total carbon uptake in unstirred and stirred samples for the 13 experiments. Values are the mean of three replicate incubations ± standard deviation. The total carbon uptake in stirred samples is given as percentage of uptake in unstirred samples. It is stated whether the two means are significantly different at the 0.05 level (t-test).

Date	Unstirred series (μg C l^{-1} h^{-1})	Stirred series (μg C l^{-1} h^{-1})	% of unstirred	Significant
Dynamic incubation				
9 June '94	31.1 ± 3.48	39.1 ± 3.94	125	no
17 June '94	81.0 ± 1.45	76.7 ± 4.47	95	no
23 June '94	63.5 ± 0.86	70.8 ± 3.45	112	yes
8 July '94	13.7 ± 1.43	11.2 ± 0.60	82	no
21 July '94	249.3 ± 19.25	265.7 ± 3.90	107	no
12 Aug. '94	124.4 ± 11.8	113.9 ± 4.6	92	no
19 Aug. '94	24.3 ± 1.0	25.4 ± 3.1	104	no
16 Sept. '94	20.4 ± 1.1	18.4 ± 2.1	90	no
23 Sept. '94	25.0 ± 2.3	22.7 ± 2.5	91	no
20 Apr. '95	24.4 ± 2.6	25.3 ± 3.1	104	no
27 Apr. '95	61.3 ± 1.7	66.2 ± 1.7	108	no
3 May '95	86.6 ± 2.2	81.1 ± 3.5	94	no
17 May '95	5.89 ± 0.21	5.55 ± 0.22	93	no
			mean: 100% ± 12%	
Static incubation				
9 June '94	37.0 ± 7.6	49.6 ± 6.2	134	no
23 June '94	81.1 ± 1.6	89.5 ± 2.3	110	yes
8 July '94	69.3 ± 5.5	83.3 ± 3.6	120	yes
21 July '94	353.7 ± 21.2	445.0 ± 21.6	126	yes
12 Aug. '94	251.8 ± 10.2	259.9 ± 4.1	103	no
19 Aug. '94	67.0 ± 8.1	55.6 ± 2.7	83	no
16 Sept. '94	37.6 ± 3.0	38.1 ± 3.6	101	no
23 Sept. '94	23.5 ± 0.4	26.0 ± 2.0	111	no
20 Apr. '95	64.7 ± 4.1	76.0 ± 4.5	118	no
27 Apr. '95	105.9 ± 4.1	99.7 ± 6.1	94	no
3 May '95	103.3 ± 8.9	103.3 ± 3.2	100	no
17 May '95	7.60 ± 0.41	7.24 ± 0.13	95	no
			mean: 108% ± 15%	

0.2 μm (Sartorius). Filters were then washed with 2 ml of filtered, non-radioactive sample water, placed on a filter paper soaked with 0.01 N HCl solution for 10 min, dried and dissolved in 4 ml scintillation cocktail (Beckman Ready Solve or Packard Filter Count). Only in the lake experiments carbon exudation was measured: after acidifying and bubbling the 0.2 μm filtrate (Gächter et al., 1984), it was mixed 1:1.5 with the scintillation cocktail Ultima Gold XR (Canberra Packard). Activity was measured with the liquid scintillation counter Tricarb 1900 TR (Canberra Packard).

Mean primary production was calculated from three replicate filters or filtrates per bottle and two or three replicate bottles according to Wetzel & Likens (1991). Carbon fixation >0.2 μm and <0.2 μm was added up to total carbon fixation. The integral production of the static incubation was estimated by integrating the vertical profile of production according to DIN 3841 2-L 13 in Deutsche Einheitsverfahren (1993). The integral production of the dynamic incubation was calculated by multiplying the production measured in the moved bottles with the amplitude of their movement.

Initial and final oxygen concentrations were measured with a Clark type temperature-compensated electrode (OXI 530, TriOxmatic EO200; WTW, Weilheim). The correct performance of the electrode was checked with the Winkler method (EN 25813 in Deutsche Einheitsverfahren 1993).

Table 2. Primary production in the dynamically incubated samples (linear and circular movement) as percentage of primary production in the static incubation. The means and standard errors of 8 lake experiments are given in each case.

Method	Amplitude of movement: 2 m		Amplitude of movement: 5 m	
	Linear movement	Circular movement	Linear movement	Circular movement
Oxygen	119 ± 2	110 ± 4	102 ± 4	115 ± 6
Radiocarbon	112 ± 5	112 ± 4	90 ± 5	106 ± 5

Results

Small scale turbulence

The operation of the bottle stirrers guaranteed a homogeneous distribution of seston in the samples during incubation. While in the unstirred bottles, particles tended to aggregate, in the stirred bottles neither the sedimentation of seston nor the accumulation of buoyant plankton in the upper part of the bottles was observed.

In the comparison of stirred and unstirred bottles, carbon fixation was increased in the stirred bottles in 14 incubations and decreased in 11 incubations (Table 1). The differences between the two methods were statistically significant, however, only in four incubations (Table 1). The ratio of production in stirred to production in unstirred bottles showed no significant correlation with underwater PAR (between 134 μ E m^{-2} s^{-1} and 1060 μ E m^{-2} s^{-1} in static incubation), attenuation coefficient (between 0.72 m^{-1} and 2.56 m^{-1}), absolute amount of production (between 6 μg C l^{-1} h^{-1} and 445 μg C l^{-1} h^{-1}), chlorophyll concentration (between 3.0 μg l^{-1} and 43 μg l^{-1}) or the contribution of diatoms (between 22% and 91%) or buoyant cyanobacteria (between 0% and 52%) to total phytoplankton biomass. On average, stirring in static incubations showed a higher impact than in dynamic incubations (108% to 100%, Table 1). When a statistically significant difference in carbon uptake between the two methods was observed, stirring always resulted in a higher production (Table 1). This was observed in the static incubation on three days in June and July 1994 when cyanobacteria (*Aphanizomenon flos-aquae* Ralfs ex Born. et. Flah.) had a larger share of phytoplankton biomass (24% to 52%).

Large scale mixing

Mesocosm experiments

The experiments succeeded in maintaining a growing population of planktonic algae under conditions of a controlled state of vertical mixing, temperature, light intensity and nutrient concentration. Chlorophyll-concentration varied between 4.7 μg l^{-1} and 87 μg l^{-1}, the vertical attenuation coefficient was between 1.00 m^{-1} and 1.65 m^{-1}.

In all primary production measurements, the dynamic incubation resulted in higher production compared to the static incubation (Figure 3). The oxygen production in the dynamic incubation amounted to a mean value of 131% ± 21% SD of the production in the static incubation. The corresponding mean value for carbon uptake was 134% ± 12% SD.

The P-I curves for the static incubated bottles did not indicate photoinhibition as it is traditionally proved, i.e. they did not show a negative slope at supersaturating irradiance (Henley, 1993). A P-I curve showing no negative slope cannot exclude photoinhibition, however (Henley, 1993). As in all mesocosm experiments the PAR above 50 cm water depth was between 240 μ E m^{-2} s^{-1} and 550 μ E m^{-2} s^{-1} it is assumed that photoinhibition of primary production took place near the water surface because independent of temperature and species photoinhibition becomes detectable above 200 μ E m^{-2} s^{-1} (Harris, 1980).

Lake experiments

The static incubation resulted in surface photoinhibition of primary production in all but one of the experiments. This was measured with the radiocarbon method (Figure 4) as well as with the oxygen method (not shown). All except one type of experiment showed an increase in primary production in the dynamic incubation compared with the static incubation (Table 2). The percentage of carbon uptake in linear dynamic incubation compared to static incubation tended to decrease as the ratio of mixed depth to euphotic depth increased (Figure 5).

By moving the incubation bottles in a linear or circular fashion, they were exposed to a different light climate (Figure 6). On the days primary production measurements were carried out, a mean attenuation coefficient of underwater PAR between 0.70 m^{-1} and 1.61 m^{-1} was recorded in lake Müggelsee. In this range of attenuation coefficients, results of theoretical calculations show that algae moving in a circular

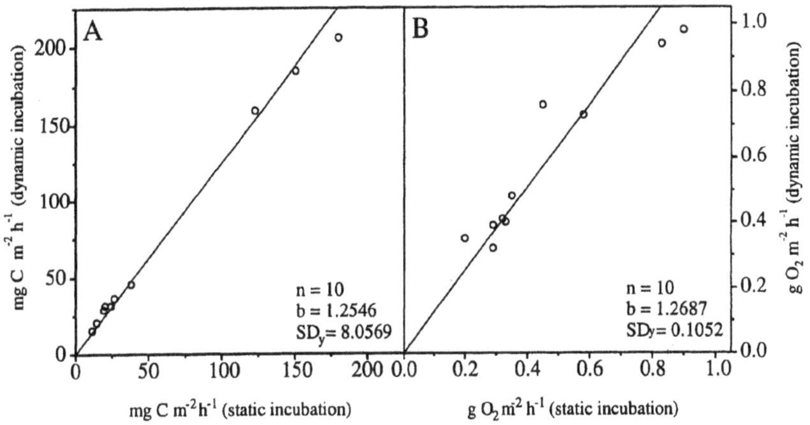

Figure 3. Primary production in dynamic incubation versus primary production in static incubation from mesocosm experiments. A – carbon uptake (>0.2 μm). B – gross oxygen production.

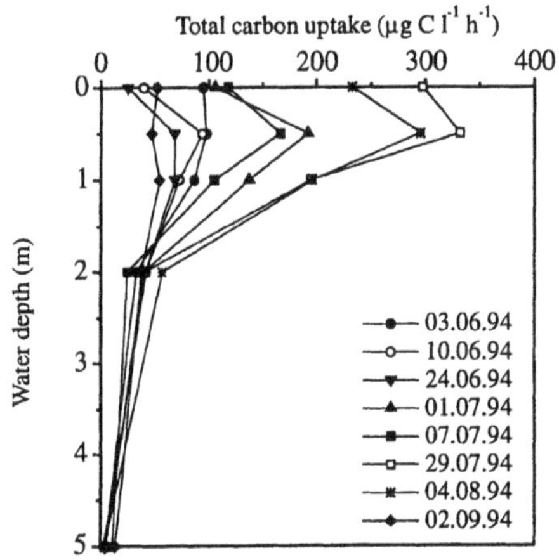

Figure 4. Primary production (total carbon uptake) in static bottles suspended in lake Müggelsee on 8 days.

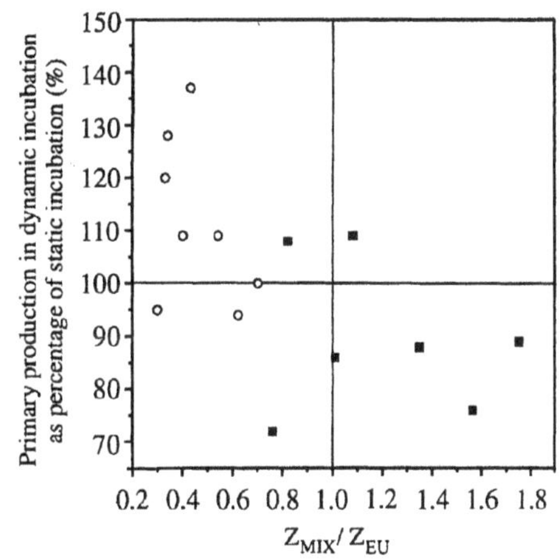

Figure 5. Primary production (total carbon uptake) in linear dynamic incubation as percentage of primary production in static incubation versus the ratio of mixed depth (Z_{mix}) to euphotic depth (Z_{eu}) measured in lake Müggelsee (circles: mixing depth = 2 m; squares: mixing depth = 5 m).

fashion receive between 17% and 4% more light than algae moving in a linear fashion if the amplitude is 2 m (Figure 7). With an amplitude of 5 m these values amount to 61% and 20% (Figure 7). The higher light supply during circular movement in comparison to linear movement resulted in higher primary production in the circulating bottles compared to the bottles that were moved in a linear fashion (Figure 8). This coherence was more evident with an amplitude of 5 m because here larger differences in light climate exist compared to an amplitude of 2 m (Figure 7).

Discussion

Small scale turbulence

The enclosure of phytoplankton in incubation bottles confines the algal cells in an artificial environment without any water motion. Under these conditions buoyant algae tend to accumulate in layers in the upper part of the bottle and fast sinking diatoms and suspended solids form a bottom layer. The accumulation of seston alters the light climate in the incubation bot-

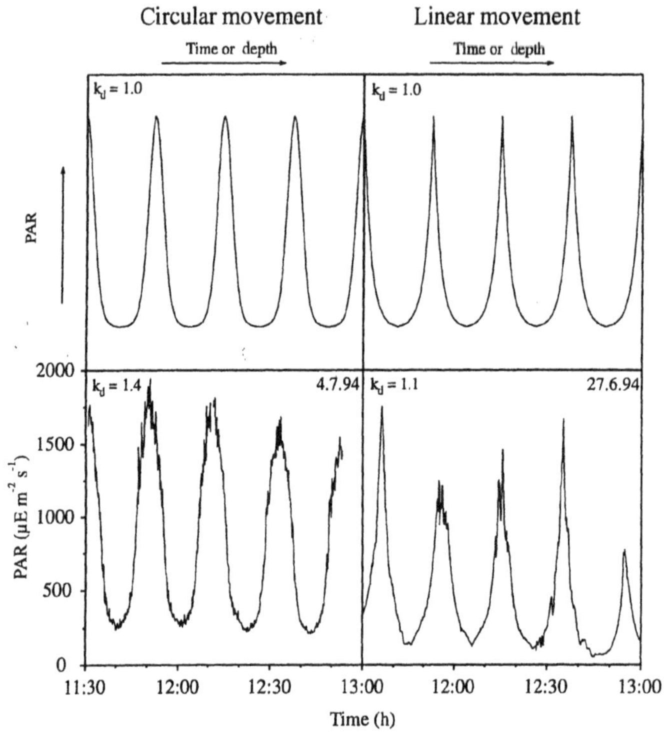

Figure 6. Photosynthetically active radiation (PAR) that is received by a bottle moved up and down in the water column with a bottle lift. The upper graphs give the theoretical values for circular and linear movement (attenuation coefficient $k_d = 1.0$). The lower graphs show the observed PAR on two different days (left: circular movement, right: linear movement).

tles especially in turbid waters (Madden & Day, 1992). The lack of turbulence results in the accumulation of waste products and the exhaustion of nutrients and carbon dioxide in the medium around the algal cells leading to a decrease in the photosynthetic rate (Olofsson, 1980; Grobbelaar, 1989, 1994). This effect is dependent on species composition as the development of extracellular microenvironments changes with cell size (Richardson & Stolzenbach, 1995) and species differ in the efficiency of e.g. carbon uptake (Olofsson, 1980). Published reports on the influence of stirring on primary production show various results. Talling (1960) observed no effect of shaking on oxygen production of a marine diatom in culture. Olofsson (1980) increased the photosynthetic rate of two algal species in short-term experiments by stirring. Savidge (1981) revealed an inverse relationship between the rates of agitation and carbon uptake of phosphorus-limited cultures of *Phaeodactylum tricornutum* Bohlin. Doty & Oguri (1958) measured a 30% to 50% increase in carbon uptake by agitating marine phytoplankton in a simulated *in situ* measurement. Madden & Day (1992) recorded significant (up to 80%) productivity differences (oxygen method) in rotating and non-rotating treatments of plankton from a turbid estuary in simulated *in situ* measurements. At low light levels, rotation enhanced productivity as a result of an increase in average irradiance; at high light levels, rotation reduced productivity because rotating algae became photoinhibited as they were nearer to the light source than the algae in the sediment of the incubation bottles (Madden & Day, 1992).

The present study does not show strong effects of stirring on carbon uptake by the phytoplankton in lake Müggelsee in *in situ* incubation. Reduction (up to 18%) or enhancement (up to 34%) are only fairly high. On average, there was no influence of stirring in dynamic incubations whereas three of four statistically significant increases in carbon uptake were observed in static incubations. This points to the hypothesis that light climate rather than the supply of nutrients or dissolved gases was influenced by the stirring of Müggelsee phytoplankton because the only difference between static and dynamic incubation was in the light climate of the exposed samples. In dynamic incubation, photoinhibition of algal photosynthesis will rarely occur because algae are near the water surface only for short periods (Harris, 1980). The static incubation in our experiments fixed phytoplankton near to the lake surface and exposed them to inhibiting PAR intensities during the whole incubation period. Especially when buoyant cyanobacteria were present in larger amounts was there a significant increase in carbon uptake (110% to 126%) observed in the statically incubated, stirred samples. While in unstirred samples bundles of *Aphanizomenon flos-aquae* concentrated in the upper region of the bottles, they were homogeneously distributed in the stirred samples. Compared to the unstirred bottles stirring might have caused an overall reduction in light intensity reaching the algal filaments in the bottles. This could have resulted in the reduced photoinhibition of algal photosynthesis explaining the higher carbon uptake in stirred samples. Future experiments during periods of cyanobacterial dominance should address this hypothesis.

Large scale turbulence

Both the mesocosm experiments and the lake experiments in this study confirmed the results of Marra (1978a), Nixdorf et al. (1990, 1992) and Nixdorf & Behrendt (1991) that the dynamic *in situ* incubation can result in higher primary production measurements compared to the classic static incubation method.

The mean increase of primary production by linear dynamic incubation was at least 10% higher in the mesocosm studies than in the lake experiments (cf. Figure 3 and Table 2). This might have been due to the fact that algal populations incubated in the mesocosm experiments were physiologically adapted to the simulated mixing conditions that were the same *in situ* and in incubation. In all lake experiments samples taken from 1 m depth contained populations with an unknown and varying mixing history and physiological state.

Referring to the results of Marra (1978a), Harris (1980) pointed out that an increase in productivity by dynamic incubation is only likely when photoinhibition of the static surface samples occurs. In line with this, photoinhibition seemed to be an important cause of the underestimation of productivity in static incubation measurements in the present study. Harris (1980) also considers the concept of column compensation

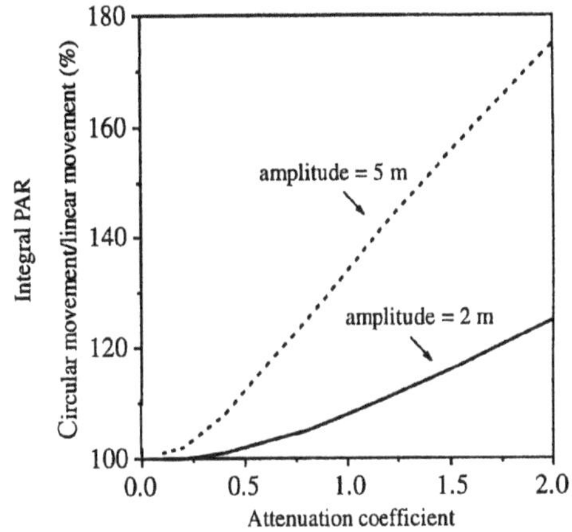

Figure 7. Integral PAR resulting from circular movement in relation to integral PAR resulting from linear movement versus attenuation coefficient for two mixing depths.

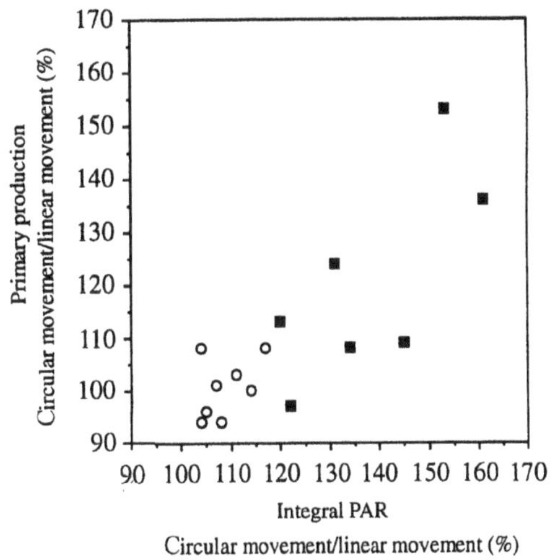

Figure 8. Primary production (total carbon uptake) versus integral PAR. For both parameters the amount resulting from circular movement is given as percentage of the amount resulting from linear movement. Primary production was measured in lake Müggelsee, PAR percentages result from theoretical calculations considering measured attenuation coefficients (circles: mixing depth = 2 m; squares: mixing depth = 5 m).

on the basis of the results of Jewson & Wood (1975). This concepts states 'that if the mixing was continued below the photic depth then the overall productivity in the column was reduced by a greater amount than was made up by the reduction (or elimination) of photoinhibition at the surface' (Harris, 1980). The results of the present study point in the same direction because with an increase in the ratio of Z_{mix}/Z_{eu}, the ratio of dynamic/static incubation fell below 100% (Figure 5).

The present study shows that with increasing mixing depth and increasing turbidity the form of the vertical motion becomes more and more important for primary production estimates. Almost all investigations of the influence of vertical mixing on primary production did not circulate their samples but moved them linearly up and down (e.g. Jewson & Wood, 1975; Marra, 1978a; Joiris & Bertels, 1985; Nixdorf et al., 1992). This linear form of water movement is no realistic simulation of natural mixing conditions. The circular motion simulated in our experiments is only slightly better because in real Langmuir circulation velocity of upwelling and downwelling water is different (Denman & Gargett, 1983) and because discontinuous fluctuations in mixing conditions occur (MacIntyre, 1993). As MacIntyre (1993) pointed out, a knowledge of real trajectories and mixing rates of phytoplankton in the surface mixed layer is necessary for the design of future experiments and models.

Conclusion

The bottle stirrers tested in this investigation are capable of simulating a natural turbulence condition in bottles that are incubated *in situ*. In the eutrophic lake studied, artificial turbulence in the incubation bottles significantly increased primary production only in periods of dominance of buoyant cyanobacteria. Working with these stirrers in other ecosystems e.g. shallow, turbid lakes or estuaries, where a larger influence of small scale turbulence on primary production has been proved (Madden & Day, 1992), would allow *in situ* measurements where up to now only simulated *in situ* measurements have been performed.

The Plankton Towers of the Max Planck Institute for Limnology (Lampert & Loose, 1992) are well suited to study primary production under controlled conditions of vertical mixing. Photosynthetic characteristics of single species or natural communities can be studied at a varying range of light intensities, mixing intensities and ratios of mixed depth to euphotic depth on a mesocosm scale.

Vertical mixing cannot be neglected when measuring primary production in shallow lakes. The form of the movement in the course of vertical mixing influences algal photosynthesis especially in more turbid systems. Therefore it is important to learn more about the real mixing conditions in lakes and oceans.

Acknowledgments

This work was supported by a DFG grant to H. Behrendt. Bottle stirrers and lifts were designed and built by Thomas Hintze and Rüdiger Biskupek. Erika Nemrava, Andreas Lorke and Andreas Irrgang helped during sampling, *in situ* incubation and primary production measurement. Marianne Graupe and Ingrid Hoffmann carried out chlorophyll-measurements, DIC-analysis and Winkler-titrations. Phytoplankton analysis was done by Sigrid Hoeg. Without the help of all these people our project could not have been carried out. Thus their help is gratefully acknowledged. Special thanks go to Professor Winfried Lampert, Karl-Otto Rothhaupt and the staff of the Max Planck Institute for Limnology for giving us the opportunity to work with the Plankton Towers in their institute.

References

Behrendt, H., 1989a. Devices for turbulence simulation during in-situ measurements of biological processes in aquatic ecosystems, I. A simple stirrer for bottles of different volume. Acta hydrophys. Berlin 33: 5–8.

Behrendt, H., 1989b. Devices for turbulence simulation during in-situ measurements of biological processes in aquatic ecosystems, II. A bottle lift for simulation of vertical movements in natural waters. Acta hydrophys. Berlin 33: 9–14.

Behrendt, H., B. Nixdorf & W. G. Pagenkopf, 1993. Phenomenological description of polymixis and influence on oxygen budget and phosphorus release in lake Müggelsee. Int. Revue ges. Hydrobiol. 78: 411–421.

Benndorf, J. & H. Baumert, 1981. Modellvorstellungen zum Einfluß der Durchmischung des Wasserkörpers auf die phytoplanktische Primärproduktion. In Unger, K. & G. Stöcker (eds), Biophysikalische Ökologie und Ökosystemforschung. Akademie-Verlag, Berlin: 333–345.

Denman, K. L. & A. E. Gargett, 1983. Time and space scales of vertical mixing and advection of phytoplankton in the upper ocean. Limnol. Oceanogr. 28: 801–815.

Deutsche Einheitsverfahren, 1983–1993. Deutsche Einheitsverfahren zur Wasser-, Abwasser- und Schlammuntersuchung. Verlag Chemie GmbH, Weinheim.

Dokulil, M., L. Hammer & D. H. Jewson, 1978. Vergleichende Untersuchungen zur Primärproduktion des Phytoplanktons im Neusiedlersee. O_2, ^{14}C und Experimente mit künstlicher Zirkulation. Biol. Forsch. Inst. Burgenland 29: 60–73.

Doty, M. S. & M. Oguri, 1958. Selected features of the isotopic carbon primary productivity technique. Rapp. Cons. Explor. Mer 144: 47–55.

Driescher, E., H. Behrendt, G. Schellenberger & R. Stellmacher, 1993. Lake Müggelsee and its environment – natural conditions and anthropogenic impacts. Int. Revue ges. Hydrobiol. 78: 327–343.

Falkowski, P. G. & C. D. Wirick, 1981. A simulation model of the effects of vertical mixing on primary productivity. Mar. Biol. 65: 69–75.

Farmer, D. M. & M. Takahashi, 1982. Effects of vertical mixing on photosynthetic responses. Jap. J. Limnol. 43: 173–181.

Franks, P. J. S. & J. Marra, 1994. A simple new formulation for phytoplankton photoresponse and an application in a wind-driven mixed-layer model. Mar. Ecol. Progr. Ser. 111: 143–153.

Gächter, R., A. Mares & M. M. Tilzer, 1984. Determination of phytoplankton production by the radiocarbon method: a comparison between the acidification and bubbling method (ABM) and the filtration technique. J. Plankton Res. 6: 359–364.

Gallegos, C. L. & T. Platt, 1982. Phytoplankton production and water motion in surface mixed layers. Deep Sea Res. 29: 65–76.

Gelbrecht, J., G. Henrion & R. Henrion, 1987. Zur Bestimmung des gesamten anorganischen Kohlenstoffes in natürlichen Gewässern durch Titration mit Salzsäure. Acta hydrochim. hydrobiol. 15: 19–28.

Grobbelaar, J. U., 1989. Do light/dark cycles of medium frequency enhance phytoplankton productivity? J. appl. Phycol. 1: 333–340.

Grobbelaar, J. U., 1994. Turbulence in mass algal cultures and the role of light dark fluctuations. J. appl. Phycol. 6: 331–335.

Guillard, R. R. L. & C. J. Lorenzen, 1972. Yellow-green algae with chlorophyllide c. J. Phycol. 8: 10–14.

Harris, G. P., 1980. The measurement of photosynthesis in natural populations of phytoplankton. In Morris, I. (ed.), The physiological ecology of phytoplankton, Blackwell Scientific Publications, Oxford: 129–187.

Henley, W. J., 1993. Measurement and interpretation of photosynthetic light-response curves in algae in the context of photoinhibition and diel changes. J. Phycol. 29: 729–739.

Jewson, D. H. & R. B. Wood, 1975. Some effects on integral photosynthesis of artificial circulation of phytoplankton through light gradients. Verh. int. Ver. Limnol. 19: 1037–1044.

Joiris, C. & A. Bertels, 1985. Incubation under fluctuating light conditions provides values much closer to real *in situ* primary production. Bull. mar. Sci, 37: 620–625.

Kromkamp, J., F. Schanz, M. Rijkeboer, E. Berdalet, B. Kim & H. J. Gons, 1992. Influence of the mixing regime on algal photosynthetic performance in laboratory scale enclosures. Hydrobiologia 238: 111–118.

Lampert, W. & C. J. Loose, 1992. Plankton Towers: Bridging the gap between laboratory and field experiments. Arch. Hydrobiol. 126: 53–66.

Lohrenz, S. E., 1993. Estimation of primary production by the simulated *in situ* method. ICES mar. Sci. Symp. 197: 159–171.

MacIntyre, S., 1993. Vertical mixing in a shallow, eutrophic lake: Possible consequences for the light climate of phytoplankton. Limnol. Oceanogr. 38: 798–817.

Mackereth, F. J. H., J. Heron & J. F. Talling, 1989. Water Analysis. Freshwater Biological Association, Ambleside, 120 pp.

Madden, C. J. & J. W. Day, Jr., 1992. Induced turbulence in rotating bottles affects phytoplankton productivity measurements in turbid waters. J. Plankton Res. 14: 1171–1191.

Maestrini, S. Y., A. Sournia & A. Herbland, 1993. Measuring phytoplankton production in 1992 and the coming years: a dilemma? ICES mar. Sci. Symp. 197: 244–259.

Mallin, M. A. & H. W. Paerl, 1992. Effects of variable irradiance on phytoplankton productivity in shallow estuaries. Limnol. Oceanogr. 37: 54–62.

Marra, J., 1978a. Phytoplankton photosynthetic response to vertical movement in a mixed layer. Mar. Biol. 46: 203–208.

Marra, J., 1978b. Effect of short-term variations in light intensity on photosynthesis of a marine phytoplankter: A laboratory simulation study. Mar. Biol. 46: 191–202.

Nixdorf, B. & H. Behrendt, 1991. Discrepancies between O_2- and ^{14}C-method resulting from measurements of primary production under consideration of vertical mixing in a shallow eutrophic lake. Verh. int. Ver. Limnol. 24: 1268–1271.

Nixdorf, B., H. Behrendt & R. Stellmacher, 1990. Comparison of methods for estimation of integral primary production in shallow aquatic ecosystems with special regard to turbulent mixing. Limnologica 20: 53–56.

Nixdorf, B., W. G. Pagenkopf & H. Behrendt, 1992. Diurnal patterns of mixing depth and its influence on primary production in a shallow lake. Int. Revue ges. Hydrobiol. 77: 349–360.

Nusch, E. A., 1981. Continuous chlorophyll fluorometry and integral measurement of photosynthetic oxygen production in a eutrophic impoundment. Vom Wasser 57: 297–307.

Olofsson, J. A., 1980. The role of microlayers in controlling phytoplankton primary productivity. In: Barica, J. & L. R. Mur (eds), Hypertrophic Ecosystems. Developments in Hydrobiology 2. Dr W. Junk Publishers, The Hague: 83–93.

Pahl-Wostl, C., 1992. Dynamic versus static models for photosynthesis. Hydrobiologia 238: 189–196.

Platt, T. & S. Sathyendranath, 1993. Fundamental issues in measurement of primary production. ICES mar. Sci. Symp. 197: 3–8.

Richardson, L. L. & K. D. Stolzenbach, 1995. Phytoplankton cell size and the development of microenvironments. FEMS Microb. Ecol. 16: 185–191.

Rott, E., 1981. Some results from phytoplankton counting intercalibrations. Schweiz. Z. Hydrol. 43: 34–62.

Savidge, G., 1981. Studies of the effects of small-scale turbulence on phytoplankton. J. mar. biol. Ass. U.K. 61: 477–488.

Savidge, G., 1986. Growth and photosynthetic rates of *Phaeodactylum tricornutum* BOHLIN in a cyclical light field. J. exp. mar. Biol. Ecol. 100: 147–164.

Talling, J. F., 1960. Comparative and field studies of photosynthesis by a marine planktonic diatom. Limnol. Oceanogr. 5: 62–77.

Utermöhl, H., 1958. Zur Vervollkommnung der quantitativen Phytoplankton-Methodik. Mitt. int. Ver. Limnol. 9: 1–38.

Venrick, E. L., J. R. Beers & J. F. Heinbokel, 1977. Possible consequences of containing microplankton for physiological rate measurements. J. exp. mar. Biol. Ecol. 26: 55–76.

Wetzel, R. G. & G. E. Likens, 1991. Limnological Analyses. Springer-Verlag, New York, 391 pp.

Yoder, J. A. & S. Bishop, 1985. Effects of mixing-induced irradiance fluctuations on photosynthesis of natural assemblages of coastal phytoplankton. Mar. Biol. 90: 87–93.

Hydrobiologia **342/343**: 107–115, 1997.
L. Kufel, A. Prejs & J. I. Rybak (eds), Shallow Lakes '95.

Factors controlling the dominance of *Planktothrix agardhii* and *Limnothrix redekei* in eutrophic shallow lakes

Jacqueline Rücker, Claudia Wiedner & Paul Zippel
Faculty Environmental Sciences, Water Protection, Brandenburg Technical University Cottbus, Seestraße 45, D-15526 Bad Saarow, Germany

Key words: cyanobacteria, shallow lakes, *Planktothrix agardhii*, *Limnothrix redekei*

Abstract

Lakes in the Scharmützelsee-region (East-Brandenburg, Germany) were found to be dominated either by *Limnothrix redekei* or *Planktothrix agardhii*. These dominance regimes were stable over the whole investigation period from 1993 to 1995. Six lakes (maximum depth between 2 and 12 m) were compared in regard to physical and chemical conditions to find reasons for the superiority of the first or the second species. All investigated lakes are polymictic but show, nevertheless differences in their mixing-behaviour. In lakes dominated by *Limnothrix redekei* stratification events occur more often than in the lakes dominated by *Planktothrix agardhii*. The latter are more windexposed due to their greater size and/or their surroundings. In combination with polymixis light and nutrient supply are discussed as the key factors for the species composition. A lower content of soluble reactive phosphorus and lower light supply are considered as possible advantages for *Limnothrix redekei* over *Planktothrix agardhii* in three lakes. A lower total nitrogen/total phosphorus ratio may be a reason for the dominance of *Planktothrix agardhii* in the other three investigated lakes.

Introduction

An intensive investigation program on lakes of the Scharmützelsee-region, East Brandenburg, Germany was started in July 1993. The lakes are situated in the catchment area of river Dahme. They are mostly shallow and in an eutrophic to hypertrophic state. They are highly turbid and mass developments of filamentous cyanobacteria like *Limnothrix redekei* Meffert and *Planktothrix agardhii* Anagn. & Kom. often occur.

In the thirties, Wundsch (1940) had already given a similar description of lakes in this region. He introduced the 'H_2S-Oscillatoria-type' of lakes which was frequently found in the region Berlin-Brandenburg. Many lakes in the studied area belong to this type of lake, characterized by high abundances of Oscillatoriales (*Planktothrix agardhii* and *Limnothrix redekei*) from spring to autumn (Czensny, 1938), by Secchi-depths to less than one meter and H_2S formation in deep water layers. This paper describes the phenomenon of long lasting dominance regimes of *Limnothrix redekei* in some of the lakes and of *Planktothrix agardhii* in others. With respect to the dominating cyanobacterial species, two types of lakes are distinguished - the 'Limnothrix-' and the 'Planktothrix-lake'. For this study, phytoplankton composition, mixing-behavior and nutrient content of six lakes (three for each type of lake) have been investigated monthly. Physiological differences of *P. agardhii* and *L. redekei* known from culture experiments are discussed in the context of physical and chemical conditions in the lakes to find out reasons for the different dominance regimes in lakes of the same area.

Description of sites studied

The investigated lakes are situated in the Scharmützelsee-region in East Brandenburg. A map is shown in Nixdorf & Deneke (1997). The lakes in this area are connected by running waters and form two lake-chains connected with the large Scharmützelsee. The so-called Glubigseenkette is characterized by five smaller lakes (area 0.06–0.63 km^2 and 4.5 to 22 m deep). The water runs into lake Scharmützelsee. The second lake-chain, the Storkower Seenkette, consists of three larger lakes (1.55–5.79 km^2, 3.1 to 12.5 m deep) which are connected by navigable channels from lake Scharmützelsee to the river Dahme.

For this contribution, lakes Melangsee and Kleiner Glubigsee both belonging to the lake-chain Glubigseenkette and lakes Wolziger See and Langer See, the two last lakes of the Storkower lake-chain were chosen. Furthermore, the study includes data obtained for lake Lebbiner See which is connected via its flowing off to the Storkower channel and lake Petersdorfer See which is located 3 km north of lake Scharmützelsee. With the exception of lake Wolziger See, these waters are very shallow in the sense of Nixdorf & Deneke (this volume). Some topographical and morphometrical properties are listed in Table 1. The three lakes Melangsee, Kleiner Glubigsee and Petersdorfer See are relatively small and surrounded by hills, forests and settlements. They are less affected by wind compared to the lakes Wolziger See, Langer See and Lebbiner See, which are much greater (Wolziger See and Langer See) and have agriculturally used flat surroundings.

Material and methods

Water temperature, oxygen concentration and saturation were measured monthly with a multiparameter probe (Hydrolab H2O) at the deepest points of the lakes. Additionally the Secchi-depth was measured. Mixed samples from the whole water column or from epilimnion during stagnation periods were taken with a 2-liter-sampler and analyzed for chlorophyll, dissolved nutrients (soluble reactive phosphorus (SRP), nitrate, ammonia), total nitrogen (TN) and total phosphorus (TP) according to standard methods (Nixdorf & Deneke, this volume). The phytoplankton biovolume was estimated from the same sample using an inverse microscope (Utermöhl, 1958). The phytoplankton composition of fresh algal material has been examined directly after sampling since 1994.

Results

In the lakes Melangsee, Kleiner Glubigsee and Petersdorfer See mass developments of *L. redekei* could be observed in summer 1994 (Figure 1). This species reached 46 to 66% of total biovolume in lake Melangsee, 33 to 62% in lake Kleiner Glubigsee and about 80% in lake Petersdorfer See. These lakes will be referred to as 'Limnothrix-lakes' in the following. Data presented here for *L. redekei* include also the species *Limnothrix planctonica* Meffert, *Limnothrix amphigranulata* Meffert and *Pseudanabaena limnetica* Meffert. In the considered period, *P. agardhii* reached only small portions of up to 7% of the total biovolume in these three lakes whereas nitrogen-fixing species like *Anabaena* spp. and *Aphanizomenon* spp. formed 6 to 19% of the total biovolume in lake Melangsee in summer 1994 and up to 33% in lake Kleiner Glubigsee (Figure 1). These species could not be found in lake Petersdorfer See.

On the other hand the lakes Langer See, Wolziger See and Lebbiner See were dominated by *P. agardhii* during the whole investigation period (Figure 2). *P. agardhii* reached up to 78% in lakes Langer See and Wolziger See. In lake Lebbiner See it represented to 22 and 42% of total biovolume. Besides *P. agardhii*, nitrogen fixing species were important in these lakes. *Aphanizomenon* spp. and *Anabaena* spp. formed 2 to 29% of total biomass in lake Langer See, up to 21% in lake Wolziger See and up to 45% in lake Lebbiner See. The occurrence of *Cylindrospermopsis raciborskii* (Wolosz.) Seenayya et Subba Raju is remarkable. Previously, it has not been recorded in these high latitudes (Padisák, pers. com.). This species appeared in all lakes of this study. It reached the maximum biomass of 5.5 mg l^{-1} in the 'Planktothrix-lake' lake Lebbiner See. *L. redekei* was able to form populations up to 16% of the total biomass in these lakes. On the contrary, the concentration of *P. agardhii* was negligible in the 'Limnothrix-lakes'.

Lake Melangsee was classified as a 'Limnothrix-lake' based on the phytoplankton composition in 1994. However, the maximum concentration of *L. redekei* reached only 9 to 10% in summer 1993 or 8% in June 1995. *P. agardhii* and N-fixing species dominated the cyanobacterial plankton in 1993 and 1995 (Figure 1). This pattern is more similar to the 'Planktothrix-lakes'. Except lake Melangsee, all the other lakes showed similar dominance regimes from summer 1993 to 1995. This is based on the qualitative investigation of the fresh algal material and justifies the differentiation between a *Limnothrix* and a *Planktothrix* dominated type of lake. Clear water periods indicated by low chlorophyll concentrations and/or high Secchi-depths (data not shown) could be observed in lake Wolziger

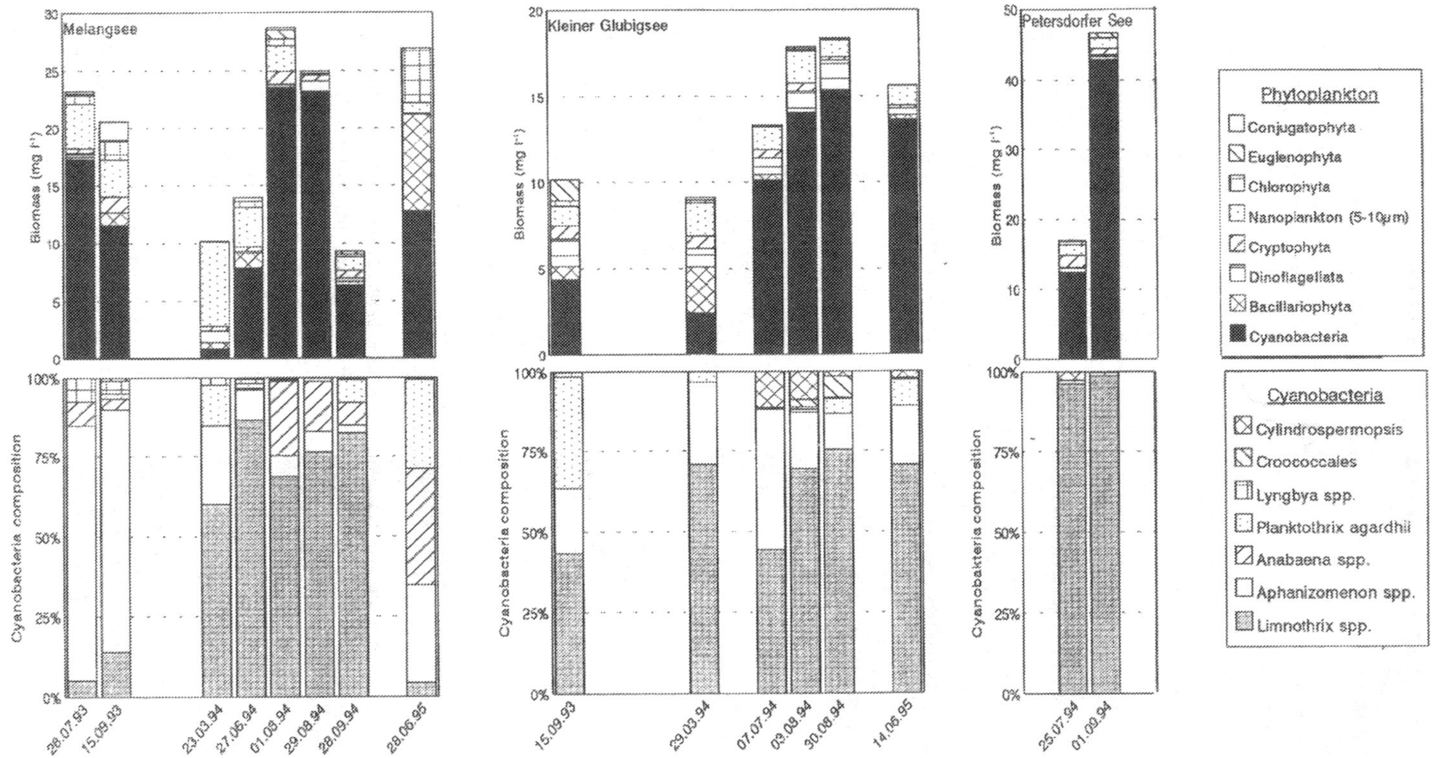

Figure 1. Biomass of phytoplankton groups (top) and cyanobacterial composition (bottom) of lakes Melangsee, Kleiner Glubigsee and Petersdorfer See (from left to right).

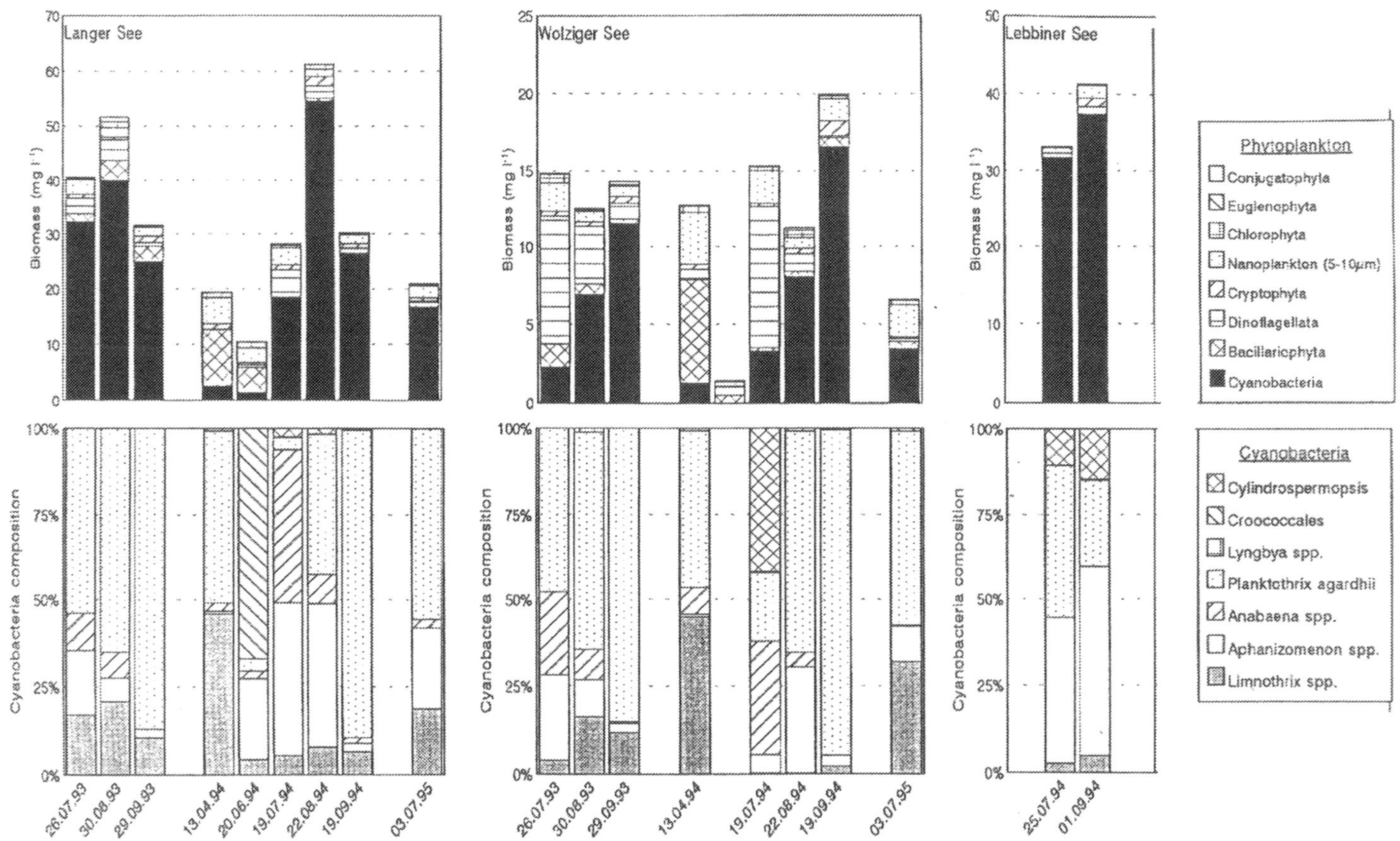

Figure 2. Biomass of phytoplankton groups (top) and cyanobacterial composition (bottom) of lakes Langer See, Wolziger See and Lebbiner See (from left to right).

Table 1. Topographic and morphometric parameters of the investigated lakes (A = area; V = volume; L_{max} = maximum length; W_{max} = maximum width; L = shore length, Z_{max} = maximum depth, Z_{mean} = mean depth (after Nixdorf & Deneke (1997))

Lake	A (km^2)	V (10^6 m^3)	L_{max} (km)	W_{max} (km)	L (km)	Z_{max} (m)	Z_{mean} (m)
Kleiner Glubigsee	0.06	0.14	0.57	0.13	1.34	4.5	2.4
Melangsee	0.11	0.17	0.51	0.31	1.38	3.3	1.6
Petersdorfer See	0.23	0.53	1.11	0.29	2.62	4.0	2.3
Lebbiner See	0.28	0.58	0.71	0.55	2.03	4.0	2.1
Langer See	1.55	3.27	2.14	1.40	6.61	3.8	2.1
Wolziger See	5.79	32.02	3.25	2.40	10.12	13.0	5.5

Table 2. Means and standard deviation (in brackets; n=4 to 12) of total phosphorous (TP), total nitrogen (TN), soluble reactive phosphorus (SRP), nitrate (NO_3-N) and ammonia (NH_4-N) of the epilimnion (during stratification) or the whole water column (unstratified conditions) of the studied lakes (MEL=Melangsee, KGL=Kleiner Glubigsee, PET=Petersdorfer See, LAN=Langer See, WOL=Wolziger See, LEB=Lebbiner See) from June to September of 1994 and 1995.

Lake	TP ($\mu g\ l^{-1}$)	TN ($\mu g\ l^{-1}$)	TN/TP	SRP ($\mu g\ l^{-1}$)	NO_3-N ($\mu g\ l^{-1}$)	NH_4-N ($\mu g\ l^{-1}$)
MEL	70.4 (±25.5)	2,687 (±780)	35.6 (±11.6)	4.9 (±1.9)	4.5 (±1.9)	82.5 (±61.8)
KGL	39.0 (±19.8)	1,990 (±877)	54.0 (±23.5)	5.6 (±2.9)	5.2 (±2.8)	201.8 (±209.8)
PET	54.4 (±9.9)	3,365 (±1,030)	58.4 (±8.7)	6.1 (±4.1)	6.4 (±4.2)	117.7 (±53.0)
LAN	116.7 (±40.0)	2,689 (±854)	23.2 (±8.4)	7.3 (±3.8)	8.7 (±3.7)	112.8 (±45.4)
WOL	103.7 (±45.0)	2,561 (±695)	22.4 (±8.3)	13.6 (±10.8)	15.0 (±7.4)	226.6 (±263.8)
LEB	91.3 (±12.4)	3,375 (±841)	35.0 (±8.0)	14.2 (±15.8)	4.7 (±1.2)	391.9 (±470.2)

See in May and June of 1994 and 1995 and in lake Melangsee between March and May 1995.

Means and standard deviations (4–12 values) of total nitrogen, total phosphorus, SRP, nitrate and ammonia of the vegetation periods (June to September) 1994 and 1995 are given in Table 2. The lowest mean SRP-concentrations of about 5 $\mu g\ l^{-1}$ was found in the three 'Limnothrix-lakes' followed by lake Langer See with 7 $\mu g\ l^{-1}$ and lakes Wolziger See and Lebbiner See with mean values above 10 $\mu g\ l^{-1}$, but with high standard deviations. The mean nitrate concentrations were lowest in the lakes Melangsee, Lebbiner See and Kleiner Glubigsee (about 5 $\mu g\ l^{-1}$). It was higher in the lakes Petersdorfer See (6 $\mu g\ l^{-1}$) and lake Langer See (7 $\mu g\ l^{-1}$) and had its maximum value of 14 $\mu g\ l^{-1}$ in lake Wolziger See again with high standard deviation. The lowest ammonia concentrations, ranging from 80 to 120 $\mu g\ l^{-1}$, were found in the lakes Melangsee, Langer See and Petersdorfer See. It was somewhat higher than 200 $\mu g\ l^{-1}$ in lakes Kleiner Glubigsee and Wolziger See and had its highest value (mean about 400 $\mu g\ l^{-1}$) in lake Lebbiner See. This could be a hint to the external load with sewage.

The analysis of temperature and oxygen profiles showed that also in very shallow lakes a certain stratification may occur. Figure 3 shows isopleths of temperature of all investigated lakes for 1994 and 1995. The most shallow lakes Melangsee (maximum depth 2 m), Lebbiner See (2.5 m) and Langer See (3 m) showed essentially no thermal stratification. Lake Langer See was always mixed in 1994. The counterpart to lake Langer See is lake Petersdorfer See which is also not deeper than 3 m. In both years, especially in 1995, the water was stratified there several times. Although only 4.5 m deep, lake Kleiner Glubigsee was stable stratified in the summers of both years. A temperature difference between surface and bottom of up to 14 °C was found, accompanied with a difference in oxygen saturation of 161%. Due to its greater depth (12.5 m), lake Wolziger See was also thermally stratified in 1994 and 1995. The average depth of the epilimnion was 7 m. Nevertheless, this lake was completely mixed in the relatively cool summer of 1993.

Discussion

The observed dominance of *L. redekei* during the whole vegetation period is remarkable and has been reported so far only in a few cases (e.g. Meffert, 1975; Rojo &

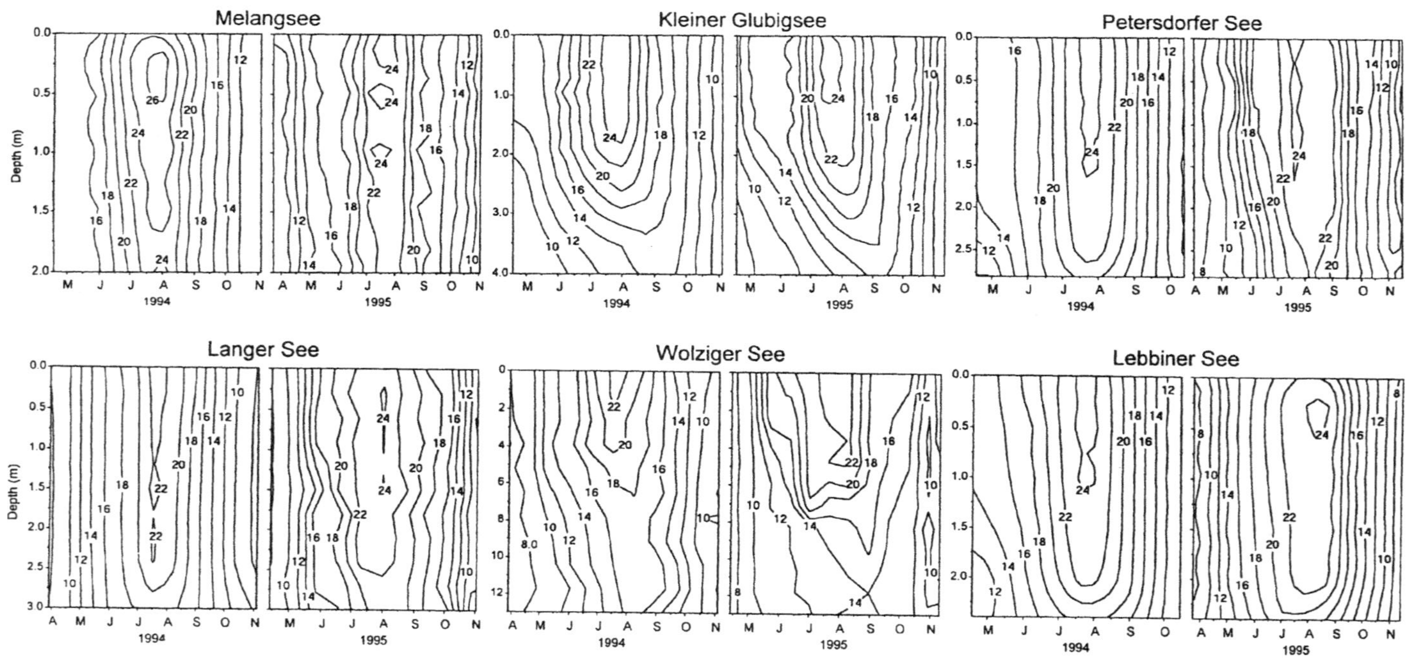

Figure 3. Isopleths of temperature of the lakes Melangsee, Kleiner Glubigsee, and Petersdorfer See (upper row from left to right) and of the lakes Langer See, Wolziger See and Lebbiner See (lower row from left to right).

Alvarez Cobelas, 1994). *Limnothrix redekei* is mostly described as a species with mass developments in spring and autumn whereas in summer after stratification of lakes *P. agardhii* may become the dominant species (e.g. Moed et al., 1984; Berger & Sweers, 1988; Chorus, 1989). The faster growth of *L. redekei* compared to *P. agardhii* at low temperature and low light supply (Foy et al. 1976; Foy, 1983; Nicklisch & Kohl, 1989; Nicklisch et al., 1991) may favor the development of *L. redekei* in spring and autumn. Long lasting dominances of *P. agardhii* are known for Dutch lakes (e.g. Berger, 1975).

All lakes included in this study are polymictic. Therefore long periods of stagnation are rare. Depending on the impact of wind, events of stagnation do more often occur in 'Limnothrix-lakes' than in 'Planktothrix-lakes'. The lakes Wolziger See and Langer See where *P. agardhii* is the dominant species are much greater than our 'Limnothrix-lakes'. They are strongly affected by wind which can be seen from homogenous distribution of temperature in lake Langer See and the great epilimnion depth (7 m) of lake Wolziger See. The latter may even be polymictic as observed in summer 1993 (compare Nixdorf & Deneke (1997)). The lakes Lebbiner See (*Planktothrix* dominated) and Petersdorfer See (*Limnothrix* dominated) have approximately the same size and morphometry. Due to their different surroundings, the impact of wind is greater at lake Lebbiner See. The small influence of wind on the 'Limnothrix-lakes' is obvious for the very small lake Kleiner Glubigsee which is stratified although it is only 4.5 m deep. For lake Melangsee it is more difficult to prove this fact on the basis of our monthly or biweekly measured temperature profiles. Because of the short watercolumn (i) the process of warming (and cooling) is relatively quick (radiation penetrates almost to the bottom) so that differences between day and night are supposed to be high and (ii) a complete mixing can occur in short time scales. This is a general problem in the description of mixing-processes in shallow lakes. Possibly, continuous measurements could give more information.

Depending on the degree of mixis, abiotic conditions may change quickly for algae drifting through the water column. The first factor we would like to discuss is the underwater light climate in combination with vertical mixing. Algae are particularly affected by the amount of photosynthetic active radiation (PAR), which might reach in more stratified lakes from photoinhibiting conditions in the upper water layer to limiting conditions in deeper water layers. In lakes with deep vertical mixing algae pass along the whole available gradient of PAR. The question whether turbulent or stratified conditions are more favorable for algal growth in the whole water column has been studied in many investigations (e.g. Dokulil, 1994; Nixdorf, 1994). Recent results indicate that vertical mixing may be advantageous for algal development. Gervais et al. (this volume) showed that primary production of algae in a defined water column under simulated turbulent conditions is higher than under non-turbulent conditions. Consequently, the utilization of the supplied light is more efficient under turbulent than under non-turbulent, i.e. stratified conditions. In this context, light limitation is more likely in the 'Limnothrix-lakes' where stratification events occur more often. Considering that *L. redekei* can outcompete *P. agardhii* under light limitation, we conclude that light is a key factor regulating the dominance of *L. redekei* and *P. agardhii* in the investigated shallow lakes. The higher concentrations of *L. redekei* in lake Langer See and Wolziger See in 1993 may be a further evidence for this theses because the summer 1993 was cool and cloudy and light limitation was more likely than in 1994 or 1995.

Next, we discuss the influence of nutrient supply on the different cyanobacterial species. Referring the competitiveness of both species under nutrient limitation, it could be shown that *L. redekei* is able to outcompete *P. agardhii* under phosphorus limitation (Nicklisch et al., 1991). In the course of competition experiments under nitrogen limitation, *L. redekei* and *P. agardhii* coexisted over six weeks at different daylength (Nicklisch et al., 1991). Only due to pulsed nitrogen addition with long term intervals, *P. agardhii* could be a successful competitor (Nicklisch, 1994). The studies of Meffert (1989a, 1993) and Nicklisch (1994) suggest that loss rates increase rapidly for *L. redekei* due to lysis under certain suboptimal growth conditions. Such a phenomenon has not been reported for *P. agardhii*. That is why a higher mechanical stability of *P. agardhii* under nutrient limitation could be an explanation for their dominance in the large, windexposed lakes in the present study. In conclusion, the greater competitiveness of *Limnothrix redekei* under phosphorus limitation could promote their dominance in the 'Limnothrix-lakes' with the lower SRP-concentrations (Table 2). On the other hand, in consequence of low TN/TP-ratios, nitrogen limitation seems to be more likely in the 'Planktothrix-lakes' which could explain the minor occurrence of *L. redekei* there. Moreover, a greater turbulence influences the nutrient budget of a lake due to intensive sediment contact and the raise of sediment particles.

Apart from light and nutrients, loss processes are of the same importance for the phytoplankton growth as production processes. Algal losses due to zooplankton grazing may cause clear water periods which could also be observed in some cases in the present study. *L. redekei* never appeared in higher concentrations after a clear water period, e.g. in lake Melangsee (1995) or in lake Wolziger See (1994). This corresponds to observations from lake Müggelsee where in years with *Oscillatoria* dominance (succession from *L. redekei* in spring to *P. agardhii* in summer) no clear water periods could be observed (Nixdorf & Hoeg, 1993). This could be a hint that *L. redekei* may be removed by zooplankton grazing. In contrast to *L. redekei*, *P. agardhii* is hardly influenced by the large zooplankton as Infante & Abella (1985) could show for *Daphnia pulicaria* and *D. thorata* of Lake Washington.

Moreover, the size of the starting population at the beginning of the vegetation period is of great importance for the phytoplankton development. Based on very slow growth rates during the winter months, *L. redekei* may dominate the phytoplankton in early spring (Foy et al., 1976; Meffert, 1989b). Post et al. (1985) showed that also *P. agardhii* is able to adapt effectively to low light and low temperature which ensure the survival of this species in winter. We found *L. redekei* as well as *P. agardhii* under ice during the winter 1995/1996.

To sum up our results, the morphometry of the lakes determines the extent and degree of polymixis in shallow lakes. *Limnothrix redekei* seems to be more successful in smaller lakes with lower water turbulence. The explanation for this preference may be the lower light supply in these lakes compared to permanently mixed lakes and their greater sensibility to mechanical stress under nutrient limitation. This is important since nitrogen limitation is more likely in the 'Planktothrix-lakes'. The greater competitiveness of *L. redekei* compared to *P. agardhii* under phosphorus limitation is the second possibility to explain their dominance in the 'Limnothrix-lakes'.

Acknowledgments

The authors are grateful to Prof. B. Nixdorf for critical discussion. Furthermore, we wish to thank G. Lippert for their careful work in the laboratory and W. Terlinden and I. Hentschke for practical work in the field.

References

Berger, C., 1975. Occurrence of *Oscillatoria agardhii* GOM. in some shallow eutrophic lakes. Verh. int. Ver. Limnol. 19: 2689–2697.

Berger, C. & H. E. Sweers, 1988. The IJsselmeer and its phytoplankton – with special attention to the suitability of the lake as a habitat for *Oscillatoria agardhii* GOM.. J. Plankton Res. 10: 579–599.

Chorus, I., 1989. Phytoplankton und Primärproduktion im Schlachtensee 1982/83, zu Beginn der Sanierung. WaBoLu Hefte 3, Berlin, 104 pp.

Czensny, R., 1938. Die Oscillatorienerkrankung unserer Seen, Biologie und Chemismus einiger märkischer Seen. Vom Wasser 8: 36–57.

Dokulil, M. T., 1994. Environmental control of phytoplankton in turbulent turbid systems. Hydrobiologia 289: 65–72.

Foy, R. H., 1983. Interaction of temperature and light on the growth rates of two planktonic Oscillatoria species under a short photoperiod regime. Br. phycol. J. 18: 267–273.

Foy, R. H., C. E. Gibson & R. V. Smith, 1976. The influence of daylength, light intensity and temperature on the growth rates of planktonic blue-green algae. Br. phycol. J. 11: 151–163.

Gervais, F., D. Opitz & H. Behrendt, 1997. Influence of small scale turbulence and large scale mixing on phytoplankton primary production. Hydrobiologia 342/343: 95–105.

Infante, A. & S. Abella, 1985. Inhibition of *Daphnia* by *Oscillatoria* in Lake Washington. Limnol. Oceanogr. 30: 1046–1052.

Meffert, M.-E., 1975. Analysis of the population dynamics of *Oscillatoria redekei* Van Goor in Lake Edeberg. Verh. int. Ver. Limnol. 19: 2682–2688.

Meffert, M.-E., 1989a. Planktic unsheated filaments (Cyanophyceae) with polar and central gas vacuoles. II. Biology, population dynamics and biotopes of *Limnothrix redekei* (Van Goor) Meffert. Arch. Hydrobiol. 116: 257–282.

Meffert, M.-E., 1989b. *In situ* and *in vitro* photosynthesis of *Limnothrix redekei* (Cyanophyta). Arch. Hydrobiol. 116: 401–414.

Meffert, M.-E., 1993. The relation between *Limnothrix redekei* (Cyanophyta) and the accompanying heterotrophic bacteria. Arch. Hydrobiol. Suppl. 100: 57–73.

Moed, J. R., J. Bloem & H. L. Hoogveld, 1984. Study of the seasonal growth of *Oscillatoria redekei* in Tjeukemeer, The Netherlands. Verh. int. Ver. Limnol. 22: 882–885.

Nicklisch, A., 1994. Does mortality by nitrogen deficiency influence the succession of *Limnothrix redekei* and *Planktothrix agardhii*? Verh. int. Ver. Limnol. 25: 2214–2217.

Nicklisch, A. & J.-G. Kohl, 1989. The influence of light on the primary production of two planktic blue green algae. Arch. Hydrobiol. Beih. Ergebn. Limnol. 33: 451–455.

Nicklisch, A., B. Roloff & A. Ratsch, 1991. Competition experiments with two planktic bluegreen algae (*Oscillatoriaceae*). Verh. int. Ver. Limnol. 24: 889–892.

Nixdorf, B., 1994. Polymixis of a shallow lake (Großer Müggelsee, Berlin) and its influence on seasonal phytoplankton dynamics. Hydrobiologia 275/276: 173–186.

Nixdorf, B. & R. Deneke, 1997. Why 'very shallow' lakes are more successful opposing reduced nutrient loads. Hydrobiologia 342/343: 269–284.

Nixdorf, B. & S. Hoeg, 1993. Phytoplankton – community structure, succession and chlorophyll content in lake Müggelsee from 1979 to 1990. Int. Revue ges. Hydrobiol. 78: 359–378.

Post, A., R. de Wit & L. R. Mur, 1985. Interactions between temperature and light intensity on growth and photosynthesis of the cyanobacterium *Oscillatoria agardhii*. J. Plankton Res. 7: 487–495.

Rojo, C. & M. Alvarez Cobelas, 1994. Population dynamics of *Limnothrix redekei*, *Oscillatoria lanceaeformis*, *Planktothrix agardhii* and *Pseudanabaena limnetica* (cyanobacteria) in a shallow hyperthrophic lake (Spain). Hydrobiologia 275/276: 165–171.

Utermöhl, H., 1958. Zur Vervollkommnung der quantitativen Phytoplanktonmethodik. Mitt. int. Ver. Limnol. 9: 1–38.

Wundsch, H. H., 1940. Beiträge zur Fischereibiologie märkischer Seen, VI. Die Entwicklung eines besonderen Seentypus (H_2S-Oscillatorien-Seen) im Flußgebiet der Spree und Havel, und seine Bedeutung für die fischereibiologischen Bedingungen in dieser Region. Z. Fischerei 38: 443–648.

Hydrobiologia **342/343:** 117–124, 1997.
L. Kufel, A. Prejs & J. I. Rybak (eds), Shallow Lakes '95.

Meiobenthos communities of some subarctic lakes

Vladimir V. Skvortsov
Institute for Lake Research of Russian Ac. Sci., Sevastyanova, 9, 196199 St. Petersburg, Russia

Key words: meiobenthos, subarctic lakes, oil pollution, regression model

Abstract

Meiobenthos communities of 19 relatively small tundra lakes (area 0.2–320 ha) were studied during summer seasons of 1986–1988 and 1993. The lakes are located in the northeastern region of European Russia. All lakes are separated into deep (with mean depth from 2.0 to 8.0 m) and shallow ones (mean depth is less than 1.5 m). Several lakes were low – or strongly polluted by crude oil. Concentrations of hydrocarbons in lake water were respectively 0.42–0.92 mg l^{-1} and 7.4–6.3 mg l^{-1}. Species composition, number and biomass of meiobenthos were investigated. Relationships between those parameters and environmental and trophic factors were analyzed.

Meiobenthos biomass ranged from 0.003 to 0.911 g m^{-2} in central zones of deep lakes and from 0.38 to 17.61 g m^{-2} in littoral zones. The biomass of littoral Cladocera, Harpacticoida and Nematoda was higher in deep unpolluted mesotrophic lakes. The biomass of Bivalvia positively correlated with primary production of phytoplankton.

Oil pollution caused structural changes of communities. Benthic Cladocera and Chironomidae are more vulnerable to the effect of oil pollution. They were only abundant at unpolluted and law-polluted localities of the lake bottom. Highest bionnass of meiobenthic Bivalvia occurred in low polluted lakes. Cyciopoida is the most resistant cornponent of meiobenthos communities. Their abundance was highest at most polluted localities.

Introduction

Meiobenthos community consists of both constant (nematods, microcrustaceans, tardigrads) and temporal (juvenile forms of oligochaets, molluscs and insect larvae) components. A significance of this community in lakes of temperate zone is reasonably well known. In particular, my previous studies have shown that meiobenthos biomasses ranged from 5 to 40% of total bottom community biomass in these lakes (Skvortsov, 1985; Skvortsov et al., 1988). Some factors such as lake trophic level, hydrochemical and hydrophysical properties of sediments and human impacts can affect structure, abundance, biomass and production of meiobenthos of temperate lakes (Holopajnen & Paasivirta, 1977; Stańczykowska, 1967; Prejs & Stańczykowska, 1972; Särkkä, 1979, Särkkä, 1987).

Meiobenthos biomass forms about 50% (in average) of biomass of the entire bottom community in subarctic lakes (Belyakov & Skvortsov, 1994). However, relationships between meiobenthos community and environmental factors was still poorly known for subarctic area.

The proposed study has main objectives: to characterize the structure, abundance, and spatial distribution of meiofauna in subarctic lakes in relation with essential environmental factors and to investigate the pollution response patterns of whole meiobenthos communities.

Study area

Nineteen lakes located in the central part of Bolshezemelskaya tundra (North Eastern region of European Russia) were investigated during summer seasons of 1986–1988 and 1993 (Table 1).

The main characteristics of lake morphometry, hydrochemical properties, productivity of phytoplankton and level of oil pollution are given in Table 2. All these data were taken from the recently published

Table 1. Location of studied lake territories and years of investigations.

Territory	N	E	Years
I	67° 50′	59° 00′	1986-1987
II	67° 30′	57° 00′	1988
III	68° 30′	58° 30′	1993

papers (Drabkova, & Bystrov, 1994; Kuznetsov, 1994; Trifonova, 1994).

The investigated lakes can be divided on deep (mean depth >2 m) and shallow (mean depth <2 m) ones. The deepest lakes were located only within the territories I and III, all lakes of the territory II were shallow. The deep lakes had funnel-shaped basins. These ones had a distinct wide littoral zone, a sublittoral and a profundal. Basins of the shallow lakes had a plane bottom surface. In the littoral of the deep lakes, sediments were presented by fine sands (M_d = 0.10–0.17 mm), to a depth of 2 meters. In the sublittoral and profundal of the deep lakes and in central zone of the shallow ones there were muddy sediments with an organic matter content from 5 to 40% (Strukova, 1994).

The investigated lakes were similar in main hydrochemical parameters. The concentrations of nutrients were ranged within limits: P_{inorg} 0.00–0.02 mg l^{-1}, P_{tot} 0.01–0.32 mg l^{-1}, N_{inorg} 0.00–0.67 mg l^{-1}, N_{tot} 0.00–2.66 mg l^{-1}. The values of parameters characterizing an organic matter content were varied within range: BOD 0.8–6.50 mg O_2 l^{-1}, COD 4.48–64.8 mg O l^{-1}, and pH 6.2–8.21. However, the whole set of lakes can be divided on two different groups at significance level of 0.05%. The division on two clusters was based on distinctions of water color (F-ratio = 73.955, p = 0.000) and concentration of major ions (F-ratio = 5.949, p = 0.016). The lakes of the first group (# # 1–12) were characterized by low colored waters and high concentration of major ions (Pt-Co units = 4.00–52.00, CMI = 19.50–88.5 mg l^{-1}), while the lakes of the second group (# # 13–19) had more colored waters and low concentration of major ions (Pt-Co units = 80.35–199.20, CMI= 13.00–29.00 mg l^{-1}.

Materials and methods

The samples of meiobenthos were taken by MB-TE-type corer (Travianko & Evdokimova, 1968) with a surface 12.6 cm in 3–5 replicates from main biotops of lakes. The collected samples were sifted through a 106 μm mesh and were preserved with a weak solution of formalin (4%). The samples were taken twice on July and August 1986 and 1988 in each lake, seven times through July to September 1987 in lakes # # 1 and 7, and ones in August 1993 in lakes # # 9–11. Special surveys were carried out on 28 sites of Mitrofan lake (# I) on July 25 and August 23, 1986 in order to study a spatial distribution of meiobenthos. The count of meiobenthos animals was made with use a dissecting microscope. Body weight of the animals was determined under appropriate formulas (Ankar & Elmgren, 1976; Tsalolichin, 1981; Balushkina, 1982, Methodical recommendations, 1982, 1983). Following statistical methods were applied to determine relationships of meiobenthos biomass to environmental characteristics: ANOVA (analysis of variance), factor analysis, simple and multiple linear regressions, and nonlinear regression (SYSTAT, 1992). Meiobenthic and environmental data were not transformed.

Results

Species composition

The determinations of the species was made only for Cladocera, Cyclopoida, Chironomidae larvae, Mollusca and predominant Nematoda. The complete list of meiobenthos species has been published in our previous paper (Belyakov & Skvortsov, 1994).

Altogether forty-three species of invertebrate animals have been found out in investigated lakes. Chironomidae were presented by the highest number of species (22). The following chironomid species were meeting the most frequently: *Clodotanytarsus mancus* (Walk.), *Tanytarsus holochlorus* Edw., *Paratanytarsus lauterbornii* (Kieff), *Limnochironomus nervosus* (Staeger), *Polypedilum nubeculosum* (Mg.), *Procladius choreus* Meig., *Psectrocladius psilopterus* Kieff.

Cladocera were presented by: *Alona affinis* Leydig, *Alona rectangula* Sars, *Alonopsis elongata* (Sars), *Camptocercus rectirostris* Shoedler, *Chydorus ovalis* Kurz., *Eurycercus lamellatus* (O.F.Müller), *Ilyocriptus acutifrons* Sars, *Acanthocyclops viridis* (Jur.) were the most usual among Cyclopoida. Bivalve molluscs were presented mainly by juvenile forms of *Euglesa* sp. and *Sphaerium amnicum* (Müll.). Nematods were presented by *Dorylaimus stagnalis* Dujardin, *Ironus tenuicaudatus* de Man, *Mononehus truncatus* Bastian and *Tobrilus* sp.

Despite essential distinctions between lakes, they distinguish insignificantly by a species structure of meiobenthos communities (calculated Sorensen's index is equal to 0.71). It is possible to note that *E. lamellatus* was found only in deep lakes on sandy sediments, and *I. acutifrons* in shallow ones with muddy bottom. Common *A. viridis* have been replaced by *A. gigas* (Claus) in the most oil polluted lake (Lake # 8). These crustaceans were almost unique components of the meiobenthos community in this lake.

Spatial distribution

Sediment-depended distribution of meiobenthos was investigated. The obtained data have shown that meiobenthos concentrated in the upper part of littoral zone. The highest number of meiobenthos (60 000 ind m^{-2} has been registered at depth about 1 meter, decreasing to 8000 ind m^{-2} at 2 m depth, in the sublittoral and profundal the meiobenthos abundance did not exceed 1000 ind m^{-2}. It is important to note, that the number of meiobenthos inhabiting a shorewater boundary (depth less than 0.5 m) reduced down to 15 000 ind m^{-2}, likely due to a regular wave-action on meiobenthos organisms dwelling in this bottom area.

Different groups of meiobenthos populations preferred sediments with a different particle size. A particle-size range of sediment preferred by meiobenthos was determined by occurrence of their abundance peaks. Bottom Cladocera, Harpacticoida, Cyclopoida and Ostracoda did not display an association with median particle size. Only chironomid larvae and Nematoda preferred more coarse littoral sands from 0.12 to 0.17 mm median particle size (t=3.607, p<0.01 and t=3.030, p<0.01 respectively).

Abundance of meiobenthos

As it has been shown above, the meiobenthos community was most abundant in a lakeshore zone in deep lakes. The number of meiobenthos organisms in central parts in these lakes ranged in limits from 200 to 4500 ind m^{-2}, and biomass from 0.003 to 0.91 g m^{-2} wet weight. The abundance of meiobenthos in center of shallow lakes was, as a rule higher, than directly at beach. At the further analysis I shall operate with data relating to littoral zone, considering central bottom biotops of shallow lakes as littoral ones.

The average values of biomass of littoral meiobenthos of investigated lakes are given in Table 3. To determine major factors, which controlled meiobenthos community in littoral, I was consistently using some statistical procedures: factor analysis, two-way ANOVA, step-wise multiple regression procedure and nonlinear regression.

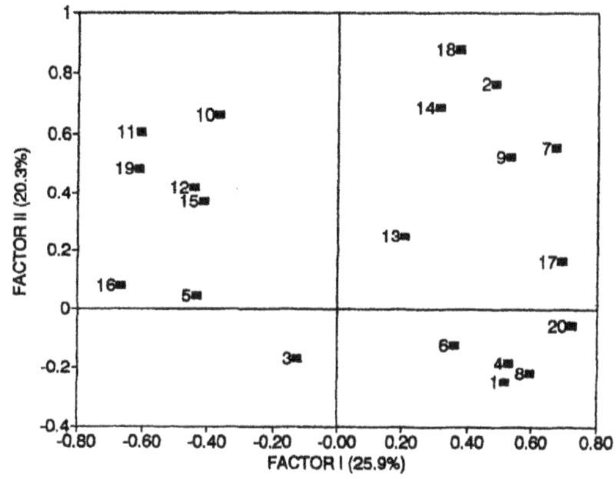

Figure 1. Relationship of meiobenthic biomass with environmental data (on the results of factor analysis). 1- Cladocera, 2-Cyclopoida; 3-Chironomidae, 4-Nematoda, 5-Ostracoda; 6-Bivalvia, 7-Oligochaeta; 8-Harpacticoida; 9-P_{inorg}; 10-P_{tot}; 11-N_{inorg}; 12-N_{tot}; 13-BOD; 14-pH; 15-COD; 16-Color; 17-CMI, 18-Oil; 19-Chlorophyll a; 20-mean depth.

The factor analysis was based on data on meiobenthos and the following variables: concentrations of inorganic and total forms of P and N, pH, water color (Pt-Co units), CMI, BOD, COD, oil-products concentration, mean lake depth, and chlorophyll a concentration.

As a result, four factors have been determined. These factors explain 66.5% of total variance of environmental and meiobenthic data (Table 4).

The first factor (25.9% explained variance) describes differences between two groups of the lakes. This factor positively correlates with such parameters, as the average lake depth and concentration of CMI. Simultaneously it negatively correlates with chlorophyll a concentration, water-color and contents of the forms of nitrogen in water. Thus, it is possible to formulate that the deep and low productive lakes are placed along the positive part of first factor axis, whereas shallow polyhumic and high productive lakes are placed along the negative part of the first factor axis. Figure 1 shows that Cladocera, Halpacticoida and Nematoda have almost similar requirements to environmental conditions. An optimal combination of these conditions for these taxonomic groups of meiobenthos occurred in deep lakes. Ostracods are placed in the negative area of this factor, that is evidence for their preference of shallow and water-colored lakes.

Table 2. Main characteristics of the lakes investigated. DEPTH- mean depth (m); CMI- concentration of major ions (mg l^{-1}); Color- Pt-Co units; OIL - oil concentration (mg l^{-1}); PP - primary production (g C m^{-2} d^{-1}); CHL - chlorophyll *a* concentration (μg l^{-1}).

##	Lake Territory names	Area	DEPTH ha	CMI	Color	OIL	PP	CHL	
1	Mitrofan	I	30.9	6.1	88.5	45.4	0.00	0.407	1.471
2	Izjato	I	334.9	2.4	34.0	20.6	0.00	0.300	1.190
3	Hasujto	I	45.5	1.7	48.0	25.0	0.00	0.275	3.643
4	Pjartavto	I	51.2	1.1	19.5	28.9	0.00	0.336	6.023
5	Nahyto	I	32.6	2.4	39.5	44.8	1.90	0.194	4.450
6	Nocmato	I	10.7	1.0	48.0	27.5	0.00	0.054	1.157
7	Talijto	I	10.5	0.9	48.0	43.8	0.00	0.295	2.407
8	Tibejto	I	27.0	2.4	67.0	52.9	6.30	0.522	8.300
9	Naulto	III	322.0	8.0		4.0	0.00	0.220	0.800
10	Lake I	III	0.7	5.8		4.0	0.00	0.390	0.640
11	Lake 2	III	0.2	2.2		4.0	0.00	0.280	0.530
12	Kyvtanhasyrej	II	120.0	0.9		38.5	0.43	1.950	46.000
13	N1	II	3.3	1.0	13.0	98.8	0.01	0.127	4.300
14	N2	II	8.8	0.9	29.0	119.2	0.01	0.942	84.150
15	N3	II	21.6	0.6	13.0	199.2	0.01	0.215	3.900
16	N4	II	6.4	1.3	23.0	94.2	2.40	1.480	71.550
17	N5	II	10.0	0.5	24.0	98.0	0.01	0.184	3.450
18	N6	II	2.1	1.5	14.0	80.4	0.01	0.222	1.500
19	N7	II	3.5	1.1	23.0.	86.4	0.64	0.070	1.100

Table 3. Average biomass of the littoral meiobenthos (g m^{-2}, wet weight). CLAD - Cladocera, CYCL - Cyclopoida, CHIR - Chironomidae, NEM - Nematoda, OSTR -Ostracoda, BIV - Bivalvia, OLIG - Oligochaeta, HARP - Harpacticoida; lake ## are the same as in Table 2.

Lake ##	CLAD	CYCL	CHIR	NEM	OSTR	BIV	OLIG	HARP	Total
1	0.576	0.211	0.298	0.242	0.131	0.203	0.038	0.260	1.960
2	1.058	0.121	1.306	0.431		0.415	0.032	0.025	3.388
3	0.013		3.164	0.124	0.044				3.345
4	0.002	0.277	2.777	0.161		2.890	0.005	0.001	6.113
5	0.445	0.888	1.360	0.764	0.216	4.679	0.186	0.118	8.654
6	0.026	0.064	0.130	0.080	0.059	0.438	0.011		0.808
7			0.133			1.218			1.352
8		3.715	0.075				0.214		4.004
9	0.474	0.183	0.254	0.030			2.110	0.266	3.315
10		0.145	0.069	0.006	0.357	0.349	0.842		1.768
11		1.000	0.117	0.004	0.136	8.238	0.817		10.312
12	0.047	0.157	1.090	0.036	0.254	15.607	0.422		17.613
13	0.008	0.191	0.896	0.028	0.108	0.498			1.729
14	0.013	0.011	1.669	0.003	0.461				2.156
15	0.092	0.113	1.212	0.117	0.483	0.036			2.051
16		0.266	0.049	0.030	0.040				0.384
17	0.008	0.135	1.124	0.071	0.050	0.071			1.459
18	0.004	0.035	0.479	0.005	0.008	0.036			0.567
19	0.004	0.132	0.459	0.004	0.004				0.602

Table 4. Loadings in factor analyses for some environmental characteristics and littoral meiobenthos biomass.

Variable	Factor I	II	III	IV
Cladocera	0.516	−0.247	0.389	−0.125
Cyclopoida	0.484	0.765	−0.163	0.17
Chironomidae	−0.124	−0.168	0.181	0.405
Nematoda	0.528	−0.182	0.543	0.285
Ostracoda	−0.434	0 047	0.698	0.192
Bivalvia	0.359	−0.121	0.275	0.556
Oligochaeta	0.674	0.556	0.168	0.302
Harpacticoida	0.594	−0.218	0.582	−0.315
P_{inorg}	0.532	0.525	−0.248	0.166
P_{tot}	−0.365	0.663	0.416	0.146
N_{inorg}	−0.607	0.607	0.169	−0.379
N_{tot}	−0.442	0.418	0107	−0.318
BOD	0.207	0.25	−0.37	0.13
pH	0.313	0.688	0.012	−0 211
COD	−0.414	0.372	0.276	0.464
Color	−0.666	0.08	0.215	0.052
CMI	0.695	0.164	0.16	−0.374
Oil	0.374	0.878	−0.141	0.059
CHL	−0.613	0.485	0.36	−0.24
Depth	0.726	−0.054	0.391	−0.413
% of variance	25.9	20.3	11.5	8.8

The second factor (20.3% explained variance) describes conditions, caused by oil pollution. This factor positively correlates with oil concentration, nutrients content and pH. Distribution of Cyclopoida and Oligochaeta is associated with the second factor (Figure 1).

The third factor (11.5% explained variance), I think, reflects the tendency of increase of an abundance Cladocera, Harpacticoida and Nematoda with increase of lake productivity within limits of their optimum conditions.

The fourth factor (8.8% explained variance) shows, that the optimum environmental and trophic conditions for Chironomidae and Bivalvia occurred in shallow lakes with high contents of dissolved organic matter.

ANOVA has been conducted with the same data set (Table 5). Basically, the conclusions received earlier have been confirmed. Also, additional information has been received. So, reliable influence of pH to Nematoda abundance has been marked. A positive influence of oil pollution on biomass of Bivalvia, and a weak negative one of high concentrations of major ions on Chironomidae has been revealed.

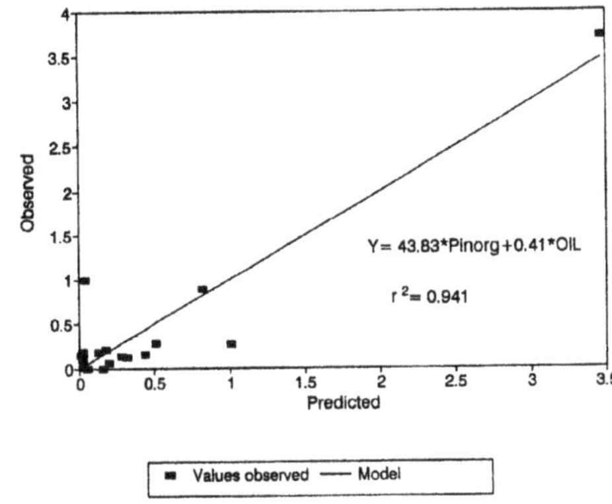

Figure 2. Predicted and observed values of cyclopoid biomass.

Factor analysis and ANOVA permitted only to evaluate an influence of the factors at a qualitative level, but they are not suitable for creation of predictive models. For this purpose can serve simple and multiple regression procedures.

I have constructed a set of empilical regression models (Table 6), which if they do not describe connections of meiobenthos with the environmental and trophic factors with high accuracy, then demonstrate them. Equations 1 and 2 show, that the abundance of Chironomidae larvae was conditioned mainly by the trophic factor. The Cyclopoida abundance on 73% was explained by the factor, reflecting trophic conditions in lake (inorgorganic phosphorus content) too (Equation 3). Only the equation 3 it is possible to use as a predictive model (Figure 2), other are only descriptive ones.

In my opinion, a more realistic approach to a description of interaction of meiobenthos community with environmental factors is to use nonlinear regression models. For approximation of field observations, I have chosen a function; that has the form:

$$Y = \text{Constant} + a * X^b * e^{cX},$$

whose a graphic representation depends on signs of its coefficients (Bronshtein; Semendyaev, 1953).

Figure 3 displays the dependence of biomass of juvenile Bivalvia on oil concentration (r^2=0.725). It is clear, that optimum conditions for this group of meiobenthic animals have been occurred in moderate oil pollution area. Besides the nonlinear interaction of Bivalvia witl daily primary production of phytoplankton was found out too (r^2=0.352). The creation of two-factor model was a further step of a nonlinear regression analysis (r^2=0.913), whose graphic repre-

Table 5. Analyses of variance for the effect of some environmental characteristics on littoral meiobenthos biomass. Color -Pt-Co units; OIL - oil concentration (mg l^{-1}) CMI-concentration of major ions (mg l^{-1}); DEPTH - mean depth (m)

Group	Parameter	F-ratio	Significance level	% of variance explained	Effect
Nematoda	pH	4.72	0.015	16	negative
	DEPTH	18.96	0.0001	33	positive
Bivalvia	Color	3.12	0.056	12	negative
	OIL	5.22	0.001	20	positive
Cladocera	CMI	3.85	0.03	16	positive
	DEPTH	5.00	0.031	10	positive
Chironomidae	CMI	3.75	0.032	16	negative
Cyclopoida	OIL	6.32	0.005	25	positive
Ostracoda	Color	3.24	0.053	26	positive

Table 6. Regression equations predicting meiobenthos biomass. PP -primary production (g C m^{-2} d^{-1}); B_{zoo} - zooplankton biomass (g m^{-3}) COD - chemical oxigen demand (mg O l^{-1}); OIL -oil concentration (mg l^{-1})

##	Group	Predictor	Coefficient	SE	t-value	Significance level	r^2
1	Chironomidae	PP	0.576	0.189	3.045	0.005	0.416
		B_{zoo}	0.126	0.045	2.770	0.010	
2	Chironomidae	COD	0.029	0.006	4.650	0.004	0.603
3	Cyclopoida	P_{inorg}	43.83	14.38	3.048	0.009	0.941
		OIL	0.41	0.052	7.725	0.000	

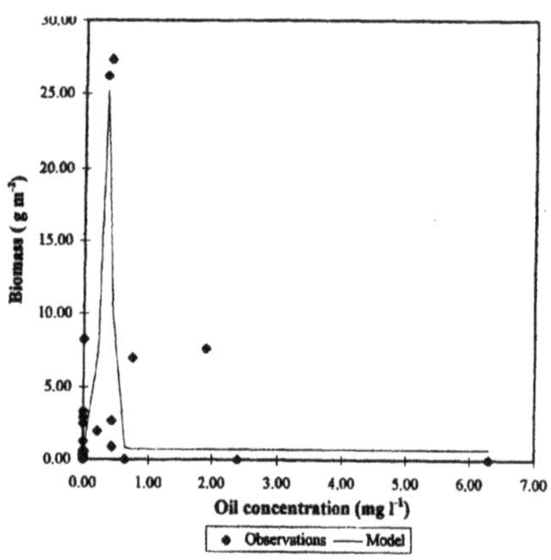

Figure 3. Dependence of bivalves biomass on oil concentration (r^2=0.725).

sentation is given in Figure 4. This model also shows, that the abundance of meiobenthic molluscs was controlled by the trophic factors. The optimum conditions for Bivalvia occur in productive (2.1 g C m^{-2} d^{-1} and moderately oil polluted (0.49 mg l^{-1}) lakes.

Discussion

Meiobenthos of freshwater lakes of moderate latitudes of Russia is rather well-known (Babitsky, 1980; Skvortsov, 1985, Skvoltsov et al., 1988; Kurashov & Belyakov, 1987, Kurashov, 1994). Information about a species structure of meiobenthos, its abundance, distribution and reaction on environmental factors is widely available (Holopajnen & Paasivirta, 1977; Stańczykowska, 1967, Prejs & Stańczykowska, 1972; Särkkä, 1979; Särkkä, 1987). At the same time, the meiofauna of tundra lakes was not investigated enough so far.

As have shown our studies, a species structure of meiobenthos of subarctic lakes did not practically differ from those in lakes of the temperate zone of Europe (Särkkä, 1983; Skvortsov et al., 1988; Kurashov, 1994).

The distribution of meiobenthos organisms in deep tundra lakes differed as a rule from distribution of meiofauna in deep lakes of a moderate zone. In contrast to them, the maximum abundance of meiobenthos was marked in littoral zone. The connection between an abundance of meiobenthos invertebrate and size of sed-

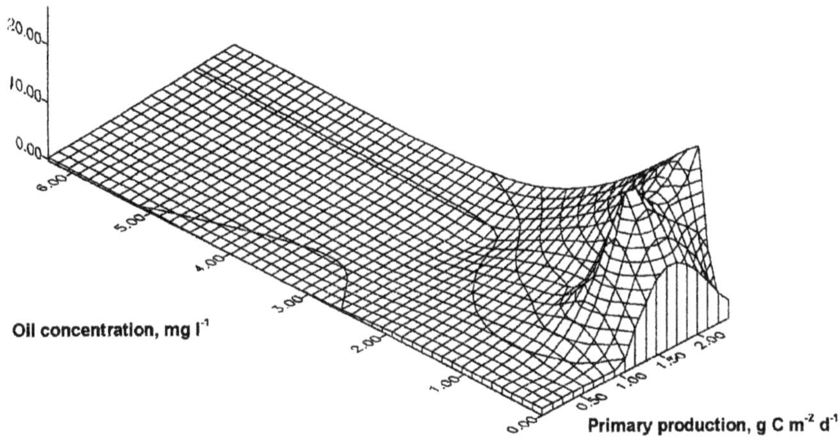

Figure 4. Bivalves biomass as the function of oil concentration and daily primary production (r^2=0.913).

iment particles was found out only for Chironomidae larvae and Nematoda. The increase of number of nematods with increase of sizes of particles was marked also in Lake Michigan (Nalepa & Quigly, 1983).

A reaction of taxonomic groups of meiobenthos of subarctic lakes on environmental factors in the majority of cases was identical with described in the literature. As well as in Finnish Lake Paijanne (Särkkä, 1979), in tundra lakes Nematoda preferred oligohumic sites and with low COD. Besides in investigated tundra lakes within the framework of the optimum combination of the environmental factors Nematoda biomass positively correlated with chlorophyll *a* concentration (the third factor, Table 4). Some authors indicated, that a heavy organic pollution caused an increase of an abundance of Cyclopoida, Harpacticoida and Nematoda (Särkkä, 1975; Kansanen, 1981). Särkkä (1979) specified, that meiobenthic Eucyclopinae and resting stages of Cyclopoida positively correlated with COD, N_{tot}, P_{tot} water color and phytoplankton biomass. Under our observations, the biomass of Cyclopoida increases with increase of P_{inorg} concentration and oil pollution (Table 6). The highest biomass (*Acanthocyclops gigas*) was found in the most oil polluted lake. However, the correlation with other variables was expressed much more poorly (Table 4, Figure 2).

In the same work it was specified, that the abundance of mollusc *Pisidium conventus* Clessin was negatively connected with such variables, as COD, P_{tot} and color, that coincides with our observations (Table 4, Figure 1).

The influence of oil pollution was investigated mainly on marine meiobenthos in field experiments. Results of experiments, indicated in those papers were inconsistent. Some researchers (Alongi et al., 1983) marked initial decreases in abundance of meiobenthos, other did not mark decrease, but registered enhanced nematod and copepod abundance (Fleeger & Chandler, 1983). The computed theoretical curve fitting the relationship between the nematod biomass and oil concentration (r^2=0.647) has shown, that the maximum abundance of nematods should be occurred in polluted lakes (oil content 1.4 mg l^{-1}). However, I can not insist on it, as there are not appropriate field data for this range of oil concentrations.

Effect of oil on bivalve *Macoma balthica* (L.) is described (Eimgren et al., 1983). The authors have shown, that *M. balthica* was more resistant organism to oil pollution. After oil spill the increase of biomass of this mollusc occurred. The authors connected this effect to a reduction of pressure of benthic predators on juvenile bivalves. According to our observations, juvenile bivalves displayed maximal biomass in oil polluted localities. Dependence of their biomass on oil pollution had complex nonlinear character (Figures 3 and 4). Probably, the increase of the bivalve biomass in moderate polluted sites connected with trophic factors and may be explained by the high resistance to toxic action of oil.

Acknowledgments

I am very grateful to *The John D. and Catherine T. MacArthur Foundation* for the financial support of my participation in 'Shallow Lakes'95' Conference (August 20–26, 1995, Mikołajki, Poland).

References

Alongi, D. M., D. F. Boesh & R. J. Diaz, 1983. Colonization of meiobenthos in oil-contaminated subtidal sands in the lower Chesapeake Bay. Mar. Biol. 72: 325–335.

Ankar, S. & R. Elmgren, 1976. The benthic macro- and meiofauna of the Asko Landsort area. A stratified random sampling survey. Contribution from the Asko Laboratory, Univ. of Stockholm. Sweden, 11, 115 pp.

Babitsky, V. A., 1980. Microzoobentos trech ozer raznogo tipa. [Microozoobenthos of the three different type lakes]. Gidrobiol. zh. 16, 1: 37–45.

Balushkina, E. V., 1982. Zavisimost massy tela litshinok chironomid ot ich dliny. [Dependence of chironomid larvae body mass on their length]. Gidrobiol. zh. 27, 3: 53–60.

Belyakov, V. P & V. V. Skvortsov, 1994. Makro- i meiobentos, ich produkcija. [Macro- and meiobenthos, their production]. In Drabkova, V. G. & I. S. Trifonova (eds), Osobennosti struktury ekosistem ozer Krajnego Severa. [Peculiarities of ecosystem structure of Far North lakes]. Nauka Publ. House, St.Petersburg: 193–202.

Bronshtein I. N. & K. A. Semendyaev, 1953. Spravotshnik po matamatike. [Mathematical reference book]. Moskva, 608 pp.

Drabkova, V. G. & S. P. Bystrov, 1994. Gidrochimitsheskaja charakteristika ozer. [Hydrochemical characteristics of the lakes]. In Drabkova, V. G. & I. S. Trifonova (eds), Osobennosti struktury ekosistem ozer Krajnego Severa. [Peculiarities of ecosystem structure of Far North lakes]. Nauka Publ. House, St.Petersburg: 48–64.

Elmgren, R., S. Hansson, U. Larsson, B. Sundeljn & P. D. Boehm, 1983. The 'Tsesis' Oil Spill: Acute and Long-Term Impact on the Benthos. Mar. Biol. 73: 51–65.

Fleeger, J. W. & G. T. Chandler, 1983. Meiofauna responses to an experimental oil spill in a Louisiana salt marsh. Mar. Ecol. Prog. Ser. 11: 257-264.

Holopajnen, I. J. & L. Paasivirta, 1977. Abundance and Biomass of the meiozoobenthos in the oligotrophic and mesohumic lake Pääjärvi, southern Finland. Ann. Zool. Fenn. 14: 124–34.

Kansanen, P. H., 1981. Effects of heavy pollution on the zoobenthos in Lake Vanajavesi, southern Finland, with special reference to the meiozoobenthos. Ann. Zool. Fenn. 18: 243–251.

Kurashov, E. A., 1994. Mejobentos kak komponent ozernoj ekosistemy. [Meiobenthos as a component of lake ecosystem]. 'Alga-Fond' Publ. House, St. Petersburg: 224 pp.

Kurashov, E. A. & V. P. Belyakov, 1987. Rol meiofauny v bentosnom soobshestve raznotipnych ozer Latgalii. [Role of meiofauna in benthic communities of lakes of various types in the Lattgalian Hills]. Gidrobiol. Z. 23: 46–50.

Kuznetsov V. K., 1994. Morfometrija, vodnyj balans i vodoobmen ozer kak rezultat razvitija ozernogo landshafta. [Morphometry, water budget and lake water exchange as the result of lake landscape development]. In Drabkova, V. G. & I. S. Trifonova (eds), Osobennosti struktury ekosistem ozer Krajnego Severa. [Peculiarities of ecosystem structure of Far North lakes]. Nauka Publ. House, St. Petersburg: 23–29.

Metoditsheskie rekomendacii po sboru i obrabotke materialov pri gidrobiologitsheskich issledovanijach na presnovodnych vodoemach. Zooplankton i ego produkcija. [Methodical recommendations on collecting and treatment of materials during hydrobiological investigations on freshwater waterbodies. Zooplankton and its production]. 1982. Leningrad: 33 pp.

Metoditsheskie rekonlendacii po sboru i obrabotke materialov pri gidrobiologitsheskich issledovanijach na presnovodnyeh vodoemach. Zoobentos i ego produkcija. [Methodical recommendations on collecting and treatment of materials during hydrobiological investigations on freshwater waterbodies. Zoobenthos and its production]. 1983. Leningrad: 52 pp.

Nalepa, T. F. & M. A. Quigley, 1983. Abundance and biomass of the meiobenthos in nearshore Lake Michigan with comparisons to the macrobenthos. J. Great Lakes Res. 9: 530–547.

Preis, K. & A. Stańczykowska, 1972. Spatial differentiation and changes in time of zoomicrobenthos in three Masurian lakes. Ecol. pol. 20: 733-745.

Särkkä, J., 1975. Effects of the pollution on the profundal meiofauna of Lake Päijäne, Finland. Aqua Fenn. 1975: 3–11.

Särkkä, I., 1979. The zoobenthos of Lake Päijänne and its relations to some environmental factors. Acta Zool. Fenn. 160: 1–46.

Särkkä, J., 1983. A quantitative ecological investigation of littoral zoobenthos of an oligotrophic Finnish lake. Ann. Zool. Fenn. 20: 157–178.

Särkkä, J., 1987. Meiobenthos of the profundal depths of lake Päijänne. Biol. Res. Rep. Univ. Jyvaskyiä 10: 95–104.

Skvortsov V. V., 1985. Kolitshestvennaja ocenka utshastija mejobentosa raznotipnych ozer Karelskogo pereshejka v processach transformacii organitsheskogo vestshestva donnymi soobstshestvami. [The estimation of role of the meiobenthos of lakes of different types of Karelian Isthmus in organic matter transformation by bottom communities]. Ph.D. thesis, Leningrad, 21 pp.

Skvortsov, V. V., P. J. Lavrentyev & V. V. Maslevtsov, 1988. Soobshestva mejo- i mikrobentosa v ozerach raznogo tipa. [The meio- and microbenthos communities of the lakes of different types]. In Drabkova, V. G & M. J. Prytkova (eds), Izmenenie struktury ekosistem ozer v uslovijach vozrastajushchej biogennoj nagruzki. [Changes of lake ecosystem structures under influence of increasing anthropogenic impact]. Nauka Publ. House, Leningrad: 267–282.

Stańczykowska, A., 1967. Comparison of the zoomicrobentkos occurring in the profundal of several lakes in Northern Poland. Bull. Acad. Pol. Sci. Cl. TI, 15: 349–353.

Strukova, R. Yu., 1994. Chimitsheski; sostav donnych otlozhenij ozer. [Chemical composition of lake bottom deposits]. In Drabkova, V. G. & I. S. Trifonova (eds), Osobennosti struktury ekosistem ozer Krajnego Severa. [Peculiarities of ecosystem structure of far North lakes]. Nauka Publ. House, St.Petersburg: 65–71.

SYSTAT. Statistics, Version 5.2 Edition. Evanston, IL. Systat, Inc. 1992.

Travianko, V. S. & L. V. Evdokimova, 1968. Bentometr MB-TE. [Corer MB-TE]. Gidrobiol. Z 4: 94–96.

Trifonova, I. S., 1994. Soderzhanie chlorofilla, intensivnost firtosintesa fitoplanktona i sootnoshenie produkcionno- destrukcionnych processov v ozerach i ich trofitsheskij stahls. [Chlorophyll content, intensity of phytoplankton photosynthesis, ratio of production- destruction processes in lakes and their trophic status]. In Drabkova, V. G. & I. S. Trifonova (eds), Osobennosti struktury ekosistem ozer Krajnego Severa. [Peculiarities of ecosystem structure of Far North lakes]. Nauka Publ. House, St. Petersburg: 109–119.

Tsalolichin, S. J., 1981. Opredelenie vesa presnovodnych nematod. Evolucija, sistematika, morfologia i ekologija svobobnozhivushchich nematod. [Determination of freshwater nematod weight. Evolulion, systematics, morphology and ecotogy of free-living nematods]. Leningrad, 80–85.

Hydrobiologia **342/343:** 125–132, 1997.
L. Kufel, A. Prejs & J. I. Rybak (eds), Shallow Lakes '95.

Significance of subfossile chironomid remains in classification of shallow lakes

Klaus P. Brodersen & Claus Lindegaard
Freshwater Biological Laboratory, University of Copenhagen, 51 Helsingørsgade, DK-3400 Hillerød, Denmark

Key words: Chironomidae, macroinvertebrate, paleoecology, shallow lakes, benthic communities

Abstract

The relative species composition of subfossile chironomid headcapsules in surface sediments from ten shallow Danish lakes of different trophic status were analysed through similarity indices and multidimensional scaling. It was possible to distinguish between high and low productivity lakes by means of subfossile remains. The low productivity shallow lakes (TP<150 μg P l^{-1}) were characterised by the taxa *Psectrocladius, Pseudochironomus, Microtendipes* and Pentaneurini. The high productivity lakes (TP>150 μg P l^{-1}) show similarities to littoral communities in eutrophic stratified lakes and are represented by genera such as *Cricotopus, Procladius, Microchironomus* and *Chironomus*. In surface samples from Lake Stigsholm subfossile species composition was compared with species composition in samples of trapped adult midges and samples of living larvae. The relative species composition between the three sample types varied considerably. Samples of living larvae showed a high degree of variation within the lake while surface sediment samples of subfossile chironomid remains seemed to be a persistent and reliable tool in classification of shallow lakes.

Introduction

Despite the long tradition in biological monitoring and zoobenthic community studies of freshwater systems, very few attempts have been made to classify or assess the quality of shallow standing waters by means of chironomid larvae. One of the reasons could be the lack of a clear stated hypothesis of how environmental factors influence shallow water communities. The traditional lake typology initiated by Thienemann (1921) and later modified several times (e.g. Brundin, 1949, 1958; Sæther, 1979; Wiederholm, 1980) is limited to profundal communities in (deep) stratified lakes. Thus, primarily the degree and extent of the hypolimnetic oxygen depletion and food availability are considered as main factors controlling the zoobenthic species composition. For paleolimnological studies of chironomid stratigraphy in deep lakes, the theory from these classification systems has been the basis for evaluation and interpretation of lake history (Warwick, 1980; Hofmann, 1988; Walker, 1993). In shallow unstratified lakes and in the upper littoral zone of deep lakes, factors other than oxygen regimes (e.g. structure and organic content of the sediment, submersed vegetation, wind induced resuspension, interspecific interactions and fish communities) could be expected as factors influencing the composition of the macrozoobenthic communities. Thus, no lucid scheme of chironomid distribution for shallow lakes of different trophic state is available for interpretation of chironomid assemblages in core studies from nonstratified standing waters.

In this study, surface sediment samples of subfossile chironomid headcapsules from the shallow Lake Stigsholm are compared with species composition of trapped adult midges and samples of living chironomid larvae to verify the headcapsule identifications and to examine the comparability of midge communities in three different sample types. Surface samples from Lake Stigsholm are compared to similar samples of chironomid communities from nine additional shallow lakes with different total phosphorus concentrations, and the feasibility of constructing a multivariate training dataset to evaluate water quality and trophic status

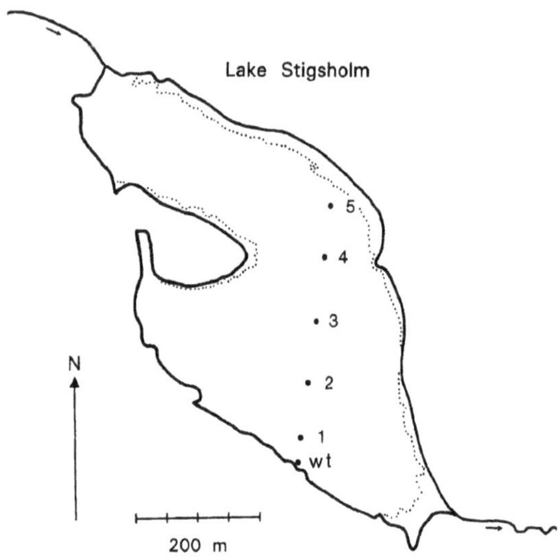

Figure 1. Map of Lake Stigsholm showing the stations for larvae samples (subfossile and living larvae) 1–5 and the location of the window trap (WT). Dotted line indicates the extension of the reed belt.

Table 1. Mean and maximum depth (m) and mean summer total phosphorus concentration (TP μg P l^{-1}) for Lake Stigsholm and the nine additional lakes.

	Lake	Mean depth (m)	Max depth (m)	TP (μg P l^{-1})
1	Holm	0.7	1.4	17
2	Madum	2.9	8.0	26
3	Hornum	1.5	2.6	69
4	Kvie	1.2	2.6	76
5	Stigsholm	0.8	1.1	129
6	Hinge	1.2	2.6	164
7	Kilen	2.9	6.5	305
8	Søbygaard	1.0	1.8	310
9	Lemvig	2.0	3.7	377
10	St. Søgaard	2.7	6.6	387

Table 2. Number of taxa identified in the three sample types. Values in brackets are the number of taxa used in similarity analysis after adjustment to larvae identification levels.

	Window traps		Subfossiles	Living larvae
Tanypodinae	8	(7)	3	1
Diamesinae	1	(1)	0	0
Orthocladiinae	27	(20)	2	0
Chironomini	20	(14)	9	7
Tanytarsini	12	(6)	4	2
Total	68	(48)	18	10

of shallow Danish lakes from subfossile chironomid headcapsules is discussed.

Localities

Lake Stigsholm is a small (0.27 km^2) shallow lake with a maximum depth of 1.1 m (Figure 1). The lake has one inlet stream and one outlet and the water retention time is only 43 days. The summer mean total phosphorus and nitrogen concentrations ranged from 0.10–0.15 mg P l^{-1} and 0.95–1.66 mg N l^{-1} in 1988–1992. The lake is surrounded by farmland and varying extent of *Phragmites* belt characterises the lake margin.

Samples from nine additional shallow lakes of different nutrient status were compared with subfossile samples from Lake Stigsholm (Table 1).

Methods

Fauna samples

Sediment samples of chironomid headcapsules and living larvae were collected at five stations along a south/north transect in Lake Stigsholm, October 1994. (Figure 1). One surface sample of headcapsules from a core study (station 5-B, May 1993 = station 5, Figure 1) was included in the material.

At each station, three sediment surface samples were taken with a core sampler (21.2 cm^2) for headcapsule analysis. The uppermost 1 cm of each core was sliced off and the three samples pooled. In the laboratory, the procedures of Hofmann (1986) were followed with deflocculation in hot 10% KOH, acid treatment and sieving with a 112 μm screen. Chironomid headcapsules were sorted for a minimum of 200 headcapsules, dehydrated in 99% ethanol, and mounted in euparal. Samples from the nine additional lakes were treated with the same procedure. The headcapsules were identified according mainly to Hofmann (1971), Wiederholm (1983) and Moller Pillot (1984a, 1984b).

Living larvae were sorted from five core samples (21.2 cm^2) taken at each station along the transect (Figure 1) in March 1995.

Adult chironomids were collected throughout the emergence period (April–October) in 1993 and 1994 (Lindegaard & Brodersen, unpubl. data) using a window trap constructed and handled as described by Jonsson et al. (1986). The trap was placed next to the shore in the southern end of the lake near station 1 (Figure 1).

Data analysis

The community similarities within and between sampling methods and lakes were measured using nonmetric multidimensional scaling (NMDS) (Field et al., 1982; Clarke, 1993). The distance measure used was the Bray-Curtis dissimilarity coefficients (Bray & Curtis, 1957) following the program package PRIMER (Clarke & Warwick, 1994). All data were standardised to percentage data and 4th root transformed ($X^{0.25}$) before analysis. Standardisation is necessary to cope with large differences in sample size and the transformation gives more weight to rare species in the analysis.

Results

Imagines

Sixty-eight adult chironomid taxa were identified (Table 2) and a total of 14 000 and 23 000 chironomids were trapped in 1993 and 1994, respectively (Lindegaard & Brodersen, unpubl. data). Among the orthocladiines, 14 taxa were considered terrestrial or semi-terrestrial in the data analysis: *Camptocladius stercorarius, Chaetocladius* spp., *Limnophyes* spp., *Metriocnemus* spp., *Paraphaenocladius impensus, Pseudorthocladius curtistylus, Smittia aterrima* and some unidentified genera, comprising 3.3% of the total. Three species were considered as running water species: *Diplocladius cultriger, Micropsectra apposita* and *M. atrofasciata*, making up only 0.2% of the total. *Cladotanytarsus* spp. and *Tanytarsus* spp. (mainly *T. mendax* Kieffer) made up 85% of the total catch in both years, although the individual proportions of these two taxa varied notably.

For some of the adult species, the taxonomic level has been reduced and adjusted to generic level to make the data sets comparable in the similarity analysis (Tables 2 and 3).

Subfossile surface samples

In Lake Stigsholm, 18 taxa of subfossile headcapsules were identified with *Tanytarsus* spp. and the non-flying midge *Corynocera ambigua* as the most dominating taxa (60%) (Table 3). *Cladopelma* gr. *lateralis* was not caught in the window trap.

Living larvae

Only ten taxa of living larvae were identified in the core samples (Table 3). Species of *Procladius* dominated and made up 45% of the total number of chironomid larvae ranging from 10 000 m^{-2} to 18 000 m^{-2} at the five stations. As in the headcapsule samples, *Corynocera ambigua* contributed 27% to the total number of chironomids.

Multivariate analysis

Two-dimensional ordinations (NMDS) were carried out on different species/sample subsets. The distance between points display the ranked dissimilarity between communities and the NMDS plots can be arbitrary scaled, rotated and inverted. The stress values in all presented ordinations are <0.08, indicating almost excellent representation (Clarke & Warwick, 1994).

Figure 2A shows the similarities between the three types of samples when using the complete species record (Table 3). The long distance (dissimilarity) between the window trap samples and the larvae samples is not surprising. The presence of terrestrial and running water species (although the numbers are low) and the absence of the abundant non-flying *Corynocera ambigua* in the window trap samples can account for this large difference.

By removing the terrestrial and the running water species and adding 25% *Corynocera ambigua* to the window trap samples, a new ordination diagram arises (Figure 2B). The window trap samples becomes less dissimilar to the larvae samples and the deviation between subfossile samples and living larvae samples becomes distinct. The headcapsule samples are located between the two other groups of sample types because of the moderate number of species registered. The within-sample-type community variation is very low compared to the between-sample-type variation.

In an ordination of only the larvae samples (living larvae and headcapsules), a relatively high dissimilarity is found among the samples of living larvae, while the dissimilarity between the headcapsule samples is

Table 3. The percentage composition of chironomids in window traps 1993 and 1994, surface samples of subfossile headcapsules (mean ± 95% confidence limits (CL), n=6), and samples of living larvae (mean ± 95% CL, n=5).

	Imagines 1993	Imagines 1994	Subfossil samples	95% CL	Living larvae	95% CL
TANYPODINAE						
Ablabesmyia sp.	0.06	0.02	0.36	0.70	–	–
Anatopynia plumipes (Fries)	0.01	–	–	–	–	–
Conchapelopia melanops (Meigen)	0.01	–	–	–	–	–
Monopelopia tenuicalcar (Kieffer)	0.01	0.01	–	–	–	–
Pentaneurini	0.08	–	0.08	0.22	–	–
Procladius spp.	5.48	1.69	9.91	1.54	44.65	16.54
Xenopelopia nigricans (Goetghebuer)	–	0.41	–	–	–	–
DIAMESINAE						
Diamesa tonsa (Walker)	0.01	–	–	–	–	–
ORTHOCLADIINAE						
Acricotopus lucens (Zetterstedt)	0.88	0.79	–	–	–	–
Camptocladius stercorarius (De Geer)	0.04	0.01	–	–	–	–
Chaetocladius spp.	0.11	0.01	–	–	–	–
Corynoneura sp. gr. *scutellata*	0.98	1.48	–	–	–	–
Cricotopus spp.	1.60	0.74	3.24	3.10	–	–
Diplocladius cultriger Kieffer	0.01	0.01	–	–	–	–
Limnophyes minimus (Meigen)	0.46	0.90	–	–	–	–
Limnophyes pentaplastus (Kieffer)	0.04	–	–	–	–	–
Limnophyes sp. cf. *habilis* type	1.26	1.35	–	–	–	–
Metriocnemus fuscipes (Meigen)	0.04	–	–	–	–	–
Metriocnemus sp. a	0.06	0.07	–	–	–	–
Metriocnemus sp. b	0.07	0.60	–	–	–	–
Nanocladius sp.	0.33	0.42	–	–	–	–
Orthocladius sp.	0.01	0.07	–	–	–	–
Paraphaenocladius impensus (Walker)	0.06	–	–	–	–	–
Psectrocladius spp.	0.16	0.14	1.48	1.54	–	–
Pseudorthocladius curtistylus (Goetghebuer)	0.03	–	–	–	–	–
Smittia aterrima (Meigen)	0.34	0.06	–	–	–	–
Orthocladiinae gen. indet. a	0.61	0.57	–	–	–	–
Orthocladiinae gen. indet. b	0.09	–	–	–	–	–
CHIRONOMINAE						
CHIRONOMINI						
Camptochironomus tentans Fabricius	0.27	0.46	–	–	–	–
Chironomus plumosus Linnaeus	0.03	0.03	2.21	1.77	3.53	6.05
Chironomus riparius Meigen	0.02	0.03	–	–	–	–
Chironomus spp.	0.03	0.10	–	–	–	–
Cladopelma gr. *lateralis*	–	–	0.25	0.29	–	–
Cryptochironomus sp.	0.01	–	0.81	0.52	2.81	3.54
Demicryptochironomus vulneratus Zetterstedt	0.03	–	–	–	–	–
Dicrotendipes spp.	0.49	0.04	7.79	1.52	0.50	0.88
Einfeldia gr. B	–	–	–	–	0.20	0.55
Endochironomus gr. *albipennis*	–	–	0.92	0.70	–	–
Endochironomus albipennis (Meigen)	0.17	0.19	–	–	–	–
Glyptotendipes spp.	0.24	3.31	1.39	1.20	–	–
Microtendipes gr. *pedellus*	0.30	0.61	1.68	1.61	0.18	0.49
Polypedilum spp.	1.37	0.29	1.82	1.51	2.06	0.76
Parachironomus spp.	0.01	0.07	–	–	–	–
Pentapedilum sordens (van der Wulp)	–	0.09	–	–	–	–
Stictochironomus sticticus (Fabricius)	0.08	0.24	0.10	0.27	–	–
PSEUDOCHIRONOMINI						
Pseudochironomus prasinatus (Staeger)	–	–	–	–	0.16	0.43

Table 3. Continued

	Imagines 1993	Imagines 1994	Subfossil samples	95% CL	Living larvae	95% CL
TANYTARSINI						
Cladotanytarsus spp.	60.99	79.01	7.58	2.13	–	–
Corynocera ambigua Zetterstedt	–	–	25.59	2.32	27.48	23.72
Micropsectra apposita (Walker)	0.04	0.01	–	–	–	–
Micropsectra atrofasciata (Kieffer)	0.19	0.01	–	–	–	–
Paratanytarsus sp.	0.10	–	0.36	0.62	–	–
Tanytarsini gen. indet.	0.06	–	–	–	–	–
Tanytarsus spp.	22.80	6.15	34.42	2.50	18.45	9.28

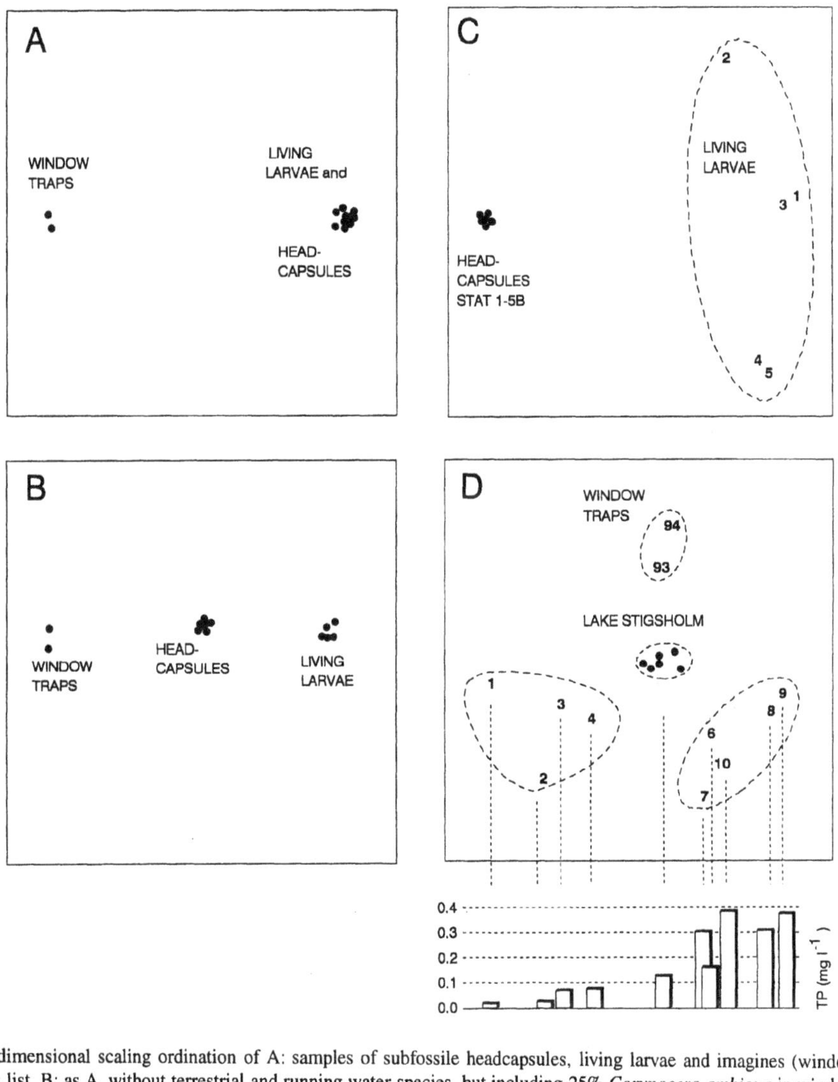

Figure 2. Multidimensional scaling ordination of A: samples of subfossile headcapsules, living larvae and imagines (window traps) on the complete species list. B: as A, without terrestrial and running water species, but including 25% *Corynocera ambigua* in window trap samples. C: ordination of larvae samples only. D: ordination of samples from Lake Stigsholm, nine additional lakes (subfossile) and imagines from Lake Stigsholm 1993 and 1994. Bars indicating the mean summer total phosphorus concentration (mg P l^{-1}). Numbers of additional lakes corresponds to Table 1.

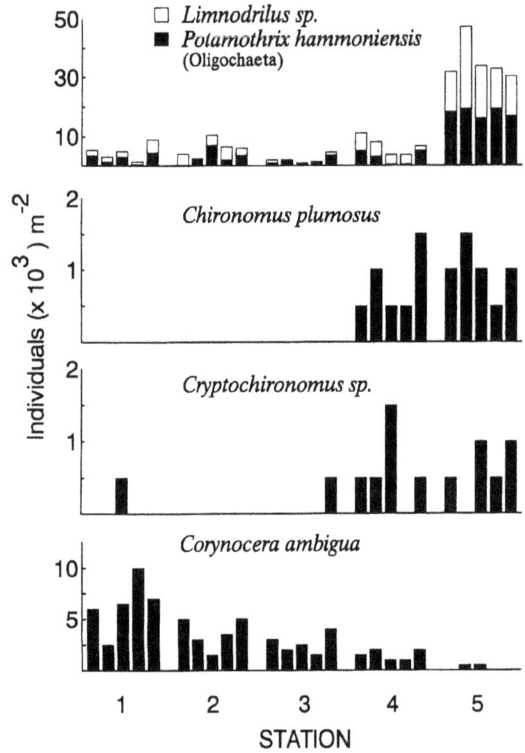

Figure 3. Abundance of selected species at station 1–5 in March 1995.

still very low (Figure 2C). For the living larvae samples, the two northern stations (4 and 5) are different from the other stations in having high numbers of *Chironomus plumosus* and *Cryptochironomus* sp., while the abundance of *Corynocera ambigua* is decreasing along the transect (Figure 3). The most northern station (5) is also characterised in having high numbers of the oligochaetes *Potamothrix hammoniensis* (Michaelsen) and *Limnodrilus* sp. (Figure 3), but only the chironomids are included in the ordination.

The between-lake variation in subfossile chironomid communities from surface sediments is presented in Figure 2D. The similarities of window trap samples and replicate headcapsule samples from Lake Stigsholm are ordinated with headcapsule samples from the nine additional shallow lakes. A horizontal separation in lakes corresponding to a gradient of total summer phosphorus concentration is suggested ($r=0.65$, Spearman rank correlation, Clarke & Ainsworth, 1993) and a cluster analysis divides the data into two groups of lakes, those less than 150 μg TP l^{-1} and those above this level. The difference between these two groups in the NMDS is significant ($P<0.01$) when tested by a nonparametric permutation procedure, ANOSIM (Clarke & Green, 1988). Again, the variation within Lake Stigsholm is low (Figure 2D), and although window trap samples deviate from all other samples, they are still in the same perpendicular projection along the horizontal TP 'gradient' as the headcapsule samples. Chironomid taxa separating the two groups are: *Psectrocladius* spp., *Pseudochironomus* sp., *Microtendipes* sp., and Pentaneurini with high frequencies in lakes of low TP ($<150\ \mu g\ l^{-1}$); and *Chironomus plumosus, Procladius* sp., *Microchironomus* sp., and *Cricotopus* spp. in the more phosphorus rich lakes ($TP>150\ \mu g\ l^{-1}$).

Discussion

Window trap samples

When analyzing the window trap samples together with the larvae samples, the most striking difference is the relatively large number of taxa registered among the adult midges. The presence of terrestrial and running water species not arising from Lake Stigsholm will of cause contribute to this high richness. Additionally, continuous collection of flying midges during the season and the large number of individuals caught in the window traps will increase the probability of finding the more infrequent and uncommon species from the lake. Species level identification is possible for a high number of adult (male) chironomids. This feature increases the number of taxa recorded and should be regarded as a major strength of this type of sample. The within-lake variation in window trap samples is expected to be low when data are adjusted for exclusion of terrestrial and running water species from nearby locations. The seasonal variations in window trap samples can be considerable, and data integration over the entire flight period is probably required. An abundant peak of *Cladotanytarsus* spp. (9000 individuals in early July 1994) in Lake Stigsholm is an example of notable variation which could be due to either year-to-year population fluctuations, local meteorological conditions at the time of swarming, or both.

Living larvae samples

The spatial variation in samples of living larvae with increasing numbers of the deposit feeding oligochaetes and *Chironomus plumosus* at the two northern stations in Lake Stigsholm, might be related to changes in sediment structure and organic content in the sediments as found by e.g. Dévai & Moldován (1983) in the shallow Lake Balaton, and in experimental ponds (Francis & Kane, 1995). The variation is considerable and clearly highlights the complications in defining a representative chironomid fauna by means of living larvae for even a small lake. The living larvae populations and communities are affected by physical and chemical stress such as wind induced resuspension, fluctuations in oxygen content, pH and temperature; and by biological interactions such as inter- and intraspecific competition, predation, swarming success, initial population size, etc. Thus, as with window trap samples, frequent sampling through out the year and multiple sampling stations are required to obtain a representative picture of the chironomid community in the lake.

Headcapsule samples

In the surface sediment samples of subfossile remains, only more or less complete chironomid larval headcapsules are preserved. Consequently, usually only generic level or species group identifications are possible for subfossile headcapsules (Hoffmann, 1988). The number of examined headcapsules in paleoecological studies is usually a few hundred or less per sample, but the abundance can vary from zero to several thousand per cubic centimeter of sediment (Walker, 1987) depending on lake sedimentation rate and hydrological conditions. Thus, limitations in identification levels and a moderate number of individuals reduces the obtainable 'richness and diversity' in the subfossile samples. The headcapsule samples (1993–94) from Lake Stigsholm show very little variation between the five sampling stations. This is because the 1 cm surface layer will integrate information on the chironomid community structure for the period of sedimentation, and because resuspension of the surface layers will mix and equalize the spatial variation within the lake. Headcapsule samples are, therefore, not affected by short-term events and in-lake variation in the same way as living larvae samples and window trap samples.

Although most of the subfossile material is identified only to genus level, the window trap samples provide a verification of the subfossile results. In this study, the total percentage abundance of taxa (identifications) in the window traps which are not registered in headcapsule samples, is only 6–8% but divided on 30 and 23 taxa in 1993 and 1994, respectively, including 11 terrestrial and 3 running water species. Within Tanypodinae and Chironomini there is good correspondance between window trap samples and headcapsule samples. However, among the Tanytarsini, lack of imagines of the non-flying *Corynocera ambigua* and an abundance peak in *Cladotanytarsus* spp. provide notable differences between the two sample types. Verification of subfossile species identification through imagines collection would in this study suggest an overestimation of subfossile *Tanytarsus* spp. at the expense of *Cladotanytarsus* spp., while the deviation in *Corynocera ambigua* is easily explained by flying behaviour.

When comparing the headcapsule samples from Lake Stigsholm with the nine other shallow lakes, the variation within Lake Stigsholm is still low. Lake Stigsholm falls between the two high/low total-phosphorus lake-group divisions, as would be expected from its measured phosphorus concentration. The distinct separation of lakes along a TP gradient in the horizontal ordination plane offers promising possibilities for constructing a multivariate training dataset on chironomid assemblages. Multivariate nonparametric methods, such as nonmetric multidimensional scaling, are used in meta-analysis of community stress in marine zoobenthic communities (Warwick & Clarke, 1993), and parametric methods (e.g. weighted average regression, calibration and ordination) are already widely used to reconstruct pH and TP by means of subfossile diatom assemblages (e.g. Stevenson et al., 1989; Anderson, 1993; Bennion, 1994) and profundal chironomid communities (Johnson, 1989). Assuming that most shallow Danish lakes have not changed notably within the time frame represented by 1 cm of sediment, similar ordinations of headcapsule samples may probably be a very steady and reliable method for defining chironomid communities in shallow lakes. However, Anderson & Odgaard (1994) illustrate the problems in dating and diatom-inferred TP reconstruction of sediments from shallow lakes due to intense mixing. In shallow Lake Væng (mean depth 1.2 m), the diatom record failed to trace four years of biomanipulation despite a substantial change in the phytoplankton communities. Nevertheless, even with coarse time resolution, the results of our study suggest that analyses of subfossile remains can be a reliable tool in classification of shallow lakes, and that samples of subfossiles should be compared to ('recent') subfossile headcapsule samples rather than communities of living larvae and adults, if possible.

Conclusion

Exact comparison of chironomid communities between samples of living and subfossile larvae, and window trap samples of imagines can be difficult. Samples of adult midges make high-level species identification possible and can be used for verification of subfossile results. Window trap samples and samples of living larvae are sensitive to short-term events that affect living populations, and local in-lake variation is considerable within samples of living larvae from Lake Stigsholm. The variation in the subfossile chironomid communities along the transect in Lake Stigsholm was very low and subfossile surface samples have an advantage in classification of shallow lake chironomid communities because of a lower sensitivity to short-term events.

Acknowledgments

The authors wish to thank the technical staff of the National Environmental Research Institute, Silkeborg, for assistance in collection of imagines 1993–94. Many thanks to Julie Wolin for English corrections and to Peter C. Dall and two anonymous referees for criticising the manuscript.

References

Anderson, N. J., 1993. Natural versus anthropogenic change in lakes: The role of the sediment record. Trends Ecol. Evol. 10: 356–361.

Anderson, N. J. & B. V. Odgaard, 1994. Recent palaeolimnology of three shallow Danish lakes. Hydrobiologia 275/276: 411–422.

Bennion, H., 1994. A diatom-phosphorus transfer function for shallow, eutrophic ponds in southeast England. Hydrobiologia 275/276: 391–410.

Bray, J. R. & J. T. Curtis, 1957. An ordination of the upland forest communities of Southern Wisconsin. Ecol. Monogr. 27: 325–349.

Brundin, L., 1949. Chironomiden und andere Bodentiere der sudschwedischen Urgebirgsseen. Rep. Inst. Freshwat. Res. Drottningholm 30: 1–914.

Brundin, L., 1958. The bottom faunistic lake type system and its application to the southern hemisphere. Moreover a theory of glacial erosion as a factor of productivity in lakes and oceans. Verh. int. Ver. Limnol. 13: 288–297.

Clarke, K. R., 1993. Non-parametric multivariate analyses of change in community structure. Aust. J. Ecol. 18: 117–143.

Clarke, K. R. & M. Ainsworth, 1993. A method of linking multivariate community structure to environmental variables. Mar. Ecol. Prog. Ser. 92: 205–219.

Clarke, K. R. & R. H. Green, 1988. Statistical design and analysis for a 'biological effect' study. Mar. Ecol. Prog. Ser. 46: 213–226.

Clarke, K. R. & R. M. Warwick, 1994. Change in marine communities: an approach to statistical analysis and interpretation. Nat. Envir. Res. Council, UK, 144 pp.

Dévai, G. & J. Moldován, 1983. An attempt to trace eutrophication in a shallow lake (Balaton, Hungary) using chironomids. Hydrobiologia 103: 169–175.

Field, J. G., K. R. Clarke & R. M. Warwick, 1982. A practical strategy for analysing multispecies distribution patterns. Mar. Ecol. Prog. Ser. 8: 37–52.

Francis, D. R. & T. C. Kane, 1995. Effect of substrate on colonization of experimental ponds by Chironomidae (Diptera). J. Freshwat. Ecol. 10: 57–63.

Hofmann, W., 1971. Zur Taxonomie und Palökologie subfossiler Chironomiden (Dipt.) in Seesedimenten. Arch. Hydrobiol. Beih. Ergeb. Limnol. 6: 1–50.

Hofmann, W., 1986. Chironomid analysis. In Berglund, B. E. (ed.), Handbook of Holocene Palaeoecology and Palaeohydrology. John Wiley, Chichester, England: 715–727.

Hofmann, W., 1988. The significance of chironomid analysis (Insecta: Diptera) for paleolimnological research. Paleogeography, Paleoclimatology, Paleoecology 62: 501–509.

Johnson, R. K., 1989. Classification of profundal chironomid Communities in oligotrophic/humic lakes of Sweden using environmental data. Acta Biol. Debr. Oecol. Hung. 3: 167–175.

Jónsson, E., A. Gardarsson & G. M. Gislason, 1986. A new window trap used in the assessment of the flight periods of Chironomidae and Simuliidae (Diptera). Freshwat. Biol. 16: 711–719.

Moller Pillot, H. K. M., 1984a. De larven der Nederlandse Chironomidae (Diptera) (Tanypodinae, Chironomini). Nederlandse Faunistische Mededelingen, 1A: 1–277.

Moller Pillot, H. K. M., 1984b. De larven der Nederlandse Chironomidae (Diptera) (Orthocladiinae). Nederlandse Faunistische Mededelingen, 1B: 1–175.

Sæther, O. A., 1979. Chironomid communities as water quality indicators. Holarct. Ecol. 2: 65–74.

Stevenson, A. C., H. J. B. Birks, R. J. Flower & R. W. Batterbee, 1989. Diatom-based pH reconstruction of lake acidification using canonical correspondence analysis. Ambio 18: 228–233.

Thienemann, A., 1921. Biologische Seetypen und die Gründung einer hydrobiologischen Anstalt am Bodensee. Arch. Hydrobiol. 13: 347–370.

Walker, I. R., 1987. Chironomidae (Diptera) in paleoecology. Quat. Sci. Rev. 6: 29–40.

Walker, I. R., 1993. Paleolimnological biomonitoring using freshwater benthic macroinvertebrates. In Rosenberg, D. M. & V. H. Resh (eds). Freshwater Biomonitoring and Benthic Macroinvertebrates. Chapman & Hall. New York: 306–343.

Warwick, W. F., 1980. Palaeolimnology of the Bay of Quinte, Lake Ontario: 2800 years of cultural influence. Can. Bull. Fish. aquat. Sci. 206: 1–117.

Warwick, R. M. & K. R. Clarke, 1993. Comparing the severity of disturbance: a meta-analysis of marine macrobenthic community data. Mar. Ecol. Prog. Ser. 92: 221–231.

Wiederholm, T., 1980: Use of benthos in lake monitoring. J. wat. Pollut. Cont. Fed. 52: 537–547.

Wiederholm, T. (ed.), 1983: Chironomidae of the Holarctic region. Part 1, Larvae. Ent. Scand. Suppl. 19: 1–457.

Hydrobiologia **342/343:** 133–142, 1997.
L. Kufel, A. Prejs & J. I. Rybak (eds), Shallow Lakes '95.

Chironomus plumosus – what regulates its abundance in a shallow reservoir?

Zdzisław Kajak
Department of Hydrobiology, Institute of Ecology, Dziekanów Leśny, 05-092 Łomianki, Poland

Key words: *Chironomus plumosus*, benthos abundance, shallow reservoir, field experiment

Abstract

A high increase of the abundance of the main dominant benthic species, *Chironomus plumosus*, was reached by elevating the local mud with benthos in trays 40 and 70 cm above the bottom in 6 m deep site of a shallow polymictic, highly eutrophic reservoir. The differences of numbers between the trays and the bottom varied during the season (May–August). The maximum ratio of Chironomus numbers trays: bottom (almost 8), was found in early summer (June/July), when abundance in the bottom was the lowest. Young larvae were especially more numerous in trays. The main causes of much higher *Chironomus* numbers in trays than in the bottom in early summer could be differences in oxygen deficit. In August–September numbers of larvae in the bottom were moderate (several times lower than maximal in spring, but several times higher than minimal in early summer). In trays in summer numbers were only 1.5 times higher than in the bottom. The reasons for these relatively low numbers in summer, despite prolonged occurrence of imagos and young larvae, could be insufficient numbers of eggs due to heavy predation by swallows on egg laying females, or the negative influence of big larvae on the survival of youngest ones.

Introduction

The aim of this paper was to determine the main causes of *Chironomus plumosus* (L.) abundance in a shallow, lowland reservoir. The essential principle applied was the comparison of the density of this species in the bottom with its density in the mud above the bottom level in a field experiment. Zegrzyński reservoir situated near Warsaw, and fed by the Narew and Bug rivers, was chosen as the study area. The reason for this choice was the extremely high abundance of *Chironomus plumosus*, which is the dominant in the benthos of the central part of the reservoir (Dusoge, 1989; Dusoge et al., 1985; Dusoge et al., in press; Kajak, 1990a, 1990b, 1991; Kuklińska, 1989, 1992). We wanted to check whether the abundance at the bottom, although usually high, and periodically very high, could be increased by elevating the mud and whether it is possible to avoid strong fluctuations of abundance during the year.

The main idea behind this work was to find out whether moving the mud a bit above the bottom level, with all other factors unchanged, would essentially modify the situation for benthos (in this case – its main component – *Chironomus plumosus*) survival and in consequence – its abundance. The hope was that benthos would change, and that this will help to understand a bit better what is most important for its abundance in these circumstances. *Chironomus* does sometimes reach very high abundance in various environments (Kajak, 1987, 1988; Konstantinov, 1958; Marian & Pandian, 1985; Sokolova, 1983).

Study area and methods

The reservoir, fed by two polluted rivers, is polymictic, very eutrophic, with average concentrations of P_{tot} 150–200 $\mu g\ l^{-1}$ and N_{tot} 2.7 mg l^{-1}. Organic matter in the bottom sediments is relatively low – about 9.0% of the dry weight. Average depth of the reservoir is 3.5 m, maximal one, near the dam – 9 m. The detailed description of the reservoir and its environmental conditions are given by Kajak (1990a, 1991) and Kajak & Dusoge (1989). The work was carried out in the central part of the reservoir at 6 m depth, which was the previous bed of the river. A location with slight water flow was chosen to avoid strong oxygen deficits.

The proportion of *Chironomus* in total benthos biomass was on the average 80% in the study site, and that of the other dominant group – *Tubificidae* was about 20%. The main species of *Tubificidae* were *Potamotrix hammoniensis* (Mich.), *Limnodrillus hoffmeisteri* (Clap.), *Tubifex tubifex* (O. F. Müll.); within Chironomidae, apart from *Chironomus*, only *Procladius Skuze* and *Glyptotendipes ex grege gripekoveni* Kieff. sporadically occurred, usually not more than few hundred ind. m^{-2} each (Dusoge, 1989; Dusoge et al. 1985, 1990).

Square plastic trays (30 × 30 cm and 12.5 cm deep) were located in special stands, at 40 and 70 cm above the bottom level and also at the bottom (Kajak & Dusoge, 1996). Similar trays, but only in the bottom, were also applied by Soster & McCall (1990). In the various experiments trays were filled with mud which was taken either with an Ekman grab, or by a diver directly at the bottom near the stand, with mud from other water bodies or mud without *Chironomus* and most of other benthos; some trays were set empty, to be filled with resuspending mud and sedimenting tripton. This method has been described in more detail by Kajak & Dusoge (1996). Trays were kept in the reservoir for 2–4 weeks. Bottom samples near the stands were taken every week or every few weeks during spring and early summer, later in summer samples were taken every month, in 1993 (Figures 1 and 2). They were taken from the boat with a 20 cm^2 Kajak plexiglass corer or by the diver with the plexiglass tube of same sampling area, and then stoppered at both ends. Samples from trays were taken by the diver. As a rule series of 5 (sometimes 10) samples were taken each time from the bottom and from every tray. All samples were split into surface (0–3 cm) and deeper (3 to about 12 cm) layer, the first being sieved using a 0.2 mm mesh sieve (to keep youngest larvae) the second – using a 0.4 mm mesh sieve. Samples were preserved in 4% formalin, then sorted in the laboratory. Larvae were measured with 1 mm accuracy. Total numbers of larvae and their numbers in the following size classes were analysed: ≤ 6 mm (the youngest), ≤ 12 mm (young; this class includes of course that ≤ 6 mm) and ≥ 18 mm. The abundance of the first class informs about hatching from eggs; the abundance of class ≤ 12 mm – about survival of young larvae, and the abundance of the oldest class – ≥ 18 mm – about numbers of larvae 'ready for emergence'. The class $\geq 7 \leq 13$ mm was not considered.

The four series of trays were examined, exposed during the following periods: 7–28 May, 7 May–11 June, 11–30 June and 30 July–30 August 1993, that is at varying abundance and size composition of *Chironomus* population at the bottom near the stands (Figures 1 and 2; Table 1). There are no data for July, because strong winds emptied the trays or even threw them out of stands; on June 11th thickness of mud in three trays was only 1–2 cm; so only one tray, with 5 cm layer of mud was taken into account, but also here surface layer of mud with some *Chironomus* could have been resuspended. Numbers of larvae in trays would probably have been higher, if there was not washing out of mud.

Numbers and size composition of the *Chironomus* population in the serious of trays located in the higher and lower stand, and between particular trays within one stand, differed. However, this differentiation was relatively small, when compared with differences between the average values for series of trays in both (low and high) stands and the data for the bottom near the trays (Table 1, Kajak & Dusoge, 1996).

General differences between series of trays in stands and the nearby bottom will be analysed, neglecting relatively smaller differences between trays within one stand and between stands; these differences are discussed in detail in other papers (Kajak & Dusoge, 1996 and in preparation).

Temperature and oxygen vertical profile were measured once to twice a month, between April and September, using oxygen Meter JSI model 57, Yellow Springs Instr. Co. Inc., also Secchi disc visibility was registered.

A Mann-Whitney U-test was used to check the significance of differences between variants and between consecutive bottom series.

Results

The bottom

In the bottom at the end of April the abundance of *Chironomus* larvae was very low and the oldest (≥ 18 mm) class was dominant. Only few larvae of the youngest (≤ 6 mm) and young (≤ 12 mm) size classes were present. These probably were the ones which overwintered in this stage; the young larvae do occur till late autumn, and they do not grow during low winter temperatures.

Table 1. Average numbers of *Chironomus* larvae in the bottom (B) and in trays (T) over the bottom, and the share (%) of size classes in total numbers at the end of the exposition period of particular tray series. In brackets - accordingly number of trays and range of average *Chironomus* abundance for particular trays in their series.

Period of experiment	Trays - T bottom - B	Numbers ind. 20 cm^{-2}				% of size classes in total abundance		
			Size classes (in mm)					
		Total	≤6	≤12	≥18	≤6	≤12	≥18
	T(1)	77.8	32.0	63.5	6.5	41.1	81.6	8.4
7–28.05	B	17.8	4.6	13.4	2.4	26.0	75.3	13.5
	T(1)	26.2	3.0	7.4	18.0	11.5	28.2	68.7
7.05–11.06	B	8.6	0.6	2.6	4.6	7.0	30.0	53.5
	T(6)	21.2	1.9	8.4	8.7	8.0	39.6	41.0
11–30.06		(18.6–25.8)	(0.6–2.7)	(4.2–10.8)	(6.5–11.2)	(3.9–13.5)	(22.6–50.0)	(32.5–58.1)
	B	2.8	0.0	0.6	1.2	0.0	21.4	42.9
	T(7)	21.0	3.4	7.1	12.6	16.2	33.3	60.0
30.07–30.08		(19.0–25.8)	(2.2–4.4)	(5.0–8.6)	(10.6–15.2)	(10.8–21.6)	(24.5–42.6)	(55.1–70.9)
	B	13.8	2.4	7.0	6.0	17.4	50.7	43.5

All differences between the average values for trays and the bottom (for total numbers and numbers in size classes) within particular series are statistically significant at P=0.05 level (Mann-Whitney U-test)

Such pattern was observed also in previous years (Dusoge, 1989; Dusoge et al., 1990; Dusoge & Kajak, in preparation).

The oldest larvae of the overwintered population metamorphosed and emerged in April–beginning of May. This resulted in the appearance of new, very abundant generation in May. Its peak numbers were found on May the 19th–36.7 ind. 20 cm^{-2}, that is 18.4 thousands ind. m^{-2} (Figure 1). The youngest class made up almost 80% of total numbers, and all young larvae – ≤12 mm – almost 100%. The oldest class (≥18 mm) was absent (Figure 2).

Later during May and till the end of June (Figures 1 and 2) paralelly to the growth of larvae, very abrupt reduction of their numbers occurred. Already 10 days after the maximum – numbers decreased by half and during the following week numbers were almost halved again.

On June the 11th only few youngest larvae (≤ 6 mm) occurred, and they were absent on June the 30th, when the total numbers dropped to the minimum - 2.8 ind. 20 cm^{-2}, that is 1400 ind. m^{-2}.

During July numbers were increasing with all size classes coexisting. The next, much lower, peak of abundance was observed at the end of July. No abrupt decline of numbers, like in May–June, was observed here. During following 2 months the population was abundant, numbers of larvae were very stable (about 13–14 ind. 20 cm^{-2} that is about 6.5–7.0 thousands ind. m^{-2}). The absolute numbers and the share of oldest larvae in total numbers was increasing since end of July till end of September, and paralelly those of younger classes of larvae – was slightly decreasing. However the share of youngest larvae was relatively high (about 20–55%) till the end of September (Figures 1 and 2; Table 1).

The summer population differed essentially from the spring one by the lack of a distinct, high peak of numbers, stable level of abundance, and the prolonged, quite high proportion of the youngest (≤6 mm) larvae throughout the August and September (Figure 1, Table 1). There undoubtedly was permanent emergence of imagos and parallel hatching of larvae, which however did not result in the very high abundance of the population. No any border between generations (no distinct cohorts) can be found during this period, and thus the duration of generations can not be estimated.

Trays versus the bottom

In the mud located above the bottom, in trays, the abundance of *Chironomus* was always higher than in the bottom. However, the ratio trays/bottom differed much – it was high in May – early June (4.4–3.0 for total numbers, and up to 7.0 for the youngest class). It was the highest at the end of June (7.6 for total numbers and 14.1 for young – ≤ 12 mm) larvae; the ratio for the youngest (≤ 6 mm) larvae can not be given because this class was absent in the bottom when total numbers in the bottom were minimal.

At the end of August when the total numbers in the bottom were very similar to those at the end of

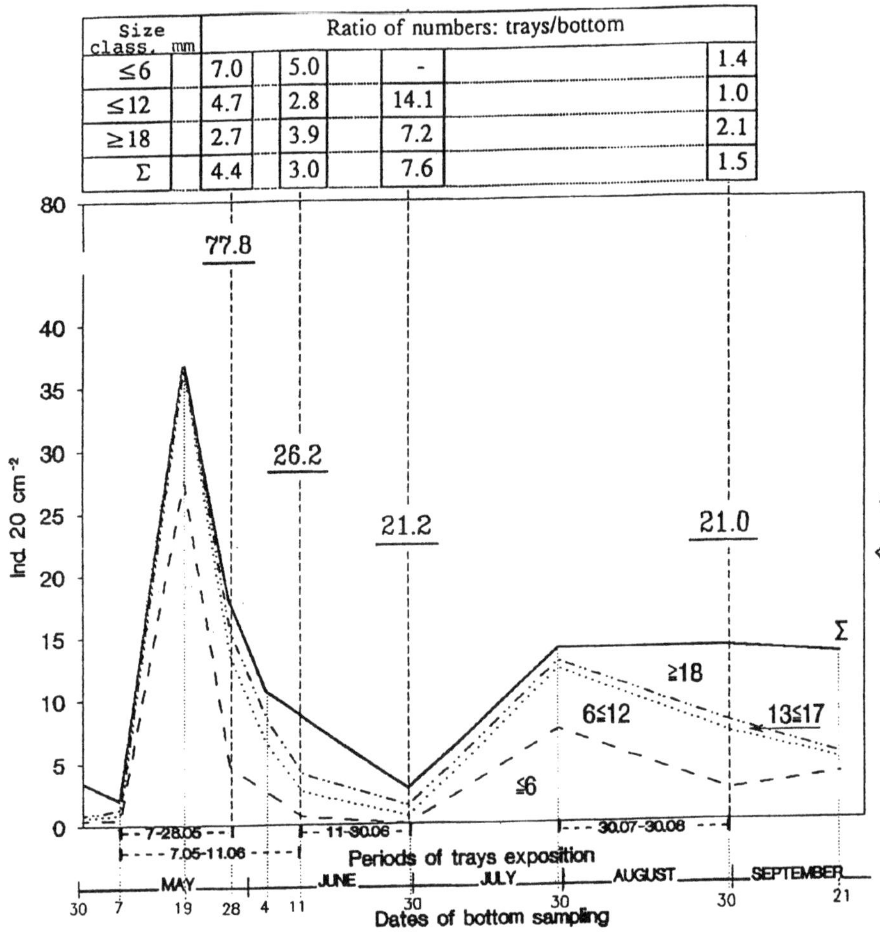

Figure 1. Average for both total numbers and that in size classes, (mm) of *Chironomus* in the bottom (curves) and in trays (fat, underlined figures on vertical dashed lines) at the end of their exposure period. In table over the graph – ratio: numbers of the larvae in trays to those in the bottom. Statistically notsignificant differences between consecutive dates for total numbers and size classes: 30.04–7.05 – N, ≤ 6, ≤ 12. 7–19.05 – ≥ 18. 28.05–4.06 – ≤ 6, ≥ 18. 4–11.06 – N, ≤ 6, ≥ 18. 11–30.06 – ≤ 6. 30.06–30.07 – ≥ 18. 30 07–30.08 – N. 30.08–21.09 – all classes. All other differences are statistically significant.

May (about 15 ind. 20 cm^{-2}) numbers in trays, were only 1.5 times higher (much less than in May) than in the bottom. Absolute numbers in trays in August were similar to that in June (Figure 1). So the lower ratio trays/bottom results from higher numbers of larvae in the bottom.

The smallest size class (≤ 6 mm) was always more or less abundant in trays (1.4–7.0 times more abundant than in the bottom) whilst it was completely absent in the bottom at the end of June (Figures 1 and 2; Table 1 and 2). Besides of higher absolute numbers, the percentage of smallest larvae was usually also higher in trays than in the bottom, excluding August 30., when it was roughly equal (Table 1).

The proportion (%) of oldest class (≥ 18 mm) was lower in the trays than in the bottom only in May; in other periods it was equal, or higher in trays (Table 1).

During August–September the numbers of oldest (≥ 18 mm) class were increasing, and paralelly that of the youngest class (≤ 6 mm) decreasing (Figure 1).

In May, when total *Chironomus* numbers in the raised tray were 4.4 times higher than in the bottom the youngest (≤ 6 mm) size class was as much as 7 times, whilst the oldest one (≥ 18 mm) was only 2.7 times

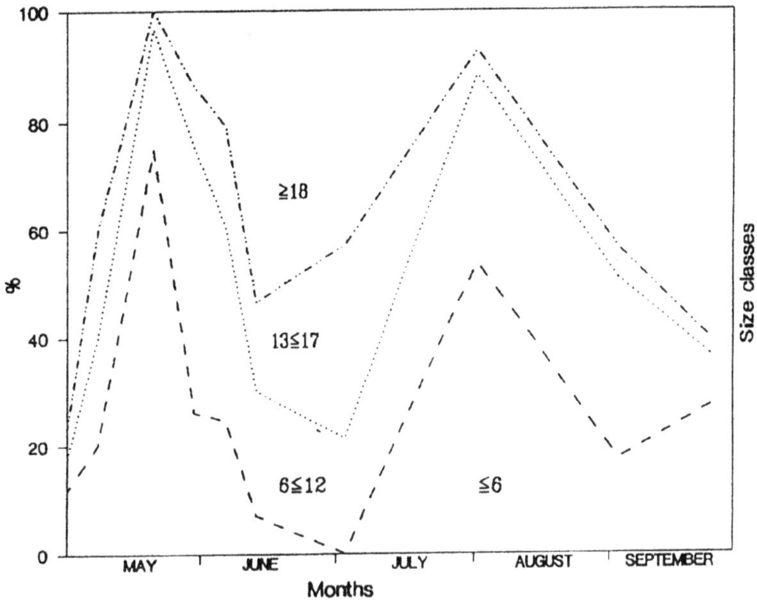

Figure 2. Proportion (%) of *Chironomus* size classes (in mm) in its total numbers in the bottom.

Table 2. Oxygen concentration (mg O_2 dm^{-3}, [%]) and temperature (°C) at the study site during the season.

Date		Depth (m)						
		0	1	2	3	4	5	6
	t	18.7						17.7
30.04	O_2	17.3 [187]						8.5 [90]
	t	20.7	20.7	20.7	20.7	20.2	20.2	20.2
14.05	O_2	14.8 [166]	14.7 [165]	14.7 [165]	14.4 [162]	13.9 [154]	13.9 [154]	13.7 [153]
	t	21.7						20.7
19.05	O_2	12.5 [144]						10.7 [120]
	t	23.7	22.0	21.7	21.2	21.0	20.9	20.9
11.06	O_2	6.7 [80]	5.6 [64]	3.1 [36]	0.8 [9]	0.2 [2]	0.2 [2]	0.2 [2]
	t	18.3	18.0	17.7	17.0	17.0	16.9	16.9
30.06	O_2	12.5 [132]	12.4 [132]	10.9 [116]	8.3 [86]	7.5 [78]	6.6 [69]	1.8 [19]
	t	23.7	22.7	22.7	22.7	22.7	20.7	20.7
6-08	O_2	13.2 [157]	6.0 [70]	3.6 [42]	2.9 [34]	2.6 [30]	2.5 [28]	2.5 [28]
	t	16.7					16.7	16.7
30.08	O_2	9.5 [99]						8.0 [83]

more abundant than in the bottom. The proportion (%) of smallest larvae (class ≤ 6 mm) was 1.6 times higher, whilst that of the oldest one almost twice lower than in the bottom (Table 1).

At the end of June, when the abundance in the bottom was minimal, and total numbers in trays exceeded those in the bottom much more (7.6 times) than in May, the youngest class of larvae was present in trays although in low numbers; it was completely absent in the bottom at that time; young (≤ 12 mm) larvae numbers in trays surpassed that in the bottom by as much as 14.1 times. Both the total numbers and the oldest (≥ 18 mm) class were in trays about 7 times more abundant than in the bottom sediment.

In August the abundance of the youngest class exceeded that in the bottom in the same proportion (about 1.5) as the total numbers of larvae; the oldest class was in trays slightly more abundant than the total numbers (Figure 1; Table 1).

Environmental conditions

Surface temperature fluctuated from 17–19 °C in April, late August and also beginning of July, whereas it varied between 21–24 °C in May, June, and at the beginning of August; it was as a rule 0.5–1.5 °C lower near the bottom (at 5–6 m) than at the surface. Only during and after strong winds it was the same at the surface and near the bottom (Table 2); the usual values of temperature in various years see also Kajak (1990a, 1990b), Kajak & Dusoge (1989).

Oxygen concentration between the surface and the bottom differed much more than temperature (Table 2). A strong vertical oxygen gradient existed usually during calm weather. At these times the oxygen concentration at 1 m depth was about 50% of that just below the surface and dropped strongly towards the bottom.

Oxygen concentration near the bottom is undoubtedly one of the main factors influencing benthos (including *Chironomus*) abundance. In the study area it was high in May, very low in June, and higher (but much lower near the bottom than in surface waters) in June/July and September (Table 2). These oxygen deficits were surprising and unexpected in this place which is shallow, with some water flow, and moderate BOD_5 – on the average 3.7, permanganate COD – 10.9 and dichromate COD – 27.9 mg O_2 l^{-1}; also organic matter in bottom deposits is relatively low (usually about 9% dry weight) (Kajak, 1990a; 1990b; Kajak & Dusoge, 1989). Oxygen depletion was sometimes surprisingly strong; e.g. on June the 11th, and on August the 6th when there was a clear oxygen deficit already at 1–2 m under the surface!

These data of course provide insufficient information. To understand the influence of oxygen on benthos in shallow, highly eutrophic water bodies, we should measure it continuously; this is hardly possible; even frequent measurements (e.g. every few days) would not help much, as the concentration is very changeable. We also have no accurate methods available to measure oxygen (and other substances) concentration in the field at the very surface of the bottom, and these are the concentrations which are important for benthos!

A good indication of oxygen conditions was the colour of the mud. It was yellow in the bottom (due to Fe in oxidized form) down to several cm depth on 19th May, but yellow-brown on 28th May, when a strong decline of *Chironomus* numbers was found. It was again yellow on June the 4th, but uniform only to 2 cm; below this the sediment was only yellow few mm around *Chironomus* tubes.

The yellow colour and the depth it reached was clearly dependent on larval abundance. The yellow colour was always lighter and reached greater depth (to about 5–6 cm; deeper – only few mm along *Chironomus* tubes) in the trays where numbers of larvae were much higher than in the bottom.

Secchi disc visibility was low and stable – at 0.65 m, from April to August.

Permanganate and dichromate COD values in both rivers, at the very top of the reservoir, for the period May–August were within 10.6–22.6 and 33.6–71.7 mg O_2 l^{-1} respectively; the values were slightly lower in spring than in summer. Other factors like N_{tot}, N_{org} NH_4-N, P_{tot} PO_4-P were very changeable, but on the average – similar in particular months of the season. However the flow of water was almost twice greater in May than in the rest of the season (data of the Inst. of Meteorology and Water Management).

To sum up:

– The numbers of *Chironomus* larvae in trays were always higher than in the bottom, although in various degree.

– The highest absolute numbers in trays (almost 80 ind. 20 cm^{-2}, that is 40 000 ind. m^{-2} occurred at the end of May, while the lowest (about 20 ind. cm^{-2}, 10 000 ind. m^{-2}) occurred at the end of June/beginning of July and at the end of August. So the numbers in trays, always higher than in the bottom, were less higher in late summer than in spring and early summer. There was no strong decline of numbers in trays at the end of June/beginning of July, which was observed in the bottom.

– The highest ratio (almost 8.0) of total *Chironomus* numbers in trays to those in the bottom occurred in the period of minimal numbers in the bottom (the end of June), but it was high (more than 4.0) also in May when there were relatively high numbers in the bottom. It was lowest (about 1.5) at the end of August when the bottom population was relatively abundant. An extremely high ratio (14.0 on the average) during minimal numbers in the bottom, at the end of June was observed for young ($\leq$12 mm) larvae.

– The youngest class ($\leq$ 6 mm) in trays was abundant throughout the season; its percentage of total numbers in trays was higher than in the bottom in May and June; in August the percentage was similar (slightly lower, probably due to quicker growth of larvae in trays).

Discussion

Always higher (although in various degree) *Chironomus* numbers in elevated trays than in the bottom must be due to better conditions in trays. The question is – which conditions. Usually one of the first thoughts in case of the bottom fauna is – oxygen concentration. Let us consider this.

The strongest decline of *Chironomus* numbers in the bottom during late May and June, and the lowest *Chironomus* numbers found there at the end of June (Figure 1, Table 1) resulted probably from oxygen deficits registered during June (Table 2). These registered deficits were however the same at the bottom and at the elevated trays level, while *Chironomus* numbers, especially young ones, in trays were much higher than in the bottom. These differences were greater in trays located a bit higher (70 cm) than lower (Figure 1, Table 1).

Higher *Chironomus* numbers (although only twice, not as much as in the elevated trays) were found also in trays located in the bottom; also the youngest larvae, absent in the bottom, were present in bottom trays (Kajak, in press). The mud in these trays could be just 1–2 cm higher than in the nearby bottom. This small distance obviously made the difference for *Chironomus* survival. Of course there could be complete oxygen deficit at the very bottom (while there were traces of oxygen measured just above the bottom and at elevated trays level)(Table 2). The deficit could also last longer at the bottom than at elevated trays level. It is however doubtful the oxygen concentration in the bottom tray (where *Chironomus* numbers were about twice greater than at the bottom) was higher than in the bottom nearby. So probably there is something else, besides the lack of oxygen which matters for *Chironomus* survival. This can be the strongly higher concentration of some toxic substances like H_2S and NH_4^+, at the very bottom, than just above it (e.g. in the bottom tray).

There is not any reason to think there were no oxygen deficits during July–August–September (although they were not found in periodical measurements, Table 2). The conditions for oxygen deficits - the concentrations of O_2 consuming substances, and also of nutrients in main inflowing rivers, as well as water temperature were similar in summer and in spring. Nevertheless the *Chironomus* numbers during this summer period were stable and the youngest larvae were always present.

Larval tubes stretching above the mud surface are thought to be an adaptation to poor oxygen conditions (and thus – the index of oxygen deficit) (Sokolova, 1983; Thienemann, 1954). In this study the tubes were low – protruding to a maximum of about 3 mm, but usually they were shorter – also during registered oxygen deficits in May–June. Perhaps in this study site (contrary to the profundal of lakes, where oxygen concentration declines gradually) the oxygen deficit appeared suddenly, so that the larvae had not enough time to build up the tubes?

In any case, whether the differences between the trays and the bottom were mainly due to oxygen, or also other causes, it is certain that conditions for survival in the trays (which were located not very high over the bottom,at the same registered oxygen conditions), are better than at the very bottom. This is proved by greater total numbers and especially by greater numbers of the youngest ($\leq$6 mm) *Chironomus* larvae in trays.

Apart of the shown differences between *Chironomus* population in the bottom and in trays there were some similarities – namely higher numbers in spring and lower numbers in summer. Numbers in trays during spring were much higher than in the bottom; differences are small in summer (August). It may be that the main cause of high numbers of larvae in spring was high number of eggs which developed in good oxygen conditions. The oxygen concentrations at the bottom measured in May twice (May 14th and 19th) during mass appearance of young larvae were high (although temperature at the bottom was not lower than later in summer) (Table 2). In June oxygen concentration decreased to almost zero.

Anyhow – relatively low numbers of large larvae (Fig. 1) and consequently - of adults (imagos) in early spring – April (because of low numbers of larvae which survived winter – Figure 1) resulted in high abundance of larvae of the spring generation.

Numbers of large larvae (the basis for emerging adults) in early summer (30.06) were not lower than in spring (7.05). So, this could not be the reason of much lower abundance of larvae (especially young ones) in summer (since July) than at their peak abundance in May. Besides, probably many adults emerged before 30 June. Emerging adults, and egg laying females were observed throughout the summer – and as late as 31.08. So, contrary to the spring period, emergence of adults and egg laying in summer was very prolonged! It did not however result in very high larval abundance. The maximal number of larvae in the bottom during summer were much lower than in spring (this was observed also in previous years – Dusoge & Kajak, in preparation; Dusoge et al. 1990 and in press). On the other hand, the numbers in summer were much more stable than in spring (no sharp decline of numbers was observed), and the youngest class was always present and relatively abundant. The oxygen conditions measured three times during July and August were good (the lowest value was 2.5 mg l^{-1}), so perhaps they were not limiting for *Chironomus* larvae (we know however nothing about oxygen between dates of measurements).

The interpretation of the influence of oxygen deficit on benthos in shallow, eutrophic or polytrophic water bodies is much more difficult than in deeper, stagnant ones, where oxygen concentrations have clear regular patterns and are much more stable.

One of the possible causes of relatively low summer *Chironomus* abundance could be strong predation pressure on adults by birds, especially swallows. Swallows were not seen at the reservoir during the *Chironomus* emergence in spring, but masses of them (obviously a new abundant generation) hunting for egg-laying *Chironomus* females were observed in summer (also late – last days of August).

Although the importance of eggs number for the abundance of larvae is often overestimated (as far as eggs number laid is usually very high, and oviposition prolonged) there can be special situations, like in this study, that the number of eggs which can develop may be limiting the abundance of larvae.

The percentage of young larvae (both ≤ 6 and ≤ 12 mm) during the season was usually higher in trays than in the bottom, but not very much; the ratio of this percentage trays/bottom never exceeded 2.0, usually it was about 1.0–1.6 (in August – below one) (Table 1).

The percentage of large larvae (≥ 18 mm) in May was lower in trays than in the bottom. This was probably due to quicker development of larvae in the trays in better oxygen conditions; this resulted in the pupation of some of large larvae which then left the reservoir as adults. Probably a faster growth was also the reason for the smaller percentage of young larvae (≤ 6 and ≤ 12 mm) in trays than in the bottom at the end of August. Despite the small distance (30 cm) between trays in the low and high stands the development of larvae was clearly quicker in high stand (Kajak & Dusoge, 1996).

Generally, locating the substrate a small distance above the bottom (not only as much as in this paper – 40–70 cm, but also just milimeters or few cm above the bottom level – in 'bottom trays'), resulted in better survival and consequently – higher abundance of larvae (Kajak & Dusoge, 1996). The newly hatched small larvae are the most sensitive to an oxygen deficit (Sokolova, 1983; Thienemann, 1954). After reaching some size they can already survive some oxygen deficits. It is therefore very probable that locating some substrate (e.g. old fishing nets) over the bottom, to support good oxygen conditions for development of eggs and hatching larvae, could lead to an increase of larval abundance in the bottom as a result of the immigration of grown up larvae from the raised substrate to the bottom (Kajak & Dusoge, 1996).

As discussed here the essential factors determining *Chironomus* abundance in the environment studied, seem to be oxygen (or may be more generally – environmental conditions at the bottom) and the mortality of females due to swallow predation. It is improbable that the higher numbers of *Chironomus* in trays are due to lower fish exploitation in trays than in the bottom; in fact the opposite scenario is probable (Kajak & Dusoge, 1996). Also immigration of larvae from the bottom to trays (both actively and by resuspension–resedimentation) does not seem to be important (Kajak & Dusoge, 1996). As far as resedimentation is concerned, we would expect more larvae to reach the lower trays whereas in fact numbers are greater in those located higher (Kajak & Dusoge, 1996). There are of course other factors and mechanisms, like competition and food conditions, discussed in other papers (Kajak, 1977, 1988; Lellak, 1965; Rasmussen, 1985; Resh & Rosenberg, 1984; Sokolova, 1983) which influence *Chironomus* abundance.

There is no any reason to think the food conditions are limiting. The guts of larvae were always full (Kajak, unpublished), the transport of nutritious sedimenting tripton, mainly from inflows, is stable.

The negative influence of big larvae on small ones must be taken into account. While peak abundance of youngest larvae, in May, the big, grown up ones were practically absent. On the other hand, they were always abundant during summer. During August–September increasing numbers of oldest (≥ 18 mm) larvae coincided with decreasing numbers of youngest (≤ 6 mm) ones. Big larvae do reduce numbers of youngest ones. This was proved experimentally (Kajak, 1988; Kajak & Dusoge, 1967) at much lower densities of larvae. Perhaps the moderate densities of *Chironomus* both in trays and in the bottom during July–September, much lower than in May, result both from insufficient amount of eggs (due to their exploitation by swallows) and from the mortality of the youngest larvae, caused by big, old ones.

Conclusions

1. The changes in *Chironomus* population numbers were very different in spring and in summer.
2. Low numbers of overwintered population of big larvae resulted in high numbers of new generation in spring. Much higher numbers of big larvae in summer resulted in relatively smaller abundance of young larvae.
3. The spring peak of abundance in the bottom composed almost exclusively of the youngest size class, dropped strongly and rapidly. There was no any peak of abundance in summer; the numbers were stable and moderate; during August–September, all size classes were present.
4. The numbers in trays elevated 40–70 cm over the bottom were always higher than in the bottom, but up to 8 times in spring (even in the tray 1–2 cm over the bottom they were twice higher) and only 1.5 time in summer.
5. The reduction of numbers in spring – very strong in the bottom (14 times) was relatively small (4 times) in trays.
6. The spring decline of numbers resulted from oxygen deficit. However measured oxygen concentration was the same at the bottom and at trays level. Obviously the consequences of oxygen deficit are different in 'islands of mud' – in trays, and in vast areas of the bottom.
7. The survival of the youngest larvae was higher and the development quicker in trays than in the bottom. This was more strongly pronounced in trays located higher. Obviously the conditions for larvae are improving with the increasing distance from the bottom. This can be due to differences in the period of the oxygen deficit and its other chemical consequences in the bottom and above it.
8. Relatively small difference of numbers trays/bottom in summer can be explained by two mechanisms:
   - heavy predation of swallows on egg-laying females of *Chironomus*
   - negative influence of big larvae (absent in spring) on the survival of freshly hatched ones.

Acknowledgments

I am obliged to Dr K. Dusoge for help in field work and cooperation in the preparation of data, MSc D. Rutkowski for diving and other help in field work, Mrs J. Nawotka for sorting samples and help in processing data, Prof. M. Grzybkowska for checking of taxonomic identification of adult *Chironomus plumosus*, two anonymous referees for helpful and constructive criticism, Miss D. Stephen for correcting my English.

References

Dusoge, K., 1989. Distribution and structure of benthos in the lowland Zegrzyński Reservoir. Ekol. pol. 37: 281–298.

Dusoge, K., L. Bownik-Dylińska, J. Ejsmont-Karabin, I. Spodniewska & T. Węgleńska, 1985. Plankton and benthos of man-made Lake Zegrzyńskie. Ekol. pol. 33: 455–479.

Dusoge, K., B. K. Lewandowski & A. Stańczykowska, 1990. Liczebność i biomasa fauny dennej w różnych ś rodowiskach Zbiornika Zegrzyńskiego (Benthos abundance and biomass in various environments of Zegrzyński dam – reservoir). In Kajak, Z. (ed.), Funkcjonowanie ekosystemów wodnych, ich ochrona i rekultywacja. Cz.I. Ekologia zbiorników zaporowych i rzek (Functioning of freshwater ecosystems, their protection and recultivation. I. Ecology of dam – reservoirs and rivers). SGGW AR, Warszawa, 50: 57–85.

Dusoge, K., B. K. Lewandowski & A. Stańczykowska. Benthos of various environments in the Zegrzyński reservoir. Acta Hydrobiol. (in press)

Kajak, Z., 1977. Factors influencing benthos biomass in shallow lake environments. Ekol. pol. 25: 421–429.

Kajak, Z., 1987. Determinants of maximum biomass of benthic Chironomidae (Diptera). Ent. scand. Suppl. 29: 303–308.

Kajak, Z., 1988. Considerations on benthos abundance in freshwaters, its factors and mechanisms. Int. Revue ges. Hydrobiol. 73: 5–19.

Kajak, Z., 1990a. Ecology of Lowland Zegrzyński reservoir near Warsaw. Arch. Hydrobiol. Beih. 33: 841–850.

Kajak, Z., 1990b. Główne cechy ekosystemu Zegrzyńskiego zbiornika zaporowego na Narwi (Main features of the ecosystem of Zegrzyński dam reservoir at Narew river). In Kajak, Z. (ed.), Funkcjonowanie ekosystemów wodnych, ich ochrona i rekultywacja. Cz.I. Ekologia zbiorników zaporowych i rzek (Functioning of freshwater ecosystems, their protection and recultivation. I. Ecology of dam – reservoirs and rivers). SGGW AR, Warszawa, 50: 188–200.

Kajak, Z., 1991. Limnology of lowland shallow impoundment near Warsaw, Poland. Verh. int. Ver. Limnol. 24: 1344–1348.

Kajak, Z. & K. Dusoge, 1967. Influence of artificially increased abundance of *Chironomus plumosus* on the benthos. Bull. Ac. Pol. Sc. Cl. II, 15: 27–33.

Kajak, Z. & K. Dusoge, 1989. Temporal and spatial diversity of trophy-indicators in a lowland dam reservoir. Ekol. pol. 37: 211–233.

Kajak, Z. & K. Dusoge, 1996. Substantial increase of *Chironomus* abundance obtained in the field experiment. Int. Revue ges. Hydrobiol. 81: 469–480.

Konstantinov, A. S., 1958. Biologia chironomid i ich razvedenie (Biology of chironomids and their cultivation). Tr. Sarat. Otd. VNIORCh. 5: 1–362.

Kuklińska, B., 1989. Zoobenthos communities of near-shore zone in the Zegrzyński reservoir. Ekol. pol. 37: 299–318.

Kuklińska, B., 1992. Chironomidae communities of the near-shore zone in Zegrzyński Reservoir, Poland. J. aquat. Ecol. 26: 385–392.

Lellak, J., 1965. The food supply as a factor regulating the population dynamics of bottom animals. Mitt. int. Ver. Limnol. 13: 128–138.

Marian, M. P. & T. J. Pandian, 1985. Interference of *Chironomus* in an open culture system for *Tubifex tubifex*. Aquaculture 44: 249–251.

Rasmussen, J. B., 1985. Effects of density and microdetritus enrichment on the growth of Chironomid larvae in a small pond. Can. J. Fish. aquat. Sci. 42: 1418–1422.

Resh, V. & M. Rosenberg, 1984. The ecology of aquatic insects Praeger Pbl-s, 625 pp.

Sokolova, N. Ju. (ed.), 1983. Motyl *Chironomus plumosus* L. (Diptera: Chironomidae). Izd. Nauka, Moskva, 309 pp.

Soster, F. M. & P. L. McCall, 1990 Benthos response to disturbance in western lake Erie: Field experiments. Can. J. Fish. aquat. Sci. 47: 1970–1985.

Thienemann, A., 1954. *Chironomus*. Leben, Verbreitung und wirtschaftliche Bedeutung der Chironomiden. Binnengewässer, 20: 1–834.

Hydrobiologia **342/343**: 143–150, 1997.
L. Kufel, A. Prejs & J. I. Rybak (eds), Shallow Lakes '95.

Macroinvertebrate communities in relation to submerged vegetation in two *Chara*-dominated lakes

Marcel S. van den Berg[1], Hugo Coops[2], Ruurd Noordhuis[2], John van Schie[2] & Jan Simons[1]
[1]*Free University, Department of Ecology and Ecotoxicology, De Boelelaan 1087, 1081 HV Amsterdam, The Netherlands*
[2]*Institute for Inland Water Management and Waste Water Treatment, P.O. Box 17, 8200 AA Lelystad, The Netherlands*

Key words: macrophytes, macroinvertebrates, *Chara, Potamogeton*, Chironomidae, Gastropoda

Abstract

Relationships between macroinvertebrates and the presence of submerged vegetation were studied in two shallow eutrophic lakes in The Netherlands, Lake Veluwemeer and Lake Wolderwijd. A shift from turbid water with sparse macrophyte cover (*Potamogeton perfoliatus, Potamogeton pectinatus*) towards clear water with a dense cover of submerged vegetation (*Chara* spp.) has been observed in the lakes over the past 10 years. Relatively large *Chara* meadows (300–500 ha) have recently developed in both lakes. The composition of macroinvertebrate fauna was determined at sites varying in cover and dominant vegetation type by sampling sediment and water during 1992 and 1994. Macrophyte biomass, sampling year and vegetation type were the major determinants of macroinvertebrate community composition. *Valvata piscinalis, Bithynia tentaculata, Gammarus tigrinus* and *Chironomus* sp. characterized the sites with high charophyte biomass, whereas *Potamopyrgus antipodarum, Cladotanytarsus* sp., *Stictochironomus* sp. dominated the samples with low charophyte biomass. *Chara* vegetation was different from *Potamogeton* sp. by showing lower densities of the midge larvae *Einfeldia dissidens* and *Cricotopus* gr. *sylvestris*.

Seasonal variations in densities of the dominant mollusc species (*V. piscinalis, P. antipodarum*) were closely related to the development of *Chara* biomass as well as to periphyton cover on charophytes. Thus, changes of the light climate in both lakes, which have led to an increase in colonization by submerged vegetation (particular *Chara* meadows), indirectly had a large impact on macroinvertebrate communities.

Introduction

Macroinvertebrates form an important component of shallow lake ecosystems; they are important as detritus decomposers (McQueen et al., 1986) and constitute a food source for fish (Keast, 1985) and birds (Batzer et al., 1993). Macroinvertebrates occur in the sediment and the water column as well as among submerged plants. The communities living on or between submerged vegetation are characterized by epiphyton grazers (Cyr & Downing, 1988b; Vermaat, 1994). Biomass, species composition and seasonal duration of aquatic vegetation determine the composition and abundance of macroinvertebrates (Lodge, 1985; Rooke, 1986; Kornijów, 1989).

Over the past decades eutrophication has negatively affected the submerged macrophyte vegetation of shallow lakes in the Netherlands. Restoration measures in lakes, such as the reduction of the nutrient load or biomanipulation, often result in an increased cover of macrophyte vegetation. Such a shift to the dominance of aquatic vegetation was reported to have changed the macroinvertebrate composition in Lake Zwemlust, a small lake in the Netherlands where gastropod density increased (Kornijów et al., 1990; Van Donk, 1990). After restoration measures, Lake Wolderwijd (2700 ha) and Lake Veluwemeer (3400 ha) changed from *Potamogeton* dominated to charophyte dominated lakes (Coops & Doef, 1996). The area covered by charophytes increased from absence in 1980 to

c. 400 ha in 1993, whereas *Potamogeton* cover (200 ha) showed no remarkable changes during this period. The related changes of the macroinvertebrate community in Lakes Wolderwijd and Veluwemeer has not been reported so far. The present paper describes macroinvertebrate communities and their relationship to the type of vegetation and its biomass in these lakes.

Materials and methods

Study area

Lake Veluwemeer (3400 ha) and Lake Wolderwijd (2700 ha) are shallow (mean depth 1.5 m), eutrophic lakes situated in the central part of The Netherlands. Both lakes are the remains of the former brackish Zuiderzee estuary. The sediments are sandy and contain marine shells. Submerged vegetation was reduced due to eutrophication in the late 1960's; *Potamogeton pectinatus* L. and *Potamogeton perfoliatus* L. have been present throughout the 1970's and 1980's. In 1979 measures were taken to reduce eutrophication (Hosper & Meijer, 1986), whereafter *Chara* sp. developed strongly (*C. aspera* Deth. ex Willd. in Lake Veluwemeer; *C. contraria* A. Braun ex Kütz. and *C. vulgaris* L. in Lake Wolderwijd). During this development *P. pectinatus* slightly decreased, whereas *P. perfoliatus* showed a slight increase.

Sampling

Macroinvertebrates were sampled in July 1992 and September 1994 at 86 sites in Lake Veluwemeer. A 10 cm diameter plexiglass core was used at sites without *Potamogeton* species. Sediment from each sample was sieved through a 0.5 mm mesh sieve, whereafter macrophytes and macroinvertebrates were collected. At sites with *P. pectinatus* and *P. perfoliatus* a 0.4 × 0.4 m stainless steel box was used. The plants were cut off at the sediment surface, whereafter macroinvertebrates were collected by filtering all the water trough a 0.5 mm mesh sieve. The sediment in each box was sampled using a 0.15 × 0.15 m Birge Ekman grab.

All macroinvertebrates were collected in jars and fixed in 96% alcohol. For each sample, plant species were separated and dry weight biomass (48 h at 105 °C) was determined. Gastropods and gammarids were identified to species level, chironomids at least to genus level and other taxa were identified to higher taxonomical groups.

Between April–October 1994 macroinvertebrates and plant biomass were investigated with two to four week intervals. In each of the lakes, a site in which a dense *Chara* vegetation had developed was compared to a neighbouring site without vegetation. Periphyton coverage on charophytes was estimated by using an image analyzer (Leica, Quantimet 500+). On the images, the heavily covered areas appeared darker than the lightly covered ones. From each lake and sampling date 25 fixed parts (alcohol 70%; 2 cm pieces) of charophytes were scanned. The area of 8 different grey intervals, ranging from white to black, was determined. The grey value with 50% black filled area was estimated by interpolation. This grey value was used as a parameter representing periphyton development. The data were calculated relative to the first date to compensate for the colour differences between macrophyte species.

Numerical analysis

Sampling data from Lake Veluwemeer 1992 and 1994 (35 taxa with a density higher than 150 ind. m^{-2} in one sample) were used in a two-way indicator species cluster analysis, using the program TWINSPAN (Hill, 1979). The abundance of each taxon was used as pseudospecies. Abundances of taxa were cut by eight levels (150, 300, 600, 1200, 2400, 4800, 9600, 19 200 m^{-2}) into nine density groups (1 to 9). Species occurring in lower densities than 150 ind. m^{-2} were deleted from the data set to compensate for the different sample sizes. Standard default parameters were used throughout.

Results

Macroinvertebrate communities in Lake Veluwemeer

The macroinvertebrate community consisted of 37 taxa, with Mollusca, Chironomidae and Gammaridae being the most abundant. Mollusca were characterized by very high densities of the gastropods *Valvata piscinalis* (Müller) (c. 5000 ind. m^{-2}) and *Potamopyrgus antipodarum* (Gray) (c. 2000 ind. m^{-2}), whereas bivalve species were less abundant (<100 ind. m^{-2}). The Chironomidae were dominated by sediment-inhabiting species, such as *Cladotanytarsus* sp., *Chironomus* sp. and *Cryptochironomus* sp. *Gammarus*

Table 1. Sample characteristics of clusters derived after TWINSPAN. Number of samples are given for sampling year and vegetation type; biomass of macrophytes is expressed in g dw m^{-2} with SE between parentheses. Differences between clusters were tested by a Chi-square test for sampling years and plant species; a Kruskal-Wallis test was used for the biomass samples. Significant differences are marked with different symbols (P< 0.05).

	Dominant cluster species				
	Chironomus sp. (n=24)	*Endochironomus albipennis* (n=11)	*Cryptochironomus* sp. (n=13)	*Einfeldia dissidens* (n=21)	*Cladotanytarsus* sp. (n=17)
Biomass (g.m^{-2})	162 (±22)^{A}	225 (±40)^{A}	78 (±10)^{B}	46 (±11)^{B}	3 (±1)^{C}
1992–1994	0/24^{A}	11/0^{B}	2/11^{A}	0/21^{A}	17/0^{B}
Chara-Potamogeton	24/0^{A}	11/0^{AB}	9/4^{BC}	0/21^{D}	17/0^{AB}

tigrinus Sexton (Gammaridae) occurred in almost all samples.

Clustering revealed five groups, separated on the basis of chironomid and gastropod species (Figure 1). At the first level samples with *Potamopyrgus antipodarum, Stictochironomus* sp. and *Cladotanytarsus* sp. were separated from those in which *V. piscinalis* and *Bithynia tentaculata* L. occurred in high densities. The clusters in the *P. antipodarum* dominated group consistently occurred at a lower macrophyte biomass compared to the *V. piscinalis* group (Table 1). The *P. antipodarum* group was subdivided into two clusters; one with high densities of midge larvae *Cladotanytarsus* sp. and the other with high densities of *Stictochironomus* sp. The *Cladotanytarsus* sp. cluster contained a high number of samples taken in 1992. Moreover, the *Cladotanytarsus* sp. cluster contained samples without any vegetation (Table 1). The other first level group characterized by *V. piscinalis*, was divided into two groups. The *Chironomus* sp. group contained mostly samples from 1994 and the *Endochironomus albipennis* (Meigen) group contained mostly samples from 1992. At the lowest cluster level, a subdivision of the *Stictochironomus* sp. group separated the *Potamogeton* from *Chara* samples showing a low biomass (Table 1). Samples taken from *Potamogeton* vegetation were characterized by three taxa: *Einfeldia dissidens* (Walker), *Cricotopus* gr. *sylvestris* (Fabricus) and Sphaeriidae, whereas samples taken from open *Chara* vegetation were characterized by *Cryptochironomus* sp.

Species correlation with charophyte biomass

All dominant taxa were correlated with the biomass of *Chara*, either positively or negatively (Table 2). The gastropods *V. piscinalis* and *B. tentaculata* were positively correlated with *Chara* biomass, whereas *Potamopyrgus antipodarum* was negatively correlated with it. The chironomids, *Cladotanytarsus* sp., *Cryptochironomus* sp. and *Stictochironomus* were negatively related with *Chara* biomass, whereas *E. albipennis, Chironomus* sp., *Psectrocladius* sp. were positively related with *Chara* biomass. The correlations with charophytes of macroinvertebrate taxa found in Lake Veluwemeer were in general identical to those found in Lake Wolderwijd.

Comparison between Lake Wolderwijd and Lake Veluwemeer

The species composition of macroinvertebrate fauna in the two lakes was generally similar (Figure 2). The lakes shared 32 of the 33 taxa which occurred in more than one sample, but most taxa were at least twice as abundant in Lake Wolderwijd as in Lake Veluwemeer. *V. piscinalis* occurred in extremely high densities in Lake Wolderwijd (six times higher than in Lake Veluwemeer). However, *Stictochironomus* sp., *E. albipennis* and *Potamopyrgus antipodarum* occurred in similar densities in both lakes.

In Figure 3 the seasonal pattern of *V. piscinalis* and *Potamopyrgus antipodarum* abundances in 1994 is shown in relation to the development of periphyton and charophytes in Lake Wolderwijd and Lake Veluwemeer. Charophytes reached equal maximum biomasses in the two lakes. In Lake Wolderwijd the charophytes were covered by a periphyton community of seasonally increasing density. The latter was not the case in Lake Veluwemeer, where a periphyton community hardly seemed to develop. In addition, the growing season appeared to be roughly a month shorter in Lake Wolderwijd than in Lake Veluwemeer. In spring and early summer, *V. piscinalis* density inside the vegeta-

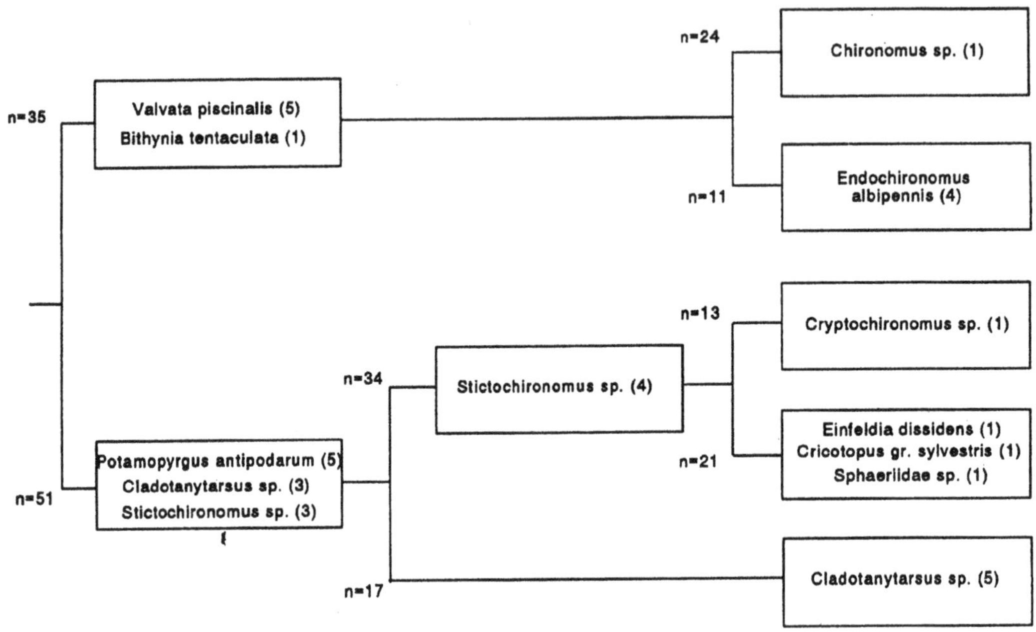

Figure 1. Indicator taxa of clusters derived from TWINSPAN. Numbers after each species represents the density class.

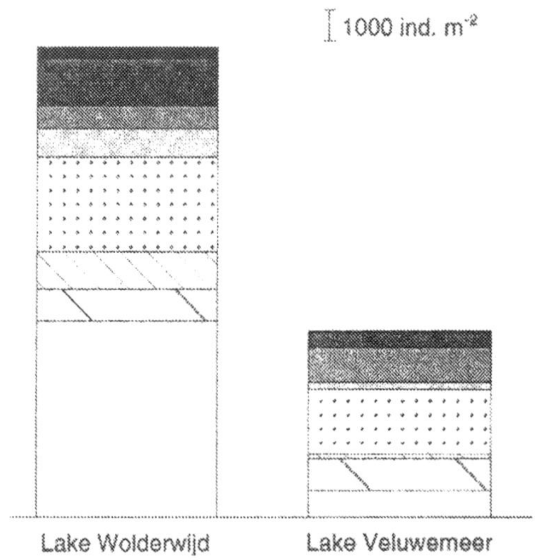

Figure 2. Comparison between Lake Wolderwijd ($n = 46$) and Lake Veluwemeer ($n = 60$) of averaged macroinvertebrate abundances in samples taken inside the Charophyte vegetation in the period August 1992 and May–September 1994. Length of bar indicates the number of individuals per square meter.

tion increased along with charophyte biomass. In Lake Wolderwijd however, peak density was not reached after the onset of the decline of charophyte biomass. The peak density of *V. piscinalis* inside the vegetation contrasted with a low density outside, suggesting active migration. In Lake Veluwemeer, *V. piscinalis* density remained lower and reached its peak a month earlier than in Lake Wolderwijd, more or less coin-

Table 2. Spearman correlation coefficients of the ten most abundant taxa of macroinvertebrates (average density higher than 100 ind. m^{-2}) with charophyte biomass in Lake Veluwemeer.

Taxon	Correlation coefficient
Valvata piscinalis	+ 0.77 ***
Bithynia tentaculata	+ 0.35 **
Potamopyrgus antipodarum	− 0.58 ***
Gammarus tigrinus	+ 0.64 ***
Cladotanytarsus sp.	− 0.72 ***
Chironomus sp.	+ 0.41 **
Cryptochironomus sp.	− 0.66 ***
Endochironomus albipennis	+ 0.50 ***
Stictochironomus sp.	− 0.50 ***
Psectrocladius sp.	+ 0.38 **

n=86, * P<0.05, ** P<0.01, *** P<0.001.

ciding with the maximum biomass of the vegetation. *Potamopyrgus antipodarum* showed a different pattern. In Lake Veluwemeer *P. antipodarum* occurred at very low densities inside the charophyte vegetation, whereas in Lake Wolderwijd densities of *P. antipodarum* were similar in the vegetated and unvegetated locations. In Lake Wolderwijd *P. antipodarum* increased at both locations until October, when the charophytes disappeared completely.

Discussion

Like in many other eutrophic lakes the composition of the macroinvertebrate fauna in Lake Wolderwijd and Lake Veluwemeer is characterized by the dominance of Gastropoda, Chironomidae and Gammaridae. Most of the taxa found are widely distributed and are characteristic of eutrophic lakes. The densities of pelagic organisms may have been underestimated, because they were not sampled effectively by the method used. Since our main goal was to sample the taxa which are directly associated with macrophytes or the sediment, this probably did not affect the results.

The observed communities of macroinvertebrates varied in relation to plant biomass, year of sampling and plant species. Vegetation type exerts a strong effect on both macroinvertebrate composition (Cyr & Downing, 1988a) and density (Hanson, 1990). Differences between plant species may be due to differences in the structure of leaves (Dvorak & Best, 1982; Rooke, 1986) or differences in the coverage by periphyton (Rooke, 1984; Lodge, 1985). Our study suggests that the difference between plant species is of less importance to macroinvertebrates, since no notable differences were found between *Potamogeton perfoliatus* and *Potamogeton pectinatus*. *Chara* vegetation, however, showed different macroinvertebrate densities. A similar distinction between charophyte and rooted plant vegetation was found by Hanson (1990). In our study the greatest differences were found between dense *Chara* vegetation and *Potamogeton* spp., while low-biomass *Chara* vegetation showed an intermediate position. Hence, an effect of plant biomass rather than vegetation type may be involved.

Different chironomid species were found in the two sampling years at identical locations. This is probably due to differences between sampling date in relation to the seasonal development of the larvae: in 1992 sampling took place five weeks earlier than in 1994.

The biomass of charophytes showed a strong correlation with macroinvertebrate species composition and density. Both negative and positive relations were observed. The relationships between the biomass of macrophytes and macroinvertebrates might be explained by considering food preferences of the organisms involved (Dvorak & Best, 1982). *Potamopyrgus antipodarum* prefers deposit feeding, but is also capable of scraping periphyton directly from plants (Haynes & Taylor, 1984). Contradictions in literature exist concerning the preference of *P. antipodarum* for aquatic macrophytes. High densities of *P. antipodarum* in macrophyte beds as compared to densities outside vegetation were found by Dorgelo (1987), whereas Dvorak & Best (1982) showed lower densities on plants. In our study *P. antipodarum* seemed to avoid very dense *Chara* beds in Lake Veluwemeer, while in Lake Wolderwijd no obvious avoidance or attraction was observed. The gastropods species positively related to *Chara* biomass, viz. *Valvata piscinalis* and *Bithynia tentaculata*, scrape periphyton from aquatic plants (Cleland, 1954; Vermaat, 1994) and may benefit from a higher food availability in a dense vegetation of macrophytes. Moreover, higher densities of *V. piscinalis* in Lake Wolderwijd compared to Lake Veluwemeer could be explained by higher food availability for *V. piscinalis* due to the strong periphyton development on charophytes in Lake Wolderwijd. However, the correlation between macroinvertebrates and charophyte biomass may also indicate at an indirect relationship. Macrophytes may act as shelter preventing predation of smaller organisms by fish (Crowder & Cooper, 1982), particular charophytes have been suggested to be a refuge to avoid fish predation (Hanson, 1990).

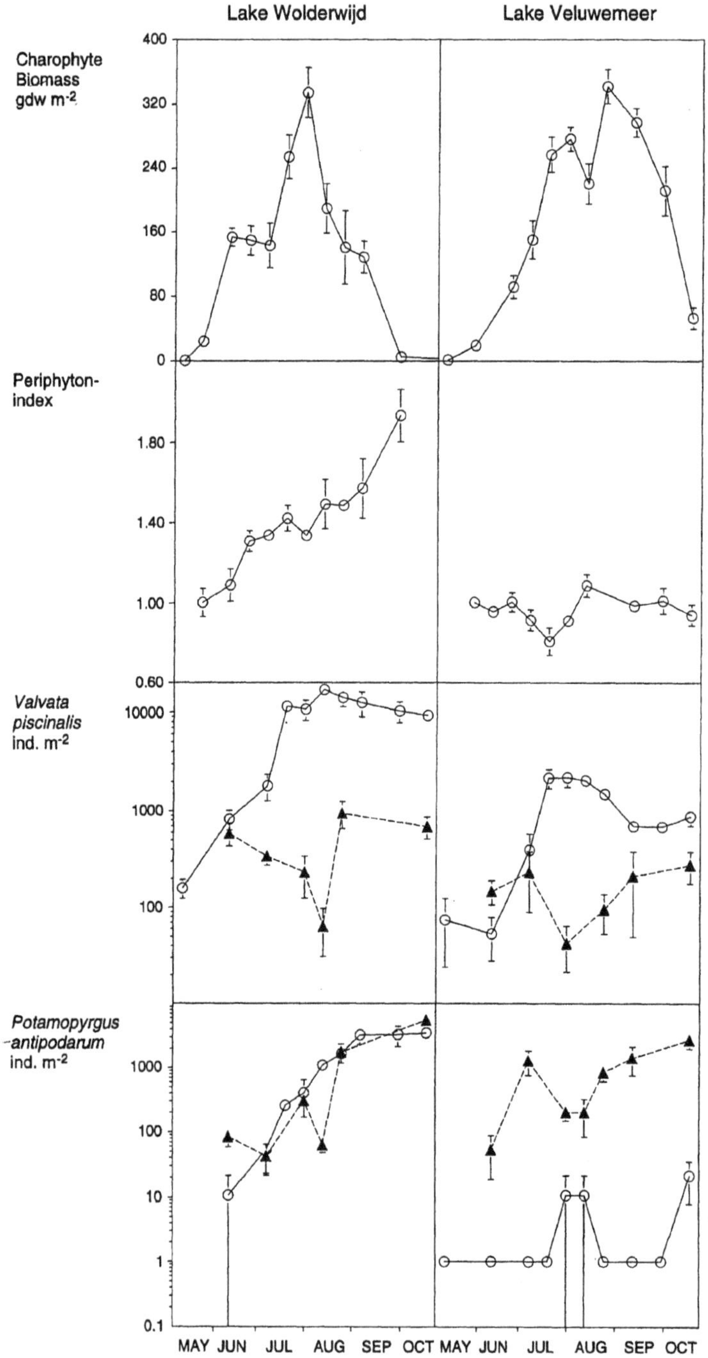

Figure 3. Seasonal course of biomass of charophytes, relative periphyton index, densities of *Valvata piscinalis* and densities of *Potamopyrgus antipodarum* in Lake Wolderwijd (left) and Lake Veluwemeer (right). Black symbols represent samples taken outside vegetation, open circles inside vegetation. Standard errors are shown.

In agreement with Hershey (1985) and Kornijów (1992) *Psectrocladius* sp. and *Endochironomis albipennis*, which occur on the plants themselves, prefer macrophyte vegetation. Sediment-related Chironomidae showed either positive or negative correlation with *Chara* biomass, suggesting sediment differences due to the presence or absence of vegetation. On mineral sediments such in Lake Wolderwijd and Lake Veluwemeer, the oxidized top layer (redox potential c. +50 mV, unpubl. result) is replaced by a reduced sediment (redox potential c. −250 mV) if dense vegetation is present. Oxygen-stress tolerance of involved Chironomidae has been suggested to be related with size and to a lesser extent with hemoglobin content (Heines et al., 1994). In agreement to this, chironomids negatively related to *Chara* (*Cladotanytarsus* sp. and *Stictochironomus* sp.) are of small size and are intolerant for anoxic conditions, whereas *Chironomus* sp., which prefers the sediment covered by charophytes, is of a large size and is tolerant for anoxic conditions.

Apart from the impact of macrophyte biomass on composition of macroinvertebrate fauna, the two lakes showed differences in macroinvertebrate densities. This could be explained by differences in the *Chara*-species composition, growth-season duration or the periphyton development. The charophyte vegetation in Lake Wolderwijd was characterized by the dominance of *Chara contraria*, a dense vegetation cover in June and July, and a strong overgrowth by periphyton. In contrast, the vegetation in Lake Veluwemeer was characterized by the dominance of *Chara aspera*, a longer duration of the growing season (from June until September) and less cover by periphyton. It is not fully understood whether these factors were interrelated, but they seem to affect macroinvertebrate density related with charophytes.

In conclusion, a new habitat for macroinvertebrates has developed due to the establishment of dense *Chara* vegetation. Plant biomass rather than the dominant macrophyte species appears to determine the characteristics of the macroinvertebrate community.

Acknowledgments

We would like to thank the Ministry of Public Works and Transport, Directorate Lake IJsselmeer area for technical and financial support to this study. We acknowledge the assistance by Laura Markus during the research.

References

Batzer, D. P., M. McGee, V. H. Resh & R. R. Smith, 1993. Characteristics of invertebrates consumed by mallards and prey response to wetland flooding schedules. Wetlands 13: 41–49.

Cleland, D. M., 1954. A study of the habits of *Valvata piscinalis* (Müller) and the structure and functions of the alimentary canal and reproductive system. Proc. Malacol. Soc. Lond. 30: 167–203.

Coops, H. & R. W. Doef, 1996. Submerged vegetation development in two shallow, eutrophic lakes. Hydrobiologia 340: 115–120.

Crowder, L. B. & W. E. Cooper, 1982. Habitat structural complexity and interaction between bluegills and their prey. Ecology 63: 1802–1813.

Cyr, H. & J. A. Downing, 1988a. Empirical relationships of phytomacrofaunal abundance to plant biomass and macrophyte bed characteristics. Can. J. Fish. aquat. Sci. 45: 976–984.

Cyr, H. & J. A. Downing, 1988b. The abundance of phytophilous invertebrates on different species of submerged macrophytes. Freshwat. Biol. 20: 365–374.

Dorgelo, J., 1987. Density fluctuations in populations (1982–1986) and biological observations of *Potamopyrgus jenkinsi* in two trophically differing lakes. Hydrobiol. Bull. 21: 95–110.

Dvorak, J. & E. P. H. Best, 1982. Macroinvertebrate communities associated with the macrophytes of Lake Vechten: structural and functional relationships. Hydrobiologia 95: 115–126.

Hanson, J. M., 1990. Macroinvertebrate size distributions of two contrasting freshwater macrophyte communities. Freshwat. Biol. 24: 481–491.

Haynes, A. & B. J. R. Taylor, 1984. Food finding and food preference in *Potamopyrgus jenkinsi* (E. A. Smith) (Gastropoda: Prosobranchia). Arch. Hydrobiol. 100: 479–491.

Heines, F., P. J. Sweerts & E. Loopik, 1994. The micro-environment of chironomid larvae in the littoral and profundal zone of Lake Maarsseveen I, The Netherlands. Arch. Hydrobiol. 130: 53–67.

Hershey, A. E., 1985. Littoral chironomid communities in an arctic Alaskan (USA) lake. Holarct. Ecol. 8: 39–48.

Hill, M. O., 1979. TWINSPAN A FORTRAN program for arranging multivariate data in an ordered two-way table by classification of individuals and attributes. Cornell University Ithaca, New York, 90 pp.

Hosper, S. H. & M. L. Meijer, 1986. Control of phosphorus loading and flushing as restoration methods for Lake Veluwe, The Netherlands. Hydrobiol. Bull. 20: 183–194.

Keast, A., 1985. Planktivory in a littoral-dwelling lake fish association: prey selection and seasonality. Can. J. Zool. 62: 1289–1303.

Kornijów, R., 1989. Seasonal changes in the macrofauna living on submerged plants in two lakes of different trophy. Arch. Hydrobiol. 117: 49–60.

Kornijów, R., 1992. Seasonal migration by larvae on an epiphytic chironomid. Freshwat. Biol. 27: 85–89.

Kornijów, R., R. D. Gulati & E. Van Donk, 1990. Hydrophyte-macroinvertebrate interactions in Zwemlust, a lake undergoing biomanipulation. Hydrobiologia 200/201: 467–474.

Lodge, D. M., 1985. Macrophyte gastropod associations: observations and experiments on macrophyte choice by gastropods. Freshwat. Biol. 15: 695–708.

McQueen, D. J., J. R. Post & E. L. Mills, 1986. Trophic relations in freshwater pelagic ecosystems. Can. J. Fish. aquat. Sci. 43: 1571–1581.

Rooke, B., 1984. The invertebrate fauna of four macrophytes in a lotic system. Freshwat. Biol. 14: 507–513.

Rooke, B., 1986. Macroinvertebrates associated with macrophytes and plastic imitations in the Eramosa River, Ontario, Canada. Arch. Hydrobiol. 106: 307–325.

Van Donk, E., 1990. Necessity for aquatic plant management after lake restoration by biomanipulation. Proc. 8th Symp. Aquatic Weeds, Uppsala 1990: 91–96.

Vermaat, J., 1994. Periphyton removal by freshwater micrograzers. In Van Viersen, W., M. Hootsmans & J. Vermaat (eds), Lake Veluwe, a macrophyte-dominated system under eutrophication stress. Kluwer Academic Publishers, Dordrecht: 213–249.

Hydrobiologia **342/343:** 151–164, 1997.
L. Kufel, A. Prejs & J. I. Rybak (eds), Shallow Lakes '95.

Top-down control in freshwater lakes: the role of nutrient state, submerged macrophytes and water depth

Erik Jeppesen, Jens Peder Jensen, Martin Søndergaard, Torben Lauridsen, Leif Junge Pedersen & Lars Jensen
Dept. of Lake and Estuarine Ecology, National Environmental Research Institute, Vejlsøvej 25, P.O. Box 314, DK-8600 Silkeborg, Denmark.

Key words: top-down control, shallow lakes, trophic structure, trophic cascade, macrophytes, zooplankton, fish, biomanipulation

Abstract

Based on data from 233 Danish lakes, enclosure experiments, full-scale experiments and published empirical models we present evidence that top-down control is more important in shallow lakes than in deep lakes, excepting lakes with a high abundance of submerged macrophytes. The evidence in support is: (1) That at a given epilimnion total phosphorus concentration (TP) the biomass of fish per m^2 is independent of depth, which means that biomass per m^3 is markedly higher in shallow lakes. (2) That the biomass of benthic invertebrates is higher in shallow lakes, which means that the benthi-planktivorous fish are less dependent on zooplankton prey than in deep lakes. By their ability to shift to zooplankton predation their density can remain high even in periods when zooplankton is scarce and they can thereby maintain a potentially high predation pressure on zooplankton. (3) That the possibilities of cladocerans to escape predation by vertical migration are less. (4) That the zooplankton:phytoplankton mass ratio per m^2 is lower and presumably then also the grazing pressure on phytoplankton. (5) That nutrient constraints appear to be weaker, as evidenced by the fact that at a given annual mean TP, summer TP is considerably higher in shallow lakes, especially in eutrophic lakes lacking submerged macrophytes. (6) That negative feedback on cladocerans by cyanobacteria is lower as cyanobacterial dominance is less frequent in shallow lakes and more easily broken (at least in Northern temperate lakes), and (7) That top-down control by benthi-planktivorous fish is markedly reduced in lakes rich in submerged macrophytes because the plants serve as a refuge for pelagic cladocerans and encourage predatory fish at the expense of prey fish. We conclude that manipulation of fish and submerged macrophytes may have substantial impact on lake ecosystems, in particular in shallow eutrophic lakes. On the contrary, if the conditions for more permanent changes in plant abundance or fish community structure are lacking the feed-back mechanisms that endeavour a return to the original turbid state will be particularly strong in shallow lakes.

Introduction

Until recently most lake research concentrated on deep lakes, even in Denmark (Hunding, 1977) in which the majority of lakes are shallow (Jeppesen et al., 1991). During the last ten-twenty years, however, shallow lakes and the littoral zone of deeper lakes have come into focus (e.g. Gulati et al., 1990; Mortensen et al., 1994). This may due to the increasing attention now paid to trophic interactions and top-down versus bottom-up control: Since alteration of the nutrient balance and biological system of shallow lakes often results in marked structural changes, they are particularly suitable for identifying interactions and feedback mechanisms (e.g. Gulati et al., 1990; Mortensen et al., 1994). Moreover, shallow lakes are easier to manipulate than deep lakes. A second reason for the increasing interest in shallow lakes is that lake restoration by means of biological manipulation seems more effective than in deep lakes (Lammens et al., 1990; Jeppesen et al., 1990b; McQueen, 1990). This is partly attributable to the fact that submerged macrophytes are able to colonize relatively large areas in shallow lakes and the potential for obtaining clearwater effects of macrophytes is consequently greater in these lakes (e.g. Scheffer et al., 1993). However, biomanipulation of shallow lakes may have marked effects even without the establishment of submerged macrophytes (Jeppesen et al., 1997). In the present report based on experience and data from Danish lakes and published empirical equations (Hanson & Leggett, 1982; Zdanowski, 1982; Hanson & Peters 1984, Pridmore et al., 1985; Downing et al., 1990), we provide evidence that top-down control is in fact potentially higher in shallow lakes. How this can be exploited in lake management is discussed, special emphasis being placed on the role of fish and submerged macrophytes.

Top-down control: shallow versus deep lakes

Is the importance of fish higher in shallow lakes?

At increasing nutrient levels significant changes in fish abundance, biomass and composition take place in temperate lakes. Fish biomass and density per unit of lake area increase with increasing total phosphorus concentration (TP) (Hanson & Leggett, 1982; Quiros, 1990; Table 1) and the percentage contribution of piscivorous fish decreases (Persson et al., 1988; Jeppesen et al., 1990b, 1994; Figure 1). At the same time, the predation pressure on zooplankton increases as evidenced by a reduction in the zooplankton:phytoplankton biomass ratio (Figure 1), which means that the grazing pressure on phytoplankton most likely decreases. There are several indications suggesting that the top-down effect of these changes is greatest in shallow lakes. Cross analysis of data from 29 lakes covering a wide range of phosphorus levels (TP = 0.008-0.54 mg P l^{-1}) and mean depth (2.4-148 m) revealed that fish biomass per unit area is negatively related to mean depth (Hanson & Leggett, 1982): Fish biomass decreased from 239 to 46 kg ww ha^{-1} when mean depth increased from 1 to 10 m corresponding to a difference as high as 48-fold in mean biomass per unit volume (24 and 0.5 mg ww fish l^{-1}, respectively). The reduction in biomass per unit area may be a coincidence related to a difference in the nutrient state of the lakes included in the analysis, since depth dependence disappears if TP is included in the analysis (Table 1). Downing et al. (1990) also found that the fish biomass could be related to TP and that depth did not contribute significantly to the relationship. The two studies thus

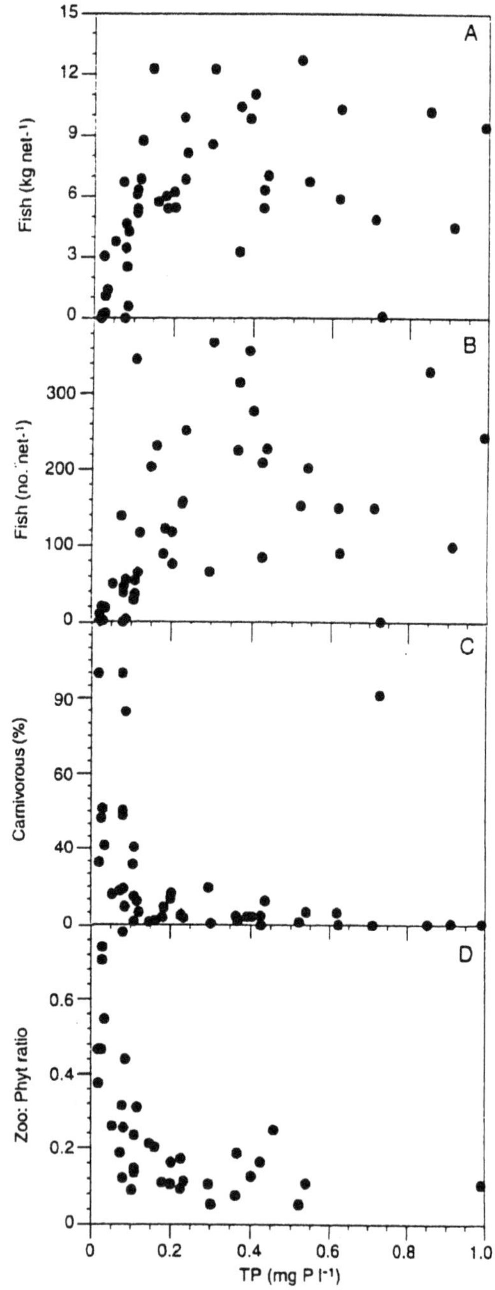

Figure 1. Biomass (A) and density (B) of planktivorous fish and contribution of carnivorous fish (%) (C) to the total number of fish caught in multiple mesh sized gill-net surveys conducted between 15 August and 15 September, and the ratio of zooplankton to phytoplankton biomass (D) versus the concentration of total phosphorus in the epilimnion. Each point represents one lake. The outlier in panel C represents a lake in which TP is high due to former sewage discharge, but where external N and P loading is presently low (from Jeppesen, unpubl. data).

Table 1. Regression equations relating fish biomass (FISHBIO, kg ww ha^{-1}) to total phosphorus concentration (TP, μg P l^{-1}) and mean depth (Z_{mean}, m) (From Hanson & Leggett, 1982).

	Intercept	Log_{10} TP	Log_{10} Z_{mean}	r^2	P[a]
Log_{10} FISHBIO	2.38	–	−0.72	0.27	<0.02
Log_{10} FISHBIO	0.774	0.708	n.s.	0.75	<0.001

[a] *T*-test

show that fish biomass per unit volume at fixed TP reduces proportionately with mean depth. In very shallow lakes the biomass can be considerably higher than the above-mentioned maximum of 24 mg l^{-1}. Thus, in bream and carp dominated lakes, biomass has been reported to be as high as 600-1,000 kg ww ha^{-1} (Meijer et al., 1990; Jeppesen et al., 1990b; Grimm & Backx, 1990), corresponding to 60-100 mg l^{-1}.

Although it is likely, a higher density of fish per unit volume in shallow lakes does not necessarily indicate a higher predation pressure on zooplankton, as the biomass (and then probably also production) of prey zooplankton per unit volume is also higher in shallow lakes. Thus using data covering a broad spectrum of water depth (maximum 1.8-397 m) and TP levels (0.004–0.2 mg P l^{-1}), Hanson & Peters (1984) found that zooplankton biomass increased with increasing TP and decreased with maximum depth (Table 2). At a TP of 0.1 mg P l^{-1}, for example, biomass decreases from 1.23 to 0.86 mg dw l^{-1} (or 30%) when maximum depth increases from 1 to 10 m and accordingly zooplankton biomass per unit area markedly increases with depth. It is a prerequisite for these calculations that the data in Hanson & Peters (1984) investigation represents the entire water column, which cannot be evaluated. A similar slope on mean depth similar to the one of Hanson & Peters on maximum depth has, however, been obtained with survey data from 35 Danish lakes sampled fortnightly during summer for six years in a way that integrates the entire water column (Table 3).

The lower biomass of zooplankton per unit area in shallow lakes may indicate higher fish predation pressure than in deep lakes. Accordingly, the grazing pressure on phytoplankton can be expected to be lower, as in fact seems to be the case. Like zooplankton, chlorophyll *a* in the photic zone of lakes increases with increasing TP and decreases with increasing mean depth. The slope of the depth relationship is higher than was the case with zooplankton, however, the estimate for Danish lakes, for instance, being −0.26 (Table 3). Using this equation we find that at a TP of 0.1 mg P l^{-1} chlorophyll *a* decreases from 46 μg l^{-1} at a mean depth of 1 m to 25 μg l^{-1} at 10 m (54%). Accordingly phytoplankton biomass per unit area only increases 5-fold as compared with a 7-fold increase for zooplankton (Figure 2, Table 3). We obtained a similar slope on chlorophyll *a* using regression analysis of data from Polish lakes published by Zdanowski (1982) (Table 3), while Pridmore et al. (1985) found a somewhat higher slope using data on New Zealand lakes (Table 4). The generally higher slope of the depth relationship for chlorophyll *a* than for zooplankton thus shows that the zooplankton:phytoplankton per unit area ratio – and hence most likely also the grazing pressure on phytoplankton – increases with increasing mean depth (Figure 2). In practice the zooplankton:phytoplankton ratio will most likely increase more than shown in Figure 2, as the calculations imply that chlorophyll a measurements represent the whole water column. This is often not the case in stratified lakes, in which chlorophyll *a* is often lower in the hypolimnion. As samples usually cover the epilimnion only, the increase of chlorophyll-*a* per unit of area towards depth will therefore be less significant, implying a lower zooplankton:phytoplankton ratio.

Several factors may explain a higher fish predation in shallow lakes. Firstly, the possibilities for crustacean zooplankton to employ vertical migration as a means of avoiding predation – a phenomenon known to be important in deep lakes (Ringelberg, 1991; Lampert, 1993) – are probably considerably less in shallow lakes. Secondly, benthiplanktivorous fish such as bream (*Abramis brama* (L.)), roach (*Rutilus rutilus* (L.)), and rudd (*Scardinius erythrophthalmus* (L.)) may rely more on benthic feeding in shallow lakes since benthic invertebrate biomass (Hanson & Peters, 1984) (Table 2, Figure 2) and production (Lindegaard, 1994) at a given TP level are both higher than in deep lakes. Thus plankti-benthivorous fish are less likely to be as sensitive to variations in zooplankton abundance as in deep lakes. Accordingly, their density may remain comparatively high even in periods when zooplankton

Table 2. Regression equations relating zooplankton biomass (ZOOPL, μg dw l^{-1}) and zoobenthos biomass (ZOOBEN, g ww m^{-2}) to total phosphorus concentration (TP, μg l^{-1}), and maximum or mean depth (Z_{max} and Z_{mean}, m). The equations are developed from 38–49 lakes from North Amenca, Europe and Africa (from Hanson & Peters, 1984).

	Intercept	$\log_{10}$ TP	$\log_{10} Z_{max}$	$\log_{10} Z_{mean}$	r^2
$\log_{10}$ ZOOPLB	1.13	0.98	−0.16	–	0.75
$\log_{10}$ ZOOBEN	−0.38	0.65	−0.22	–	0.54
$\log_{10}$ ZOOBEN	+ 0.38	0.65	–	−0.22	0.54

Table 3. Regression equations relating zooplankton biomass (ZOOPLB, mg dw l^{-1}), chlorophyll *a* (CHLA, μg l^{-1}) and Secchi depth (SECCHI, m) to total phosphorus (TP, mg P l^{-1}), total nitrogen (TN, mg N l^{-1}) and mean depth (ZMEAN, m) in Danish lakes (TP-range: 0.017–1.91 mg P l^{-1}, mean depth range: 0.7–16.5 m) and Polish lakes (TP-range 0.020–0.94 mg P l^{-1}, mean depth range: 11–39 m). Also are shown r^2, root mean square error (RMSE) and the number of lakes, n. Both data sets concurrently show that chlorophyll *a* markedly reduces and Secchi depth markedly increases with increasing mean depth.

	Intercept	Log_e TP	Log_e TN	$\text{Log}_e Z_{mean}$	R^2	RMSE	n
Danish lakes							
Log_e ZOOPLB	0.65 ± 0.02****	0.47 ± 0.08****	–	−0.14 ± 0.10$^{n.s.}$	0.59	0.46	35
Log_e ZOOPLB	0.39 ± 0.005****	0.39 ± 0.10***	0.17 ± 0.1$^{n.s.}$	−0.16 ± 0.10$^{n.s.}$	0.60	0.46	35
Log_e CHLA	5.78 ± 0.17****	0.85 ± 0.08****	–	−0.26 ± 0.10**	0.72	0.61	60
Log_e CHLA	5.16 ± 0.26****	0.65 ± 0.10****	0.46 ± 0.16**	−0.26 ± 0.10**	0.74	0.58	59
Log_e SECCHI	−1.23 ± 0.11****	−0.45 ± 0.05****	–	0.42 ± 0.06****	0.76	0.38	59
Log_e SECCHI	−0.83 ± 0.16****	−0.33 ± 0.06****	−0.31 ± 0.10**	0.42 ± 0.06****	0.79	0.35	59
Polish lakes◊							
Log_e CHLA	5.38 ± 0.26****	0.84 ± 0.12****	–	−0.34 ± 0.13*	0.62	0.74	67
Log_e CHLA	4.56 ± 0.44****	0.67 ± 0.14****	0.77 ± 0.33*	−0.25 ± 0.14$^{n.s.}$	0.65	0.71	65
Log_e SECCHI	−1.36 ± 0.18****	−0.41 ± 0.08****	–	0.50 ± 0.08****	0.67	0.48	65
Log_e SECCHI	−0.98 ± 0.30**	−0.33 ± 0.10**	−0.34 ± 0.22$^{n.s.}$	0.46 ± 0.10****	0.66	0.66	64

◊ Data from Zdanowski, 1982
* $p<0.05$, ** $p<0.01$, *** $p<0.001$, **** $p<0.0001$, $^{n.s.}$ = not significant

is scarce; and being able to shift to zooplankton predation, they can maintain a continuously higher capacity to control zooplankton than in deep lakes. Thirdly, as zoobenthos biomass generally decreases and zooplankton biomass per unit area conversely increases with depth, the zooplankton:zoobenthos biomass ratio increases markedly with increasing mean depth (Figure 2). The same seems to apply to production, Lindegaard (1994) having found that zoobenthic production constituted as much as 86% of the total zoobenthic + zooplankton production in two shallow (mean depth 1.9–2.3 m) lakes, but only 47% and 32% in two deep lakes (mean depth 13.5 m and 34.1 m, respectively) with comparable nutrient levels. A shift from benthic to pelagic feeding – for example during times when prey zooplankton density is high – is therefore likely to have a substantially greater impact on zooplankton in shallow lakes. The higher zoobenthos biomass and production in shallow lakes may reflect higher sedimentation mediated by the lower settling distance (Jeppesen et al., 1990a), as well as the lower grazing pressure on phytoplankton due to the higher fish predation. Moreover, the nutritive value of the settled material is probably higher in shallow lakes since the retention time in the water column is lower. An exception may be large shallow lakes in which frequent resuspension enhances retention and decomposition in the water, thereby reducing the availability of settled material for deposit-feeding zoobenthos.

While it can be surmised that fish predation pressure on zooplankton is highest in shallow lakes, experiments involving fish manipulation indicate that if predation pressure is reduced, the capacity of zooplankton to control phytoplankton will be especially high in shallow lakes (Lammens et al., 1990; Jeppesen et al., 1990b, Moss, 1990). One reason may be that

Table 4. Regression equations relating chlorophyll *a* (CHLA, $\mu g\ l^{-1}$) to total phosphorus (TP, $\mu g\ l^{-1}$), total nitrogen (TN, $\mu g\ l^{-1}$), and mean depth (Z_{mean}, m) in 16 New Zealand lakes. TP ranged from 10-153 μg P l^{-1}, mean depth from 2–97 m and chlorophyll *a* from 0.6–54 $\mu g\ ^{-1}$ (from Pridmore et al., 1985).

	Intercept	$\log_{10}$ TP	$\log_{10}$ TN	$\log_{10} Z_{mean}$	r^2
$\log_{10}$ CHLA	−1.26	1.50	–	–	0.73
$\log_{10}$ CHLA	−1.43	1.40	0.11	–	0.73
$\log_{10}$ CHLA	−0.11	1.00	–	−0.50	0.80

large *Daphnia* spp. like fish, use the settled seston and epipelon as alternative food sources in shallow lakes. If that is the case, the zooplankton will be able to maintain a potentially high grazing pressure on the phytoplankton even during periods in which the phytoplankton density is so low that starvation followed by a decline in zooplankton biomass would otherwise have occurred. In support of this view typical zooplankton-phytoplankton predator-prey oscillations do not often occur in nutrient-rich shallow lakes in the absence of fish. Algal biomass remains continuously low during the major part of the summer, while the biomass of large *Daphnia* (*D. magna* Straus, *D. pulex* Baird) – although fluctuating – remains high (Barica, 1975; Fott et al., 1980; Carvalho, 1994; Jeppesen, unpubl. data). In addition, enhanced phytobenthos growth, due to *Daphnia* grazing-mediated improved light conditions, may directly by phytobenthos uptake and indirectly by changing the chemical environment in the sediment reduce the nutrient release from the sediment as observed in laboratory experiments by Hansson (1989) and Jansson (1989). This will further help *Daphnia* in controlling phytoplankton because of the consequent greater nutrient constraints on phytoplankton (increased bottom-up control).

Fish may influence both bottom-up and top-down control via their search for food at the bottom. Benthivorous fish, in particular carp (*Cyprinus carpio* (L.)) and bream, stir up sediment when searching for food; at high fish density the concentration of suspended organic and inorganic matter increases markedly. Thus, in some shallow fish-manipulated lakes (Meijer et al., 1990) and from pond experiments (Breukelaar et al., 1994) with varying densities of benthivorous fish revealed a highly significant linear relationship between the abundance of inorganic suspended matter and carp and bream biomass, Secchi depth thus being considerably lower when the biomass of these fish was high. At reduced transparency the food capture efficiency of visually hunting predators such as perch (*Perca fluviatilis* L.) most likely decreases, thus reducing their capacity to control planktivorous prey fish with a negative feedback on the large-bodied zooplankton grazers, i.e. the daphnids. In addition, a high concentration of suspended matter may negatively influence the growth of *Daphnia* spp. (and thus top-down control of phytoplankton) as their assimilation efficiency is reduced (Arruda et al., 1983; Hart, 1988). Moreover, it may lower their competitiveness compared to rotifers (Kirk & Gilbert, 1990; Kirk 1991), which are less efficient grazers on large phytoplankton (Bogdan & Gilbert, 1984; Arndt, 1993). As consequence, the capacity to control phytoplankton is further reduced.

The impact of benthivorous fish on resuspension is believed to increase with decreasing mean depth, partly because the sediment stirred up by fish is distributed within a smaller water volume, and partly because wind forces in shallow lakes more easily hold resuspended sediment in suspension. In addition, wind-induced resuspension is more likely to occur in shallow lakes. That water depth and benthivorous fish are important is supported by empirical relationships developed for Danish lakes (Windolf et al., 1993): In shallow lakes the concentration of suspended matter increased with increasing TP concentration and lake area (the latter being a substitute for wind impact) and decreases with mean depth and increasing grazing pressure on phytoplankton. Even when these variables were included the bream biomass still made a significant contribution to the relationship. In deep lakes, however, bream and lake area did not make a significant contribution to the variation in suspended matter. Fish-induced resuspension of sediment not only has a negative impact on top-down control forces, it probably also reduces the nutrient constraint on phytoplankton growth as a result of nutrient release (Havens, 1991; Søndergaard et al., 1992). On the other hand, the reduction in light conditions may negatively affect the phytoplankton growth

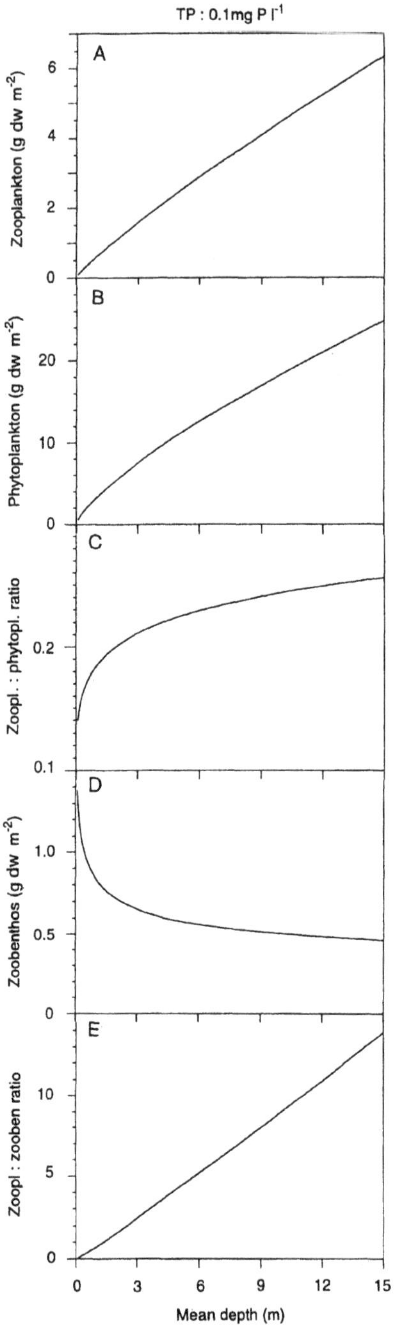

Figure 2. Biomass per unit area of zooplankton (A) and phytoplankton (B), zooplankton:phytoplankton per unit area ratio (C), zoobenthos biomass (D) and the zooplankton:zoobenthos ratio (E) in lakes with an epilimnion total phosphorus concentration of 0.1 mg P l^{-1} and different mean depths. The curves in A and B are based on the equations from Danish lakes shown in Table 3, while that in D is based on the equation by Hanson & Peters (1984), shown in Table 2. The equations used are those including total phosphorus and mean depth.

(Hoyer & Jones, 1983). Accordingly, while high resuspension does not necessarily imply high phytoplankton growth, it always results in lower water transparency.

How important are YOY fish? In order to facilitate the development of methods to combat top-down control by fish it is important to know the relative contribution of small and large fish to predation on zooplankton. In recent years it has been established that young-of-the year (YOY) fish play a far more important structuring role in lake ecosystem than previously believed (Mills & Forney, 1983; Cryer et al., 1986; Mills et al., 1987; Whiteside, 1988; Gliwicz & Pijanowska, 1989, Hewett & Steward, 1989). Some studies indicate that YOY are responsible for the mid-summer decline in zooplankton (Luecke et al., 1990), a phenomenon previously attributed to the increased density of inedible phytoplankton such as cyanobacteria (e.g. Bernardi & Guissani, 1990). The importance of YOY seems, however, to decrease with increasing TP as indicated by the seasonal dynamics in zooplankton grazing pressure on phytoplankton (Figure 3). Thus, in mesotrophic lakes there is a gradual change from a high pre-summer and, in some of the lakes, a high autumn grazing pressure on phytoplankton to a low grazing pressure all year round in the most hypertrophic lakes (Figure 3). The latter probably reflects that due to a high biomass planktivorous fish maintain a high predation pressure on zooplankton resulting in a low grazing pressure on phytoplankton throughout the season irrespective of the number of YOY. The low grazing pressure in hypertrophic lakes cannot be due to lack of edibility or palatibility of the phytoplankton as most of these lakes were dominated by edible green algae (Jeppesen et al., 1990a, 1990b; Jensen et al., 1994). Likewise, the low grazing pressure during mid-summer at lower TP levels cannot be primarily due to negative feedback from cyanobacteria since the decline occurs every year irrespective of whether or not cyanobacteria are present (Jeppesen, unpubl. data). The potential role og zooplankton grazing may be higher than indicated in Figure 3, however, as the zooplankton was sampled during the day; because even though the samples integrate the whole water column, zooplankton seeking refuge at the sediment surface may have escaped sampling (Jeppesen, unpubl. data). Nevertheless, grazing pressure on pelagic phytoplankton seems very low during summer, at least during the day.

Although the question of whether or not YOY fish are particularly important in shallow lakes remains to be clarified, the findings indicate that YOY-fish play

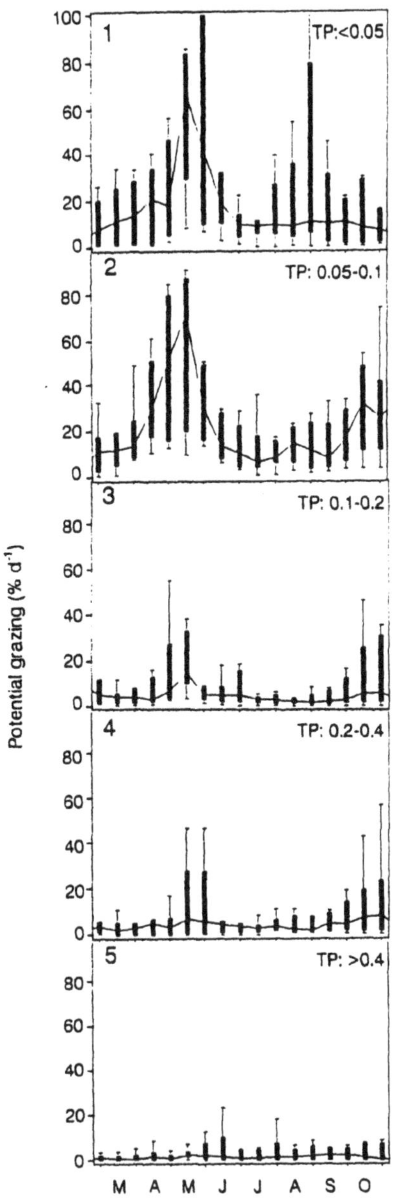

Figure 3. Seasonal variation in the zooplankton potential grazing pressure on phytoplankton (% of the phytoplankton biomass ingested per day) at five different phosphorus categories (mg P l^{-1}). The curve indicates the median, the thick bars the 25–75% percentiles, and the thin bars the 10–90% percentiles. Each phosphorus category includes 6–10 lakes. The potential grazing pressure is calculated assuming that cladocerans and copepods ingest phytoplankton corresponding to 100% and 50%, respectively, of their biomass per day (from Jeppesen, unpubl. data).

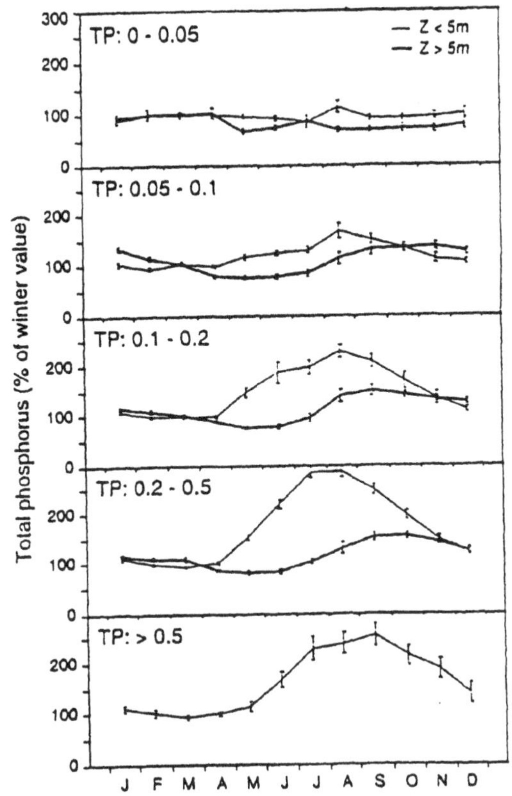

Figure 4. Seasonal variation in epilimnion total phosphorus concentration, expressed in per cent of winter values (monthly mean ± SD) in Danish lakes in different categories of mean annual TP (mg P l^{-1}). The data set is divided into lakes with a mean depth < 5 m (n = 191) and > 5 m (n = 42).

an important role for the overall predation pressure in mesotrophic and slightly eutrophic shallow lakes. In hypertrophic shallow lakes YOY-fish are probably less important because older fish seem to maintain a high predation pressure throughout the year anyway.

Are nutrient constraints less important in shallow lakes? The importance of top-down control is also influenced by the strength of bottom-up factors, and these seem to be weaker in shallow lakes. The shallow lake is typically fully mixed throughout the year, whereas most deep lakes are summer-stratified. This affects nutrient availability for the phytoplankton. Due to nutrient loss to the hypolimnion through sedimentation TP typically decreases in stratified lakes in late spring, thereafter to gradually increase to an autumn peak in connection with the thermocline being forced downwards and eventually broken (Figure 4). In shallow mesotrophic and macrophyte-rich Danish lakes (<0.05 mg P l^{-1}) there is also a tendency towards a slight decrease in TP in late spring followed by a minor peak in August. In the more nutrient-rich shallow lakes, in contrast, there is a marked increase in TP during summer that peaks in August. The summer–winter difference in TP levels increases markedly with increasing nutrient level, TP being 2.5–3-fold greater during summer than winter in hypertrophic lakes. The increasing amplitude may reflect increased internal loading during summer. In a number of the lakes with short residence time, however, it also reflects increased TP concentration in the water inlets because of the reduced groundwater input to inflowing rivers in summer and the resultant increased proportion of sewage in the inlet water (Kristensen et al., 1991). As phosphorus is often the growth-limiting nutrient in lakes, this suggests that nutrient constraint (bottom-up control) is comparatively weak in shallow lakes.

Is the competitive capacity of cyanobacteria lower in shallow lakes? Although fish thus seem to be the main cause of the decrease in zooplankton grazing pressure from mesotrophic to eutrophic lakes, large and in particular filamentous algae can also have a negative impact on grazing capacity and thereby weaken the top-down control of phytoplankton. This is particularly the case with cyanobacteria in freshwater lakes (Bernardi & Giussani, 1990). Thus large cyanobacteria may affect the filtration capacity of zooplankton by interference and be less edible, less nutritious, and sometimes toxic (Bernardi & Guissani, 1990). Cyanobacteria are often dominant during summer in eutrophic shallow lakes (Sas, 1989) but appear to contribute less to biomass at relatively low and high TP in shallow non-stratified lakes than in deep stratified lakes – at least in the Northem temperate region. Thus, analysis of data from 178 Danish lakes (Jeppesen et al., 1990b; Jensen et al., 1994) showed that cyanobacteria dominate at intermediate TP levels (0.1–0.5 mg P l^{-1}), while green algae usually dominate in the most hypertrophic lakes and a mixed community is found at TP levels below 0.1 mg P l^{-1}. Green algal dominance of shallow hypertrophic lakes has also been observed elsewhere (Nygaard, 1949; Pavoni, 1963; Leah et al., 1980). In contrast, the majority of deep lakes are dominated by cyanobacteria at TP levels above some 0.02–0.05 mg P l^{-1} (Sas, 1989). Cyanobacteria are S-strategist and have a low specific settling rate, among other reasons, because of a high buoyancy (Reynolds, 1984). This makes them competitive in stratified systems where loss by sedimentation is critical because, as discussed above, the nutrients do not immediately return to epilimnion. In fully mixed lakes, in contrast, nutrients are rapidly returned from the sediment (Søndergaard et al., 1990a) and TP is often higher during summer (Figure 4), thereby rendering nutrient loss by sedimentation less critical. In addition, due to temporal fluctuations in temperature on a seasonal and diurnal scale as well as wind-induced resuspension nutrient release to the photic zone probably fluctuates more in shallow lakes, thus enabling R-strategists to compete more easily with S-strategists (Sommer, 1985) such as cyanobacteria. It therefore seems reasonable to assume that the overall negative influence which cyanobacteria may have on the top-down control forces is more important in deep lakes.

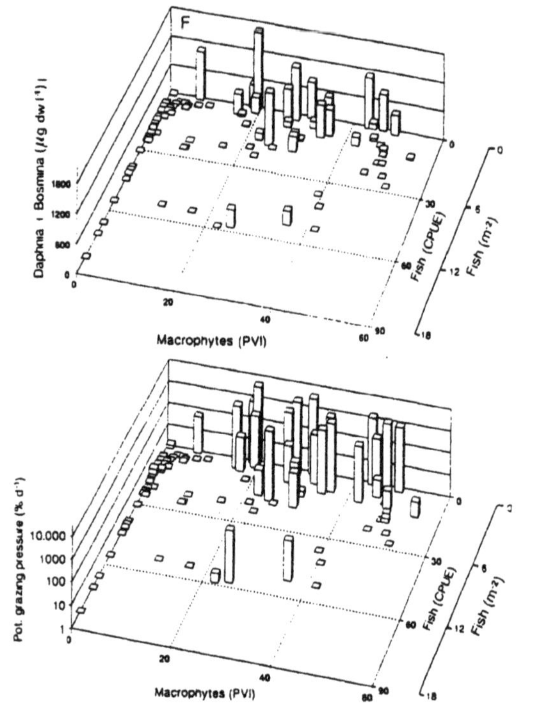

Figure 5. Biomass of the dominant pelagic cladocerans (*Bosmina* + *Daphnia*; upper panel) and their potential grazing pressure on phytoplankton (estimated 24 h ingestion by *Bosmina* + *Daphnia* in % of phytoplankton biomass; lower panel) versus the abundance of 0+ and 1+ roach and three-spined sticklebacks (CPUE in traps) and macrophyte plant volume ingested, PVI (%) in Lake Stigsholm enclosure experiments involving manipulation of plants (mainly *Potamogeton* species) and fish density (modified from Schriver et al., 1995).

How is the impact of submerged macrophytes on top-down control? Several studies have shown that submerged macrophytes have a stabilizing effect on lake ecosystems and, if abundant, contribute to maintaining a clearwater state even at high nutrient concentrations (Canfield et al., 1984; Jeppesen et al., 1990b; Scheffer et al., 1993; Moss et al., 1994; Meijer et al., 1994). The reasons for this are numerous, as discussed by e.g. Moss (1990), Scheffer (1990), Jeppesen et al. (1991) and Scheffer et al. (1993). We will concentrate on macrophytes as a refuge for pelagic cladocerans as submerged macrophytes protect pelagic and plant-associated cladocerans from fish predation (Timms & Moss, 1984), thereby helping to maintain a higher grazing pressure on phytoplankton produced in the plant-covered areas or which have drifted in from the pelagic with the water current (Schriver et al., 1995). The efficacy of the macrophyte as a refuge for pelagic cladocerans depends on plant density (Crowder & Cooper, 1979; Winfield, 1986; Diehl, 1988) and type (Irvine et al., 1990), as well as on the density and composition of the planktivorous fish present in the vegetation (Persson, 1991). For example, Schriver et al. (1995) found a poor refuge effect of macrophytes even at low densities of 0+ and 1+ planktivorous fish (three-spined sticklebacks (*Gasterosteus aculeatus* (L.)) and roach if the plant volume infested (PVI) (Canfield et al., 1984) of filamentous *Potamogeton* species was lower than 10-15%, but a high refuge effect at higher PVI provided that fish density was low (Figure 5). In the latter case the estimated grazing pressure on phytoplankton was extremely high, phytoplankton biomass consequently being low and dominated by grazing-tolerant forms. If fish density was higher than about 4 fish per m^2 the grazing pressure on phytoplankton again fell drastically to values so low that the zooplankton could no longer control the phytoplankton. Although the experiments were based on calculated grazing pressure (see legend to Figure 5), subsequent in vitro grazing experiments with ^{14}C-labelled phytoplankton have confirmed the results (Jeppesen, unpubl. data).

A factor that contributes to the stabilization is that pelagic cladocerans primarily use the macrophytes as a daytime refuge against predation from fish. During night they, migrate into the open water where they contribute to control of pelagic phytoplankton (e.g. Timms & Moss, 1984). Macrophyte refuges are therefore comparable to vertical migration and compensate for the poor possibilities for cladocerans for using the vertical alternative in shallow lakes. This daytime migration to the plant beds takes place despite the fact that

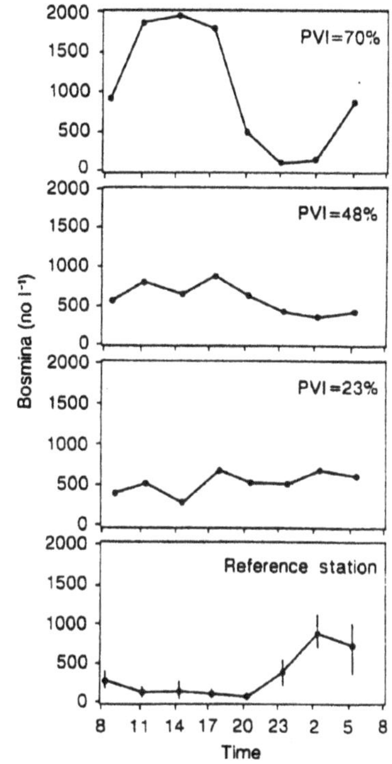

Figure 6. Diel variation in the density of *Bosmina* spp. (mainly *B. longirostris*) in exclosures with various macrophyte densities (PVI) and at a reference station (including SD, $n = 3$) located in a plant-free part of Lake Stigsholm. Average density and diel amplitude (high during day and low during night) are highest at high PVI. At the reference station, in contrast, the highest density occurred at night (Pedersen, unpubl. data).

submerged macrophytes may have a repellent effect on pelagic cladocerans which may be attributable to chemical cues (Hasler & Jones, 1949; Pennak, 1966, 1973; Quade, 1969). Regarding zooplankton this seems reasonable as the open water among the plants in dense macrophyte beds is frequently poor in phytoplankton and microorganisms due to high grazing pressure by plant-associated cladocerans, etc. (Jeppesen et al., unpubl. data). Pelagic cladocerans are also influenced by chemical cues from planktivorous fish (Dodson, 1988; De Meester, 1993; Loose et al., 1993; Lauridsen & Lodge, 1996), however, and the net result of these two contradictory cues is that migration to macrophyte beds occurs during the day – at least in shallow eutrophic lakes with high fish densities (Lauridsen et al., 1996; Lauridsen & Buenk, 1996). Aggregation of pelagic cladocerans seems to be especially important in the case of small macrophyte beds (Lauridsen et al., 1996) since pelagic cladocerans mainly appear in

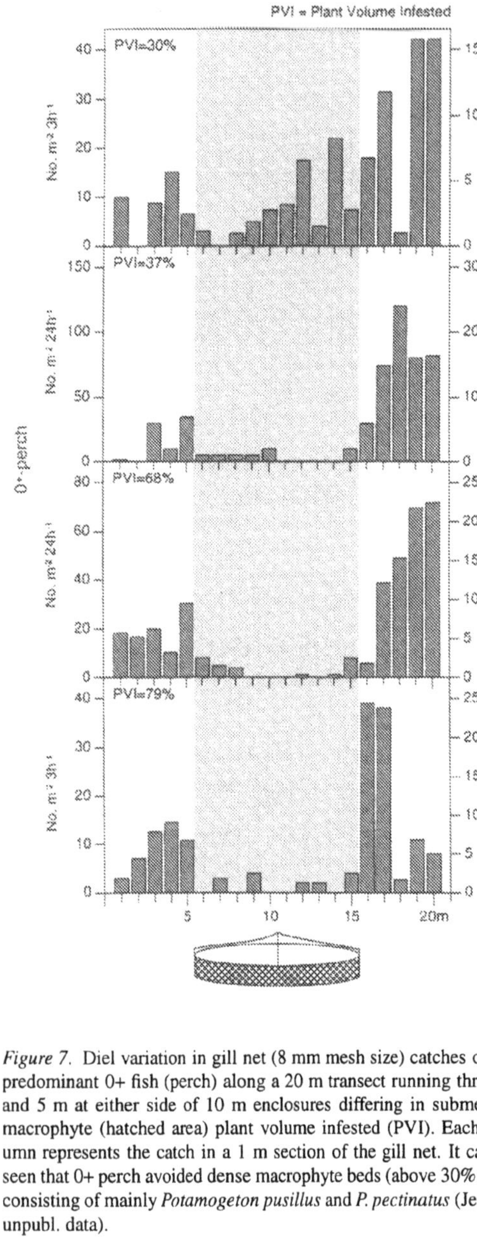

Figure 7. Diel variation in gill net (8 mm mesh size) catches of the predominant 0+ fish (perch) along a 20 m transect running through and 5 m at either side of 10 m enclosures differing in submerged macrophyte (hatched area) plant volume infested (PVI). Each column represents the catch in a 1 m section of the gill net. It can be seen that 0+ perch avoided dense macrophyte beds (above 30% PVI) consisting of mainly *Potamogeton pusillus* and *P. pectinatus* (Jensen, unpubl. data).

the transitional zone between the macrophyte belt and open water (Lauridsen & Buenk, 1996). Aggregation also appears to be especially high in dense macrophyte beds (Figure 6 and Stansfield et al., 1997), partly because the zooplanktivorous fish tend to avoid these areas (Engel, 1988; Phillips et al., 1995; Figure 7), and partly because they provide better protection against fish predation (Winfield, 1986; Engel, 1988; Schriver et al., 1995; Figure 5).

Apart from influencing the phytoplankton in the macrophyte beds during the day, night-time migration to the open water may also have a significant effect on the pelagic system. On the basis of migration data from densely vegetated 2 m beds Lauridsen et al. (1996) calculated that only 3% coverage of lake area was needed for an increase of the average night-time concentration of cladocerans 2-fold in the entire water column of shallow lake Stigsholm (mean depth: 1 m; surface area: 21 ha). Establishment of even small numbers of small, dense macrophyte beds may therefore markedly enhance the capacity of the pelagic zooplankton to control phytoplankton in the open water. This is further strengthened by the fact that submerged macrophytes favour piscivorous fish such as perch and pike (*Esox lucius* L.) at the expense of planktivorous fish such as roach and bream (Winfield, 1986; Grimm & Backx, 1990) and thereby indirectly the large pelagic cladocerans and grazing pressure on phytoplankton. Submerged macrophytes therefore compensate for the overall higher predation risk in shallow lakes.

Management implications (biomanipulation)

Fish and submerged macrophytes seem to play an important role in shallow lakes, and lake manipulation of these two compartments may thus have substantial impact on the ecosystem. Conversely, it also implies that if the conditions for more permanent changes in plant density or fish stock are not present the feed-back mechanisms that endeavour to return lake systems to their original turbid state will be particularly strong in shallow lakes. It is important to bear this in mind when undertaking biomanipulation in shallow lakes.

Several management-oriented biological restoration methods have been developed (e.g. Moss, 1990). Selective removal of planktivorous fish is one of the methods that may enhance top-down control of phytoplankton, and several experiments have been conducted with varying degrees of success (e.g. Shapiro & Wright, 1984; Benndorf, 1987; Gulati et al., 1990; De Melo et al., 1992; Reynolds, 1994; Meijer et al., 1994). An alternative or supplementary method is stocking of predatory fish, e.g. pikeperch (Benndorf et al., 1988), perch (Riemann et al., 1990) or YOY pike (Prejs et al., 1994). In Denmark we have experimented with the stocking of large numbers of YOY pike, the idea being that they would forage on newly hatched roach and bream since the YOY of these two species seem to maintain a high predation pressure on pelagic zooplankton during summer (Figure 2). The Danish studies (Berg et al., 1997) and those of a Polish lake (Prejs et al., 1994) have shown that at high density YOY pike may have a highly significant negative impact on the density of YOY prey fish, and that this cascades to the lower trophic levels resulting in a shift to large-bodied zooplankton, a decrease in chlorophyll *a* and TP and an increased Secchi depth (Søndergaard et al., 1997). Due to cannibalism among pike during winter the long-term effect is poor, however, although repeated stocking of pike can be used during a transitional period following nutrient loading reduction in order to maintain a high transparency and thereby stimulate growth of submerged macrophytes (Søndergaard et al., 1997).

Since the first experiments involving fish manipulation in Denmark were initiated in 1986 (Riemann et al., 1990; Søndergaard et al., 1990b; Jeppesen et al., 1990a, 1990b), 14 additional experiment have been undertaken. The results indicate that the method will have a long-term effect in shallow lakes if nutrient loading has been reduced to levels so low that the TP concentration in the future state of equilibrium will be below 0.05-0.1 mg P l^{-1} (Jeppesen & Søndergaard, unpubl. data) which is somewhat lower than the 0.08-0.15 mg P l^{-1} given by Jeppesen et al. (1991). However, if nitrogen loading is low a positive result may also be obtained at higher phosphorus concentrations. Although it has been claimed that fish manipulation is most likely to be successful in lakes with a surface area <0.04 km^2 (Reynolds, 1994), this is not in accordance with our experience. Thus a marked trophic cascade has taken place in lake Arreskov (3 km^2) after natural fish kill, resulting in a shift to a clearwater stage (Fyns Amtskommune, 1995), and following fish manipulation of lake Christina (16 km^2) (Hanson & Butler, 1994) and lake Finjasjön (11 km^2) (Annedotter et al., personal comments) the lakes became clear with extensive growth of submerged macrophytes. Also cyanobacteria have been reported to render it difficult to initiate a shift to the clearwater state as they negatively influence the zooplankton in various ways (Bernardi & Guissani, 1990). Again, however, this is not in accordance with our experience (e.g. Søndergaard et al., 1990b; Jeppesen et al., 1997). Further evidence is given by cross-analyses of survey data from Danish lakes showing that low grazing pressure on phytoplankton during summer in eutrophic lakes rather reflects a high fish predation pressure on zooplankton (especially by YOY) than it is a result of negative cyanobacterial feedback (Jeppesen, unpubl. data). Fish manipulation methods still need to be further developed, but will undoubtedly become an important lake restoration tool, especially in the case of shallow lakes as a follow-up to a reduction in external nutrient loading.

Plant refuges, protecting submerged macrophytes against waterfowl grazing may be an alternative or supplementary method to increase the top-down control of phytoplankton, especially in cases where grazing by waterfowl, such as coot and mute swans, delays the establishment of submerged macrophytes such as has frequently been observed at low plant densities (Lauridsen et al., 1993, 1994; Van Donk et al., 1994; Søndergaard et al., 1996, but see Perrow et al., 1997). The refuges serve two purposes: They enable the plants to grow in a predation-free environment from where they can spread vegetatively or via seeds from to the remaining part of the lake, and they increase the day-time refuge possibilities for pelagic cladocerans. As described above, the migration intensity of pelagic cladocerans is probably greatest in small, dense macrophyte beds. The refuges must therefore be adapted accordingly. They can be simple, consisting of wooden poles covered by open mesh wire fencing, etc.. The use of plant refuges as a restoration tool is probably limited to the same nutrient interval as fish manipulation because it is in this nutrient interval that colonization of submerged macrophytes is most likely to succeed aeppesen et al., 1990b, 1991). We predict that the method will be most efficient in small lakes, because coot density is relatively higher in these lakes (Brøgger)(Jensen & Jørgensen, 1992; Søndergaard et al., 1996).

Acknowledgements

The assistance of the technical staff of the National Environmental Research Institute, Silkeborg, is gratefully acknowledged. Field and laboratory assistance was provided by L. Hansen, J. Stougaard, B. Laustsen, J. Glargaard, K. Jensen, and L. Nørgaard. Technical assistance was provided by K. Møgelvang, A. M. Poulsen and A. D. Matharu. D. I. Barry provided valuable editorial comments. The study was supported by the Centre for Freshwater Environmental Research.

References

Arndt, H., 1993. Rotifers as predators on components of the microbial web (bacteria, heterotrophic flagellates, ciliates) – a review. Hydrobiologia 255/256: 231–246.

Arruda, J.,A., G. R. Marzolf & R. T. Faulk, 1983. The role of suspended sediments in the nutrition of zooplankton in turbid reservoirs. Ecology 65: 1225–1235.

Barica, J., 1975. Collapses of algal blooms in prairie pothole lakes: Their mechanisms and ecological impact. Verh. int. Ver. Limnol. 19: 606–615.

Benndorf, J., 1987. Food web manipulation without nutrient control: A useful strategy in lake restoration? Schweiz. Z. Hydrol. 49: 237–248.

Benndorf, J., H. Schultz, A. Benndorf, R. Unger, E. Penze, H. Kneschke, K. Kossatrz, R. Dumke, U. Hornic, R. Kruspe & S. Reichels, 1988. Food-web manipulation by enhancement of piscivorous fish stocks: Long-term effects in the hypertrophic Bautzen Reservoir. Limnologica 19: 97–110.

Berg, S., E. Jeppesen & M. Søndergaard, 1997. Pike (*Esox lucius* L.) stocking as a biomanipulation tool. 1. Effects on the fish population in Lake Lyng (Denmark). Hydrobiologia 342/343: 311–318.

Bernardi, R. de & G. Guisanni, 1990. Are blue-green algae suitable food for zooplankton? An overview. Hydrobiologia 200/201: 29–41.

Bogdan, K. G. & J. J. Gilbert, 1984. Body size and food size in freshwater zooplankton. Proc. natn. Acad. Sci. U.S.A. 81: 6427–6431.

Breukelaar, A. W., E. H. R. Lammens, J. P. G. Klein Breteler & I. Tatrai, 1994. Effects of benthivorous bream (*Abramis brama* L.) and carp (*Cyprinus caprio* L.) on sediment resuspension and concentration of nutrients and chlorophyll *a*. Freshwat. Biol. 32: 113–121.

Brøgger-Jensen, S. & H. E. Jørgensen, 1992. Vandfugle og søers miljøtilstand [The environmental state of waterfowl and lakes]. Miljøprojekt nr. 200. Danish Environmental Protecton Agency, Copenhagen, 64 pp.

Canfield, D E., J. V. Shireman, D. E. Colle, W. T. Haller, C. E. Watkins, & M. J. Maceina, 1984. Prediction of chlorophyil *a* concentrations in Florida lakes: Importance of aquatic macrophytes. Can. J. Fish. aquat. Sci. 44: 497–501.

Carvalho, L., 1990. Top-down control of phytoplankton in a shallow hypertrophic lake: Little Mere (England). Hydrobiologia 200/201: 53–64.

Crowder, L. B. & W. E. Cooper, 1979. Structural complexity and fish-prey interactions in ponds: A point of view. In Johnson, D. L. & R. A. Stein (eds), Response of fish to habitat structure in standing water. North Central Division. Am. Fish. Soc. Spec. Pub. 6: 1–10.

Cryer, M., G. Pierson & C. R. Townsend, 1986. Reciprocal interactions between roach *Rutilus rutilus*, and zooplankton in a small lake: Prey dynamics and fish growth and recruitment. Limnol. Oceanogr. 31: 1022–1038.

De Meester, L., 1993. Genotype, fish-mediated chemicals, and planktonic behaviour in *Daphnia magna*. Ecology 74: 1467–1474.

De Melo, R., R. France & D. J. McQueen, 1992. Biomanipulation: Hit or myth? Limnol. Oceanogr. 37: 192–207.

Diehl, S., 1988. Foraging efficiency of three freshwater fishes: Effects of structural complexity and light. Oikos 53: 207–214.

Dodson, S., 1988. The ecological role of chemical stimuli for the zooplankton: the predator-avoidance behaviour in *Daphnia*. Limnol. Oceanogr. 33: 1431–1439.

Downing, J. A., C. Plante & S. Lalonde, 1990. Fish production correlated with primary productivity and the morphoedaphic index. Can. J. Fish. aquat. Sci. 47: 1929–1936.

Engel, S., 1988. The role and interactions of submersed macrophytes in a shallow Wisconsin lake. J. Freshwat. Ecol. 4: 329–340.

Fott., J., L. Pechar & M. Prazakowa, 1980. Fish as a factor controlling water quality in ponds. In Barica, J. & L. R. Mur (eds), Hypertrophic Ecosystems, Developments in Hydrobiology 2. Dr W. Junk Publishers, The Hague: 255–261.

Fyns Amtskommune, 1995. Vandmiljøovervågning - Arreskov Sø 1994 [Environmental Survey - Lake Arreskov Sø 1994]. Fyns Amt, 123 pp.

Gliwicz, Z. M. & J. Pijanowska, 1989. The role of predation in zooplankton succession. In Sommer, U. (ed.), Plankton Ecology. Springer Verlag, London: 253–296.

Grimm, M. P. & J. J. G. M. Backx, 1990. The restoration of shallow eutrophic lakes and the role of northern pike, aquatic vegetation and nutrient concentration. Hydrobiologia 200/201: 557–566.

Gulati, R. D., E. H. R. R. Lammens, M.-L. Meijer & E. van Donk, 1990. Biomanipulation, tool for water management. Hydrobiologia, 200/201: 1–628.

Hanson, J. M. & M. G. Butler, 1994. Responses of plankton, turbidity and macrophytes to biomanipulation in a shallow prairie lakes. Can. J. Fish. aquat. Sci. 51: 1180–1188.

Hanson, J. M. & W. C. Leggett, 1982. Empirical prediction of fish biomass and weight. Can. J. Fish. aquat. Sci. 39: 257–263.

Hanson, J. M. & R. H. Peters, 1984. Empirical prediction of crustacean zooplankton biomass and profundal macrobenthos biomass in lakes. Can. J. Fish. aquat. Sci. 41: 439–445.

Hansson, L. A., 1989. The influence of a periphytic biolayer on phosphorus exchange between substrate and water. Arch. Hydrobiol. 115: 21–26.

Hart, P. C., 1988. Zooplankton feeding rates in relation to suspended sediment content: Potential influences on community structure in a turbid reservoir. Freshwat. Biol. 19: 123–139.

Hasler, A. & F. Jones, 1949. Demonstration of the antagonistic action of large aquatic plants on algae and rotifers. Ecology 30: 359–364.

Havens, K. E., 1991. Fish-induced sediment resuspension: Effects on phytoplankton biomass and community structure in a shallow hypereutrophic lake. J. Plankton. Res. 13: 1163–1176.

Hewett, S. W. & D. J. Stewart, 1989. Zooplanktivory by alewives in Lake Michigan: Ontogenetic, seasonal and historical patterns. Trans. am. Fish. Soc. 118: 581–596.

Hoyer, M. V. & J. R. Jones, 1983. Factors affecting the relation between phosphorus and chlorophyll *a* in Midwestern reservoirs. Can. J. Fish. Aquat. Sci. 40: 192–199.

Hunding, C., (ed.) 1977. Danish Limnolgy. Reviews and Perspectives. Folia limnol. scand. 17, 136 pp.

Irvine, K., B. Moss & J. Stansfield, 1990. The potential of artificial refugia for maintaining a community of large-bodied cladocera against fish predation in a shallow eutrophic lake. Hydrobiologia 200/201: 379–389.

Jansson, M., 1989. Role of benthic algae in transport of nitrogen from sediment to lake water in a shallow clearwater lake. Arch. Hydrobiol. 89: 101–109.

Jensen, J. P., E. Jeppesen, K. Olrik & P. Kristensen, 1994. Impact of nutrients and physical factors on the shift from cyanobacterial to chlorococcal green algal dominance in shallow Danish lakes Can. J. Fish. aquat. Sci. 51: 1692–1699.

Jeppesen, E., J. P. Jensen, P. Kristensen, M. Søndergaard, E. Mortensen, O. Sortkjaer & K. Olrik, 1990b. Fish manipulation as a lake restoration tool in shallow, eutrophic, temperate lakes 2: Threshold levels, long-term stability and conclusions. Hydrobiologia 200/201: 219–227.

Jeppesen, E., P. Kristensen, J. P. Jensen, M. Søndergaard, E. Mortensen & T. Lauridsen, 1991. Recovery resilience following a reduction in external phosphorus loading of shallow, eutrophic Danish lakes: Duration, regulating factors and methods for overcoming resilience. Mem Ist. ital. Idrobiol. 48: 127–148.

Jeppesen, E. M. Søndergaard, E. Kanstrup, B. Petersen, R. B. Henriksen, M. Hammershøj, E. Mortensen, J.P. Jensen & A. Have, 1994. Does the impact of nutrients on the biological structure and function of brackish and freshwater lakes differ? Hydrobiologia 275/276: 15–30.

Jeppesen, E., M. Søndergaard, B. Kronvang, J. P. Jensen, L. M. Svendsen & T. Lauridsen. Lake and catchment management in Denmark, 1997. In Harper, D., B. Brierley, A. Ferguson, G. Phillips & J. Madgwick (eds), The ecological basis for lake and reservoir management. J. Wiley & Sons.

Jeppesen, E., M. Søndergaard, E. Mortensen, P. Kristensen, B. Riemann, H. J. Jensen, J. P. Müller, O. Sortkjaer, J. P. Jensen, K. Christoffersen, S. Bosselmann & E. Dall, 1990a. Fish manipulation as a lake restoration tool in shallow, eutrophic temperate lakes 1: Cross-analysis of three Danish case-studies. Hydrobiologia 200/201: 205–218.

Kirk, K. L., 1991. Inorganic particles alter competition in grazing plankton: the role of selective feeding. Ecology 72: 915–923.

Kirk, K. L. & J. J. Gilbert, 1990. Suspended clay and the population dynamics of planktonic rotifers and cladocerans. Ecology 71: 1741–1755.

Kristensen, P., J. P. Jensen & E. Jeppesen, 1991. Simple empirical lake models. Nitrogen and phosphorus in fresh and marine waters. NPo forskning fra Miljøstyrelsen. C-abstracts. Danish Environmental Protection Agency, Copenhagen, 125–145.

Lammens, E. H. R. R., D. Gulati, M-L. Meijer & E. van Donk, 1990. The first biomanipulation conference: A synthesis. Hydrobiologia 200/201: 619–627.

Lampert, W., 1993. Ultimate causes of diel vertical migration of zooplankton: New evidence for the predator-avoidance hypothesis. Arch. Hydrobiol. Beih. Ergebn. Limnol. 39: 79–88.

Lauridsen, T., E. Jeppesen & M. Søndergaard, 1994. Colonization and succession of submerged macrophytes in shallow Lake Væng during the first five years following fish-manipulation. Hydrobiologia 275/276: 233–242.

Lauridsen T. & D. Lodge, 1996. Avoidance by *Daphnia magna* Straus of fish and macrophytes: Chemical cues and predator-mediated use of macrophyte habitat. Limnol. & Oceanogr. 41: 794–798.

Lauridsen, T. & I. Buenk, 1996. Diel changes in the horizontal distribution of zooplankton in the littoral zone of two eutrophic shallow lakes. Arch. Hydrobiol. 137: 161–176.

Lauridsen, T., L. Junge Pedersen, E. Jeppesen & M. Søndergaard, 1996. The importance of macrophyte bed size for composition and horizontal migration of cladocerans in a shallow lake. J. Plankton Res. 18: 2283–2294.

Lauridsen. T. L., E. Jeppesen & F. Ø. Andersen, 1993. Colonization of submerged macrophytes in shallow fish manipulated Lake Væng: Impact of sediment composition and water fowl grazing. Aquat. Bot. 46: 1–15.

Leah, R. T., B. Moss & D. E. Forrest, 1980. The role of predation in causing major changes in the limnology of a hyper-eutrophic lake. Int. Rev. ges. Hydrobiol. 65: 223–247.

Lindegaard, C., 1994. The role of zoobenthos in energy flow in two shallow lakes. Hydrobiologia 275/276: 313–322.

Loose, C. J. E., von Elert & P. Dawidowicz, 1993. Chemically induced diel vertical migration in *Daphnia*: A new bioassay for kairomones exuded by fish. Arch. Hydrobiol. 126: 329–337.

Luecke, C., M. J. Vanni, J. J. Magnuson, J. F. Kitchell & P. J. Jacobson, 1990. Seasonal regulation of *Daphnia* populations by planktivorous fish: Implications for the clearwater phase. Limnol. Oceanogr. 35: 1718–1733.

McQueen, D. J., 1990. Manipulating lake community structure: Where do we go from here? Freshwat. Biol. 23: 613–620.

Meijer, M. L, W. de Haan, A. W. Breukelaar & H. Buiteveld, 1990. Is reduction of the benthivorous fish an important cause of high transparency following biomanipulation in shallow lakes? Hydrobiologia 200/201: 303–316.

Meijer, M. L., E. Jeppesen, E. van Donk., B. Moss, M. Scheffer, E. H. R. R. Lammens, E. Van Nes, J. A. Berkum, G. J. de Jong, B. A. Faafeng & J. P. Jensen, 1994. Long-term responses to fish-stock reduction in small shallow lakes: interpretation of five year results of four biomanipulation cases in the Netherlands and Denmark. Hydrobiologia 275/276: 457–466.

Mills, E. L. & J. L. Forney, 1983. Impact on *Daphnia pulex* of predation by yellow perch in Oneida Lake, New York. Trans. am. Fish. Soc. 112: 154–161.

Mills, E. L., J. L. Forney & K. J. Wagner, 1987. Fish predation and its cascading effect on the Oneida Lake food chain. In Kerfoot, W. C. & A. Sih (eds), Predation: Direct and indirect effects on aquatic communities. University Press of New England, Hanover, New Hampshire: 118–131.

Mortensen, E., E. Jeppesen, M. Søndergaard & L. Kamp Nielsen (eds), 1994. Nutrient Dynamics and Biological Structure in Shallow Freshwater and Brackish Lakes. Developments in Hydrobiology 94. Kluwer Academic Publishers, Dordrecht, 507 pp. Reprinted from Hydrobiologia 275/276.

Moss, B., 1990. Engineering and biological approaches to the restoration from eutrophication of shallow lakes in which aquatic plant communities are important components. Hydrobiologia 200/201: 367–378.

Moss, B., S. McGowan & L. Carvalho, 1994. Determination of phytoplankton crops by top-down and bottom-up mechanisms in a group of English lakes, the West Midland Meres. Limnol. Oceanogr. 39: 1020–1029.

Nygaard, G., 1949. Hydrobiological studies on some Danish ponds and lakes. Biol. Skr. K. Dan. Vidensk. Selsk. 7: 1–239.

Pavoni, M., 1963. The importance of nanoplankton compared to netplankton. Schweiz. Z. Hydrol. 25: 219–341.

Pennak, R. W., 1966. Structure of zooplankton populations in the littoral macrophyte zone of some Colorado lakes. Trans. am. microsc. Soc. 85: 329–349.

Pennak, R. W., 1973. Some evidence for aquatic macrophytes as repellents for a limnetic species of *Daphnia*. Int. Revue. ges. Hydrobiol. 58: 569–576.

Perrow, M. R., J. Schutten, J. R. Howes, T. Holzer, F. J. Madgwick & A. J. D. Jowitt, 1997. Interactions between coot (*Fulica atra*) and submerged macrophytes: the role of birds in the restoration process. Hydrobiologia 342/343: 241–255.

Persson, L., 1991. Behavioural response to predators reverses the outcome of competition between prey species. Behav. Ecol. Sociobiol. 28: 101–105.

Persson, L., G. Anderson, S. F. Hamrin & L. Johansson, 1988. Predation regulation and primary production along the productivity gradient of temperate lake ecosystems. In Carpenter, S. R. (ed.), Complex interactions in lake communities. Springer Verlag, New York: 45–65.

Phillips, G. L., M. Perrow & J. Stansfield, 1996. Manipulating the fish-zooplankton interaction in shallow lakes: a tool for restoration. In S. P. R. Greenstreet & M. L. Tasker (eds), Aquatic predators and their prey. Blackwell Scientific Publications, Oxford: 174–183.

Prejs, A., A. Martyniak, S. Boron, P. Hliwa & P. Koperski, 1994. Food web manipulation in a small eutrophic Lake Wirbel, Poland: Effect of stocking with juvenile pike on planktivorous fish. Hydrobiologia 275/276: 65–70.

Pridmore, R. D., W. N. Vant & J. C. Rutherford, 1985. Chlorophyll-nutrient relationships in North Island lakes (New Zealand). Hydrobiologia 121: 181–189.

Quade, H. W., 1969. Cladoceran faunas associated with aquatic macrophytes in some lakes in northwestern Minnesota. Ecology 50: 170–179.

Quiros, R., 1990. Predictors of relative fish biomass in lakes and reservoirs of Argentina. Can. J. Fish. aquat. Sci. 47: 928–939.

Reynolds, C. S., 1984. The ecology of freshwater phytoplankton. Cambridge University Press, Cambridge, 384 pp.

Reynolds, C. S., 1994. The ecological basis for the successful biomanipulation of aquatic communities. Arch. Hydrobiol. 130: 1–33.

Riemann, B., K. Christoffersen, H. J. Jensen, J. P. Müller, C. L. Lindegaard & S. Bosselmann, 1990. Ecological consequences of a manual reduction of roach and bream in a eutrophic, temperate lake. Hydrobiologia 200/201: 241–250.

Ringelberg, J., 1991. Enhancement of the phototactic reaction in *Daphnia hyalina* by a chemical mediated by juvenile perch (*Perca fluviatilis*). J. Plankton Res. 13: 17–25.

Sas, H. (Ed.), 1989. Lake restoration by reduction of nutrient loading. Expectation, experiences, extrapolation. Acad. Ver. Richardz Gmbh. 497 pp.

Scheffer, M., 1990. Multiplicity of stable states in freshwater systems. Hydrobiologia 200/201 (Dev. Hydrobiol. 61): 475–486.

Scheffer, M., S. H. Hosper, M.-L. Meijer, B. Moss & E. Jeppesen, 1993. Alternative equilibria in shallow lakes. Trends in ecology and evolution (TREE) 8: 275–279.

Schriver, P., J. Bøgestrand, E. Jeppesen & M. Søndergaard, 1995. Impact of submerged macrophytes on the interactions between fish, zooplankton and phytoplankton: Large-scale enclosure experiments in a shallow lake. Freshwat. Biol. 33: 255–270.

Shapiro, J. & D. I. Wright, 1984. Lake restoration by biomanipulation. Round Lake, Minnesota – the first two years. Freshwat. Biol. 14: 371–383.

Sommer, U., 1985. Comparison between steady state and non-steady state competition: Experiments with natural phytoplankton. Limnol. Oceanogr. 30: 335–346.

Søndergaard, M., E. Jeppesen & S. Berg, 1997. Pike (*Esox lucius*) stocking as a biomanipulation tool, 2. Effects on lower trophic levels in Lake Lyng, Denmark. Hydrobiologia 342/343: 319–325.

Søndergaard, M., E. Jeppesen, P. Kristensen & O. Sortkjær, 1990a. Interactions between sediment and water in a shallow hypertrophic lake: A study on phytoplankton collapses in Lake Søbygard, Denmark. Hydrobiologia 191: 149–164.

Søndergaard, M., E. Jeppesen, E. Mortensen, E. Dall, P. Kristensen & O. Sortkjær, 1990b. Phytoplankton biomass reduction after planktivorous fish reduction in a shallow, eutrophic lake: A combined effect of reduced internal P-loading and increased zooplankton grazing. Hydrobiologia 200/201: 229–240.

Søndergaard, M., P. Kristensen & E. Jeppesen, 1992. Phosphorus release from resuspended sediment in the shallow and wind exposed Lake Arresø, Denmark. Hydrobiologia 228: 91–99.

Søndergaard, M., L. Olufsen, T. Laundsen, E. Jeppesen & T. Vindbæk Madsen, 1996. The impact of grazing waterfowl on submerged macrophytes: In situ experiments in a shallow eutrophic lake. Aquat. Bot. 53: 73–84.

Stansfield, J. H., M. R. Perrow, L. D. Tench, A. J. D. Jowitt & A. A. L. Taylor, 1997. Submerged macrophytes as refuges for grazing Cladocera against fish predation: observations on seasonal changes in relation to macrophyte cover and predation pressure. Hydrobiologia 342/343: 229–240.

Timms, R. M. & B. Moss, 1984. Prevention of growth of potentially dense phytoplankton populations by zooplankton grazing in the presence of zooplanktivorous fish, in a shallow wetland ecosystem. Limnol. Oceanogr. 29: 472–486.

Van Donk, E., E. De Deckere, J. G. P. Klein Breteler & J. Meulemans, 1994. Herbivory by waterfowl and fish on macrophytes in a biomanipulated lake: Effects on long-term recovery. Verh. int. Ver. Limnol. 25: 2139–2143.

Whiteside, M. C., 1988. 0+ fish as major factors affecting abundance patterns of littoral zooplankton. Verh. int. Ver. Limnol. 23: 1710–1714.

Windolf, J., E. Jeppesen, M. Søndergaard, J. P. Jensen & L. Sortkjær, 1993. Vandmiljøplanens Overvågningsprogram 1992. Ferske vandområder - Søer. [The Action Plan on the Aquatic Environment 1992. Freshwater area - Lakes]. National envir. Res. Inst., 129 pp.

Winfield, L. J., 1986. The influence of simulated aquatic macrophytes on the zooplankton consumption rate of juvenile roach, *Rutilus rutilus*, rudd, *Scardinius erythrophthalmus*, and perch, *Perca fluviatilis*. J. Fish. Biol. 29: 37–48.

Zdanowski B., 1982. Variability of nitrogen and phosphorus contents and lake eutrophication. Pol. Arch. Hydrobiol. 29: 3–4.

Hydrobiologia **342/343:** 165–174, 1997.
L. Kufel, A. Prejs & J. I. Rybak (eds), Shallow Lakes '95.

Seasonal successions and trophic relations between phytoplankton, zooplankton, ciliate and bacteria in a hypertrophic shallow lake in Vienna, Austria

Jutta Mayer[1], Martin T. Dokulil[1], Monika Salbrechter[1], Martina Berger[1], Thomas Posch[1], Gerald Pfister[1], Alexander K. T. Kirschner[2], Branko Velimirov[2], Andrea Steitz[2] & Til Ulbricht[2]
[1]*Institute for Limnology, Gaisberg 116, A-5310 Mondsee, Austria*
[2]*Institute of General Biology, University of Vienna, Währingerstraße 11, A-1090 Wien, Austria*

Key words: eutrophication, trophic relations, filamentous cyanobacteria, ciliates, zooplankton, bacteria

Abstract

Alte Donau nowadays is an eutrophic urban lake within the city of Vienna. Increasing nutrient concentrations and massive blooms of cyanobacteria mainly caused by *Limnothrix redekei* Van Goor and *Cylindrospermopsis raciborskii* (Wołsz.) Seenayya et Subba Raju were recently registered. As a consequence Secchi depth was significantly reduced especially during the summer season (minimum: 0.25 m). An investigation including water chemistry, phytoplankton, macrophytes, and sediment was initiated in 1993 and extended to metazooplankton, ciliates and bacteria in 1994. The first half of the year 1994 was characterised by relatively clear water and a high diversity of the phytoplankton composition due to flushing of the lake with water of better quality by the end of 1993. Ciliates and metazooplankton held about 10% of the total biomass of all the investigated trophic levels. The vanishing of the remaining macrophytes enlarged the nutrient supply during summer 1994 and favoured the development of cyanobacteria. The high water temperatures which excluded certain zooplankton species, and the inedibility of the filaments further increased the dominance of cyanobacteria. In November, when the algal bloom finally ceased, the highest bacterial numbers of the investigation period occurred. Thereafter, other algal groups, bacteria and metazooplankton gained more importance.

Interactions are possible because of close overlap in space and time due to the turbulent mixed conditions of the water body and the change from the macrophyte dominated to the algal dominated stable state. Planned restoration measures must aim to re-establish the previous macrophyte dominated clear-water stage.

Introduction

The open water zone of lakes are characterised by complex interactions. Community interactions in pelagic food webs are affected by small- to large scale physical, chemical and biological processes, are trophically dynamic and are governed by nutrient limitation, competition, predation and other ecological forces (Carpenter, 1988). In shallow waters, trophic level interactions are complicated by detrital pathways and influences from the sediments (Herzig, 1979). Community structure, biomass, and production are influenced by both producers and consumers which function simultaneously. If bottom up effects by producers or top down control by consumers dominate is largely a function of the trophic level of the lake (McQueen et al., 1986).

In this paper we describe the seasonal development of phytoplankton, algal picoplankton, bacteria, ciliates and metazooplankton from a shallow hypertrophic urban lake in 1994 and try to elucidate some of the interactions between food web components.

Table 1. Morphometric data for Alte Donau.

Altitude	157 m a.s.l.
Area	1.583 km^2
Volume	3.697×10^6 m^3
Maximum depth	6.8 m
Mean depth	2.33 m
Min. theoret. retention time	20 days

Site description

Alte Donau, formerly the main course of the River Danube, was formed as an ox-bow lake in the years 1870–1875 when the river has been regulated. Today, Alte Donau is a shallow lake with no superficial connection with the river, fed only by subterranean flow from the ground-water and precipitation. Location and morphometry are indicated in Figure 1. Morphometric data are summarised in Table 1. As an urban lake, Alte Donau is of high recreational importance.

In the past, Alte Donau was always characterised by clear water, concentrations of chlorophyll *a* below 5 μg l^{-1}, and a domination of various macrophytes covering almost the whole area. In 1992, a remarkable decline of the macrophytes was observed leaving only remnants of *Charophyceae*-stands. The massive growth of phytoplankton at the same time caused a significant reduction of the Secchi-depth to less than 0.25 m during the summer season. This shift to a new stable state (Dokulil & Janauer, 1995) was caused by altered hydrological conditions which resulted in an increase in the water level by 0.4 m and largely reduced level fluctuations (Donabaum et al., 1996).

Limnological investigations, including macrophytes, phytoplankton, water chemistry and sediment chemistry, were initiated in 1993 to develop restoration measures. The project was extended to include metazooplankton, rotifers, ciliates and bacteria in 1994.

As an immediate short-term improvement of water quality, about half of the lake water was drained in December 1993 and replaced with water of better quality from the nearby Neue Donau (Figure 1).

Methods

The investigation period was from April to November 1994. During the summer season the samples were taken at fortnight intervals, while in October and November samples were taken monthly. Because of the narrow and shallow connection between the two basins of Alte Donau and the irregular morphometry in the south (Figure 1), three sampling stations were selected in the southern basin, two in the northern basin. Bacteria were investigated only at one station in each basin.

Previous results from 1993 indicated that vertical differences were negligible. Therefore integrated water samples for chemical analysis and phytoplankton biomass were taken with an acrylic plastic tube from the top 2 m. In addition, near-bottom water was collected with a Schindler sampler. Standard analytical techniques were used for all the chemical and physical variables.

Samples for the phytoplankton enumeration were fixed with Lugol's solution. Quantification of the phytoplankton was carried out with an inverted microscope (NIKON). About 400 individuals of the dominating species were counted (Utermöhl, 1958).

Lengths and widths of phytoplankton, zooplankton and ciliates were measured using an image analysing system (Lucia M., NIKON) to determine mean biovolumes. For calculating the cell volumes of phytoplankton and ciliates cell shapes were approximated by simple geometric forms.

Abundance and biomass of algal picoplankton and bacteria were investigated at two sites of Alte Donau Samples were collected from four depths in the northern basin and from two depths in the southern basin. Depending on season, 10 to 50 ml of water were fixed with formaldehyde solution (4% f.c.) and stored at 4 °C. Cells were counted by their fluorescence with an excitation at 515 nm–565 nm and at 1250 × magnification. Cell size of picocyanobacteria was measured and average cell volume was calculated and converted to carbon assuming a conversion factor of 200 fg C mm^{-3} (Weisse, 1991). Carbon biomass was estimated by multiplying the average cell carbon content with the total algal picoplankton cell counts.

For bacterial numbers and biomass 10 ml samples were fixed with buffered formaldehyde solution (4% f.c.; pH = 8) in sterile 20 ml plastic scintillation vials and stored at 4 °C in the refrigerator prior to the microscopic examination. Bacteria were counted and their cell volumes measured according to the method of Hobbie et al. (1977) Bacterial carbon biomass was calculated by converting cell volumes (CV) into cellular carbon content (CC) using the allometric relation of Norland (1993), where the CC (in f.w. C cell^{-1}) amounts to 120 CV$^{0.72}$. The average CC was then multiplied by the total number of bacteria.

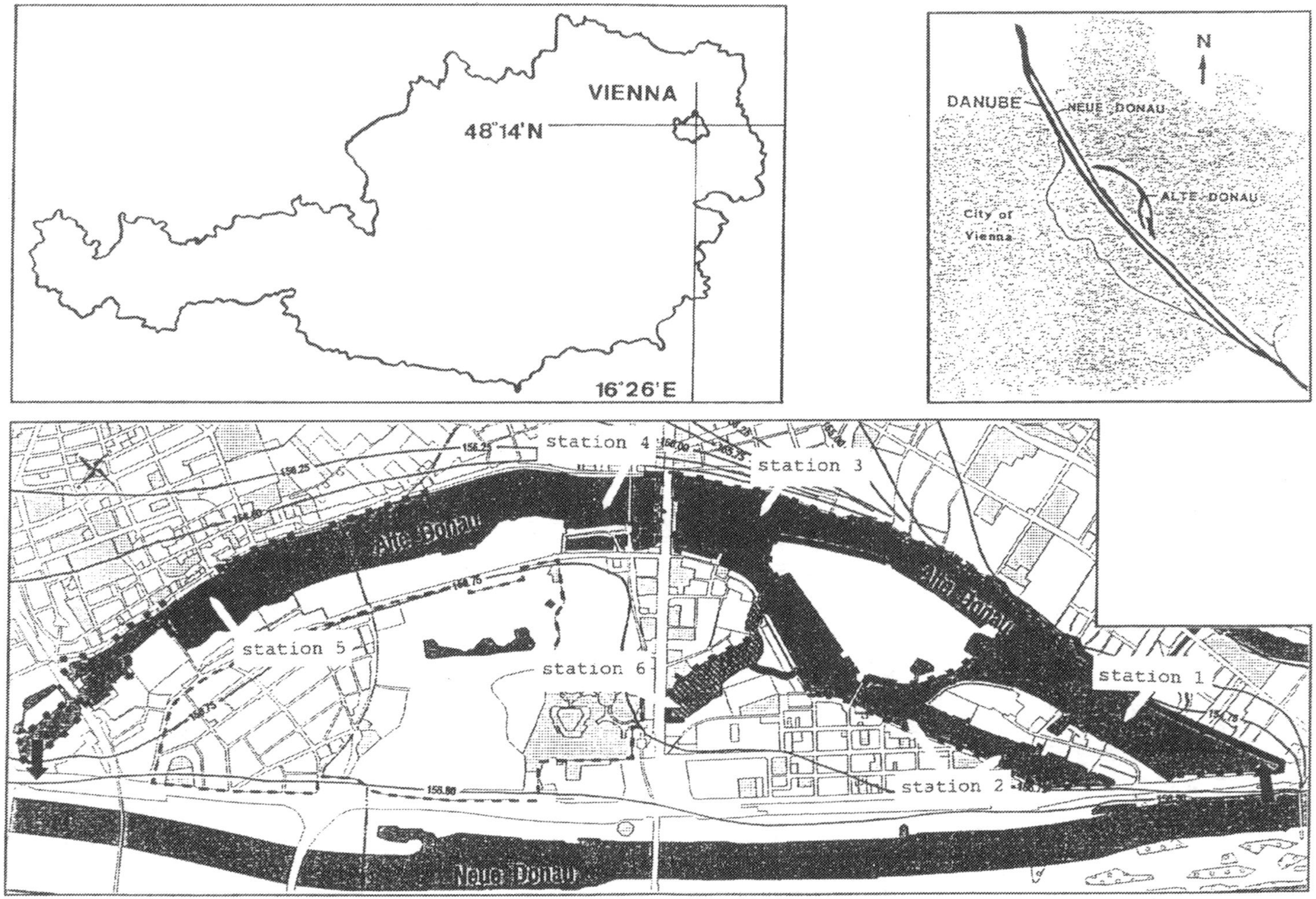

Figure 1. Map of Austria, location of Alte Donau in Vienna and map of the study site with sampling stations.

The metazooplankton and ciliates samples were taken with a 5 l Schindler sampler from three different depths (surface, middle, bottom) to account for vertical inhomogeneities and then combined to an integrated sample. For the metazooplankton, the whole water volume (15 l) was filtered through a net of 44 μm mesh size, resuspended in 50 ml lake water and fixed with formaldehyde (4% f.c.). Subsamples were analysed by the Utermöhl method under an inverted microscope (Utermöhl, 1958). Crustacean biovolumes were estimated using length/weight regressions published by McCauley (1984) and Bottrell et al. (1976) assuming a dry weight/wet weight ratio of 0.15. Rotifer biovolumes were calculated using measurements of all three dimensions according to Ruttner-Kolisko (1977).

Unconcentrated subsamples were fixed with Bouin's fluid as described by Skibbe (1994) respectively with Lugol's solution for the Ciliates. One litre of untreated water was taken for living examination. The quantitative protargol stain (QPS) and living observation were used for the taxonomical identification based on publications of Kahl (1930–1935), Krainer (1988), Foissner et al. (1991, 1992 & 1994) and Foissner (1994). The applied techniques generally allowed the identification of ciliates to genera, in some cases to species. Ciliate abundances were determined using protargol preparations described in detail by Skibbe (1994). An obvious shrinkage of stained ciliates was noticed due to the protargol preparation. Therefore calculated cell volumes were multiplied with a correcting factor of 0.4 (Jerome et al., 1993). Conversion to fresh weight (F.W.) assumed a specific density of one for all species.

Results

Physical and chemical variables

Concentrations of total phosphorus (TP) total nitrogen (TN) and chlorophyll *a* increased significantly from 1987 to 1994 while Secchi-depth decreased correspondingly. Concentrations of soluble reactive phosphorus (SRP) were sometimes below detection limits during the summer of 1994 and oxygen depletion occurred at times near the sediment. Flushing Alte Donau with water of better quality in December 1993 resulted in increased Secchi-depth, lower chlorophyll *a* and lower nutrient concentrations during the first half of 1994, and greater variability between basins (Table 2).

Phytoplankton

The filamentous cyanobacteria *Limnothrix redekei* and *Cylindrospermopsis raciborskii* were first detected in 1992 and dominated the phytoplankton from May to November 1993 (Dokulil & Mayer, 1996). As a result of the flushing of Alte Donau in December 1993, phytoplankton species composition was significantly altered in the first half of 1994. Diatoms, represented by *Fragilaria ulna* var. *acus* (Kützing) Lange-Bertalot and centric diatoms (*Cyclotella* spp., *Stephanodiscus* spp.) as well as Chrysophytes (*Dinobryon* ssp.) became more prominent and cyanobacterial contribution became largely suppressed (Figure 2). Green algae, such as *Tetraedron minimum* (A. Br.) Hansg., *Scenedesmus quadricauda* (Turp.) Bréb and *Scenedesmus acuminatus* (Lag.) Chod., gained importance in June/July. Despite of low abundance, the dinoflagellates *Ceratium hirundinella* (O. F. Müller) Dujardin, *Glenodinium* ssp., and *Peridinium* ssp. contributed significantly to total biomass.

Phytoplankton biomass ranged from 4.22 mg l^{-1} to 53.92 mg l^{-1} in the northern basin and from 4.85 mg l^{-1} to 60.87 mg l^{-1} in the southern basin. Although cyanobacterial contribution in April/May was greater in the southern basin when compared to the northern basin, total algal biomass remained below 10 mg l^{-1} until mid-July. More green algae developed in the northern part and biomass started to increase by mid-June (Figure 2).

Limnothrix redekei and *Cylindrospermopsis raciborskii* began to increase exponentially in both basins in August. These species formed algal blooms during September and early October when maximal development of both total phytoplankton and cyanobacterial biomass was reached.

Algal picoplankton

Cell numbers of algal picoplankton ranged from 0.5×10^8 (2.9.94) to more than 30×10^8 cells l^{-1} (13.7.94). A first maximum was observed in May and a larger one in July, when small colonies were the dominating picoalgal forms. Thereafter numbers of algal picoplankton decreased rapidly to minimum values in late summer. During autumn and winter the determination of cell densities was not possible due to high amounts of detritus in the water column. The fluctuation of the biomass of the algal picoplankton followed the pattern of the cell numbers, with maximum values

Table 2. Annual mean values and ranges for selected variables for the years 1987, 1993 and 1994. For 1994 values are shown for both basins. n.d. = not determined.

Variable (unit)	1987	1993	1994 southern basin	1994 northern basin
Secchi-depth (m)	1.65	0.4 (0.3-0.5)	0.7 (0.25-1.90)	0.8 (0.25-1.80)
pH-value	n.d.	n.d.	8.4 (7.52-9.14)	8.5 (7.9-9.32)
Oxygen (mg l^{-1})	n.d.	8.5 (0-14.8)	8.1 (5.2-14.4)	7.8 (4.1-14.3)
SRP (μg l^{-1})	n.d.	n.d.	3 (0-22)	4 (0-16)
Total phosphorus (μg l^{-1})	35 (25-50)	55 (23-131)	67 (24-98)	86 (41-145)
Total nitrogen (mg l^{-1})	0.56	1.96 (0.5-7.76)	1.7 (0.49-2.28)	1.8 (0.62-4.25)
DOC (mg l^{-1})	n.d.	n.d.	5.9 (3.15-9.61)	6.0 (3.9-8.59)
Chlorophyll-a (μg l^{-1})	5	41 (6-102)	40 (1-114)	52 (10-143)

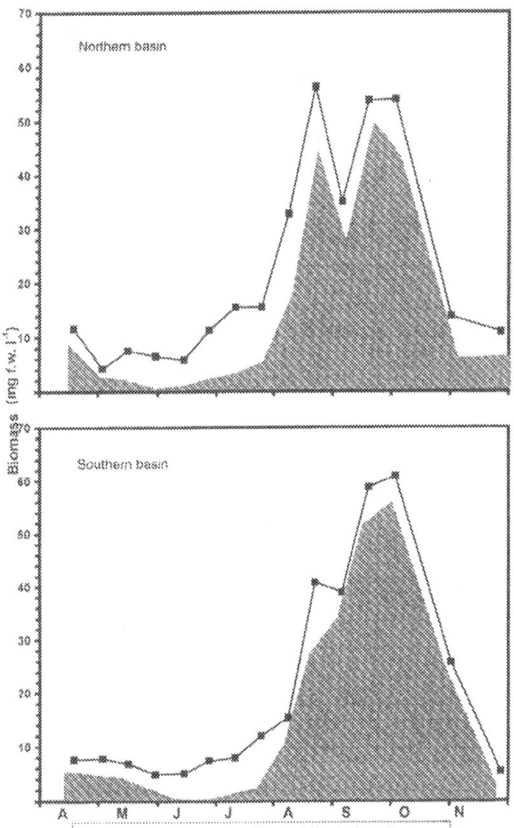

Figure 2. Seasonal development of total phytoplankton and cyanobacterial biomass in the northern basin (upper panel) and southern basin (lower panel).

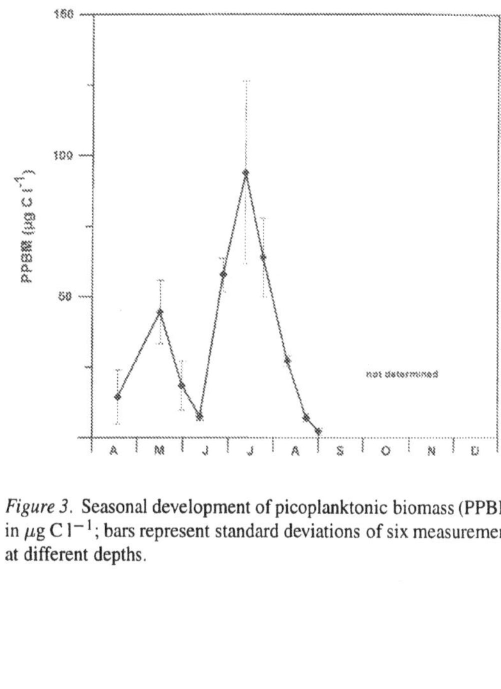

Figure 3. Seasonal development of picoplanktonic biomass (PPBM) in μg C l^{-1}; bars represent standard deviations of six measurements at different depths.

of about 100 μg C l^{-1} in July and minimum values of less than 10 μg C l^{-1} in late summer (Figure 3).

Heterotrophic bacterioplankton

At both sampling sites bacterial numbers continuously increased from densities of about 15×10^9 in early spring to an average maximum in December of 15×10^9 cells l^{-1}. A significant difference between the two basins of Alte Donau were registered in the beginning of May, when bacterial numbers in the northern basin were about 2.5 times higher than in the southern basin. Bacterial biomass varied between 30 and 115 μg C l^{-1} steadily increasing towards the end of the year (Figure 4).

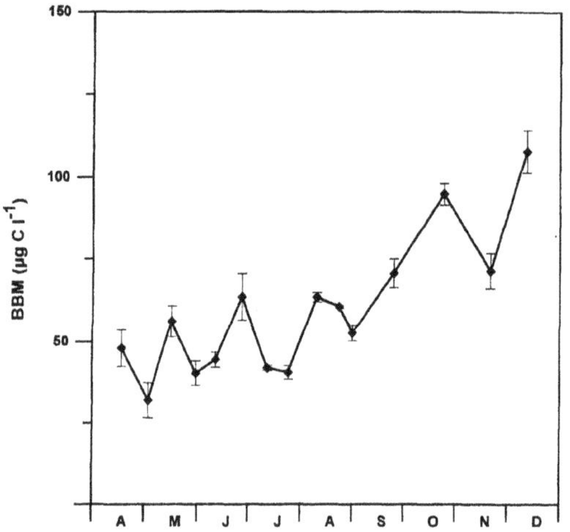

Figure 4. Seasonal variation of bacterial biomass (BBM); values are given in μg C l^{-1}; bars represent standard deviations of six measurements at different depths.

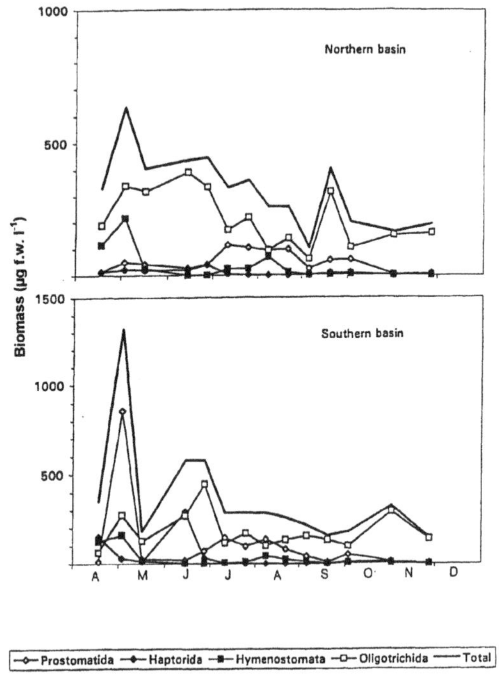

Figure 5. Seasonal development of total biomass of the ciliates and the four most important groups in the northern basin (upper panel) and in the southern basin (lower panel).

Ciliates

Cell abundance ranged from 9 to 135 ind. ml^{-1} during the sampling period. The mean ciliate number was 54 ind. ml^{-1} in the northern basin and 46 ind. ml^{-1} in the southern basin. There were no significant differences in the mean abundance between the southern and the northern basin. In both basins the highest abundances were found at the end of June after a smaller peak in the first days of May. However, seasonal succession differed between basins. A second peak developed in the northern basin in early autumn, whereas cell abundances were decreasing in the southern part.

Biomass ranged from a minimum of 58 mg to a maximum of 2472 mg F.W. m^{-3}. The mean value was 323 mg in the northern basin and 367 mg F.W. m^{-3} in the southern basin. Similar like abundance, ciliate biomass was different in the two basins. The peak value in the southern basin at the beginning of May was mainly caused by the large prostomatid *Bursellopsis spumosa* (Schmidt) Corrliss. After this species decreased, haptorids strongly contributed to total ciliate biomass (Figure 5). In the northern basin ciliates of the group Hymenostomata significantly contributed to total biomass at the beginning of May (Figure 5). During the rest of the year biomass values of both basins were very similar with an exception in the northern basin in early autumn. This biomass peak corresponded with higher abundances of the Oligotrichida at that time (Figure 5).

Genera of the Oligotrichida (*Halteria* spp., *Strobilidium* spp., *Pelagostrombidium* spp.) were the dominant components of ciliate plankton both in numbers and biomass during the whole sampling period. Their percentage share ranged from 61% to 63%. Second in importance were the prostomatids *Urotricha* spp., *Coleps* spp. and *Balanion* sp. (16%–18%). During summer, small hymenostomes (*Cyclidium* spp., Scuticociliata) became important to total ciliate abundance with relative contributions of about 12%. Haptorids (especially *Askenasia* spp., *Mesodinium* sp. and *Monodinium* sp.) were relatively abundant only until the end of July. Peritrichs and other ciliate groups were not abundant in the plankton during the investigated period (relative abundances <2%).

Metazooplankton and rotifers

The seasonal changes in biomass of rotifers, cladocerans and copepods are shown in Figure 6. The seasonal development of metazooplankton was very similar in the northern and southern basin, with maximum biomass of copepods in summer. Rotifers reached high

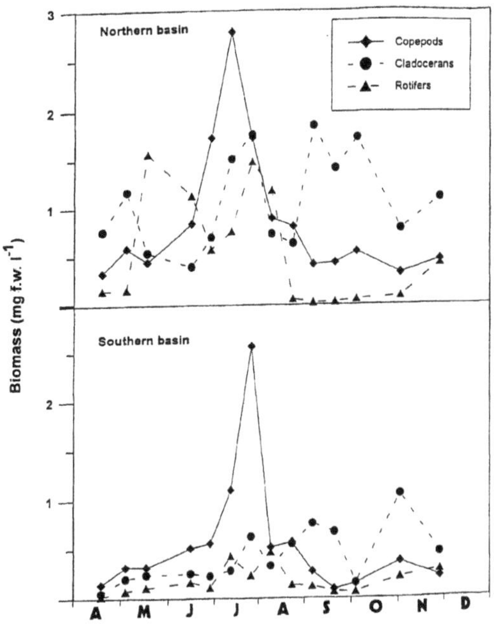

Figure 6. Seasonal development of the metazooplankton and rotifer biomass in the northern basin (upper panel) and in the southern basin (lower panel).

biomass in May and August, cladocerans increased in autumn. Although the seasonal changes were similar, the annual mean for total zooplankton differed between the northern (1425 ind. l^{-1}; 1236 g l F.W. l^{-1}) and southern basin (3511 ind. l^{-1}; 3462 μg F.W. l^{-1}).

Metazooplankton was dominated by cyclopoid copepods, which contributed 44% in the northern and 25% in the southern basin to the biomass. The herbivorous calanoid copepod *Eudiaptomus gracilis* (Sars) gains in significance during winter time. The most important cladoceran species was *Eubosmina coregoni* Baird (60% of number). With numbers up to 1000 individuals per litre, this was the dominating species at the end of the investigated period in both basins. *Daphnia cucullata* Sars and *Chydorus sphaericus* (O. F. Müller) got important during spring and summer.

The seasonal cycle of rotifers showed two development peaks in the southern basin. The first was composed of a very diverse spring community, while the summer maximum consisted only of one species. In the northern basin, a real spring maximum could not be observed but in both basins the reduced development of rotifers in summer appeared at the same time when cyclopoids were dominant. The rotifer fauna consisted of 31 species. Mainly the genera *Brachionus, Keratella* and *Polyarthra* were responsible for the obtained densities. Among the registered rotifers the abundant species belong to the group of unselective microfiltrators and have to be taken into account as important herbivores.

Discussion

The phytoplankton composition, especially the bloom of filamentous cyanobacteria in the second half of the year is an indication of the eutrophic to hypertrophic state of Alte Donau. *Cylindrospermopsis raciborskii* is adapted to low light conditions, is able to fix atmospheric nitrogen (Lampert & Sommer, 1993) and can store phosphorus through luxury uptake (Zaiss, 1985). However, filamentous algae are less edible than more smaller cells. The feeding pressure from metazooplankton is therefore negligible (Kurmayer, unpubl. data) and does not significantly influence the development of the cyanobacteria.

Bacterial numbers are comparable with the abundance in other eutrophic waters (e.g. Chrost & Faust, 1983; Riemann & Søndergaard, 1986). Biomass seems to be somewhat influenced by ciliate and metazooplankton grazing. The increase in bacterial numbers during the second half of the year may be a reaction to better food supply due to the expected release of easily utilisable dissolved organic matter during the phytoplankton bloom and after its breakdown.

The high number of different ciliate species (27) found during the investigation and the annual mean abundance of 57 640 individuals are characteristic features of eutrophic waters. In late summer, when the highest nutrient concentrations were measured and algae blooms occurred, Hymenostomata of the genus *Cyclidium* sp. appeared indicating an increase of the trophic situation. The decline of ciliate biomass in the second half year seems to be a result of the change in the spectrum of food particles because of the altered phytoplankton composition (Pfister, 1995). However, ciliates can feed on fine particulate detritus (Posch, 1995).

The species composition of metazooplankton also indicates the eutrophic state. The high abundances of cyclopoid copepods (*Thermocyclops*) and the low numbers of calanoid copepods (*Eudiaptomus gracilis*) are characteristic for waters with increased eutrophication (Gliwicz, 1969). The high contribution of cladocerans (McNaught, 1975) to metazooplankton

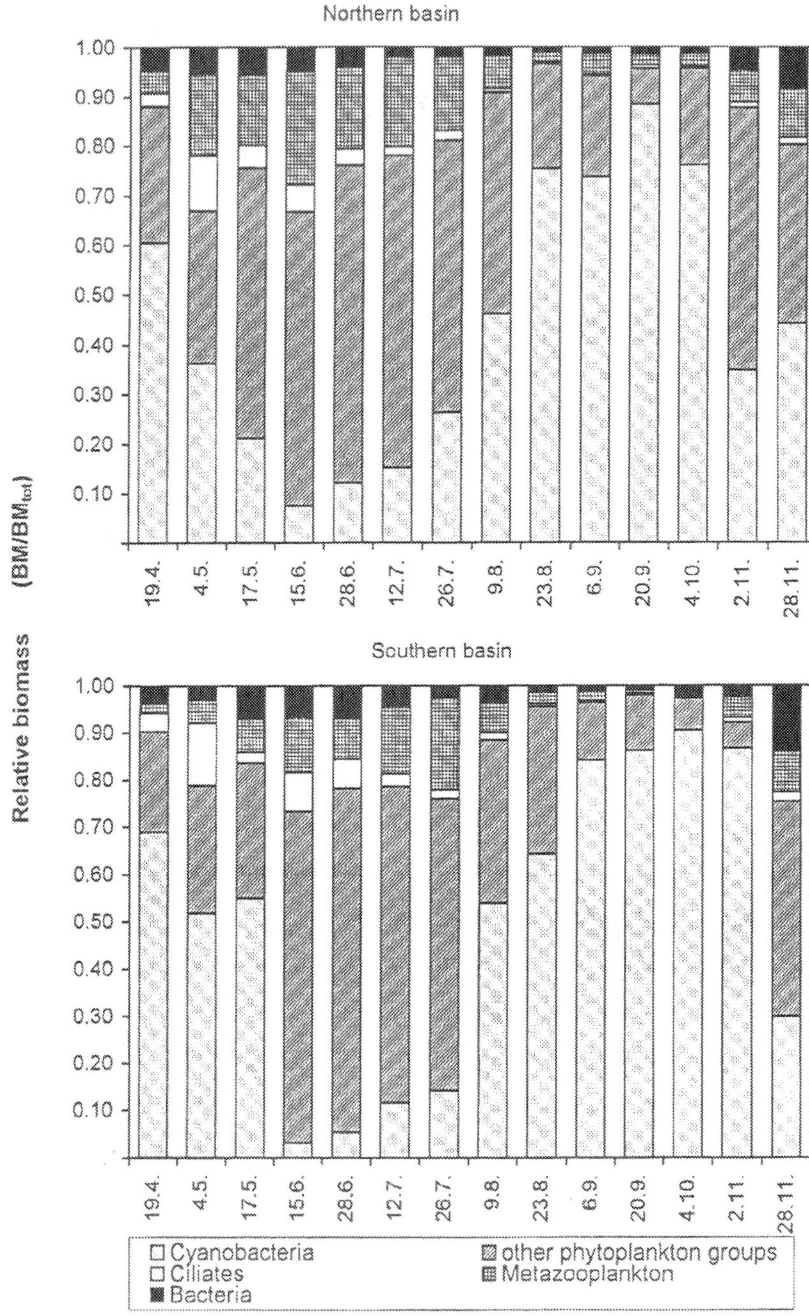

Figure 7. Seasonality of the relative biomass proportion of all investigated trophic levels in the northern basin (upper panel) and in the southern basin (lower panel).

abundance and biomass also associated with nutrient enrichment has been recorded at all sampling stations. Rotifers, like *Brachionus angularis*, *B. calyciflorus*, *Anureopsis* sp and *Filinia longiseta* as well as the cladocera *Daphnia cucullata* are indicators for eutrophic conditions. The high abundance of small cladocerans (*Eubosmina coregoni*) compared to big daphnids (*Daphnia cucullata*) is in agreement with literature, that small cladocerans are favoured through eutrophication (Beaver & Crisman, 1982). The feeding pressure of planktivorous fish seems to be responsible for that, as unpublished preliminary data suggest.

The algal feeding zooplankton and the ciliates had a significantly higher biomass in the southern basin. Feeding pressure on the edible phytoplankton species must have been therefore stronger in the southern basin than in the northern basin.

As a consequence of flushing the Alte Donau with water of better quality in December 1993, good light conditions in early spring lower nutrient concentrations and a mixed phytoplankton assemblage retarded cyanobacterial development. High water temperatures favoured growth of cyanobacteria in summer. At the same time, zooplankton biomass decreased (Figure 7) because of the shift in food supply to domination of inedible algae and water temperatures which are too high for certain species. The phytoplankton biomass gains importance during summer and amounts to almost 100% of the total biomass of all investigated trophic levels in September and October (Figure 7). Because of this shift in food supply more bacteria feeding daphnides and ciliates appeared.

In both basins the ciliates peaked in May and in June. The predatory *Bursellopsis spumosa* together with the metazooplankton caused high feeding pressure on ciliates. As a consequence, ciliate biomass declined. A second peak was observed in the northern basin at the end of September which might be an outcome of better food supply, possibly by fine particulate detritus (Posch, 1995; Posch & Arndt, 1996).

In autumn, the structure of the zooplankton community is characterised by the high abundance of cyanobacteria and increasing bacterial numbers. Large filter feeders, inhibited by filamentous inedible algae, gained significance (Gliwicz, 1977). Abundance and biomass of metazooplankton is dominated by *Eubosmina coregoni*, which is believed to be a 'low efficiency bacteria feeder' by Geller & Müller (1981). The available food composition, with bacteria as main food source, and the handicap of big filamentous cyanobacteria (length of the filaments between 70 and 400 μm) for other zooplankton species, offered best requirements for *Eubosmina coregoni*.

Rotifers and nauplii played an important role as herbivorous components in spring and early summer, when the phytoplankton community was composed of smaller forms (e.g. chlorococcal green algae, centric diatoms). Microfiltrators like *Keratella* spp. and *Brachionus* sp. feeding on particles in the range of 0.5 to 20 μm, including detritus and bacteria (Pourriot, 1977), had high abundance in spring and contributed to the decline of the algal biomass. *Polyarthra* sp. building large peak values in May, prefers food in a somewhat larger size range from 1–40 μm (Bogdan et al., 1980; Bogdan & Gilbert, 1987).

To summarise, physico-chemical structuring forces of the environment, such as underwater light conditions and temperature, are mainly responsible for the seasonal succession and the growth of phytoplankton species. Bacterial numbers are controlled by the availability of organic substances and affected by predation of ciliates and rotifers. Bacterial availability as food, algal composition and predation imposed by zooplankton are factors structuring the ciliate assemblage. The metazooplankton community is largely structured by food supply from algae and bacteria and the feeding pressure by planktivorous fish.

All these interactions are possible because of close overlap in space and time due to the turbulent mixed conditions of the water body and the change from the macrophyte dominated to the algal dominated stable state (Dokulil & Janauer, 1995). Any restoration measures must aim to re-establish the previous macrophyte dominated clear-water stage.

References

Beaver, J. R. & T. L. Crisman, 1982. The trophic response of ciliated protozoans in freshwater lakes. Limnol. Oceanogr. 27: 246–253.

Bogdan, K. G., J. J. Gilbert & P. L. Starkweather, 1980. *In situ* clearance rates of planktonic rotifers. In Dumont, H. J. & J. Green (eds), Rotatoria. Developments in Hydrobiology 1. Dr W. Junk Publishers, The Hague: 73–77. Reprinted from Hydrobiologia 73.

Bogdan, K. G. & J. J. Gilbert, 1987. Quantitative comparison of food niches in some freshwater zooplankton. A multitracer approach. Oecologia 72: 331–340.

Bottrell, H. H., A. Duncan, Z. M. Gliwicz, E. Grygierek, A. Herzig, A. Hillbricht-Ilkowska, H. Kurasawa, P. Larsson & T. Weglenska, 1976. A review of some problems in zooplankton production studies. Norw. J. Zool. 24: 419–456.

Carpenter, S. R. (ed.), 1988. Complex Interactions in Lake Communities. Springer Verlag, 283 pp.

Chróst, R. J. & M. A. Faust, 1983. Organic carbon release by phytoplankton: its composition and utilisation by bacterioplankton. J. Plankton Res. 5: 477–493.

Dokulil, M. T. & G. A. Janauer, 1995. Alternative stable states during eutrophication of a shallow urban lake in Vienna, Austria. Proc. 6th Int. Conf. Conservation and management of lakes-Kasumigaura 95 Univ. Tsukuba, Japan 2: 730–733.

Dokulil, M. T. & J. Mayer, in press. Population dynamics and photosynthetic rates of a *Cylindrospermopsis-Limnothrix* association in a highly eutrophic urban lake, Alte Donau, Vienna, Austria. Algol. Stud.

Donabaum, K., M. Schagerl & M. T. Dokulil. Integrated lake management for the restoration of a stable equilibrium. In Ferguson, A., D. Harper, B. Brierley & G. Phillips (eds), The ecological Basis for River Management. John Wiley Publ., in press.

Foissner, W., 1994. Progress in taxonomy of planktonic freshwater ciliates. Mar. Microbiol. Food Webs 8: 9–35.

Foissner, W., H. Berger & F. Kohmann, 1992. Taxonomische und ökologische Revision der Ciliaten des Sabrobiensystems. Band 2: Peritrichia, Heterotrichida, Odontostomatida. Informationsberichte des Bayer. Landesamts für Wasserwirtschaft, Heft 5/92, 502 pp.

Foissner, W., H. Berger & F. Kohmann, 1994. Taxonomische und ökologische Revision der Ciliaten des Sabrobiensystems. Band 3: Hymenostomata, Prostomatida, Nassulida. Informationsberichte des Bayer. Landesamts für Wasserwirtschaft, Heft 1/94, 548 pp.

Foissner, W., H. Blatterer, H. Berger & F. Kohmann, 1991. Taxonomische und ökologische Revision der Ciliaten des Sabrobiensystems. Band 1: Cyrtophorida, Oligotrichida, Hypotrichia, Colpodea. Informationsberichte des Bayer. Landesamts für Wasserwirtschaft, Heft 1/91, 478 pp

Geller, W. & H. Müller, 1981. The filtration apparatus of Cladocera: Filter mesh-sizes and their implications on food selectivity. Oecologia 49: 316–321.

Gliwicz, Z. M., 1969. Studies on the feeding of pelagic zooplankton in lakes with varying trophy. Ekol. pol. 17A: 663–708.

Gliwicz, Z. M., 1977. Food size selection and seasonal succession of filter-feeding zooplankton in an eutrophic lake. Ekol. pol. 25: 179–225.

Herzig, A., 1979. The zooplankton of the open lake. In Löffler, H. (ed.), Neusiedlersee. Limnology of a Shallow Lake in Central Europe. Dr W. Junk Publishers, The Hague: 281–336.

Hobbie, J. E., R. J. Dayley & S. Jasper, 1977. Use of nuclepore filters for counting bacteria by epifluorescense microscopy. Appl. envir. Microbiol. 33: 1225–1228.

Jerome, C. A., D. J. S. Montagnes & F. J. R. Taylor, 1993. The effect of the quantitative protargol stain and Lugol's and Bouin's fixatives on cell size: A more accurate estimate of ciliate species biomass. J. Euk. Microbiol. 40: 254–259.

Kahl, A., 1930–1935. Urtiere oder Protozoa I. Wimpertiere oder Ciliata (Infusoria). In Dahl, F. (ed.), Die Tierwelt Deutschlands 26 (18, 21, 25, 30). Fischer, Jena, 1886 pp.

Krainer, K. H., 1988. Alpha-Taxonomie und ökologie neuer sowie mehrerer wenig bekannter pelagischer Ciliaten (Protozoa: Ciliophora aus den Klassen Kinetophragminophora, Oligohymenophora, Polyhymenophora) einiger Grundwasserbaggerteiche des nördlichen Leibnitzer Feldes (Steiermark, Österreich). Dissertation an der Karl-Franzens-Universität Graz, 209 pp.

Lampert, W. & U. Sommer, 1993. Limnoökologie. Thieme Verlag, Stuttgart-New York, 440 pp.

McCauley, E., 1984. The estimation of the abundance and biomass of Zooplankton in samples. In Downing, J. A. & F. H. Riegler (eds), A Manual on Methods for the Assessment of Secondary Productivity in Fresh Waters, Blackwell Scientific Publications, Oxford, London: 228–265.

McNaught, D. C., 1975. A hypothesis to explain the succession from calanoids to cladocerans during eutrophication. Verh. int. Ver. Limnol. 19: 724–731.

McQueen, D. J., J. R. Post & E. L. Mills, 1986. Trophic relationships in fresh-water pelagic ecosystems. Can. J. Fish. aquat. Sci. 43: 1571–1581.

Norland, S., 1993: The relationship between biomass and volume of bacteria. Curr. Meth. aquat. microbiol. ecol.: 303–307.

Pfister, G., 1995. Untersuchungen zur Nahrungsselektivität von Protozoen am Beispiel des omnivoren Ciliaten *Stylonychia mytilus*. Diplomarbeit, Univ. Innsbruck, 73 pp.

Posch, Th., 1995. Über die Aufnahme von feinpartikulärem Detritus durch Ciliaten. Diplomarbeit, Univ. Innsbruck, 64 pp.

Posch, Th. & H. Arndt, 1996. Uptake of sub-micrometre- and micrometre-sized detrital particles by bacterivorous and ominvorous ciliates. Microbiol. Ecol. 10: 45–53.

Pourriot, R., 1977. Food and feeding habits of rotifera. Arch. Hydrobiol. Beih. Ergebn. Limnol. 8: 243–260.

Riemann, E. & M. Søndergaard, 1986. Regulation of bacterial secondary production in two eutrophic lakes and in experimental enclosures. Plankt. Res. 8: 519–536.

Ruttner-Kolisko, A., 1977. Suggestions for biomass calculation of plankton rotifers. Arch. Hydrobiol. Beih. Ergebn. Limnol. 8: 71–76.

Skibbe, O., 1994. An improved quantitative protargol stain for ciliates and other planktonic protists. Arch. Hydrobiol. 130: 339–347.

Utermöhl, H., 1958. Zur Vervollkommnung der quantitativen Phytoplankton-Methodik. Mitt. int. Ver. Limnol. 9: 1–38.

Weisse, T., 1991. Ecological Characteristics of autotrophic picoplankton in a prealpine Lake. Int. Revue ges. Hydrobiol. 76: 493–504.

Zaiss, U., 1985. Physiologische und ökologische Untersuchungen zur Regulation der Phosphatspeicherung bei *Oscillatoria redekei*. Arch. Hydrobiol./Suppl.72: 166–219.

Hydrobiologia **342/343**: 175–183, 1997.
L. Kufel, A. Prejs & J. I. Rybak (eds), Shallow Lakes '95.

Zooplankton–phytoplankton interactions in lakes Võrtsjärv, Peipsi (Estonia) and Yaskhan (Turkmenia)

Tiina Nõges
Võrtsjärv Limnological Station of the Institute of Zoology and Botany, Rannu, Tartu County, Estonia EE2454 & Institute of Zoology and Hydrobiology, Tartu University, Vanemuise 46, EE 2400, Tartu, Estonia

Key words: zooplankton, phytoplankton, interactions, grazing, nutrient regeneration, shallow eutrophic lake

Abstract

Zooplankton-phytoplankton interactions have been studied in three shallow lakes of different trophic state. In strongly eutrophic, large and very shallow Lake Võrtsjärv the grazing does not play a leading role in controlling phytoplankton production and its standing stock. Small-size zooplankton can not eat filamentous blue-greens. The nutrient regeneration by zooplankton has a weak impact on phytoplankton, the latter being limited rather by underwater light than by nutrient availability. In large moderately eutrophic Lake Peipsi the presence of concentrated zooplankton in the experimental vessel mostly stimulated steady-state phytoplankton growth and negative grazing values were measured. Most probably the nutrients (N, P), excreted by zooplankton in grazing chamber stimulated the growth of larger phytoplankton which dominated because of heavy grazing pressure on edible forms. In Lake Peipsi phytoplankton seems to be nutrient-limited and heavily controlled by zooplankton community. In eutrophic, macrophyte-dominated Lake Yaskhan zooplankton in grazing chamber also mainly stimulated the growth of phytoplankton community which seemed to be nutrient-limited because of strong competition with macrophytes.

Introduction

Herbivores exert their control over plant communities both directly, by selective or nonselective feeding, and indirectly, by altering resource availability and/or growth conditions. They can consume considerable quantities of plant nutrients and store them in their own bodies or release them in forms that are more or less available for subsequent uptake by plants (Elser, 1992).

The abundance of different species in phytoplankton is the result of the balance of reproduction on one hand and selective grazing, mortality and sedimentation on the other hand. In general, low community grazing rates would favor edible phytoplankton species, at high grazing rates inedible species would be favored by regenerated from edible species nutrients (Sterner, 1989).

As herbivorous zooplankton occupies the key position in the food chain of lakes determining its type (grazing or detrital), interactions between zooplankton and phytoplankton are a central topic in plankton ecology.

In this paper the results of grazing measurements in shallow eutrophic lakes Võrtsjärv, Peipsi and Yaskhan are presented, and the mechanisms of zooplankton-phytoplankton interactions in these waterbodies are evaluated.

Study sites, material and methods

Three shallow lakes of different surface area were studied (Table 1). They are located in Estonia (L. Võrtsjärv, L. Peipsi) and Turkmenia (L. Yaskhan).

In Lake Võrtsjärv 15 experiments were performed in 1984 and 1985 (May–November). Grazing of the dominant zooplankton species (Method 1 as described below) was measured during May–September 1984, and that of the *in situ* zooplankton community (Method 2) in September 1984–November 1985. Experiments in Lake Peipsi were carried out in

Table 1. Characteristics of studied lakes.

Lake	Location	Area	Mean depth	Max depth	Trophic state	Mean Secchi disc transparency
L. Vortsjärv	Estonia	270 km^2	2.8 m	6 m	highly eutrophic	0.9 m
L. Peipsi	Estonia/Russia	2611 km^2	8.3 m	12.9 m	moderately eutrophic	1.6 m
L. Yaskhan	Turkmenia, Kara-Kum desert		2.5 m	6.4 m	eutrophic, macrophyte-dominated	3 m

1985–1986 (twice a month from May to November), and in Lake Yaskhan in 1988 during three expeditions (in April, July and November) applying Method 2.

Method 1: experiments with the initial handling of animals

Zooplankton was collected from the lake with a plankton net. A fixed number of specimens of the dominant cladocerans (*Chydorus sphaericus* or *Bosmina coregoni* were introduced into 4 experimental bottles (20, 250 or 500 ml) which contained 150 μm screened water with either natural or ^{14}C-labeled lake phytoplankton. Two other bottles remained devoid of zooplankton and served as final control for estimating changes in the absence of zooplankton. The bottles were placed on a rotating wheel and exposed for 24 hours. The temperature was regulated according to that of the lake. After exposition the phytoplankton biomass and/or the concentration of chlorophyll *a* (chl *a*) were measured in both experimental and control bottles, and the ingestion rate (I) was calculated using the formula (Umnov, 1983):

$$I = \frac{1}{t} * \frac{V}{n} * \ln\frac{Nk}{No} * \frac{No * (Nk - N)}{Nk - No};$$

where

- I = zooplankton ingestion rate; (mg phytoplankton WW grazed per animal (or per mg zooplankton WW) per day,
- V = volume of the experimental vessel; ml,
- t = exposition time; days,
- n = zooplankton amount in one experimental bottle; individuals or mg WW ml^{-1},
- No = initial phytoplankton biomass (mg WW ml^{-1}; direct or calculated from the concentration of chl *a*),
- N and Nk = phytoplankton biomass (mg WW ml^{-1}) in experimental and control bottles at the end of experiment.

The clearance rate (C; ml animal^{-1} or ml mg^{-1} day^{-1}) was calculated using the formula:

$$C = I/B_{\text{phyto}};$$

where B_{phyto} is the average phytoplankton biomass (mg WW ml^{-1}).

In experiments with ^{14}C-labeled algae, the clearance rate was calculated initially by the formula (Tackx & Van de Vrie, 1985):

$$C = \frac{1}{t} * \frac{V}{n} * \ln\frac{Rk}{R};$$

where R and Rk = radioactivity of particles (DPM ml^{-1}) in experimental and control bottles at the end of experiment.

The ingestion rate was calculated as

$$I = C * B_{\text{phyto}}$$

Method 2: experiments with in situ *zooplankton community without handling*

These experiments were carried out in a twin water sampler (see more in detail in Nõges, 1992), a special device designed for *in situ* grazing experiments on the natural zooplankton community. Zooplankton was concentrated in the grazing chamber, the lower opening of which was covered with a plankton net (pore size of 150 μm), by pulling the device through the water column from the bottom to the surface (5 m). Both openings of the control chamber were covered with a plankton net, which prevented the intrusion of zooplankton in the vessel. The volume of each chamber was 2 l. After filling, the grazing device was fastened on a rotating wheel and exposed, for 24 h, to temperature similar to that of the lake. Samples for chl *a* as well as for phytoplankton biomass analyses were taken from both chambers through lower taps before and after exposition. At the end of experiments, the amount and species composition of zooplankton in the grazing chamber was estimated. The grazing rate was calculated according to the formula (Geller, 1984):

$$I = \frac{1}{t} * \frac{V}{n} * \ln\frac{Nk}{Nko} * \frac{Nk * No - N * Nko}{Nk - Nko};$$

where

I = zooplankton ingestion rate, mg phytoplankton WW grazed per animal (or per mg zooplankton WW) per day,

V = volume of the experimental vessel; ml,

t = exposition time; days,

n = zooplankton amount in the experimental vessel, individuals or mg WW ml^{-1}

No and Nko = phytoplankton biomass (mg WW ml^{-1}; direct or calculated from the concentration of chl *a*) in experimental and control vessels at the beginning of experiment;

N and Nk = phytoplankton biomass (mg WW ml^{-1}) in the experimental and control vessels at the end of experiment.

The concentrations of chlorophyll's (*a, b, c*) and pheopigment's were determined spectrophotometrically (Recommendations..., 1979). The excretion of phosphorus by zooplankton was calculated using the formula worked out by Gutelmakher (1986):

$$E = 1.047 * \mathrm{DW}^{0.801},$$

where

E = excretion of phosphorus at 20 °C, mg P per individual per hour;

DW = dry weight of one individual. For temperatures other than 20 °C, correction was applied.

The following coefficients were used in calculations: for phytoplankton, 1 mg wet weight = 0.1 mg C = 0.002 mg chl *a* (Nauwerk, 1963; Vollenweider et al., 1974; Bulyon, 1976); for zooplankton, 1 mg wet weight = 0.167 mg dry weight (Waters, 1977). For carbon-phosphorus recalculations the Redfield ratio: C/P = 106 mol mol^{-1} was applied (Lynch & Hobbie, 1988).

Results

Lake Võrtsjärv

The algal community is represented by an association of filamentous algae, the species of *Aulacoseira* in spring, *Oscillatoria amphibia* f. *tenuis* (Anissim.) Elenk., *Planktolyngbya limnetica* (Lemm.) Kom. et Cronb. and *L. redekei* (van Goor) Meffert in summer and autumn are accompanied by a low variable biomass of small algae, mostly chlorophytes and chrysophytes. During the last decades phytoplankton biomass has seldom exceeded 30 g WW m^{-3}. The yearly average particulate primary production is 243 g C m^{-2} and chlorophyll *a* concentration 29 mg m^{-3}. The dominating zooplankton species are *Anuraeopsis fissa* (Gosse), *Keratella cochlearis* (Gosse), *Keratella quadrata frenzeli* (Eckst.), *Polyarthra dolichoptera* Idels (in winter), *Polyarthra luminosa* Kut., *Synchaeta verrucosa* Nipk. (in winter), *Trichocerca rousseleti* (Voigt), *Bosmina c. coregoni* Baird, *Bosmina longirostris* (Müller), *Chydorus sphaericus* (Müller), *Daphnia cucullata* Sars and species of the genus *Mesocyclops*. The number of zooplankton varies from 168 to 4048 thousand individuals per m^3, biomass from 0.1 to 2.6 g WW m^{-3} (Haberman et al., in press).

The average total number of bacteria is 3.2 millions cells ml^{-1}. The average production of heterotrophic bacteria 90 g C m^{-2} $year^{-1}$. About 500 species and forms of bottom animals have been found in Lake Võrtsjärv while Chironomidae hold the first position with 136 species and species groups, being followed by Mollusca (70 species), and Oligochaeta (54 species). The most abundant fish in Lake Võrtsjärv is bream (*Abramis brama* (L.)), its mean biomass in 1981–1990 was 47.9 kg ha^{-1}, production 31.1 kg ha^{-1}. Pike-perch (*Stizostedion lucioperca* (L.)) and pike (*Esox lucius* L.) occupy the leading position among predators, their biomasses are 11.5 kg ha^{-1} and 7.8 kg ha^{-1}, respectively. (Nõges et al., in press).

The clearance rate (C) on phytoplankton was 0.35–0.62 ml $animal^{-1}$ day^{-1} (57–102 ml mg WW day^{-1}), the ingestion rate (I) 5.8–10 mg WW $animal^{-1}$ day^{-1} (1.14–1.60 mg WW mg WW day^{-1} or 1.43–2 mg C mg C^{-1} day^{-1}). C and I were the highest in June, at the end of August and in the beginning of September. The lowest or negative values were measured in May, July, at the beginning of August, and at the end of September. Negative values indicate the stimulation of phytoplankton growth by the presence of zooplankton.

For calculating the grazing pressure (GR) of the dominant cladoceran community, the cladocerans' biomass (Haberman, 1984), phytoplankton biomass and integral primary production PPint were used. Maximum GR up to 2.5 g WW m^{-3} day^{-1} occurred in September forming 12% of the phytoplankton biomass and 133% of primary production per day (Figure 1).

The average value of GR was 0.84 g WW m^{-3} day^{-1} which was 2.5% of Bphyto per day and 28.5% of PPint per day.

Lake Peipsi

In phytoplankton diatoms, mostly the *Aulacoseira* species are dominant during the whole vegetation period. The abundance of smaller centric diatoms. *Stephanodiscus* and *Cyclotella* reach their peak densities in June. In summer and autumn blue-green algae *Aphanothece saxicola* Näg., *Microcystis aeruginosa* Kütz. supplement the species list. In zooplankton rotifers *Synchaeta verrucosa*, *Polyarthra dolichoptera*, *Keratella quadrata* dominate in spring, cladocerans *Bosmina* sp., *Daphnia* sp., juvenil and adult copepods *Eudiaptomus gracilis* (G. O. Sars), *Mesocyclops leukarti* Claus in summer and autumn (Nõges et al., 1993). Total count of bacteria is on the average 2.5 million cells ml^{-1}. The profundal zoobenthos is typical for eutrophic lakes, dominated by *Chironomus plumosus*. Among big clams, *Dreissena polymorpha* Pallas introduced in 1930s has become the dominating animal in the lake (Timm et al., 1990). Main commercial fishes are vendace (*Coregonus albula* (L.)) which abundance has sharply decreased in recent years, pike-perch, burbot (*Lota lota lota* (L.)), lake whitefish (*Coregonus lavaretus maraenoides* Poljakow), roach (*Rutilus rutilus* (L.)) and ruff (*Acerina cermia* (L.)) (Pihu, 1996).

The presence of zooplankton in the experimental vessel frequently stimulated phytoplankton growth: in many cases negative grazing values were found (Figure 2) and it was impossible to calculate average grazing.

The average values of production per vegetation period were: phytoplankton 204 g C m^{-2}, bacterioplankton 38 g C m^{-2}; herbivorous zooplankton 20.6 g C m^{-2} and predatory zooplankton 1.5 g C m^{-2}. The average ratio of filter-feeding zooplankton production (Pfzp) and primary production (PPint) during the vegetative period was 10% (Nõges et al., 1993).

Lake Yaskhan

The species composition of phytoplankton is characteristic of eutrophic lakes. Plankton diatoms, particularly *Synedra acus* var. *angustissima* Grun. dominate in spring. The bottom-living blue-greens (*Microcystis grevillei* (Hass.) Elenk., *Aphanothece stagnina* (Spreng.) A. Br., *A. salina* Elenk. et Danil). are quite numerous in water throughout the whole vegetation period. Blue-greens (*Cyanodictyon imperfectum* Cronb. et Weibull, *Gomphosphaeria* spp.) and green algae (*Cosmarium* spp., *Monoraphidium* spp., *Oocystis* spp., *Planktonema lauterbornii* Schmidle) dominate in summer. In autumn diatoms, blue-greens and green algae are represented more or less equally. The biomass is rather small, fluctuating from 0.2 to 7.2 g m^{-1}, the content of chlorophyll a is 2.4–4.4 mg m^{-3} in the surface layer and reaches up to 46.7 mg m^{-3} near the bottom. The scantiness of phytoplankton is connected with abundant macrophyte vegetation as a competitor for nutrients and a producer of plentiful detritus. The number of free-living bacteria ranges from 3.1 to 6.5 millions per ml. Small rotifers, (*Keratella quadrata*, *Synchaeta oblonga* Ehr. and *Brachionus plicatilis* Müll.) dominate in zooplankton in spring and summer with a modest biomass (0.3–5.5 g WW m^{-3}) but having a great production capacity (5–280 mg C m^{-2} day^{-1}). The macrozoobenthos of Lake Yaskhan is rather poor as a result of strong pressure by benthivorous fish. The mean value of biomass in 1988 was 0.5 g WW m^{-2}. The prevailing group is Chironomidae. Seven fish species have been recorded in the lake, five being the most abundant: bream, roach, mosquito fish (*Gambusia affinis* Baird et Girard), perch (*Perca fluviatilis* L.) and wild carp (*Cyprinus carpio* L.) (Kirsipuu et al., 1993).

The stimulation of the formation of chlorophyll *a* and *b* occurred in the presence of concentrated zooplankton in all experiments. The concentration of chlorophyll *c* in the experimental vessel decreased slightly in July and November, while the concentration of pheopigment's increased in spring and summer but decreased in autumn (Figure 3).

The amount of phosphorus excreted by zooplankton was calculated on the basis of the number and biomass of zooplankton (Haberman, 1993). The consumption of phosphorus by phytoplankton was calculated from daily integral primary production (Nõges, 1993) using the Redfield ratio. In spring and autumn the amount of phosphorus consumed by phytoplankton exceeded the amount of that element recycled by zooplankton, while in summer the situation was opposite (Figure 3).

Discussion

The situation in nature reflects more or less steady-state conditions: the formation of any compound or organ-

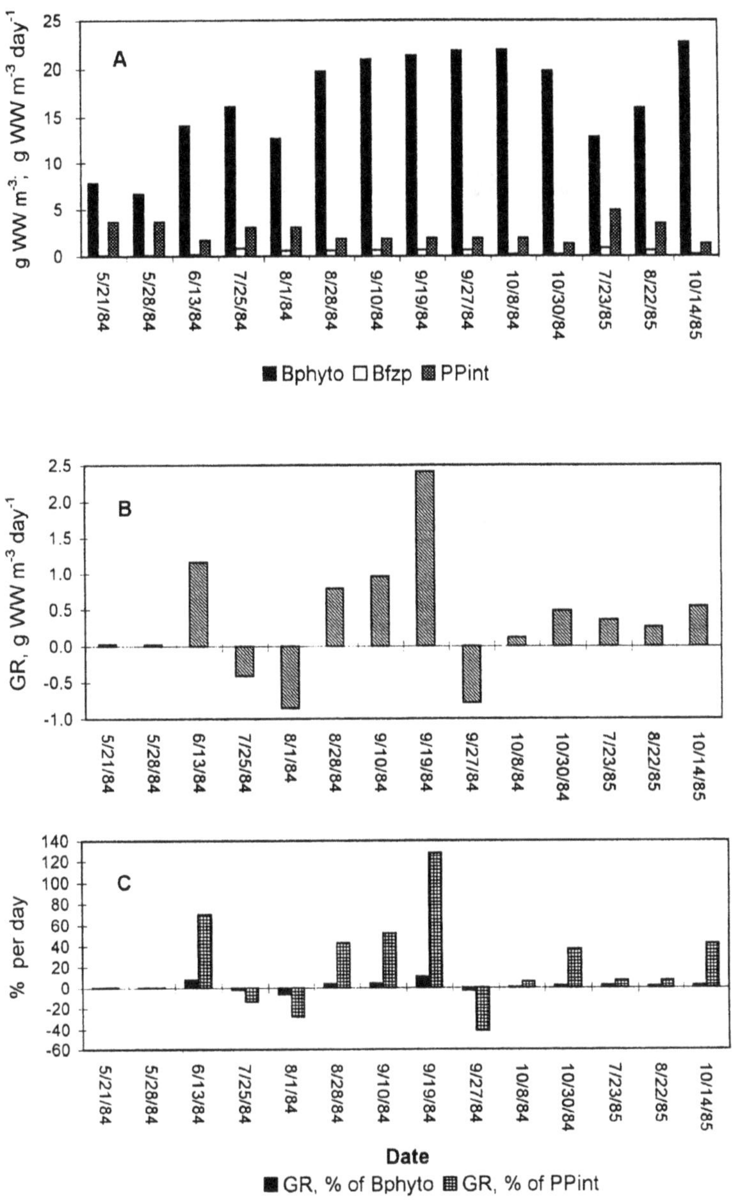

Figure 1. A: biomass of phytoplankton (Bphyto; g WW m^{-3}) and filter-feeding zooplankton (Bfzp; g WW m^{-3}) and primary production (PPint; g WW m^{-3} day^{-1}); B and C: grazing (GR) on Bphyto (g WW m^{-3} day^{-1} and % per day) and on PPint (% per day) in Lake Võrtsjärv in 1984 and 1985.

ism is balanced with its consumption. The accumulation indicates the lack of consumers or the disturbance of the steady state owing to rapid changes in physico-chemical conditions. In experiments the steady state is sometimes disturbed artificially. Changes which take place for achieving a new steady state can give information about processes which are well balanced and therefore hidden under undisturbed conditions.

Two different feedback mechanisms are permanently coexisting in zooplankton-phytoplankton interactions:

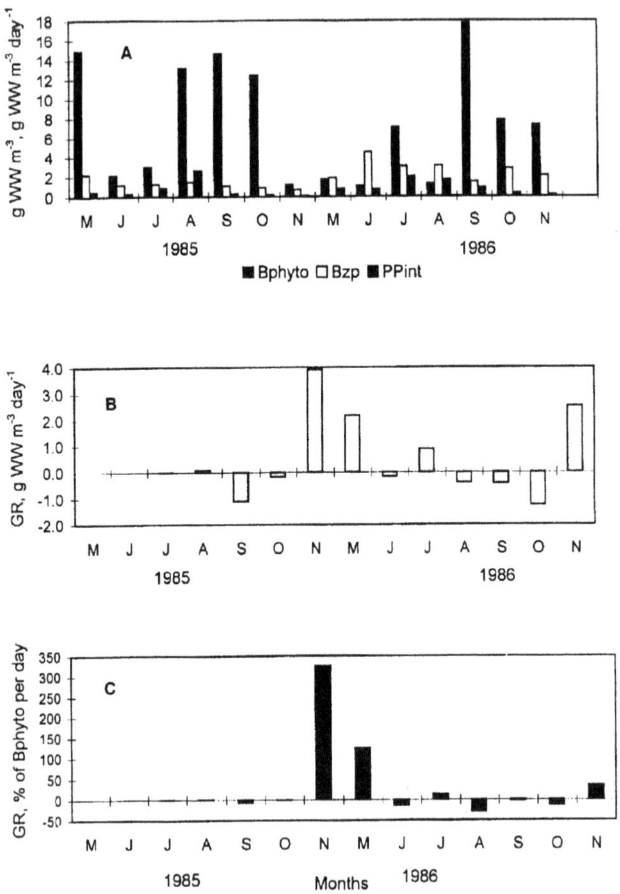

Figure 2. A: biomass of phytoplankton (Bphyto; g WW m^{-3}) and zooplankton (Bzp; g WW m^{-3}) and primary production (PPint; g WW m^{-3} day^{-1}); B and C: grazing (GR) on Bphyto (g WW m^{-3} day^{-1} and % per day) in Lake Peipsi in 1985 and 1986.

1. Grazing on edible phytoplankton species and decreasing of their amount.
2. Nutrient regeneration by zooplankton and stimulation of phytoplankton growth.

The prevailing of the first effect over second one is revealed in positive community grazing value, and *vice versa*. In our experiments zooplankton was eliminated from one vessel and its concentration was increased in the other. If phytoplankton amount was mainly determined by grazing mortality, the increase of zooplankton abundance resulted in the decrease of phytoplankton biomass and the elimination of zooplankton favored algal development. In the case of nutrient limitation the stimulation of phytoplankton occurred at the increased zooplankton concentration due to higher nutrient regeneration rate.

The first-type feedback prevailed quite frequently in Lake Võrtsjärv where concentrated zooplankton rather eliminated phytoplankton than stimulated its growth (Figure 1B and C) In Lake Peipsi grazing was positive only in early spring and late autumn (Figure 2B) while in L. Yaskhan stimulation (negative grazing) was overwhelmingly dominating.

The prevalence of positive grazing values in Lake Võrtsjärv is most probably caused by the fact that phytoplankton growth in this lake is generally not nutrient-limited. Only in late summer and autumn in periods of relatively low nitrogen concentration, nutrient limitation might appear causing negative grazing values. As also the other investigations proved primary production in that lake being rather light- than nutrient-limited (Nõges & Nõges, in press), no strong stimu-

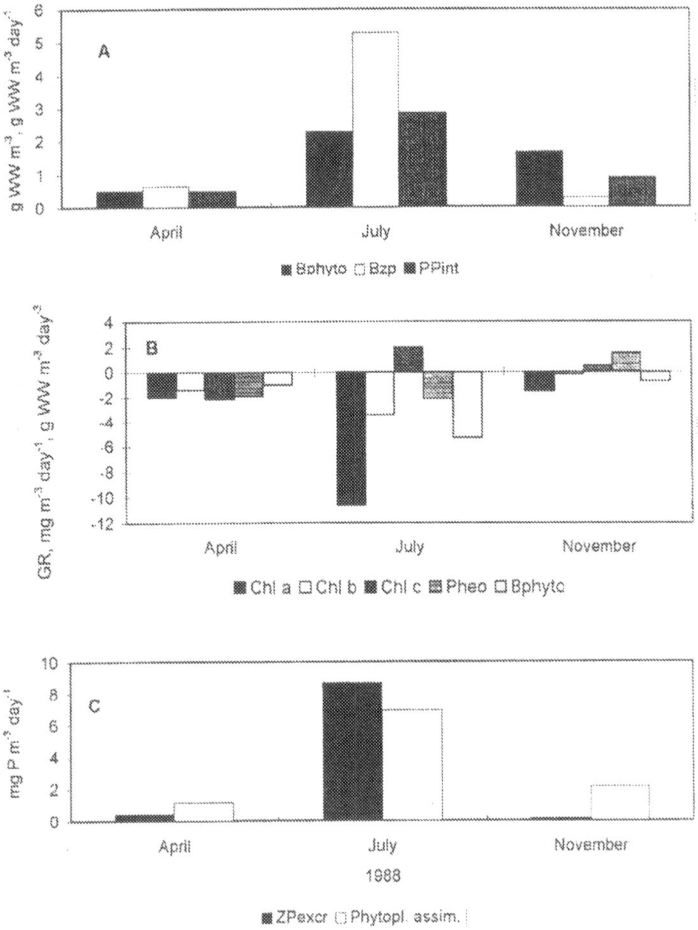

Figure 3. A: biomass of phytoplankton (Bphyto; g WW m^{-3}) and zooplankton (Bzp; g WW m^{-3}) and primary production (PPint; g WW m^{-3} day^{-1}); B: grazing (GR) on Bphyto (g WW m^{-3} day^{-1}), and different chlorophyll forms and pheopigments (mg m^{-3} day^{-1}); C: phosphorus excretion by zooplankton (ZPexcr; mg P m^{-3} day^{-1}) and its assimilation by phytoplankton (C; mg P m^{-3} day^{-1}) in Lake Yaskhan in 1988.

lation of phytoplankton by zooplankton was assumed. On the other hand, considering the species composition of phyto- and zooplankton (great filamentous algae as *Aulacoseira* sp., *Limnothrix redkei, Planktolyngbya limnetica* and small zooplankton as rotifers, *Bosmina coregoni coregoni, Chydorus sphaericus*) one can not expect high grazing pressure on phytoplankton. If to compare C and I in L. Võrtsjärv with some values from the literature (Peters & Downing, 1984; Lampert & Taylor, 1985; Gulati, 1984; James & Forsyth, 1990; Jarvis, 1986; Kankaala, 1988; Gutelmakher, 1988) quite modest grazing intensity on phytoplankton could be noticed in Lake Võrtsjärv, as it was expected. This provides also an explanation to the fact that no distinct zooplankton-caused clear-water phase occurs in Lake Võrtsjärv (Nõges & Nõges, in press). According to Lampert (1988) the grazing rate of 80% of phytoplankton biomass per day should be sufficient to balance the growth rate of algae. In Lake Võrtsjärv the highest grazing rate was 11.7% of phytoplankton biomass per day (average 2.46%) which means that zooplankton can not control phytoplankton biomass in this lake. Grazing losses (2.5% of biomass per day) were 10 times less than the daily increase of phytoplankton standing stock (P/B = 25%). The detrital food chain is evidently prevalent in Lake Võrtsjärv. This assumption is also supported by low ratio of yearly herbivorous zooplankton production and primary production (Pfzp/PPint = 2%) in that lake indicating low efficiency of energy transformation from zooplankton to phytoplankton.

The second effect predominated in Lakes Peipsi and Yaskhan being considered as nutrient-limited environments. Zooplankton grazing was intensive and the proportion of edible phytoplankton low due to high grazing pressure. In such cases the concentrated zooplankton caused rather an increase than a decrease of phytoplankton biomass in experimental vessel.

Considering the yearly ratio of filter-feeding zooplankton production and phytoplankton primary production (10%) in Lake Peipsi, the herbivorous zooplankton in that lake seems to feed mostly on living algae, while the detrital chain is of less importance. However, the large filamentous or colonial algae such as *Aulacoseira* ssp. and *Aphanothece saxicola*, which are not suitable food for zooplankton, dominate in phytoplankton during the major part of the vegetation period (Nõges et al., 1992). It is likely that phytoplankton species consumable for zooplankton, are eliminated very intensively and therefore their share in the algal biomass remains permanently low. The study revealed the stimulative effect of the presence of zooplankton on phytoplankton. Lake water in which the content of small algae is low due to high grazing pressure, does not provide abundant food for concentrated zooplankton in the grazing chamber and, therefore, the elimination of phytoplankton is low. The stimulative effect of nutrients (N, P), excreted by zooplankton leads to the intensification of non-grazed phytoplankton production.

In Lake Yaskhan rotifers (*Keratella quadrata, Synchaeta oblonga*) dominated in zooplankton in April while colonial cyanophytes (*Cyanodictyon* spp.) and quite large diatoms (*Synedra acus* var. *angustissima* Grun.) were numerous in phytoplankton (Laugaste & Ott, 1993). It is evident that small rotifers can not graze on large phytoplankton. The requirements of phytoplankton for mineral phosphorus were higher than the recycling ability of zooplankton (Figure 3C); therefore, the increased amount of zooplankton in experiment gave an additional portion of phosphorus and stimulated phytoplankton growth.

In summer concentrated zooplankton also stimulated phytoplankton growth, but the decrease of chlorophyll *c* concentration in experimental vessel, may indicate the grazing of zooplankton on diatoms, which contain chl *c* (Wetzel, 1975). The rather high proportion (32%) of the macrofiltrator *Diaphanosoma brachyurum* in zooplankton biomass in July (Haberman, 1993) as well as that of small chlorophytes e.g. (*Cosmarium* spp. and *Oocystis* spp. in phytoplankton biomass (42%) (Laugaste & Ott, 1993) gives reason to assume the important role of phytoplankton in zooplankton food. It is hard to believe, however, that algae could provide all food for zooplankton whose biomass exceeded three times the biomass of phytoplankton (Figure 3A). Therefore, the important role of bacteria and detritus in zooplankton food was presumed. This assumption is supported by the intensive excretion of phosphorus by zooplankton which even exceeded its consumption by phytoplankton (Figure 3C).

Zooplankton biomass as well as its phosphorus regeneration ability was low in November. Primary production was evidently based on nutrients released by bacteria rather than zooplankton. The intensity of bacterial development, however, was not as high as in spring and summer (Nõges, 1993), and phytoplankton was suffering from nutrient deficiency. The increase in chl *a* concentration and decrease in pheopigment's in the experimental water sampler (Fig. 3B) might indicate that concentrated zooplankton released phosphorus mainly from detritus and therefore stimulated phytoplankton growth. High phyto- to zooplankton biomass ratio in November (5.8) (Figure 3A) conditioned the accumulation of 'edible' chlorophytes (*Oocystis* spp., *Crucigenia tetrapedia* (Kirchn.) W. et G. S. West) (Laugaste & Ott, 1993). It is hard to understand why zooplankton did not consume these algae. Changes toward the winter community could have been taking place in zooplankton, the high proportion of cyanophytes in phytoplankton (45% of biomass) or rather low water temperature (14 °C) in comparison with summer (30 °C) could have a suppressive influence on zooplankton activity etc.

Conclusions

The experiments of zooplankton-phytoplankton interactions gave basis to conclude that phytoplankton growth in lakes Peipsi and Yaskhan is limited by nutrients and depends on their recycling by herbivorous zooplankton. In Lake Võrtsjärv the linkage between primary producers and herbivores is weak. In this lake grazing does not control algal growth and biomass while phytoplankton is not much dependent of zooplankton nutrient regeneration.

References

Bulyon, V. V., 1976. Soderzhaniye organicheskogo veshchestva v fotosinteticheskih pigmentah v planktone ozera Baikal. [Organic matter content in photosynthetic pigments of the plankton in Lake Baikal]. In Winberg, G. G. (ed.), Gidrobiologicheskiye osnovy samoochishcheniya vod [Hydrobiological basis of water self-purification.] Nauka, Leningrad (St. Petersburg).

Elser, J. J., 1992, Phytoplankton dynamics and the role of grazers in Castle Lake, California. Ecology 73: 887–902.

Geller, W., 1984. A device for the *in situ* measurement of zooplankton food selection, grazing and respiration rate. Verh. int. Ver. Limnol. 22: 1425–1431.

Gulati, R. D., 1984. The zooplankton and its grazing as measures of trophy in Loosdrecht lakes. Verh. int. Ver. Limnol. 19: 1202–1210.

Gutelmakher, B. L., 1986. Metabolizm planktona kak edinogo tselogo: trofometabolicheskiye vzaimodeistviya zoo- i fitoplanktona [Metabolism of the plankton community as a whole: trophic relations of zoo- and phytoplankton.] Leningrad (St. Petersburg), 155 pp.

Haberman, J., 1984. Võrtsjärve zooplankton aastail 1964–1982. Võrtsjärve ökosüsteemi seisund [Zooplankton of Lake Võrtsjärv in 1964–1982. The Ecosystem of Lake Võrtsjärv]. Tartu: 65–78.

Haberman, J., 1993. Zooplankton of Lake Yaskhan. Limnologica 23: 215–225.

Haberman, J., P. Nõges, E. Pihu, T. Nõges, K. Kangur & V. Kisand. Characterization of L. Võrtsjärv. Limnologica (in press).

James, M. R. & D. J. Forsyth, 1990. Zooplankton-phytoplankton interactions in a eutrophic lake. J. Plankton Res. 12: 455–472.

Jarvis, A. C., 1986. Zooplankton community grazing in a hypertrophic lake (Hartbeespoort Dam, South Africa). J. Plankton Res. 8: 1065–1078.

Kankaala, P., 1988. The relative importance of algae and bacteria as food for *Daphnia longispina* (Cladocera) in a polyhumic lake. Freshwat. Biol. 19: 285–296.

Kirsipuu, A., T. Kallejärv, R. Laugaste, J. Haberman, H. Timm, A. Järvalt & P. Nõges, 1993. Brief characterization of Lake Yaskhan. Limnologica 23: 179–182.

Lampert, W., 1988. The relationship between zooplankton biomass and grazing: a review. Limnologica 19: 11–20.

Lampert, W. & B. E. Taylor, 1985. Zooplankton grazing in a eutrophic lake: implications of diel vertical migration. Ecology. 66: 62–82.

Laugaste, R. & I. Ott, 1993. Phytoplankton of Lake Yaskhan. Limnologica 23: 195–214.

Lynch, J. M. & I. E. Hobbie, 1988. Micro-organisms in Action: Concepts and Applications in Microbial Ecology. Oxford–Melbourne, 363 pp.

Nauwerk, A., 1963. Die Beziehung Zwishen Zooplankton and Phytoplankton im See Erken. Symbolae Botanicae Uppsaliensis 17: 1–163.

Nõges, T. 1992. Comparison of two methods of zooplankton grazing measurements. Int. Revue ges. Hydrobiol. 77: 665–672.

Nõges, P., 1993. Hydrochemical characteristics, productivity of phytoplankton, bacterioplankton and amphiphytes of the freshwater part of L. Yaskhan. Limnologica 23: 183–193.

Nõges, T., J. Haberman, H. Tammert, M. Timm, I. Ott & P. Nõges, 1992. Ecological relations of main plankton components in the pelagial of Lake Peipsi. Proc. Estionian Acad. Sci. Ecol. 2: 137–155.

Nõges, T., J. Haberman, M. Timm & P. Nõges, 1993. The seasonal dynamics and trophic relations of the plankton components in Lake Peipsi (Peipus). Int. Revue ges. Hydrobiol. 77: 513–519.

Nõges, T. & P. Nõges. Primary production of L. Võrtsjärv. Limnologica. (in press)

Nõges, T., P. Nõges, J. Haberman, V. Kisand, K. Kangur, A. Kangur & A. Järvalt. Food web structure in shallow eutrophic Lake Võrtsjärv (Estonia). Limnologica (in press).

Peters, R. H. M. & J. A. Downing, 1984. Empirical analysis of zooplankton filtering and feeding rates. Limnol. Oceanogr. 29: 763–784.

Pihu, E., 1996. Fishes, their biology and fisheries management in Lake Peipsi. Hydrobiologia 338: 163–172.

Recommendations for marine biological studies in the Baltic Sea. Phytoplankton and Chlorophyll. 1979. BMB Publications No 5, 38 pp.

Sterner, R. W., 1989. The role of grazers in phytoplankton succession. In Sommer, U. (ed.), Plankton Ecology. Succession in plankton communities. Brock/Springer Series in Contemporary Bioscience. Springer Verlag, 369 pp.

Tackx, M. L. M. & E. M. van de Vrie, 1985. Calculations of results in grazing experiments using the counting method. Hydrobiol. Bull. 19: 29–36.

Timm, H., K. Kangur, T. Timm & V. Timm, 1990. Peipsi makrozoobentose seisund [State of macrozoobenthos of Lake Peipsi]. In Timm, T. (ed.), Peipsi järve seisund [State of Lake Peipsi.] Tartu: 123–126.

Umnov, A. A., 1983. Otsenka tochnosti rascheta skorosti filtratsij vody i ratsiona vodnyh zhivotnyh s filtratsionnym tipom pitaniya [Precision of filtration activity and ration calculations of filtrative-feeding animals]. In Winberg, G. G. (ed.), Troficheskie svyazi i ih rol' v produktivnosti prirodnyh vodoemov [Trophic relations and their role in production of natural water bodies]. Nauka. Leningrad (St. Petersburg): 90–99.

Vollenweider, R. A, M. Munavar & P. Stadelman, 1974. A comparative review of phytoplankton and primary production in the Laurential Great Lakes. J. Fish. Res. Bd Can. 31: 739–762.

Waters, T. F, 1977. Secondary production in inland waters. In McFayden, A. (ed.), Advances in Ecological Research. London–San Francisco: 91–164.

Wetzel, R. G. 1983. Limnology. Saunders College Publishing, Philadelphia–Toronto, 767 pp.

Hydrobiologia **342/343:** 185–195, 1997.
L. Kufel, A. Prejs & J. I. Rybak (eds), Shallow Lakes '95.

Biotic interactions of the zooplankton community of a shallow, humic lake

Teresa Węgleńska[1], Jolanta Ejsmont-Karabin[2] & Jan Igor Rybak[3]
[1]*Department of Hydrobiology, Institute of Ecology, Polish Academy of Sciences, Dziekanów Leśny, 05-092 Łomianki, Poland*
[2]*Hydrobiological Station, Institute of Ecology, Polish Academy of Sciences, 11-730 Mikołajki, Poland*
[3]*Department of Hydrobiology, University of Warsaw, Stefana Banacha 2, 02-097 Warszawa, Poland*

Key words: humic lake, zooplankton, biotic interactions, vertical and horizontal migrations

Abstract

The objective of this paper was to recognize the predator-prey relations and competitive interactions in the zooplankton community of a shallow, humic Lake Flosek (Great Masurian Lake District). The year to year changes in the seasonal pattern and sequence of peak abundance of herbivore and predatory Crustacea and Rotifera were observed. Zooplankton in this lake showed also a distinct vertical distribution and more or less visible diel vertical migrations as well as the horizontal migration of Crustacea species between pelagial and near-shore zones of lake in summer period. The changes of demographic events like the decrease of mean body size of adults as well as decrease of fecundity along the inshore-offshore transect were also observed. This distinct habitat partition between crustaceans and rotifers can reflect a combined effect of food scarcity and predator pressure.

Zooplankton of the shallow Lake Flosek is controlled by the complicate net of feed-back relations, resulting in a rather stable community structure which can explain its rather unsignificant changes in the period of about 20 years after the liming of water in this lake.

Introduction

The food suspension for plankton filtrators in humic lakes is usually scarce due to low primary production, inadequate composition (Dinophyceae dominate) and low abundance of planktonic algae (Kwiatkowski & Roff, 1976; Olem, 1991). In this situation one may expect rather strong 'bottom-up' control which, in turn, enhances considerably the food competition between different species of rotifers and crustaceans (Roff & Kwiatkowski, 1977; Stenson, 1985). At the same time the shallowness of the lake, high water transparency and lack of typical littoral (with its submerged and emergent macrophytes) do not supply sufficient space to accommodate strong predator-prey relations. Therefore, a strong 'top-down' control of planktonic herbivores in these lakes is expected (Salonen & Lehtovaara, 1992; Stenson & Svensson, 1994).

Several life strategies which enable the coexistence of competing species, and the effective defense mechanisms against predation can be observed in zooplankton community. Zooplankton may avoid interactions with potential predators by their life history modifications, diurnal vertical and horizontal migration patterns, and variety of behavioral, morphological and physiological adaptations (Havel, 1987; Gliwicz & Pijanowska, 1988).

This study describes the predator-prey relations and competitive interactions in the zooplankton community of a small, shallow, humic Lake Flosek. The aim of the study was to recognize whether the defensive life strategies observed among the zooplankton species of deep and stratified lakes also recorded in a shallow, humic lake.

Study sites

The investigation was carried out for two periods: 1966–1974 (1st) and 1990–1993 (2nd) in a small (area

4 ha) and shallow (z_{max} 8 m), polyhumic, mid-forest Lake Flosek (Great Masurian Lake District). Previously (1st) this lake had a slightly acid water (pH 5.5–6.5) and a low Ca concentration (5 mg l^{-1}); it changed considerably after liming (in 1970) and finally (2nd period) the lake has pH of water 7–8 and Ca concentration about 13 mg l^{-1} being still humic and oligotrophic. The emergent macrophytes in the near shore zone are scarce and there is no submerged macrophytes (Ozimek, unpubl. data). The low concentration of P-PO_4 (0.01–0.02 mg l^{-1}) and P total (0.02–0.05 mg l^{-1}), low gross primary production (about 41.8 KJ m^{-2} d^{-1} in summer period), low chlorophyll *a* concentration (about several μg per litre in summer), high water transparency (Secchi disc 4 m in summer) and low biomass of phytoplankton (1–5 mg F.W. l^{-1}) indicate consistently the low trophic status of studied lake in 1st as well as in 2nd period (Hillbricht-Ilkowska et al., 1977, and authors unpubl. data).

Lake Flosek has also an unique thermal and mictic regime i.e. the summer stratification with a sharp thermocline on depth 4 m lasts usually from May to the end of August. In the remaining periods the lake is mixed till the bottom. Occasionally, a drop of oxygen concentration over the bottom is observed in summer (from 0.5 to 2.5 mg l^{-1}). The concentration of various chemical compounds (like phosphates, calcium, ammonium, nitrogen, chlorophyll *a* and the values of gross primary production as well as phytoplankton biomass) show occasionally the sudden increases on depth 3 to 4 m in the periods of summer stratification (Hillbricht-Ilkowska, *loc. cit.*).

The phytoplankton in Lake Flosek is not abundant and has rather simplified species composition. The share of nanoplankton in total biomass is rather low (about 20%). In 1966 (before liming) dinoflagellates (*Gymnodinium* sp., *Ceratium hirundinella* (O. F. Müller), *Peridinium* sp.) were dominating in biomass (about 90%). After liming, the dinoflagellates were replaced by chrysophytes (several species of *Dinobryon*) and chloromonads (mainly *Vacuolaria viridis* (Dang) (Spodniewska unpubl. data).

The low trophic status of Lake Flosek was mainly responsible for simplified composition and low biomass of zooplankton (1–3 mg F.W. l^{-1}). The species typical for oligotrophy and the species with a wide environmental tolerance were dominating in zooplankton. The most abundant within rotifers were: *Polyarthra vulgaris* Carlin, *P. remata* Skorikov, *P. dolichoptera* Idelson, *Asplanchna priodonta* Gosse, *Keratella cochlearis* (Gosse), *Gastropus stylifer* Imhof, *Anureopsis fissa* (Gosse), *Collotheta mutabilis* (Huds.), *Conochiloides coenobasis* (Skorikov). Crustacean communities were dominated by *Ceriodaphnia quadrangula* 0. F. Müller, *Bosmina longirostris* 0. F. Müller, *Daphnia longispina* 0. F. Müller, *Eudiaptomus graciloides* (Lillj.), *Mesocyclops leuckarti* (Claus) and *Cyclops vicinus* Uljan. The marked temporal and spatial changes in the abundance and relative composition of zooplankton during study period were observed due to strong and fluctuating pressure of invertebrate predators (*Asplanchna priodonta*, Cyclopoida copepods and *Chaoborus flavicans* Meig. larvae) as well as to temporary strong pressure of fish (mainly bleak, roach, bream and tench) (Hillbricht-Ilkowska et al., 1977, authors unpubl. data).

Material and methods

Zooplankton were sampled monthly during vegetation period in 1966–1974 and 1990–1993 from the deepest part of the lake at 1 m intervals, from surface to the bottom. Additionally, in summer 1993–1994, zooplankton were sampled during 24 hour periods at different distances from the shore to mid lake zone for the study of diel vertical and horizontal migration of zooplankton. Samples of 30 litre volume were taken using the 5 litre Bernatowicz sampler, condensed by means of plankton net of 30 μm mesh size, and fixed with Lugol's solution and formalin 4%. The material was elaborated by routine methods according to Bottrell et al. (1976). The mean depth of the population (D) was calculated using the weighted expression

$$D = \sum_i D_i N_i / \sum_i N_i,$$

where D_i is depth of the ith sample and N_i the concentration of animals at that depth.

The pressure of invertebrate predator was estimated according to Ejsmont-Karabin (1974) and Gliwicz et al. (1978), as an 'animal' part of diurnal consumption, assuming, that it averages 50% of body weight for *Asplanchna priodonta* and 25% for copepodites and adults of cyclopoid copepods at 20 °C; and within the range of 5–25 °C, it alternates in accordance to the Krogh' curve. The chlorophyll *a* content was determined with the standard spectrophotometric method after acetone extraction (Golterman, 1969). Water temperature and oxygen concentration were measured by means of an oxygen metre YSI Model 57.

Results and discussion

Seasonal succession of zooplankton

The year to year changes of *Asplanchna priodonta* abundance in Lake Flosek may explain the observed fluctuations of density of its attractive prey i.e. *Keratella cochlearis*. This last small size species occurs in Lake Flosek in high densities only in years with low *A. priodonta* abundance (Figure 1).

Many species of planktonic rotifers secrete a layer of mucus that partially or completely surrounds the body of the animal. These mucus sheaths make the rotifers effectively larger and therefore less vulnerable to tactile predators (e.g. *A. priodonta* and copepods) without making them more visible to zooplanktivorous fish (Stemberg & Gilbert, 1987).

The high population density of *Conochiloides coenobasis* and other species with mucus sheaths in Lake Flosek (Figure 1) can be considered as the result of above mentional kind of accommodation.

The decrease of density of *K. cochlearis* which followed the decrease of abundance of *C. coenobasis* (Figure 1) can indicate also the strong interspecific competition among these two detritovorous species. The different seasonal succession and the sharp sequence in time of peak numbers of the both populations seems to be the example of the temporary separation of species with similar food requirements.

The year to year variability in the seasonal pattern and sequence of peak abundances of herbivore Calanoida (*Eudiaptomus graciloides*), predatory Cyclopoida (*Mesocyclops* sp. and *Cyclops vicinus*) and predatory midge larvae of *Chaoborus flavicans* can be also explained in terms of predator avoidance and exploitative food competition (Figure 2). In 1966 (before liming), the predator *Mesocyclops* sp. as well as its prey – nauplii of *E. graciloides* maintained very low population densities during the whole year, due to food scarcity. The low impact of invertebrate predation (Cyclopoida nad larvae of *C. flavicans*) also suggest that interspecific competition might have been relative weak.

The improvement of feeding conditions in 1973 after liming i.e. changes in the phytoplankton composition (increase of Chrysophyceae) and the increase of rotifers abundance enhanced the development of predatory Cyclopoida (*Mesocyclops* sp.) as well as herbivorous Calanoida (*Eudiaptomus graciloides*). The increase of the abundance of both Copepoda groups stimulated not only the prey–predator relations between adults and copepodits of *Mesocyclops* sp. and nauplii of *E. graciloides* but also the interspecific food competition between both species. It was possibly caused by the omnivorous nature of adults and copepodits of Cyclopoida (i.e. they can feed not only on the nauplii of *E. graciloides* but also on the other animals as well as plant material) and by the fact that the nauplial stages of Cyclopoida and Calanoida show dietary overlap. In effect of these predation and competition relations the marked temporal segregation in time of peak abundance of *E. graciloides* and *Mesocyclops* sp. was observed. The peak of reproduction and development of *E. graciloides* occurred in early spring, i.e. one month before the peak of the development of *Mesocyclops* sp. population. The sudden decrease of the abundance of both Copepoda groups took place in late summer and autumn, and was correlated significantly ($p < 0.05$, $r = 0.68$) with a strong increase of abundance of *C. flavicans* (Figure 2).

In 1992 the small (<1 mm) eurytopic species from *Mesocyclops* genus were replaced by big size (> 1 mm) cold stenothermic species *Cyclops vicinus*. The peak numbers of this latter species occurred in early spring and coincided with low impact of *Chaoborus* larvae. *C. vicinus* entered the diapause and rested in lake sediments at the same time when the abundance of predatory *Chaoborus* larvae began to increase. A possible reason for this is, that a time for entering the diapause Cyclopoida could be an adaptation for predator avoidance (Hairston, 1987). In 1992 a time of peak numbers of *E. graciloides* shifted towards the summer months, i.e. the period when the predatory pressure of *C. flavicans* was the weakest, and predatory Cyclopoida did not occur (Figure 2).

Vertical distribution and diel migration of zooplankton

Zooplankton in Lake Flosek is strongly stratified, and this pattern of vertical distribution changes seasonally. In this way, a distinct habitat partition between crustaceans and rotifers can be attributed to a combined effect of food scarcity and predation pressure.

Due to the forest surroundings and wind protection, a sharp thermic and oxygen stratification occurred in Lake Flosek in summer. Oxygen was depleted (below 1 mg l^{-1}) at the bottom, transparency of water was very high (Secchi disc up to 3.5 m) and maximal concentration of chlorophyll *a* was observed at 4 m (Figure 3). Feeding pressure of invertebrates was weak and rather uniform in the whole water column. It was the effect of the disappearance of *Asplanchna priodonta* in water

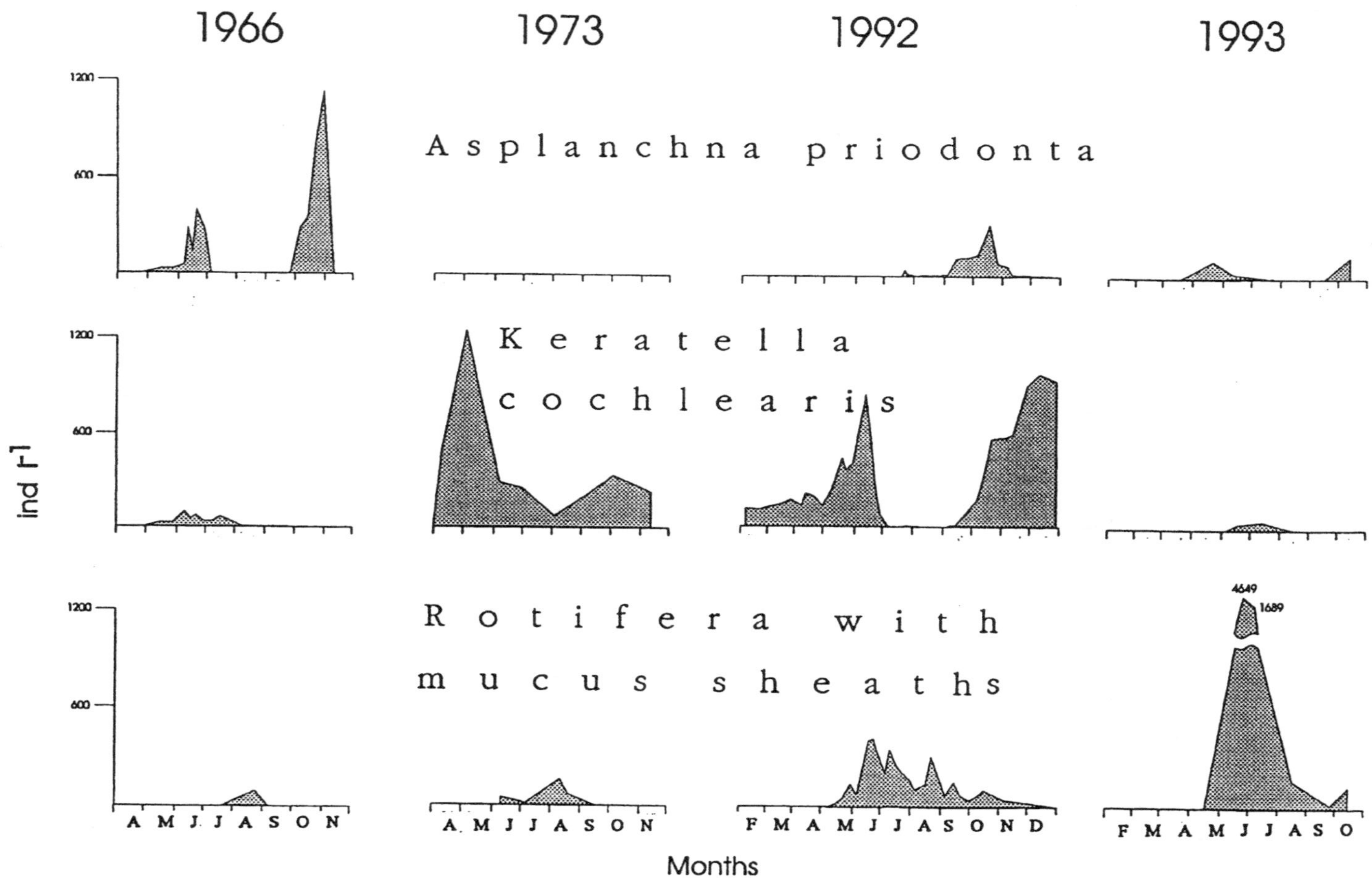

Figure 1. The pattern of seasonal dynamics of predatory rotifer (*Asplanchna priodonta*), its prey (*Keratella cochlearis*) and rotifers with mucus sheaths (*Collotheca mutabilis, Conochiloides coenobasis, Gastropus stylifer* in Lake Flosek in control year (before liming) and 3, 22 and 23 years after liming.

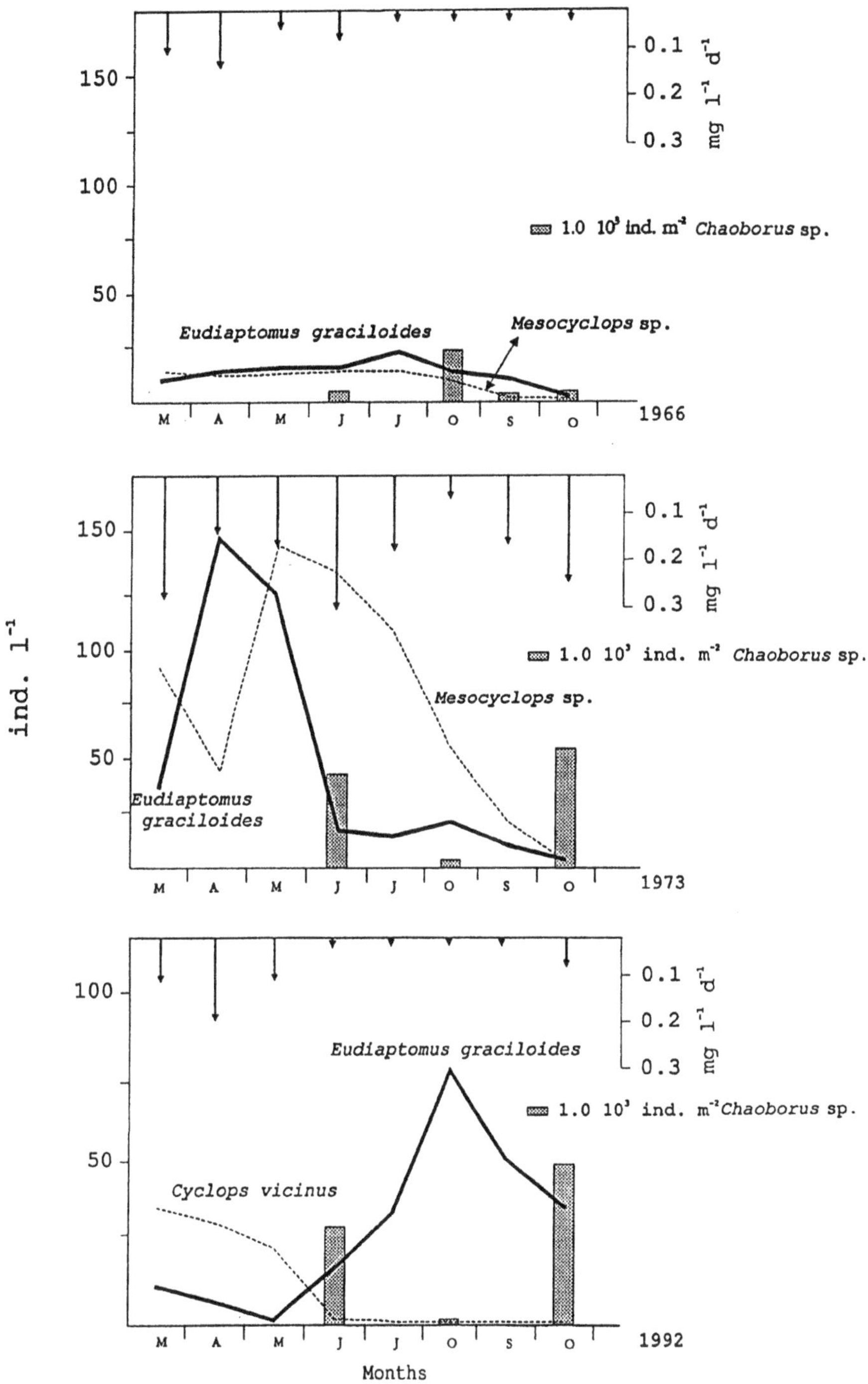

Figure 2. The pattern of seasonal dynamics of cyclopoid predators (*Mesocyclops* sp. and *Cyclops vicinus*) – dotted lines, and midge larvae *Chaoborus flavicans* – bars, and its potential prey (*Eudiaptomus graciloides*) solid lines, against changes of invertebrate predation pressure – arrows, (calculated as daily feeding rates from densities of cyclopoid copepods) in Lake Flosek in control year (before liming), and 3 and 22 years after liming.

column, and the small numbers of Cyclopoida as well as *Chaoborus flavicans* larvae. However, a rather high pressure of fish in summer 1993 was noted (Jaskóta pers. com.).

The density of zooplankton during a day was the highest near the almost anoxic bottom, close to the maximum concentration of phytoplankton (by chlorophyll *a*), and probably of bacterioplankton. The density of rotifers and crustaceans found at the bottom (at 5 m) was 8–9 times higher than near the surface. The increased crustacean number, and particularly that of big species like *Daphnia longispina*, older copepodits and adults of *Eudiaptomus graciloides* in bottom layer may be explained by the fact that due to high water transparency, predation pressure of visual fish predators had to be strong in upper layers of lake.

It is obvious that migration pattern of zooplankton in Lake Flosek is a complicated behavioral adaptation which can not be explained by a single factor. Therefore, one may not exclude other environmental factors which can affect the vertical distribution of crustaceans, like fluctuating light intensity, food distribution and interspecific competition.

Rotifers, as a group, are not seriously affected by visual predators. Therefore, it seems, that vertical distribution of food resources and exploitative interspecific competition are the factors determining the vertical distribution of rotifers in Lake Flosek in summer, i.e., in period, when the pressure of invertebrate predation is rather weak. The abundant rotifer community at bottom was composed mainly of species like *Polyarthra dolichoptera* or *Anureopsis fissa*, being absent or rare at the surface layer in this period. These species are known to feed almost exclusively on detrital and bacterial suspension (Pourriot, 1977), their accumulation at the bottom suggests their capability to exploiting the bottom food resources.

During the autumn circulation, the environmental factors like temperature, oxygen and chlorophyll *a* concentration were distributed uniformly in water column (Figure 3). The weak fish pressure and lower water transparency (below 2 m) were probably the factors which make all size classes of *Daphnia longispina* to remain in upper layers. In autumn period, the vertical distribution and, in consequence, the habitat partitioning by Cladocera species having the similar food-size requirements was remarkably stronger than in summer period. For instance, the maximum densities of *Ceriodaphnia quadrangula*, *Daphnia longispina* and *Bosmina longirostris* were found close to the surface, at 1 m and at 2–3 m depth respectively, while *Eudiaptomus graciloides* remained mostly in near-bottom layer (4–5 m). This kind of distribution pattern suggest a strong food competition among zooplankters. All above mentioned species feed on the same or similar food particles (in terms of size and shape). And this food suspension is relatively scarce and homogenous in humic Lake Flosek.

The maximum densities of rotifers in autumn period were restricted to surface layer (1 m) (Figure 3). It seems that strong pressure of invertebrate predators, due to mass occurrence of *Asplanchna priodonta* and numerous *Chaoborus* larvae, was the main factor influencing the vertical distribution of rotifers in Lake Flosek in this period. This pressure seems to be about two orders of magnitude higher than at calculated for summer period. An inverse relationship in vertical distribution of predatory *Asplanchna priodonta* and suspension feeders like *Keratella cochlearis* and *Conochiloides coenobasis* may reflect the interactions as competition and predation.

Crustacea in Lake Flosek show a classic diel migration pattern i.e. the density increases at lower layers during a day and a dispersion upwards occurs during night (Figure 4). The mean community depth of Crustacea was found at 3–4 m during day and at 2 m during night. Evidently, particular species migrate to a different depth during the 24 hours cycle (Figure 4). Small size species like *Bosmina longirostris* can migrate only on short distance, while the migration depth range for bigger species like *Ceriodaphnia quadrangula* and copepodits and adults of *Eudiaptomus graciloides* can be much larger. In case of *Daphnia longispina*, the depth range of migration is particularly large, especially for adults. It supports the hypothesis that more conspicuous individuals (like: big, pigmented or egg-carrying females) are more easily detected by predators and, for this reason, their migrations are more pronounced, especially if they do not have individual evasive capabilities (Stirling et al., 1990; Lampert, 1992).

The diel migration pattern of rotifers in Lake Flosek was not so pronounced as that of crustaceans, but the maximum density of rotifers occurred usually at 3 m at night (Figure 5), while at 1–2 m in day. The individual species did not follow any consistent pattern of movement and some forms of reverse migration (in respect to what was observed among Crustacea) was commonly observed. The most 'migrating' species among dominants were: *Conochiloides coenobasis, Polyarthra remata, Synchaeta kitina*. However each of these species was 'wandering' at different time over

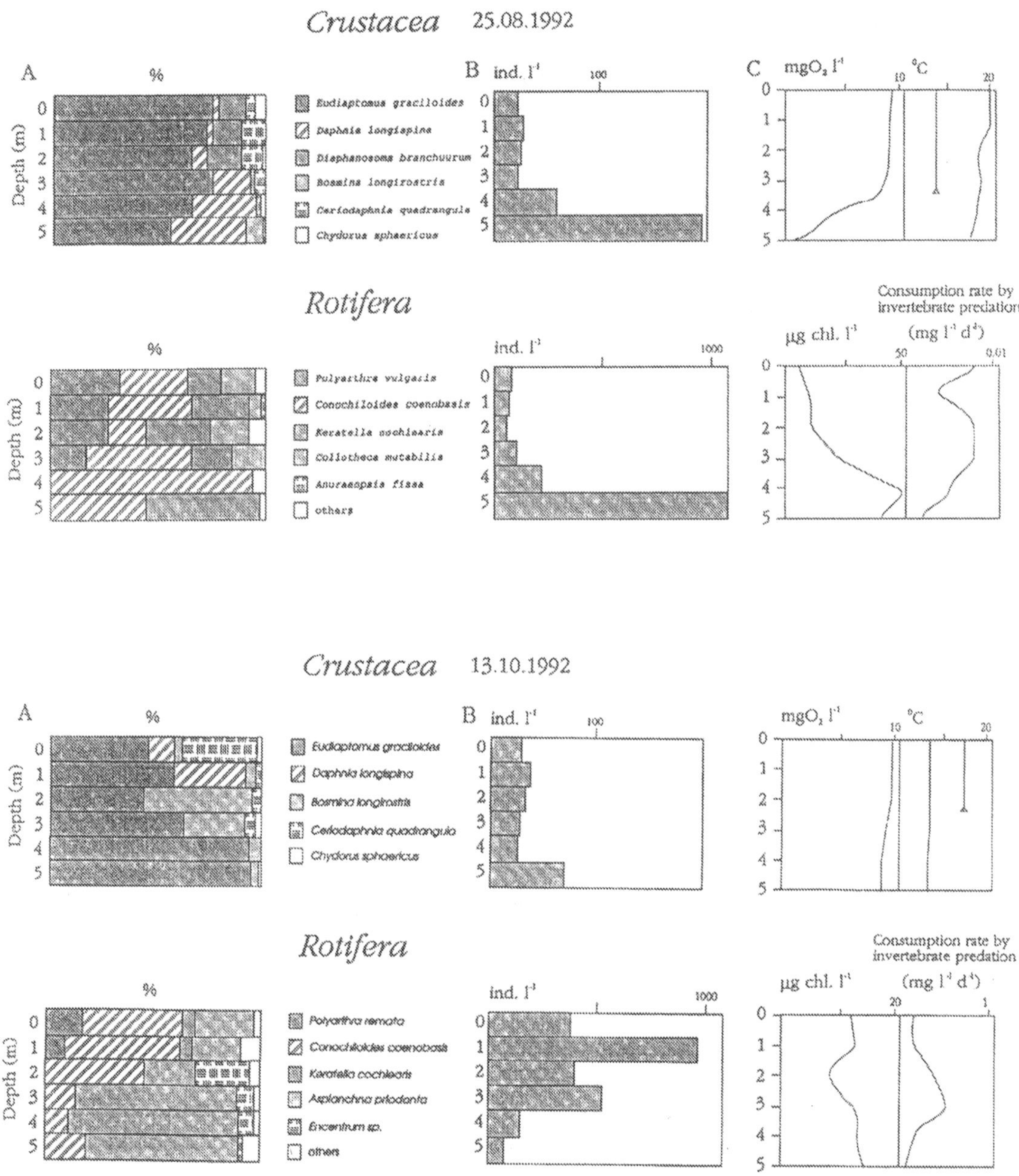

Figure 3. The vertical distribution of the relative species composition (A) and numbers of Crustacea and Rotifera (B) in relation to the depth profiles of environmental parameters (C) (temperature, dissolved oxygen, Secchi disc, chlorophyll *a* concentration and invertebrate pressure (calculated as daily feeding rates from densities of cyclopoid copepodes and rotifer *Asplanchna priodonta*) in Lake Flosek during summer stagnation (August) and autumn circulation (October) in the year of moderate planktivorous fish pressure.

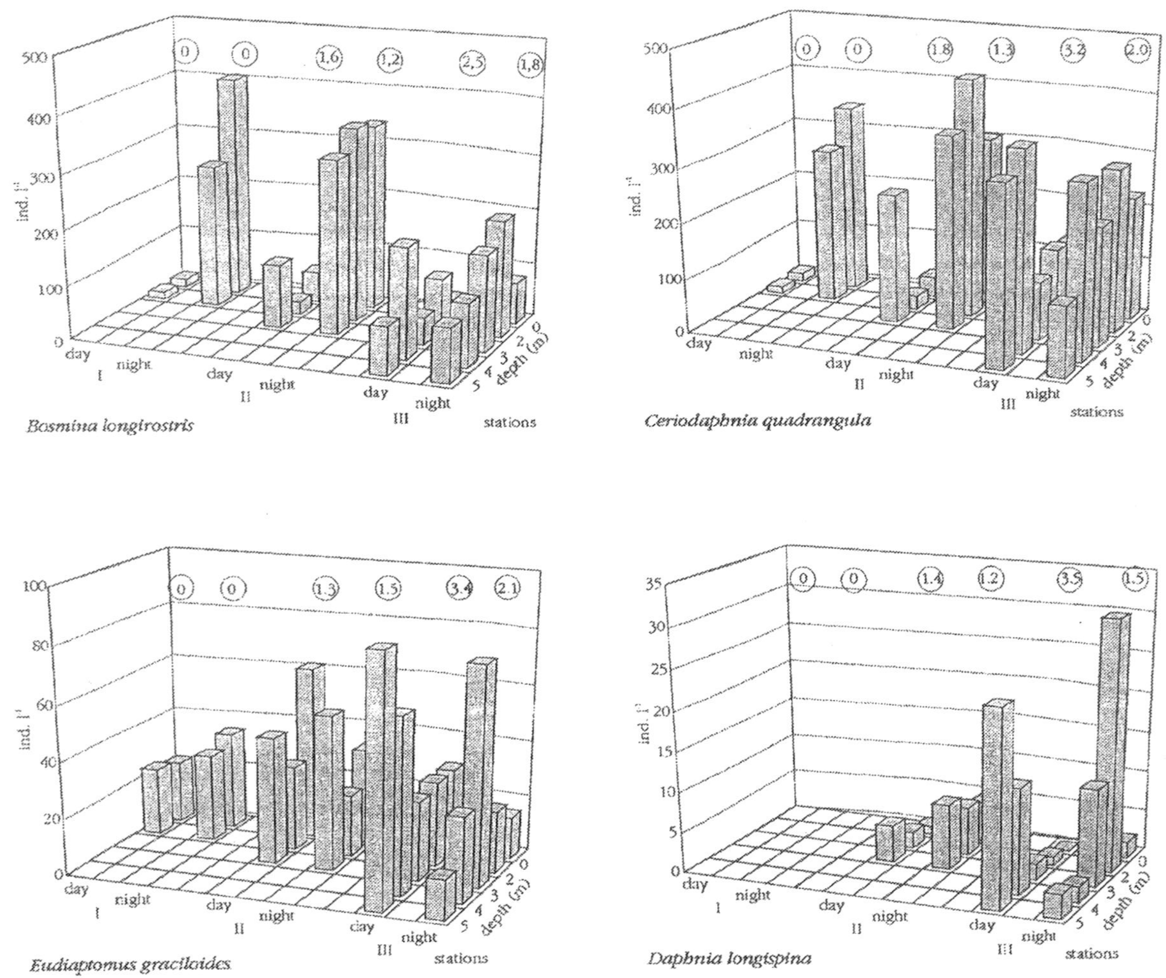

Figure 4. Daily transversal density gradient of four dominant Crustacea species in Lake Flosek during the summer 1993. In circles – mean population depth; I – station close to the shore, II – station about 10 m offshore, III – station about 50 m offshore (central part of lake).

diel cycle (Figure 5). Some field studies (Ohman et al., 1983; Stich & Lampert, 1981) suggest that rotifers may show well defined small scale movements of less than 1 m, over a time period of several hours, which are negatively and significantly correlated with the diurnal movements of predators (for instance *Chaoborus* larvae which migrate at night towards the surface and rest over the bottom during the day).

The studies revealed that the part of zooplankton community may stay below the 4 m level trough the whole 24 hours period, despite almost complete lack of light and oxygen in this water layer. A numerous species of Rotifera and Crustacea may tolerate and even thrive in water with low oxygen concentration (Ruttner-Kolisko, 1975; Lair, 1991; Hanazato, 1992).

Horizontal distribution of zooplankton

Lake Flosek is a small lake without vegetated littoral zone. Therefore one can expect rather homogenous horizontal distribution of zooplankton over the whole lake. However, a horizontal distribution of zooplankton over the 50 m distance between shore zone and pelagial zone is significantly diversified during 24 hours periods (Figure 4, Table 1).

Population density of small and medium size littoral species like *Bosmina longirostris* and *Ceriodaphnia quadrangula* from the region of shore zone to pelagial, increased significantly during the day and decreased during the night. The abundance of these both species at the shore was much higher in night than in day. *Eudiaptomus graciloides* showed during day–night period a relatively homogenous horizontal distribution along the transect inshore-offshore, while the occurrence of euplanktic large size body *Daphnia longispina* is confined exclusively to pelagial zone. Only single juvenile forms of this species were found up to 10 m from the shoreline. No significant nocturnal increase of the number of individuals of these two species was found (Table 1). All species studied showed an upward migration at night and downward one during the day. Mean population depth at night decreased with the distance from the shore, while mean depth in daytime increased towards the shoreline (Figure 4). These data point to the fact that the migrations of crustaceans in Lake Flosek are not strictly vertical, but they are rather transversal. During a day, the animals leave not only from the upper layers of the lake, but also from the shallow water zones. The studies of several authors (Boikova, 1986; Matveev, 1986) suggest that there is a possible relation between these two kinds of migrations and the well-known phenomenon of shore-avoidance (Uferflucht) by plankton.

The strong inshore-offshore horizontal variation of mean body size of adults of *B. longirostris, C. quadrangula, E. graciloides*) was also observed. The value of this demographic parameter significantly decreased ($p = 0.01$, *t*-test) during a day towards the near-shore zone (Table 1).

The significant difference of the relative (%) abundance of egg-bearing adults ($p < 0.05$, *t*-test) between pelagic and near-shore zones were also found for four Crustacea species. During the day, the egg-bearing females occurred mainly in off-shore stations while during a night their percent contribution increased considerably at close to shore stations (Table 1). The near-shore-pelagial gradients in density, mean body size and fecundity of Crustacea do not result from the differences in environmental factors and different food conditions in these both zones. In Lake Flosek biotic and abiotic factors differed only slightly between the sampling stations (chlorophyll *a*, oxygen concentrations and temperature of water). It is also unlikely that diel differences of Crustacea abundance in near shore zone were the effect of different pressure of invertebrate predators. The abundances of *Chaoborus* larvae (Dusoge, unpubl. data) and predatory Cyclopoida were in these period quite low in near-shore zone in comparison with pelagial zone. Moreover, *Chaoborus* larvae were occurring in water mass only during the night.

The most likely explanation is that during the day the fish predation is much higher in the near-shore zone than in the pelagial zone. (Gliwicz & Jachner, 1992). Some authors suggested that the low numbers of Crustacea and absence of large body size species in shallow water might result from a direct effect of high mortality of species that are more vulnerable to predation by littoral fish (Węgleńska, 1964; Gliwicz & Rybak, 1976; Gliwicz & Rykowska, 1992). The absence of large body size species like *Daphnia longispina* at near-shore zone as well as the smaller body size and low percentage of ovigerous females of all other species in near-shore zone of Lake Flosek seem to confirm the above ideas. Only one among all studied species, i.e. *Eudiaptomus graciloides* had no pronounced horizontal differentiation of population density. It is probably because this species is less vulnerable (in comparison with *Daphnia longispina*) on the pressure of visual fish predators due to relatively slow body mobility and an efficient evasive success (Drenner et al., 1988).

Although the littoral fish can exploit the near-shore zone in Lake Flosek, the phenomenon of inshore-

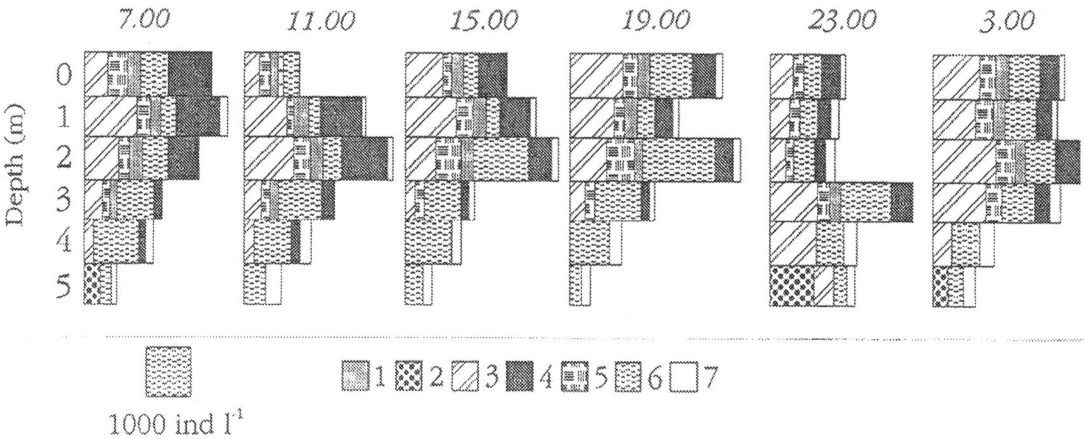

Figure 5. Daily vertical migration of Rotifera in 24–25 June 1993 1 – *Keratella cochlearis*, 2 – *Polyarthra dolichoptera*, 3 – *P. remata*, 4 – *Synchaeta kitina*, 5 – *P. vulgaris*, 6 – *Conochiloides coenobasis*, 7 – others.

Table 1. Mean population density (ind l^{-1}) (a), proportion of egg-bearing adults (%) (b), and mean body size (μm) of adults (c), in four Crustacea species across the transect inshore-offshore in Lake Flosek during 24 hours periods in July 1993.

Time		*Bosmina longirostris*			*Ceriodaphnia quadrangula*			*Eudiaptomus graciloides*			*Daphnia longispina*		
		I	II	III	I	II	III	I	II	III	I	II	III
	a	14	66	94	14	99	219	23	40	40	–	2*	18
Day	b	–	–	2	–	15	15	–	–	20	–	–	40
	c	350	350	375	450	470	510	1060	1250	1300	–	600	958
	a	384	358	139	364	312	238	37	33	32	–	5*	22
Night	b	10	3	2	18	15	7	20	20	33	–	–	45
	c	402	351	374	492	485	501	1265	1310	1302	–	745	1065

* Only juvenile forms
I - station close to the shore, II - station about 10 m offshore, III - station about 50 m offshore (central part of lake)

offshore gradient of Crustacea abundance may be attributed also to the ability of these species to 'avoid actively the near-shore regions' (Boikova, 1986). Such shore – avoidance behaviour may evolve and be sustained as an efficient mechanism to avoid areas of high predation risk (Gliwicz & Rykowska, 1992).

The big differences (as many as 20–30 times) in the abundance of *Bosmina longirostris* and *Ceriodaphnia quadrangula* (Table 1) observed over the 24 hours cycle in near-shore zone of Lake Flosek seem to confirm this phenomenon.

Conclusions

The food resource scarcity as well as strong predator-prey interactions influence the pattern of seasonal succession and sharp sequence in time of peak numbers in zooplankton community of Lake Flosek.

Vertical stratification of the abiotic and biotic factors (i.e. temperature, oxygen and chlorophyll concentration, abundance of invertebrate and vertebrate predators) observed in this lake are sufficient for enhancing spatial segregation of species with similar food requirements. Zooplankton community in Lake Flosek shows more or less visible diel vertical migration in summer period. In this shallow, steeply stratified lake with an anoxic hypolimnion, vertical migrations seem to offer a considerable advantage in resource partitioning and predator avoidance. The ability of zooplankton to stay in low-oxygen water layer during diel vertical migration may provide a temporary refuge, which in turn can mitigate invertebrate predation and filter-feeding fish impact.

Migrations of Crustacea in Lake Flosek are not strictly vertical, but rather transversal. During a day, the animals leave not only the upper layers of the lake, but also the shallow water shore zones. The diurnal transversal migrations seem to be the only possible defensive strategy for crustaceans to avoid the predators in shallow Lake Flosek. These conditions are: lack of natural refuges (like macrophytes at near-shore zone), and the lack of 'dashened' parts of water column (the transparency reaches the bottom). Although the littoral fish can direct exploit zooplankton in the near-shore zone of Lake Flosek, the phenomenon of the inshore-offshore gradient of Crustacea density may be attributed also to the ability of crustaceans to 'avoid actively the littoral regions'.

References

Boikova, O. S., 1986. Horizontal distribution of crustacean in Lake Glubokoe. Hydrobiologia 141: 113–123.

Bottrell, H. M., A. Duncan, Z. M. Gliwicz, E. Grygierek, A. Hercig, A. Hillbricht-Ilkowska, A. Kurosawa, P. Larson & T. Węgleńska, 1976. A review of some problems in zooplankton production studies. Norw. J. Zool. 24: 419–456.

Drenner, R. W. & A. McComas, 1980. The role of zooplankters escape ability and fish size selectivity in the selective feeding and impact of planktivorous fish. In Kerfoot, W. C. (ed.), Evolution and Ecology of Zooplankton Communities. The University Press of New England, Hanover (N.H.): 587–593.

Ejsmont-Karabin, J., 1974. Studies on the feeding of planktonic polyphagous *Asplanchna priodonta* (Gosse), Rotatoria. Ekol. pol. 22: 311–317.

Gliwicz, Z. M., A. Hillbricht-Ilkowska & T. Węgleńska, 1978. Contribution of fish and invertebrate predation to the elimination of zooplankton biomass in two Polish lakes. Verh. int. Ver. Limnol. 20: 1007–1011.

Gliwicz, Z. M. & A. Jachner, 1992. Diel migration of juvenile fish: a ghost of predation past or present? Arch. Hydrobiol. 124: 385–410.

Gliwicz, Z. M. & J. Pijanowska, 1988. Predation and resource depth distribution in shaping behaviour of vertical migration in zooplankton. Bull. mar. Sci. 43: 695–709.

Gliwicz, Z. M. & J. I. Rybak, 1976. Zooplankton. In Pieczyńska, E. (ed.), Selected Problems of Lake Littoral Ecology. Wyd. Uniwersytetu Warszawskiego, Warszawa: 69–96.

Gliwicz, Z. M. & A. Rykowska, 1992. 'Shore avoidance' in zooplankton: a predator – induced behavior or predator – induced mortality? J. Plankton Res. 14: 1331–1342.

Golterman, H. L. (ed.) & R. S. Clymo, 1969. Methods for chemical analysis of freshwaters. IBP Handbook, No. 8, Blackwell Scientific Publications, Oxford, 188 pp.

Hairston, N. G. Jr., 1987. Diapause as a predator-avoidance adaptation. In Kerfoot, W. C. & A. Sih (eds), Predation: Direct and Indirect Impacts on Aquatic Communities. The University Press of New England, Hanover (N.H.): 3–16.

Hanazato, T., 1992. Direct and indirect effects of low-oxygen layers on lake zooplankton communities. Arch. Hydrobiol. 35: 87–98.

Havel, J. E., 1987. Predator-induced defenses. In Kerfoot, W. C. & A. Sih (eds), Predation: Direct and Indirect Impacts on Aquatic Communities. The University Press of New England, Hanover (N.H.): 263–278.

Hillbricht-Ilkowska, A., J. I. Rybak, Z. Kajak, K. Dusoge, J. Ejsmont-Karabin, I. Spodniewska, T. Węgleńska & A. Godlewska-Lipowa, 1977. Effect of liming on a humic lake. Ekol. pol. 25: 329–420.

Kwiatkowski, R. E. & J. C. Roff, 1976. Effects of acidity on the phytoplankton and primary productivity of selected northern Ontario lakes. Can. J. Bot. 54: 2546–2561.

Lair, N., 1991. Grazing and assimilation rates of natural populations of planktonic cladocerans in an eutrophic lake. Hydrobiologia 215: 51–63.

Lampert, W., 1992. Zooplankton vertical migrations: Implications for phytoplankton-zooplankton interactions. Arch. Hydrobiol. 35: 69–78.

Matveev, V. F., 1986. Long-term changes in the community of planktonic crustaceans in relation to predation and competition. Hydrobiologia 141: 33–45.

Ohman, M. O., B. N. Frost & E. B. Cohen, 1983. Reverse diel vertical migration. An escape from invertebrate predators. Science 220: 1404–1407.

Olem, H., 1991. Liming acid surface waters. Lewis Publ., Michigan, 331 pp.

Pourriot, R., 1977. Food and feeding habits of rotifers. Arch. Hydrobiol. 8: 243–260.

Roff, J. C. & R. E. Kwiatkowski, 1977. Zooplankton and zoobenthos communities of selected northern Ontario lakes of different acidities. Can. J. Zool. 55: 899–911.

Ruttner-Kolisko, A.., 1975. The vertical distribution of plankton rotifers in a small alpine lake with a sharp oxygen depletion (Lunzen, Obersee). Verh. int. Ver. Limnol. 19: 1286–1294.

Salonen, K. & A. Lehtovaara, 1992. Migrations of haemoglobin rich *Daphnia longispina* in a small, steeply stratified, humic lake with an anoxic hypolimnion. Hydrobiologia 229: 271–288.

Stemberger, R. S. & J. J. Gilbert, 1987. Defenses of planktonic rotifers against predators. In Kerfoot, W. C. & A. Sih (eds). Predation: Direct and Indirect Impacts on Aquatic Communities. The University Press of New England, Hanover (N.H.): 3–16.

Stenson, J. A., 1985. Biotic structure and relation in the acidified Lake Gårdsjön system. Ecol. Bull. (Stockholm), 32: 319–326.

Stenson, J. A. & J. E. Svensson, 1994. Manipulations of planktivore fauna and development of crustacean zooplankton after restoration of the acidified humic Lake Gårdsjön. Arch. Hydrobiol. 131: 1–23.

Stich, H. B. & W. Lampert, 1981. Predator evasion as an explanation of diurnal vertical migration. Nature 293: 396–398.

Stirling, D. G., D. J. McQueen & M. R. S. Johannes, 1990. Vertical migration of *Daphnia galeata mendote* (Brooks) demographic responses to changes in planktivore abundance. Can. J. Fish. aquat. Sci. 47: 395–400.

Węgleńska, T., 1964. Variation of incidence, size and fertility of plankton Crustacea in space and time. Bull. Acad. Pol. Sci. Cl. II 12: 386–390.

Hydrobiologia **342/343:** 197–205, 1997.
L. Kufel, A. Prejs & J. I. Rybak (eds), Shallow Lakes '95.

Interactions between piscivores, zooplanktivores and zooplankton in submerged macrophytes: preliminary observations from enclosure and pond experiments

Lene Jacobsen[1*], Martin R. Perrow[2*+], Frank Landkildehus[2], Morten Hjørne[3], Torben L. Lauridsen[2] & Søren Berg[1]
[1]*Danish Institute for Fisheries Research, Department of Inland Fisheries, Vejlsøvej 39, 8600 Silkeborg, Denmark*
[2]*National Environmental Research Institute, Department of Lake and Estuarine Ecology, Vejlsøvej 25, 8600 Silkeborg, Denmark*
[3]*Institute of Biology, Odense University, Campusvej 55, 5230 Odense, Denmark*
**Corresponding authors*
+Present address: ECON, School of Biological Sciences, University of East Anglia, Norwich, Norfolk NR4 7TJ, UK

Key words: predator/prey interactions, refugia, indirect effects, *Esox lucius*, *Perca fluviatilis*, *Rutilus rutilus*, *Daphnia*

Abstract

The effects of piscivores upon zooplanktivore behaviour and distribution and the impact of zooplanktivores on the abundance and distribution of zooplankton are well documented. However, the potential indirect effect of piscivores reducing the predation pressure upon grazing zooplankton through behavioural changes of zooplanktivores has received little attention, even though this may be an important mechanism in enhancing the stability of submerged macrophytes in shallow lakes. Preliminary observations from an unreplicated large-scale field enclosure experiment and a replicated pond experiment suggest that this mechanism is plausible with the set of piscivores (pike *Esox lucius* and perch *Perca fluviatilis*) and the zooplanktivores (0+ roach *Rutilus rutilus* and perch) common in temperate Europe. The presence of piscivores typically changed the habitat use and the activity level of zooplanktivores and the presence of zooplanktivores typically changed the habitat selection of cladoceran zooplankton. In the case of piscivore/zooplanktivore interactions, the risk of predation was enough to generate clear responses even where the losses to predation were low. However, only in the enclosure experiment was an indirect impact of the presence of piscivores, enabling *Daphnia* spp. to utilise open water in the presence of a high density of zooplanktivorous fish observed. Whether the magnitude and direction of the effect of piscivores is sufficient to benefit zooplankton may depend on the functional group (capable of foraging within structured habitats) of the predator (both piscivore and zooplanktivore), absolute and relative densities of predator and prey and predator dietary choice.

Introduction

The role of submerged macrophytes in shaping the interactions between piscivorous fish and their fish prey has recently received much attention (Gotceitas & Colgan, 1987; Savino & Stein, 1989; Christensen & Persson, 1993; Eklöv & Diehl, 1994; Bean & Winfield, 1995; Persson & Eklöv, 1995). In comparison, a number of studies have focussed on the influence of submerged macrophytes in the zooplanktivorous fish/cladoceran zooplankton interaction (Schriver et al., 1995; Phillips et al., 1996; Stansfield et al., 1997). In both cases, increased structural complexity typically leads to a decrease in the foraging ability of the predator concerned (Winfield, 1986; Diehl, 1988; Bean & Winfield, 1995; Tatrai & Herzig, 1995). Thus, in the presence of predators, prey may switch habitat preference, perhaps abandoning the profitable open water zone for the relative safety of structured habitats (Werner et al., 1983; Timms & Moss, 1984; Gliwicz & Jachner, 1992; Lauridsen & Lodge, 1996). This may increase the chances of survival but at the expense of reduced growth rate (Persson, 1993; Persson & Eklöv, 1995).

Cascading effects of reduced predation pressure at the top of the food chain in aquatic systems are well documented (Carpenter et al., 1987; McQueen, 1990). This has led to the use of biomanipulation (reduced zooplanktivorous fish stock or enhanced piscivore stock – Lammens et al., 1990). However, the potential knock-on effect of piscivores reducing the predation pressure upon zooplankton through habitat and dietary shifts or reduced activity of zooplanktivores has received little attention (Hambright et al., 1986; Turner & Mittelbach, 1990) even in the well-studied food webs of temperate European shallow lakes. However, such a mechanism may be important in enhancing the beneficial effect of grazing Cladocera (i.e. reduction of the phytoplankton standing crop and enhanced water clarity), which ultimately may act as a stabilising mechanism for the persistence of submerged macrophytes in shallow lakes at nutrient loadings where either phytoplankton or macrophytes may exist (Irvine et al., 1989; Scheffer et al., 1993).

The purpose of this paper is to: (a) report on observations from a large-scale field enclosure experiment examining zooplanktivore/zooplankton/macrophyte interactions, where the accidental entrapment of piscivorous perch in one enclosure allowed a preliminary assessment of the influence of piscivores to be made, and (b) describe the results from subsequent replicated pond experiments designed to examine the interactions more closely. Experiments used the two principal piscivores of temperate European shallow lakes, pike (*Esox lucius* (L.)) and perch (*Perca fluviatilis* L.) and the two principal zooplanktivores, 0+ perch and roach (*Rutilus rutilus* (L)).

We try to assess whether predation risk is an important force structuring the behaviour and distribution of this set of zooplanktivores and whether this has consequences for the distribution of cladoceran zooplankton.

Materials and methods

Enclosure experiment

Experimental design

The experiment was conducted in July 1993 in Lake Stigsholm in the central Jutland in Denmark (Schriver et al., 1995). Two enclosures were placed near the southwestern shore on a sandy bottom in 60–90 cm of water. They were circular, 10 m in diameter (area 78.5 m^2) with an inner ring of 5 m (area 19.6 m^2) which contained macrophytes (*Potamogeton pectinatus* L.) at a density of 26–29% PVI (plant volume infested – Canfield et al., 1984). The enclosures were made of thick plastic sheeting with chain along the bottom edge. This was pulled up, and situated on wooden poles prior to the experiment, to ensure natural conditions in the enclosures.

Each of the two enclosures was stocked with underyearling perch (mean $\pm$ S.E. total length = 44 mm $\pm$ 0.5) and roach (mean $\pm$ S.E. total length = 39 mm $\pm$ 0.4): These had been caught in traps in the area a few days before and retained in large keepnets (4 $\times$ 4 m) prior to introduction. The unintentional retention of some fish in the enclosures as the plastic sheet was lowered into the sediment at the beginning of the experiment caused some variability in fish density. The final densities ($\pm$ S.E.) were estimated from catch-depletion using a seine net and were; 0+ perch, 10.0 $m^{-2} \pm 0.14$ and 10.3 $m^{-2} \pm 0.03$; 0+ roach, 3.2 $m^{-2} \pm 0.06$ and 5.5 $m^{-2} \pm 0.04$ respectively. This planktivorous fish density is considered to be high but has been observed locally in the littoral zone of some eutrophic lakes (Berg, unpubl. data). The first enclosure also contained a density of four fish per m^{-2} of piscivorous 1+ perch (mean $\pm$ S.E. total length = 110 mm $\pm$ 1.4) which had been trapped in the same manner. The design of the experiment was thus an unreplicated comparison of zooplanktivorous fish in the presence and absence of piscivores.

Fish sampling

Both enclosures were sampled for three days and nights (July 26–29). Transparent plexiglass Breder traps (Breder, 1960) were used to monitor fish habitat selection and activity. Ten of these were placed on the bottom and near the surface in three habitats: open water, the interface between open water and macrophytes, and macrophytes. Traps were emptied every four hours during the monitoring period. Fish were identified and counted before being released back into the enclosure.

Comparison of the activity level (expressed as total trap capture on each sample occasion) of both 0+ perch and roach in the presence/absence of piscivores was made with Wilcoxon signed ranks test for matched pairs using the sample occasion ($n = 18$) as a replicate (Siegel & Castellan, 1988). Further, habitat use (expressed as CPUE per trap) of perch and roach in the presence/absence of piscivores during day (0900–2100) and night (2100–0900) was compared using Chi-square 3×2 or 3×1 contingency tables. The location of any significant difference was determined by the partitioning procedure (Siegel & Castellan, 1988).

During the first 24-hour period, a maximum of ten 0+ fish of each species was taken for dietary analysis at six hourly intervals from each enclosure. Any 1+ perch were also taken as they became available. All fish were immediately preserved in 80% ethanol. In the laboratory, the recognizable items in the stomachs of perch and the foreguts of roach were identified and enumerated. A total of 47 0+ perch, 16 0+ roach and 18 1+ perch were analysed. The number of *Daphnia* spp. in the guts of 0+ fish in all samples in the presence and absence of piscivores were tested using Man-Whitney Wilcoxon W-tests (Siegel & Castellan, 1988).

Zooplankton sampling

The density and distribution of zooplankton in macrophytes and open water were monitored at twelve hourly intervals during day (1300) and night (0100). Five samples from each habitat were collected randomly with a core sampler (1.5 m length, 7.4 cm diameter), sampling the entire water column. The samples from each habitat on each sampling occasion were then pooled, filtered through a 20 μm mesh net and fixed in 4% lugol.

The proportion of *Daphnia* spp. in open water in the two enclosures were compared using a Wilcoxon signed ranks test using the sampling occasions ($n = 6$) as a replicate (Siegel & Castellan, 1988).

Pond experiment

Experimental design

Three experimental plastic square ponds (area 16 m^2) with a water depth of 40 cm were used. An open stainless steel framework divided each pond into two sections, and also further subdivided one of the sections into three compartments. Each of the latter was stocked with natural macrophytes (*Elodea canadensis* Michaux) at either 25%, 50% or 75% PVI. These were prevented from drifting between compartments by means of stainless rods trapping their ends onto the bottom of the pond and by nylon filament netting of a large enough mesh size (130 mm) not to hinder fish movement.

The relative position of the central division and of the different macrophytes densities were determined randomly. A green curtain (mesh 2 mm) with stainless steel rods stitched along its upper edge, was fixed to the bottom to the framework along each division. This was used to prevent the movement of fish between sections upon completion of the experiment. This was achieved with the minimum of disturbance by pulling on nylon ropes attached to each of the four ends of the curtain, quickly raising it from its resting position on the bottom of the pond.

Three treatments: 0+ roach (mean $\pm$ 1 S.E. fork length = 49.0 mm $\pm$ 0.7) alone; 0+ roach in the presence of one pike (mean $\pm$ 1 S.E. fork length = 247.0 mm $\pm$ 8.2); and 0+ roach in the presence of two 1+ (mean $\pm$ 1 S.E. fork length = 123.0 mm $\pm$ 4.8) and one 2+ (mean $\pm$ 1 S.E. fork length = 186.0 mm $\pm$ 3.0) perch: were compared with a fishless control. The use of as single pike and three perch mimicked the typical hunting strategy of these fish, single and group hunters respectively (Eklöv, 1992). The density of roach was 1 m^{-2} which may be considered to be a typical average whole-lake density in shallow eutrophic lakes (Søndergaard et al., 1990; Stansfield et al., this volume; Perrow et al., 1996). This is in contrast to the densities in the enclosure experiment, which tested shoaling density achieved by shoals of young fish (Berg, unpubl. data). The biomass of piscivores was equivalent to 75–100 kg ha^{-1} which is typical for temperate European shallow lakes (Grimm & Backx, 1990; Berg et al., 1994; Phillips et al., 1996).

A randomised block design was used to assign treatments to ponds over the experimental period (28th June–7th July 1995). Five replicates of each treatment and control were run. Each replicate used different, inexperienced fish.

Fish sampling

Fish used in the experiment were captured from nearby lakes and acclimatised in laboratory ponds for three weeks prior to experimentation and fed natural food, including live fish (trout) in the case of piscivores. Piscivores were not fed at least three days before use.

The day before use fish were transferred outside into stock ponds to acclimatize to natural ambient temperatures. Fish were then stocked in the ponds in the morning and left undisturbed for the duration of the experiment. Pilot observations had indicated that fish had developed a consistent pattern of apparently natural behaviour within two hours of being introduced. After six hours the curtains between the sections were drawn up and all fish were recaptured by electrofishing. The proportion of roach in the open water compared to macrophytes in the absence and presence of different piscivores was compared by the Kruskal-Wallis test. Where significant, post hoc comparisons between treatments were made (Siegel & Castellan, 1988).

At the end of each experiment five roach were taken for dietary analysis. All recognizable items in the foregut were identified and counted. In addition, all predatory fish were thoroughly stomach-flushed and any roach present was counted. Some fish were dissected to confirm the reliability of the stomach flushing method.

Zooplankton sampling

Daphnia spp. (principally *D. magna*) were added to the ponds prior to experimentation and maintained at a density of 80–100 l^{-1}, representative of natural densities under a low fish predation pressure (Townsend & Perrow, 1989; Søndergaard et al., 1990; Carvalho, 1994). The distribution of zooplankton was determined at the end of each experiment by means of nine pooled samples within each of the major habitats. Samples were taken with a core sampler of 7.4 cm diameter. The proportion of *Daphnia* spp. in open water were compared between treatments using the Kruskal-Wallis test.

Results

Enclosure experiment

Of the 1+ perch taken for dietary analysis 33% had roach (28%) or perch (6%) in the stomachs confirming that these fish did represent a predation threat to the zooplanktivorous fish in the enclosures.

The general level of zooplanktivorous fish activity, represented by captures in traps, was significantly lower in the presence of piscivorous perch (perch – Wilcoxon signed ranks test, $T^+ = 171$, $n = 18$; $z = 3.72$, $p < 0.001$; roach – Wilcoxon signed ranks test, $T^+ = 171$, $n = 18$; $z = 3.72$, $p < 0.001$). The total captures in absence of piscivores being 1863 perch and 721 roach, whereas in the presence of piscivores the total catch was only 401 and 42 respectively (Figure 1).

Pooled CPUE data for samples over the monitoring period, showed a significant difference in the habitat selection of 0+ perch in the presence/absence of piscivores both during the day ($\chi^2 = 82.97$, df = 2, $p < 0.001$) and night ($\chi^2 = 85.61$, df = 2, $p < 0.001$) (Figure 1). Partitioning the contingency tables revealed that during the day perch preferred open water in the absence of piscivores, but utilised macrophyte and interface zones in the presence of piscivores ($\chi^2 = 81.45$, df = 1, $p < 0.001$) (see Figure 1). At night, perch utilised macrophytes more than both open water and interface zones ($\chi^2 = 31.64$, df = 1, $p < 0.001$) in the presence of piscivores. The interface was also preferred to open water ($\chi^2 = 9.28$, df = 1, $p < 0.01$). In the absence of piscivores however, the opposite was true.

Due to extremely low captures in the presence of piscivores it was not possible to test if roach exhibited shifts in habitat preference according to predation risk. However, it is clear that in the absence of piscivores, roach preferred open water both during the day ($\chi^2 = 28.08$, df = 2, $p < 0.001$) and night ($\chi^2 = 22.87$, df = 2, $p < 0.001$) (Figure 1).

Daphnia spp. contributed a mean of 63% of the dietary items in the stomachs of individual perch and a mean of 50% in the foreguts of roach. The rest of the diet of both species consisted of other zooplankton species and occasional macroinvertebrates. This illustrates that these zooplankters were an important dietary component of both fish species and implies a significant predation risk for *Daphnia* in the enclosures. In addition, the total number of *Daphnia* spp. consumed per capita by perch, was high (157.6 ± 31.0 without piscivores and 71.0 ± 25.8 with piscivores).

The mean ($\pm$ 1 S.E.) number of *Daphnia* spp. per perch stomach, from day and night samples combined, was significantly lower in the presence of piscivores (Mann-Whitney Wilcoxon test: W = 137, $n = 21$ and 26 with and without piscivores respectively, $p < 0.01$). It was not possible to test for differences between the consumption of perch at day and night or roach between treatments, as too few fish samples were available.

The proportion of *Daphnia* spp. in open water representing samples from both day and night was significantly higher in the presence of piscivores (Wilcoxon signed ranks test, $T^+ = 20$, $n = 6$, $p < 0.05$) (Figure 2).

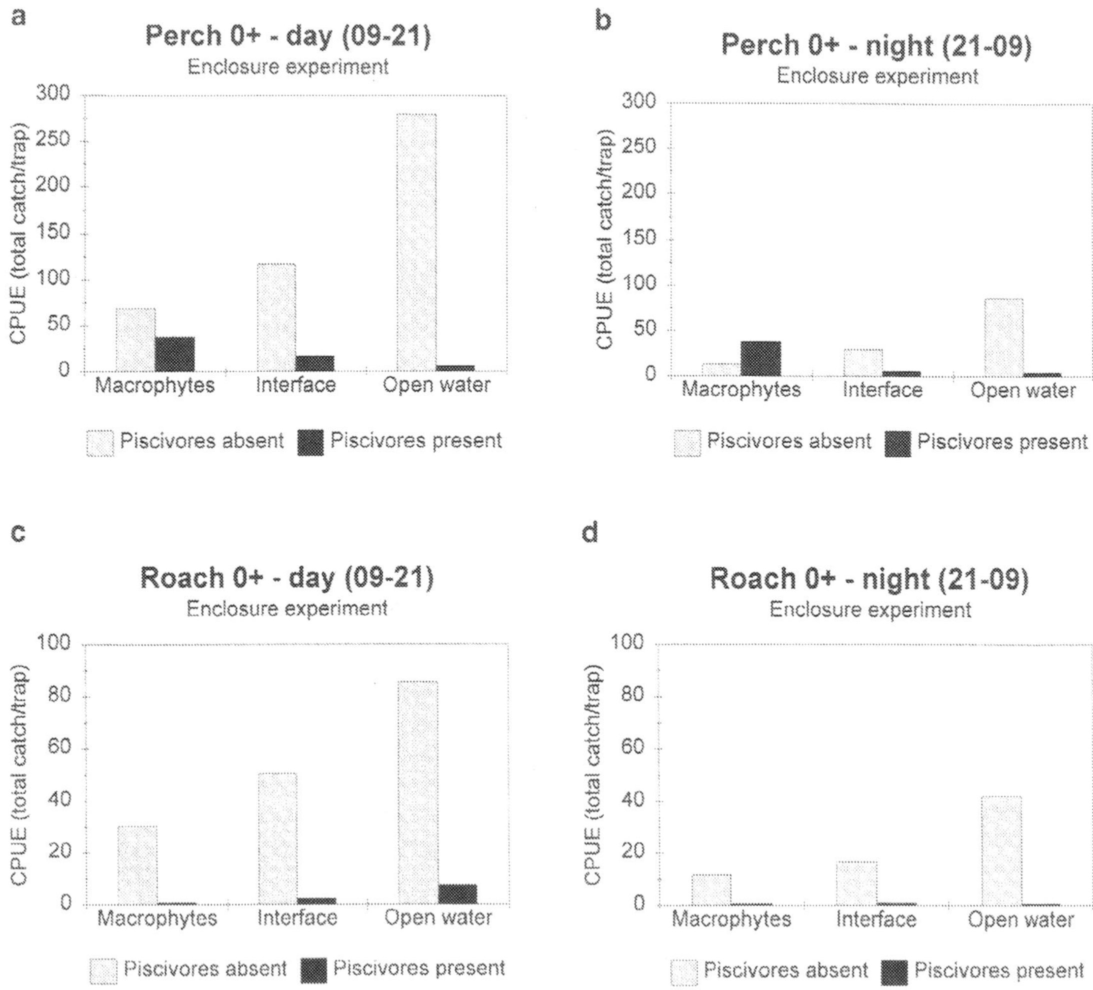

Figure 1. (a) The habitat use of perch 0+ during the day with and without piscivores. CPUE data are pooled for 3 days. Total number of catches were: Piscivores absent, $n=1465$; Piscivores present, $n=222$. (b) The habitat use of perch 0+ during the night with and without piscivores. CPUE data are pooled for 3 nights. Total number of catches were: Piscivores absent, $n=398$; Piscivores present, $n=179$. (c) The habitat use of roach 0+ during the day with and without piscivores. CPUE data are pooled for 3 days. Total number of catches were: Piscivores absent, $n=533$; Piscivores present, $n=33$. (d) The habitat use of roach 0+ during the night with and without piscivores. CPUE data are pooled for 3 nights. Total number of catches were: Piscivores absent, $n=224$; Piscivores present, $n=8$.

Pond experiment

There was a significant difference between the habitat use of roach in the three treatments (Kruskal-Wallis test, $KW=5.83$, $n=5$, $p<0.05$). Nevertheless, the post hoc comparisons between treatments failed to reveal any significant differences. Figure 3 illustrates that roach utilizes macrophytes most in the presence of piscivorous perch, and least in the presence of pike.

In contrast, both pike and perch were virtually exclusively recaptured in the macrophytes. The only exceptions were a single perch in open water in two replicates. Both pike and perch were generally unsuccessful in capturing on roach in the ponds. Only one roach was eaten by a pike on one occasion.

The diet of roach in the ponds was dominated by small chironomid larvae in 70% of the analysed fish. The mean ($\pm$ 1 S.E.) number of *Daphnia* spp.

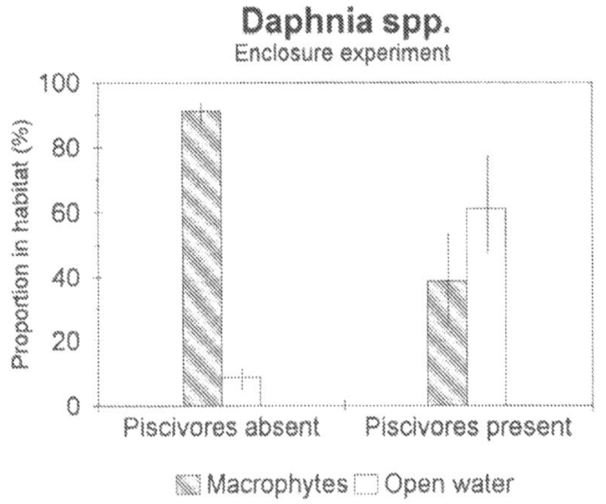

Figure 2. The distribution of *Daphnia* spp. in open water and macrophytes in the two treatments. Mean proportional habitat use (± 1 S.E.) during the three days and nights is shown.

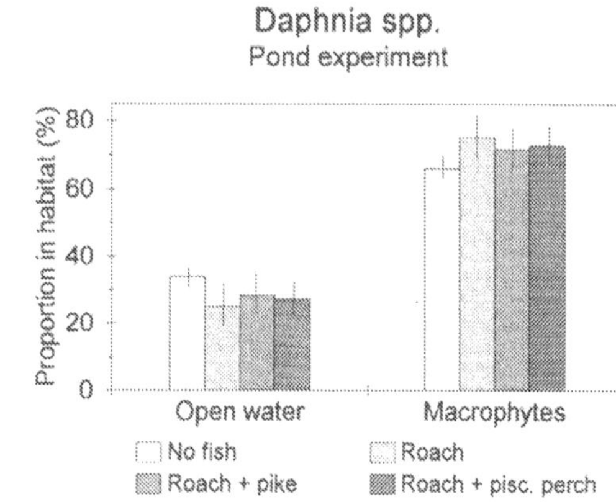

Figure 4. The habitat distribution of *Daphnia* spp. (mean ± 1 S.E.) without fish and with three different fish treatments in the pond experiment.

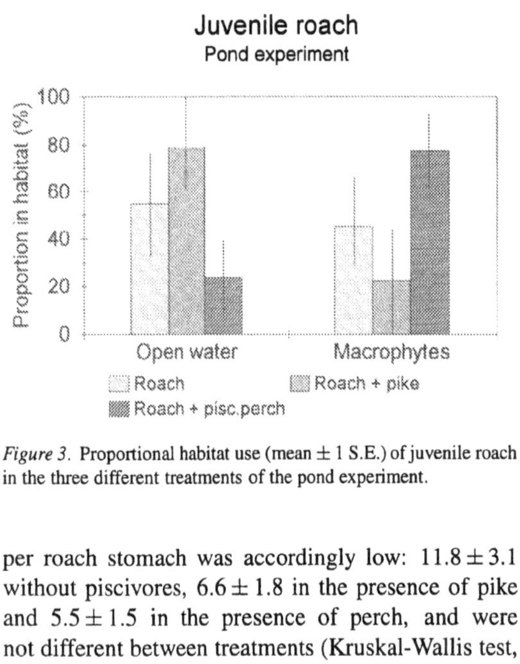

Figure 3. Proportional habitat use (mean ± 1 S.E.) of juvenile roach in the three different treatments of the pond experiment.

per roach stomach was accordingly low: 11.8 ± 3.1 without piscivores, 6.6 ± 1.8 in the presence of pike and 5.5 ± 1.5 in the presence of perch, and were not different between treatments (Kruskal-Wallis test, $KW = 0.78$; $n = 23, 14, 25$; $p = 0.67$).

Daphnia spp. showed no habitat preference between treatments (Kruskal-Wallis test, $KW = 1.3$, $n = 5$, $p = 0.73$), with a general association with macrophytes (Figure 4).

Discussion

In the unreplicated enclosure experiment, the risk of predation for zooplanktivorous fish elicited distinct behavioural responses in terms of habitat selection and activity level, that cascaded down to the habitat selection of zooplankton. This indicates some promise for the mechanism of piscivorous fish having an indirect beneficial effect on the grazing potential of zooplankton and suggests that predation risk plays an important role in structuring lake communities. However, this effect could not be detected in the subsequent replicated pond experiment, with zooplanktivorous fish showing a subdued response to the presence of predators and zooplankton showing no change in habitat selection.

The relative predator/prey densities were similar between the two experiments and, in fact, biased in favour of the piscivorous perch in the ponds (1:5 compared to 1:3 in the exclosures) suggesting that the different densities of fish used, did not contribute to the difference in the scale of the response of zooplanktivorous fish between the two experiments. Rather, it is suggested that the different predation rate on zooplanktivorous fish between the enclosures (33% of all perch contained fish) and the ponds (no perch contained fish) conferred a high risk in the enclosures and a low risk of predation in the ponds. Moreover, some of the difference in predation rates and accordingly risk may be attributable to differences in behaviour of the piscivorous perch in the two systems. Although often associated with lakes with a high coverage of macrophytes (Persson et al., 1991) perch is often reported as an open water predator (Eklöv, 1992), a fact supported in the enclosure experiment by most of the captures of 1+ perch in this habitat. However, in the pond experiment perch were typically associated with macrophytes, a pattern also observed by Eklöv & Diehl (1994). This difference may be a function of predator density. Perch has been shown to be an effective group hunter, with foraging success increasing with group size. Although the prey fish may evade one attacker, the chances of escape are reduced with multiple attacks from several predators in quick succession (Eklöv, 1992). Even though the relative difference in predator/prey density was of the same magnitude in the two experiments, the difference in predator density (4 m^{-2} compared to 0.2 m^{-2}) may have resulted in perch assuming a group foraging strategy in open water in the enclosure and a more solitary ambush-style in the ponds (see Eklöv, 1992), with consequent differences in the perception of predation risk and anti-predation strategy adopted by zooplanktivorous fish.

This is demonstrated by the different behaviour patterns exhibited by roach in the pond and enclosure experiments. When in the presence of large perch in the open water of the enclosure, roach responded by significantly reducing their activity level (as shift in habitat selection could not be detected due to the low number of fish captured), a response also observed by Bean & Winfield (1995) when roach were exposed to pike. We suggest such behaviour may be a response to intense predation risk (e.g. when predator density is abnormally high as in the enclosure experiment or in tank experiments, where the chances of escape are small) where the best strategy may be not to attract the attention of visually foraging predators. In the pond experiment however, roach showed little change in habitat preference when in the presence of large perch compared to when in isolation. It may be that roach are capable of relying on mobility, swimming fast near the surface and making evasive jumps when attacked (Christensen & Persson, 1993). As already noted, roach did not suffer any mortality in this experiment (see also Persson & Eklöv, 1995) thus underlining the lack of a significant predation threat from perch, when they apparently adopted a solitary hunting strategy by concealing themselves within the macrophytes.

What of pike? Did they represent a significant predation threat for roach, as observed by Eklöv (1992) and Bean & Winfield (1995)? The fact that only one roach was consumed by one pike during the pond experiment suggests not. This may simply be due to a low spatial overlap between predator and prey. In the presence of pike, roach continued to utilise their preferred open water habitat (Persson, 1993). Pike fulfilled their role as an ambush predator and exclusively inhabited the densest part of the macrophytes, ready to strike outwards into open water (Bean & Winfield, 1995). Thus, in the pond experiment with a relatively large open water zone, roach may always have been able to avoid being in close proximity to pike. The contradictory results of Bean & Winfield (1995) in experiments in aquaria, with roach taking refuge in the macrophytes and significantly reducing activity level in the presence of pike, may be explained by the limited escape opportunities in open water for the prey fish.

Although neither predator seemed to present a significant threat to zooplanktivorous fish in the ponds, the differential habitat selection of roach between treatments, with a tendency to select for macrophytes in the presence of perch and open water in the presence of pike, suggests that roach did perceive some risk of predation. In addition, although both predators tended to use the macrophytes, roach seemed able to distinguish between the different predators and the potential threat they posed. For example, pike were likely to attack from macrophytes at the interface with open water and perch were likely to be more successful foraging as a group in open water.

Taking results from both experiments it has been established that zooplanktivores may respond to the presence of a predator but that the magnitude and direction (in terms of type of habitat selected) of response is variable. This also appears to be the case for zooplankton. The basis for this has been shown to be the trade off between repellence (Pennak, 1973; Dorgelo & Heycoop, 1985) and refuge (Timms & Moss, 1984; Schriver et al., 1995; Phillips et al., 1996; Stansfield et al., this volume) provided by macrophytes, *Daphnia* taking refuge in macrophytes only in the presence of fish (Lauridsen & Lodge, 1996).

Some authors (e.g. Venugopal & Winfield, 1993) have argued that any refuge effect is nullified by the presence of large numbers of 0+ fish in macrophyte-dominated habitats, themselves concentrated by the risk of predation in open water (Werner et al., 1983; Naud & Magnan, 1988; Gliwicz & Jachner, 1992). In support to this, Schriver et al. (1995) demonstrate that at low PVI at least, an increase in fish density will negate any refuge effect. However, many fish species show reduced foraging efficiency on zooplankton in structured habitats (Winfield, 1986; Tatrai & Herzig, 1995) ultimately leading to a reduced predation pressure on zooplankton (Phillips et al., 1996). Further, the observations of Stansfield et al. (this volume) suggest that this is a function of PVI, denser macrophytes offering better refuges.

In the enclosure experiment, *Daphnia* spp. were associated with macrophytes in the presence of high numbers of zooplanktivorous fish, consistent with the refuge hypothesis. Whereas in the presence of piscivores and the concentration of 0+ perch (at least) in the macrophytes combined with reduced activity level of perch and roach, *Daphnia* spp. appeared to be released from requiring a refuge and moved back to the preferred open water habitat.

The question is, why was no impact on *Daphnia* observed in the pond experiment, particularly as the zooplanktivorous fish responded differently in the presence of different predators, implying differential predation risk for zooplankton in different habitats? Moreover, the foraging efficiency of roach is known to be reduced in structured habitats (Winfield, 1986), which should enhance the refuge for zooplankton within macrophytes. The answer to the question may be linked to the actual rate of predation changing the perception of risk. In the ponds, the density of roach (1 m^{-2}) was far lower than the density of planktivorous fish in the enclosure experiment (13–15 m^{-2}). The level of predation may thus be simply a function of fish density. This appears to be supported by the low consumption of *Daphnia* spp. in the pond experiment compared to the enclosures. This implies the risk of predation for zooplankton was low, thus providing no selection pressure for different patterns of behaviour between treatments. However, why *Daphnia* illustrated a preference for macrophytes in all treatments and the control, behaviour normally associated with a refuging response, is unknown.

Conclusions

There is clear potential for the indirect effects of piscivores initiating behavioural changes (habitat shifts, reduced activity levels) of zooplanktivores with beneficial consequences for grazing cladocerans, and consequently some indication that this mechanism has a role in structuring natural lake communities. However, these effects are complex, linked to the functional group of piscivore and zooplanktivore and predator density and behaviour, making predictions as to the strength and direction of any effect problematic. In particular, it is unclear whether the refuge for grazing cladocerans afforded by macrophytes will be enhanced or reduced in the presence of piscivores, or alternatively whether the open water will become available as relative predation risk from zooplanktivores declines in the presence of piscivores. We also expect any effects to be further complicated by changes in density and perhaps nature of macrophyte beds, modifying the foraging ability of the predators and consequently the escape opportunities for prey.

To unravel the mechanisms involved, there is a clear need for more unambiguous experimental testing. We stress the importance of mimicking the natural situation as closely as possible concerning fish densities and appropriate refugia. Observations on patterns of behaviour or distribution must be supported by dietary analysis to confirm predation rate and likely intensity of risk for prey species.

Acknowledgments

We gratefully acknowledge the field and laboratory assistance of J. S. Mikkelsen, J. Stougaard Pedersen and K. Jensen. We wish to thank E. Jeppesen and M. Søndergaard for valuable discussions on experimental designs of enclosure and pond experiments, and two anonymous reviewers for valuable comments on the manuscript. Two authors (LJ and MRP) were sponsored by The Danish Environmental Programme, Centre for Fresh Water Environmental Research.

References

Bean, C. W. & I. J. Winfield, 1995. Habitat use and activity patterns of roach (*Rutilus rutilus* (L.)), rudd (*Scardinius erythrophthalmus* (L.)), perch (*Perca fluviatilis* L.) and pike (*Esox lucius* L.) in the laboratory: the role of predation threat and structural complexity. Ecol. of Freshwat. Fish 4: 37–46.

Berg, S., E. Jeppesen, M. Søndergaard & E. Mortensen, 1994: Environmental effects of introducing whitefish, *Coregonus lavaretus* (L.), in Lake Ring. Hydrobiologia 275/276: 71–80.

Breder, C. M., 1960. Design for a fish trap. Zoologica, N.Y. 45: 155–159.

Canfield, D. E. Jr., J. V. Shireman, D. E. Colle, W. T. Haller, C. E. Watkins II & M. J. Maceina, 1984. Prediction of clorophyll *a* concentrations in Florida lakes: importance of aquatic macrophytes. Can. J. Fish. aquat. Sci. 41: 497–501.

Carpenter, S. R., J. F. Kitchell, J. R. Hodgson, P. A. Cochran, J. J. Elser, M. M. Elser, D. M. Lodge, D. Kretchmer, X. He & C. N. von Ende, 1987. Regulation of a lake primary productivity by food web structure. Ecology 68: 1863–1876.

Carvalho, L., 1994. Top-down control of phytoplankton in a shallow hypertrophic lake: Little Mere (England). Hydrobiologia 275/276: 53–64.

Christensen, B. & L. Persson, 1993. Species-specific antipredatory behaviours: effects on prey choice in different habitats. Behav. Ecol. Sociobiol. 32: 1–9.

Diehl, S., 1988. Foraging efficiency of three freshwater fishes: effects of structural complexity and light. Oikos 53: 207–214.

Dorgelo, J. & M. Heycoop, 1985. Avoidance of macrophytes by *Daphnia longispina*. Verh. int. Ver. Limnol. 22: 3369–3372.

Eklöv, P., 1992. Group foraging versus solitary foraging efficiency in piscivorous predators: the perch, *Perca fluviatilis*, and pike, *Esox lucius*, patterns. Anim. Behav. 44: 313–326.

Eklöv, P. & S. Diehl, 1994. Piscivore efficiency and refuging prey: the importance of predator search mode. Oecologia 98: 344–353.

Gliwicz, J. M. & A. Jachner, 1992. Diel migrations of juvenile fish: a ghost of predation past or present? Arch. Hydrobiol. 124: 385–410.

Gotceitas, V. & P. Colgan, 1987. Selection between densities of artificial vegetation by young bluegills avoiding predation. Trans. am. Fish. Soc. 116: 40–49.

Grimm, M. P. & J. J. G. M. Backx, 1990. The restoration of shallow eutrophic lakes, and the role of northern pike, aquatic vegetation and nutrient concentration. Hydrobiologia 200/201: 557–566.

Hambright, K. D., R. J. Trebatoski & R. W. Drenner, 1986. Experimental study of the impacts of bluegill (*Lepomis macrochirus*) and largemouth bass (*Micropterus salmoides*) on pond community structure. Can. J. Fish. aquat. Sci. 43: 1171–1176.

Irvine, K., B. Moss & H. Balls, 1989. The loss of submerged plants with eutrophication II. Relationships between fish and zooplankton in a set of experimental ponds, and conclusions. Freshwat. Biol. 22: 89–107.

Lammens, E. H. R. R., R. D. Gulati, M.-L. Meier & E. van Donk, 1990. The first biomanipulation conference: a synthesis. Hydrobiologia 200/201: 619–627.

Lauridsen, T. L. & D. M. Lodge, 1996. Avoidance by *Daphnia magna* of fish and macrophytes: chemical cues and predator-mediated use of macrophyte habitat. Limnol. Oceanogr. 4: 794–798.

McQueen, D. J., 1990. Manipulating lake community structure: where do we go from here? Freshwat. Biol. 23: 613–620.

Naud, M. & P. Magnan, 1988. Diel onshore-offshore migrations in northern redbelly dace *Phoxinus eos* (Cope), in relation to prey distribution in a small oligotrophic lake. Can. J. Zool. 66: 1249–1253.

Pennak, R. W., 1973. Some evidence for aquatic macrophytes as repellants for a limnetic species of *Daphnia*. Int. Revue ges. Hydrobiol. 58: 569–576.

Perrow, M. R., A. J. D. J. Jowitt & L. Zambrano Gonzalez, 1996. Sampling fish communities in shallow lowland lakes: point sample electricfishing versus electricfishing within stop-nets. Fish. Mgmt Ecol. 3(4): 303–313.

Persson, L., 1993. Predator-mediated competition in prey refuges: the importance of habitat dependent prey resources. Oikos 68: 12–22.

Persson, L., S. Diehl, L. Johansson, G. Andersson & S. F. Hamrin, 1991. The succession of fish communities along the productivity gradient of temperate lakes-patterns and the importance of size-structured populations. J. Fish. Biol. 38: 281–293.

Persson, L. & P. Eklöv, 1995. Prey refuges affecting interactions between piscivorous perch and juvenile perch and roach. Ecology 76: 70–81.

Phillips, G. L., M. R. Perrow & J. H. Stansfield, 1996. Manipulating the fish-zooplankton interactions in shallow lakes: a tool for restoration. In Greenstreet, S. P. R. & M. L. Tasker (eds), Aquatic Predators and their prey. Blackwell Scientific Publications Ltd., Oxford, England: 174–183.

Savino, J. F. & R. A. Stein, 1989. Behavioural interactions between fish predators and their prey: effects of plant density. Anim. Behav. 37: 311–321.

Scheffer, M., S. H. Hosper, M.-L. Meijer, B. Moss & E. Jeppesen, 1993. Alternative equilibria in shallow lakes. TREE 8: 275–279.

Schriver, P., J. Bøgestrand, E. Jeppesen & M. Søndergaard, 1995. Impact of submerged macrophytes on fish-zooplankton-phytoplankton interactions: large-scale enclosure experiments in a shallow lake. Freshwat. Biol. 33: 255–270.

Siegel, S. & N. J. Castellan Jr., 1988. Nonparametric statistics for the behavioural sciences. McGraw-Hill Book Co., London and New York, 399 pp.

Stansfield, J. H., M. R. Perrow, L. D. Tench, A. J. D. Jowitt & A. A. L. Taylor, 1997. Submerged macrophytes as refuges for grazing Cladocera against fish predation: observations on seasonal changes in relation to macrophyte cover and predation pressure. Hydrobiologia 342/343: 229–240.

Søndergaard, M., E. Jeppesen, E. Mortensen, E. Dall, P. Kristensen & O. Sortkjær, 1990. Phytoplankton biomass reduction after planktivorous fish reduction in a shallow, eutrophic lake: a combined effect of reduced internal P-loading and increased zooplankton grazing. Hydrobiologia 200/201: 229–240.

Tatrai, I. & A. Herzig, 1995. Effects of habitat structure on the feeding efficiency of young stages of razor fish (*Pelecus cultratus* (L.)): an experimental approach. Hydrobiologia 299: 75–81.

Timms, R. M. & B. Moss, 1984. Prevention of growth of potentially dense phytoplankton populations by zooplankton grazing, in the presence of zooplanktivorous fish, in a shallow wetland system. Limnol. Oceanogr. 29: 472–486.

Townsend, C. R. & M. R. Perrow, 1989. Eutrophication may produce population cycles in roach, *Rutilus rutilus*, by two contrasting mechanisms. J. Fish. Biol. 34: 161–164.

Turner, A. M. & G. G. Mittelbach, 1990. Predator avoidance and community structure: interactions among piscivores, planktivores, and plankton. Ecology 71: 2241–2254.

Venugopal, M. N. & I. J. Winfield, 1993. The distribution of juvenile fishes in a hypereutrophic pond. Can macrophytes potentially offer refuges for zooplankton? J. Freshwat. Ecol. 8: 389–396.

Werner, E. E., J. F. Gilliam, D. J. Hall & G. G. Mittelbach, 1983. An experimental test of the effects of predation risk on habitat use in fish. Ecology 64: 1540–1548.

Winfield, I. J., 1986. The influence of simulated aquatic macrophytes on the zooplankton consumption rate of juvenile roach, *Rutilus rutilus*, and perch, *Perca fluviatilis*. J. Fish. Biol. 29 (suppl. A): 37–48.

Hydrobiologia **342/343:** 207–213, 1997.
L. Kufel, A. Prejs & J. I. Rybak (eds), Shallow Lakes '95.

The impact of predation by perch on the size-structure of *Chironomus* larvae – the role of vertical distribution of the prey in the bottom sediments, and habitat complexity

Ryszard Kornijów
Department of Hydrobiology and Ichthyobiology, University of Agriculture, 20-950 Lublin 1, P.O. Box 158, Akademicka 13, Poland

Key words: Macrophytes, refuge, habitat complexity, predation, *Chironomus* larvae, size-structure, vertical distribution, lake littoral

Abstract

The study was performed in June–August 1993, in eighteen enclosures (1 × 2 m) built in a fertile, small (area 2.5 ha) and shallow (max. depth 1.7 m) lake, Little Mere (north-western England). Three densities of *Nuphar lutea* leaves were manipulated in enclosures to 10%, 50% and 90% coverage of the water surface. Perch (total length ca. 15 cm) was stocked in half of the enclosures in a density 3 fish m^{-2}. The vertical distribution of *Chironomus* f.l. *plumosus* larvae (classified in the size classes: 0–5, 5–10, 10–15, 15–20 and >20 mm) in the bottom sediment showed a clear pattern, with bigger larvae burrowing to the greatest depth. Densities of all larval size classes except 0–5 mm were negatively affected both by plant density and by perch in three sediment layers considered (0–2, 2–5 and 5–10 cm). The larvae formerly at the surface did not move deeper in the sediments to avoid fish predation. The size distribution of the larvae, with the prevalence of the 15–20 and 10–15 mm larvae, found in the guts of perch was the most similar to that in the upper sediments at low and medium plant densities, and to that in the intermediate, 2–5 cm, sediment layer at high plant density. The effect of the interaction between plant density and fish was significant for all the larvae except those of 10–15 and 15–20 mm. Although the results of the experiment showed size-selective feeding by perch, the direct overall effect of the fish on the size-structure of the larvae was limited by vertical distribution of the prey in the bottom sediment. In addition, it was modified by nymphaeids, which, at high density, altered food conditions and acted as a refuge for *Chironomus* larvae both directly (providing shelter) and indirectly (providing fish with the additional, epiphytic prey).

Introduction

According to the theory of optimal foraging, the most frequently eaten prey are the most abundant, largest and most easily captured organisms, which, for the predator, are energetically the cheapest (MacArthur & Pianka, 1966; Werner & Hall, 1974). This model is supported by the size-efficiency hypothesis (Brooks & Dodson, 1965) and by the general refuge concept (Peters, 1992) which claim that predators normally prefer larger prey (within the limits of predator capacities).

The importance and implications of size-selective predation have been extensively documented for zooplankton-feeding fish (review by Northcote, 1988; Walls et al, 1990). However attempts to apply the model to benthic communities have given ambiguous results. While some authors have found a profound effect of fish on the mean size or weight of benthic prey (Hall et al., 1970; Kajak et al., 1976; Post & Cucin, 1984; Mittelbach, 1988; Diehl, 1995; Kornijów & Kairesalo, unpubl. data), others have found unclear or no effects (Thorp & Bergey, 1981; Hanson & Legget, 1986; Persson & Greenberg 1990; Diehl, 1992). Finally, Hershey (1985) suggested that size-selective feeding by fish on chironomid larvae may depend on their availability, which, in turn, is connected with burrowing behaviour of the prey. This hypothesis has never been proved for there have been no studies investigating simultaneously the fish impact on in-benthos, and its vertical distribution and size structure.

In this paper I test the hypothesis that size-selective structuring by an epi-benthic predator (perch) on benthic larvae of *Chironomus* f.l. *plumosus* (L.) in the lake littoral does occur, but is limited to the depth of the bottom sediments penetrated by fish, thus being dependent on the availability of the prey.

Perch is much better adapted than cyprinid fishes like roach, rudd or bream (Winfield, 1986; Diehl, 1988) to a strongly vegetated littoral. Nonetheless, habitat complexity may affect the foraging efficiency and diet of perch (Savino & Stein, 1982; Diehl, 1988, 1992; Mattila, 1992). Therefore, the present study was carried out at various densities of vegetation (nymphaeids) to test whether macrophytes, providing in-benthic chironomids with a refuge, can act as a factor influencing size-selective feeding by perch.

Methods

The experiment was performed in 18 top- and bottomless enclosures (1 × 2 m) built in a dense and relatively uniform bed of *Nuphar lutea* (L.) Sm. in Little Mere in north-west England in Cheshire (area 2.5 ha, max. depth 1.7 m; for more details on the lake see Carvalho, 1994). The enclosures were of curtain netting (mesh size 0.5 mm) fixed onto a wooden framework, pushed into the bottom sediment at the depth of about 75 cm. The lake bottom consisted of peat covered by loose mud of about 20–25 cm thickness. Three densities of *Nuphar lutea* leaves were manipulated in enclosures to 10%, 50% and 90% coverage of water surface by removing leaves at the petiole base. On 16 June 1993, six perch [mean length (LF) 14.8 cm, SD = 0.8 cm] were added to each of three enclosures at each manipulated plant density and all the enclosures were covered by a thin wide-mesh (2 × 2 cm) netting to prevent avian predation. There were thus six treatments: low, medium and high plant density without fish and low, medium and high plant density with fish, each replicated three times with treatments randomised among the enclosures.

The enclosures were inspected every second or third day and dead fish were replaced by individuals of a similar size. The overall replacement of the fish was 12%. All the fish used in the experiment were collected by seining in Little Mere. The experiment was terminated on 15 August by removing all the fish using a net constructed from 8 mm mesh material attached to a wooden frame which fit snugly inside the enclosures. The fish were killed, measured, and their stomach contents preserved with 4% formaldehyde solution for further analyses.

Benthic macroinvertebrates were sampled from within the enclosures three times: at the start of the experiment, after 39 days, and at the end of the experiment. Samples were collected using a perspex tube (area 15 cm^2, length 120 cm). In each enclosure 18 cores were taken on each occasion. In July and August six cores from each enclosure were sliced in the field, immediately following sampling, into three depth-fractions (0–2 cm, 2–5 cm and 5–10 cm) using the method of Van de Bund & Groenendijk (1994). The material collected was sieved through a 0.4-mm mesh. Invertebrates retained were sorted with the unaided eye and preserved in 4% formaldehyde solution. Larvae of *Chironomus* were counted and classified in the following five size classes: 0–5 mm, 5–10 mm, 10–15 mm, 15–20 mm, and >20 mm.

Results

Larvae of *Chironomus* f.l. *plumosus* constituted on average about 40% of the total density and 70% of the biomass of the whole zoobenthos. The remaining important taxa were: *Limnodrilus hoffmeisteri* Clap. and *Euilyodrilus hammoniensis* Mich. (Tubificidae), *Asellus aquaticus* Racov. (Isopoda), *Crangonyx pseudogracilis* Bous. (Amphipoda), and larvae of Ephemeroptera.

Occurrence of all larval size classes, except those 0–5 and 10–15 mm long, which were distributed in the sediment relatively evenly, clearly depended on the sediment depth (one-way ANOVA, effect of the sediment layer on density of 0–5, 5–10, 10–15, 15–20, >20 mm larvae: $p<0.26$, 0.003, 0.08, 0.001, 0.001, respectively), with higher density and bigger larvae burrowing deeper (Figure 1). The 5–10 mm larvae were usually most abundant in the upper and middle layers, to a depth 5 cm. Density of the 15–20 mm larvae was lowest at 0–2 cm and highest in the deepest (5–10 cm) sediment layer. The biggest larvae (>20 mm) were rare or absent in the surface sediment layer, while their highest density was in the deepest layer.

Plant density negatively influenced density of all larval size classes except 0–5 mm (Table 1, Figure 2). Different larval size classes were affected in different sediment layers; in 0–2 cm and 2–5 cm layers these

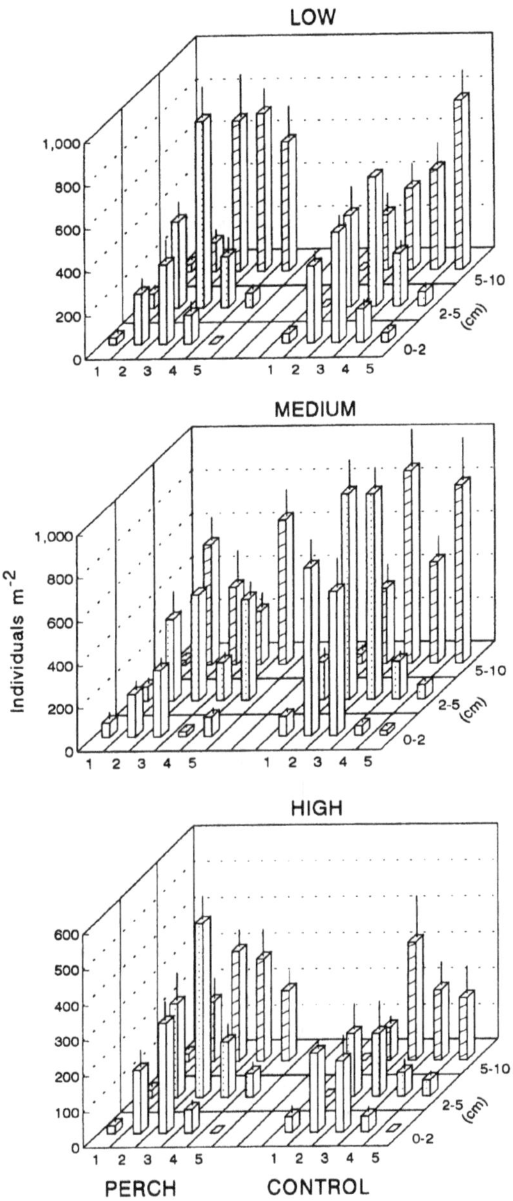

Figure 1. Vertical distribution of various size classes of *Chironomus* larvae in bottom sediments, divided into three layers (0–2, 2–5 and 5–10 cm) in enclosures with perch and in control fishless enclosures at low, medium and high densities of *Nuphar* leaves. Means ± SE are shown for samples taken from each enclosure on two occasions (in July and August). 1, 2, 3, 4, 5 are the larvae 0–5, 5–10, 10–15, 15–20 and >20 mm long, respectively.

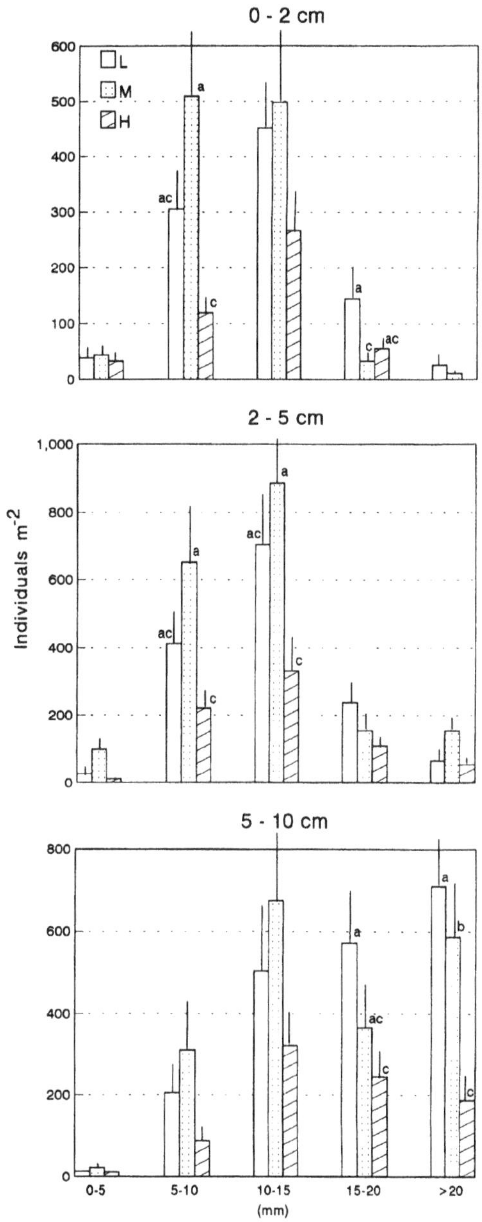

Figure 2. Density of various size classes (0–5, 5–10, 10–15, 15–20, >20 mm) of *Chironomus* larvae in three sediment layers (0–2, 2–5, 5–10 cm) at low (L), medium (M) and high (H) densities of *Nuphar* leaves. Data are means and 1 SE; $n = 4$–6 for the enclosures with perch and the controls. Means not showing a common lower case letter (a, b, c) are significantly different at the 0.05 level (one-way ANOVA).

Table 1. Three-way repeated-measures ANOVAs of the effects of *Nuphar lutea* density, and feeding by perch on density of various size classes of *Chironomus* larvae in the separate sediment layers (0–2, 2–5 and 5–10 cm) over the experimental period (time effect). * $p < 0.05$; ** $p < 0.01$; *** $p < 0.001$; NS not significant.

Source of variation	Size classes (mm) 0–5	5–10	10–15	15–20	>20
	0–2 cm				
Plant density	NS	*	NS	NS	NS
Perch	NS	**	NS	NS	NS
Plant density × perch	NS	*	NS	NS	NS
Time	NS	NS	NS	NS	NS
Time × plant density	NS	NS	NS	NS	NS
Time × perch	NS	NS	NS	**	*
Time × plant density × perch	NS	NS	NS	**	NS
	2–5 cm				
Plant density	NS	***	**	NS	NS
Perch	NS	NS	NS	NS	*
Plant density × perch	*	***	NS	NS	NS
Time	NS	**	NS	NS	NS
Time × plant density	NS	*	NS	NS	NS
Time × perch	NS	NS	NS	NS	NS
Time × plant density × perch	NS	*	NS	NS	NS
	5–10 cm				
Plant density	NS	NS	NS	*	***
Perch	NS	NS	NS	NS	NS
Plant density × perch	NS	NS	NS	NS	*
Time	NS	NS	*	NS	NS
Time × plant density	NS	NS	NS	NS	NS
Time × perch	NS	NS	NS	NS	NS
Time × plant density × perch	NS	NS	NS	NS	NS

were the 5–10 and 10–15 mm larvae, and in deeper sediments – the larvae of 15–20 and >20 mm (Figure 2). The larvae occurred the most abundantly at low or medium plant density (Figure 2).

Cumulative measures over time (July and August) analysed by ANOVA revealed the significant influence of perch only on density of the 5–10 mm larvae in the upper 0–2 cm sediment layer (Table 1). Data from single sampling dates were less variable than cumulative measures, and showed the influence of fish on density of the 5–10 mm and >20 mm larvae in 2–5 and 5–10 cm sediment layers, respectively, in July. Larvae of all size classes in the upper sediment layer tended to be more abundant than in the controls, too, but the results were not significant due to high variability among replicates. In August the impact of fish on the

Table 2. Similarity comparison of the size structure of *Chironomus* larvae in the diet of perch and in various sediment layers in the enclosures at low, medium and high *Nuphar* density at the end of the experiment (August). The results were obtained by adding the smaller or equal percentages of the various size classes of larvae (0–5 mm, 5–10 mm, 10–15 mm, 15–20 mm and >20 mm) found in the fish gut and the separate layers of the bottom sediments (0–2 cm, 2–5 cm, 5–10 cm) in the enclosures with perch. The closer the values to 100, the greater the similarity, the closer the values to 0, the smaller the similarity.

Layers of sediments (cm)	*Nuphar* density Low	Medium	High
0–2	75	64	68
2–5	43	39	83
5–10	50	33	76

larvae was stronger, and density of as many as three size classes (5–10, 10–15 and 15–20 mm) in 0–2 and 2–5 cm sediments were affected. Larvae of these size classes were at least double in the controls than in the enclosures with fish (Figure 3).

The size-structure of the larvae found in the guts of perch at the end of the experiment was the most similar to that in the upper sediments at low and medium plant densities, and to that in the intermediate sediment layer at high plant density (Table 2, Figure 4). The highest proportions in the perch guts usually were of the larvae of 15–20 and 10–15 mm (Figure 4). The characteristic feature was the lack of the biggest (>20 mm) larvae both in the perch guts as well as in the surficial sediment layers in the enclosures with fish.

The effect of interaction between plant density and perch was significant for the 5–10 mm larvae in the upper sediment layer, for the 0–5 and 5–10 mm larvae in the 2–5 cm layer, and for the >20 mm larvae in the deepest sediments (Table 1). There was no such effect on the 10–15 and 15–20 mm larvae in any sediment layer considered.

The effect of time on larval density was rather negligible, and concerned only the 5–10 and 10–15 mm larvae in the sediment layers of 2–5 and 5–10 cm, respectively (Table 1).

Discussion

Vertical distribution of *Chironomus* larvae presented in this study parallels the one found by other authors that

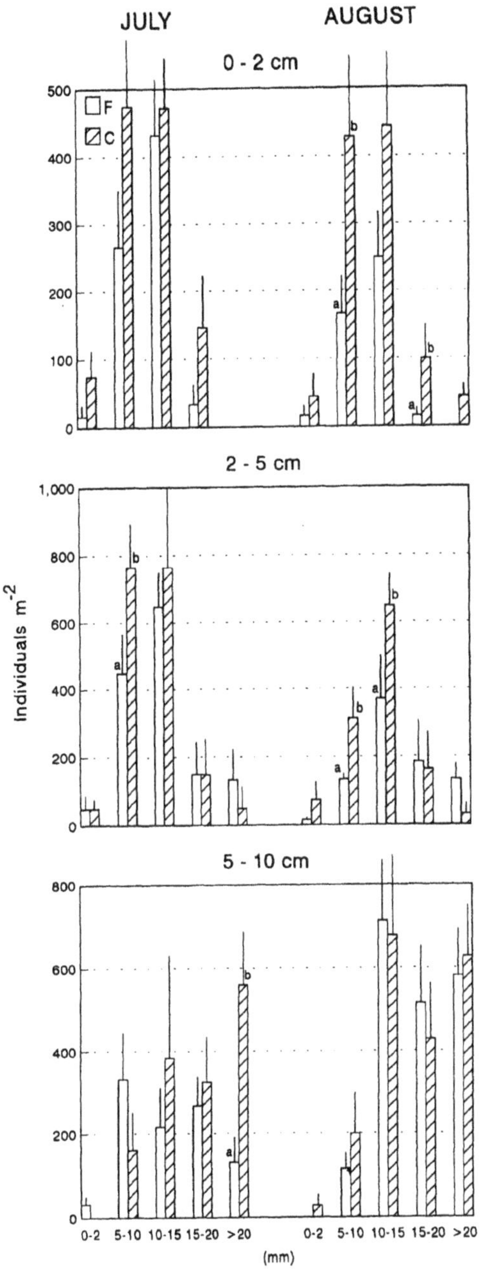

Figure 3. Density of various size classes (0–5, 5–10, 10–15, 15–20, >20 mm) of *Chironomus* larvae in three sediment layers (0–2, 2–5, 5–10 cm) in the enclosures with perch (F) and in the controls (C). Data are means and 1 SE, $n = 16$–18 for low, medium and high densities of *Nuphar* leaves. Means not showing a common lower case letter (a, b) are significantly different at the 0.05 level (one-way ANOVA).

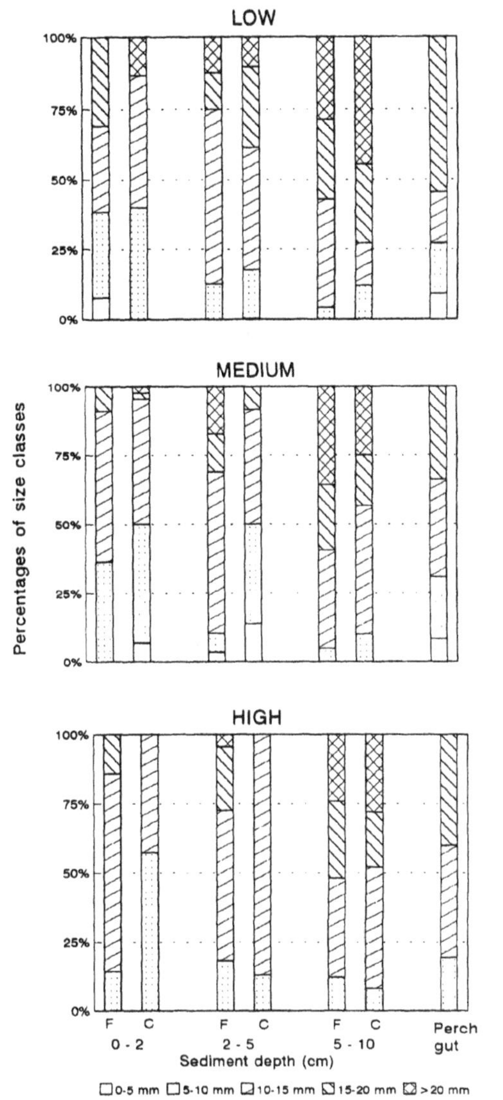

Figure 4. Percentages of various size classes (0–5, 5–10, 10–15, 15–20, >20 mm) of *Chironomus* larvae, based on their total density, found at the end of the experiment (August) in three sediment layers (0–2, 2–5, 5–10 cm) in the enclosures with perch (F), in the control fishless enclosures (C), and in the perch guts, at low, medium and high densities of *Nuphar* leaves.

larger forms burrow deeper, which, in consequence, leads to a maximum of benthos biomass a few centimeters below the sediment surface (e.g. Kajak & Dusoge, 1971; Becket et al., 1992; Van de Bund & Groenendijk, 1994). The size structure of the larvae eaten by perch, which is considered an epi-benthic predator (Mattila,

1992), was most similar to that in the upper bottom sediments (Table 2, Figure 4). However, perch was able to penetrate the bottom sediments deeper, for once the effect of its foraging on the larval density was significant even in the deepest, 5–10 cm, sediment layer (Figure 3). The reason for this might be the nature of the bottom sediments which were very loose.

The comparison between the size distribution of the larvae found in the perch guts and that in the environment at the end of the experiment suggests a preference of perch for bigger larvae (Figure 4). Especially the high contribution of the 15–20 mm larvae in the perch diet, and low in the environment is striking. The significant effect of perch on the small, 5–10 mm, larvae (Table 1, Figure 3) is also worth mentioning. Larvae of this size most abundantly occurred in the two upper sediment layers. It should be pointed that their density, as well as density of the 0–5 larvae must have been highly underestimated since a substantial part of the larvae could easily pass through the netting used for washing samples. The reason for high vulnerability of the 5–10 mm larvae, apart from their high abundance, might also be high mobility of the larvae. Yamagushi & Fukuhara (1971) found *Chironomus plumosus* larvae of the similar size (8–12 mm) as most frequently swimming above the mud surface. Migrations from the bottom sediments are usually stimulated by oxygen depletion in the near-bottom waters, which in the enclosures studied often took place, particularly at dusk (Moss & Kornijów, in press).

At the end of the experiment the biggest, >20 mm, larvae were not present in the surficial sediment layer in the enclosures with fish (Figure 4). This finding suggests that the biggest larvae were consumed by perch there as first, as confirmed by significance of interaction between time and perch (Table 1).

In consequence, consumption by perch affected size-structure and vertical distribution of the larvae in the bottom sediments. The larvae formerly at the surface did not move deeper to avoid fish predation. Otherwise larval densities in the deeper sediment layers would have been higher in the presence of fish than in the controls (Figure 3).

Nuphar negatively affected density of majority of the larval size classes (Table 1). The fact that the deeper the bottom sediments, the bigger larvae were influenced by vegetation could simply be connected with the larval vertical distribution in the bottom sediments (Figure 1). The bigger larvae occurred in the deeper sediments in sufficient numbers to make the statistical analysis of treatment effects on their abundance meaningful.

The effect of vegetation on density of the larvae could be connected with their feeding behaviour and food conditions. Larvae of *Chironomus* are tubicolous animals and can be classified functionally as collectors-gatherers or collectors-suspension feeders (Johnson et al., 1989). Their basic food consists of both planktonic and epipelic algae. Vegetation itself did not influence chlorophyll *a* concentration in the water (Moss & Kornijów, in press). Chlorophyll concentrations were inversely related to plant density in the presence of perch, which could account for deterioration of food conditions for the larvae. This was reflected by the interaction of the effect between plant density and perch on larval density, which was significant in the 5–10, 10–15 and >20 mm size classes (Table 1). Another, complementing reason for the existence of this interaction could be that dense vegetation, as a potential refuge, decreased searching effectiveness of perch and increased the chance of prey survival. An equally plausible explanation may be that at high vegetation density, fish could exploit additional food resources – epiphytic invertebrates abundantly living on the plant surface (Kornijów, Moss & Measey, unpubl. data). The impact of perch on the in-benthic community could then be softened.

The present paper supports the two first hypotheses posed: (1) feeding of perch on *Chironomus* larvae is size-selective, although above all it is detectable only in the sediment layers available to fish; (2) vegetation, particularly if occurring at high density, can modify the influence of perch on the size structure of in-benthic prey.

Feeding efficiency of fish on chironomid larvae decreases steeply with increasing depth of the larvae in the substrate (Lammens et al., 1987). Some of previous studies may have failed to detect an influence of epi-benthos-feeding fish on the prey size distribution because they sampled the whole sediment column occupied by benthos and did not make a distinction between available and non-available sediment fractions for fish. It is likely that in such cases any impact of fish could be masked due to the inclusion of big size classes occurring beyond the limit of fish penetration.

The present study shows that the theoretical assumptions of optimal foraging theory and the size efficiency hypothesis are valid also for interactions between an epi-benthic fish predator and its benthic prey. In addition, this paper suggests the existence of a habitat complexity threshold, above which predation rates, as shown in simplified laboratory conditions (Savino & Stein, 1982), are sharply reduced. However, in a complex natural environment, it can be that fish simply switch to other food patches such as epiphytic prey – as in this case (Kornijów, Moss & Measy, unpubl. data), or pelagic prey (Persson & Greenberg, 1990; Diehl, 1992). This again concords with optimal foraging theory. Thus not only the prey behaviour (Werner et al., 1983) but also that of its predator should be taken into consideration while testing or using foraging models.

Acknowledgments

I would like to thank Brian Moss for his hospitality and giving me opportunity to have a great and fruitful time in his lab at the University of Liverpool, and for his assistance with the project. I also thank John Measey and Sabri Kilinc for their help in field and laboratory work. Support was provided by the British Council.

References

Becket, D. C, T. P. Aartila & A. C. Miller, 1992. Contrasts in density of benthic invertebrates between macrophyte beds and open littoral patches in Eau Galle Lake, Wisconsin. Am. Midl. Nat. 127: 77-90.

Brooks, J. L. & S. L. Dodson, 1965. Predation, body size and composition of plankton. Science 150: 28–35.

Carvalho, L., 1994. Top-down control of phytoplankton in a shallow hypertrophic lake: Little Mere (England). Hydrobiologia 275/276: 53–63.

Diehl, S., 1988. Foraging efficiency of three freshwater fishes: effects of structural complexity and light. Oikos 53: 207–214.

Diehl, S., 1992. Fish predation and benthic community structure: the role of omnivory and habitat complexity. Ecology 73: 1646–1661.

Diehl, S., 1995. Direct and indirect effects of omnivory in a littoral lake community. Ecology 76: 1727–1740.

Johnson, K., B. Boström & W. van de Bund, 1989. Interactions between *Chironomus plumosus* (L.) and the microbial community in surficial sediments of a shallow, eutrophic lake. Limnol. Oceanogr. 34: 992–1003.

Hall, D. J.,W. E. Cooper & E. E. Werner, 1970. An experimental approach to the production dynamics and structure of freshwater animal communities. Limnol. Oceanogr. 15: 839–928.

Hanson, J. M. & W. C. Legget, 1986. Effect of competition between two freshwater fishes on prey consumption and abundance. Can. J. Fish. aquat. Sci. 43: 1361–1372.

Hershey, A. E., 1985. Effects of predatory sculpin on the chironomid communities in an arctic lake. Ecology 66: 1131–1138.

Kajak, Z. & K. Dusoge, 1971. The regularities of vertical distribution of benthos in bottom sediments of three Masurian lakes. Ekol. pol. 19: 485–499.

Kajak, Z., J. Zawisza & A. Hillbricht-Ilkowska, 1976. Effect of experimentally increased fish stock on biocenosis and recovery processes of a pond type lake. Limnologica 10: 595–601.

Lammens, E. H. R. R., J. Geursen & P. J. McGillavry, 1987. Diet shifts, feeding efficiency and coexistence of bream (*Abramis brama*), roach (*Rutilus rutilus*) and white bream (*Blicca björkna*) in eutrophicated lakes. In Kullander, S. O. & B. Fernholm (eds), Proc. Fith Congress Europ. Ichthyol. Stockholm, Swed. Mus. Nat. Hist.: 153–162.

MacArthur, R. H. & E. R. Pianka, 1966. On optimal use of patchy environment. Am. Nat. 100: 603–609.

Mattila, J., 1992. Can fish regulate benthic communities on shallow sof-bottoms in the Baltic Sea. The role of perch, ruffe and roach. Ph.D. dissertation, Abo Akademi University.

Mittelbach, G. G., 1988. Competition among refuging sunfishes and effects of fish density on littoral zone invertebrates. Ecology 69: 614–623.

Moss, B. & R. Kornijów, in press. The effects of nymphaed (*Nuphar lutea* L.) density and predation by perch (*Perca fluviatilis* L.) on the size and composition of zooplankton communities in a shallow lake. J. anim. Ecol.

Northcote, T. G., 1988. Fish in the structure and function of freshwater ecosystems: a 'topdown' view. Can. J. Fish. aquat. Sci. 45: 361–379.

Persson, L. & L. Greenberg, 1990. Optimal foraging and habitat shift of perch (*Perca fluviatilis*) in a resource gradient. Ecology 71: 1699–1713.

Peters, R. H., 1992. Lessons from the size efficiency hypothesis II. The mire of complexity. Hydrobiologia 235/236: 435–455.

Post, J. R. & D. Cucin, 1984. Changes in the benthic community of a small precambrian lake following the introduction of yellow perch, *Perca flavescens*. Can. J. Fish. aquat. Sci. 41: 1496–1501.

Savino, J. F. & R. A. Stein, 1982. Predator-prey interaction between largemouth bass and bluegills as influenced by simulated submerged vegetation. Trans. am. Fish. Soc. 111: 255–266.

Thorp, J. H. & E. A. Bergey, 1981. Field experiments on responses of a freshwater benthic community to vertebrate predators. Ecology 62: 365–375.

Van de Bund, W. J. & D. Groenendijk, 1994. Seasonal dynamics and burrowing of littoral chironomid larvae in relation to competition and predation. Arch. Hydrobiol. 132: 213–225.

Walls, M., I. Kortelainen & J. Sarvala, 1990. Prey responses to fish predation in freshwater communities. Ann. zool. fenn. 27: 183–199.

Werner, E. E. & D. J. Hall, 1974. Optimal foraging and the size selection of prey by the bluegill sunfish (*Lepomis macrochirus*). Ecology 55: 1042–1052.

Werner, E. E., G. G. Mittelbach, D. J. Hall & J. F. Gilliam, 1983. Experimental tests of optimal habitat use in fish: the role of relative habitat profitability. Ecology 64: 1525–1539.

Winfield, I. J., 1986. The influence of simulated aquatic macrophytes on the zooplankton consumption rate of juvenile roach, *Rutilus rutilus*, rudd, *Scardinius erythrophthalmus*, and perch, *Perca fluviatilis*. J. Fish Ecol. 29, suppl. A: 37–48.

Yamagushi, H. & H. Fukuhara, 1971. Note on the swimming behavior of *Chironomus plumosus* larvae in lake Suwa. Jap. J. Ecol. 20: 256–257.

Hydrobiologia **342/343**: 215–221, 1997.
L. Kufel, A. Prejs & J. I. Rybak (eds), Shallow Lakes '95.

Impact of fish predation on a small-bodied cladoceran: limitation or stimulation?

Mirosław Ślusarczyk
Department of Hydrobiology, University of Warsaw, Banacha 2, 02-097 Warszawa, Poland

Key words: Bosmina, fish predation, antipredator defenses, life history

Abstract

In seven enclosure experiments the short-term effects of exposing *Bosmina longirostris*, a small bodied cladoceran, to high levels of fish predation were analyzed. The density of *Bosmina* did not change significantly in the experimental enclosures, suggesting no direct fish impact on *Bosmina* demography. However, detailed studies revealed strong fish effects on some demographic and life history parameters of *Bosmina*. While the mean size of individual *Bosmina* decreased, reproduction enhanced (due to larger clutch sizes and a higher proportion of gravid females) in the enclosures with fishes. There is life history shift towards lower females sizes at a first reproduction in the fish enclosures.

The maintenance of *Bosmina* density in the fish enclosures may be attributed to: (1) higher reproductive rate due to elevated food level; (2) a life history shift towards smaller size reducing vulnerability to fish predation; (3) release from copepod predation caused by the presence of fish.

Introduction

Planktivorous fish prey selectively on conspicuous zooplankton, selecting large, pigmented, actively moving individuals (Hrbáček et al., 1961; Brooks & Dodson, 1965; Zaret, 1980). Elimination of easily visible, egg-carrying females affects the reproductive rate and lowers the probability of prey population recovering (Gliwicz & Pijanowska, 1989). Planktonic populations can be effectively controlled by fish; in extreme situation, fish can lead to the extinction of their prey (Gliwicz, 1985). On the other hand, fish which reduce the abundance of large-bodied individuals, protect small-sized zooplankton from competition for limited food resources (Vanni, 1987). Nutrients released to the water by the fish stimulate phytoplankton growth, thus improving food conditions for filter feeding zooplankton (Lamarra, 1975; Anderson et al., 1978; Schindler, 1993). Fish can reduce the invertebrate predation risk for small-bodied zooplankton by feeding on relatively large sized planktonic predators (Dodson, 1974).

Which effect (either suppression or stimulation) predominates, is highly dependent on prey size: bigger individuals are strongly suppressed by fish, and small bodied animals can benefit from the presence of fish.

The aim of this paper was to analyze short term responses of the small-bodied cladoceran *Bosmina longirostris* (O. F. Müller) to exposure to fish pressure in a series of in situ enclosure experiments.

The lake, its plankton and fish

Lake Wirbel (surface area 11 ha, mean depth 1.8 m, max depth 4 m) is a highly eutrophic, polimictic lake located in Mazurian Lakeland in north-eastern Poland. The phytoplankton composition changed during the experimental period from early spring (dominated by diatoms, blue-greens and green algae) to the early summer (dominated by blue-greens and diatoms).

The lake zooplankton was numerically dominated by rotifers (mainly *Keratella cochlearis* (Gosse), *Polyarthra* spp. and *Trichocerca* spp.). Cyclopoid copepods (mainly *Mesocyclops* sp.) were the most abundant group of planktonic crustacea during experiment. *Bosmina longirostris* was the most abundant cladocer-

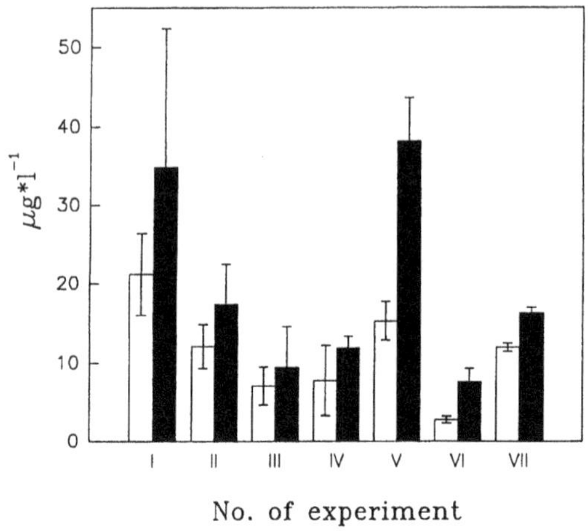

Figure 1. Mean ± 1 SD chlorophyll *a* concentration in seven experiments, in fish-free (empty bars) and fish (filled bars) enclosures.

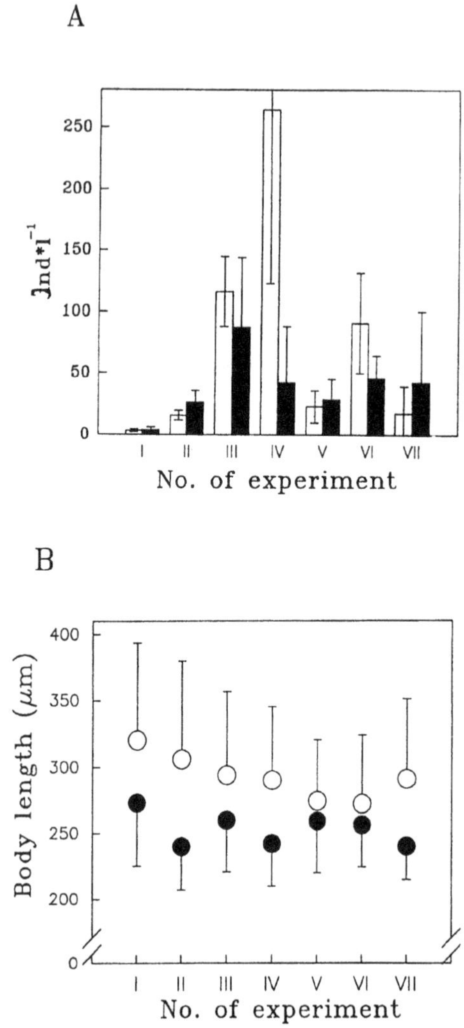

Figure 2. Mean ± 1 SD density (top) and mean ± 1 SD body length (bottom) of *Bosmina* at the end of seven experiments in fish-free (empty bars and circles) and fish (filled bars and circles) enclosures.

an until mid-July when it was replaced by *Diaphanosoma brachyurum* (Lièvin), *Ceriodaphnia quadrangula* (O. F. Müller), *Daphnia cucullata* G. O. Sars, *Polyphemus pediculus* L. and occasionally *Leptodora kindti* (Focke).

The lake was inhabited by numerous populations of various fish species that are either obligatory or temporary plankton feeders: *Leucaspius delineatus* Heckel, *Rutilus rutilus* L., *Blicca björkna* L., *Perca fluviatilis* L., and *Scardinius erythroptalmus* L. (Prejs et al., 1994).

Materials and methods

Seven 11-day experiments were performed from the beginning of May to mid-July 1988. In each of the experiments two sets of three enclosures (polyethylene bags of 300 dm^3 capacity each) were used. Each bag was filled with lake water containing natural zooplankton. Three enclosures were always left fish free while three roach *R. rutilus* (1+, 60– 100 mm *l.t.*) were added to each of the remaining enclosures. Dead fish were removed from enclosures but not replaced. Fish survival in the enclosures is shown in the Table 1.

Water samples for chlorophyll determination and zooplankton samples were taken from the lake at the beginning of each experiment and then from each enclosure after 11 days. Chlorophyll *a* concentration in the fraction of particles <50 μm was determined using a standard method (Golterman & Clymo, 1969). This gave a measure of phytoplankton biomass available for filter– feeding zooplankton. Twenty five liter zooplankton samples were randomly taken at three different depths: 0, 50, 100 cm with a 1 l Patalas sampler. Samples were filtered through 50 μm mesh and preserved with 4% sugar formalin.

Zooplankton was counted under a dissecting microscope. Seventy individuals of *B. longirostris* were measured in each sample. Gravid females and number of eggs carried in brood chambers were counted. Six size classes have been distinguished: I – 200–240 μm, II – 241–280 μm, III – 281–320 μm, IV – 321–360 μm, V – 361–400 μm, VI – >400 μm. The size distribution of

Table 1. Roach survivorship: number of fishes alive in the fish inhabited enclosures at the end of each experiment.

Exp.	Enclosure 1	2	3
I	3	3	3
II	3	3	3
III	1	1	1
IV	3	3	3
V	3	1	1
VI	2	1	3
VII	3	3	3

gravid females was obtained from additional measurements of 40–70 egg-carrying individuals, if present, in each enclosure.

The minimal size of mature individual was arbitrarily taken as the length of the smallest egg-bearing female found in each experiment. The proportion of gravid females in a population (number of egg-bearing females divided by number of adult females) and fecundity (number of eggs per gravid females – clutch size, and number of eggs per adult females) was calculated. The clutch size to body length ratio was calculated to eliminate effect of different females body size on their fecundity in the enclosures. The proportion of gravid females was additionally separately determined for each size class of females in the third experiment (where large quantity of gravid females enabled more detailed determination).

Results

Algal biomass was higher in the fish enclosures than in the fish-free ones (ANOVA, $F = 16.5, p = 0.0005$) at the end of all the experiments (Figure 1).

In the fish-free enclosures, the final density of species that were rare or absent in fish enclosures (*D. brachyurum, C. quadrangula, P. pediculus, L. kindti*) usually increased. In fish enclosures, large bodied cladocerans (*L. kindti, P. pediculus*) were exterminated while the density of rotifers was on average 5.5 times higher then in fish-free enclosures. The total density of *Cyclopoida* did not differ significantly between the two sets of experimental enclosures (ANOVA, $F = 2.14, p = 0.85$).

In the fish enclosures, *Bosmina* density was not clearly affected; in three experiments the final densities of *Bosmina* was higher in the fish as compared to the fish-free enclosures while in four others it was lower (Figure 2); all those values did not significantly differ (ANOVA, $F = 2.22$, $p = 0.14$). However, the effect of fish presence on *Bosmina* body length was clear cut. The average size of *Bosmina* in the fish-free enclosures was significantly larger than in the fish ones (Figure 2) ($p<0.0001$, Mann-Whitney test). The increase of the size of *Bosmina* in the fish-free enclosures resulted both from the increased number of large-bodied and decreased number of small-sized individuals (Figure 3).

Size structure of egg bearing females differed between fish and fish-free enclosures (Figure 3); large egg-carrying females were absent in the presence of fish, and females started reproduction at a smaller size (the minimum size of gravid females was significantly lower in the fish enclosures ($p<0.0001$, Mann-Whitney test)).

The clutch size to body length ratio, the mean number of eggs per adult female, and the proportion of egg bearing females within adults were significantly lower in the fish-free enclosures (Table 2) while the mean number of eggs per gravid females did not differ significantly.

In the third experiment (where more detailed analysis was performed), the mean clutch size in the fish-free enclosures was lower (than in the fish presence) when female size was incorporated into the model (Figure 4) (ANCOVA, $F = 172, p<0.0001$) but did not differ significantly when size was not considered (ANOVA, $F = 1.5, p = 0.22$). The proportion of egg-bearing females in separate size classes was lower in the fish-free enclosures (Figure 5).

A significant linear regression (positive slope) between chlorophyll *a* concentration and number of eggs per adult *Bosmina* was found ($r^2 = 0.33, p = 0.01$ and $r^2 = 0.26, p = 0.05$ in fish-free and fish enclosures, respectively; data from all experiments pooled).

A significant linear regression (negative slope) between density of Cyclopoida and average clutch size of adult females was found in fish-free enclosures ($r^2 = 0.38, p = 0.006$), but not in the fish ones ($r^2 = 0.14$, $p = 0.13$).

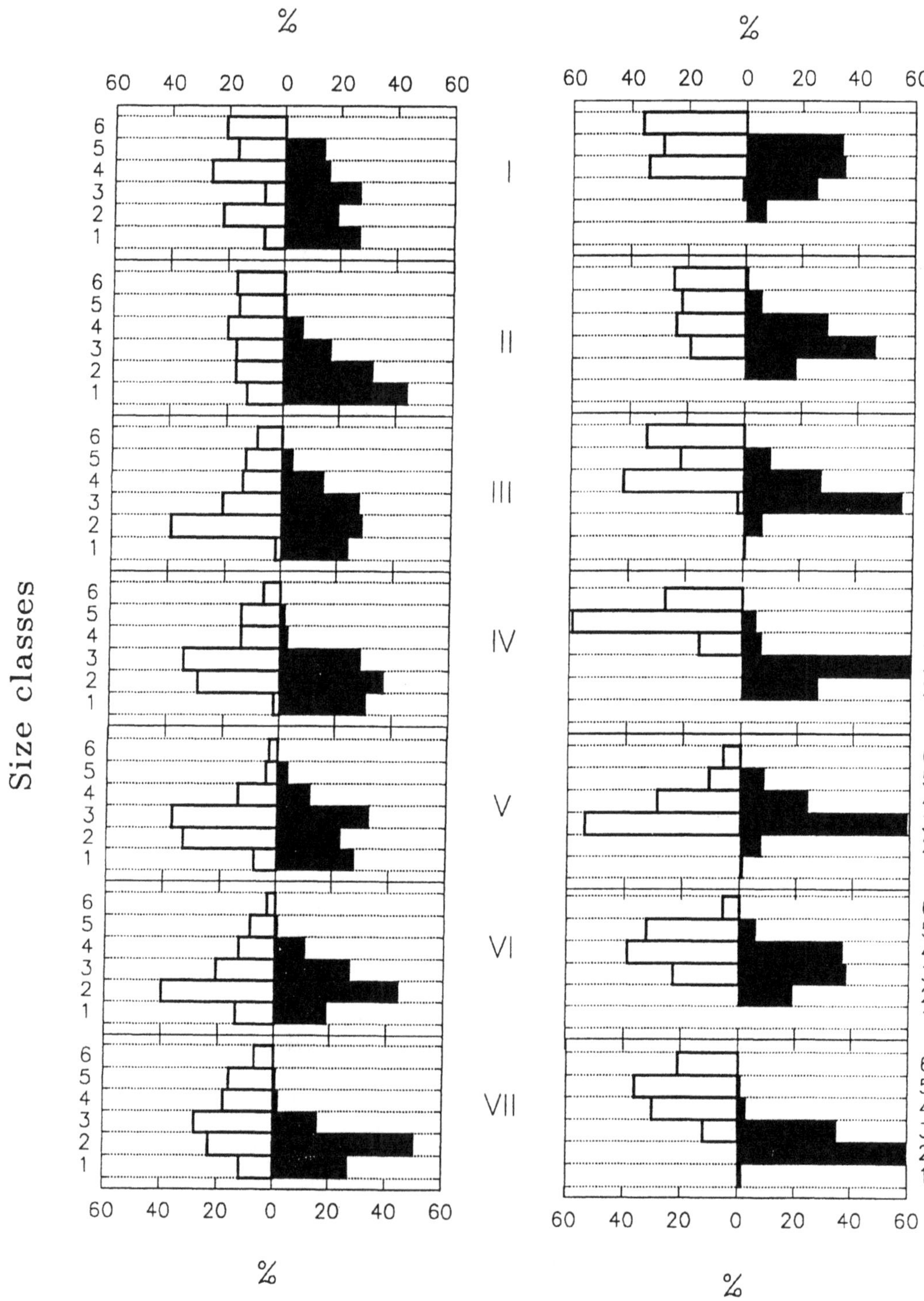

Figure 3. Size distributions of all individuals (on the left hand side) and of egg-bearing females (on the right) of *Bosmina* in fish-free (empty bars) and fish (filled bars) enclosures at the end of seven (I–VII) experiments.

Table 2. Fecundity of *Bosmina* expressed as average clutch size per body length, average clutch size and mean number of eggs per adults, respectively and proportion of gravid females within adults, in fish and fish-free enclosures, at the end of each experiment. Marked with asterix are statistically significant differences between fish and fish enclosures (ANOVA, $p<0.05$).

Exp.	Average clutch size per body length mm^{-1} *		Average clutch size		Average number of eggs in broad chambers of adults *		Average proportion of gravid females within adults *	
	no fish	fish	no fish	fish	no fish	fish	no fish	fish
I	14.7	19.7	2.94	2.61	1.45	1.58	0.49	0.62
II	16.2	19.3	2.63	1.95	1.18	0.95	0.44	0.43
III	14.0	26.6	2.69	2.47	0.85	1.35	0.32	0.55
IV	6.8	19.7	1.44	1.35	0.07	0.19	0.05	0.47
V	14.7	23.9	1.53	2.05	0.67	1.30	0.43	0.63
VI	10.5	17.6	1.56	1.67	0.41	0.64	0.25	0.37
VII	12.8	22.3	2.13	1.65	0.51	1.03	0.24	0.62

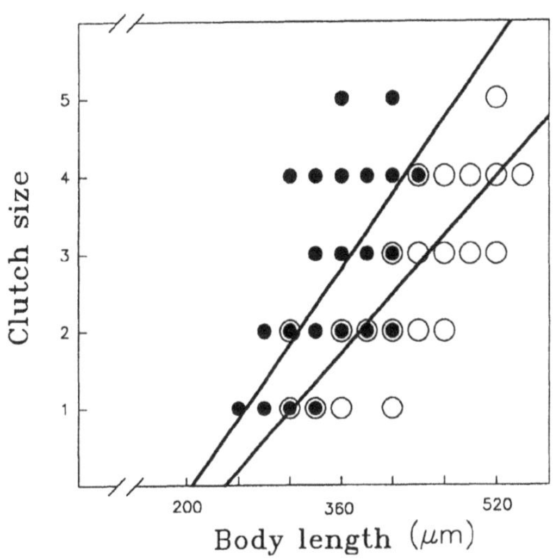

Figure 4. Clutch size vs body length in fish-free (empty circles) and fish (filled circles) enclosures at the end of the third experiment. (Both regression lines are plotted on the basis of about 180 points which overlap on the graph. Different size of circles was used to demonstrate overlapping filled and empty circles).

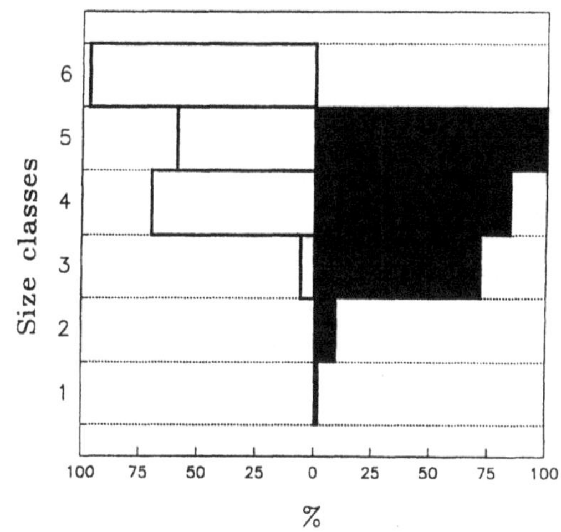

Figure 5. Proportion of gravid females in separate size classes of *Bosmina* in fish-free (empty bars) and fish (filled bars) enclosures at the end of the third experiment.

Discussion

The isolation of zooplankton in fish-free enclosures caused an increase in relative density of large-bodied species (*L. kindti, P. pediculus, D. brachyurum, C. quadrangula*) which were highly suppressed in the lake and in the fish enclosures due to their susceptibility to fish predation. No significant difference in *Bosmina* density between fish-free and fish inhabited enclosures was observed, which might indicate that *Bosmina* was not affected by fish. However, a comparison of size distributions of *Bosmina* from different treatments showed a strong effect of size-selective predation. The number of large-bodied females, usually bearing a large number of eggs was dramatically reduced in the presence of fish. However, the reproductive potential was not higher in fish-free enclosures. There are at least three plausible explanations of the observed phenomena:

(1) Despite strong selective predation, survivors could have benefited from the fish presence (e.g by increased reproduction) to an extent which compensated for losses of large and fecund females. Indeed, the higher proportion of gravid females and larger clutch sizes accounted for an increased reproductive potential of *Bosmina* in the fish enclosures. This could be

caused by an increase in food availability for herbivorous zooplankton due to the indirect effect of fish on phytoplankton growth and enhanced resuspension rate of seston particles. Phytoplankton biomass was always higher and the fecundity of *Bosmina* was positively correlated with chlorophyll *a* concentration when fish were present.

The increased production of eggs in the fish enclosures was probably offset by the absence of large individuals, potentially carrying more eggs, which were selectively removed by fish. Thus the average clutch size did not significantly differ between the two treatments when female size was not considered. Mean number of eggs per body length might be more reliable measure of reproductive effort when comparing females which differ strongly in size.

However, this hypothesis cannot explain the dramatic increase in the minimum size of gravid females in the fish-free enclosures, since the size of primiparous females in cladocerans tends to decrease with a decrease in food concentration (Węgleńska, 1971; Taylor, 1985).

(2) The observed contrasting patterns of population size structures between fish and fish-free enclosures could not have been caused merely by the selective effects of fish predation. It is likely that the different life history traits of *Bosmina* in fish and fish-free enclosures caused size structure difference. Planktonic animals grow often slower in the presence of visual predators (e.g. fish) which select for large preys. They allocate less energy to somatic growth and more to reproduction, which enables them to mature earlier at smaller size, and produce more eggs per body unit than individuals not endangered by fish (Macháček, 1991; Stibor, 1992; Weider & Pijanowska, 1993). In the present experiments a rapid shift in *Bosmina* life history was observed. Soon after the isolation of *Bosmina* from fish in the enclosures, the resource allocation pattern switched to faster growth at the expense of reproduction. Maturation was delayed, size at first reproduction increased and fecundity per unit of body length decreased. Thus animals could achieve high reproductive ratio in further instars once they grew larger.

(3) The observed differences in *Bosmina* size structures could have been influenced by the impact of fish on invertebrate predators. The pressure of invertebrate predation increased in the fish-free enclosures, mostly on either smaller, or slower (e.g. carrying eggs) individuals. Invertebrate predation could enhance differences of *Bosmina* size structures between fish and fish-free enclosures and thus affect *Bosmina* fecundity. Among invertebrate predators, only Cyclopoida were abundant during all the experiments. Though the density of cyclopoid copepods was not lower in the fish enclosures, it is possible that the mere presence of fish could stress them and 'focus' their attention on defence against predators (Boolens & Stearns, 1992) and thus release the pressure on the smaller filter feeders. The negative correlation between fecundity of *Bosmina* and Cyclopoida density in fish-free enclosures and the lack of correlation in the enclosure with fish supports this hypothesis. Selective elimination of gravid females (due to their possible reduced ability to escape) or the removal of eggs from brood pouches of gravid females (Gliwicz & Stibor, 1995) could explain these relationships.

It's likely that all three effects took place in my experiment, but it is hard to judge their relative impact.

Acknowledgments

I thank Paweł Koperski and Małgorzata Paradowska for help in performing the experiment and Joanna Pijanowska, Maciej Gliwicz and Piotr Dawidowicz for assistance in experimental design and constructive comments. Many thanks are due to Stephen Young for linguistic corrections of my manuscript.

References

Andersson, G., H. Berggren, G. Cronberg & C. Gelin, 1978. Effect of planctivorous and benthivorous fish on organisms and water chemistry in eutrophic lakes. Hydrobiologia 59: 9–15.

Boolens, S. M. & D. E. Stearns, 1992. Predator-induced changes in the diel feeding cycle of a planktonic copepod. J. exp. mar. Biol. Ecol. 156: 179–186.

Brooks, J. L. & S. I. Dodson, 1965. Predation, body size and composition of plankton. Science 150: 28–35.

Dodson, S. I., 1974. Zooplankton competition and predation: an experimental test of the size-efficiency hypothesis. Ecology 55: 605–613.

Gliwicz, Z. M., 1985. Eutrofizacja jezior tatrzanskich-użyźnienie czy zarybienie? (Eutrophication of Tatra lakes – fertilization or fish introduction?) Wiad. ekol. 31: 351–390.

Gliwicz, Z. M. & J. Pijanowska, 1989. The role of predation in zooplankton succession-W: Plankton ecology. In Sommer, U. (ed.), Succession in plankton communities. Springer-Verlag, Berlin, Heidelberg, New York, 253–296.

Gliwicz, Z. M. & H. Stibor, 1995. Egg predation by copepods in *Daphnia* brood cavities. Oecologia 95: 295–298.

Golterman, H. L. & R. S. Clymo, 1969. Methods for chemical analysis of fresh waters. IBP Handbook No. 8, Blackwell Scientific Publications, 172 pp.

Hrbáček, J., M. Dvorakova, V. Kořinek & L. Procházková, 1961. Demonstration of the effect of the fish stock on the species composition of zooplankton and the intensity of metabolism of the whole plankton association. Prir. Ved. 75: 1–65.

Lamarra, V., 1975. Digestive activities of carp as a major contributor to the nutrient loading of lakes. Verh. int. Ver. Limnol. 22: 658–665.

Macháček, J., 1991. Indirect effects of planktivorous fish on growth and reproduction of *Daphnia galeata*. Hydrobiologia 225: 193–197.

Prejs, A., A. Martyniak, S. Boroń, P. Hliwa & P. Koperski, 1994. Food web manipulation in a small, eutrophic Lake Wirbel, Poland: effect of stocking with juvenile pike on planktivorous fish. Hydrobiologia 275/276: 65–70.

Schindler, D. E., 1993. Nutrient regeneration by sockeye salmon (*Oncorhynchus nerka*) fry and subsequent effects on zooplankton and phytoplankton. Can. J. Fish. aquat. Sci. 49: 2498–2506.

Stibor, H., 1992. Predator induced life-history shifts in a freshwater cladoceran. Oecologia 92: 162–165.

Taylor, B., 1985. Effect of food limitation on growth and reproduction of *Daphnia*. Arch. Hydrobiol. 21: 285–296.

Vanni, M. J., 1987. Indirect effect of predators on age-structured prey populations: planktivorous fish and zooplankton. In Kerfoot, W. C. & A. Sih (eds), Predation: Direct and Indirect Impacts on Aquatic Communities. New England: 149–160.

Weider, L. J. & J. Pijanowska, 1993. Plasticity of *Daphnia* life histories in response to chemical cues from predators. Oikos 67: 385–392.

Węgleńska, T., 1971. The influence of various concentrations of natural food on the development, fecundity and production of planktonic Crustacean filtrators. Ekol. pol. 19: 427–471.

Zaret, T., 1980. Predation and freshwater communities. Yale University Press, New Haven, London, 187 pp.

Hydrobiologia **342/343:** 223–228, 1997.
L. Kufel, A. Prejs & J. I. Rybak (eds), Shallow Lakes '95.

Size-selective filtration and taxon-specific digestion of plankton algae by silver carp (*Hypophthalmichthys molitrix* Val.)

Lajos Vörös[1], Imre Oldal[2], Mátyás Présing[1] & Katalin V.-Balogh[1]
[1]*Balaton Limnological Research Institute of the Hungarian Academy of Sciences, P.O. Box 35, H-8237, Tihany, Hungary*
[2]*Laboratory for Environmental Protection of South Transdanubian Environmental Inspectorate, P.O. Box 15, H-7615, Pécs-Cserkút, Hungary*

Key words: silver carp, phytoplankton, size selective filtration, taxon specific digestion

Abstract

A comparison of the composition of natural phytoplankton communities with the gut content of silver carp (*Hypophthalmichthys molitrix*) showed that this fish species can not take up algae smaller than 10 μm. The species composition of algal assemblages in the gut content of the silver carp usually differs significantly from the composition of the >10 μm size fraction of the natural phytoplankton assemblages. This difference is mainly a result of the taxon selective digestion of silver carp. *In vitro* experiments with digestive enzymes of this species resulted in a very fast (1–2 minutes) disintegration of non-mucilaginous cyanobacteria. The digestion of diatoms and cryptophytes was also effective, but mucilaginous cyanobacteria and green algae (*Chlorococcales*) proved practically indigestible.

Introduction

The food and feeding of silver carp attracted significant interest in the 1970s because this fish was considered a potential tool for controlling eutrophication. Studies in lakes, ponds and experimental enclosures led to contradictory results. In experimental ponds Januszko (1974) observed a slight increase in phytoplankton biomass at a fish stock level of 1500 ind. ha^{-1}, while 3000 ind. ha^{-1} resulted in only a 10% decrease in phytoplankton biomass. In the case of a much higher fish density (up to 12 000 ind. ha^{-1}) the biomass and production of phytoplankton significantly increased (Januszko, 1978). In the enclosure experiment at 450 and 1350 kg ha^{-1} silver carp density the phytoplankton biomass decreased significantly (Kajak et al., 1975). In whole pond and lake ecosystems, however, studies showed that silver carp did not reduce the phytoplankton biomass and sometimes even inceased it. The carp reduced the biomass of zooplankton significantly (Kajak et al., 1975; Opuszyński, 1979; Spataru & Gophen, 1985).

Analysis of the food ingestion and the feeding selectivity of silver carp supports the hypothesis that its food selectivity is a mechanical, passive function of its filter morphology (Spataru & Gophen, 1985; Smith, 1989). Distances between the silver carp's gill rakers range from 12 to 26 μm (Hampl et al., 1983) which is consistent with experimental results that the lower limit for available food particles is about 10 μm (Cremer & Smitherman, 1980; Smith, 1989). A logical consequence of the above-mentioned results is that silver carp grazing can not control total algal biomass, but will modify the size structure of algal communities (Smith, 1989).

There are numerous studies on the silver carp's effect on species composition of natural phytoplankton and on the gut content of this fish. However the size structures of the free living and ingested algal assemblages have not been compared yet. The digestibility of different taxa of phytoplankters has not been investigated. The aim of this study is to compare the phytoplankton size structure in nature and in the gut content of the silver carp, and to determine the digestibility and/or resistance to digestive enzymes of various phytoplankton taxa ingested by silver carp.

Materials and methods

Study area

The Marcali reservoir was built in 1984 on the south part of the catchment area of Lake Balaton to decrease the phosphorus load to the lake. Its surface area is 4 km^2 and the average depth is 1.8 m. After the autumn drainage and fish harvesting the reservoir was filled up and stocked with 1+ fish (480 kg ha^{-1}) in the spring of 1988. At the end of this farming cycle in the autumn of 1989 the harvested fish biomass was 1400 kg ha^{-1}. The stocked and the harvested fish population consisted of about 80% of silver carp (*Hypophthalmichthys molitrix* Val.), 15% of bighead carp (*Aristichthys nobilis* Rich.), and 5% of common carp (*Cyprinus carpio* L.). During this two year period the reservoir retained 83% of soluble reactive phosphorus loading and 38% of total phosphorus loading. Approximately 85% of the retained phosphorus was bound up in the sediment while 9% was accumulated by the fish community and 6% by submerged macrophytes. The average water residence time was 4 months and the total phosphorus load was found to be 73 kg^{-1} ha^{-1} year^{-1} (Vörös, 1990).

Sampling, measurements and experiments

In 1988 material was collected on four dates: 18 May, 15 June, 14 July and 5 August from the upper (site 1), middle (site 2) and lower (site 3) parts of the reservoir. The water samples were preserved with Lugol's-solution for nano- and microplankton fraction measurements. For the quantitative determination of phototrophic picoplankton the subsamples were fixed with formaldehyde (1.5% final concentration). Parallel with the phytoplankton sampling, silver carp (age 1+) were collected with fishing nets from the three sampling sites. The fish samples were immediately dissected for collection of gut content. Five specimens of silver carp were examined at each time from each sampling area. The total contents of the intestines were homogenised. Aliquots were resuspended in filtered lake water and preserved with formaldehyde solution (5% final concentration). After this treatment the microscopic investigation of these samples was identical with that for the phytoplankton samples. Algal taxa larger than 2 μm were counted with an inverted microscope (Utermöhl, 1958). The quantity of picoplankton (<2 μm) was determined using epifluorescence microscopy (Caron, 1982). Biomass (fresh weight) of different algal taxa was calculated from cell volumes and abundances.

The index of electivity for each of the different food components (algal species) was calculated by Ivlev's formula (Ivlev, 1961):

$$E = \frac{g_i - w_i}{g_i + w_i} \tag{1}$$

where g_i = relative frequency of phytoplankton species in the gut of a fish, w_i = relative frequency of the same phytoplankton species in the surrounding water. The E values can range between −1 (total neglect) and +1 (total preference), whereas the $E = 0$ value shows an phytoplankton uptake corresponding to its frequency of occurrence in the surrounding water.

In vitro experiments on the digestion of phytoplankton by fish gut juice were carried out in August 1989. The reservoir phytoplankton in a sample of water was concentrated approximately 100 times by centrifugation. Filamentous, nitrogen fixing cyanobacteria were collected from Lake Balaton. For the same experiment starved silver carp (age 2+) were dissected and their gut juice was collected from the first 30 cm of the intestine. Freshly collected particle-free (GF/C filtered) gut juice (0.5 ml) and an equal volume of the concentrated phytoplankton suspension was mixed and incubated in test tubes during 0, 2, 4, 8, 15, 30, 60 and 120 minutes at 25 °C. The digestion process was stopped by addition of formaldehyde (35%), and the samples were diluted with particle free reservoir water. The quantity and composition of the algal mass was determined using the same method as in the case of the natural phytoplankton. The digestion process was characterised by the quantitative changes of different algal taxa during the incubation time.

Results

In May centric diatoms (*Stephanodiscus hantzschii* and *S. astraea) Trachelomonas volvocina*, one *Scenedesmus* and *Cryptomonas* species had the greatest biomass in the reservoir (Table 1). Among the most abundant species only the *S. astraea* was more frequent in the fish gut than in the water (E = 0.89). The dominant *S. hantzschii* had the same frequency in the gut and in the water (E = 0.01) while the contribution of *Scenedesmus, Trachelomonas, Cryptomonas* and *Phacus* species to total biomass signifi-

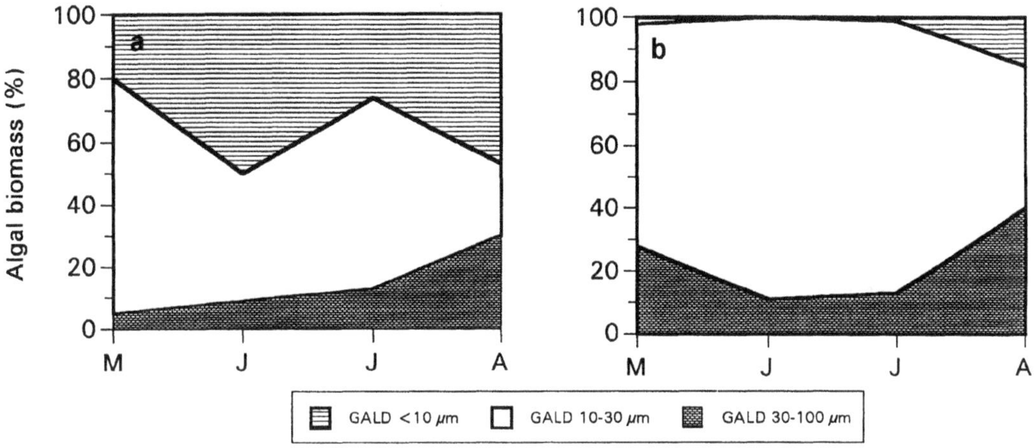

Figure 1. Temporal changes of phytoplankton size (GALD) classes (% of total biomass) in open water (a) and in gut content of silver carp (*Hypophthalmichthys molitrix*) (b) in Marcali reservoir (average values of sampling sites 1, 2 and 3). (GALD = greatest axial linear dimension).

cantly decreased in the gut content (Table 1). The contribution of picoplankton to total phytoplankton biomass in the reservoir water was 4% and it was undetectable in the intestine of fish. The total phytoplankton biomass in different parts of the reservoir ranged between 14.5 and 16.5 mg l^{-1}.

In June the total phytoplankton biomass decreased significantly varying between 6.9 mg l^{-1} and 8.2 mg l^{-1}. In the reservoir the dominant phytoplankton organisms were cryptophytes (*Rhodomonas minuta, Cryptomonas* sp.), euglenophytes (*Euglena pisciformis*) and green algae (*Scenedesmus quadricauda, Scenedesmus* sp. *Pediastrum boryanum, Phacotus lenticularis*). Picoplankton contribution to total phytoplankton biomass was about 13%. These species occurred in negligible amounts in the gut content. One centric diatom (*Cyclotella meneghiniana*) amounted to 83% of biomass in the intestine while its contribution to the free living phytoplankton community was only 4% (Table 1).

In July the total phytoplankton biomass increased reaching between 18 mg l^{-1} and 23 mg l^{-1}. Diatoms (*Melosira granulata, Cyclotella meneghiniana*), green algae (*Scenedesmus, Crucigenia*) and *Cryptomonas* species were the most abundant organisms in the water. In the gut content the percentage of *Cyclotella* significantly increased while the frequency of *Melosira, Scenedesmus* and *Cryptomonas* taxa decreased dramatically. In the total phytoplankton biomass the picoplankton contribution was about 5% and it was undetectable in the fish guts (Table 1).

In August the phytoplankton biomass varied between 20 mg l^{-1} and 26 mg l^{-1}, Centric diatoms (*M. granulata, S. hantzschii*), green algae (*Scenedesmus*), *Cryptomonas* sp., *Phacus pyrum* and picocyanobacteria were most abundant in the water. In the gut content the contribution of *S. astraea, Planktonema lauterbornii* and *Nitzschia sublinearis* to the total phytoplankton biomass increased, while the percentage of *M. granulata* and *S. quadricauda* did not change significantly. The relative frequency of *Scenedesmus* sp. decreased dramatically, while the *Cryptomonas, Peridinium, Phacus* and picocyanobacteria were undetectable in the gut content (Table 1).

Planktonic algae larger than 100 μm did not occur in the reservoir during the observed period. The contribution of algae smaller than 10 μm to the total phytoplankton biomass was low (20%) in May but became higher (30–50%) later. The 10–30 μm size algal group had a similar contribution to that observed for the previous group, and the contribution of algae larger than 30 μm in most cases was unimportant (Figure 1a). In the gut contents the amount of algae smaller than 10 μm was negligible and the frequency of larger forms was much higher than in the surrounding water (Figure 1b).

During the *in vitro* digestion experiment carried out with algae originating in the reservoir (Figure 2) the biomass of *Chlorococcales* green algae (*Scenedesmus, Tetraedron, Tetrastrum, Pediastrum* and *Oocystis*) only slightly decreased. At the end of the experiment more than 80% of their original biomass was present. Their cells were found to be intact, and they did not show any damage (Figure 2a). The biomass of *Euglenophy-*

Table 1. Biomass distribution (% of total biomass) of algal taxa in the water (W) of Marcali reservoir and in the gut contents (G) of silver carp (*Hypophthalmichthys molitrix*). The species are ranked by decreasing values of electivity index (E). Data represent the average values of sampling sites 1, 2 and 3.

Taxon	May		June		July		August		E-index
	W	G	W	G	W	G	W	G	Max
Cyclotella meneghiniana Kütz.			4.04	83.20	14.58	59.46			0.91
Stephanodiscus astraea (Ehr.) Grun.	8.76	17.27					1.60	28.38	0.89
Planctonema lauterbornii Schmidle							0.64	3.18	0.66
Other algae	8.44	31.84	11.37	16.36	7.03	17.99	4.28	17.40	0.60
Nitzschia sublinearis Hust.							1.93	6.26	0.53
Chlorella sp. (10 μm)					2.04	4.24	1.68	2.51	0.35
Nitzshia acicularis W. Smith					3.80	2.30	1.80	2.70	0.20
Melosira granulata (Ehr.) Ralfs					25.55	8.70	17.12	18.88	0.05
Stephanodiscus hantzschii Grun.	44.29	44.98					8.13	3.88	0.01
Scenedesmus quadricauda (Turp.) Bréb.	3.59	0.03	3.44	0.44	6.37	5.98	7.49	5.93	−0.03
Crucigenia quadrata Morren			3.07	0.00	1.48	0.23	1.60	0.99	−0.24
Tetrastrum staurogeniaeforme (Schröd.) Lemm.	1.02	0.58	1.19	0.00					−0.27
Gomphosphaeria lacustris Chod.					1.98	1.10			−0.28
Trachelomonas volvocina Ehr.	8.04	3.55							−0.39
Scenedesmus sp.	7.82	1.75	8.88	0.00	18.52	0.00	27.80	9.87	−0.48
Ankistrodesmus angustus Bern.			1.24	0.00					−1.00
Chlamydomonas sp. (10 μm)	1.79	0.00	1.86	0.00					−1.00
Cryptomonas sp. (25 × 15 μm)			13.09	0.00	5.01	0.00			−1.00
Cryptomonas sp. (40 × 25 μm)	4.14	0.00			2.23	0.00	6.74	0.00	−1.00
Didymocystis planktonica Kors.			1.58	0.00					−1.00
Euglena pisciformis Klebs			5.34	0.00	4.14	0.00			−1.00
Lyngbya limnetica Lemm.	1.81	0.00							−1.00
Pediastrum boryanum (Turp.) Menegh.			6.93	0.00					−1.00
Peridinium inconspicuum Lemm.			3.32	0.00			4.48	0.00	−1.00
Phacotus lenticularis Ehr.			12.35	0.00					−1.00
Phacus longicauda (Ehr.) Duj.	2.65	0.00							−1.00
Phacus pyrum (Ehr.) Stein					1.38	0.00	7.49	0.00	−1.00
Picocyanobacteria	4.00	0.00	13.00	0.00	4.47	0.00	7.22	0.00	−1.00
Rodomonas minuta	1.68	0.00	9.30	0.00					−1.00
Scenedesmus acutus Meyen	1.98	0.00			1.38	0.00			−1.00
Total phytoplankton	100	100	100	100	100	100	100	100	

ta (*Euglena, Phacus* and *Trachelomonas*) showed a decrease of 40% during the digestion experiment (Figure 2b). All the observable *Euglena* cells became spherical. The biomass of *Cryptomonas* and *Rhodomonas* cells decreased rapidly and after 30 minutes they completely disintegrated (Figure 2c). The behaviour of centric diatoms (*Cyclotella* and *Stephanodiscus*) was similar, but about 20% of them were resistant to the silver carp digestive enzymes (Figure 2d). The decrease of *Chroococcales*-type cyanobacteria showed a linear trend and at the end of the experiment less than 20% of the original biomass was detectable (Figure 2e).

Because filamentous cyanobacteria were absent in the Marcali reservoir the digestibility of these types of algae was checked with phytoplankton originated from Lake Balaton collected at the time of a mass development of nitrogen fixing cyanobacteria. The dominant species were *Aphanizomenon flos-aquae* and *Cylindrospermopsis raciborskii.* The enzymatic action was extremely rapid and after ten minutes about 90% of their biomass had disappeared. At the end of the incubation time the presence of *A. flos-aquae* and *C. raciborskii* was not detectable. Only some half-digested *Oscillatoria* filaments were found (Figure 2f). The rapid (within 30–60 seconds of enzyme treatment) and complete disintegration of the filaments of cyanobacteria was clearly visible under the microscope.

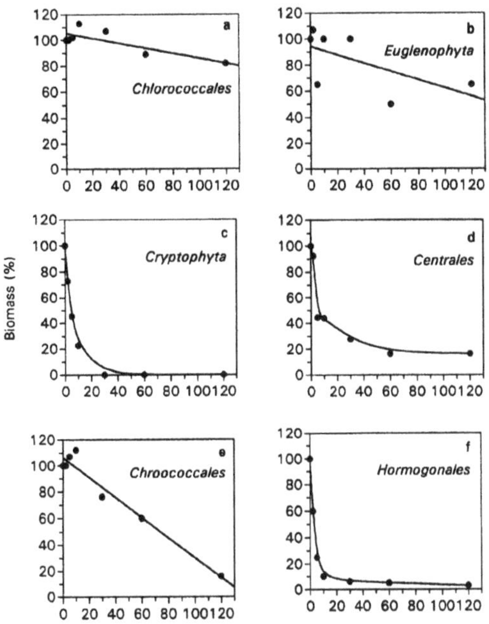

Figure 2. In vitro digestion (remained biomass in % of initial algal biomass after incubation periods) of different phytoplankton groups from Marcali reservoir (a–e) and cyanobacteria from Lake Balaton (f) by gut juice of silver carp (*Hypophthalmichthys molitrix*).

Discussion

The morphology of the silver carp's filtering apparatus predicts the lower limit (about 10 μm) of the available particle size (Hampl et al., 1983). Cremer and Smitherman (1980) found a predominance of *Selenastrum* in the plankton samples but not in the intestinal content. The absence of *Selenastrum* in the intestinal content was probably the result of its size which is generally less than 10 μm. Feeding experiments with different particle sizes (Smith, 1989) support this hypothesis.

Our simultaneous size structure analysis of natural phytoplankton and algae in the gut contents of silver carp in Marcali reservoir was found to be in accordance with these results. In spite of the permanent and significant contribution of small phytoplankters (<10 μm) to total phytoplankton biomass the carp's guts rarely contained algae smaller than 10 μm. Silver carp selected positively the >10 μm phytoplankton size fraction, but species composition of this algal size group in natural phytoplankton communities and in gut content was quite different (Table 1). These differences are partly explainable by the increasing filtration rate with the increasing particle diameter (Smith, 1989). But another factor most probably the taxon specific digestion, was also responsible for these differences.

Our *in vitro* digestion experiments gave a very good explanation for the absence of some phytoplankton taxa in the fish guts. Studying zooplankton feeding, Porter (1973) observed three groups of phytoplankton that cut across taxonomical lines: non available, available but non digestible and available and digestible. In the case of silver carp and probably for other filter feeding fish species this classification could be valid also, but the size limits and the digestibility of different taxa can vary. According to our results in the case of silver carp these phytoplankton groups were the following:

1. Non available algae (<10 μm): autotrophic picoplankton (*Synechococcus* type), *Rhodomonas minuta*, small centric diatoms, several *Chlorococcales* green algae were the dominant members of this group of phytoplankton.

2. Available (>10 μm) but non digestible algae: the most common *Chlorococcales* green algae belonged to this group. Besides the *Scenedesmus, Tetraedron, Tetrastrum, Pediastrum, Oocystis, Coelastrum* and *Chodatella* species some mucilagenous *Chroococcales*-type cyanobacteria and *Euglenophyta* were not easily digested.

3. Available (>10 μm) and digestible algae: inside this size category, non mucilaginous filamentous cyanobacteria (*Aphanizomenon flos-aquae, Cylindrospermopsis raciborskii*), *Cryptomonas* species and diatoms proved to be well digestible.

Depending on the species composition of the phytoplankton, the gut content analysis and the index of electivity may lead to the importance of group 2 being considerably overestimated, and furthermore may not be able to reveal the importance of group 3 or may lead to its being considerably underestimated. This group (available (>10 μm) and digestible algae) has high feeding significance. This means, that the analysis of gut content and the index of electivity can result only in information about the size selectivity of the fish filtration process if the sizes of the food particles are known.

These results suggest that the silver carp grazing has a complex effect on the phytoplankton structure. The size selective filtration can control the large sized digestible algae, including the nuisance blooms of cyanobacteria. There is much experimental evidence for this statement. Miura (1990) reported the effective control of *Microcystis* by silver and bighead carp filtration leading to the dominance of green algae smaller than 10 μm. Enclosure experiments in a tropical reservoir demonstrated the successful reduction of *C. raciborskii* biomass in the presence of silver carp (Starling & Rocha, 1990). Bitterlich (1985) also observed the *in vitro* digestion of a filamentous cyanobacteria (*Anabaena flos-aquae*) but his results suggest that the cell disintegration is not necessarily followed by its being well digested.

Our results are in accordance with the view that the growth of small-sized phytoplankton (<10 μm) is significantly enhanced in the presence of silver carp because the size selective filtration of fish controls their large sized algal competitors and grazers (zooplankton) (Opuszyński, 1979; Kajak, 1979; Spataru & Gophen, 1985; Drenner et al., 1986).

In deep lakes the silver carp filtration can decrease the biomass of the algae belonging to the second group (available (>10 μm) but non digestible algae) because faecal pellets sink into the non photic layer. In shallow lakes the egested faecal pellets continuously resuspend in the euphotic zone. Consequently in these lakes ingestion can be advantageous to these non digestible species because they can take up nutrients in the fish gut. Porter (1973, 1976) explained the resistance of several algal cells to the digestive enzymes of zooplankters by the presence of a gelatinous cover. Its mucopolisaccharide sheaths acted as molecular sieves, allowing small molecular and ionic nutrients to enter while keeping out large digestive enzymes. In the case of filter feeding fishes Miura & Wang (1985) also demonstrated the resistance of *Anabaena spiroides* due to their outer gelatinous sheaths. We found that not only can the mucilage protect the algal cells against the digestive enzymes, but also that (at least in the case of *Chlorococcales* green algae) their cellulose cell walls are impenetrable to the digestive enzymes of the silver carp (*Hypophthalmichthys molitrix*).

Acknowledgments

This study was supported by a grant from South Transdanubian Environmental Inspectorate.

References

Bitterlich, G., 1985. Digestive processes in silver carp (*Hypophthalmichthys molitrix*) studied *in vitro*. Aquaculture 50: 123–131.

Caron, D. A., 1983. Technique for enumeration of heterotrophic and phototrophic nanoplankton, using epifluorescence microscopy, and comparison with other procedures. Appl. envir. Microbiol. 46: 491–498.

Cremer, M. C. & R. O. Smitherman, 1980. Food habits and growth of silver and bighead carp in cages and ponds. Aquaculture 20: 57–64.

Drenner, R. W., S. T. Threlkcd & M. D. McCracken, 1986. Experimental analysis of the direct and indirect effects of an omnivorous filter-feeding clupeid on plankton community structure. Can. J. Fish. aquat. Sci. 43: 1935–1945.

Ivlev, V. S., 1961. Experimental ecology of the feeding of fish. Yale University Press, New Haven: 302 pp.

Januszko, M., 1974. The effect of three species of phytophagous fish on algae development. Pol. Arch. Hydrobiol. 21: 431–454.

Januszko, M., 1978. The influence of silver carp *Hypophthalmichthys molitrix* (Val.) on eutrophication of the environment of carp ponds. Roczn. Nauk Rol., H. T. 99: 55–79.

Hampl, A., J. Jirasek & D. Sirotek, 1983. Growth morphology of the filtering apparatus of silver carp (*Hypophthalmichthys molitrix* Val.) II. microscopic anatomy. Aquaculture 31: 153–158.

Kajak, Z., 1979. The possible use of fish, especially silver carp – *Hypophthalmichthys molitrix* (Val.) – to overcome water blooms in temperate water bodies. In Bíró, P. & J. Salánki (eds), Human Impacts on Life in Fresh Waters. Symp. Biol. Hung. 19: 77–86. Akadémiai Kiadó, Budapest.

Kajak, Z., J. I. Rybak, I. Spodniewska & W. A. Godlewska-Lipowa, 1975. Influence of the planktonivorous fish *Hypophthalmichthys molitrix* (Val.) on the plankton and benthos of the eutrophic lake. Pol. Arch. Hydrobiol. 22: 301–310.

Miura, T., 1990. The effects of planktivorous fishes on the plankton community in an eutrophic lake. Hydrobiologia 200/201: 567–579.

Miura, T. & J. Wang, 1985. Chlorophyll *a* found in faeces of phytoplanktivorous cyprinids and its photosynhetic activity. Verh. int. Ver. Limnol. 22: 2636–2642.

Opuszyński, K.,1979. Silver carp, *Hypophthalmichthys molitrix* (Val.) in carp ponds III. Influence on ecosystem. Ekol. pol. 27: 117–133.

Porter, K. G., 1973. Selective grazing and differential digestion of algae by zooplankton. Nature 244: 179–180.

Porter, K. G., 1976: Enhancement of algal growth and productivity by grazing zooplankton. Science 192: 1332–1334.

Smith, D. W., 1989. The feeding selectivity of silver carp, *Hypophthalmichthys molitrix* Val., J. Fish. Biol. 34: 819–828.

Spataru, P. & M. Gophen, 1985. Feeding behaviour of silver carp *Hypophthalmichthys molitrix* (Val.) and its impact on the food web in Lake Kinneret, Israel. Hydrobiologia 120: 53–61.

Starling, F. L. R. M. & A. J. A. Rocha, 1990. Experimental study of the impacts of planktivorous fishes on plankton community and eutrophication of a tropical Brazilian reservoir. Hydrobiologia 200/201: 581–591.

Utermöhl, H., 1958. Zur Vervolkommung der quantitativen Phytoplankton-Methodik. Mitt. int. Ver. Limnol. 9: 1–38.

Vörös, L., 1990. A Marcali-tározó hidrobiológiai vizsgálata, Tudományos jelentés. Hydrobiological investigation of Marcali reservoir. Scientific report. Balaton Limnological Research Institute, Tihany, Hungary. 75 pp.

Hydrobiologia **342/343:** 229–240, 1997.
L. Kufel, A. Prejs & J. I. Rybak (eds), Shallow Lakes '95.

Submerged macrophytes as refuges for grazing Cladocera against fish predation: observations on seasonal changes in relation to macrophyte cover and predation pressure

Julia H. Stansfield[1], Martin R. Perrow[2]*, Louise D. Tench[1], Adrian J. D. Jowitt[2] & Ayesha A. L. Taylor[1]
[1]*National Rivers Authority, Biology Laboratory, Station Road, St Olaves, Great Yarmouth, NR31 9JA, UK*
[2]*ECON, School of Biological Sciences, University of East Anglia, Norwich NR4 7TJ, UK*
(**author for correspondence*)

Key words: zooplankton, zooplanktivorous fish, macrophyte, refugia, predator/prey interactions, lake restoration

Abstract

The provision of a refuge from fish predation for large-bodied Cladocera among stands of submerged macrophytes is thought to be an important stabilising mechanism against nutrient-induced phytoplankton increases in clear water shallow lakes. The occurrence of any refuge effect in relation to the seasonal impact of both macrophyte development and recruitment of zooplanktivorous fish was monitored over the summer months (May to September) in three lakes (Cromes Broad, Hoveton Little Broad/Pound End, and Upton Broad) of variable trophic status, fish community structure and nature and extent of macrophyte cover, in the Norfolk Broads in Eastern England. At all sites, *Daphnia* spp. exhibited a early summer peak of abundance but had declined rapidly by July probably as a result of predation from underyearling fish. In extensive macrophyte stands (Cromes Broad) *Daphnia* spp. persisted after its elimination in open water, indicating some refuge effect. At sites with macrophyte cover and/or low fish predation pressure, *Daphnia* spp. was replaced by *Ceriodaphnia* spp., both with and without *Simocephalus* sp., thus maintaining large populations of grazing Cladocera, apparently capable of exerting a high grazing pressure on phytoplankton. This occurred, even under high predation pressure from a relatively high density of zooplanktivorous 0+ fish (Cromes Broad), consistent with the refuge hypothesis. In addition, at this site, significant positive associations of cladoceran abundance with increasing macrophyte cover were apparent throughout the summer. From information on the distribution of fish it is suggested that dense macrophytes offer the most suitable refuge through provision of predator-free space.

Introduction

Eutrophication of shallow freshwater lakes frequently results in increased standing crops of phytoplankton or perhaps filamentous algae. This in turn leads to the decline or loss of submerged macrophytes, typically through shading (Phillips et al., 1978; Scheffer, 1990), with consequent gross changes in biotic structure and function (Moss, 1983; De Nie, 1987; Bronmark & Weisner, 1992). However, there is growing empirical and experimental evidence that stabilising mechanisms associated with the presence of stands of macrophytes, may prevent or reduce increases in algal growth up to a threshold level of nutrient loading (Howard-Williams, 1981; Timms & Moss, 1984; Mitchell et al., 1988; Balls et al., 1989; Andersson et al., 1990; Blindlow et al., 1993; Van Donk et al., 1993; Schriver et al., 1995). Consequently, within a range of nutrient concentration, either macrophytes of phytoplankton may dominate; the alternative stable state hypothesis (Irvine et al., 1989; Scheffer et al., 1993).

Several stabilising mechanisms which perpetuate the macrophyte-dominated state have been suggested. These include: allelopathy, the production of compounds limiting algal growth by the macrophytes themselves (Wium-Anderson et al., 1982); competition for N, with macrophytes maintaining a competitive advantage over algae through luxury uptake (Balls et al., 1989) and acting as a sink for available nutrients (in both water column and sediments) (Ozimek et al., 1990; Van Donk et al. , 1993); prevention of resuspension of sediment and recirculation of fast sinking algae by wind and wave action through calming of the water column (James & Barko, 1990); provision of refuges for grazing zooplankton (Timms & Moss, 1984) and refuges for piscivorous fish acting upon the production (Spencer & King, 1984; Grimm & Backx, 1990) or the distribution (Werner et al., 1983; Persson & Eklov, 1995) of zooplanktivorous fish thereby reducing predation pressure upon the zooplankton (Hambright et al., 1986, Turner & Mittelbach, 1990).

Detailed knowledge of the mechanisms responsible for maintaining clear water and low algal populations in lakes dominated by macrophytes is critical for those lake managers concerned with the sustainable restoration (i.e. reversal of anthropogenic eutrophication) of shallow lakes, particularly if this is to be achieved in a predictable and cost-effective manner (Shapiro, 1990). In restoration programmes, the refuge effect for zooplankton is particularly appealing (Shapiro, 1990). This is due to two reasons. First, after nutrient loading has been reduced to the range where macrophyte dominance is likely to be stable (TP<0.15 mg l^{-1} or higher if N is strongly limiting-Hosper & Jagtman, 1990; Jeppesen et al., 1990), the removal of zooplanktivorous fish, often termed biomanipulation (Lammens et al., 1990), is often required to enhance populations of grazing zooplankton to reduce algal crops to provide a suitable light climate for macrophyte growth. If this mechanism operates, then it is likely that grazing has some importance in maintaining clear water within macrophyte beds. Second, during the development of macrophyte stands following biomanipulation which may or may not be rapid (Meijer et al., 1994; Perrow et al., 1994; Moss et al., 1996), any stabilising mechanism is of increased value if it also preserves water clarity in the open water surrounding macrophyte stands. The diel migration of large-bodied Cladocera from macrophyte stands to open water to feed provides such a mechanism (Timms & Moss, 1984).

Although macrophyte stands have been demonstrated to provide refuges for zooplankton (Schriver et al., 1995; Phillips et al., 1996) the timing of such effects relative to the seasonal development of macrophytes and recruitment of 0+ fish is poorly understood. The aim of this study was to describe the seasonal pattern of use of submerged macrophytes by grazing Cladocera in relation to potential predation pressure from fish and to determine whether the refuge effect is plausible as a mechanism in promoting clear water in beds of macrophytes. This adopted a comparative approach by monitoring zooplankton abundance and community structure, fish abundance and biomass, chlorophyll a concentration (as a index of algal abundance) and macrophyte cover during the macrophyte growing season (May–September) in three lakes of varying trophic status, fish community structure and nature and extent of macrophyte cover.

Description of study sites

The three lakes studied (Hoveton Little/Pound End, Cromes and Upton Broads) are among the Norfolk Broads, a series (around 50) of small (1–124 ha), shallow (0.5–2 m) lakes in eastern England, formed from flooded medieval peat diggings. Most may now be classed as hypertrophic and only four retain their original macrophyte populations (Moss, 1983).

The study sites represent a continuum from phytoplankton to macrophyte dominance. The first, Hoveton Little Broad, has been divided into two sections as part of a long term management strategy aimed at evaluating the effects of different restorative actions (Table 1). For the purposes of this study, Hoveton Little Broad, acts as a control section with turbid phytoplankton dominated water and few submerged macrophytes, whereas the other biomanipulated section, Pound End, has clear water and some macrophytes, although these are restricted to bird exclosures. The second, Cromes Broad, retains a relatively high nutrient loading but has recently developed an extensive cover of *Ceratophyllum demersum* L. with *Potamogeton freisii* Rupr. (Perrow et al., 1997). The third, Upton Broad, being isolated from riverine inputs and surrounded by wet woodland, has largely retained its original character and has a large cover of macrophytes, particularly *Najas marina* L., with *Chara* spp., *P. pectinatus* L. and *Zannichellia palustris* L. (Perrow et al., 1997). The physical characteristics, management measures employed and relevant biological data are shown in Table 1.

Table 1. Characteristics of the study sites.

	Hoveton Little Broad	Pound End	Cromes Broad	Upton Broad
Size (ha)	14.3	5.5	2.0	5.6
Mean depth (m)	1.7	1.1	1.0	1.2
Riverine connection	River Bure	River Bure via Hoveton Little	None	None
Management	Dredged 1990	Dredged 1990 Biomanipulated 1990-	Dredged 1988 Isolated 1992	None
Macrophyte cover				
% occurrence	–	6%	80%	91%
mean % cover	–	37%	66%	23%
(point survey Sept '94)				
Chlorophyll a (μg l^{-1}) (mean ± S.E. Apr-Sept '94)	58.2 ± 14.0	12.2 ± 2.2	14.2 ± 2.3	11.5 ± 2.2
Total phosphorus (μg l^{-1}) (mean ± S.E. Apr-Sept '94)	74.0 ± 9.0	47.0 ± 5.0	115.0 ± 19.0	40.0 ± 5.0

Materials and methods

Zooplankton, macrophytes and chlorophyll a

Sampling for zooplankton, macrophytes, and chlorophyll *a* was carried out monthly from May to September 1994. At each site, ten zooplankton samples were taken from both open water and from amongst submerged macrophytes, where present. Where macrophytes occurred the density in a 0.25 m^2 quadrat centred on each sampling point was expressed as PVI (per cent volume infested-Canfield et al., 1984). This was estimated by multiplying the estimated % cover by the height of the stand and dividing by the depth of water. At Pound End, five samples were taken from each of the two large (a. 500–1000 m^2) bird exclosures. In the open water and when macrophyte stands reached the water surface, a tube sampler was used (length 90 cm, diameter 10 cm) in order to sample the whole water column. When macrophyte stands did not reach the water surface a Patalas sampler (10 litre capacity) was used to obtain a sample from among the macrophyte stand. Water was strained through a 64 μm net and the animals retained were preserved in 70% alcohol. Samples were counted under a stereo microscope using 25 × magnification.

Differences between mean abundance of important cladoceran zooplankton between open water and macrophytes were tested using Mann-Whitney U-tests on each sampling occasion. In addition, the relationships between abundance of particular cladoceran species and % macrophyte cover were explored using the Spearman rank correlation coefficient (r_s). Chlorophyll *a* concentrations were determined according to the methods of Moss et al. (1986).

Fish

Fish populations were sampled on two occasions, in May and September. This corresponded to the period prior to the recruitment of cyprinids and after the late summer period in which 0+ fish are likely to exert a depressive effect on zooplankton (Townsend & Perrow, 1989), respectively. In addition, sampling in May provided a baseline of the adult fish stock prior to the recruitment of underyearlings, whereas sampling in September allowed the relative recruitment strength of the 0+ fish to be estimated.

Point abundance sampling (PAS) (Copp & Penáz, 1988) conducted with high frequency (600 Hz) pulsed DC electrofishing equipment was used to sample all habitats in proportion to their abundance. During sampling the 3 m fibreglass dinghy, with one electrofishing operator and one person controlling the boat using 'push' (facing the direction of travel) rowing, was rowed systematically along transect lines (8–10 oar strokes apart). At the equivalent distance along the transect line, a point was sampled by rapidly immersing the large (45 cm diameter) anode and swinging it through a 2 m arc in one motion. Any stunned fish were removed and transferred to individual buckets. The area sampled was quantified as 2.4 m^2. This was achieved by using a voltmeter to determine the distance at which the voltage gradient was reduced to 0.12 V which may be used as the threshold limit for effective electrofishing (Copp & Penáz, 1988). At the end of each transect all fish were identified, counted, measured to the nearest mm fork length and weighed to the nearest gram, before being returned to a stretch of water that had been sampled previously. In addition, the % of the water column occupied by submerged macrophytes was estimated visually at each sampling point.

The entire area of Pound End, Upton and Cromes Broads were sampled, whereas only part of Hoveton Little Broad (c. 6 ha-comparable to the size of Pound End) was covered.

Results

Fish

The spring biomass of adult cyprinid and percid fishes was variable but low (up to 110 kg ha^{-1} but typically <15 kg ha^{-1}) (Table 2). The increase in biomass and especially, density, of potentially zooplanktivorous fish (all except pike, *Esox lucius* L. and eel, *Anguilla anguilla* (L.)) over the sampling period underlines the important contribution the recruitment of 0+ fish makes to fish production in the study lakes (Table 2). The relative recruitment strength of the different species in the different lakes is illustrated in Figure 1.

Zooplankton

Three genera of large-bodied grazing Cladocera zooplankton were found at significant density (>50 l^{-1}) in some or all of the study sites. These were *Daphnia hyalina/longispina* (Leydig/O.F.M.) *Ceriodaphnia reticulata/quadrangula* (Jurine/O.F. M.) and *Simocephalus vetulus* (O.F.M.).

Daphnia spp. had an early summer peak of abundance, albeit of varying magnitude, at all sites, but had declined to low populations by July. At Cromes Broad, where extensive macrophyte stands developed from June onwards (Figure 1), *Daphnia* spp. persisted amongst the macrophytes after its elimination in open water (Figure 2). At Cromes and Pound End *Daphnia* spp. was replaced by *Ceriodaphnia* spp. both with (Cromes and Pound End) and without (Upton) *Simocephalus* sp., thus maintaining large populations of grazing Cladocera. This was not the case in Hoveton Little Broad, where cladoceran populations remained extremely low throughout the summer.

The patterns of habitat use by *Ceriodaphnia* spp. varied between the different sites (Figure 2). At Pound End, macrophyte beds were favoured in early season, although once populations had built up by mid-summer, significant numbers (>100 l^{-1}) were found in open water. At Upton, prior to the peak in macrophyte populations, *Ceriodaphnia* spp. was more abundant in open water, thereafter there was no significant difference between the different habitats. In contrast, in Cromes, *Ceriodaphnia* spp. was significantly more abundant in macrophyte beds compared to open water on all occasions (Figure 2). A similar pattern was shown by *Simocephalus* sp. where it occurred, as it was virtually always significantly more abundant in macrophyte beds (Figure 2).

For sites with macrophytes, the relationships between PVI and the abundance of the different genera as well as all grazing Cladocera combined are shown in Table 3. At Cromes, there is a consistent pattern of increasing abundance of all genera, including *Daphnia* spp. while it was still present, with increasing PVI. At Upton no such pattern was found, with either no significant relationships, or in July, a significant negative relationship with *Ceriodaphnia* spp. Pound End represented an intermediate situation, with different species exerting different preferences: *Daphnia* spp. tended to be negatively associated with macrophytes; *Ceriodaphnia* spp. displayed positive associations early in the summer but no relationship in subsequent samples and *Simocephalus* sp. showed significant positive correlations with PVI on all occasions.

Discussion

Seasonal interactions between macrophytes, fish, zooplankton and phytoplankton

There is clearly some interaction between extent of macrophyte development, intensity of predation by 0+ fish measured simply as a function of density and the structure and abundance of the cladoceran community and its potential impact on algal abundance (measured as chlorophyll *a*) (Figure 1). Before considering the seasonal patterns of these interactions in more detail, it

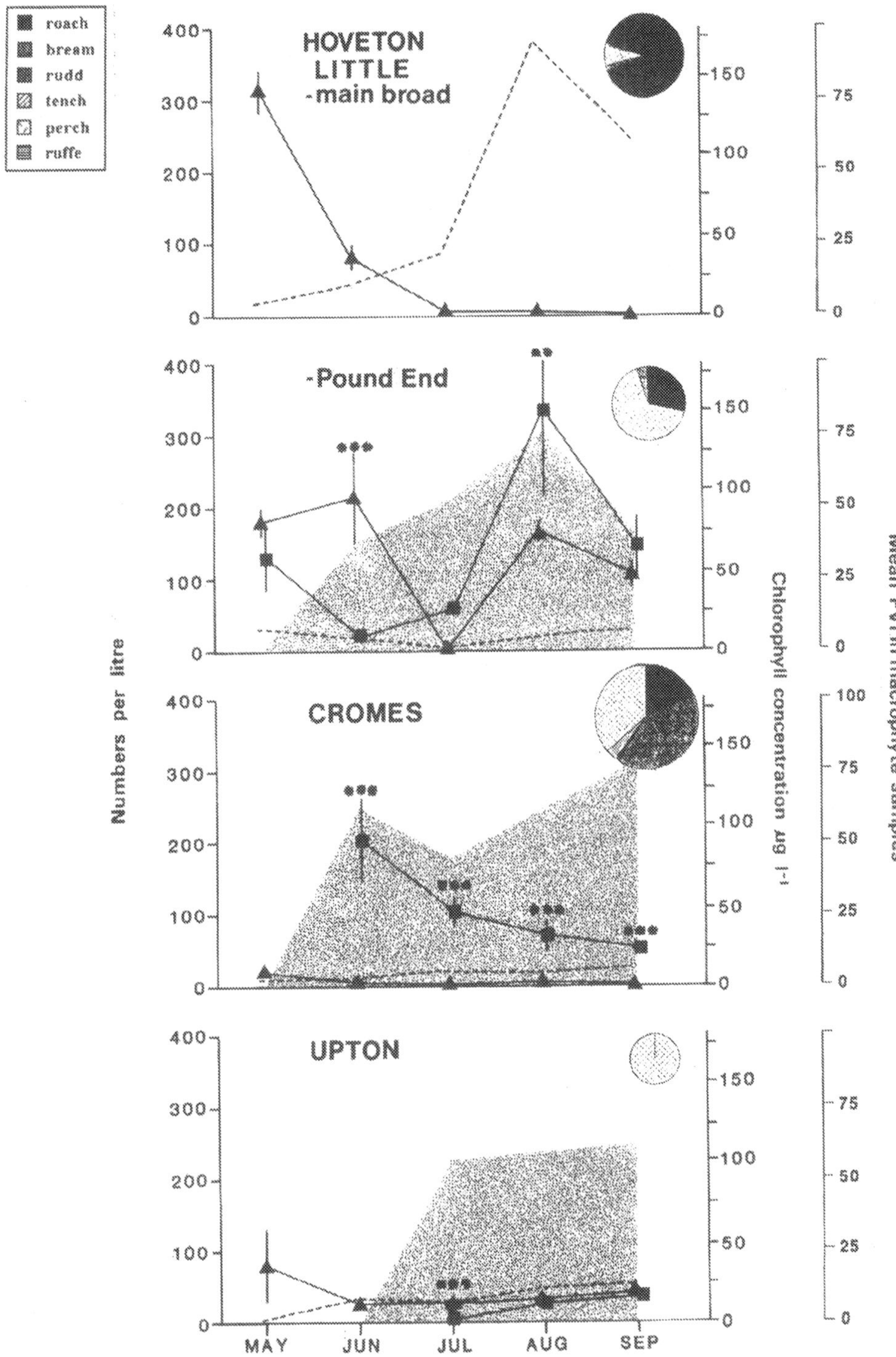

Figure 1. Relationships between abundance of grazing Cladocera (numbers l^{-1}) in macrophytes (squares) and open water (triangles), chlorophyll *a* concentration ($\mu g\ l^{-1}$) (dotted line), relative recruitment strength of 0+ fish (relative abundance in September) and mean PVI in macrophyte samples (shaded area), at all sites over the study period. For Cladocera, significant differences in abundance between the different habitats according to Mann-Whitney U-tests are shown where $^{**}p<0.01$ and $^{***}p<0.001$. For fish recruitment strength, pies are proportional in size to mean number of fish present.

Table 2. Mean ($\pm$ 1 S.E.) numbers (n m^{-2}) and biomass (g m^{-2}) of fish caught in electrofishing surveys in all sites (with number of sampling points) in May and September.

	Hoveton Little (180)		Pound End (158)		Cromes (149)		Upton (160)	
	n m^{-2}	g m^{-2}	n m^{-2}	g m^{-2}	n m^{-2}	g m^{-2}	n m^{-2}	g m^{-2}
May								
Roach	<0.01	<0.01	0.15	1.25	0.02	0.22	–	–
	±<0.01	±0.01	±0.11	±0.76	±0.02	±0.15		
Bream	–	–	–	–	<0.01	10.34	–	–
					±0.01	±10.34		
Tench	–	–	0.02	0.08	–	–	0.02	0.02
			±0.01	±0.07			±0.01	± 0.01
Perch	–	–	<0.01	0.03	0.02	0.24	0.07	0.25
			±<0.01	0.12	±0.01	±0.24	±0.02	±0.25
Ruffe	–	–	0.01	0.12	0.01	0.15	0.01	0.12
			±0.01	±0.10	±0.01	±0.08	±0.01	±0.06
Pike	<0.01	3.91	0.02	8.21	0.01	5.61	0.01	10.50
	±<0.01	±3.91	±0.01	±4.78	±0.01	±3.08	±<0.01	±8.48
Eel	0.04	4.94	0.09	9.24	0.03	4.94	0.02	1.47
	±0.01	±1.76	±0.02	±3.75	±0.01	±1.76	±0.01	±0.70
Total	0.03	6.30	0.28	18.92	0.10	21.49	0.13	12.36
	±0.01	±4.02	±0.11	±6.19	±0.03	±10.84	±0.03	±8.50
September	(163)		(160)		(126)		(169)	
Roach	0.52	1.70	0.16	1.23	0.24	2.40	–	–
	±0.21	±0.77	±0.12	±0.63	±0.07	±0.73		
Bream	0.01	0.01	–	–	0.27	0.23	–	–
	±0.01	±0.01			±0.12	±0.12		
Rudd	–	–	–	–	0.01	0.18	–	–
					±0.01	±0.14		
Tench	–	–	–	–	0.03	0.02	0.01	1.24
					±0.01	±0.01	±<0.01	±1.12
Perch	0.04	0.19	0.38	1.81	0.40	1.46	0.24	0.92
	±0.01	±0.07	±0.11	±0.53	±0.08	±0.26	±0.04	±0.14
Ruffe	<0.01	0.01	0.02	0.05	–	–	–	–
	±<0.01	±0.01	±0.01	±0.03				
Pike	<0.01	8.84	0.01	3.65	0.04	4.15	<0.01	0.32
	±<0.01	±8.84	±<0.01	±2.60	±0.01	±2.17	±<0.01	±0.32
Eel	0.04	0.88	0.02	3.32	0.01	0.81	0.02	2.80
	±<0.01	±0.62	±0.01	±1.51	±<0.01	±0.49	±0.01	±1.23
Total	0.58	11.63	0.58	10.06	1.00	9.24	0.27	5.28
	±0.22	±8.91	±0.16	±3.06	±0.17	±2.53	±0.04	±1.68

Bream - *Abramis brama* (L.) Rudd - *Scardinius erythrophthalmus* (L.), Tench - *Tinca tinca* (L.), Ruffe - *Gymnocephalus cernuus* (L.).

is important to establish whether the decline of *Daphnia* spp. after the early summer peak of abundance is related to fish predation and not, for example, by food limitation.

First, such a decline corresponds closely to the recruitment of 0+ fish. This is well documented in the Broads particularly from underyearling roach predation (Cryer et al., 1986; Townsend & Perrow, 1989).

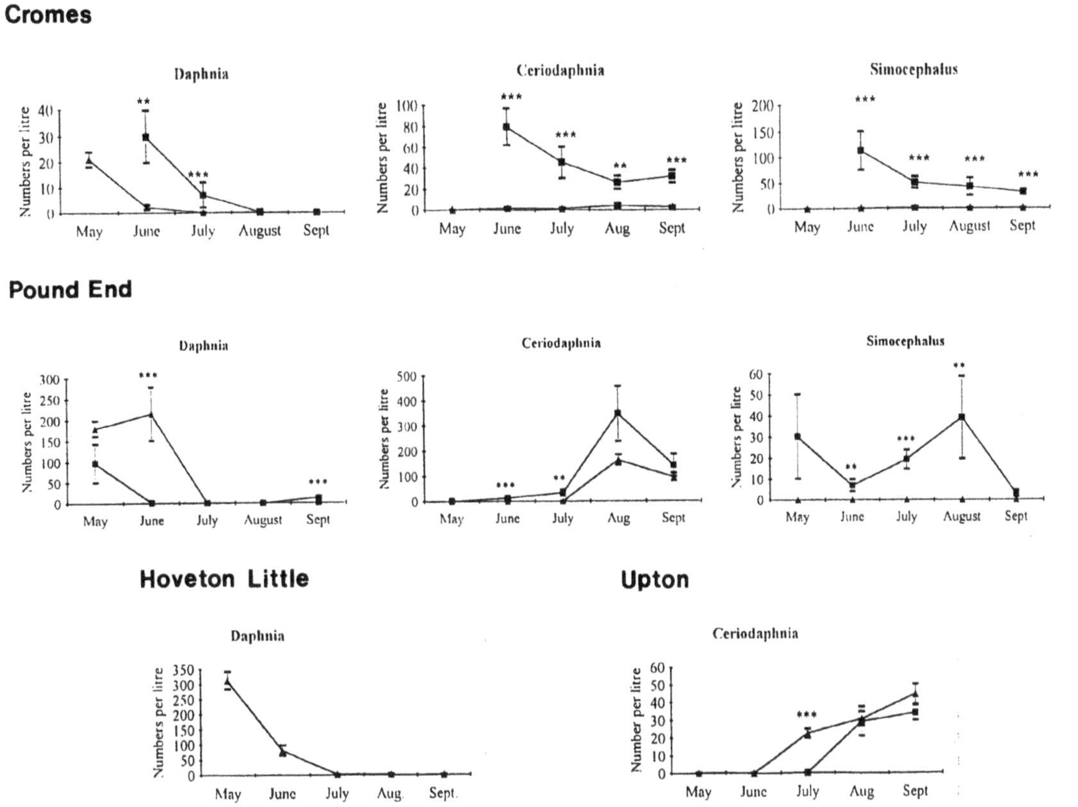

Figure 2. Seasonal changes in the abundance (numbers l^{-1}) of the principal cladoceran genera amongst macrophytes (squares) and in the open water (triangles) of the study sites. Significant differences in abundance between the different habitats according to Mann-Whitney U-tests are shown where $^{**}p<0.01$ and $^{***}p<0.001$.

Indeed, where fish recruitment is poor or absent, *Daphnia* spp. populations typically persist throughout the summer months (Perrow et al., 1994). Conversely, it is clear that the relatively low overwintering stock of older fish exerted little influence on the development of the *Daphnia* spp. population. Second, in all sites, *Daphnia* spp. was quickly replaced by high densities of other species which is not consistent with food limitation and indeed at no site was the abundance of edible algae (expressed as chlorophyll *a*) apparently limiting. It therefore seems reasonable to conclude that predation by 0+ fish is responsible for the decline in *Daphnia* spp. Moreover, this appears to be readily achieved by a relatively low number of 0+ fish in Upton (prior to the development of macrophytes) and relatively moderate recruitment in Pound End and Hoveton Little Broads. However, *Daphnia* spp. persisted in Cromes Broad for a further month even in the presence of a suite of zooplanktivorous fish at relatively high density. This seems related to the early development of an extensive cover of macrophytes at this site. *Daphnia* spp. showed a positive association with macrophytes (Figure 2) and increasing macrophyte cover (Table 3) at this time, quite unlike the situation in any other site. Such use of macrophytes is consistent with the refuge hypothesis.

However, *Daphnia* spp. appeared unable to exploit macrophytes as a refuge over an extensive period. This is in keeping with the behavioural ecology of the group which are typically pelagic and have been shown to avoid macrophytes (Pennak, 1973; Dorgelo & Heykoop, 1985). Recent experiments in the laboratory have shown that a clear trade-off exists, *Daphnia magna* Straus preferring to stay in open water until exposed to the risk of predation from zooplanktivorous fish whereupon it migrates to macrophytes, predation risk clearly outweighing repellency (Lauridsen & Lodge, 1996). It is likely that *Daphnia* spp. prefers

Table 3. Spearman rank correlations (r_S) between abundance of different genera and total Cladocera and PVI (per cent volume infested) of macrophytes in sites and occasions with macrophytes.

	Daphnia spp.	*Ceriodaphnia* spp.	*Simocephalus* sp.	Total grazers
Pound End				
June	−0.86***	+0.82***	+0.84***	−0.75***
July	−0.19	+0.87***	+0.91***	+0.79***
August	–	−0.46	+0.68**	−0.40
September	−0.67***	+0.10	+0.53*	+0.10
Cromes				
June	+0.59**	+0.89***	+0.88***	+0.88***
July	+0.74***	+0.76***	+0.83***	+0.81***
August	–	+0.80***	+0.91***	+0.86***
September	–	+0.92***	+0.84***	+0.88***
Upton				
July	−0.21	−0.86***	–	−0.74***
September	–	−0.19	–	−0.16

* $p<0.05$, ** $p<0.01$, *** $p<0.001$

open water in which to feed and Timms & Moss (1984) provide evidence that this occurs at night. However, many common fish species, particularly 0+ cyprinids, are capable of feeding in darkness (Townsend & Risebrow, 1982) and so the risk of predation, although reduced, is still evident. In addition, *Daphnia* spp. populations may suffer some predation from the fish concentrated in a low density of macrophytes (Figure 3). Therefore, in the face of intense predation pressure from a high density of zooplanktivorous fish, as at Cromes Broad, *Daphnia* spp. may not be able to persist through time, even with the provision of suitable refuges.

Following the decline of *Daphnia* spp., at those sites retaining at least some macrophyte cover *Ceriodaphnia* spp. dominated, both with (Pound End) and without *Simocephalus* sp. (Upton). At Pound End, there was even some partial recovery of *Daphnia* populations in late summer. In contrast, the connected Hoveton Little Broad, with relatively high numbers of 0+ roach, displayed no replacement of *Daphnia* spp. with other species or any recovery of populations. The consequence of the presence of *Ceriodaphnia* spp. and *Simocephalus* was the maintenance of large populations (typically $>50\ l^{-1}$, Figure 1) of grazing Cladocera throughout the summer months in Cromes and Pound End.

The patterns of habitat use of these species also appeared to be related to the potential predation pressure from fish, thus also consistent with the refuge hypothesis. For example, where predation pressure was presumably low from a low density of 0+ perch as in Upton, *Ceriodaphnia* spp. was present in both open water and macrophyte beds (Figure 2). Under moderate fish predation pressure in Pound End (0.45 m^{-2} of 0+ perch and roach), *Ceriodaphnia* spp. and particularly *Simocephalus* sp. were more abundant in macrophytes, although *Ceriodaphnia* spp. also occurred at high density in open water (Figure 2). In contrast, where predation pressure was likely to be intense from a suite of 0+ fish at high density (0.8 m^{-2}) in Cromes, large populations of both *Ceriodaphnia* spp. and *Simocephalus* sp. were only maintained through association with macrophytes. Furthermore, under intense predation pressure the abundance of cladocerans was almost always positively correlated with increasing PVI suggesting that increasing cover offered a better refuge (Table 3).

What is the mechanism behind the refuge effect?

Schriver et al. (1995) demonstrate that a refuge effect may exist at low macrophyte density (15–20% of volume) but that this is dependent on fish density. In the current study, it is suggested that macrophytes may act as refuges for zooplankton at relatively high fish

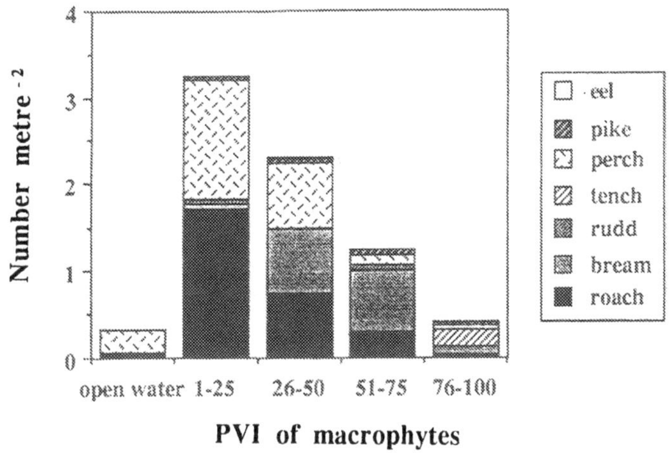

Figure 3. Relative distribution of all fish species in different levels of macrophyte PVI in Cromes Broad in September.

density (Cromes) where macrophyte coverage is extensive and the density of macrophytes is high (Table 1). However, the situation at Pound End suggests that even where the distribution of macrophytes is limited, providing patches of dense macrophytes are available (Table 1), a refuge against moderate predation pressure may be provided. The reasons for this may be that dense macrophyte stands contain relatively few fish (Figure 3). By implication, this suggests that predation risk for Cladocera declines with increasing macrophyte cover. Evidence for this is provided by Phillips et al. (1996), who demonstrated that predation pressure, derived from known numbers of fish with known diet, upon zooplankton declined with increasing macrophyte cover. It is therefore suggested that the formation of predator-free space is the mechanism responsible for the refuge effect. The question is, why do zooplanktivorous fish avoid dense macrophytes? (Engel, 1988).

Several plausible mechanisms exist. First, oxygen concentrations (Davis, 1975) and pH (Beklioglu & Moss, 1995) may be unsuitable in macrophyte stands, although there is no evidence that either was limiting in the samples collected during routine monitoring of the study sites. Second, and perhaps more applicable, the foraging efficiency of some fish may decline with increasing habitat complexity (Winfield, 1986; Diehl, 1988). However, when foraging on zooplankton, this is clearly species and size dependent. For example, 0+ perch (*Perca fluviatilis* L.) may show high feeding efficiency in structured environments unlike 0+ roach (*Rutilus rutilus* (L.)) (Winfield, 1986), and vegetation may serve as a competitive refuge for perch (Persson, 1993). In support of this, several authors (Hall & Werner 1977; Carpenter et al., 1987; Venugopal & Winfield, 1993) have demonstrated that small fish may associate with macrophytes. However, this is typically a behavioural response to potential predation risk (Savino & Stein, 1982; Werner et al., 1983; Persson & Eklov, 1995) and not because of enhanced foraging opportunity. Whether small fish take refuge in macrophytes in response to predation risk is dependent on a number of factors including the functional type of predator and prey (Jacobsen et al., 1997). It is perhaps significant that in this study the relative proportion of piscivorous pike increased with increasing macrophyte cover (Figure 3). In addition, the individuals present in these habitats were 0+, the size class most likely to consume 0+ zooplanktivores (Van Densen, 1994). Therefore, a third explanation for the avoidance of dense macrophytes by fish in this study may be that they were exposed to a greater risk of predation in these habitats. This in turn may have led to a reduced predation risk for zooplankton (Turner & Mittelbach, 1990) thereby enhancing the refuge effect.

Are refuging Cladocera able to control phytoplankton populations?

The presence of large numbers of cladoceran grazers exerting an influence on phytoplankton populations is essential to the value of any refuge effect in explaining the presence of clear water in macrophyte-dominated lakes. From Figure 1, it appears that in sites with rea-

Table 4. Relative filtering rate (ml by mean number of animals l^{-1}) of different cladocerans in open water and amongst submerged macrophytes in Cromes and Upton Broads.

	May		June		July		August		September	
	open	macro.	open	macro.	open	macro.	open	macro.	open	macro.
Cromes										
Daphnia spp.	56.3	27.7	–	143.1	–	–	–	–	–	–
Ceriodaphnia spp.	–	–	–	201.8	–	119.2	–	77.2	–	49.7
Simocephalus sp.	–	57.0	–	633.4	–	421.8	–	468.1	–	158.1
Total	56.3	84.7	–	978.3	–	541.0	–	545.3	–	207.8
Upton										
Daphnia spp.	372.0	–	254.1	–	–	–	–	–	–	–
Ceriodaphnia spp.	–	–	–	–	54.8	67.4	51.1	61.2	98.0	65.7
Total	372.0	–	254.1	–	54.8	67.4	51.2	61.2	98.0	65.7

sonable numbers of cladoceran zooplankton ($>30\,l^{-1}$), chlorophyll *a* concentrations remained low ($<25\ \mu g\ l^{-1}$). But is this attributable to grazing? Particularly to species other than *Daphnia* spp.? The relationship between grazing rate and body length are very similar for *Ceriodaphnia* spp. and *Simocephalus* sp. compared to *Daphnia* spp. (Peters & Downing, 1984; Knoechel & Holtby, 1986; Bern, 1987; Mourelatos & Lacroix, 1990) and if at high enough density it seems reasonable to conclude that these species may be able to graze out phytoplankton crops, as observed by Turner & Mittelbach (1992). The interconnected Pound End and Hoveton Little Broads, with the same nutrient concentration, but radically different chlorophyll *a* values, appear to provide a convincing example of the grazing potential of these Cladocera (Figure 1). Moreover, in Cromes Broad, the filtering capacity of *Simocephalus* spp. and *Ceriodaphnia* spp., which are maintained at densities in excess of $50\ l^{-1}$ throughout the summer months, appears to more than compensate for that of *Daphnia* spp. (Table 4). In Upton, the situation is less clear, where *Simocephalus* sp. is absent and the filtering rate of *Ceriodaphnia* spp. in midsummer does not compensate the loss of *Daphnia* spp. However, it is still comparable with that of *Daphnia* spp. at Cromes in early season, before the development of macrophytes when the only likely explanation for the control of phytoplankton populations is grazing. Therefore, it is reasonable to conclude that grazing by *Ceriodaphnia* spp. is at least a contributory factor in limiting phytoplankton development in Upton in mid- to late summer.

From the patterns of habitat use of *Ceriodaphnia* spp. and *Simocephalus* sp., it is likely that only *Ceriodaphnia* spp. may have the potential to graze in the open water between stands of macrophytes, *Simocephalus* sp. always being limited to the macrophytes themselves (Lauridsen et al., in press). Thus in developing macrophyte populations at the beginning of the growing season or following a restoration attempt using control of zooplanktivorous fish, it seems that large populations of *Daphnia* spp. are required to provide the clear water necessary for macrophyte development. Following the establishment of macrophyte cover and/or recruitment of 0+ fish, *Daphnia* may be expected to decline, with *Ceriodaphnia* assuming the role of grazing in open water between as well as within macrophyte stands, with *Simocephalus* contributing to control of phytoplankton standing crop in the macrophyte beds themselves (Table 4).

Conclusions

Several results of the study are consistent with the refuge hypothesis:

- The elimination of *Daphnia* spp. by 0+ fish is delayed where extensive stands of macrophytes are present in early summer.
- Large populations ($>50\,l^{-1}$) of grazing Cladocera may be maintained throughout the summer at relatively high fish density.
- The level of use of macrophyte stands is linked to fish predation pressure.
- Cladoceran abundance is positively associated with increasing macrophyte cover.

Essential to the value of the refuge effect as a mechanism promoting the stability of submerged macrophyte populations in shallow lakes, the presence of significant populations of grazing Cladocera seems to contribute to the maintenance of a low phytoplankton standing crop and the attainment of clear water. Moreover, macrophyte-associated species such as *Ceriodaphnia* spp. and *Simocephalus* sp. appear to make an important contribution to this effect where *Daphnia* spp. is reduced by fish predation and/or disfavoured by extensive macrophyte cover. The mechanism for the refuge effect in this study appears to be the provision of predator-free space as zooplanktivorous fish appear to avoid dense macrophytes.

Acknowledgments

This study was undertaken as part of the Broads Authority/National Rivers Authority EC LIFE programme project (LIFE 92-3/UK/031 & NRA Project 475) on the Restoration of the Norfolk Broads: Development of Biomanipulation Techniques & Control of Phosphorus Release from Sediments (Project Managers Jane Madgwick & Dr Geoff Phillips) with further financial support from the Soap & Detergents Industry Association. We thank Dominic Harvey for assistance in counting some of the zooplankton samples and Luis Zambrano González and Laurie Cartwright for assistance with electrofishing.

References

Andersson, G., I. Blindlow, A. Hargeby & S. Johansson, 1990. Det våras for Krankesjon [The recovery of Lake Krankesjon]. Anser 29: 53–62.

Balls, H., B. Moss & K. Irvine, 1989. The loss of submerged plants with eutrophication I. Experimental design, water chemistry, aquatic plant and phytoplankton biomass in experiments carried out in ponds in the Norfolk Broadland. Freshwat. Biol. 22: 71–87.

Beklioglu, M. & B. Moss, 1995. The impact of pH on interactions among phytoplankton algae, zooplankton and perch (*Perca fluviatilis*) in a shallow, fertile lake. Freshwat. Biol. 33: 497–509.

Bern, L., 1987. Zooplankton grazing on (methyl-^{3}H) Thymidine-labelled natural particle assemblages: determination of filtering rates and food selectivity. Freshwat. Biol. 17: 151–159.

Blindlow, I., G. Andersson, A. Hargeby & S. Johansson, 1993. Long-term pattern of alternative stable states in two shallow eutrophic lakes. Freshwat. Biol. 30: 159–167.

Bronmark, C. & S. E. B. Weisner, 1992. Indirect effects of fish community structure on submerged vegetation in shallow, eutrophic lakes: an alternative mechanism. Hydrobiologia 243/244: 293–301.

Canfield, D. E. Jr., J. V. Shireman, D. E. Colle, W. T. Haller, C. E. Watkins II & M. J. Maceina, 1984. Prediction of chlorophyll *a* concentrations in Florida lakes: importance of aquatic macrophytes. Can. J. aquat. Sci. 41: 497–501.

Carpenter, S. R., J. F. Kitchell, J. R. Hodgson, P. A. Cochran, J. J. Elser, M. M. Elser, D. M. Lodge, D. Kretchmer, X. He & C. N. von Ende, 1987. Regulation of lake primary productivity by food web structure. Ecology 68: 1863–1876.

Copp, G. H. & M. Peňáz, 1988. Ecology of fish spawning and nursery zones in the flood plain, using a new sampling approach. Hydrobiologia 169: 209–224.

Cryer, M., G. Peirson & C. R. Townsend, 1986. Reciprocal interactions between roach *Rutilus rutilus*, and zooplankton in a small lake: Prey dynamics and fish growth and recruitment. Limnol. Oceanogr. 31: 1022–1038.

Davis, J. C., 1975. Minimal dissolved oxygen requirements of aquatic life with emphasis on Canadian species: A review. J. Fish. Res. Bd Can. 32: 2295–2232.

De Nie, A. W., 1987. The decrease in aquatic vegetation in Europe and its consequences for fish populations. EIFAC Occasional Paper 19. FAO, Rome, 88 pp.

Diehl, S., 1988. Foraging efficiency of three freshwater fish: effects of structural complexity and light. Oikos 53: 207–214.

Dorgelo, J. & M. Heykoop, 1985. Avoidance of macrophytes by *Daphnia longispina*. Verh. int. Ver. Limnol. 22: 3369–3372.

Engel, S., 1988. The role and interactions of submerged macrophytes in a shallow Wisconsin lake. J. Freshwat. Ecol. 4: 229–241.

Grimm, M. P. & J. J. G. M. Backx, 1990. The restoration of shallow eutrophic lakes, and the role of northern pike, aquatic vegetation and nutrient concentration. Hydrobiologia 200/201: 557–566.

Hall, D. J. & E. E. Werner, 1977. Seasonal distribution and abundance of fishes in the littoral zone of a Michigan lake. Trans. am. Fish. Soc. 106: 545–555.

Hambright, K. D., R. J. Trebatoski & R. W. Drenner, 1986. Experimental study of the impacts of Bluegill (*Lepomis macrochirus*) and Largemouth Bass (*Micropterus salmoides*) on pond community structure. Can. J. Fish. aquat. Sci. 43: 1171–11

Hosper, S. H. & E. Jagtman, 1990. Biomanipulation additional to nutrient control for restoration of shallow lakes in The Netherlands. Hydrobiologia 200/201: 523–534.

Howard-Williams, C., 1981. Studies on the ability of a *Potamogeton pectinatus* community to remove dissolved nitrogen and phosphorus compounds from lake water. J. appl. Ecol. 18: 619–637.

Irvine, K. B. Moss, H. Balls, 1989. The loss of submerged plants with eutrophication II. Relationships between fish and zooplankton in a set of experimental ponds, and conclusions. Freshwat. Biol. 22: 89–107.

Jacobsen, L., M. R. Perrow, F. Landkildehus. M. Hjørne, T. L. Lauridsen & S. Berg, 1997. Interactions between piscivores, zooplanktivores and zooplankton in submerged macrophytes: preliminary observations from enclosure and pond experiments. Hydrobiologia 342/343: 197–205.

James, W. F. & J. V. Barko, 1990. Macrophyte influences on the zonation of sediment accretion and composition in a north-temperate reservoir. Arch. Hydrobiol. 2: 129–142.

Jeppesen, E., J. P. Jensen, P. Kristensen, M. Søndergaard, E. Mortensen, O. Sortjkœr & K. Olrik, 1990. Fish manipulation as a lake restoration tool in shallow, eutrophic, temperate lakes 2: threshold levels, long-term stability and conclusions. Hydrobiologia 200/201: 219–227.

Knoechel, R. & L. B. Holtby, 1986. Construction and validation of a body-length-based model for the prediction of cladoceran community filtering rates. Limnol. Oceanogr. 31: 1–16.

Lammens, E. H. R. R., R. D. Gulati, M.-L. Meijer, E. van Donk, 1990. The first biomanipulation conference: a synthesis. Hydrobiologia 200/201: 619–628.

Lauridsen, T. L. & D. M. Lodge, 1996. Avoidance by *Daphnia magna* of fish and macrophytes: chemical cues and predator-mediated use of macrophyte habitat. Limnol. Oceanogr. 22: 805–810.

Lauridsen, T. L., L. J. Pedersen, E. Jeppesen & M. Søndergaard (in press). The importance of macrophyte bed size for composition and horizontal migration of cladocerans in a shallow lake. J. Plankton Res.

Meijer, M.-L., E. Jeppesen, E. van Donk, B. Moss, M. Scheffer, E. Lammens, E. van Nes, J. A. van Berkum, G. L. de Jong, B. A. Faafeng & I. P. Jensen, 1994. Long-term responses to fish-stock reduction in small shallow lakes: interpretation of five-year results of four biomanipulation cases in The Netherlands and Denmark. Hydrobiologia 275/276: 457–466.

Mitchell, S. F., D. P. Hamilton, W. S. MacGibbon, P. K. B. Nayar & R. N. Reynolds, 1988. Interactions between phytoplankton, submerged macrophytes, black swans and zooplankton in a shallow lake. Int. Revue ges. Hydrobiol. 73: 145–170.

Moss, B., 1983. The Norfolk Broadland: experiments in the restoration of a complex wetland. Biol. Rev. 58: 521–526.

Moss, B., H. Balls, K. Irvine & J. H. Stansfield, 1986. Restoration of two lowland lakes by isolation from nutrient-rich water sources with and without removal of sediment. J. appl. Ecol. 23: 391–414.

Moss, B., J. H. Stansfield, K. Irvine, M. R. Perrow & G. L. Phillips, 1996. Progressive restoration of a shallow lake- a twelve-year experiment in isolation, sediment removal and biomanipulation. J. appl. Ecol. 33: 71–86.

Mourelatos, S. & G. Lacroix, 1990. *In situ* filtering rates of Cladocera: effect of body length, temperature, and food concentration. Limnol. Oceanogr. 35: 1101–1111.

Ozimek, T., R. D. Gulati & E. van Donk, 1990. Can macrophytes be useful in biomanipulation of lakes? The Lake Zwemlust example. Hydrobiologia 200/201: 399–407.

Pennak, R. W., 1973. Some evidence for aquatic macrophytes as repellents for a limnetic species of *Daphnia*. Int. Revue ges. Hydrobiol. 58: 569–576.

Perrow, M. R., B. Moss & J. H. Stansfield, 1994. Trophic interactions in a shallow lake following a reduction in nutrient loading: a long term study. Hydrobiologia 275/276: 43–52.

Perrow, M. R., J. Schutten, J. R. Howes, T. Holzer, F. J. Madgwick & A. J. D. Jowitt, 1997. Interactions between coot (*Fulica atra*) and submerged macrophytes: the role of birds in the restoration process. Hydrobiologia 342/343: 241–255.

Persson, L., 1993. Predator-mediated competition in prey refuges: the importance of habitat dependent prey resources. Oikos 68: 12–22.

Persson, L. & P. Eklov, 1995. Prey refuges affecting interactions between piscivorous perch and juvenile perch and roach. Ecology 76: 70–81.

Peters, R. H. & J. A. Downing. 1984. Empirical analysis of zooplankton filtering and feeding rates. Limnol. Oceanogr. 29: 763–784.

Phillips, G. L., D. F. Eminson & B. Moss, 1978. A mechanism to account for macrophyte decline in progressively eutrophicated freshwaters. Aquat. Bot. 4: 103–126.

Phillips, G. L., M R. Perrow & J. H. Stansfield, 1996. Manipulating the fish-zooplankton interaction in shallow lakes: a tool for restoration. In Greenstreet, S. P. R. & M. L. Tasker (eds), Aquatic Predators and their Prey. Blackwell Scientific Publications Ltd., Oxford, England: 174–183.

Savino, J. & R. A. Stein, 1982. Predator-prey interaction between Largemouth Bass and Bluegills as influenced by simulated, submersed vegetation. Trans. am. Fish Soc. 111: 255–266.

Scheffer, M., 1990. Multiplicity of stable states in freshwater systems. Hydrobiologia 200/201: 475–486.

Scheffer, M., S. H. Hosper, M.-L. Meijer, B. Moss & E. Jeppesen, 1993. Alternative equilibria in shallow lakes. TREE 8: 275–279.

Schriver, P., J. Bøgestrand, E. Jeppesen & M. Søndergaard, 1995. Impact of submerged macrophytes on fish-zooplankton-phytoplankton interactions: large-scale enclosure experiments in a shallow lake. Freshwat. Biol. 33: 255–270.

Shapiro, J., 1990. Biomanipulation: the next phase-making it stable. Hydrobiologia 200/201: 13–27.

Spencer, C. N. & D. L. King, 1984. Role of fish in regulation of plant and animal communities in eutrophic Ponds. Can J. Fish. aquat. Sci. 41: 1851–1855.

Timms, R. M. & B. Moss, 1984. Prevention of growth of potentially dense phytoplankton populations by zooplankton grazing, in the presence of zooplanktivorous fIsh, in a shallow wetland system. Limnol. Oceanogr. 29: 472–486.

Townsend, C. R. & M. R. Perrow, 1989. Eutrophication may produce population cycles in roach, *Rutilus rutilus*, by two contrasting mechanisms. J. Fish. Biol. 34: 161–164.

Townsend, C. R. & A. J. Risebrow, 1982. The influence of light level on the functional response of a zooplanktivorous fish. Oecologia 53: 293–295.

Turner, A. M. & G. G. Mittelbach, 1990. Predator avoidance and community structure: interactions among piscivores, planktivores, and plankton. Ecology 71: 2241–2254.

Turner, A. M. & G. G. Mittelbach, 1992. Effects of grazer community composition and fish on algal dynamics. Can J. Fish. aquat. Sci. 49: 1908–1915.

Van Densen, W. L. T., 1994. Predator enhancement in freshwater fish communities. In Cowx, I. G. (ed.), Rehabilitation of Freshwater Fisheries. Fishing News Books, Blackwell Scientific Publications Ltd., Oxford, England: 102–119.

Van Donk, E., R. D. Gulati, A. Iedema & J. T. Meulemans, 1993. Macrophyte-related shifts in the nitrogen and phosphorus contents of the different trophic levels in a biomanipulated shallow lake. Hydrobiologia 251: 19–26.

Venugopal, M. N. & I. J. Winfield, 1993. The distribution of juvenile fishes in a hypereutrophic pond: Can macrophytes potentially offer a refuge for zooplankton? J. Freshwat. Ecol. 8: 389–396.

Werner, E. E., J. F. Gilliam, D. J. Hall & G. G. Mittelbach, 1983. An experimental test of the effects of predation risk on habitat use in fish. Ecology 64: 1540–1548.

Winfield, I. J., 1986. The influence of simulated aquatic macrophytes on the zooplankton consumption rate of juvenile roach, *Rutilus rutilus*, rudd, *Scardinius erythrophthalmus*, and perch, *Perca fluviatilis*. J. Fish Biol. 29: 37–48.

Wium-Anderson, S., U. Anthoni, C. Christophersen & G. Houen, 1982. Allelopathic effects on phytoplankton by substances isolated from aquatic macrophytes (Charales). Oikos 39: 187–190.

Hydrobiologia **342/343**: 241–255, 1997.
L. Kufel, A. Prejs & J. I. Rybak (eds), Shallow Lakes '95.

Interactions between coot (*Fulica atra*) and submerged macrophytes: the role of birds in the restoration process

Martin R. Perrow[1*], J. Hans Schutten[2*], John R. Howes[1], Tim Holzer[1], F. Jane Madgwick[3] & Adrian J. D. Jowitt[1]
[1]*ECON, School of Biological Sciences, University of East Anglia, Norwich NR4 7TJ, UK*
[2]*ARISE, Aquatic Ecotoxicology Section, University of Amsterdam, Kruislan 320, 1098 SM Amsterdam, The Netherlands*
[3]*Broads Authority, Thomas Harvey House, 18 Colegate, Norwich NR3 1BQ, UK*
**Authors for correspondence*

Key words: herbivory, bird grazing, bird diet, macrophyte colonisation, macrophyte growth, seasonal population trends, shallow lakes

Abstract

Grazing by herbivorous birds is often cited as an important factor in suppressing macrophyte development in shallow lakes undergoing restoration, thus delaying the attainment of the stable clear water state. Development and succession of macrophyte communities and size, diet and grazing pressure of coot (*Fulica atra*) populations upon macrophytes, were monitored over the seasonal cycle at ten shallow lakes of varying nutrient status, in the Norfolk Broads in eastern England. In spring, territorial breeding birds were at relatively low density and included only a small proportion of macrophytes in their diet, resulting in low grazing pressure on macrophytes. In summer, there was a significant relationship between macrophyte cover and bird density, illustrating the importance of macrophytes in the dispersion phase for birds following breeding. Macrophytes comprised the bulk of bird diet where they were available and the consumption of macrophytes was up to 76 fold higher than in spring. However, losses to grazing in both periods were negligible when compared to potential growth rates documented in the literature. Grazing experiments at two biomanipulated lakes confirmed that birds were not responsible for limiting macrophytes during the spring colonisation phase or in the summer growth period. During the period of autumnal senescence and over the winter months where some macrophyte species remain available, e.g. as developed individuals or dormant buds, grazing by birds may conceivably have an impact on the development and structure of macrophyte populations in subsequent growing seasons.

The relative importance of bird grazing compared to other factors limiting the development of macrophytes in shallow lakes is discussed in the light of other experimental studies.

Introduction

Much scientific and technical effort has recently been directed at the restoration of shallow lakes in temperate Europe and North America as a consequence of their huge ecological and sociological importance (National Research Council, US, 1992; Broads Authority, 1994; National Environmental Research Institute, Denmark, 1994). In most cases, the reversal of anthropogenic eutrophication is the target of this investment (Ryding & Rast, 1990). In simple terms, clear water dominated by submerged macrophytes is preferred over turbid, phytoplankton-dominated water (Moss, 1990). Further, the simple reduction of nutrient loading has rarely been successful as a single measure (Sas, 1989). It is widely accepted that restructuring the biological community of shallow lakes, usually through biomanipulation of the fish community (Lammens et al., 1990) is important in counteracting buffering mechanisms tending to preserve algal domination even where the nutrient concentration is suitable for stable macrophyte populations (Irvine et al., 1989; Scheffer et al., 1993).

Once clear water, the essential prerequisite for a submerged macrophyte community, is attained, there are two scenarios. The first is the spontaneous development of an extensive and dense macrophyte community (Meijer et al., 1994). However, even where this has occurred, it may prove to be unstable in the long term (Blindlow et al., 1993; Perrow et al., 1994) with a concurrent increase in algal populations. Alternatively, macrophytes may be slow to colonise (Lauridsen et al., 1994; Moss et al., 1996; Strand, unpubl. data).

The impact of grazing birds, principally coot (*Fulica atra*) and swans (especially Mute swan *Cygnus olor*) has been widely discussed as the limiting factor in the colonisation, development and subsequent stability of macrophytes (Jupp & Spence, 1977; Lauridsen et al., 1993, 1994; Perrow et al., 1994; Moss et al., 1996; Søndergaard et al., in press). However, information relating to the impact of grazing by birds frequently relies on experiments using macrophytes transplanted into the site (Lauridsen et al., 1993; Moss et al., 1996; Søndergaard et al., in press), rather than the impact on naturally occurring macrophytes (Anderson & Low, 1976; Esler, 1989). Alternatively, a measure of macrophyte production is compared with losses due to grazing often derived by multiplying the number of bird days against some estimate of dietary input derived from the literature (Kiørboe, 1980; Schutten et al., 1994). In addition, there is often little attempt to determine macrophyte/bird interactions in relation to population parameters, including seasonal fluctuation, macrophyte phenology or actual bird diet.

This paper reports on a two-pronged study of macrophyte/bird interactions. First, detailed monitoring over the seasonal cycle was undertaken on macrophyte and coot populations, including dietary selection and consumption of macrophytes, at a range of sites in an attempt to assess likely grazing effects according to seasonal patterns of macrophyte development and bird populations and dietary intake. Second, short-term exclosure experiments were conducted at two biomanipulated sites, both of which had shown a lag in the development of macrophyte populations under apparently suitable conditions. These tested hypotheses developed during the monitoring phase.

Background information and site descriptions

The Norfolk Broads in eastern England, is an internationally important wetland complex of around 50 shallow (<2 m) lakes (broads) many of which are connected by rivers and interspersed by areas of fen and marsh. As a result of intense anthropogenic eutrophication, principally through P-rich sewage effluent, many of the lakes are now classed as hypertrophic and only 4 retain their original macrophyte populations (Broads Authority, 1994). Restoration attempts have been undertaken involving (a) reduction of external nutrient supply through installation of works at sewage treatment plants,(b) isolation (Moss et al., 1986, 1990; Perrow et al., 1994), (c) control of internal loading primarily through suction dredging (Moss et al., 1986, 1996) and (d) biomanipulation (Perrow, 1990; Phillips et al., 1996; Moss et al., 1996). Clear water conditions have been attained in many cases. Development of stable macrophyte populations in lakes with a suitable light climate has been retarded however, and grazing by birds, particularly coot, has been cited as a possible negative factor (Moss et al., 1986; Perrow et al., 1994, Moss et al., 1996). Experiences in attempting to establish macrophyte inocula over the last ten years by the water managers, the Broads Authority, also implicated bird grazing (unpubl. data). The management need to understand the role of bird grazing led the Broads Authority to commission the current project.

Coots are a common, widespread species in the UK with the resident population being supplemented by continental birds on autumnal passage (e.g. in 1990/91 peak numbers of over 85 000 birds were recorded in September, Ferns & Kirby, 1992) and throughout winter (Cramp & Simmons, 1980). The Norfolk Broads are known to harbour considerable populations e.g. the wintering population in the 1960's at Hickling Broad alone was estimated at 2500 birds (Seago, 1967). Coots occurred at all broads studied in all seasons (Howes & Perrow, 1994) and frequently outnumbered other waterfowl (Perrow, Howes & Holzer, unpubl. data), particularly during the breeding season. With a known high dietary intake of macrophytes (see Cramp & Simmons, 1980, for review), coots were considered to be the principal herbivorous bird in the Broads.

Coots exhibit considerable flexibility in their foraging behaviour (Cramp & Simmons, 1980; Draulans & Vanherck, 1987; Howes & Perrow, 1994); upending in shallow water or diving in deeper water (generally 1–2 m but to 6.5 m) to forage on macrophytes, algae, detritus and benthic invertebrates; cropping emergent

Table 1. Characteristics of the study sites.

Lake	Area (ha)	Direct Riverine connection	Management	Submerged macrophytes '92
Alderfen	5.1	None	Isolated 1979 Suction-dredged 1992/93 Biomanipulated 1993-	2 species 50% *C. demersum*
Belaugh	1.9	River Bure	Dredged 1987	4 species <5% *C. demersum*
Cockshoot	5.5	None	Isolated 1981 Suction-dredged 1981 Biomanipulated 1989-	9 species 25% *C. demersum*
Cromes	2.0	None	Suction-dredged 1988 Isolated 1992	1 species 50% *C. demersum*
Hickling	129.8	None	None	11 species 75% *M. spicatum*
Hoveton Great	36.2	River Bure	1 ha Exclosure Biomanipulated 1992-	5 species 25% *P. crispus*
Martham North	6.7	River Thurne	None	16 species 75% *N. marina*
Ormesby	56.4	None	None	10 species <5% *C. globularis*
Pound End	5.5	River Bure via Hoveton Little	Suction-dredged 1990 Biomanipulated 1990-	5 species <5% *C. demersum*
Upton	5.6	None	None	3 species 75% *N. marina*

Macrophyte data is taken from Kennison (1993). The number of species and maximum % cover of the dominant species are shown.

or bankside vegetation from on the water or on land; gleaning insects, seeds and fruits from the water surface or vegetation; or taking handouts or leftovers from human visitors. Such behavioural plasticity and ability to take a wide range or plant and animal material enabled coots to persist at all broads studied, even where macrophytes were absent. The continuous presence of coots was thought to restrict the colonisation and establishment of macrophytes at low density (Howes & Perrow, 1994).

The primary monitoring phase of the study was undertaken at ten sites, which ranged from phytoplankton to macrophyte dominated and encompassed those lakes subjected to recent restoration measures. Characteristics of the study lakes are shown in Table 1. During the subsequent experimental phase, two small biomanipulated sites, Pound End and Cockshoot Broad were selected (Table 1).

Methods and materials

Monitoring study

Bird monitoring

Coot populations were monitored monthly at all sites from July 1993–June 1994. The preliminary study of Howes & Perrow (1994) recorded foraging activity throughout the hours of daylight with birds feeding in bouts interspersed by preening and resting (Brown, unpubl. data). Birds then roosted at night (Draulans & Vanherck, 1987; Brown, unpubl. data). Observations

Table 2. Mean (+ 1 S.E.) wet weight (g) per coot bill length of the submerged plant taxa consumed.

Taxon	Wet weight (g)
Callitriche spp.	0.37 ± 0.02
Ceratophyllum demersum	1.14 ± 0.19
Ceratophyllum submersum	0.13 ± 0.01
Chara spp.	0.57 ± 0.06
Elodea canadensis	0.40 ± 0.03
Hippurus vulgaris	1.00 ± 0.03
Myriophyllum spicatum	0.23 ± 0.03
Najas marina	0.85 ± 0.03
Nitellopsis obtusa	0.50 ± 0.03
Potamogeton pectinatus	0.26 ± 0.02
Potamogeton crispus	0.80 ± 0.03
Zannichellia palustris	0.25 ± 0.01
Filamentous algae	1.44 ± 0.12

conducted at any time of day were thus thought to be representative of foraging activity and grazing pressure on macrophytes at the site, particularly as bird numbers fluctuated little over the course of the day. Two sites a day were monitored and observations were made over the entire lake, except at Ormesby and Hoveton Great Broad, where observations were restricted to a 12 ha section and the 1 ha fish exclosure respectively. Bird numbers in the latter area could not be related directly to macrophyte abundance in the main broad and this site was therefore excluded from that part of the analysis.

Observations were made from land (usually bird hides) or from a boat using a 30 magnification × 80 mm Optolyth telescope or 8 × 20 Leica binoculars. If birds were disturbed by the observers taking position, monitoring was delayed for at least twenty minutes until birds had repositioned and resumed natural activity. Birds were counted at intervals to determine peak numbers. Food intake and selection was determined through continuous observation for fifteen minutes each on a minimum of four randomly selected individual birds.When foraging underwater, coots bring items to the surface to handle and swallow. This allowed the majority of dietary items to be identified and their size to be estimated relative to bill length before they were consumed.

Daily consumption of each dietary item at each site was calculated by extrapolating the consumption in fifteen minutes to one hour and then multiplying by the number of daylight hours and the number of birds present.

For macrophytes and filamentous algae, the mean wet weight (g) per bill length was measured by taking ten replicate one bill-length equivalent samples of each species and weighing them to the nearest 0.1 g (Table 2). The consumption per occasion in bill lengths at each location of known size was then converted to wet weight (g m^{-2}) for each species.

Macrophyte monitoring

Macrophyte densities were monitored at all sites at bimonthly intervals from April to October 1993. At each site, 3 permanent transects (25–50 m in length) were sampled by a snorkel diver. The vertical projected sediment coverage of each macrophyte species (all vascular plants and *Chara* spp.) and filamentous algae as well as total community cover along the line and in the area up to 2 m (depending on visibility) on either side of the line were estimated using an adapted Tansley scale (see Schutten et al. (1994) for details).

Grazing experiments

The impact of coot grazing on a natural communities of macrophytes and filamentous algae at Cockshoot and Pound End was tested using ten replicate pairs of cylindrical exclosures (1 m height, 113 cm diameter enclosing a total surface area of 1 m^2) made of black, plastic 'Netlon Tensar SS3' mesh (71 × 51 mm). Lids of the same mesh were fitted to exclude birds from the treatment exclosures, whereas controls had no lids. Each exclosure was pushed into the sediment, leaving 75 cm above the sediment surface, but still under water, and anchored in place with a 2.5 m wooden stake.

The ten pairs, one treatment and its control, were placed within the known territory of a pair of coots at each of the sites, in early May 1994, the season after the monitoring study. The members of each pair of exclosures were within 2 m of each other, whilst pairs were at least 7 m apart. Macrophytes were monitored every 2–3 weeks from late June to mid September. The % cover of each macrophyte species and/or filamentous algae was estimated visually through a perspex-bottomed bucket from a stationary boat. In late September, the plants in five randomly-selected pairs were harvested using a long-handled rake. Species were separated and wet weighed to the nearest 0.5 g.

Differences between treatments and controls on each occasion (total cover and cover of common plant species, were tested using Wilcoxon signed ranks tests.

At each broad, coot diet and macrophyte intake were monitored using the methods described above, within a few days of the macrophyte sampling.

Results

Monitoring study

Macrophyte populations

The seasonal changes in macrophyte and filamentous algae density are variable (Figure 1). Turbid lakes known to be dominated by phytoplankton (Broads Authority, 1994), including Hoveton Great, Ormesby and Belaugh Broads, were characterised by low cover of macrophytes (<5%) with representation of only a few species, typically *Zannichellia palustris* L. and several *Potamogeton* species in spring to summer, and *Ceratophyllum demersum* L. in late summer and autumn. Filamentous algal cover was typically an early season phenomenon, with densities up to 25%.

The clear-water biomanipulated sites, Alderfen and Pound End (Table 1) had very few macrophytes with the same pattern as above. In contrast, Cockshoot had more macrophyte species with a combined cover of up to 30%, and some seasonal succession from *Zannichellia/Potamogeton* spp. to *C. demersum* and *Lemna trisulca* L. Filamentous algal cover was high (to 45%) at both Cockshoot and Pound End with a peak in mid-summer.

Cromes, Upton and Hickling and Martham North were all dominated by macrophytes. The former three were characterised by slow spring development, with a peak in macrophyte cover (>50%) in late summer and autumn by only a few species; *C. demersum* at Cromes, *Chara* spp. and *Myriophyllum spicatum* L. at Hickling and *Najas marina* L. at Upton. Martham North, on the other hand, had a large number of species/genera of macrophytes (12), with pronounced seasonal shifts in abundance, starting with *Hippuris vulgaris* L. and *M. spicatum*, to *Chara* spp. and finally to *N. marina* in late season. Macrophyte cover was always high, increasing from 40% in spring to 80% in late summer.

Bird populations

In general, birds were at relatively low density in spring (mean + S.E. density in April = 1.58 ± 0.30 individuals ha^{-1}) (Figure 2), as a result of the intense territoriality of breeding pairs (Cramp & Simmons, 1980). At Broads with little plant cover, such as Alderfen, Belaugh and Hoveton Great Broad, the population consisted almost solely of breeding pairs and their offspring resulting, although with some decline over winter, in relatively stable densities. In contrast, bird density fluctuated markedly (the pattern is damped at Hickling on account of its large size) at broads with significant plant cover (with the exception of Ormesby), with the incursion of juvenile and non-breeding/post-breeding birds to supplement the resident population. The timing of this immigration was usually between July and September possibly related to the attainment of peak abundance of the dominant plant(s). For example, high densities of birds corresponded to the peak occurrence of *Z. palustris/Potamogeton* spp. in Ormesby in July, maximum cover of *N. marina* and *Chara* spp. in Martham North in August and high coverage of filamentous algae in Pound End in August and Cockshoot in September. Only Cromes and Upton received an incursion of birds in winter, where the late developing *C. demersum* and *N. marina* dominated respectively. In the increase phase, bird numbers peaked between 8 ha^{-1} (Cromes) and 18 ha^{-1} (Martham North). Breeding densities were resumed between January and March.

Bird diet

The diet of birds varied considerably between site and season, although several general trends were identified (Figure 3). In spring (March to May) the diet was dominated by invertebrates and filamentous algae. This is explained by these 'soft' foods being fed to chicks (Howes & Perrow, 1994). Consequently, macrophytes made up a low proportion of the diet at this time, even where they were present. In contrast, in summer (June–August) and autumn (September–November) where present, macrophytes formed the bulk of the diet, although filamentous algae was often also important. In autumn and especially winter (December–February) the diet of birds was generally more variable with invertebrates, emergent plants, detritus and filamentous algae all dominant at different sites. However, in lakes with a peak macrophyte cover of >50%, macrophytes were still the major dietary item.

Interactions between birds and macrophytes

If all sites on all occasions are included, there is no relationship between coot density (birds ha^{-1}) and % total plant cover (Spearman rank correlation, $r_s = 0.04$, $n = 23$, $p = 0.87$). However, if this is teased apart in separate analyses, coot density is not related to %

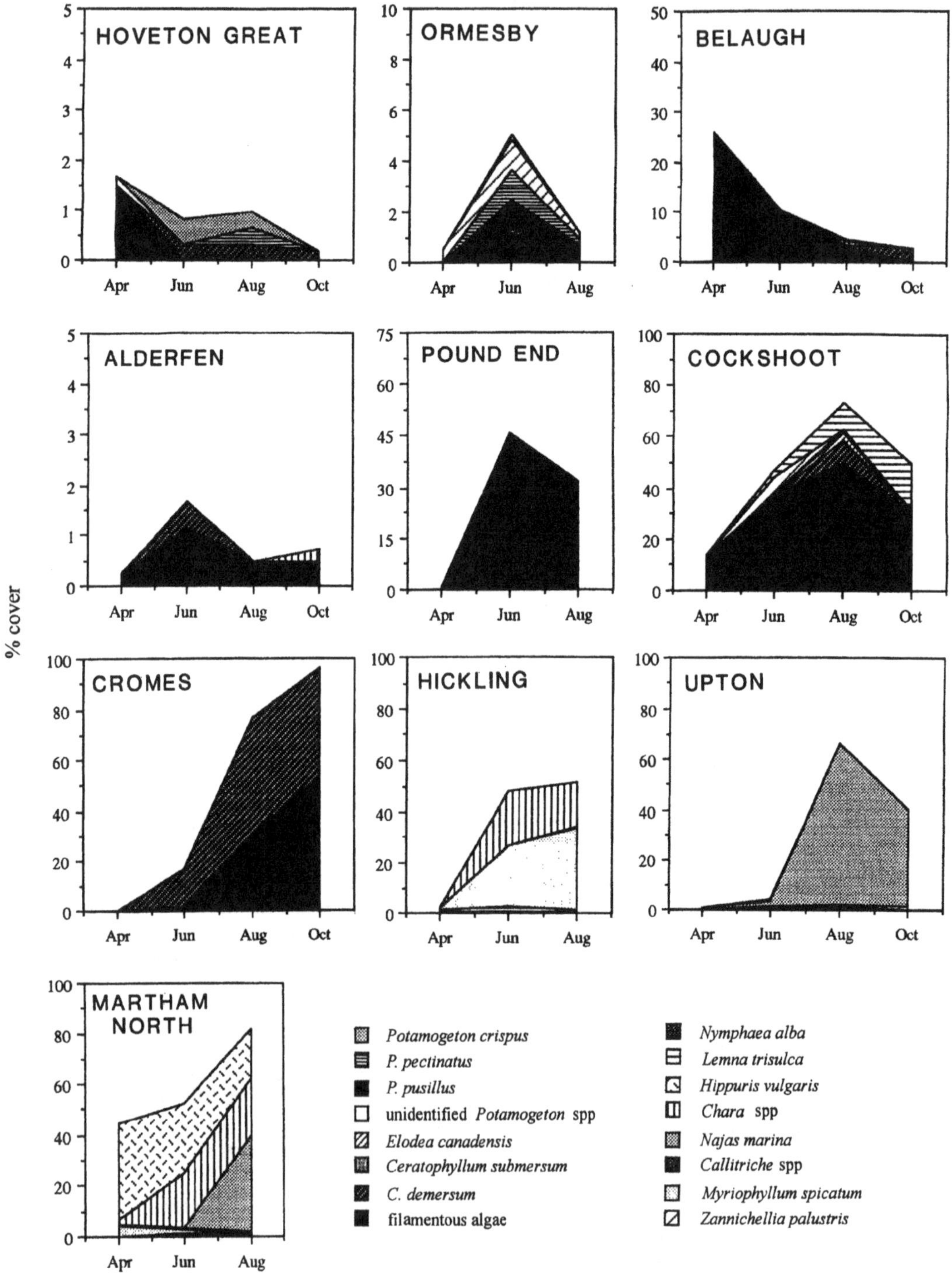

Figure 1. Seasonal changes in the cover (%) of macrophytes and filamentous algae during 1993, in the ten study sites.

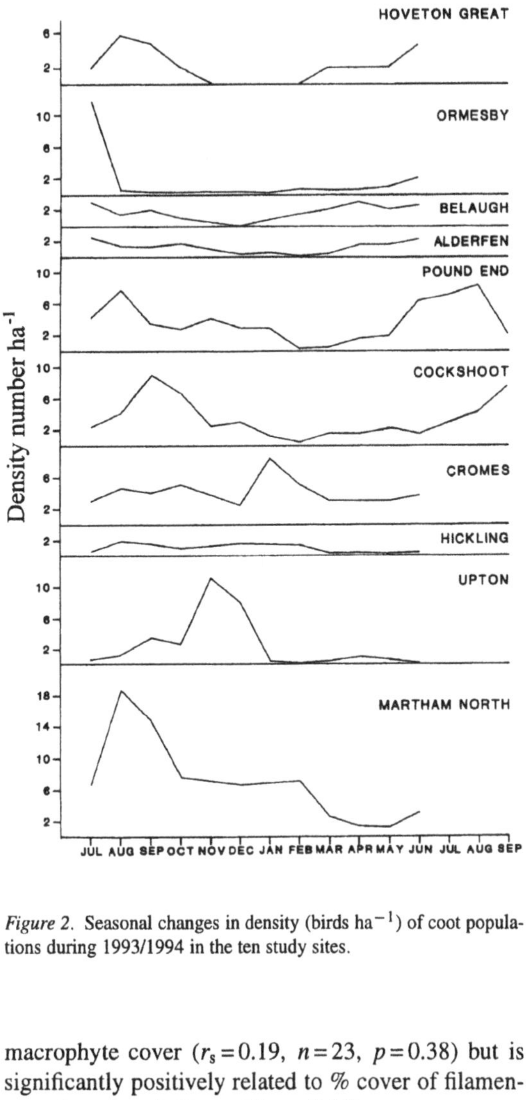

Figure 2. Seasonal changes in density (birds ha^{-1}) of coot populations during 1993/1994 in the ten study sites.

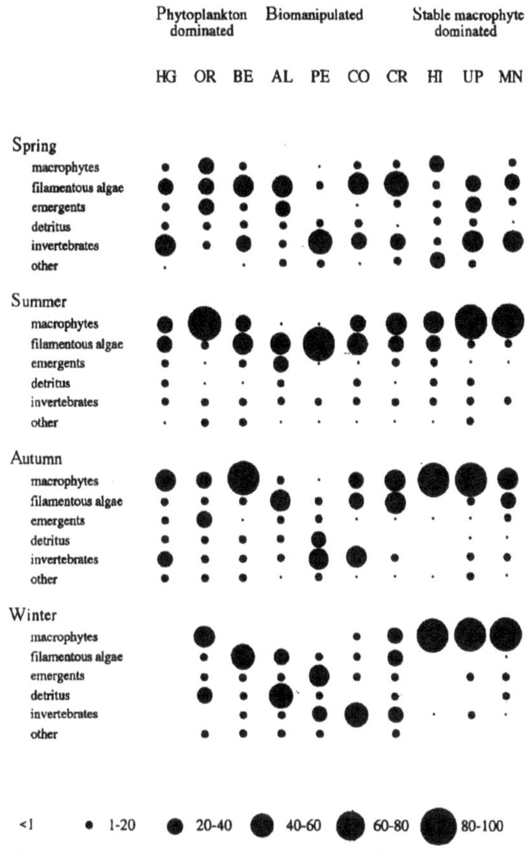

Figure 3. Seasonal dietary composition of coot populations at all study sites where: HG = Hoveton Great, OR = Ormesby, BE = Belaugh, AL = Alderfen, PE = Pound End, CO = Cockshoot, CR = Cromes, HI = Hickling, UP = Upton, MN = Martham North. The size of circle is proportional to the contribution (%) made by each group to the total bill lengths consumed in each season (spring March–May, summer June-August, autumn September–November, winter December–February).

macrophyte cover ($r_s = 0.19$, $n = 23$, $p = 0.38$) but is significantly positively related to % cover of filamentous algae ($r_s = 0.58$, $n = 23$, $p < 0.01$).

Taking the population fluctuations in late summer and autumn into account, by expressing the densities in August and October as a % of the breeding density in June, the relationships between bird density and plant cover change markedly. There was a significant positive relationship between % total cover and birds ($r_s = 0\ 55$ $n = 14$, $p < 0.05$), and it is macrophyte cover ($r_s = 0.67$, $n = 14$, $p < 0.01$) and not algae ($r_s = 0.08$, $n = 14$, $p = 0.79$) that is the important factor.

Furthermore, there were significant positive relationships between the proportion of macrophytes or filamentous algae in the diet of birds and % cover ($r_s = 0.73$, $n = 23$, $p < 0.001$ and $r_s = 0.72$, $n = 23$, $p < 0.001$ respectively). In lakes with macrophytes, an increase in the number of birds with an increasing proportion of macrophytes in the diet, results in an increase in grazing pressure (g m^{-2} day^{-1} consumed) ($r_s = 0.34$, $n = 80$, $p < 0.01$).

Grazing experiments

Cockshoot

The total plant cover and cover of the seven different species in both protected treatments and grazed controls over the experimental period are illustrated in Figure 4. *P. berchtoldii & Z. palustris* made an important contribution to early season cover, whereas *C. demersum* tended to dominate later in the summer and into

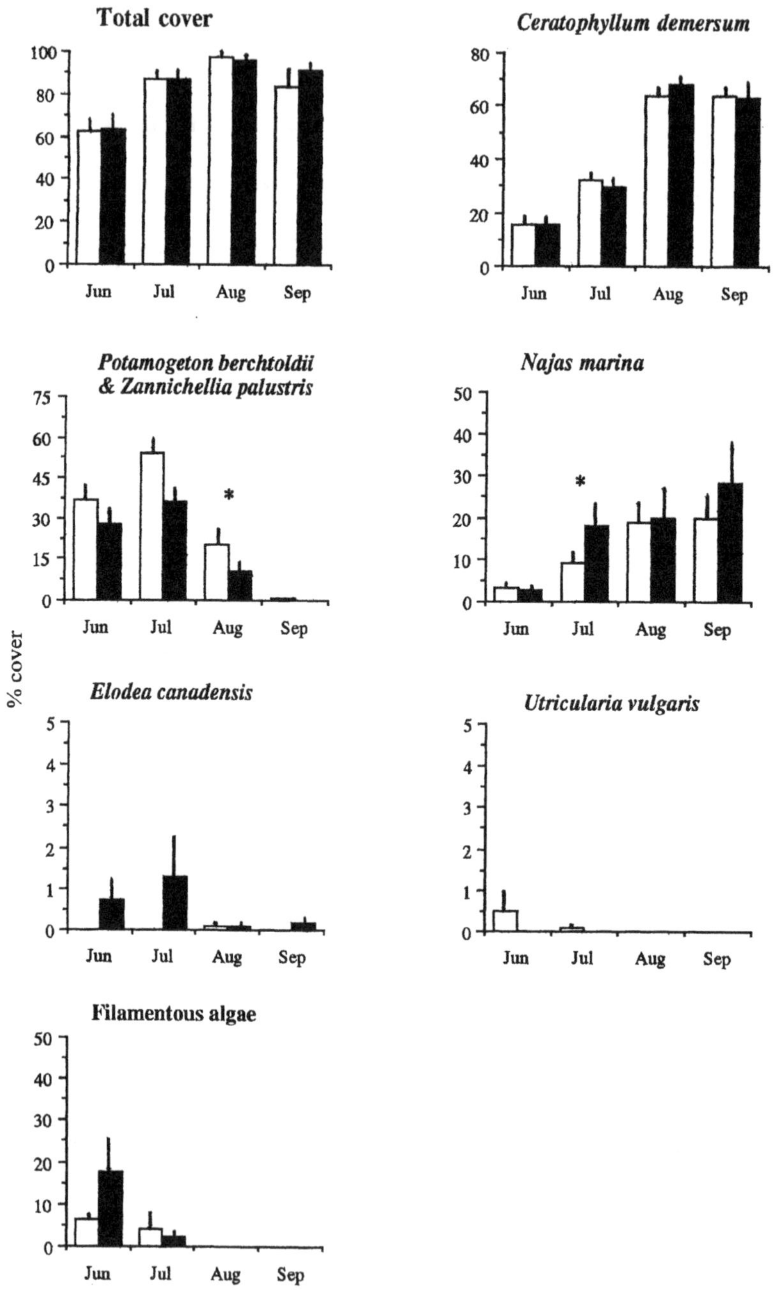

Figure 4. Mean (± 1S.E, $n = 10$) total plant cover and cover of different plant species in both grazed controls (open bars) and protected treatments (shaded bars) over the experimental period at Cockshoot Broad. Statistical tests were only conducted on total cover, *C. demersum, P. berchtoldii/Z. palustris* & *N. marina*. Significant differences at the $p<0.05$ level are indicated (*).

autumn. At peak, macrophyte cover (with some filamentous algae) approached 100%.

Overall, there were few significant differences between treatments and controls illustrating potential positive and negative effects of bird grazing. Any differences were only apparent after at least two months of the experiment had elapsed and never occurred in a consistent fashion. That no differences occurred in the early stages of the experiment is consistent with the negligible consumption of macrophytes by the coot population in this period (Figure 5). Even though consumption by coots increased markedly during the experimental period, this did not lead to any significant differences in the biomass of any macrophyte taxon at the end of the experiment (Table 3).

Pound End
In contrast to Cockshoot, Pound End was dominated by filamentous algae, particularly *Enteromorpha* spp. This gradually declined over the course of the experiment (Figure 6) until only a small biomass was present in early October (Table 3). Only two macrophyte species were found, *Elodea canadensis* Michaux and *C. demersum*, in two separate treatment exclosures and *C. demersum* in one control on one occasion (Figure 6). There were no significant differences in the cover of algae or macrophytes on any occasion.

Discussion

To critically evaluate the potential impact of grazing coots upon submerged macrophyte populations it is necessary to divide the cycle of seasons into distinct phases according to macrophyte ecology and coot behaviour and ecology. In simple terms, these are the spring germination or initial growth phase for macrophytes and the nesting season for birds; the summer growing and reproduction (for sexually reproducing species such as *Potamogeton* spp.) phase for macrophytes and the recruitment and dispersal phase for birds; the autumn period of senescence for most species of macrophytes and the peak passage of birds at many sites; and the winter period in which macrophytes are either unavailable, coinciding with the lowest bird populations, or occur either as senescent material or as overwintering stages (most notably in the case of a few asexually reproducing species such as *Ceratophyllum* and *Elodea*) where high coot populations are maintained.

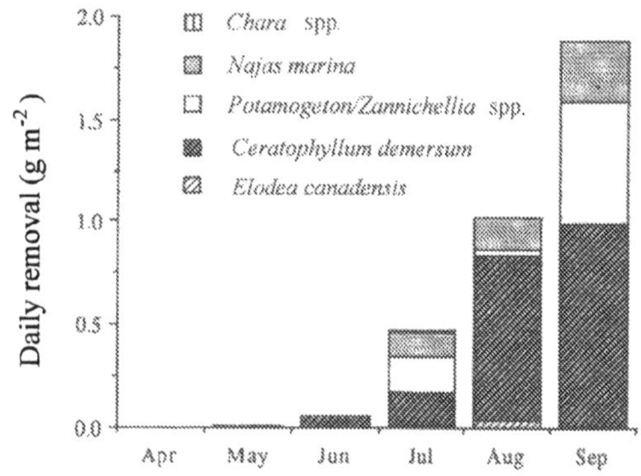

Figure 5. The relative daily consumption (g m^{-2} WW) of different macrophyte species by coot at Cockshoot Broad from April–September 1994, covering the period of the grazing experiment.

Is colonisation of macrophytes in spring affected by bird grazing?
Germination of macrophyte seedlings from seed or overwintering buds/turions takes place in spring. Limitation during this establishment phase (sensu Harper, 1977) determines successful colonisation. A number of factors are prerequisites for the establishment of macrophytes, prior to any possible limiting effect of bird grazing. These include the presence of a viable seed bank; suitable conditions for germination e.g. the presence of reducing sediments and consequent high production of sulphides and ammonia may inhibit seeds germinating (Perrow et al., 1994); and suitable conditions for subsequent growth and survival e.g. Jupp & Spence (1977) and Lauridsen et al. (1993) illustrated the importance of sediment structure and organic content.

What of the impact of birds? Sandsten (1995), in Lake Krankesjon, showed that were there was a viable seed bank, under suitable conditions for germination and growth, macrophytes colonised successfully in the presence of birds. In contrast, Lauridsen et al. (1993), Søndergaard et al. (in press) and Moss et al. (1996) concluded that bird grazing was responsible for limiting colonisation and establishment. These studies used transplanted macrophytes and therefore we suggest it is inappropriate to draw conclusions from these studies on the response of naturally occurring macrophyte populations. In addition, transplants represent unusual concentrations of food which may attract birds and induce unnaturally high estimates of grazing pressure.

In the current study, the density of breeding birds was considerably lower than peak populations. Birds

Table 3. The mean (+ 1 S.E.) plant biomass (g m^{-2} WW) in exclosures in Cockshoot and Pound End at the end of the experimental period. No significant differences were recorded between ungrazed treatments and grazed contro]ls (each n=5) for any taxon.

Taxon	Cockshoot – Grazing	Cockshoot + Grazing	Pound End – Grazing	Pound End + Grazing
Total macrophytes	600.6 ± 88.7	461.5 ± 74.7	0.3 ± 0.2	0.0 ± 0.0
C. demersum	371.0 ± 97.0	348.0 ± 71.9	0.2 ± 0.2	0.0 ± 0.0
E. canadensis	0.1 ± 0.1	1.2 ± 1.0	0.1 ± 0.1	0.0+ 0.0
L. trisulca	1.2 ± 0.8	0.5 ± 0.3		
N. marina	222.3 ± 128.8	101.4 ± 6.8		
P. berchtoldii/ Z. palustris	6.0 ± 4.5	10.4 ± 6.8		
Filamentous algae	103.6 ± 52.2	138.4 ± 76.0	0.8 ± 0.5	0.2 ± 0.1

are strictly territorial, with factors such as shoreline length (as habitat for nesting) and food resource availability determining territory size and consequently the density of breeding pairs at a site (Cramp & Simmons, 1980). The diet of birds at this time, even where macrophytes are available, was dominated by invertebrates and filamentous algae. This seems largely due to the requirements of chicks for such foods, invertebrates in particular, dominating in the first ten days of life (Cramp & Simmons, 1980). The presence of filamentous algae in the diet in spring, as well as throughout the summer may explain the overall relationship between coot density and filamentous algal cover.

A combination of a low number of birds, with a low proportion of macrophytes in the diet resulted in a low consumption of macrophytes. When compared to the published growth rates of selected macrophyte species, it is clear that consumption by birds was typically only a fraction of the potential growth rate during spring (Table 4). The negligible impact of birds upon colonisation by macrophytes is also supported by the virtual lack of macrophyte consumption by the coot population in Cockshoot in the spring (April/May) period (Figure 5). Moreover, in Pound End, few macrophytes colonised even where they were protected from birds (Figure 6). Here, as in some other biomanipulated sites, filamentous algae was prevalent. Phillips et al. (1978) suggested that mats of benthic and epiphytic algae were the principal cause of the decline of macrophyte populations in the Broads.

In the current study, there was a significant negative relationship between the abundance of macrophytes (expressed as % cover) and filamentous algae (Spearman rank correlation, $r_s = 0.57$, $n = 18$-those occasions and sites >10% cover, $p = 0.01$). However, the extent to which this is the result of an interaction between the two groups or simply the tendency of filamentous algae to colonise clear water sites before macrophytes (re: Cockshoot between 1993 and 1994) is unknown.

The effects of bird grazing in summer and autumn

Macrophytes are clearly important to coots in late summer and autumn, as populations increased dramatically at sites with significant macrophyte cover. The proportion of macrophytes in the diet was also high, including that of young birds which switch to an adult diet within 5 weeks of birth (Cramp & Simmons, 1980; Howes & Perrow, 1994). The apparent preference for macrophytes late in the season may be linked to improved nutritional content (e.g. carbohydrates) at this time (Best & Dassen, 1987; Best & Visser, 1987). But does bird grazing have an impact?

As above, the potential growth rate of macrophytes should typically easily outstrip any losses to grazing birds during the summer period (Table 4). The only exception to this being *Potamogeton* spp./*Z. palustris* in Ormesby in July when up to 30% of potential net growth was consumed.

The lack of impact was confirmed by the experiments at Cockshoot, where macrophyte cover or biomass were not greatly influenced by bird grazing, effects being species-specific and creating positive and negative consequences perhaps as a result of changing interspecific competitive interactions (Lodge, 1991). Similar results were reported by Sandsten (1995), with *Chara* spp. apparently benefiting from an impact of

Table 4. Removal (g m^{-2} d^{-1} WW) by birds of selected macrophyte species expressed as a % of published growth rates from similar water bodies in north-western Europe.

Species		March	April	May	June	July
C. demersum[1]	mean	0.51	0.10	0.48	0.83	0.37
	max.	3.02	0.51	2.77	2.91	0.99
	min.	0.00	0.00	0.00	0.00	0.00
	n	6	6	6	6	8
C. demersum[2]	mean	0.29	0.05	0.25	1.03	0.41
	max.	1.53	0.26	1.40	3.23	1.10
	min.	0.00	0.00	0.00	0.00	0.00
	n	6	6	6	6	8
C. demersum[3]	mean	0.29	0.05	0.25	0.38	0.17
	max.	1.53	0.26	1.40	1.32	1.32
	min.	0.00	0.00	0.00	0.00	0.00
	n	6	6	6	6	8
E. canadensis[4]	mean	0.00	0.09	0.01	0.06	0.13
	max.	0.00	0.47	0.06	0.29	0.56
	min.	0.00	0.00	0.00	0.00	0.00
	n	5	5	5	5	7
N. marina[5]	mean	–	–	–	0.00	0.07
	max.	–	–	–	0.00	0.35
	min.	–	–	–	0.00	0.00
	n	–	–	–	4	5
Potamgeton spp.	mean	<0.01	<0.01	<0.01	0.03	0.92
/*Z. palustris*[6]	max.	0.03	0.04	0.05	0.22	6.50
	min.	0.00	0.00	0.00	0.00	0.00
	n	8	8	8	8	9
Potamogeton spp.	mean	0.02	0.03	0.04	0.13	4.29
/*Z. palustris*[7]	max.	0.17	0.26	0.33	1.03	30.33
	min.	0.00	0.00	0.00	0.00	<0.01
	n	8	8	8	8	9
Potamogeton spp.	mean	0.11	0.16	0.20	0.10	3.06
/*Z. palustris*[8]	max.	0.85	1.30	0.73	0.73	21.66
	min.	0.00	0.00	0.00	0.00	0.00
	n	8	8	8	8	9
P. crispus[3]	mean	0.00	0.00	0.00	0.00	0.20
	max.	0.00	0.00	0.00	0.00	0.41
	min.	0.00	0.00	0.00	0.00	0.00
	n	2	2	2	2	2
P. pectinatus[10,11,12]	mean	0.00	0.00	0.00	0.00	0.01
	max.	0.00	0.00	0.00	0.00	<0.01
	min.	0.00	0.00	0.00	0.00	0.00
	n	5	5	5	5	5

References: [1,2] Best & Visser (1987); [3,5] Phillips (1976); [4,9] Lauridsen et al. (1994); [6,10] Schutten et al. (1994); [7,8,11,12] Doef et al. (1994).

grazing on *Potamogeton* spp. and Kiørboe (1980) who found no statistical differences in the biomass of several species including *Chara* spp. and *Potamogeton* spp.

In contrast, many authors (e.g. Anderson & Low, 1976; Jupp & Spence, 1977; Esler, 1989; Lauridsen et al., 1993, 1994; Moss et al., 1996); Søndergaard et al., in press) have described effects on growth and/or

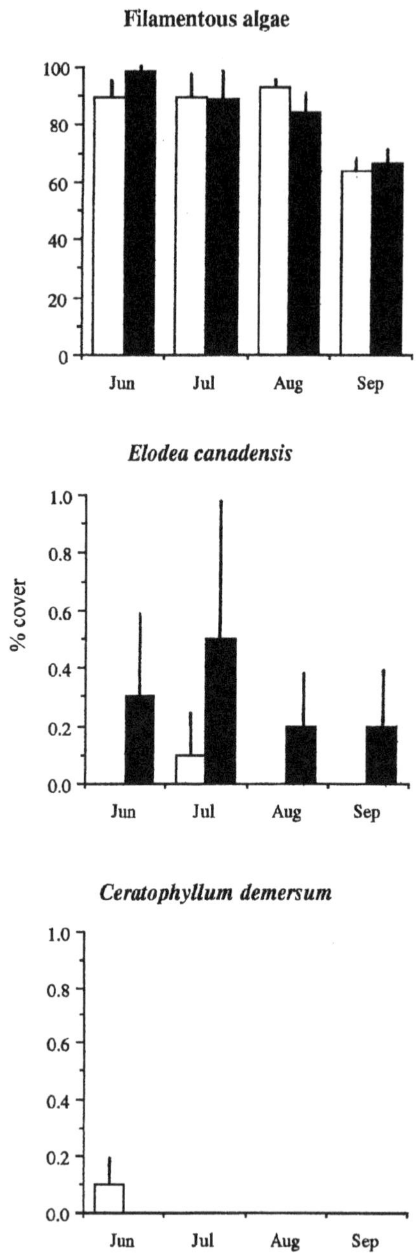

Figure 6. Mean (±S.E., $n = 10$) cover of filamentous algae and macrophyte species in both grazed controls (open bars) and protected treatments (shaded bars) over the experimental period at Pound End.

biomass of macrophytes in protected exclosures versus unprotected controls. There were many differences in experimental technique between the studies; such as the use of transplanted inocula (see above), variable numbers of replicates, comparison of treatments with true controls or adjacent areas, measurements of different variables including cover, biomass, growth (e.g. shoot length) and shoot number; but also in the abundance and type of herbivorous birds and macrophytes.

One element was relatively consistent. That is, any effect on macrophytes was only apparent from summer onwards when macrophytes reached a maximum standing crop or began to senesce, i.e. Lauridsen et al. (1993); mid July onwards; Søndergaard et al. (in press) early July and late July onwards; Moss et al. (1996) August; Anderson & Low (1976) August and September; Esler (1989) beginning in November with a consistent pattern by December.

The timing of any effect is likely to be critical. For example, for sexually reproducing flowering species, if grazing removes whole plants or sexual parts (seeds, fruits) before reproduction is complete, then grazing may be a limiting factor. Although this must be tempered against the fact that some species have shown improved seed germination after consumption by birds (Agami & Waisel, 1986).

In the Broads, as in other temperate European and North American shallow lakes *Potamogeton* spp. are widespread and common and an important component of macrophyte communities. They tend to dominate in early season, flower followed by vegetative reproduction and tuber formation in July/August and senesce in autumn. In Cockshoot at least, the peak in grazing pressure coincided with the decay of the above sediment biomass (Figure 5), birds never inducing a negative impact throughout the growing and flowering period (Figure 4, Table 3). In support of Kiørboe (1980), we suggest that significant removal of macrophytes tends to occur in autumn during the period of senescence of many species, during the peak passage period for waterfowl on migration and therefore any impacts on growth, biomass etc. are of little consequence for many of the macrophyte populations concerned. Rather, it is the abundance of birds that is controlled by the abundance/biomass of macrophytes (Mitchell et al., 1988; Giles, 1992; Hargeby et al. 1994; Lillie & Evrard, 1994; McKinnon & Mitchell, 1994).

Table 5. Maximum summer (August) standing crop (g DW m^{-2}) of *Ceratophyllum demersum* in waterbodies in the Norfolk Broads, compared with the maximum recorded consumption by a coot population (0.114 g DW m^{-2} d^{-1} in Cromes Broad), expressed as % of standing crop consumed/day.

Site	g DW m^{-2}	Source	% day^{-1}
Alderfen Broad (1982)	60.0	Perrow et al. (1994)	0.19
Alderfen Broad (1983)	38.0	Perrow et al. (1994)	0.30
Alderfen Broad (1984)	25.0	Perrow et al. (1994)	0.46
Alderfen Broad (1985)	1.5	Perrow et al. (1994)	7.60
Alderfen Broad (1987)	2.1	Perrow et al. (1994)	5.43
Alderfen Broad (1988)	16.5	Perrow et al. (1994)	0.69
Alderfen Broad (1989)	22.5	Perrow et al. (1994)	0.51
Alderfen Broad (1990)	54.0	Perrow et al. (1994)	0.21
Alderfen Broad (1991)	24.0	Perrow et al. (1994)	0.48
Belaugh Broad (1993)	1.4	Stansfield (1994)	8.14
Cockshoot Broad (1993)	12.1	Stanstield (1994)	0.94
Cockshoot Dyke (1993)	2.9	Stansfield (1994)	3.93
Pound End-bird exclosure (1993)	2.5	Stansfield (1994)	4.56
Woodbastwick Fen Dyke (1972)	248.0	Phillips (1976)	0.05
Woodbastwick Fen Dyke (1974)	338.0	Phillips (1976)	0.03

The effects on subsequent macrophyte development

There is a lack of compelling evidence to show that bird grazing limits macrophyte populations within a season, but it is possible that any impact is only manifested in subsequent seasons (Lodge, 1991). This is relevant for species such as *Elodea* or *Ceratophyllum* that may overwinter as developed plants or buds respectively (Van Donk et al., 1994) and perhaps even for tuber-forming species such as *P. pectinatus* (Anderson & Low, 1976), all of which may be available throughout the autumn and winter months.

Elodea and *Ceratophyllum* are an important component of the flora in restored lakes (Lauridsen et al., 1994; Perrow et al., 1994, Van Donk et al., 1994, Moss et al., 1996). This is probably a consequence of their rapid colonising ability, aided by dispersion from fragmentation (Best & Dassen, 1987; Best & Visser, 1987) and their tolerance to relatively eutrophic conditions (Palmer, 1992). However, they are also prone to rapid die-back (Krzyżanek, et al., 1986; Perrow et al., 1994; Moss et al., 1996) with bird grazing being implicated in some studies (Perrow et al., 1994; Moss et al., 1996). In support of this, Van Donk et al. (1994) demonstrated that herbivory by birds in the winter months was partly responsible for the decline of the *Elodea* population and restructuring of the macrophyte community in favour of *Ceratophyllum*.

Any impact of birds is obviously determined by the size of the initial standing crop. For example, the standing crop of *C. demersum* recorded in the broads is highly variable between sites and years (Table 5). Using the maximum recorded grazing pressure by a coot population, simple calculations reveal that the time taken for a coot population to remove the standing crop of *C. demersum* may be as little as 12 days or as long as the equivalent of 9 years. For sites with a low standing crops at the end of the growing season, it is conceivable that birds may quickly remove the entire overwintering stock, thereby having a major impact on the population in subsequent seasons.

The reliance on monocultures of potentially vulnerable species such as *Elodea* and *Ceratophyllum* in lake restoration programmes, underlines the need to establish a diverse macrophyte community, which may buffer any effect from bird grazing or other factors. This may only be possible where nutrient levels are reduced significantly, aiding stability (Hosper & Jagtman, 1990).

Conclusions

There is little evidence that grazing by coots birds limits colonisation and establishment of macrophytes in spring in the shallow lakes of the Norfolk Broads. This is related to the low grazing pressure from a low density of territorial breeding birds and the low prevalence of macrophytes in their diet. There is little conclusive experimental evidence in the literature to support the hypothesis that birds directly influence colonisation and establishment of naturally occurring macrophytes. Indeed, in shallow lakes undergoing restoration, any lag in the response of macrophytes is likely to be caused by a number of factors other than bird grazing. However, losses to birds may contribute to a low abundance of macrophytes.

In general, grazing by birds tends to peak in late summer and autumn after the principal growing season and after reproduction, i.e. flowering and setting of seed has occurred. Any impact of grazing may therefore be considered to be unimportant for many existing macrophyte populations. However, grazing at this time and into the winter months, where some species of macrophytes remain available as developed plants, dormant buds or even tubers, may be important in determining macrophyte abundance and community structure in subsequent growing seasons.

Overall, it is clear that further long-term replicated exclosure experiments are required in a variety of circumstances (e.g. with different start-up inocula of macrophytes and different species of macrophytes and herbivorous birds) to reach a consensus of opinion on the impact of bird grazing. If limiting, additional measures controlling bird grazing (e.g. through supplementary feeding, scaring or culling) or enhancing macrophyte populations (e.g. transplantation and initial protection) may be needed to accelerate macrophyte colonisation and development in lakes undergoing restoration.

Acknowledgments

Bird monitoring and grazing experiments were commissioned by the Broads Authority (BA) and undertaken as part of the BA/National Rivers Authority (NRA) EC LIFE programme project (LIFE 92-3/UK/031 & NRA Project 475) on the Restoration of the Norfolk Broads (Project managers Jane Madgwick & Dr Geoff Phillips) Macrophyte data was collected by J. Schutten for the Macrophyte Recovery Project, a joint project of the BA, RIZA (Institute for Inland Water Management and Wastewater Treatment, the Netherlands) and University of Amsterdam with further financial contribution by the Soap & Detegent Industry Association and collaboration of the NRA and University of East Anglia. We gratefully acknowledge those organisations and individuals, too numerous to mention, for access to sites and Anita van der Vet, Ruth Nightingale and Marcus Rowcliffe for assistance with fieldwork. Two anonymous referees provided stimulating comments on the manuscript.

References

Agami, M. & Y. Waisel, 1986. The role of mallard ducks (*Anas platyrhynchos*) in distribution and germination of seeds of the submerged hydrophyte *Najas marina* L. Oecologia 68: 473–475.

Anderson, M. G. & J. B. Low, 1976. Use of Sago pondweed on the Delta Marsh, Manitoba. J. Wildl. Mgmt 40: 233–242.

Best, E. P. H. & J. H. A. Dassen, 1987. A seasonal study of growth characteristics, and the levels of carbohydrates and proteins in *Elodea nuttallii, Polygonum amphibium* and *Phragmites australis*. Aquat. Bot. 28: 353–372.

Best, E. P. H. & H. C. W. Visser, 1987. Seasonal growth of the submerged macrophyte *Ceratophyllum demersum* L. in mesotrophic Lake Vechten in relation to insolation, temperature and reserve carbohydrates. Hydrobiologia 148: 231–243.

Blindlow, I., G. Andersson, A. Hargeby & S. Johansson, 1993. Long-term pattern of alternative stable states in two shallow eutrophic lakes. Freshwat. Biol. 30: 159–167.

Broads Authority, 1994. The Broads Plan-No Easy Answers. Broads Authority Consultation Document, Broads Authority, Norwich, England.

Cramp, S. & K. E. L. Simmons, 1980. Handbook of the Birds of Europe the Middle East and North Africa. The Birds of the Western Palearctic, Volume II. Hawks to Bustards. Oxford University Press, Oxford, UK.

Doef, R., H. Coops, M. L. Streekstra & L. H. C. A. Hector, 1994. Waterplanten in het Wolderwijden Veluwemeer (1990–1993). RIZA nota-94.046, RIZA, Lelystad, Netherlands. ISBN 9036903440.

Draulans, D. & L. Vanherck, 1987. Food and foraging of Coot *Fulica atra* on fish ponds during autumn migration. Wildfowl 38: 63–69.

Esler, D., 1989. An assessment of American coot herbivory of Hydrilla. J. Wildl. Mgmt 53: 1147–1149.

Ferns, J. & J. Kirby, 1992. Waterfowl counts in the UK. 1990–91. Wildfowl 43: 235–240.

Giles, N., 1992. Wildlife after gravel: twenty years of practical research by the Game Conservancy and ARC. Game Conservancy, Fordingbridge, UK.

Hargeby, A., G. Andersson, G. Blindlow & S. Johansson, 1994. Trophic web structure in a shallow eutrophic lake during a dominance shift from phytoplankton to submerged macrophytes. Hydrobiologia 279/280: 83–90.

Harper, J. L., 1977. Population biology of plants. Academic Press Inc., London, UK.

Hosper, S. H. & E. Jagtman, 1990. Biomanipulation additional to nutrient control for restoration of shallow lakes in The Netherlands. Hydrobiologia 200/201: 523–534.

Howes, J. R. & M. R. Perrow, 1994. Macrophyte grazing by coot (*Fulica atra*) in the broads: bird-resource interactions. In Pitt, J.-A. & G. L. Phillips (eds), The Development of Biomanipulation Techniques & Control of Phosphorus Release from Sediments, EC LIFE project 92-3/UK/031, NRA Report No. No. 475121A. National Rivers Authority/Broads Authority, Bristol, UK.

Irvine, K., B. Moss & H. Balls, 1989. The loss of submerged plants with eutrophication II. Relationships between fish and zooplankton in a set of experimental ponds, and conclusions. Freshwat. Biol. 22: 89–107.

Jupp, B. P. & D. H. N. Spence, 1977. Limitations of macrophytes in a eutrophic lake, Loch Leven. II. Wave action, sediments and waterfowl grazing. J. Ecol. 65: 431–446.

Kennison, G. C. B., 1993. Aquatic macrophyte surveys of the Norfolk Broads. Unpubl. report, Broads Authority, Norwich, UK.

Kiørboe, T., 1980. Distribution and production of submerged macrophytes in Tipper Grund (Ringkøbing Fjord, Denmark), and the impact of waterfowl grazing. J. appl. Ecol. 17: 675–687.

Krzyżanek, E., H. Kasza, W. Krzanowski, T. Kuflikowski & G. Pająk, 1986. Succession of communities in the Goczałkowice Dam Reservoir in the period 1955–1982. Arch. Hydrobiol. 106: 21–43.

Lammens, E. H. R. R., R. D. Gulati, M.-L. Meijer & E. van Donk, 1990. The first biomanipulation conference: a synthesis. Hydrobiologia 200/201: 619–628.

Lauridsen, T. L., E. Jeppesen & F. Østergaard Andersen, 1993. Colonization of submerged macrophytes in shallow fish manipulated Lake Vaeng impact of sediment composition and waterfowl grazing. Aquat. Bot. 46: 1–15.

Lauridsen, T. L., E. Jeppesen & M. Søndergaard, 1994. Colonization and succession of submerged macrophytes in shallow Lake Vaeng during the first five years following fish manipulation. Hydrobiologia 275/276: 233–242.

Lillie, R. A. & J. O. Evrard, 1994. Influence of macroinvertebrates and macrophytes on waterfowl utilization of wetlands in the Prairie Pothole Region of northwestern Wisconsin. Hydrobiologia 279/280: 235–246.

Lodge, D. M., 1991. Herbivory on freshwater macrophytes. Aquat. Bot. 41: 195–224.

McKinnon, S. L. & S. F. Mitchell, 1994. Eutrophication and black swan (*Cygnus atratus* Latham) populations: tests of two simple relationships. Hydrobiologia 279/280: 163–170.

Meijer, M.-L., E. Jeppesen, E. van Donk, B. Moss, M. Scheffer, E. Lammens, E. van Nes, J. A. van Berkum, G. L. de Jong, B. A. Faafeng & J. P. Jensen, 1994. Long-term responses to fish-stock reduction in small shallow lakes: interpretation of five-year results of four biomanipulation cases in The Netherlands and Denmark. Hydrobiologia 275/276: 457–466.

Mitchell, S. F., D. P. Hamilton, W. S. MacGibbon, P. K. B. Nayar & R. N. Reynolds, 1988. Interactions between phytoplankton, submerged macrophytes, black swans and zooplankton in a shallow lake. Int. Revue ges. Hydrobiol. 73: 145–170.

Moss, B., 1990. Engineering and biological approaches to the restoration from eutrophication of shallow lakes in which aquatic plant communities are important components. Hydrobiologia 200/201: 367–377.

Moss, B., H. Balls, K. Irvine & J. H. Stansfield, 1986. Restoration of two lowland lakes by isolation from nutrient-rich water sources with and without removal of sediment. J. appl. Ecol. 23: 391–414.

Moss, B., J. H. Stansfield & K. Irvine, 1990. Problems in the restoration of a hypertrophic lake by diversion of a nutrient-rich inflow. Verh. int. Ver. Limnol. 24: 409–425.

Moss, B., J. H. Stansfield, K. Irvine, M. R. Perrow & G. L. Phillips, 1996. Progressive restoration of a shallow lake – a twelve-year experiment in isolation, sediment removal and biomanipulation. J. appl. Ecol. 33: 71–86.

National Environmental Research Institute, Denmark, 1994. Report and activities 1993–94. Ministry of the Environment, Roskilde, Denmark.

National Research Council (US), 1992. Restoration of Aquatic Ecosystems-Science, Technology and Public Policy. National Academy Press, Washington D.C, USA: 552 pp.

Palmer, M., 1992. A botanical classification of standing waters in Great Britain. JNCC Publications, Peterborough, England. ISBN 1 873701 27 6.

Perrow, M. R., 1990; Biomanipulation in Broadland. In O'Grady, K. T., A. J. B. Butterworth, P. B. Spillett & J. C. J. Domaniewski (eds), Fisheries in the Year 2000, Proceedings of the 21st anniversary conference of the Institute of Fisheries Management. IFM, Nottingham, England: 335–337.

Perrow, M. R., B. Moss & J. H. Stansfield, 1994. Trophic interactions in a shallow lake following a reduction in nutrient loading: a long term study. Hydrobiologia 275/276: 43–52.

Phillips, G. L., 1976. An Investigation of the Distribution and Growth of Aquatic Plants in some of the Norfolk Broads. Unpubl. PhD thesis, University of East Anglia, Norwich, England.

Phillips, G. L., D. F. Eminson & B. Moss, 1978. A mechanism to account for macrophyte decline in progressively eutrophicated freshwaters. Aquat. Bot. 4: 103–126.

Phillips, G. L., M. R. Perrow & J. H. Stansfield, 1996. Manipulating the fish-zooplankton interaction in shallow lakes: a tool for restoration. In Greenstreet, S. P. R. & M L. Tasker (eds), Aquatic Predators and Their Prey. Blackwell Scientific Publications Ltd., Oxford. England: 174–183.

Ryding, S.-O. & W. Rast (eds), 1989. The Control of Eutrophication of Lakes and Reservoirs. UNESCO, Man & the Biosphere Series, Vol. 1, Parthenon Publishing Group, Lancs. England.

Sandsten, H., 1995. Can waterfowl grazing trigger a shift from clear to turbid water in shallow Lake Krankesjon? Master thesis, University of Lund, Sweden.

Sas, H., 1989. Lake restoration by reduction of nutrient loading: Expectations, experiences, extrapolations. Academia Verlag 497 pp.

Scheffer, M., S. H. Hosper, M.-L. Meijer, B. Moss & E. Jeppesen, 1993. Alternative Equilibria in Shallow Lakes. TREE 8: 275–279.

Schutten, J., J. A. van der Velden & H. Smit, 1994. Submerged macrophytes in the recently freshened lake system Volkerak-Zoom (The Netherlands), 1987–1991. Hydrobiologia 275/276: 207–218.

Seago, M. J., 1967. Birds of Norfolk. Jarrold & Sons, Norwich, England.

Søndergaard, M., L. Bruun, T. L. Lauridsen, E. Jeppesen & T. Vindbaek Madsen, in press. The impact of grazing waterfowl on submerged macrophytes: *in situ* experiments in a shallow eutrophic lake. Aquat. Bot.

Stansfield, J. H., 1994. The extent of macrophyte recovery at biomanipulated sites, and experimental evidence that bird grazing and/or water chemistry affect macrophyte establishment. In Pitt, J-A. & G. L. Phillips (eds), The Development of Biomanipulation Techniques & Control of Phosphorus Release from Sediments, EC LIFE project 923/UK/031, NRA Report No. No. 475/2/A. National Rivers Authority/Broads Authority, Bristol, UK.

Van Donk, E., E. De Deckere, J. G. P. Klein Breteler & J. T. Meulemans 1994. Herbivory by waterfowl and fish on macrophytes in a biomanipulated lake: effects on long-term recovery. Ver. int. Ver. Limnol. 25: 2139–2143.

Hydrobiologia **342/343:** 257–267, 1997.
L. Kufel, A. Prejs & J. I. Rybak (eds), Shallow Lakes '95.

Vertically-challenged limnology; contrasts between deep and shallow lakes

Brian Moss, Meryem Beklioglu*, Laurence Carvalho**, Sabri Kilinc, Suzanne McGowan & Deborah Stephen
Department of Environmental ad Evolutionary Biology, University of Liverpool, Liverpool L69 3BX, UK
**Biology Department, Middle East Technical University, 06531, Ankara, Turkey*
***Present address: Environmental Change Research Unit, University College, London, 26 Bedford Way, London WC1H 0AP, UK*

Key words: lake, depth, eutrophication, restoration

Abstract

Previous work on a set of small lakes, of varying depth, the meres of North West England, has shown that nitrogen availability controls the summer phytoplankton populations in the deeper ones (max depth>3 m) and zooplankton grazing in shallow ones. The meres have generally high total phosphorus concentrations and this may be a natural phenomenon dependent on the local geochemistry. Some anthropogenic eutrophication has occurred, however, and from a chain of three meres, sewage effluent was diverted in 1991. The upper lake, Mere Mere, lying above the point of discharge, has not changed in any systematic way since effluent diversion. The middle lake, the very shallow Little Mere, has changed markedly in water chemistry but not fundamentally in ecosystem structure. It was and remains a clear-water, macrophyte dominated lake. The third lake, the deep Rostherne Mere, has shown no response in chlorophyll *a* concentrations in four years since effluent diversion though in the past two years there appears to be a downward trend in total phosphorus. The reasons for this are explored in terms of our understanding of lake eutrophication. Comparisons are made with White Mere, a deep groundwater fed lake with a long retention time and a very high total phosphorus concentration. The deep meres may add a new dimension to our understanding of natural and anthropogenic eutrophication.

Introduction

Prejudice is always based on ignorance. Its consequences can be devastating. For centuries in Britain, it was believed that the condition of hare-lip, now known medically as cleft palate, was caused in babies by the crossing of the path of a pregnant woman by a hare. The sufferer, if a woman, was thought to be a witch. Prue Sarne, the 17th century heroine of Mary Webb's (1924) novel, *Precious Bane*, being so afflicted, eventually escapes the consequences of her condition, but not before almost succumbing to yet more ignorance. She believed that to enter the waters of a local lake, or mere, on a certain date, when the waters had 'broken', dressed in a white smock and in full view of the assembled public would cure the affliction. The 'breaking of the waters' was a well known local phenomenon which we now understand to be the rising to the surface of buoyant cyanophytes – an algal bloom.

The story is a curious one but perhaps illustrates our continued human propensity to create myths. Myths are devices for the explanation of natural phenomena when complete information is lacking. One such, deeply held for many years, was that lakes naturally became more eutrophic with time. Cyanophyte blooms were thus eventual and inevitable natural phenomena. Partly this idea resulted from a confusion between the separate processes of nutrient loading and of filling-in of lakes with sediment, the latter being almost self-evident. But on confinement of the meaning of eutrophication to increased nutrient loading, the idea of inevitable 'natural' eutrophication, as opposed to a purely anthropogenic phenomenon, has declined. This is because catchments contain a finite stock of nutrients which can be delivered to a lake and because phosphorus and nitrogen do not accumulate in the water but are lost to sinks such as the sediment, the atmosphere and the outflow.

Cyanophyte blooms were thus not 'natural' in most lakes but caused by human activities. For some limnologists, particularly palaeolimnologists, whose increasingly sophisticated studies have demolished the former myth, the new paradigm now holds as powerful a sway (Anderson, 1995) as the former. Yet a long traditional knowledge of algal blooms in the area in which 'Precious Bane' was set, the Meres and Mosses district of Shropshire (Sinker, 1962; Reynolds, 1979), might suggest that the entire truth has yet to be told.

We may also be in danger of relying too strongly on another apparently firm plank in the structure of our understanding – that where there has been anthropogenic eutrophication, and restoration of a former state is required, reduction of nutrient loading will be the solution for deep lakes, whilst it is unlikely to be effective alone in shallow lakes (Sas, 1989; Moss, 1989, 1991). The same meres of the North West Midlands are providing instructive challenges to this new paradigm before it becomes too ensconced. This paper summarises the results of five-years' monitoring of a system of three linked lakes, Mere Mere, Little Mere and Rostherne Mere. A major source of nutrients from sewage effluent was diverted, in 1991, from the system which includes a very shallow lake (Little Mere) and a comparatively deep one (Rostherne Mere). Some reference is also made to another deep mere, White Mere which is fed solely by ground water and has a very long hydraulic retention time (about 2.5–5 yr).

The North West Midland Meres

The meres (Figure 1) lie on a glacial outwash plain, roughly rectangular in shape, and bordered to the north by the valley of the River Mersey, and to the east, west and south by the hills of the Pennines, the Welsh Borders and South Shropshire. The terrain is flat or hummocky with moraines. A large number of small basins was formed on the plain in kettle holes, through moraine damming of streams, or by subsidence (Tallis, 1973; Reynolds, 1979). Many of the shallower basins have filled in with peat and sediment (the mosses) whilst about sixty retain open water (the meres) in areas from about 2 ha to about 50 ha. Their waters are mostly base-rich with high conductivities (Moss et al., 1994) though a few, based on sands, are low in ionic content. In general they have high total phosphorus concentrations (of several hundred $\mu g\ l^{-1}$ to over 1 mg l^{-1}) and low inorganic nitrogen concentrations (Figure 1) with consequently low N:P ratios. Some show signs of recent eutrophication (Reynolds & Bellinger, 1992; Nelms, 1984; Livingstone, 1979) based on palaeolimnological evidence or monitoring carried out on a discontinuous basis over a long period, but there is also a hint that others have been very eutrophic, with cyanophyte blooms, for a very long period (Phillips, 1884; Griffiths, 1925).

Reynolds (1979) noted low nitrogen and high phosphorus concentrations in the meres. Recent, more detailed, monitoring (Moss et al., 1994) has confirmed this, and revealed a correlation between winter nitrogen concentrations and growth season phytoplankton chlorophyll *a* in a group of twelve meres with maximum depths greater than 3 m and strong summer thermal stratification. In a second group, of shallower meres, no links between nutrients and phytoplankton chlorophyll *a* were found, but there was a significant inverse correlation between growth season chlorophyll *a* and zooplankton (or *Daphnia*) biomass. In the shallow meres that had low chlorophyll *a* concentrations there was a significant macrophyte coverage; in those with high chlorophyll *a* and low zooplankton biomass, macrophytes were scarce or absent. The shallow lakes were also generally affected by surface streams carrying high phosphate and ammonium concentrations (Moss et al., 1992), probably derived mostly from stock-farming, and even when they had clear water, had high nutrient concentrations. The deeper lakes tend to be groundwater fed and have longer water retention times than the shallower ones (Moss et al., 1992).

Underlying the northern part of the plain are soft, Triassic, evaporite deposits, exploited since Roman times as salt sources. They lie close to the surface and may have been disturbed by the glaciation, such that mineral rich deposits could have been carried southwards. They could thus provide, as apatite, the source of phosphorus required to maintain the very high concentrations of total phosphorus found, in the absence of obvious eutrophicating sources, in the deep meres (Moss et al., 1994). However, analysis of ground waters in the area has so far revealed concentrations of total phosphorus two orders of magnitude lower than those found in the lakes. The system of three lakes central to this paper, including Mere Mere (National Grid Reference SJ 732815), Little Mere (NGR 733823) and Rostherne Mere (NGR SJ745844), is shown in Figure 1. A detailed catchment map is given in Car-

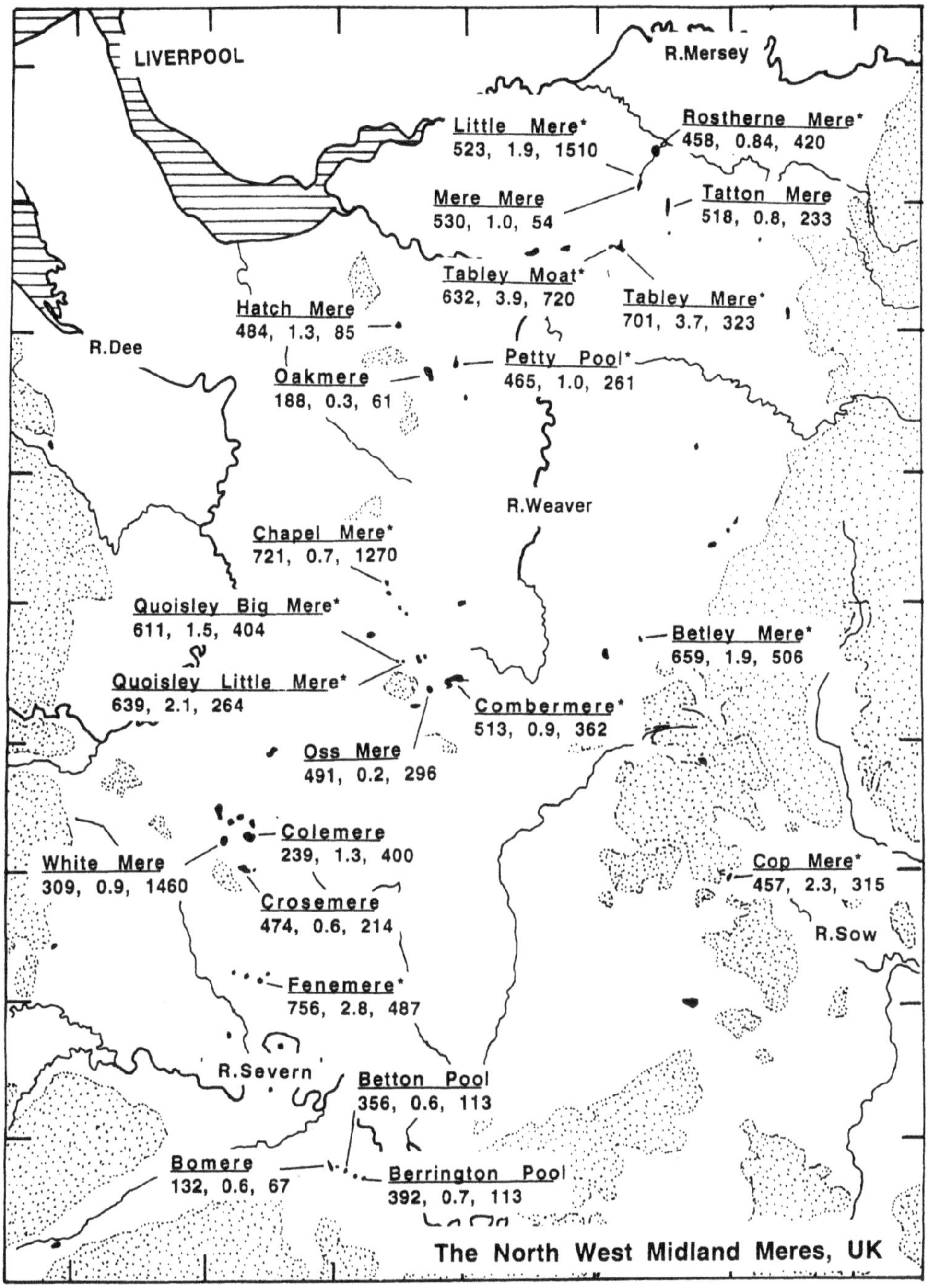

Figure 1. Map of the location of some of the North West Midland Meres. The estuary of the River Mersey lies to the top (north) and the latitude and longitude of the centre of the map are 53° 02′N and 2° 36′W. Mean annual conductivities ($\mu S\ cm^{-1}$), mean winter dissolved inorganic nitrogen concentrations (nitrate-N plus ammonium-N) ($mg\ l^{-1}$) and mean annual total phosphorus concentrations ($\mu g\ l^{-1}$) are shown, in that order, below the names of a representative selection of lakes sampled in 1991. Asterisks indicate lakes known to be polluted by stock or sewage effluent.

valho et al., 1995). It occupies a catchment, of agricultural land, small villages and a golf course, which has undergone increases in agricultural intensification (hectarage of arable and temporary grassland, and more cattle stocked) in the last fifty years, but has been more seriously affected by sewage effluent. A small stream feeds Mere Mere, which is moderately large and deep (15.8 ha, z_{max} 8.1 m) and in many ways is a 'typical' mere, with well developed littoral macrophytes, modest chlorophyll *a* concentrations, and seasonal cyanophyte blooms. At the north end of Mere Mere, the outflow stream was dammed, probably in the last century, to form a much smaller (2.5 ha) lake, Little Mere, which is very shallow (z_{max} 2.6 m). From the 1930s until June 1991, sewage effluent was discharged directly to Little Mere, from which water then drained through Rostherne Brook to the much larger (31 ha) and deeper (z_{max}, 31 m) Rostherne Mere.

Rostherne Mere is a National Nature Reserve, because of the richness of its winter bird populations. Indeed it was one of the first lakes to be called guanotrophic (Brinkhurst & Walsh, 1967) though the birds then, and certainly not now (Carvalho et al., 1995), are unlikely to have contributed more than a tiny fraction of the phosphorus budget. It has shown evidence of recent eutrophication, however (Livingstone, 1979; Nelms, 1984; Reynolds & Bellinger, 1992), which might sensibly be attributed to the nutrients provided by the upstream source of sewage effluent and more intensive agriculture since World War 2. It was felt desirable to reduce the incidence of cyanophyte blooms which were common in summer and, in view of the large literature supporting such action, it was deemed appropriate to divert the effluent away from the catchment in 1991. Nutrient budgeting has shown the effluent to have dominated the external nutrient load entering the lake by the major Rostherne Brook inflow (Table 1).

Table 1. Phosphorus loads (kg) for Rostherne Mere for an annual period prior to diversion of sewage effluent from its Rostherne Brook inflow and for an annual period immediately following diversion. values in parentheses are percentages of the total load.

Source	1990/91	1991/92
Sewage effluent	1622 (72)	0
Diffuse surface drainage	564 (25)	511 (87)
Groundwater	52 (2)	40 (7)
Direct rainfall	20 (1)	19 (3)
Bird roost	22 (1)	20 (3)
TOTAL	2280	590

Methods

Methods were standard ones and are documented variously in Carvalho (1993, 1994); Carvalho et al., 1995; Moss et al., 1994; Beklioglu, 1995; Beklioglu & Moss, 1995.

Results

Changes in Mere Mere

Mere Mere lies above the pre-1991 source of sewage effluent and although there have been year-to-year changes in the water chemistry and phytoplankton, these have been unsystematic and unrelated to the diversion of effluent (Table 2). Some of these changes, for example in nitrate concentrations, probably reflect fluctuations in weather and have been experienced through the system as a whole, but they have been relatively minor compared with the changes wrought in Little Mere by diversion of the effluent.

Changes in Little Mere

Little Mere has responded in major ways to the diversion of effluent, yet, paradoxically, is fundamentally unchanged in terms of its ecosystem structure. Before diversion (Carvalho, 1993, 1994) it had substantial stands of water lilies *Nuphar lutea* (L.) Sm and *Nymphaea alba* L.), and the floating *Lemna minor* L.. There were also small clumps of the submerged *Elodea canadensis* Michx. and *Potamogeton berchtoldii* Fieb.. There was a substantial spring growth of diatoms, but in summer the water was very clear with only prominent colonies of *Volvox* that contributed little to the chlorophyll *a* concentration. The water had very high concentrations of ammonium, soluble reactive and total phosphorus (Figure 2), but little nitrate, and low oxygen concentrations even by day (Carvalho, 1994). This might be expected from a small lake receiving a comparatively large volume of sewage effluent (on average 40%, at least, of the inflow, with 2–10% in winter and spring and >90% in summer) from a sewage works that

Table 2. Annual means of some chemical variables and of chlorophyll *a* in Mere Mere. Values for SRP and ammonium-N are given for the period October-January, when, following autumnal mixing, they are most abundant in the surface water. Those for nitrate-N are for the corresponding period (January to March) when incoming water has increased them to their maxima for the year and chlorophyll *a* concentrations are for the growth season (March–October). Total P values are annual averages.

	1990	1991	1992	1993	1994
Sol. react P ($\mu g\ l^{-1}$)	11 ± 7	13 ± 5	17 ± 16	16 ± 11	26 ± 7
Total P ($\mu g\ l^{-1}$)	79 ± 25	53 ± 19	68 ± 55	95 ± 72	75 ± 43
Ammonium-N ($\mu g\ l^{-1}$)	206 ± 31	174 ± 29	84 ± 82	103 ± 64	196 ± 45
Nitrate-N ($mg\ l^{-1}$)	1.4 ± 0.3	2.1 ± 0.3	1.5 ± 0.4	1.1 ± 0.1	0.8 ± 0.3
Chl *a* ($\mu g\ l^{-1}$)	29 ± 16	25 ± 19	19 ± 16	45 ± 40	23 ± 26

was greatly overloaded and probably operating inefficiently even where removal of organic matter was concerned. Fish were very scarce and large populations of prominently red-coloured *Daphnia magna* Sars were present during summer and autumn. With such a high algal growth potential in the water, the lake phytoplankton was controlled by grazing in summer, and the lake corresponded to the plant-dominated state of the alternative stable states model of shallow lake systems (Moss, 1989, 1991; Scheffer et al., 1993).

Possible outcomes for the lake following diversion of effluent were (i) that it would remain plant-dominated rather than phytoplankton-dominated in summer because various stabilising (buffer) mechanisms would mitigate against change. Alternatively, (ii) it was possible that with increasing oxygen concentrations, fish would recolonise from Mere Mere, just upstream, and that progressive oxygenation of organic matter, deposited with the previous sewage effluent, would lead to eventual reduction in carbon dioxide concentrations and increases in pH which would favour cyanophyte growth. Inocula are readily available in incoming Mere Mere water. Thus, a possible scenario was that the zooplankton community, under predation pressure from fish, might be unable to exert substantial grazing pressure and that the system would switch to a phytoplankton-dominated state.

Hypothesis (i) has hitherto been supported. There have been substantial changes in water chemistry (Figure 2) with marked falls in total and soluble reactive phosphorus concentrations, though little change in nitrate concentration, increases in oxygen concentrations and a recolonisation of fish, particularly perch (*Perca fluviatilis* L.). The zooplankton community progressively lost *Daphnia magna* in the two years following diversion but by 1993, the smaller *Daphnia hyalina* Leydig had formed substantial summer populations (Beklioglu, 1995). Meanwhile the submerged plant population had increased and by 1994 more or less covered the entire bottom of the lake. In turn the *Daphnia hyalina* population declined in 1994 (*D. Stephen*, unpubl. data) and it is not yet clear whether the phytoplankton community in summer is now controlled by the grazing of plant-associated Cladocera or by availability of nitrogen compounds (Ozimek et al., 1990). Despite the marked reduction in total phosphorus concentrations, from close to 2 mg l^{-1} prior to diversion, substantial concentrations of available soluble reactive phosphorus still remain (>60 $\mu g\ l^{-1}$) but ammonium has declined and both it and nitrate are close to undetectable for much of the summer.

Experiments in mesocosms, in 1992 and 1993 (Beklioglu, 1995), have shown a strong effect of grazing in controlling the phytoplankton crops and also that, despite manipulation of conditions to favour cyanophytes (Beklioglu & Moss, 1995), their crops did not increase or ever dominate the water in the mesocosms. The relative importance of grazing and nutrient limitation, now that *Daphnia* is much scarcer than previously, is being tested in a mesocosm experiment in 1995.

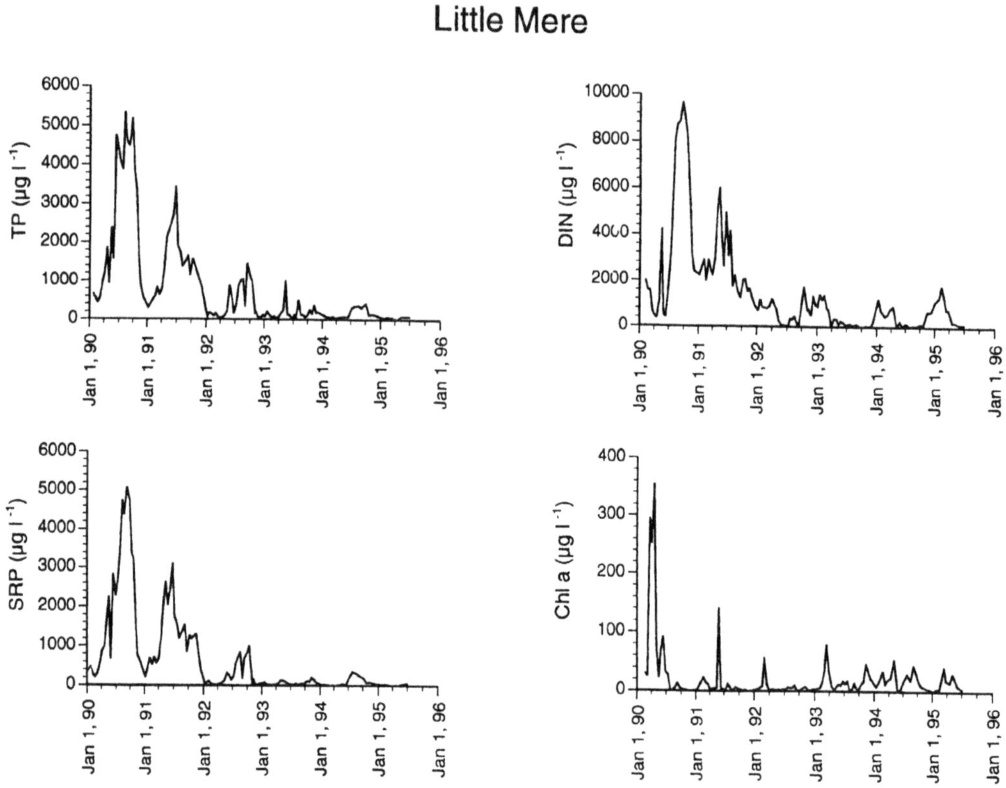

Figure 2. Concentrations of total phosphorus, soluble reactive phosphorus, dissolved inorganic-N and chlorophyll *a* in Little Mere between 1990 and 1994. Effluent was diverted from the mere in June 1991.

Changes in Rostherne Mere

Rostherne Mere is a substantial lake compared with the other meres in the series. Studies prior to diversion (Table 1) suggested that 72% of its external phosphorus load came from the sewage effluent draining through Little Mere and diversion of this effluent resulted in a reduction in total phosphorus load from at least 2280 kg per year before diversion to 590 kg afterwards. There was also a substantial reduction in the load of total inorganic nitrogen from 12 300 to 7500 kg per year as a result of the loss of the ammonium ions that had come from the effluent. The reduction due to effluent diversion was marginally enhanced by the effects of a relatively dry year in 1991/92.

Despite these differences, there is yet no change in chlorophyll *a* concentration but perhaps the start of a downward trend in total phosphorus concentration in Rostherne Mere, in the four years following effluent diversion (Figure 3). Its total phosphorus concentrations remain very high despite the major reduction in external load, though there is a slight downward trend and comparison with theoretical calculations suggests that external load reduction will eventually have some effect. There have been no systematic changes in the zooplankton community and only minor changes in the phytoplankton community but it is not yet possible to relate these to effluent diversion because comparable changes have also occurred in Mere Mere.

White Mere

White Mere is a 25 ha, relatively deep (16.5 m max.) lake lying to the south of the meres discussed above (National Grid Reference SJ 434333). It has no surface inflows or outflows and its water level remains relatively constant. With a small catchment area (around 8 ha) it has only a modest flowthrough of ground water and a retention time estimated, from net catchment precipitation and volume data, at between 2.5 and 5 years. Most notably it has very high total phosphorus concentrations (Kilinc, 1995) (Figure 4). These concentrations

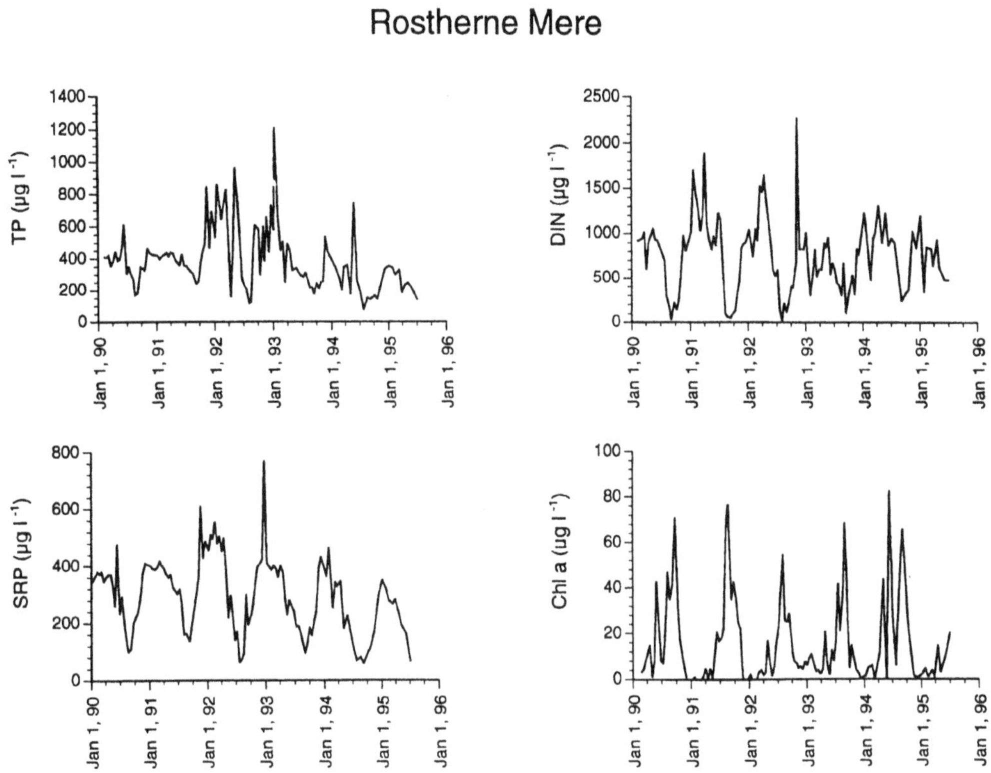

Figure 3. Concentrations of total phosphorus, soluble reactive phosphorus, dissolved inorganic-N and chlorophyll *a* in Rostherne Mere between 1990 and 1994. Effluent was diverted from the inflow stream to the mere in June 1991.

(400–3200 μg l^{-1}) contrast greatly with considerably lower concentrations (20–30 μg l^{-1}) in the inflowing groundwater. There is considerable release of phosphate from the marginal sediments into the epilimnion in summer (mean 110 mg P m^{-2} day^{-1}) and the possibility of migratory transfer of phosphorus by the algae. Physical mixing from the hypolimnion in summer appears to be a minor source. In contrast to Rostherne Mere, White Mere receives insignificant amounts of effluent (from a group of cottages, whose residents, we calculate might contribute about 19 μg $1l^{-1}$ of a total phosphorus concentration two orders of magnitude greater). White Mere has had abundant cyanophytes for at least 6000 years (S. McGowan, unpubl. palaeopigment data) and the phosphorus chemistry of the lake appears largely independent of the nature of the catchment.

Discussion

Reduction in nutrient loading has generally been unsuccessful in bringing about improvements in water clarity in shallow lakes (Marsden, 1989; Sas, 1989) and a whole hypothesis of alternative stable states in such lakes predicts that nutrient control alone over a very wide range of concentrations is unlikely to be effective in decreasing phytoplankton crops. The marked changes in Little Mere might seem to contradict the hypothesis, but this is not so. Little Mere, even before diversion of sewage effluent, was in a plant-dominated state, with clear water in summer.

The effects of diversion were to extend the coverage of submerged plants but not to increase the clarity of the water by decreasing the phytoplankton crops in summer. The size of the spring peak of phytoplankters, which develops prior to the annual seasonal establishment of the cladoceran community, was markedly reduced, however, and this may be linked with reduction in loading. It is also possible that the reduced

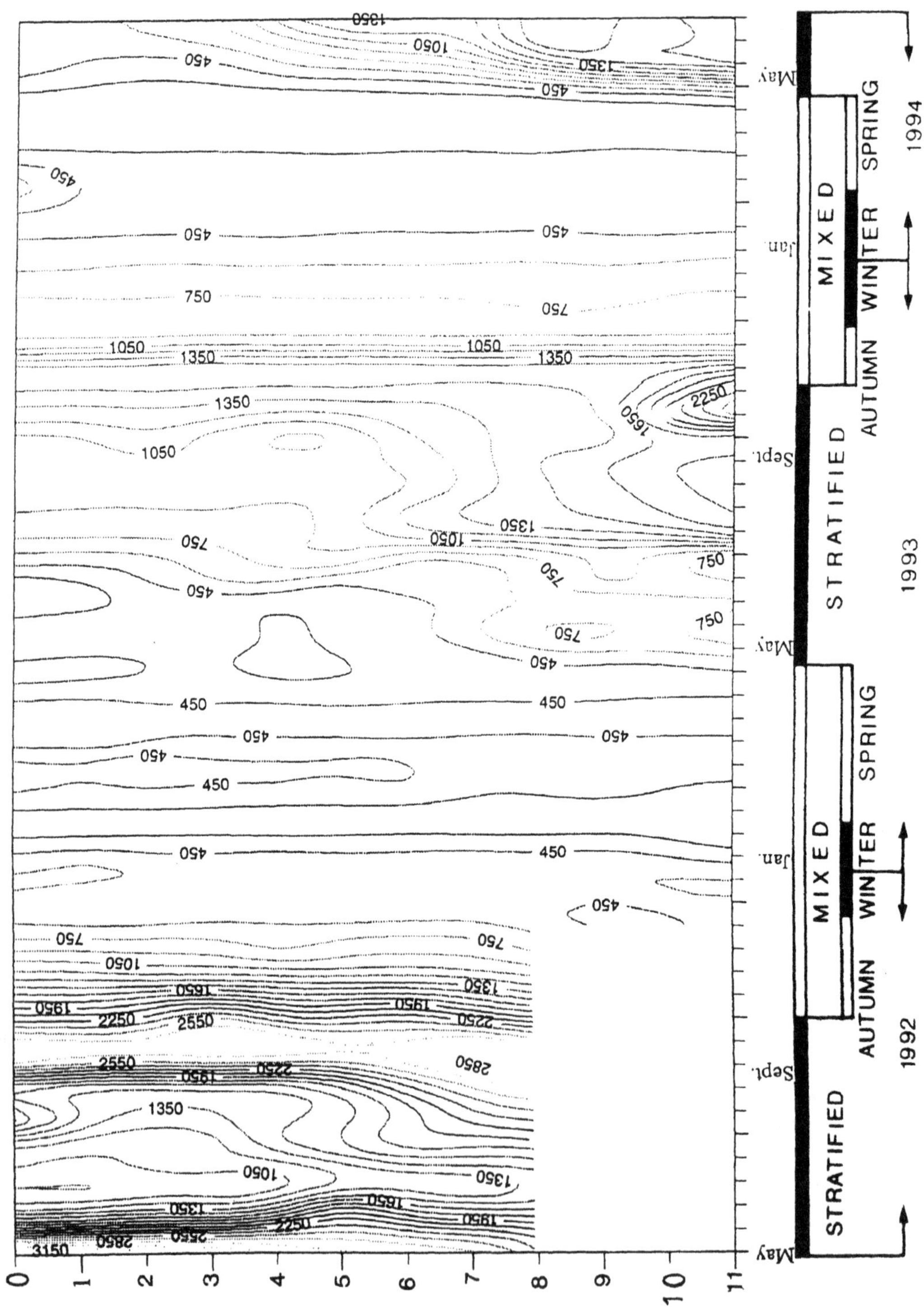

Figure 4. Depth-time diagram for total phosphorus in White Mere, Shropshire. Isopleths are given in $\mu g\ l^{-1}$ and the depth scale is in m.

nutrient concentrations had effects through reduction of periphyton on the developing plant surfaces (Phillips et al., 1978). However, even after diversion, nutrients remained abundant and it is unlikely that the phytoplankton was nutrient limited at any time up to the heavy plant growth of 1994 when submerged plants entirely covered the lake.

The lake may have been able to maintain a clear-water state prior to effluent diversion even without the deoxygenation which led to an extreme scarcity of fish, but that is unknowable. The results from the lake confirm that a plant-dominated state in summer can exist over a very wide range of nutrient concentrations (Figure 2) and that grazing by zooplankters is an important component in maintaining this state. Indeed the results suggest a considerable resilience of the community.

The behaviour of Rostherne Mere is equally interesting. The lake at first appeared not to respond at all to a major fall in external loading of phosphorus on it. It may be argued that this is inevitable if the phytoplankton is limited by some other factor such as light (Reynolds & Bellinger, 1992) or nitrogen (Moss et al., 1994). In the latter circumstances a response in phytoplankton would require a very severe fall in phosphorus loading to establish P limited conditions. Equally it can be argued that the relatively long retention time of the lake (1.6–1.8 yr) may mitigate against rapid fall in total phosphorus concentrations in the water. The existence of nitrogen limitation may reflect mechanisms which maintain the abundance of phosphorus. Nitrogen concentrations in the lake are comparatively low but not absolutely so. The existence of a surface inflow and outflow in Rostherne Mere result in its having a shorter hydraulic retention time than some solely groundwater-fed deep meres.

In most deep lakes, a number of mechanisms favours net loss of phosphorus from the water, such that a reduction in loading is followed by a response in the lake. These include sedimentation of phosphorus-containing particles (phytoplankton cells and zooplankton faeces) through the thermocline to the hypolimnion, and then washout or reprecipitation in autumn of any such phosphorus returned to the surface when the lake mixes. During the mixed period, phosphorus is bound into the sediment as oxidised complexes of iron and manganese. It is possible that one or more of these three mechanisms (sedimentation, washout, reprecipitation) is not operating in White Mere in contrast to an apparent functioning of them in Rostherne Mere.

The thermal stratification in Rostherne Mere is typical of warm monomictic lakes (Carvalho, 1993) and sedimentation of particles must be inevitable. Water flows do increase in autumn and washout must occur, but retention time is high and the washout may not be vigorous. In such a case, reprecipitation of P from the oxygenated water must be expected. A shortage of iron could inhibit this but there is no reason to expect an abnormally low input to the lake. Even if all the available iron is tied up in complexes by the large existing phosphorus concentrations, even in winter, this would be a consequence rather than a cause of the high phosphorus concentrations. Concentrations remain high also in summer and thus, despite a slow downward trend following effluent diversion, a mechanism preventing loss to the hypolimnion or one returning P to the epilimnion must be sought. One such is aerobic release of phosphorus from the marginal sediments in contact with the epilimnion and another the abundant populations of vesiculate cyanophytes and the dinoflagellate, *Ceratium hirundinella* (Muell.) Duj., which might transport P during their movements in the water column (Ganf & Oliver, 1982).

A case might thus be made that the slow response of phosphorus concentrations in Rostherne Mere to reduction in external loading, has been due to efficient recycling of sediment and hypolimnion P in summer, and low washout and chemical reprecipitation in autumn and winter. In such circumstances, the lake may never achieve very low total phosphorus concentrations and if a reduction in chlorophyll concentration is required, reduction in the nitrogen load may be needed. This is potentially likely to be frustrated by the presence of nitrogen fixing cyanophytes in the lake (Schindler, 1977).

The hypothesis that light has been a major limiting factor to phytoplankton production in Rostherne Mere, (due to deep mixing at least in spring), is challenged by the palaeolimnological evidence that the algal crops have increased in recent decades. Light may limit growth rates but is unlikely to limit the biomass except when this is so high as to be self-shading. Light limitation will then have become a consequence rather than a cause of the size of the algal crops. It may thus be that light is now sometimes limiting but only because nutrients have increased. Early data (Grimshaw & Hudson, 1970) suggest that phosphorus has been abundant for many years and thus the likely driving nutrient is currently nitrogen from increased intensification of agriculture in the catchment.

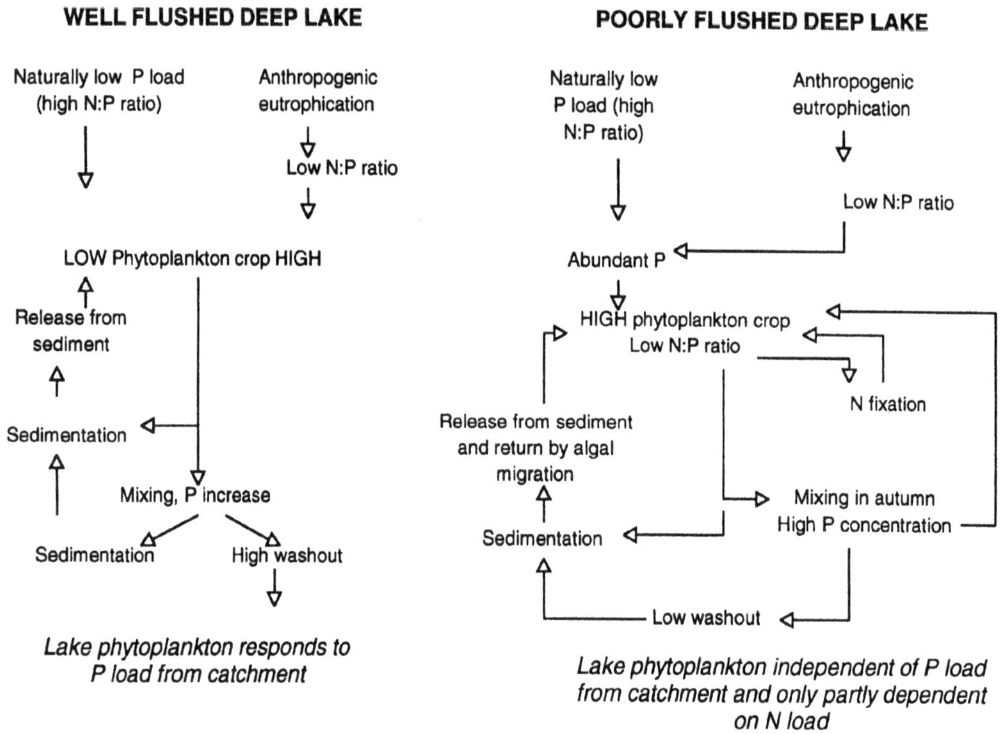

Figure 5. A model for phosphorus flux in deep stratified lakes with a short retention time and similar lakes with long retention times.

Contrasts between poorly- (White Mere) and well-flushed (Rostherne Mere) deep lakes, suggested by these results, are illustrated in Figure 5. The implication of Figure 5 is that in poorly-flushed, deep, stratifying lakes, mechanisms may exist that concentrate phosphorus and lead to maintenance of high concentrations. Central to these mechanisms is the existence of migratory algae or effective mixing from hypolimnion to epilimnion during summer, aerobic sediment release in the sediments in contact with epilimnion water, and low flushing rates. This combination of circumstances may, but need not, involve anthropogenic eutrophication. A low natural load from an undisturbed catchment could be progressively concentrated over many years. Many of the meres have similar circumstances to White Mere (Reynolds, 1971, 1979; Kilinc, 1995; Moss et al., 1994), with very high phosphorus concentrations, direct thermal stratification in summer, abundant vesicuolate cyanophytes or *Ceratium*, and often long retention times because they are fed mostly by ground water. With its shorter retention time, Rostherne Mere appears slightly unusual and this may explain its relatively lower total P concentrations.

Perhaps in the meres, therefore, there is a challenge to the orthodoxies that temperate lakes always respond to the phosphorus loading from their catchments and that eutrophication is always anthropogenic. In the deep meres, mechanisms for the concentration of phosphorus may have led to a switch from a former natural phosphorus limitation to a natural nitrogen limitation. This may have been mitigated to some extent by production of nitrogen-fixing cyanophytes. Thence there could have been a change from determination of the state of the lake by the catchment to an internal determination resilient to changes in the catchment. This is not to say that the orthodoxies do not apply to a wide range of other lakes, but in the vast glacial plains of the northern hemisphere, there must be many lakes deep enough to stratify and with long retention times, the apparent key features for establishment of the peculiarities of the deeper North West Midland meres.

Acknowledgments

We are grateful to English Nature, the Natural Environment Research Council (UK), and the Turkish Government for support, in various ways of this work.

References

Anderson, N. J., 1995. Naturally eutrophic lakes: reality, myth or myopia? T.R.E.E. 10: 137–138.

Beklioglu, M., 1995. Whole lake and mesocosm studies on the roles of nutrients and grazing in determining phytoplankton crops in a system of shallow and deep lakes. PhD Thesis, University of Liverpool, 425 pp.

Beklioglu, M. & B. Moss, 1995. The impact of pH on interactions among phytoplankton algae, zooplankton and perch (*Perca fluviatilis*) in a shallow, fertile lake. Freshwat. Biol. 33: 497–509.

Brinkhurst, R. O, & B. Walsh, 1967. Rostherne Mere, England: a further instance of guanotrophy. J. Fish. Res. Bd Can. 24: 1299–1309.

Carvalho, L. R., 1993. Experimental limnology on four Cheshire meres. PhD thesis University of Liverpool, 385 pp.

Carvalho, L. R., 1994. Top-down control of phytoplankton in a shallow, hypertrophic lake Little Mere, England. Hydrobiologia 275/276: 53–63.

Carvalho, L. R., M. Beklioglu & B. Moss, 1995. Changes in a deep lake following sewage diversion – a challenge to the orthodoxy of external phosphorus control as a restoration strategy. Freshwat. Biol. 34: 399–410.

Ganf, G. G. & R. L. Oliver, 1982. Vertical separation of light and available nutrients as a factor causing replacement of green algae by blue green algae in the plankton of a stratified lake. J. Ecol. 70: 829–844.

Griffiths, B. M., 1925. Studies on the phytoplankton of the lowland waters of Great Britain. III The phytoplankton of Shropshire, Cheshire and Staffordshire. Bot. J. Linn. Soc. Lond. 47: 75–92.

Grimshaw, H. M. & M. J. Hudson, 1970. Some mineral nutrient studies of a lowland mere in Cheshire, England. Hydrobiologia 36: 329–341.

Kilinc, S., 1995. Limnological studies on the North West Midland meres, with special reference to Whitemere. PhD Thesis, University of Liverpool, 264 pp.

Livingstone, D., 1979. Algal remains in recent lake sediment. PhD thesis, University of Leicester, 157 pp.

Marsden, M. W., 1989. Lake restoration by reducing external phosphorus loading: the influence of sedimentary phosphorus release. Freshwat. Biol. 21: 139–162.

Moss, B., 1983. The Norfolk Broadland: Experiments in the restoration of a complex wetland. Biol. Rev. 58: 521–561.

Moss, B., 1989. Water pollution and the management of ecosystems: a case study of science and scientist. In Grubb, P. J. & R. H. Whittaker (eds), Toward a More Exact Ecology. Thirtieth Symposium of the British Ecological Society. Blackwell Scientific, Oxford; 401–422.

Moss, B., 1991. The role of nutrients in determining the structure of lake ecosystems and implications for the restoring of submerged plant communities to lakes which have lost them. International Conference on N, P and Organic matter. Contributions by invited international experts. National Agency for Environmental Protection, Copenhagen, Denmark: 75–86.

Moss, B., S. McGowan & I. Carvalho, 1994. Determination of phytoplankton crops by top-down and bottom-up mechanisms in a group of English lakes, the West Midland meres. Limnol. Oceanogr. 39: 1020–1029.

Moss, B., S. McGowan, S. Kilinc & L. R. Carvalho, 1992. Current limnological condition of a group of the West Midland meres that bear SSSI status. English Nature, Peterborough, 320 pp.

Nelms, R., 1984. Palaeolimnological studies of Rostherne Mere (Cheshire) and Ellesmere (Shropshire). PhD Thesis, Liverpool Polytechnic, 128 pp.

Ozimek, T., E. van Donk & R. Gulati, 1990. Can macrophytes be useful in biomanipulation of lakes. Hydrobiologia 200/201: 399–409.

Phillips, G. L., D. F. Eminson & B. Moss, 1978. A mechanism to account for macrophyte decline in progressively eutrophicated freshwaters. Aquat. Bot. 4: 103–126.

Phillips, W., 1884. The breaking of the Shropshire meres. Trans Shrops. Arch. nat. Hist. Soc. 7: 277–300.

Reynolds, C. S., 1971. The ecology of the planktonic blue green algae in the North Shropshire meres, England. Field Stud. 3: 409–432.

Reynolds, C. S., 1979. The limnology of the eutrophic meres of the Shropshire – Cheshire plain – a review. Field Stud., 5: 93–173.

Reynolds, C. S. & E. G. Bellinger, 1992. Patterns of abundance and dominance of the phytoplankton of Rostherne Mere, England: evidence from an 18-year data set. Aquat. Sci. 54: 10–36.

Sas, H., 1989. Lake restoration by reduction of nutrient loadings: expectations, experiences, extrapolations. Academia Verlag Richarz, Sant Augustin, 497 pp.

Scheffer, M., S. H. Hosper, M.-L. Meijer, B. Moss & E. Jeppesen, 1993. Alternative equilibria in shallow lakes. T.R.E.E. 8: 275–279.

Schindler, D. W., 1977. The evolution of phosphorus limitation in lakes. Science 195: 897–898.

Sinker, C. A., 1962. The North Shropshire Meres and Mosses; a background for ecologists. Field Stud. 1: 101–138.

Tallis, J. H., 1973. The terrestrialisation of lake basins in North Cheshire with special reference to the development of a 'Schwingmoor' structure. J. Ecol 61: 537–567.

Webb, M., 1924. Precious Bane. Jonathan Cape, London, 320 pp.

Hydrobiologia **342/343**: 269–284, 1997.
L. Kufel, A. Prejs & J. I. Rybak (eds), Shallow Lakes '95.

Why 'very shallow' lakes are more successful opposing reduced nutrient loads

Brigitte Nixdorf & Rainer Deneke
Technical University Cottbus, Faculty Environmental Sciences, Water Protection, Seestr. 45, D-15526 Bad Saarow, Germany

Key words: shallow lakes, eu-hypertroph, Cyanobacteria, morphometry, mixing, euphotic depth

Abstract

There are different approaches for classifying deep and shallow waters using physically and ecologically derived parameters. Nevertheless, transition states make it difficult to define border crossing points between the two types of limnetic ecosystems and to distinguish more precisely between different types of shallow, especially highly eutrophicated lakes. We contribute a detailed analysis of different characteristics of shallow waters from lakes in the Berlin/Brandenburg-region. In the catchment area of the river Dahme in Eastern Brandenburg (Scharmützelsee-region) we find mainly shallow and highly eutrophicated lakes, dominated by Cyanobacteria. 'Very shallow' lakes of different morphometry and topography are compared with 'medium shallow' or deeper lakes in the region with similar loading characteristics for the following properties: morphometry, topography, theoretical retention time, mixing intensity, nutrient dynamics, external and internal loading, underwater light climate, z_{eu}/z_{mix}, phytoplankton development and oxygen budget. We found that 'very shallow' lakes in the region are more efficient in converting the available phosphorus into phytoplankton biomass because of the constant and sufficient underwater light climate due to the favourable relation of z_{eu} and z_{mix}. We conclude that the regular mixing regime guarantees a stable and near optimum light/dark rhythm as well as higher heterotrophic activities, stimulating primary production up to the upper limit of algal development.

Introduction

North Germany and Northern Poland have landscapes dominated by lakes, and lowland rivers connecting these lakes. Most of these waters are shallow and especially endangered by eutrophication, because of their relatively large catchment areas in an agricultural region. There have been several reports about the high degree of eutrophication, the reasons and consequences in the Berlin and Brandenburg region since the beginning of this century (Thienemann, 1932; Czensny, 1938; Wundsch, 1940; Ohle, 1953; Scharf, 1971; Klein, 1989; Behrendt et al., 1990; Behrendt & Nixdorf, 1993a; Chorus, 1995)

The lakes described in this paper are situated in Eastern Brandenburg (Germany, Scharmützelsee-region), where lakes cover about 12% of the territory. They belong to a chain of lakes in the catchment area of the River Dahme, southeast of Berlin (Figure 1). Most waters in Brandenburg and Berlin are eutrophic or hypertrophic (Mietz, 1994; Klapper, 1992), but although these lakes have similar loads of nutrients, they show quite different symptoms of eutrophication. Most are dominated by Cyanobacteria during summer and partly during autumn and winter (Nixdorf et al., 1995). Only one lake, at the beginning of the so-called Glubigsee-chain, is mesotrophic. The succeeding eight lakes are eutrophic to hypertrophic and different in topography and morphometry. In shallow lakes with a high trophic state the development of filamentous Cyanobacteria has been often recorded (Barica, 1980; Berger, 1984; Berger & Sweers, 1988; Rojo & Cobelas, 1994).

The in-lake TP-concentration in the investigated area is now between 50 and 120 μg P l^{-1}, a range in which lakes are reaching a trophic threshold and

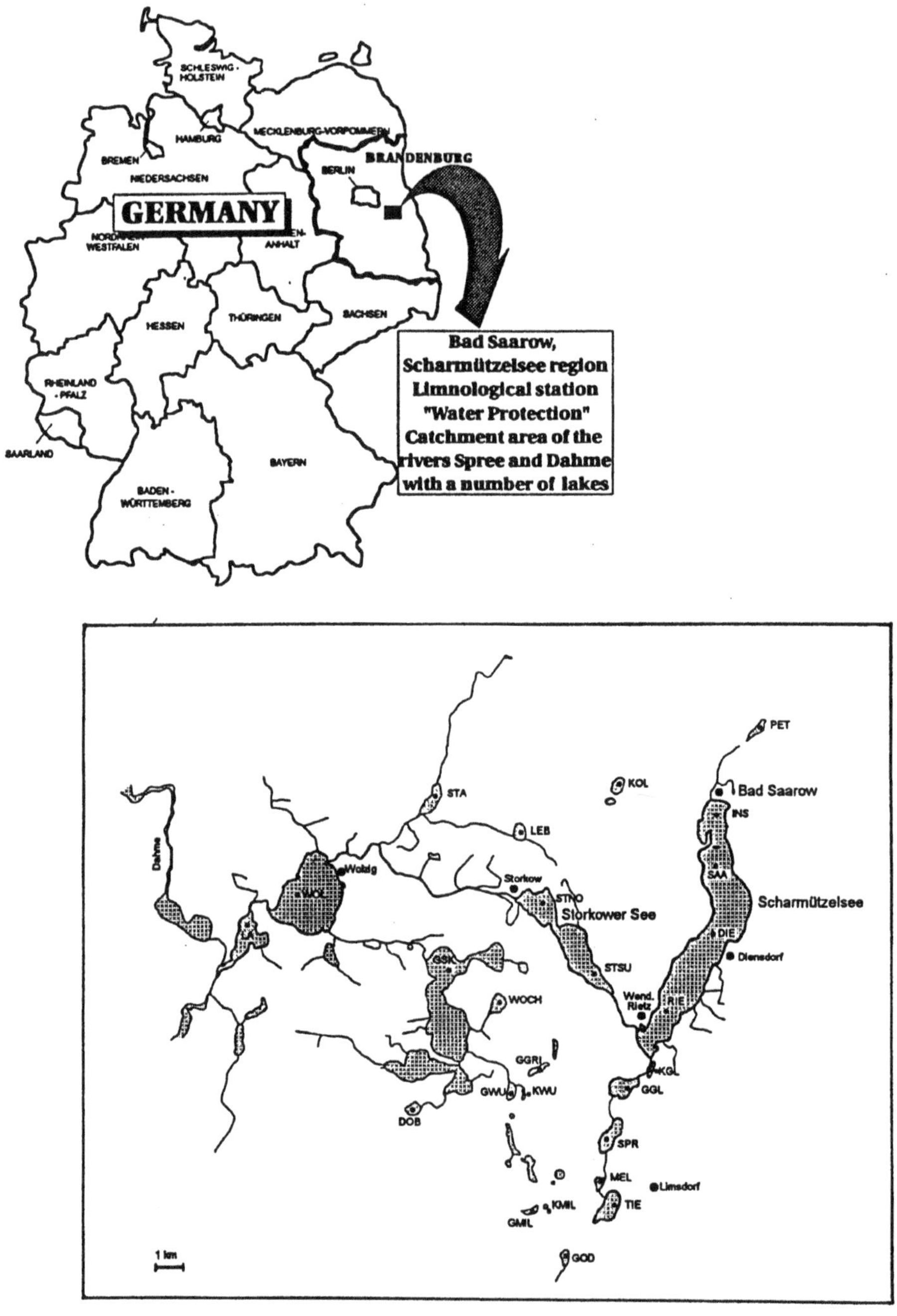

Figure 1. Map of the lakes in the Scharmützelsee-region (Brandenburg, Germany).

a shift in the planktonic succession can be expected (Sas, 1989; Chorus, 1995). Within this range, a great variability of phytoplankton responses can be observed depending on the degree of nutrient or light resource limitation of photosynthesis, respectively (Reynolds, 1992a). Several influences like internal load by phosphorus release from the sediment or unknown matter input by groundwater are the reasons for an apparent unchanged trophic situation despite restorative measures or catchment changes not only in this region.

In eutrophication management, limitation of phytoplankton biomass is of concern (Sommer, 1993). The amount of biomass per unit incorporated nutrient can be quantified by the chlorophyll/phosphorus relationship. A ratio around one means that the algal biomass is limited by phosphorus and no further development of more biomass is possible (Reynolds, 1992b; Chorus, 1995). The lakes, especially the shallow ones, presented in this paper are classified on the basis of in-lake nutrient concentration. Nevertheless, the intensity and time course of algal development and the utilization of the available phosphorus present complications. One aim of this study is to investigate the structural basis of lake productivity by comparison of morphometry and topography of different lakes.

Description of sites studied

The Scharmützelsee-region is named after one of the greatest and deepest lakes in Brandenburg, Lake Scharmützelsee. It covers about 400 km^2 (57° 20′N, 54° 35′E) in Eastern Brandenburg (Figure 1). On the Northern shore of the Scharmützelsee in Bad Saarow, the Water Protection Research Station is located about 60 km East from Berlin. The landscape around Bad Saarow is characterized by a number of lakes, which are connected by rivers or channels. Some of the lakes get a high portion of their water from groundwater (e.g. Springsee). The catchment area is dominated by forests and agriculture with increasing interest as a centre of tourism and for navigation. Most of the lakes originate from the Weichsel glacier and are located at the border of the southern plateau of the Berlin glacial valley. A detailed description of the geological background is given by Wundsch (1940) and Müller (1952). Wundsch (1940) was the first scientist to investigate a number of lakes in the region on the basis of a complex limnological approach. He conceived a specific type of lake in the catchment area of the rivers Spree and Havel, the 'H_2S-*Oscillatoria*-lakes', because of the dominance of the bluegreens *Oscillatoria* and their propensity to lyse. The consequences of this type of eutrophication have been described as the '*Oscillatoria* disease' by Czensny (1938) and Schäperclaus (1941).

From the 1960s to the end of the 1970s the nutrient load on the lakes increased due to intensification of agriculture and fish management and insufficient waste water treatment. The external nutrient loads have been decreasing since the beginning of this decade. Nevertheless, no significant changes in phytoplankton development have been yet observed because of the high internal P-load in almost all lakes.

The investigated lakes differ in maximum depth from 2.5 to 30 m, and nine out of twelve lakes are shallower than 13 m (Table 1). Area differences between lakes are much greater, ranging from 0.06 to 12.09 km^2. Nine lakes have a simple basin shape, three others (TIE, GGL, SCH) are structurally more complex. Other than the Scharmützelsee, the shape of the shoreline is mostly quite simple and the shore development (Håkanson, 1981) is very close to a circle or ellipse ranging from 1.08 to 1.56. Eight lakes are polymictic, and only four of them are dimictic with a long-lasting summer stratification. The intensity of polymixis varies between lakes. All lakes are now dominated by plankton, even the very small ones (MEL, KGL). The transparency at the beginning of the growth season is too low to allow the establishment of a macrophyte regime. The lake features presented here can be considered as typical for the situation in Eastern Brandenburg, where most of the lakes are in a similar size and depth range and are also highly eutrophicated.

Materials and methods

Morphometric and topographic estimations

Since summer 1993 following lakes have been investigated: Tiefer See (TIE), Melangsee (MEL), Springsee (SPR), Großer and Kleiner Glubigsee (GGL, KGL), Scharmützelsee (SCH), Storkower See (Southern and Northern basin: STSU, STSN), Wolziger See (WOL), Langer See (LAN), Petersdorfer (PET) and Lebbiner See (LEB) (Figure 1).

All topographical and morphometrical parameters were calculated from bathymetric maps, taken from the literature (Storkower See, Wolziger See, Langer See, Scharmützelsee are based on Wundsch, 1940; Großer Glubigsee, Springsee were drawn by Müller, 1952). With the exception of the Lebbiner See and

Table 1. Topographical and morphometrical parameters of 12 lakes in the Scharmützelsee-region (A: total lake area, V: total lake volume, L_{max} B_{max}: maximum length/width, l: shoreline length, F: shore development, t_r: theoretical retention time, C_A: catchment area, for other parameters see Table 2).

Lakes	Abbrv.	Mixis	A (km²)	V (10^6 m³)	L_{max} (km)	B_{max} (km)	l (km)	F	z_{max} (m)	z_{mean} (m)	z_{red}	t_r (d, a)	C_A (km²)
Tiefer See	TIE	di	0.63	–	1.39	0.59	3.60	1.37	22.9	–	0.81	–	–
Melangsee	MEL	poly	0.11	0.17	0.51	0.31	1.38	1.18	3.3	1.6	0.18	22 d	5
Springsee	SPR	di	0.59	5.56	1.50	0.67	3.64	1.34	18.1	9.5	0.65	1.1 a	12
Großer Glubigsee	GGL	di	0.64	2.64	1.31	0.80	3.57	1.26	12.0	4.2	0.42	0.5 a	16
Kleiner Glubigsee	KGL	(poly)/di	0.06	0.14	0.57	0.13	1.34	1.56	4.4	2.4	0.28	8 d	17
Scharmützelsee	SCH[1]	di	12.09	108.23	10.00	1.50	26.93	2.18	30.0	9.0	0.51	16 a	112
Storkower See (South)	STSU	poly	2.33	8.79	3.27	1.15	8.33	1.54	8.0	3.8	0.20	1.5 a	168
Storkower See (North)	STSN[2]	(poly)/di	1.39	6.59	1.63	1.12	5.45	1.30	11.4	4.7	0.33		
Wolziger See	WOL	poly/di	5.79	32.02	3.25	2.40	10.12	1.19	13.0	5.5	0.27	–	382
Langer See	LAN	poly	1.55	3.27	2.14	1.40	6.61	1.50	3.8	2.1	0.11	–	395
Petersdorfer See	PET	poly	0.23	0.53	1.11	0.29	2.62	1.54	4.0	2.3	0.18	–	–
Lebbiner See	LEB	poly	0.28	0.58	0.71	0.55	2.03	1.08	4.0	2.1	0.17	–	–

[1] With 3 sampling stations (RIE, DIE, SAA). [2] Equivalent to STNO.

Petersdorfer See, the remaining lakes (Melangsee and Kleiner Glubigsee) were surveyed by ourselves. For the Tiefer See a bathymetric map is not yet available, but some parameters (Table 1) could have been derived from the topographic map (1:10 000, Nr. 0909-422) and knowledge of z_{max} Volumes were calculated with a layer model based on isobaths, further parameters are calculated according to the formulae in Table 2.

Physical, chemical and biological analyses

Water samples for chemical and biological analysis were taken monthly or biweekly as mixed samples from the completely mixed waters or from epi- and hypolimnion with a 2.3 l-LIMNOS-sampler. Sampling stations were above the deepest points and if necessary also at other points (Nixdorf et al., 1995), where also Secchi depth was estimated. Depth profiles of temperature, oxygen (concentration and percentage saturation), pH, redox and conductivity were measured at 0.5 m intervals by a HYDROLAB 20 probe connected to a field computer (HUSKY Hunter). The attenuation was calculated from light measurements every 50 cm under water with two spheric quantum sensors (Licor) with a distance of 39 cm between them to minimize the effect of disturbance of the incident light by clouds and waves.

Euphotic depth (z_{eu}) was calculated from the attenuation coefficient assuming its lower limit to be at 1% of incident radiation.

Chemical analysis of different fractions of nutrients, further anions and cations, DOC and metals were determined according to standard methods (Deutsche Einheitsverfahren, 1986–1993). Phytoplankton was counted after fixation with Lugol's solution by the conventional Utermöhl technique (Utermöhl, 1958). Biovolume was calculated by measuring the corresponding dimensions and size classes of algae (Nixdorf & Hoeg, 1993). Chlorophyll *a* and chemical parameters were determined according to the standard methods. Since 1995 the community respiration (20–24 hours) has been measured in situ (oxygen bottles) using Winkler titration.

Results

Morphometry, mixing regime and criteria to distinguish deep and shallow lakes

A simple approach to distinguish between deep and shallow lakes is to group them according to their maximum depth (z_{max}). As a first coarse classification, the lakes considered here can be divided into 3 groups: first, a group of shallow lakes ($N=5$) with z_{max} smaller than 5 m, second, an intermediate group ($N=4$) with z_{max} ranging between 5 and 15 m, and a third group of deep lakes ($N=3$) with z_{max} deeper than 15 m. Mixing regimes between the first and last group differ substantially compared to the variability within groups. The first group is usually polymictic with the exception of

Table 2. Parameters used for topographical, morphometrical and hydrological characterisation of the investigated lakes (mostly according to Håkanson, 1981).

Parameter	Abbreviation	Criterion/Formula
Effective axis length	D_a (m)	$(L_e + B_e) * 0.5$
Maximum effective length or width	L_e, B_e (km)	maximum length/width not interrupted by land
Maximum depth	z_{max} (m)	–
Mean depth	z_{mean} (m)	V/A
Mixing depth	z_{mix} (m)	observed depth, where $\Delta T \leq 0.5\ °C\ m^{-1}$
Depth of summer thermocline	z_{th} (m)	mean depth of the thermocline in July - September
Median depth	z_{med} (m)	depth of 50% lake surface area, taken from the percentage hypsographic curve
Depth of 90% lake volume	z_{90} (m)	depth of 90% lake water volume, taken from the percentage volume curve
Theoretical epilimnion depth (Ventz, 1974)	z_{epi} (m)	$(D_a*10^{-3})^{0.28}*5.81$
Reduced depth	z_{red}	$z_{max}*[(A*10^6)^{0.25}]^{-1}$
Euphotic depth	z_{eu} (m)	depth of 1% surface irradiance

the Lake Kleiner Glubigsee, which is relatively stable stratified. All lakes of the third group are dimictic. For the lakes of the intermediate group z_{max} as a parameter is crucial, because one is polymictic (STSU), two are dimictic (GGL, STSN), and one lake (WOL) showed interannual variation between mixing regimes. As many lakes in Northern Germany must be classified as 'intermediate' according to z_{max}, it is worth seeking better criteria to differentiate more precisely between the different types of lakes. A classification of the lakes according to their mean depth (z_{mean}) yields no further information. This is due to the fact that z_{mean} and z_{max} are closely related to each other, as has been shown for 60 lakes in Canada (Chow-Fraser, 1991).

A new approach must consider differences in lake area of more than one magnitude as well as differences in lake basin morphometry. The basic relationship between lake area and mixing depth or depth of the thermocline in summer is well established (Hanna, 1990). Ventz (1974) calibrated this relationship for the Brandenburg lakes region (z_{epi}, Table 2). Calculated values of the theoretical epilimnion depth (z_{epi}), which are based only on length measurements (effective axis length, D_a), are in good agreement with the depth of the thermocline (z_{th}) recorded from July to September in 1993 and 1994 (Figure 2). Lake basin morphometry is introduced in this new classification by comparing the predicted value of z_{epi} to z_{90}, the depth which contains 90 percent of the lake water volume. The depth

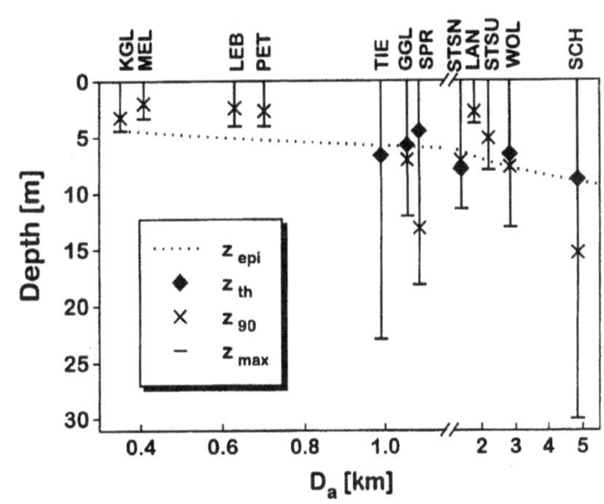

Figure 2. Different depths characterising morphometric and topographic properties of the lakes (for explanation see Table 2).

z_{90} can be derived from the percentage volume curve (Håkanson, 1981), based on the bathymetric map.

Applying these criteria to the four lakes of the intermediate group, it seems reasonable to separate the Storkower See (South) from the others, because z_{max} is very close to z_{epi} (Figure 2) and wind induced polymixis seems likely, which parallels its higher degree of polymixis in the last few years. For the Wolziger See (WOL) the z_{90} value is very close to z_{epi}, because of its particular basin shape. Thus wind-induced water movements probably can easily mix up to 90 percent of the lake volume (Figure 2). This may be the reason why a stable stratification in summer in the Wolziger See may only take place under certain climatic circumstances. Thus, mixing regime can vary between years, especially as the Wolziger See is known to be exposed to strong winds. For the two other lakes (STSN, GGL) differences between z_{90} and z_{epi} increase. The small Großer Glubigsee always shows a long-lasting summer stratification. The doubly large Storkower See (North) is also dimictic but tends to be more unstable. Lake features not quantified here may be responsible for the higher stability of the GGL, especially exposure to wind, which depends on the placement of the lake in the main wind direction (STSN) and on protection against wind by the landscape or shore vegetation (GGL). The stronger influence of wind on the STSN is reflected by the deeper position of the depth of the thermocline in respect to z_{epi} and z_{90}.

For a quick and simple classification of lakes the reduced depth (z_{red}), which is derived from z_{max} weighted by area, was used as an alternative parameter (Table 2). Scaling the lakes according to their reduced depth results in a better match with the observed patterns of mixing than the classification used so far (Table 2). There seems to be a line between poly- and dimictic lakes between 0.33 and 0.42 reduced depth.

Algal development in the investigated lakes during the growing seasons 1993 and 1994

The plankton regime of almost all investigated lakes was dominated by Cyanobacteria at least during middle and late summer until autumn in 1993/94 (Table 3). During 1993/94 a shift in the phytoplankton composition was observed from the *Oscillatoria* regime to the *Planktothrix* dominated community (Rücker et al., 1997). There were two cases when algae other than Cyanobacteria were dominant. In the Storkower See (North) and the Wolziger See, a shift from dominant Cyanobacteria to dominant Dinophyceae and vice versa was observed. One exception is the mesotrophic Lake Tiefer See at the beginning of the chain of lakes with a comparatively low algal biomass dominated by nanoplankton and diatoms. In spring 1994 Cyanobacteria (*Planktothrix agardhii*) constituted about 14% of algal biomass.

Lake Melangsee, is highly eutrophic and was dominated by Cyanobacteria, mainly *Aphanizomenon* spp. (1993), which occurred as filaments, not colonies, and *Oscillatoria* spp. (1994). In Lake Springsee the phytoplankton was dominated by *Oscillatoria limnetica* and other nonidentified *Oscillatoria* species (Rücker et al., 1997). The next lake of the chain is Lake Großer Glubigsee. Although the lake is eutrophic, it was not always dominated by Cyanobacteria but by dinoflagellates (*Ceratium*) and nanoplanktonic algae. Lake Kleiner Glubigsee has a similar algal community to Lake Springsee (*Limnothrix redekei* and *Planktothrix agardhii*). In Lake Scharmützelsee a dominance of *O. limnetica* and other very thin filamentous species of *Oscillatoria* (1.2 to 1.7 μm in diameter) occurred accompanied by *L. redekei* and *P. agardhii*. In Lake Storkower See (South) we observed large developments of diatoms (*Aulacoseira* and *Melosira*) and dinoflagellates (*Ceratium*) comprising 60 to more than 90% of the total biomass. Contrary to the phytoplankton composition in the south basin the algal community in the north basin was characterized by *Microcystis* as the main Cyanobacteria and different species of *Peridinium*. At the end of the chain of lakes, Lake Wolziger See and Langer See are occupied by Cyanobacteria with the dominant species *P. agardhii* and *L. redekei*. Dinophyceae (*Ceratium*) also had relatively high biomass in the Wolziger See. In the Langer See some species of *Anabaena* occurred but were infrequent.

Although our interest was focused on the phytoplankton composition during the main growth period, samples from late autumn and winter were also analysed. In some lakes considerable amounts of algae were found during the cold season, sometimes exceeding the summer algal biomass (on a water column basis). Values were in the range of 70 to 250 mg Chl *a* m^{-2} in Lake Großer and Kleiner Glubigsee, Springsee, Melangsee, Petersdorfer See, and both basins of Storkower See.

Trophic characterization

The concentration of total phosphorus (annual mean 1994, Figure 3) in the investigated lakes was between 27 and 124 μg P l^{-1}. The mass ratio of TN to TP was always >7. Thus, no nitrogen limitation of the biomass production was expected (Table 4). The concentration of SRP in the mixed column or in the epilimnion was very low. During the main growth period in 1994 the concentration was less than 5 μg P l^{-1} in some lakes (Tiefer See, Scharmützelsee, Springsee, Großer Glubigsee, Melangsee and Langer See). After a more or less marked clear-water phase the SRP content was below 0.5 and 1 μg P l^{-1} in some of the lakes. The SRP content in the waste water loaded Lake Lebbiner

Table 3. Dominant phytoplankton groups and species and the range of algal biomass in the lakes during summer and late summer 1993/94

Lakes	Range of algal biomass	Dominant phytoplankton groups	Dominant taxons
MEL (Melangsee)	11–30 mg l^{-1}	Cyanobacteria (50–86%) Nanoplankton (7–23%)	*Aphanizomenon* spp. (1993: 72–80%) *Oscillatoria* spp. (1994: 69–88%)
SPR (Springsee)		Cyanobacteria	*Limnothrix redekei* *Oscillatoria limnetica*
KGL (Kleiner Glubigsee)	13–18 mg l^{-1}	Cyanobacteria (75–83%)	*Limnothrix redekei* and related species *Aphanizomenon* spp.
SCH (Scharmützelsee)	2–15 mg l^{-1}	Cyanobacteria (76–83%) Dinophyceae (7–10%) Nanoplankton (5–10%)	*Oscillatoria limnetica* *Ceratium* spp.
STSU (Storkower See, South)	9–21 mg l^{-1}	Cyanobacteria (32–40%) Dinophyceae (14–60%) Diatoms (1–32%)	*Planktothrix agardhii* *Aphanizomenon* spp. *Anabaena* spp. *Peridinium* spp.
STSN (Storkower See, North)	7–13 mg l^{-1}	Cyanobacteria (32–44%) Dinophyceae (26–52%) Diatoms (3–13%)	*Planktothrix agardhii* *Aphanizomenon* spp. *Anabaena* spp. *Ceratium* spp.
WOL (Wolziger See)	11–20 mg l^{-1}	Cyanobacteria (16–83%) Dinophyceae (1–60%) Nanoplankton (5–14%)	*Planktothrix agardhii* *Limnothrix redekei* *Aphanizomenon* spp. *Ceratium* spp. *Peridinium* spp.
LAN (Langer See)	28–61 mg l^{-1}	Cyanobacteria (65–89%) Nanoplankton (2–11%) Dinophyceae (2–18%)	*Planktothrix agardhii* (1993: 52–87%) *Aphanizomenon* spp. *Limnothrix redekei*
PET (Petersdorfer See)	17–47 mg l^{-1}	Cyanobacteria (73–92%) Cryptophyceae (2–10%) Nanoplankton (3–9%)	*Limnothrix redekei* and related species
LEB (Lebbiner See)	33–41 mg l^{-1}	Cyanobacteria (90–96%) Nanoplankton (2–4%)	*Aphanizomenon* spp. *Planktothrix agardhii* *Cylindrospermopsis* spp.

Table 4. Annual means of total and soluble reactive phosphorus (italic data indicate SRP concentration during the growth period lower than 5μg P l^{-1}), total nitrogen, chlorophyll, Secchi depth and ratios between Chl*a* and TP and z_{eu} and z_{mix} in the lakes investigated in 1994 (abbreviations see Table 1).

Lake	TP (μg P l^{-1})	SRP (μg P l^{-1})	TN (mg N l^{-1})	Chl *a* (μg l^{-1})	SD (m)	Chla/TP	z_{eu}/z_{mix}
TIE	27	*4*	1.37	5	6.1	0.19	2.1
SCH	60	*16*	1.66	21	1.7	0.35	0.6
SPR	50	*11*	1.98	25	1.6	0.50	0.6
GGL	70	*9*	2.43	30	1.2	0.43	0.4
STSU	77	9	2.18	37	1.0	0.48	0.4
STSN	121	8	2.35	37	1.0	0.31	0.4
WOL	122	15	3.20	46	1.1	0.38	0.45
KGL	64	7	2.23	47	1.1	0.73	0.9
MEL	68	*4*	2.63	57	0.9	0.84	1.1
LEB	111	27	3.83	71	0.9	0.64	0.8
PET	54	7	4.04	83	0.6	1.54	0.6
LAN	124	*5*	2.81	111	0.5	0.90	0.55

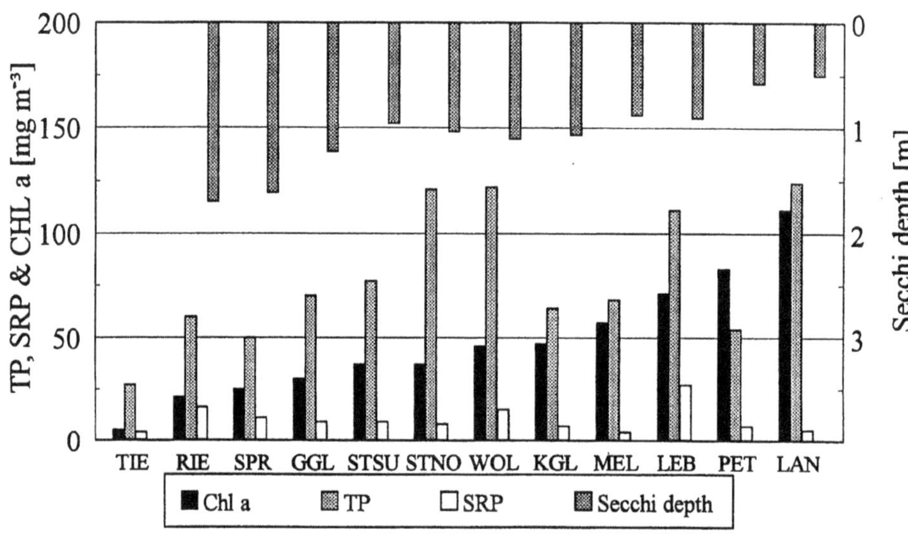

Figure 3. Annual means for total and soluble reactive phosphorus concentration, chlorophyll *a* content and Secchi depth in the lakes of the Scharmützelsee-region (1994).

See was the highest of all with 27 μg P l^{-1} (annual mean).

Monthly averages of total phosphorus and chlorophyll *a* for the investigated lakes in the Scharmützelsee-region are shown in Figure 3. Considering either, all data or annual means, a distinct trend is obvious: the in-lake TP concentrations are in the eutrophic or at the beginning of the hypertrophic range for the majority of the lakes. Nevertheless, there are some lakes with a higher chlorophyll concentration and a lower Secchi depth. This is especially obvious for two pairs of lakes: (a) both basins of Storkower See and (b) Wolziger See and Langer See (Figure 4). The first two have the same mass development of algae (37 μg Chl *a* l^{-1}), although the TP content in the Southern basin (STSU) is only 77, but in the deeper Northern basin 121 μg P l^{-1}. Lake Wolziger See and Langer See have the same TP content as the north basin of Storkower See. Wolziger

See has a little higher phytoplankton biomass (46 μg Chl *a* l^{-1}) but the algal development in Lake Langer See is more than twice as high as in Wolziger See and Storkower See (North). This phenomenon is illustrated in Figure 3 by the increase in algal development (annual means 1994) for all lakes arranged according to the chlorophyll concentration. Two facts are obvious:

(a) within the range of the observed TP concentrations the phytoplankton response is irregular and nonlinear reflecting the high fluctuation in the range from eutrophic to hypertrophic systems

(b) the efficiency of converting phosphorus (as TP) into phytoplankton biomass (as chlorophyll) depends on the shallowness of the lakes. There is a tendency to a higher efficiency (expressed as the Chl *a*/TP ratio, Table 4) in the more shallow lakes. The annual means for Chl *a*/TP relationship in the mesotrophic Lake Tiefer See is 0.19, in the 'medium shallow' lakes (MS: GGL, STSU, STSN, WOL) is 0.41 and in the 'very shallow' lakes (VS: KGL, MEL, LEB, PET, LAN) 0.93. In Figure 5 these relationships are demonstrated by the regression lines for the 'very shallow' (VS) lakes and the 'medium shallow' (MS) lakes. Because the reasons for this phenomenon are not to be sought in nutrient supply, the analysis will now focus on a further necessary resource – the light climate in the different lakes.

Underwater light climate and relation of z_{eu} and z_{mix}

During the main growth period of 1994 (April to October) there were only two lakes (Tiefer See and Melangsee) with a euphotic depth layer equal or even greater than the mixed layer ($z_{eu}/z_{mix}=2.1$ or 1.1, respectively). All the other lakes had a worse underwater light climate. Summarizing all data on the relation between euphotic and mixed depth distinction in two groups, according to the z_{eu}/z_{mix} ratios, is possible as previously shown. The lake with the clearest water (Tiefer See) is very well supplied with light. The same is true for the turbid and 'very shallow' Lake Melangsee which receives more than 1% of the surface radiation at the bottom. The z_{eu}/z_{mix} ratio in the 'medium shallow' lakes (MS) is between 0.4 and 0.5 (averaged during the growth season) and the mean of these values is 0.41. The equivalent ratio in the 'very shallow' lakes is between 0.55 and 1.1 (mean = 0.79) indicating a better under water light supply in these lakes.

Table 5. Community respiration during summer 1995 from epilimnion, hypolimnion or mixed samples of lakes in the Scharmützelsee-region (abbrevations see Table 1),

Lake	Pelagic R_{com} (mg O_2 l^{-1} d^{-1})	Average	R_{com} in hypolimnion (mg O_2 l^{-1} d^{-1})
TIE	0.28	0.28	0.47
SCH	1.04		0.55
SPR	0.93		0.80
GGL	1.13	1.04	2.22
STSU	0.98		1.08
STSN	0.95		1.49
WOL	1.20		1.60
KGL	1.38		4.92*
MEL	2.15		
LEB	1.81	1.81	
PET	1.93		
LAN	1.80		

* These values were obtained from samples taken 1 m above the sediment

Community respiration and oxygen budget

During warm and calm periods, almost all lakes may stratify slightly and as a result, oxygen is often consumed in the hypolimnion or above the sediment, causing high internal loading of phosphorus. From March 1995 on we were able to measure the oxygen consumption by the pelagic community in all the lakes. Because of the close relationship between temperature and metabolic activities we calculated 'early summer respiration' (from May to August) as an indicator for heterotrophic activity. The results of these measurements are given in Table 5.

The 'very shallow' lakes obviously have much higher respiration rates (1.81 mg O_2 l^{-1} d^{-1} on average) compared to the deeper lakes (1.04 mg O_2 l^{-1} d^{-1}). One exception is the Lake Kleiner Glubigsee which has a relatively low oxygen consumption in the upper layer of the lake, but a very high R_{com} in the deeper layers (about 5 mg O_2 l^{-1} d^{-1}) in this case about one meter above the sediment.

Discussion

The permanent dominance of *Oscillatoria* during summer and autumn is an often observed phenomenon of eutrophic lakes in middle Europe (Berger, 1984). It was described in detail by Wundsch (1940) and Czen-

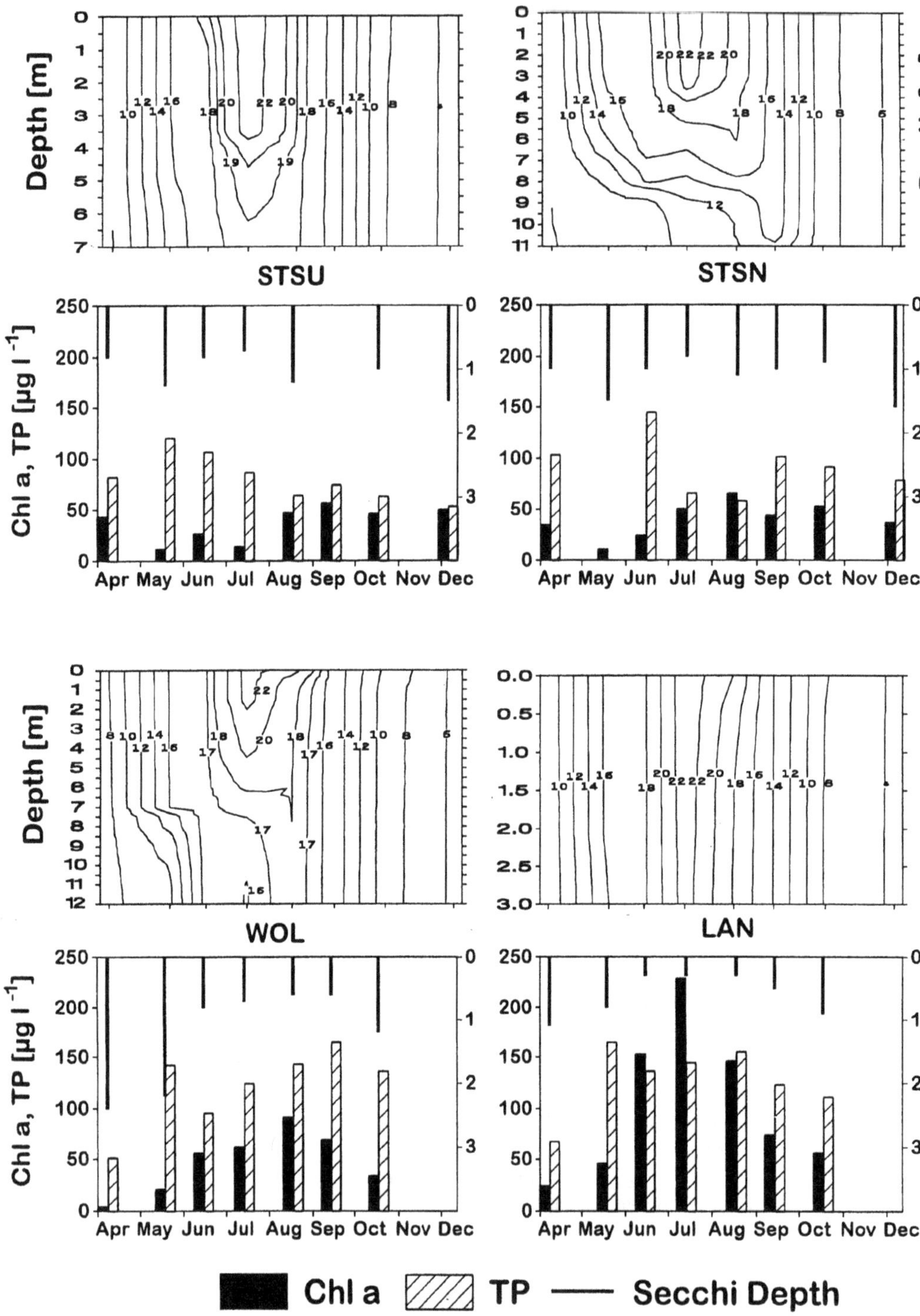

Figure 4. Mixing regime (isopleths of temperature in °C) and trophic properties in the Wolziger See, Langer See, Storkower See (South) and Storkower See (North) during the growth period 1994.

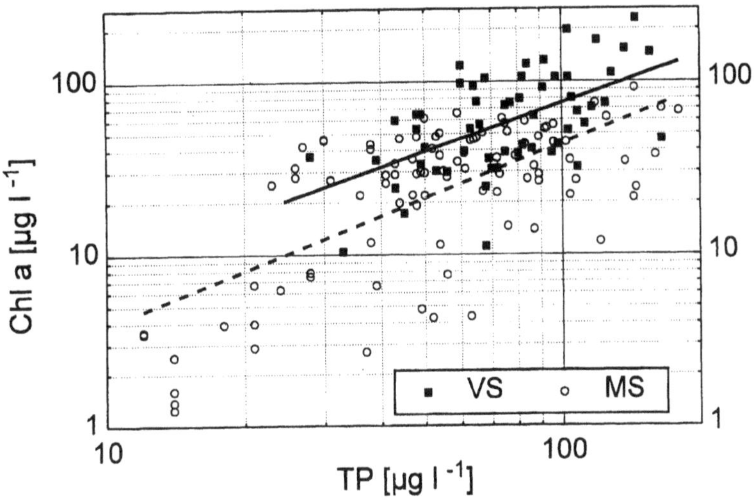

Figure 5. Regressions of monthly data of chlorophyll *a* and TP of lakes in the Scharmützelsee region, divided into 'very shallow' (VS) and 'medium shallow' (MS) categories.

sny (1938) for the Berlin/Brandenburg-region several decades ago. The development of Cyanobacteria in this region is expected because of the shallowness of the waters and the high nutrient load due to the large catchment areas (Behrendt et al., 1990; Behrendt & Nixdorf, 1993a). Discussion will focus on the following topics:

Relation between mixing regime and lake topography and morphometry

In order to determine features related to shallowness which may enhance the eutrophication process of lakes, we discuss some structural presuppositions based on size and basin shape. There is a great variability in patterns of mixing and stratification between the investigated lakes. Transient stages between shallow polymictic and deep dimictic lakes are regularly observed, as has been shown for the intermediate group of lakes with z_{max} ranging from 5 to 15 m. Lakes with a mictic instability, fluctuating between regimes, can be identified by analysis of size and morphometry. A further refinement in this direction is possible by consideration of the exposure to wind.

Within the shallow lakes, we can define three subsets, which reflect their potential mixing intensity during summer (Table 6). First, there is a group of 'very shallow' lakes (VS, $N=4$) with a probably high intensity of polymixis, containing lakes with z_{max} being much smaller than the predicted area-related z_{epi}-value, and a second group of 'medium shallow' lakes (MS1, $N=2$) with a higher potential for occasional stratification, characterized by a maximum depth deeper than z_{epi} and a z_{90}-value much smaller. For a third group of 'medium shallow' lakes (MS2, $N=3$) with comparable high maximum depth and z_{90} slightly deeper than z_{epi}, exposure to wind rather than lake morphometry seems to rule their mixing regime and in the case of the Storkower See (North) and the Wolziger See allow facultative holomixis in summer (Table 6).

For the group of 'very shallow' lakes (VS) morphological variability is limited because of the low depth. Thus, they tend to be more uniform in structural components. With increasing depth, lakes become structurally more diverse and thus more sensitive to the size-related influence of the wind on the depth and type of mixing. If the maximum depth increases, the ratio between z_{90} and the depth of the thermocline depends on lake size and on basin shape. That is why the intermediate group varies substantially: too shallow for stability, and too deep for intense polymixis. This classification is in accordance with the observed mixing patterns and may probably serve as a predictive tool.

Sas (1989) proposed a classification of highly eutrophic lakes according to physical properties in order to predict the response of lake ecosystems to reduced nutrient loading based on only 2 categories. He assumed 'deep' to be equivalent to 'not completely mixed' and 'shallow' to 'completely mixed'. As has been shown here such a coarse classification does

Table 6. Classification proposed for shallow and deeper lakes of different size and morphometry in relation to mixis regime.

Classification	Criterion	Intensity of mixis	Lakes
Very shallow (VS)	$z_{max} < z_{epi}$	Intensive polymixis	MEL, LAN, LEB, PET
Medium shallow 1 (MS1)	$z_{max} \geq z_{epi}$; $z_{90} < z_{epi}$	polymixis with short stratified periods	STSU, (KGL)
Medium shallow 2 (MS2)	$z_{max} > z_{epi}$; $z_{90} \geq z_{epi}$	Dimixis or polymixis with longer stratified periods, sensitive to exposure to wind	GGL, STSN, WOL
Deep	$z_{max} >> z_{epi}$; $z_{90} >> z_{epi}$	Stable summer stratification	TIE, SPR, SCH

not meet the actual needs for a better differentiation between mixing regime of lakes of a narrow depth range. Furthermore, the use of the median depth (z_{med}, Table 2) and the mixing depth (z_{mix}) as criteria for mixing regime has some disadvantages. The median depth is less directly related to mixing intensity as it contains much less than 90% of lake water volume, and the mixing depth cannot be predicted by the theoretical epilimnion depth as precisely as the depth of the thermocline, because it is more fluctuating during the growth season.

Nutrient conditions, uptake and storage

In 1994 the annual concentration of SRP was low in all lakes in the Scharmützelsee-region. It never exceeded 5 μg P l^{-1}, and was only 0.5 to 2 μg P l^{-1} for 4–12 weeks in many lakes during the seasonal or annual course, respectively (Rücker et al., 1997) whereas the TP content has been decreasing reaching a level between 50 and 120 μg P l^{-1}. According to Reynolds (1992b, Figure 3a p. 7) this might be comparable with the stage 2 of his model where lakes with reduced nutrient loads show a behavioural response in algal biomass development. The relatively low amount of phosphorus limits algal growth. Considering the total phosphorus, however, and the results of converting it into algal biomass in the different groups of lakes the limitation hypothesis becomes questionable because the 'very shallow' lakes appear to be able 'to make more' with the resources.

The k_s-value for phosphorus is very low for *P. agardhii* (1 or 0.2–1.1 μg P l^{-1}, Ahlgren, 1978; Zeevenboom et al., 1982). Dissolved inorganic P is a scarce resource in the lakes of the Scharmützelsee-region, but sufficient to maintain the growth of Cyanobacteria and enable them to outcompete other algal groups which are not as successful in poor underwater light climate. No excess of SRP in the mixed layer of almost all 'very shallow' lakes and in the epilimnion of the 'medium shallow' lakes, is present to allow phytoplankters other than Cyanobacteria favourable starting conditions. Only one group of algae interrupts the permanent summer and autumn dominance of Cyanobacteria: dinoflagellates occurred in Storkower and Wolziger See in 1993 and 1994 in considerable amounts (Table 3). One reason for this may be the clear-water phases due to nutrient limitation and herbivory effects on phytoplankton (Klemer & Barko, 1991; Sarnelle, 1993). In consequence, algal succession during summer can be favoured by grazing-resistant species (dinoflagellates, filamentous or colony forming Cyanobacteria) as has been observed in Lake Storkower See (North) and Wolziger See in early summer from 1993 to 1995. Both lakes are 'medium shallow' (MS2) with anoxic water and considerable phosphorus accumulation in the layer above the sediment for several weeks during summer (Figure 4). The high grazing resistance and the ability of dinoflagellates to migrate vertically and to take up SRP in deeper layers of the lakes enables at times their dominance over Cyanobacteria.

Mixing and underwater light

Most water bodies are more or less mixed, either within the epilimnion of stratified lakes or within the whole water column in polymictic lakes. Vertical mixing controls not only the distribution of matter in a water column, but also the availability of light and nutrients (Cochlan et al., 1991). Depending on the turbidity of water the intensity of light and on the light/dark rhythm may be important. In our investigations the mixing depth was estimated from the vertical measured profiles of temperature and oxygen. Sometimes it was difficult to distinguish between seasonal and diurnal thermocline. This was demonstrated by Reynolds (1992a) who showed clearly the diurnal and seasonal variability of the mixed layer. The extent of diurnal variations in the mixed depth and the consequences for algal primary production were also investigated in Lake Müggelsee (Nixdorf et al., 1992; Behrendt et al, 1993). Profiles estimated early in the morning gave the best results for the identification of the seasonal mixed layer. This fact is especially important when sampling intervals are greater than one week as in our investigations.

When mixing extends beyond the euphotic zone, algae spend a part of the light day in darkness. The consequences for the underwater light supply due to a vertical or circular transport of algae to different depths of Lake Müggelsee was clearly shown by Gervais et al. (1997). Following analysis of the OECD lakes light limitation for eucaryotic algae is expected at z_{eu}/z_{mix} below 0.25 and for Cyanobacteria below 0.15 (Sas, 1989). These values were never exceeded in our investigations. Does that mean that no light limitation exists in the lakes considered in this paper? To answer this question other possible limitation mechanisms of algal growth have to be taken into account. Limitation by nitrogen deficiency was excluded because of the high ratio of TN/TP and of the fact that N-fixing Cyanobacteria could be neglected. Following the results of the OECD study (Sas, loc. cit.), the very low annual and seasonal concentrations of SRP in most of the lakes would indicate phosphorus limitation. Periods of more than 10 weeks during the growth period with SRP below 10 $\mu g\ P\ l^{-1}$ are assumed to reduce algal development by limiting the uptake rates for phosphorus. Such situations were found in both groups of our lakes. Even the 'very shallow' Lake Langer See, with the highest algal biomass, never or seldom had more than 5 $\mu g\ P\ l^{-1}$ during the year. If P-limitation was the primary key for the regulation of algal development the question arises why lakes with the same range of phosphorus supply produce less than half of the biomass (e.g. Wolziger See) compared with Langer See. Our results show very clearly the differentiation into two groups of lakes which mainly differ in morphometry and hydrography. Thus, we focus our attention on the influence of mixing on algal growth. Beside the relation between photic and mixing depth further light induced influences on primary production and growth of phytoplankton have to be taken into account. One of these is the frequency and stability of the passive algal transport.

Assuming that phytoplankton will be transported through the mixed layer more or less continuously, algae in the different types of lakes perceive different daily photoperiods. In the clear Lake Tiefer See the photoperiod corresponds to the length of the solar day. The same can be assumed for the 'very shallow' and turbid Lake Melangsee. But there are considerable differences between the rhythms of the light/dark cycles in both lakes.

Experiments have shown that there is an optimum for planktonic primary production lasting 20 to 40 minutes per cycle in the mixed layer, so long as movement is according to the velocities and forms of Langmuir circulation (Nixdorf & Behrendt, 1991). Cyanobacteria responded to these simulations of transport with a significant increase in primary production. Kronkamp et al. (1992) found that filamentous Cyanobacteria are able to respond rapidly to changes in the mixed layer by adapting their photosynthetic performance. Chromatic adaptation was proved for *L. redekei* and *P. agardhii* very clearly by Kohl & Nicklisch (1981). On the other hand, Berger & Sweers (1988) tried to explain the dominance of *O. agardhii* Gom. by the stability of change in light/dark cycle: 'The shallower the lake, the more regular the light-dark rhythm'.

That might also be one of the reasons favouring *Oscillatoria* dominance during persistent mixing episodes in summer in Rostherne Mere (Reynolds & Bellinger, 1992). Moderate mixing favours a high biomass of Cyanobacteria by effective light harvesting for photosynthesis on the one hand and by low maintenance energy at low mean irradiance (Tilzer, 1987). Deep mixing in turbid waters causes high respiratory losses and low or negative net growth of phytoplankton. This is especially important during late autumn and winter in deeper waters (Tilzer & Beese, 1988).

Low growth rates of *Oscillatoria* may be the response to the ability of these algae to be successful under low light intensities, whereas different strategies between *Planktothrix* and *Limnothrix* concerning mixing and underwater light regime seem to exist. The growth of *Limnothrix* in Lake Müggelsee was independent of the mixing regime, whereas *Planktothrix* was favoured by stratification events in the polymictic lake (Nixdorf, 1994). It was assumed that *P. agardhii* prefers to grow in 'the shadow' of other algae, mainly *L. redekei*.

Role of heterotrophic turnover of matter

The community respiration (R_{com}) is the measure of oxygen consumption of all respiring organisms in the pelagial of the investigated lakes. The contributions of different planktonic groups to total oxygen consumption varies in a wide range and is difficult to estimate directly because the interactions of oxygen consuming and producing processes (Le Cren & Lowe-McConnell, 1980). The differences between oxygen consumption in the 'very shallow' and 'medium shallow' lakes obtained in our measurements are high. The question arises why do such differences occur and if they can be explained by algal respiration due to the higher phytoplankton biomass alone in the 'very shallow' lakes. The Chl *a* concentration in the 'very shallow' lakes exceeds the content in the 'medium shallow' waters by 10 up to 40 μg Chl *a* l^{-1}. Hamm (1991) calculated an additional BOD_5 of 20 to 60 $\mu gO_2\ l^{-1}$ per unit Chl *a* in eutrophicated running waters. Assuming that R_{com} is about one third of the BOD_5 the additional oxygen consumption by algal induced respiration is between 0.1 and 0.6 mg O_2 l- $^{-1}$. These values are too low to explain the differences in R_{com} in the lakes of varying depths.

A further approach to estimate the contribution of algal respiration confirms these results: Considering the portion of algal respiration losses on primary production. Assuming a balance between production and respiration in the pelagial of the investigated lakes, a primary production of 3 to 6 g O_2 m^{-2} d^{-1} can be calculated for the 'medium shallow' or 'very shallow' lakes, respectively. 10 to 13% of primary production is lost by phytoplankton respiration on average (Smith, 1977; Ganf, 1980; Nixdorf & Fulda, 1989; Behrendt & Nixdorf, 1993b). Thus, there must be an additional contribution of nonalgal respiration in the 'very shallow' lakes due to the bacterial community or the microbial food web. Sporadic investigations of bacterial cell numbers and morphometric properties proved a dominance of small bacteria (<1 μm) in highly eutrophicated waters (Köcher & Nixdorf, 1994; Nixdorf et al., 1995). If this phenomenon is induced by high grazing pressure or limitation by nutrients (Jost et al., 1992) will be a subject of further investigations. High turnover rates within the microbial components were reported for eutrophic and hypertrophic systems, also (Nixdorf & Arndt, 1993; Mathes & Arndt, 1994). As a result of these activities an intensive release of nutrients in very short term cycles is possible, supporting the development of higher phytoplankton biomass especially in 'very shallow' lakes under nutrient conditions near to limitation.

Summary

1. The lakes investigated allow classification into three groups of shallow lakes (very, medium (MS1) and medium (MS2) shallow) considering maximum depth, theoretical depth of epilimnion, and depth below 90% of the lake volume as well as duration and intensity of mixing.
2. All lakes are eutrophic to hypertrophic with a comparable range of in-lake TP concentration. With the exception of one lake the apparent concentration of dissolved phosphate is very low (several weeks during the growth period at or below the detection level).
3. These nutrient conditions favour the development of Cyanobacteria during summer and autumn in almost all lakes with a marked dominance of either *Planktothrix* or *Limnothrix*. After clear-water phases dinoflagellates may occur at high abundances in 'medium shallow' or deeper lakes during summer.
4. Total phytoplankton biomass in the 'very shallow' lakes was about twice as high as in the 'medium shallow' or deeper lakes although the nutrient concentration was in the same range.
5. The reasons for this differentiation and for the dominance of Cyanobacteria are to be looked for in the constant and sufficient underwater light climate due to the favourable relation of z_{eu} and z_{mix} in the 'very shallow' lakes. The regular mixing regime guarantees a stable and near optimum change in the light/dark rhythm in these lakes.
6. Cyanobacteria are very well adapted to conditions of nutrient deficiency and low underwater light climate. They are more efficient of converting phosphorus into biomass in the 'very shallow' lakes because production of algal biomass in the 'medium shallow' and deeper lakes is limited by the unfavourable light supply.
7. Measurements of community respiration in the lakes suggest that 'very shallow' lakes have an additional advantage due to the increased metabolic processes within the microbial components. R_{com} is about twice as high in the 'very shallow' lakes compared to the remaining lakes. Because the differences are only partly to be explained by algal respiration a higher bacterial turnover is supposed

to support algal growth under low ambient nutrient conditions.

Acknowledgments

We would like to thank Jacqueline Rücker and Gudrun Lippert who were responsible for the field measurements and hydrochemical program. Furthermore, we wish to thank Paul Zippel, who carried out all of the plankton analyses, and Ingo Henschke and Wolfgang Terlinden for their careful work in the field.

References

Ahlgren, G., 1978. Response of phytoplankton and primary production to reduced nutrient loading in Lake Norvikken. Verh. int. Ver. Limnol. 20: 840–845.

Barica, J., 1980. Why hypertrophic ecosystems? In Barica, J. & L. R. Mur (eds), Hypertrophic ecosystems. Developments in Hydrobiology 2, Dr W. Junk Publishers, The Hague.

Behrendt, H., E. Driescher & G. Schellenberger, 1990. Lake Müggelsee – the use of lake water and its consequences. GeoJournal 22: 175–183.

Behrendt, H. & B. Nixdorf, 1993a. Changes of Secchi depth and seston content of the Lake Müggelsee since 1931 – an example for eutrophication and feedbacks. In Guissiani, G. & C. Callieri (eds), Proceedings of the 5th Int. Conf. on the Conservation and Management of Lakes 'Strategies for Lakes Ecosystems beyond 2000': 30–33.

Behrendt, H. & B. Nixdorf, 1993b. The carbon balance of phytoplankton derived from growth, primary production and losses at different levels in Lake Müggelsee. Int. Revue ges. Hydrobiol. 78: 439–458.

Behrendt, H., B. Nixdorf & W.-G. Pagenkopf, 1993. Phenomenological description of polymixis in Lake Müggelsee. Int. Revue ges. Hydrobiol. 78: 411–421.

Berger C., 1975. Occurrence of *Oscillatoria agardhii* Gom. in some shallow eutrophic lakes. Verh. int. Ver. Limnol. 19: 2689–2697.

Berger, C., 1984. Consistent blooming of *Oscillatoria agardhii* Gom. in shallow hypertrophic lakes. Verh. int. Ver. Limnol. 22: 910–916.

Berger, C. & H. E. Sweers, 1988. The IJsselmeer and its phytoplankton – with special attention to the suitability of the lake as a habitat for *Oscillatoria agardhii* Gom. J. Plankton Res. 10: 579–599.

Chorus, I., 1995. Müssen in der Seensanierung Gesamtphosphat-Schwellenwerte unterschritten werden, bevor das Phytoplankton eine Reaktion zeigt? In Jäger, D. & R. Koschel (eds), Verfahren zur Sanierung und Restaurierung stehender Gewässer. Limnologie aktuell 8: 21–28.

Chow-Fraser, P., 1991. Use of the morphoedaphic index to predict nutrient status and algal biomass in some Canadian lakes. Can. J. Fish. aquat. Sci. 48: 1909–1918.

Cochlan, W. P., N. M. Price & P. J. Harrison, 1991. Effects of irradiance on nitrogen uptake by phytoplankton: comparison of frontal and stratified communities. Mar. Ecol. Progr. Ser. 69: 103–116.

Czensny, R., 1938. Die Oscillatorienerkrankung unserer Seen, Biologie und Chemismus einiger märkischer Seen. Vom Wasser 8: 36–57.

Deutsche Einheitsverfahren zur Wasser-, Abwasser- und Schlammuntersuchung 1986–1993. Verlag Chemie GmbH Weinheim.

Ganf, G. G., 1980. Community respiration. In Le Cren, E. D. & R. H. Lowe-McConnell (eds), The Functioning of Freshwater Ecosystems. Cambridge University Press, Cambridge: 188–194.

Gervais, F., D. Opitz & H. Behrendt, 1997. Influence of small scale turbulence and large scale mixing on phytoplankton primary production. Hydrobiologia 342/343: 95–105.

Håkanson, L., 1981. A Manual of Lake Morphometry. Springer-Verlag, Berlin, Heidelberg, 78 pp.

Hamm, A. (ed.), 1991. Studie über Wirkungen und Qualitätsziele von Nährstoffen in Fließgewässern. Academia Verlag, Sankt Augustin, 830 pp.

Hanna, M., 1990. Evaluation of models predicting mixing depth. Can J. Fish. aquat. Sci. 47: 940–947.

Jost, G, G. Klinkenberg & P. Spittler, 1992. Bacteria between grazing pressure and organic carbon limitation. Arch. Hydrobiol. Beih. Ergebn. Limnol. 37: 233–240.

Klapper, H., 1992. Eutrophierung und Gewässerschutz. – Gustav Fischer Verlag, Jena, Stuttgart, 277 pp.

Klein, G., 1989. Anwendbarkeit des OECD-Vollenweider-Modells auf den Oligotrophierungsprozeß an eutrophierten Gewässern. Vom Wasser 73: 365–373.

Klemer, A. & J. Barko, 1991. Effects of mixing and silica enrichment on phytoplankton seasonal succession. Hydrobiologia 210: 171–181.

Köcher, B. & B. Nixdorf, 1994. Bakterien und autotrophes Picoplankton in natürlichen und künstlichen Seen der Region Berlin/Brandenburg – erste Ergebnisse. Deutsche Gesellschaft für Limnologie e. V., Erweiterte Zusammenfassungen der Jahrestagung 1993 in Coburg: 284–288.

Kohl, J. G. & A. Nicklisch, 1981. Chromatic adaptation of the planktonic blue-green alga *Oscillatoria redekei* van Goor and its ecological significance. Int. Revue ges. Hydrobiol. 66: 83–94.

Kronkamp, J., F. Schanz, M. Rijkeboer, E. Berdalet, B. Kim & H. J. Gons, 1992. Influence of the mixing regime on algal photosynthetic performance in laboratory scale enclosures. Hydrobiologia 238: 111–118.

Le Cren, E. D. & R. H. Lowe-McConnell, 1980. The functioning of freshwater ecosystems. Cambridge University Press, Cambridge, 588 pp.

Mathes, J. & H. Arndt, 1994. Biomass and composition of protozooplankton in relation to lake trophy in north German lakes. Mar. Microb. Food webs, 8: 357–375.

Mietz, O., 1994. Aufbau und Entwicklung eines Seenkatasters als Grundlage zur Lösung wichtiger gewässerökologischer Fragestellungen im Land Brandenburg dargestellt am Beispiel des Teupitzer Seengebietes Brandenburg. Beiträge zur angewandten Gewässerökologie Norddeutschlands 1: 10–39.

Müller, H., 1952. Die produktionsbiologischen Verhältnisse märkischer Seen in der Umgebung Storkows. Z. Fisch. 1: 95–246.

Nixdorf, B., 1994. Polymixis of a shallow lake (Großer Müggelsee, Berlin) and its influence on seasonal phytoplankton dynamics. Hydrobiologia 275/276: 173–186.

Nixdorf, B. & H. Arndt, 1993. Seasonal dynamics of plankton components including the microbial web in Lake Müggelsee. Int. Revue ges. Hydrobiol. 78: 403–410.

Nixdorf, B. & H. Behrendt, 1991. Discrepancies between O_2- and ^{14}C-method resulting from measurements of primary production under consideration of vertical mixing in a shallow eutrophic lake. Verh. int. Ver. Limnol. 24: 1268–1271.

Nixdorf, B., H. Behrendt & W.-G. Pagenkopf, 1992. Diurnal patterns of mixing depth and its influence on primary production in shallow lakes. Int. Revue ges. Hydrobiol. 77: 349–360.

Nixdorf, B. & S. Fulda, 1989. A method for estimation of phytoplankton dark carbon losses by application of ^{14}C method. Arch. Hydrobiol. Beih. Ergebn. Limnol. 33: 445–450.

Nixdorf, B. & S. Hoeg, 1993. Phytoplankton – community structure, succession and chlorophyll content in Lake Müggelsee from 1979 to 1990. Int. Revue ges. Hydrobiol. 78: 359–377.

Nixdorf, B., J. Rücker, R. Deneke & P. Zippel, 1995. Limnological analysis of standing waters in the Scharmützelsee-region, part 1. BTUC – UW 1/95, 52 pp.

OECD (1982). Eutrophication of waters. OECD report, Paris, 154 pp.

Ohle, W., 1953. Der Vorgang rasanter Seenalterung in Holstein. Die Naturwissenschaften 40: 152–162.

Reynolds, C. S., 1992a. Dynamics, selection and composition of phytoplankton in relation to vertical structure in lakes. Arch. Hydrobiol. Beih. Ergebn. Limnol. 35: 13–31.

Reynolds, C. S., 1992b. Eutrophication and the management of planktonic algae: What Vollenweider couldn't tell us. In Sutcliffe, D. W. & J. G. Jones (eds), Eutrophication: Research and Application to Water Supply. Freshwat. Biol. Ass., Cumbria, 217 pp.

Reynolds, C. S. & E. G. Bellinger, 1992. Patterns of abundance and dominance of the phytoplankton of Rostherne Mere, England: Evidence from an 18-years data set. Aquat. Sci. 54: 10–36.

Rojo, C. & M. A. Cobelas, 1994. Population dynamics of *Limnothrix redekei, Oscillatoria lanceaeformis, Planktothrix agardhii* and *Pseudoanabaena limnetica* (Cyanobacteria) in a shallow hypertrophic lake (Spain). Hydrobiologia 275/276: 165–171.

Rücker, J., C. Wiedner & P. Zippel, 1997. Factors controlling the dominance of *Planktothrix agardhii* and *Limnothrix redekei* in eutrophic shallow lakes. Hydrobiologia 342/343: 107–115.

Sarnelle, O., 1993. Herbivory effects on phytoplankton succession in a eutrophic lake. Ecol. Monogr. 63: 129–149.

Sas, H., 1989 Lake restoration by reduction of nutrient loading – Expectations, Experiences, Extrapolations. Academia Verlag Richard, 497 pp.

Schäperclaus, W., 1941. Seenverschlechterung. Z. Fisch. 38: 345–375.

Scharf, R., 1971. Nährstoff- und Sauerstoffverhältnisse der Seen Ostbrandenburgs. Limnologica 8: 393–414.

Smith, W. D., 1977. The respiration of photosynthetic carbon in eutrophic areas in the ocean. J. mar. Res. 35: 781–794.

Sommer, U., 1993. The scientific basis of eutrophication management: Reconciling basis physiology and empirical biomass models. In Guissiani, G. & C. Callieri (eds), Proceedings of the 5th Int. Conf. on the Conservation and Management of Lakes 'Strategies for Lakes Ecosystems beyond 2000': 6–9.

Thienemann, A., 1932. Schwankungen des Grundwasserzustandes in Norddeutschland während der letzten Jahrzehnte, ihre Ursachen und ihre limnologische, geologische und wirtschaftliche Bedeutung. Arch. Hydrobiol. 24: 345–428.

Tilzer, M. M., 1987. Light-dependence of photosynthesis and growth in Cyanobacteria: Implications for their dominance in eutrophic lakes. N. Z. J. Mar. Freshwat. Res. 21 : 401–412.

Tilzer, M. M. & B. Beese, 1988. The seasonal productivity cycle of phytoplankton and controlling factors in Lake Constance. Schweiz. Z. Hydrol. 50: 1–39.

Utermöhl, H., 1958. Zur Vervollkommnung der quantitativen Phytoplanktonmethodik. Mitt. int. Ver. Limnol. 9: 1–38.

Ventz, D., 1974. Die Einflußnahme von Umgebungsfaktoren und morphometrischen Faktoren auf den Stoffhaushalt von Seen. Diss. A, Dresden, 111 pp.

Wundsch, H. H., 1940. Beiträge zur Fischereibiologie märkischer Seen, VI. Die Entwicklung eines besonderen Seentypus (H_2S-Oscillatorien-Seen) im Flußgebiet der Spree und Havel, und seine Bedeutung für die fischereibiologischen Bedingungen in dieser Region. Z. Fisch. 38: 443–648.

Zeevenboom, W., A. Bij de Vaate & L. Mur, 1992. Assessment of factors limiting growth rate of *Oscillatoria agardhii* in hypertrophic Lake Wolderwijd, 1978, by use of physiological indicators. Limnol. Oceanogr. 27: 39–52.

Hydrobiologia **342/343**: 285–295, 1997.
L. Kufel, A. Prejs & J. I. Rybak (eds), Shallow Lakes '95.

Changes in ecological relationships in a group of eight shallow lakes in the Polesie Lubelskie region (eastern Poland) over forty years

Tadeusz J. Chmielewski, Stanisław Radwan & Beata Sielewicz
Department of Hydrobiology and Ichthyobiology Academy of Agriculture, Akademicka 13, 20-033 Lublin, Poland

Key words: shallow lakes, ecosystem function

Abstract

The changes in 33 factors in 8 shallow lakes and their catchment areas were studied over a 40 year period. Large-scale changes in ecological conditions were present in the region studied: a disappearance of about 30% of wetlands, a reduction in the surface area of lakes by about 15%, and an increased forest cover in the region by about 26%. Six lakes showed increasing transparency and nutrient concentration, while two changed in the opposite direction. Various reactions were observed in zooplankton and zoobenthos.

Generally, patterns for the reactions of lake ecosystems were found to vary individually, even in very similar lakes. The character of some changes might indicate a modular type of biocoenosis reaction. The character and rate of the overgrowing of the open waters of lakes depend to a high degree on the species composition and the predominating structure of the plants in the lake and its drainage basin.

To avoid an incorrect prediction of the reactions of lake ecosystems, it is recommended that multifactor analysis be used for drainage basins over long periods of time.

Introduction

The shallow lakes of Polesie Lubelskie (eastern Poland) show a great morphological, trophic and biocoenotic differentiation (Radwan et al., 1974; Chmielewski (ed.), 1989; Wilgat et al., 1992, Radwan (ed.), 1995).

The ecological and use characterisation of the drainage basins of these lakes is also varied (Chmielewski & Radwan, 1993), with lakes reacting in different ways to changes occurring in their drainage basins over a longer period of time (Chmielewski et al., 1995).

By addressing the nature and scale of the changes in particular features in different trophic types of lakes and their drainage basins during the last few decades, it may be possible to gain a better understanding of the functioning of the shallow Polesie Lakes. This may also allow a forecasting of the ecological consequences of the changes that occur in the water environment and the working out of effective protective measures.

Description of sites studied

Eight shallow lakes, varying in trophic status but situated close to each other, were studied. Their surface areas varied between 1.5 ha and 284.1 ha and their depths between 0.8 and 6.6 m (Figure 1). The ecological character and changes in land use of the catchment areas of these lakes were studied in parallel.

Materials and methods

Among the factors studied, there were seven physical-chemical and twenty biocoenotic features, as well as the rate of macrophyte succession in the years 1968–1994. In the drainage basins of these lakes, five features illustrating changes in the character of the cover and utilisation of the area were estimated for the years 1952–1994, on the basis of a comparative analysis of 1:25000 scale aerial photographs.

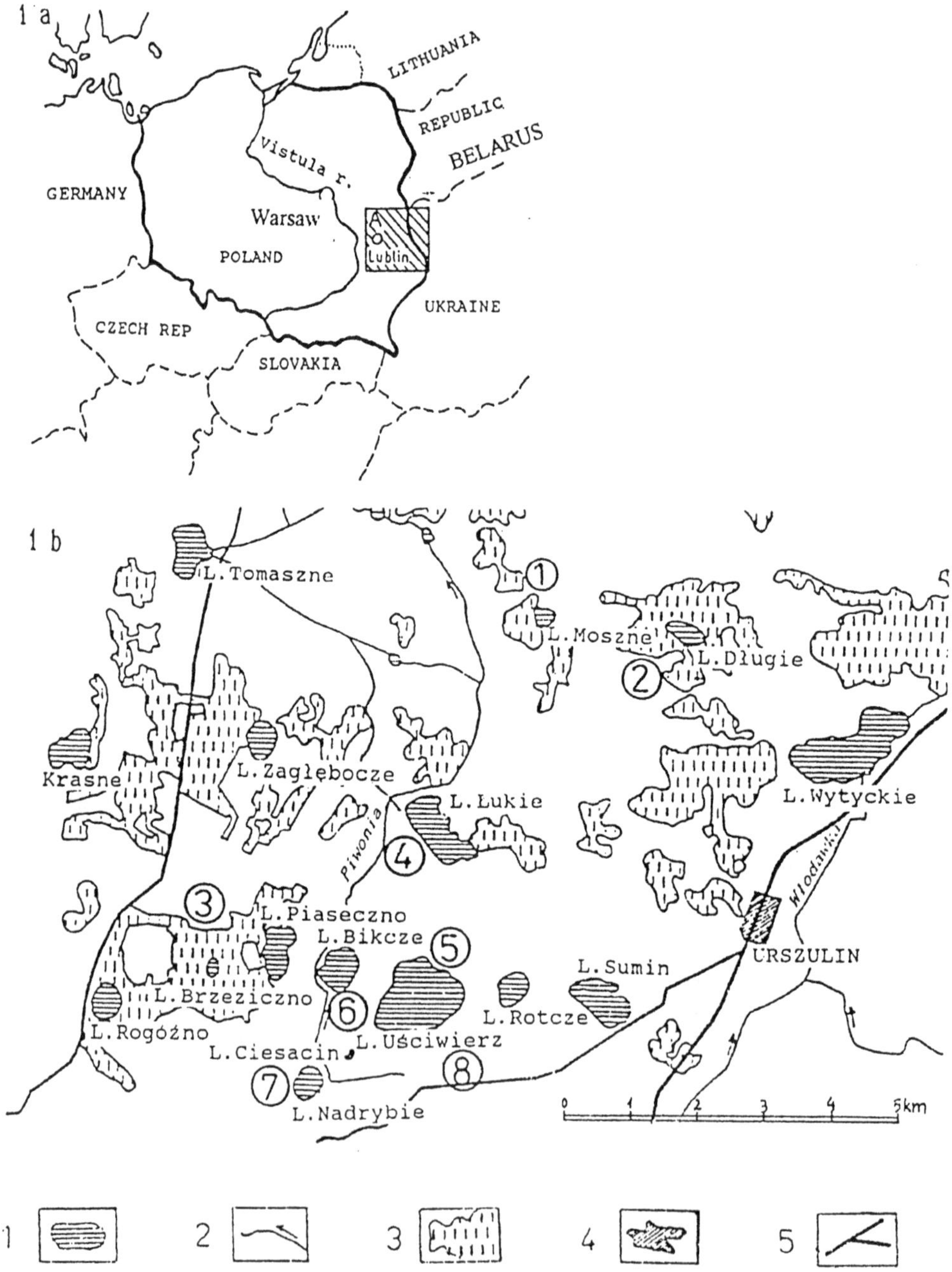

Figure 1. Location of study sites: 1a – In the country, 1b – In the Łęczna-Włodawa Lakeland, 1 – lakes, 2 – rivers, 3 – forests, 4 – localities 5 roads, 1–8 – the studied objects.

In an attempt to elucidate causal relationships, the character and scale of changes and the occurrence of particular features were compared.

The measurement of all the physical-chemical and biocoenotic features was performed in the pelagic zone of the lakes, in late spring and at the end of summer, i.e. the seasons of the greatest, quantitative and qualitative differentiation of the water biocoenosis. Standard chemical and biocoenotic methods were applied (Golterman, 1972; Edmondson & Winberg, 1971).

Out of several physical-chemical features, the following were chosen for the comparative analysis: transparency, pH (near the surface), content of the oxygen at the bottom of the lake and concentrations of the nutrients: NO_3^-, NH_4^+ and P-PO_4^-.

The biocoenotic features subjected to further analysis were the quantitative and qualitative composition of the zooplankton (Rotifera, Cladocera, Copepoda), and the quantitative and qualitative composition of the zoobenthos (Oligochaeta, Hirudinea, Crustacea, Chironomidae, Chaoborus, Mollusca and others).

The scale of changes in all the features, except those in the concentration of biogenic substances was estimated using a percentage indicator applying the formula:

$$\text{IFC} = \frac{100(C_2 - C_1)}{C_2}$$

IFC = Indicator of Feature's Change;

C_1 = value of the feature in 1966;

C_2 = value of the feature in 1993.

This methodical approach is similar to the one adopted by Johnes et al. (1993).

The large amplitude of oscillations, and the scale of changes in the concentration of biogenic substances estimated, required use of a Synthetic Indicator of the Change of Biogene Contents (SICBC):

$$\text{SICBC} = \frac{|\Delta C\text{NO}_3| + |\Delta C\text{NH}_4| + |\Delta C\text{PO}_4|}{3},$$

where ΔC is the multiplicity of the change in the concentration of particular biogenic substances during the period 1966–1993.

In addition, for chosen groups of related factors (e.g. a group of physicochemical features, the quantitative structure of particular ecological formations, methods of drainage basin exploitation etc.), a Synthetic Indicator of Feature Variability (SIFV) was calculated using the formula

$$\text{SIFV} = \frac{|\Delta C_1| + |\Delta C_2 + | \ldots + |\Delta C_n|}{n},$$

where C_1 is the absolute value for the variability of a particular feature and n = number of analysed features.

The correlation analysis of the variables examined was carried out using the Statgraphics 6.0 - C,5 programme.

To eliminate the intensity of the agricultural utilisation of the drainage basins and its influence on lake ecosystems, an Agricultural Use Index (AUI) was applied, after the formula

$$\text{AUI} = a \times i\Delta a,$$

where a = percentage of ploughed lands in the drainage basins in 1992; Δa = changes in the percentage of ploughed land in the drainage basin in the years 1952–1992 which indicate value (i); i = indicator of intensity of change in agricultural utilisation in the years 1952–1992, where:

i = 0.9 for $\Delta a > -1 < -10\%$;

i = 0.8 for $\Delta a > -10 < -25\%$;

i = 0.7 for $\Delta a > -25\%$;

i = 1.1 for $\Delta a > 1 < 10\%$;

i = 1.2 for $\Delta a > 10 < 25\%$;

i = 1.3 for $\Delta a > 25\%$.

Results

General direction and scope of changes

The investigations indicate that the last forty years witnessed some significant changes in ecological relations in the drainage basins of the lakes studied (Table 1). The most important of these are:

- the drying out of peatbogs and other wetlands by 30% on average,
- the shrinking of lake surface area by 15.4% on average,
- overgrowing of the lake surface with macrophytes by 23.0% on average,
- increase of forestation of the area by 26.4% on average.

These changes in the landscape were accompanied by different reactions of the lake ecosystems. Lakes recognized in the 1950s as dystrophic (Lakes Moszne, Długie, Brzeziczno and Nadrybie) (Fijałkowski, 1960) later became dystrophic (Moszne, Długie), highly eutrophic (Nadrybie), or hard to define, trophically

Table 1. Changes of drainage basins land-use from 1952 to 1992.

Drainage basin	Area of basin (ha)	% of basin area: Surface water 1952	1992	changes	Peatbogs 1952	1992	changes	Other wetlands 1952	1992	changes	Forests 1952	1992	changes	Agriculture and urban areas 1952	1992	changes
Moszme	711.5	2.4	2.3	−0.1	16.03	6.0	−9.7	16.2	0	−16.2	16.5*	54.9*	38.4	48.6	36.2	−2.4
Długie	490.9	5.5	5.2	−0.3	10	6.7	−3.3	19.9	0	−19.9	25.2	71.5	46.3	39.4	16.6	−22.8
Brzeziczno	593.6	1.5	1.2	−0.3	8.9	3.3	−5.6	0	0	0	45.2**	51.6**	6.4	44.4	43.9	−0.5
Łukie	1874.8	8.2	8.0	−0.2	27.6	7.6	−20.0	0	0	0	3.4	29.7	26.3	60.8	54.7	−6.7
Uściwien	926.4	30.7	28.4	−2.3	26.2	8.9	−17.3	5.5	0	−5.5	0	11.4	11.4	37.6	51.3	13.7
Bikcze	330.6	25.1	24.8	−0.9	43.7	5.9	−37.8	3.0	0	−3.0	0	4.8	4.8	27.6	64.5	36.9
Nadrybie	981.7	4.8	3.3	−1.5	30.1	4.6	−25.5	2.9	0	−2.9	1.1	8.9	7.8	61.1	83.2	22.1
Ciesacin	39.9	3.8	1.7	−2.1	96.2	28.6	−67.6	0	0	0	0	69.7	69.7	0	0	0

* Including 11.6% wet forest, drainaged after 1952
** Including 10.0% wet forest, drainaged after 1952

(Brzeziczno). Similarly, some lakes recognized in the 1950s as eutrophic are now difficult to assign to this category (e.g. Bikcze) (Table 2). During the last forty years the rate at which the surface of the lakes became overgrown with macrophytes varied between 0.7% and 62.9%.

As to physical-chemical factors, six lakes displayed a decrease in transparency as well as a considerable increase in the concentration of biogenic substances (7.2 times on average).

However, two lakes have seen a transparency increase and the level of biogenic substances declined (7 times on average). The pH of water increased considerably in four lakes (by 6.7% on average), and decreased in three (by 7.4% on average) (Table 2).

Significant structural changes have taken place in biocoenotic components. In the zoobenthos the number of Oligochaeta decreased (by 22.4% on average) while the number of Chironomidae increased considerably (by 47.4% on average). Where zooplankton is concerned, six lakes have seen the qualitative composition of Rotifera become impoverished (by 0.4% on average) while the number of Cladocera species increased. In only two lakes has the number of rotifer species increased (by 19.5%).

Where the quantitative composition of zooplankton is concerned, Rotifera species dominate in all lakes, accounting for 53 to 87% of all zooplankton (Kornijow et al., 1995; Radwan et al., in press).

The functioning of the lakes and the utilization of drainage basins

The drying of the wetlands in their drainage basins has had a great impact on the nature and functioning of the lakes. A decrease in lake size occurred. This process has been more strongly marked in the shallowest lakes and in the ones that occupy smaller parts of the drainage area (Figure 2). Forest eventually covered these dry basin areas. However, a decrease in forest cover has been observed, where agricultural utilisation of drainage basins was intensive (Lakes Uściwierz, Nadrybie and Bikcze). Only in the drainage basin of Lake Brzeziczno a rather different reaction was recorded (Figure 3). It has been noted that increased forest cover in a drainage basin may slow down ecological variability and eutrophication in the lake if succession takes place in the part of the drainage basin adjacent to the lake and mainly involves conifers (Lake Moszne). In context, a succession involving broad-leaved trees does not have a major effect on slowing down eutrophication (Lake Długie). However, if the lake is in the last stages of decline, even an area increase in forest cover of the drainage basins will not stop significant further degradation (Lake Ciesacin) (Figure 4).

The rate at which a lake surface is overgrown with macrophytes depends mainly on the depth of the littoral area and on the rate and scope of dehydration of the drainage basin (Figure 4). Trophic character of the lake seems to be of minor significance. The highest rate of growth at the surface takes place in lakes with intense ecological destruction (Ciesacin, Nadrybie). The lowest rate characterizes moderately eutrophic and dys-

Table 2. Changes in physical-chemical factors in the group of shallow lakes (1966–1993).

Lake	Year	Surface (ha)	Max deep (m)	Type of trophy	Secchi disc transpa-rency (m)	pH	O_2 mg dm^{-3}	NO_3 mg N dm^{-3}	NH_4 mg N dm^{-3}	P-PO_4 mg dm^{-3}	Ca mg dm^{-3}						
						depth of 0.5m	above bottom	depth of 0.5m	above bottom	depth 0.5m	above bottom	depth 0.5m	above bottom	depth 0.5m	above bottom	depth 0.5m	above bottom
Mozne	1985	17.5	1.1	dystr.-like	1.15	8.2	8.2	9.6	9.6	0.046	0.046	0.199	0.199	0.001	0.001	32.2	25.6
	1993	17.0			1.00	8.2	8.2	9.8	9.8	0.313	0.313	0.190	0.220	0.013	0.002	408	382
Długie	1985	28.4	1.3	dystr.-like	1.30	8.7	8.2	10.8	10.8	0.069	0.093	0.199	0.174	0.001	0.001	23.4	21.8
	1993	27.1			1.15	8.5	8.5	9.7	9.8	0.398	0.384	0.220	0.230	0.013	0.005	250	219
Brzezizno	1968	8.7	2.54	dystr.-like	2.31	6.4	6.4	8.6	9.0	–	–	–	–	–	–	3.7	–
	1993	8.6		degradated	0.90	6.2	6.6	9.5	9.2	0.313	0.142	0.300	0.330	0.002	0.005	4.4	387
Łukie	1968	152.6	6.5	middle	2.25	8.5	9.6	11.6	9.6	0.020	–	0.220	–	0.035	–	377	–
	1993	150.0		eutrophic	1.90	8.1	8.1	9.4	8.7	0.071	0.085	0.170	0.275	0.004	0.004	386	387
Uściwierz	1968	284.0	6.6	low	1.70	8.2	7.6	11.2	9.3	0.024	–	0.240	–	0.015	–	50.1	–
	1993	262.5		eutrophic	1.15	8.0	8.2	9.5	8.2	0.042	0.062	0.135	0.062	0.002	0.002	33.0	35.0
Bikcze	1968	85.0	3.3	dystr.-like	0.54	7.7	7.2	10.2	10.2	–	–	–	–	–	–	384	–
	1993	81.9		or eutrophic	0.80	8.5	7.9	10.1	8.3	0.100	0.114	0.380	0.365	0.002	0.001	366	364
Nadrybie	1966	47.0	1.95	eutrophic	1.12	8.4	–	13.2	12.4	–	–	–	–	–	–	–	–
	1993	32.7		degradated	0.50	8.6	–	9.8	9.8	0.142	0.142	0.300	0.300	0.001	0.001	–	139
Ciesacin	1966	1.5	1.8	eutrophic,	1.80	7.5	–	14.0	5.2	–	–	–	–	0.005	0.010	27.2	–
	1993	0.7	0.8	degradated	0.50	8.2	8.2	11.0	9.2	0.123	0.123	0.880	0.880	0.071	0.076	27.2	50.2

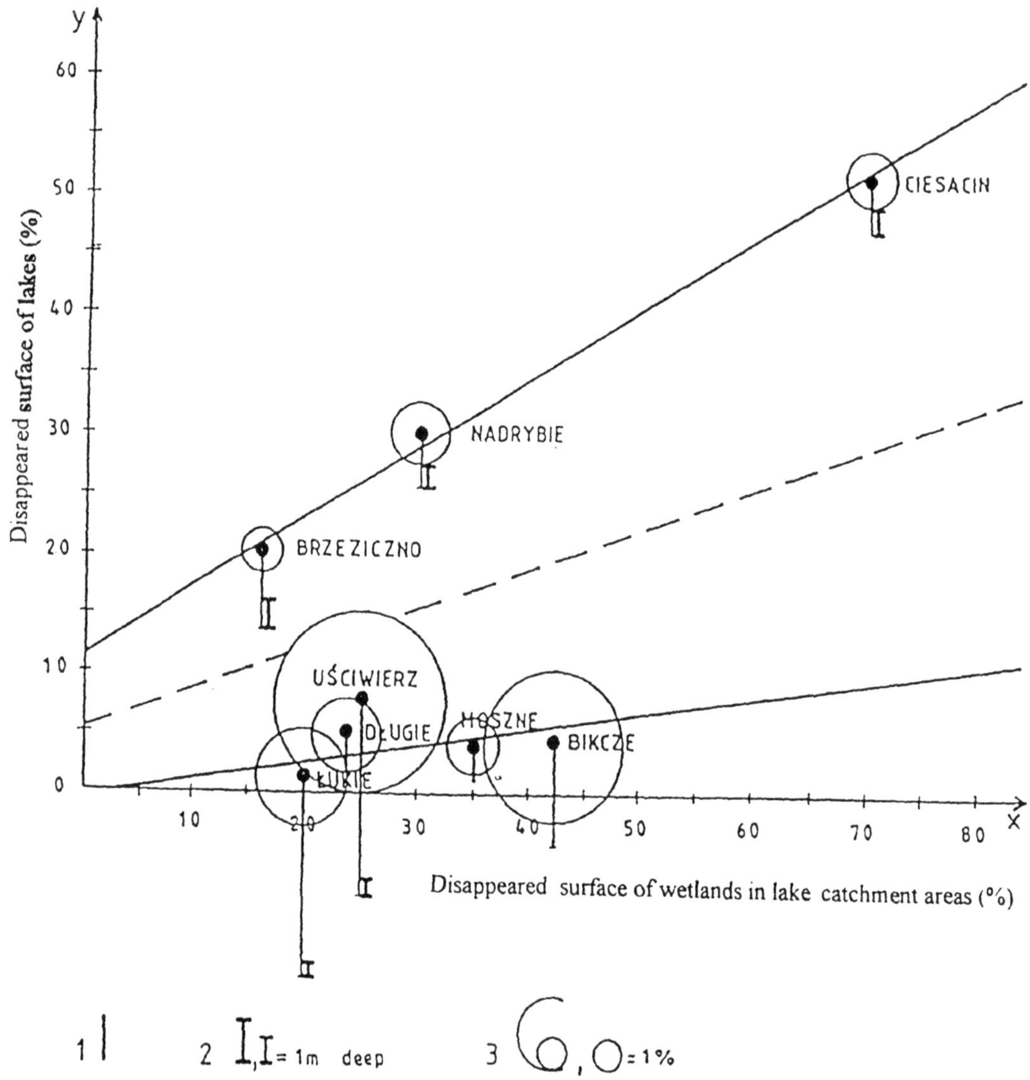

Figure 2. The concordance between the disappearance of shallow lakes and the disappearance of wetlands in their catchment areas. 1 – max. depth of lakes in 1954, 2 – decrease in depth of lakes 1954–1993 (No data for lakes Moszne, Długie and Bikcze), 3 – percentage relationship between area of lake and that of its catchment area.

trophic lakes (Łukie, Moszne, Bikcze). Great changes in the utilization of a drainage basin are usually accompanied by an increase in concentration of nutrients in the lakes, especially in smaller ones (the bigger the changes, the greater the increase in concentrations) (Figure 4).

A comparison of the results of all studies led to the elaboration of a general ideogram of correlations between the main ecological processes occurring in the drainage basins of shallow lakes (Figure 5).

Drying of an area is usually a consequence of increased agricultural utilization of drainage basins and is associated with increased eutrophication. Increase in forest cover is negatively correlated with the process of water eutrophication, and with the development of agriculture. It is, however, correlated positively with the process of drying of drainage basins. The number of species in a lake's biocoenosis is usually greatest in water with moderate trophic status, a moderate percent-

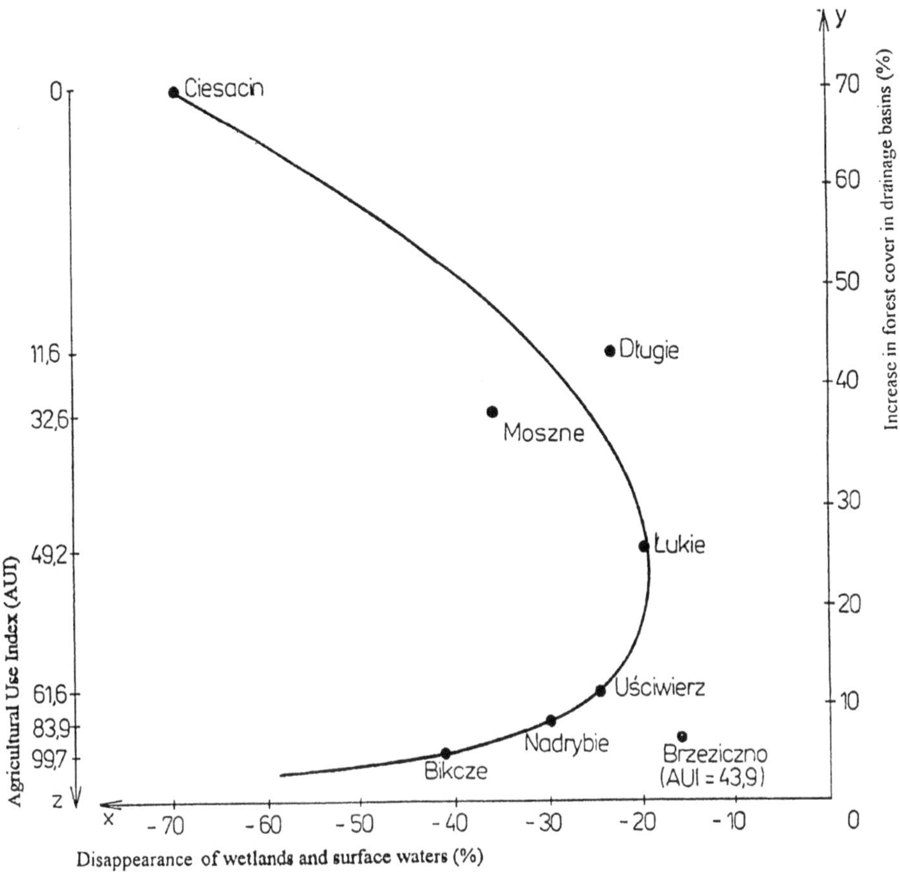

Figure 3. Correlation between drainage of an area, natural forest succession, and increased agricultural usage in the drainage basins of eight shallow lakes in the Polesie Lubuskie Region.

Table 3. Qualitative structure of phytoplankton, zooplankton and zoobenthos in a group of shallow lakes (1966–1993).

	PHYTOPLANKTON (number of species)					ZOOPLANKTON (number of species)								ZOOBENTHOS (number of species)											
	1993					1966–69				1993				1966–69						1993					
Tax. group / Lake	Cy	Chl	Bac	Oth		R.	Cl.	Co		R.	Cl.	Co		Ol	Chi	Cha	Mol	Oth		Ol	Chi	Cha	Mol	Oth	
Moszne	3	8	–	4	15	18	7	3	28	13	6	3	22	1	6	–	2	3	12	1	7	1	2	1	14
Długie	7	16	–	8	31	12	14	5	31	19	9	4	32	–	8	–	2	2	12	1	8	–	–	2	11
Brzeziczno	6	3	–	5	14	13	5	4	22	17	2	3	22	1	3	1		1	6	–	3	–	–	–	3
Łukie	7	25	2	9	43	23	10	4	37	27	15	5	47	1	1	–	–	1	3	1	10	–	1	3	15
Uściwierz	7	21	2	9	39	14	7	3	24	13	8	3	24	1	6	1	–	1	9	1	13	1	1	4	20
Bikcze	6	19	4	9	38	19	8	5	2	15	14	5	34	1	4	1	–	1	7	1	17	1	–	7	26
Nadrybie	5	12	–	6	23	21	8	3	32	7	4	3	14	–	5	–	–	1	6	1	1	–	–	–	2
Ciesacin	–	4	–	6	10	16	6	4	26	7	6	3	16	n.d.	n.d.	n.d.	n.d.	n.d.	n.d.	1	3	–	–	2	

Cy - Cyanophyta, Chl -Chlorophyta, Bac - Bacillariophyceae, R - Rotifera, Cl - Cladocera, Co - Copepoda, Ol - Oligochaeta, Chi -Chironomidae, Cha - Chaoborus, Mol - Mollusca, 0th - Others, n.d. no data

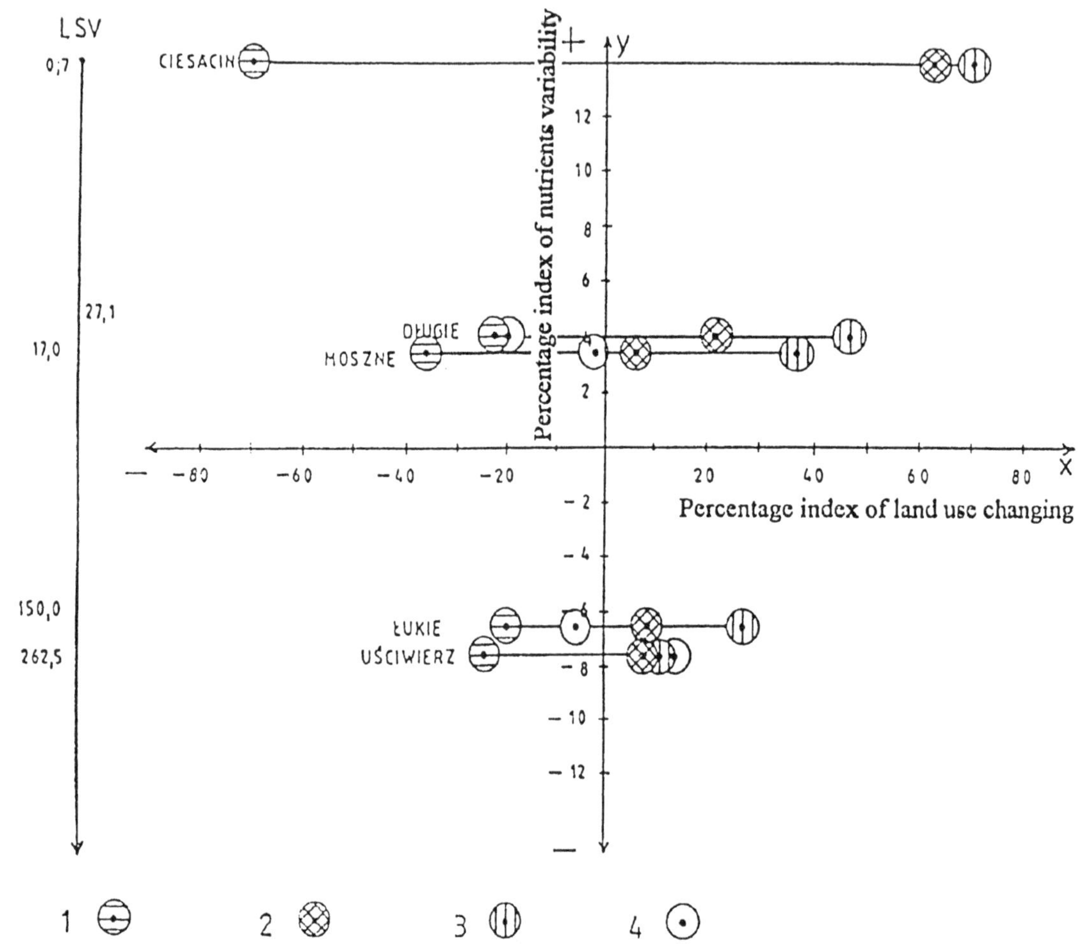

Figure 4. Relationship between variability in nutrient concentration in shallow lakes and land use changes in their drainage basin 1952(66)–1992. LSV – Lake size vector (ha), 1 – changes in water and wetland areas (%), 2 – changes in cover of the surface of lakes by macrophytes (%), 3 – changes in catchment area forest cover, 4 – changes in agriculture and urban areas (%).
There are no complete data for 1966 for nutrients in Lakes Brzeziczno, Bikcze and Nadrybie.

age of wetlands in the drainage basin and an average level of forest cover (Table 3, Figure 5).

Changes in the use of drainage basins were followed by changes in water transparency and oxygen concentration. A decrease in area of peatbogs in the drainage basins was accompanied by an increased acidity of lake water.

The process by which a surface was overgrown with macrophytes was positively correlated with decreased lake area, changes in physical-chemical features of the water, and an increase in the number of Copepod species. There was an inverse correlation between Copepoda and Rotifera. The numbers of Rotifera correlated with those of Oligochaeta and Hirudinea, while numbers of Hirudinea correlated negatively with the processes of drying in drainage basins, but positively with increased forest cover.

Changes in the physical-chemical features of lakes influence the functioning of water biocoenoses to different degrees. However, in general, limited variability in physical-chemical features is accompanied by limited changes in biocoenotic structure. This trend seems to be independent of the size of a lake. The smallest changes have been observed in lakes of low trophy (Uściwierz, Moszne). Changes of average magnitude lead to eutrophic lakes (Bikcze, Nadrybie) and those of

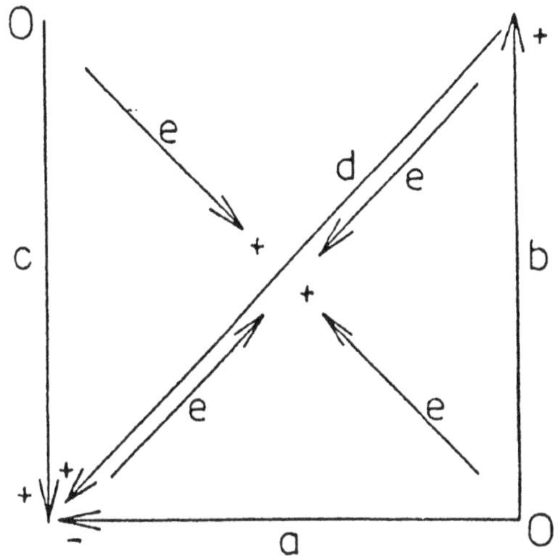

Figure 5. Ideogram of correlations between the main ecological processes which occur in the drainage basins of the shallow lakes of Polesie Lubuskie Region.
a – decreasing gradient of humidity in drainage basins, b – increasing gradient of forest cover in drainage basins, c – increasing gradient of agricultural usage in drainage basins, d – increasing gradient eutrophication, e – increasing gradient of number of species in water biocoenosis.

greatest magnitude cause a clear degradation (Ciesacin, Brzeziczno). It seems, however, that this dependence is modular rather than linear. It is possible that there are several levels of reaction of a lake biocoenosis to physical-chemical changes.

Qualitative and quantitative changes in water zoocoenoses indicate that the study area includes three groups of lakes with different degrees of variability:

- the most degraded lakes (Ciesacin, Nadrybie), which have become markedly shallow and have a small variety of species.
- the biologically richest lakes (Uściwierz, Moszne, Długie, Łukie), which are of low or moderate trophic status. Here, the high quality of water and large species diversity provide a relative ecological stability (Radwan et al., in press).
- lakes with most unstable ecological relations (Bikcze, Brzeziczno), which are eutrophic (Bikcze), or dystrophic and in the process of destruction (Brzeziczno). These lakes have a low diversity of species (Radwan et al., in press).

The results above indicate a high degree of individuality in the reaction of lake ecosystems to change in exterior factors, even where several features of these lakes are similar. To study the lake ecosystem functioning, the authors therefore suggest using a multifactor analysis.

Discussion

The study and comparison of the ecological changes that occur in the drainage basins of different lakes over longer periods may be helpful not only in gaining knowledge of the structure and function of lake ecosystems, but also in studying their functional connections with terrestrial ecosystems (O'Sulivan, 1992; Hillbricht-Ilkowska, 1984). They may also be useful in the creation of models of landscape functioning (Chmielewski, 1992; Chmielewski, et al., 1994).

The results of this research may be of great importance where the practical preservation of nature and the utilisation of the water ecosystem are concerned. (Wali, 1992; Reynolds, 1994; Radwan (ed.), 1995). The results of the present work add to the study of these problems and underlie the complexity of the issue. They also suggest, among other things, that shallow lakes may react in a very specific and complicated way to changes in their drainage basin.

The modular way in which a lake biocoenosis reacts corresponds well with e.g. analogous observations concerning the ecology of terrestrial plants (Czarnowski, 1978), and with the general concept of ecosystems and their tendency to evolve to a homeostatic state (Odum, 1972; Reichle et al., 1975; Trojan, 1980).

In turn, the fact that lake ecosystems differing greatly in size, trophic status and drainage basin management can sometimes respond in a similar, although modular way to changes of some factors in the ambient environment, may perhaps contribute to a better understanding of lake ecosystem functioning.

It may, in particular, lead to verification of methods for the evaluation of the susceptibility of shallow lakes to degradation (Bajkiewicz-Grabowska, 1987).

Conclusions

1. The ecosystems of the shallow lakes react in various ways to changes in the environment. An estimation of the directions and rate of these reactions on the basis of selected features of lakes drainage basins only may lead to a totally false picture.

It emerged that similar reactions to change were observed in such different lakes as Lake Uściwierz (low eutrophic) and Lake Moszne (dystrophic-like) while seemingly more similar lakes, like Moszne and Długie (dystrophic-like), or Lake Bikcze (dystrophic-like/eutrophic) and Lake Nadrybie (highly eutrophic) had some drastically different reactions. To avoid incorrect interpretations it is recommended that multifactor analysis of the changes be applied to lake and drainage basin features over long periods of time.

2. Larger lakes, occupying more (15–30%) of the overall area of their drainage basin (Uściwierz and Łukie) showed a higher stability in physicochemical features than smaller lakes situated in relatively large drainage basins (like lakes Brzeziczno and Nadrybie).

3. The character of some changes might show a modular type of biocoenosis reaction.

4. The character and rate of the overgrowing of open water depends to a high degree on the species composition and predominating structure of the plants in the lake and its drainage basin.

References

Bajkiewicz-Grabowska, E., 1987. Evaluation of the natural susceptibility of shallow lakes to degradation. GeoJournal 14: 367–371.

Chmielewski, T. J. (ed.), 1989. Poleski Park Narodowy. Dokumentacja naukowa. [Polesie National Park. Scientific documentation.] IGPiK, TWWP, Lublin-Warszawa: 151 pp.

Chmielewski, T. J., 1992. Próba modelowania funkcjonowania fizjocenozy jako dynamicznego układu poliekosystemowego. [An attempt of modelling of physiocoenosis considered as a dynamic polyecosystem structure.] In Chmielewski, T. J., A. Richling & K. H. Wojciechowski (eds), Funkcjonowanie i waloryzcja krajobrazu. [Functioning and evaluation of a landscape.] Int. Assotiation for Landscape Ecology, Polish Geographical Society, Lublin: 25–38.

Chmielewski, T. J. & S. Radwan, 1993. Zmiany stosunków ekologicznych w rejonie Poleskiego Parku Narodowego w ostatnich 75 latach. [Changes in ecological relations in Polesie National Park during the last 75 years.] In Radwan, S., Z. Karbowski & M. Sołtys (eds), Ekosystemy wodne i torfowiskowe w obszarach ochranianych. [Freshwater and peat-bag ecosystems in the protected areas.]. Pol. Tow. Hydrobiol., AR, TWWP, PPN, Lublin: 13–25.

Chmielewski, T. J., S. Radwan & L. Pawłowski, 1994. Using research into paths for the movement of heavy metals within the environment for the purpose of building a theory of landscape ecology, In Richling, A., E. Malinowska & J. Lechnio, (eds), Landscape research and its applications in environmental management. Warsaw University, Pol. Assoc. Landsc. Ecol., Warsaw: 165–182.

Chmielewski, T. J., B. Sielewicz & I. Butrym, 1995. Zmiany strukturalne i funkcjonalne ekosystemów jeziornych na tle zmian struktury ekologicznej rejonu Poleskiego Parku Narodowego w ciągu ostatnich 40 lat. [Structural and functional changes of lake ecosystems in the connection with the change of ecological structure of the Polesie National Park during last 40 years.] In Radwan, S. (ed.), Ochrona ekosystemów wodnych w Poleskim Parku Narodowym i jego otulinie. [Protection of the water ecosystem in Polesie National Park and its protection zone.] AR, TWWP, Lublin: 96–110.

Czarnowski, M. S., 1978. Zarys ekologii roślin lądowych. [Outline of the ecology of the terrestrial plants.]. PWN, Warszawa, 458 pp.

Edmondson, W. T. & G. G. Winberg, 1971. A manual on method for the assessment of secondary productivity in fresh waters. IBP Handb. 17: 1–358.

Fijałkowski, D., 1960. Szata roślinna jezior Łę czyńsko-Włodawskich i otaczających je torfowisk. (Wschodnia Polska.) [Plant associations of lakes and surrounding them peat-bogs in Łęzna-Włodawa Lakeland (Eastern Poland.)]. Ann. UMCS, sec. B, 14: 131–206.

Golterman, H. L., 1972. Methods for Chemical Analysis of Fresh Waters. Blackwell Scientific Publication, Oxford and Edinburgh, J.B.P. London, 166 pp.

Hillbricht-Ilkowska, A., 1984. Współczesne kierunki badawcze w ekologii wód słodkich. [Zmienność antropogenna oraz procesy wewnątrzekosystemowe [Contemporary research trends in freshwater ecology. I. Human impact on ecosystems and their internal processes.] Wiad. ekol. 30: 331–351.

Johnes, P., B. Moss & G. Phillips, 1993. Lake classification and monitoring scheme. Brampton Meeting, June 28, 1993. University of Liverpool and National Rivers Authority: 1–10.

Kornijów, R., S. Radwan, W. Kowalik, C. Kowalczyk & W. Wojciechowska, 1995. Zmiany jakościowe i ilościowe w biocenozach jezior Poleskiego Parku Narodowego i jego otuliny w latach 1955–1994. [Changes quantitative and qualitative in the lakes' biocoenosis in Poleski National Park and its protection zone in 1955–1994 years.] In Radwan, S. (ed.), Ochrona ekosystemow wodnych w Poleskim Parku Narodowym i jego otulinie. [Preservation of water ecosystems in Polesie National Park and its protection zone.]. AR, TWWP, Lublin, 69 pp.

Odum, E. P., 1972. Ecosystem theory in relation to man. In Odum, E. P. (ed.), Ecosystem Structure and Function. Oregon: 11–24.

O'Sulivan, P. E., 1992. The eutrophization of shallow costal lakes in Southwest England – understanding and recommendations for restoration, based on paleolimnology, historical records, and the modelling of changing phosphorus loads. Hydrobiologia 243/244: 421–434.

Radwan, S. (ed.), 1995. Ochrona ekosystemów wodnych w Poleskim Parku Narodowym i jego otulinie. [Protection of the water ecosystems in Polesie National Park and its protection zone.]. AR, TWWP, Lublin, 131 pp.

Radwan, S., W. Podgórski & C. Kowalczyk, 1974. Charakterystyka jezior Pojezierza Łęczynsko-Włodawskiego na podstawie abiotycznych czynników środowiskowych. [The characteristic of the lakes in the Łęczna-Włodawa Lakeland based on the abiotic environmental factors.]. Ann. UMCS, sec. C, 29: 231–246.

Radwan, S., R. Kornijów, C. Kowalczyk, W. Kowalik, B. Jarzynowa, B. Popiołek, A. Paleolog & W. Wojciechowska, (in press). Structure of biocenoses in the lakes situated in the Polesie National Park and its protection zone (Eastern Poland). Verh. int. Ver. Limnol.

Reichle, D. E., R. V. O'Neill & W. F. Harris, 1975. Principles of energy and material exchange in ecosystems. In Van Dobben, W. H., R. H. Lowe & W. McConnel (eds), Unifying Concepts in Ecology. The Hague: 27–43.

Reynolds, C. S., 1994. The ecological basis for the successful biomanipulation of aquatic communities. Arch. Hydrobiol. 130, 1: 1–33.

Trojan, P., 1980. Homeostaza ekosystemów. [Homeostasis of the ecosystems.] Wszechnica PAN, Ossolineum, Wrocław-Warszawa-Kraków-Gdańsk, 149 pp.

Wali, M. K. (ed.), 1992. Ecosystem Rehabilitation. Ecosystem Analysis and Synthesis. SPB Academic, Publishing, The Hague, 388 pp.

Wilgat, T., Z. Michalczyk, M. Turczyński & K. Wojciechowski, 1992. Jeziora Łęczyńsko-Włodawskie. [The Lakes of Łęczna-Włodawa Lakeland]. Studia Ośrodka Dokument. Fizjograf. PAN, Wrocław-Warszawa-Kraków, 19: 23–140.

Hydrobiologia **342/343**: 297–304, 1997.
L. Kufel, A. Prejs & J. I. Rybak (eds), Shallow Lakes '95.

Why does Druzno Lake (Poland) still exist? On the conditions of the pond ecosystem sustainability

Andrzej Giziński, Andrzej Kentzer & Marian Rejewski
Nicolaus Copernicus University, Department of Hydrobiology, 87-100 Toruń, Gagarina st. 9, Poland

Key words: shallow lakes, phosphorus balance, sediment's resuspension, biological production

Abstract

Druzno Lake is a semi-natural, hypertrophic basin located on the Żuławy polders (delta of Vistula). Its area is 13–29 km^2, with a mean depth of 1.25 m. The lake is extremely polymictic and of throughflow character: the water retention time is very short (20 days!). Almost 90% of the bottom is overgrown by macrophytes. The productivity, especially of phytoplankton, is very high. The main factors of ecosystem stability are: There are very good oxygen conditions in the lake. Organic matter undergoes quick mineralization. Due to frequent resuspension and intensive throughflow a considerable part of the mineral and organic seston is 'exported' out of the lake.

By dredging the navigable channel, additional amounts of suspended matter have been exported out of the system. Therefore these technical treatments slow down the processes of shallowing and aging of the lake and support the lake's existence.

Introduction

Druzno lake is a semi-natural water basin. It was formed by its separation from the southern part of Vistula Bay (Figure 1), as a result of natural development of the eastern part of the delta of the Vistula River, and the process of polderization and melioration initiated in the 12th century (Drwal, 1991). A system of polders and embankments existing presently was formed in the 16th and 17th centuries – mainly by Dutch immigrants. The lake is the main component of the melioration system of Żuławy – the region of the most fertile soils in northern Poland. In addition, it constitutes an important ornithological reserve (Nitecki et al., 1991).

At the beginning of the fifties the lake was intensively studied: hydrochemically and hydrobiologically (Mikulski, 1955; Gromadska, 1956). The results of these studies pointed out that the lake is in the last stage of overgrowing and ecological succession. Mikulski (*op. cit.*) was even afraid that the lake could disappear in 30–40 years.

The results of our studies on the lake Druzno, performed between 1986 and 1990, have turned out to be very similar to those obtained 40 years ago. The lake still exists and seems 'not to tend to disappear' in a few further centuries (Błędzki et al., 1991, Giziński et al., 1991).

The aim of this work was to answer the question raised in the title: Why does Druzno Lake still exist?, describing the surprising stability of its ecosystem.

Drainage basin, morphometry, and hydrology of Druzno Lake

The area of the drainage (catchment) basin of Druzno Lake equals ca 1100 km^2 (Cyberski & Mikulski, 1976). Żuławy territory constitutes about 24% of the area of the catchment, and Elbląg moraine plateau 76%. According to our assessments the output of phosphorus from the Żuławy territory is relatively small (0.7 kg ha^{-1} year^{-1}), but from the moraine plateau it is rather high (2.2 kg ha^{-1} year^{-1}). Correct evaluation of the input of nitrogen compounds has turned out to be impossible due to temporary pumping (to the lake) of water from polders, in which the load of this nutrient remained unknown.

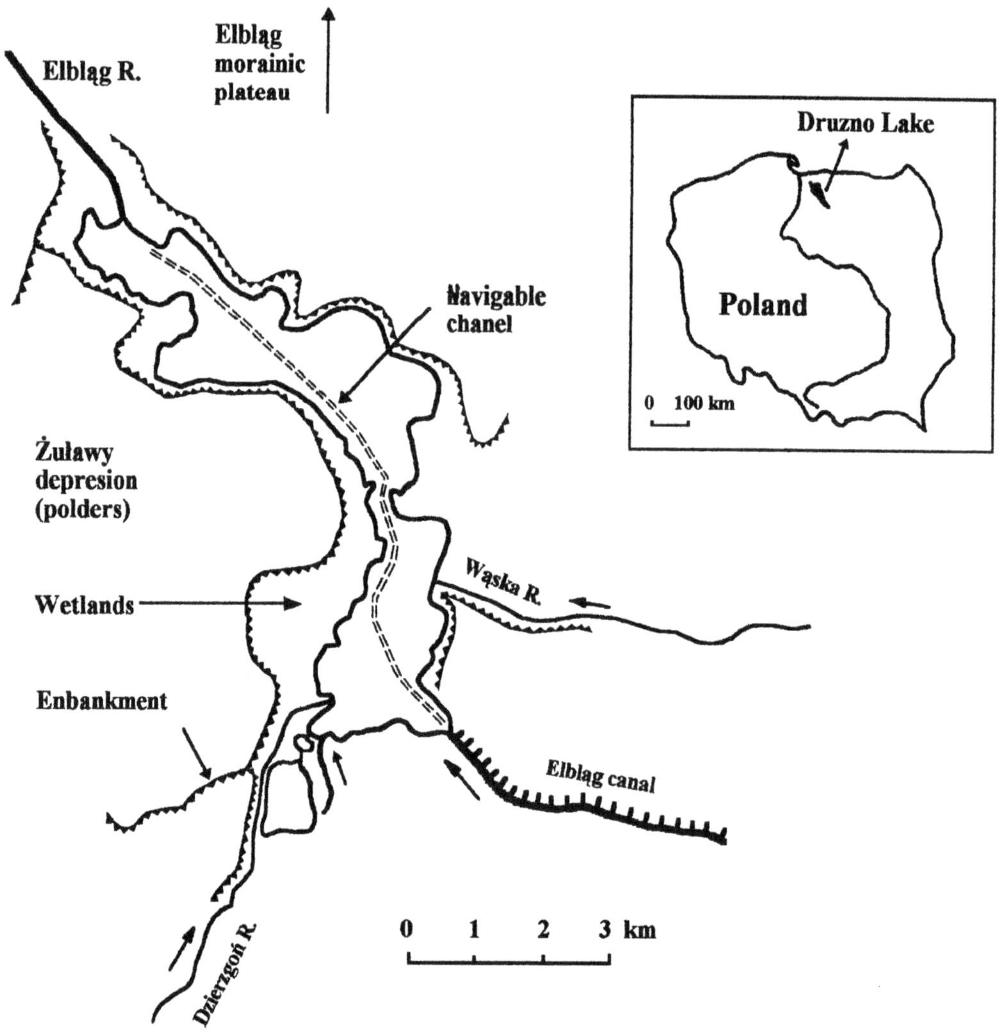

Figure 1. Localization of Druzno Lake.

The closest surroundings of the lake form timberless plains. Coastal trees do not constitute an efficient protection against winds.

The surface of the lake's water is about 1.8 m above the adjacent polder terrains. The lake and its tributaries are surrounded by artificial embankments. The level of water in the lake is determined by climatic factors and also by melioration needs.

The most important morphometric features (Cyberski & Mikulski, *op. cit.*) of Druzno Lake are variable, depending on the water level:

- Area: 12.6–29.8 km^2; we assumed the value of area mentioned most frequently by various authors, i.e. 17.9 km^2.
- Max. depth: 3 m; mean depth: ca. 1.25 m.
- Volume: 22.4 10^6 m^3.
- Max. length: 10 km; max. width: 2.2 km.
- Thickness of sediments: 6–10 m.

Due to the large area of the lake, small depth, and exposure to the action of wind, Druzno Lake is an extremely polymictic basin. Water dynamics is enlarged by considerable throughflow (Q), and dimin-

Table 1. Water balance of Druzno Lake. (after Cyberski J. & Z. Mikulski, 1976, with our modification)

Input	10^6 m^3	% of input	Output	10^6 m^3	% of output
Precipitation	15.5	4	Evaporation	14.8	~4
			Plant transpiration*	45.2	11
Inlet from Vistula Lagoon	89.2	22			
Inflow			Outflow		
– surficial	246.4	62	–surficial	342.6	85
– ground	49.0	12	–ground	–	–
			$\sqrt{\text{summ}}$	402.6	100
Total	400.1	100	Total	400.1**	100.0

* Our evaluation. ** Lowering of water level; minus $2.5 \cdot 10^6$ m^3.

ished by the high degree of overgrowth by macrophytes.

We have made a correction to the water balance presented in Table 1 – considering plant transpiration: 90% of the lake's area is overgrown by emergent plants (helophytes) and by plants with floating leaves (nymphaeids). It is known that plant transpiration is many times higher than evaporation from a water surface. We have assessed that the total plant transpiration in Druzno Lake equals 45 10^6 m^3 per year. Total output remains unchanged, because we have appropriately diminished the surface outflow.

The correction we have made has an essential importance for the proper hydrochemical characterization of the lake. Evaporation and transpiration constitute as much as 15% of throughflow (Q). This means that, due to evaporation and transpiration, concentration of substances dissolved in lake water increases on average by 15%. The total exchange of water takes place as many as 18 times per year, i.e. the retention time equals only 20 days. The above values are rather more typical of dam reservoirs than lakes.

The lake is used for recreational activities on a large scale. From the Elbląg Canal to the Elbląg River runs a navigable canal, which is periodically (every few years) deepened (Figure 1). There is a zone of intensified sedimentation after periods of resuspension of sediments.

Chemical characteristics of water and bottom sediments of the lake studied

Hydrochemistry

Druzno Lake is definitely hypertrophic. Nutrient concentrations are high for the whole year:

$$TP - 0.9 \text{ mg l}^{-1}$$
$$N - 4.8 \text{mg l}^{-1}$$

A high share of orthophosphates in total phosphorus ($P\text{-}PO_4$ = ca. 40% of TP) and very low value of N:P ratio 5-6 (Forsberg et al., 1978) are typical for polytrophic lakes. The results of these studies on nitrogen compounds point out the occurrence of large amount of organic matter in lake water. Organic nitrogen constitutes more than 90% of total nitrogen. Due to intensive decomposition of organic nitrogen compounds ammonia definitely predominates:

$$N\text{-}NO_3 = 0.029;\ N\text{-}NO_2 = 0.014;$$
$$N\text{-}NH_4 = 0.392 \text{ mg l}^{-1}$$

It is worth emphasizing that the oxygen conditions are good in the lake studied, especially in the zones of open water (without macrophytes), where, during sunny weather, a hypersaturation is noted reaching as much as 200%. Oxygen saturation rarely decreases below 60%. After the vegetation season, when macrophytes disappear, oxygen conditions in the lake are even better. Only at periods immediately after intense resuspension of sediments were short-term decreases of oxygen content observed up to values close to zero in the layer of water just above the bottom. Good oxygen conditions in Druzno Lake are mainly determined by the intense phytoplankton photosynthesis and processes of mixing. Even in the near bottom layer of water, oversaturation of oxygen was often observed (Wiśniewski, 1995). Water is additionally oxygenated during pumping of water from polders to the lake and to its tributaries (Cebulak, 1991).

The remaining properties of Druzno Lake are typical of 'normal', strongly eutrophic lakes. It is worth mentioning that the content of chlorides is high, and equals on average 50 mg l^{-1}, which is implicated by the periodical inlet of brackish waters from Vistula Lagoon.

Bottom deposits

Hydration of the bottom deposits reached approximately 95%, and organic matter content was unexpectedly low (17%). It was caused by intensive mineralization of organic matter (good oxygen conditions, frequent resuspension of bottom deposits).

Phosphorus content (TP) in bottom sediments of different parts of the lake was different: in the southern part, where tributaries are richest in phosphorus, concentration of TP was equal to 2–3 mg g^{-1} DW, whereas in the northern part it was considerably lower, ca. 1 mg g^{-1} DW. Average value for the whole lake equals ca. 1.5 mg g^{-1} DW (Figure 2).

Processes of phosphorus accumulation in bottom deposits of the lake are intensive, which is implicated by the above-mentioned good oxygen conditions. Release of phosphorus from deposits to water is negligible; it takes place only in the case of metal-containing compounds and humic substances (Hieltjes & Lijklema,1980). According to Nurnberg (1988), this form of phosphorus is soluble only at low concentrations of oxygen, but these are both rather rare and short-lasting in the lake studied. Studies on sedimenting tripton in open-water zones and at places with few macrophytes have shown (Żytkowicz, unpubl. date) that tripton chemical composition is almost identical to the composition of bottom deposits. Żytkowicz calculated that resedimentation is about 40–50-fold higher than that theoretically calculated, the maximal net sedimentation. It means that ca. 95–97% of sedimenting matter originates from the resuspension. The 'statistical' particle of bottom sediment is probably subjected 40–50 times to resuspension and mineralization in the well-oxygenated lake water. The regulatory role of resuspension was described by Wiśniewski (1995).

Biological characteristics

Plankton

The occurrence of 183 species of phytoplankton was noted, among them 112 diatoms species. Also zooplankton (Błędzki et al., 1991) was very rich, comprising 84 species; among them are 43 species of Rotatoria. Species composition of plankton was typical of lowland dam reservoirs and of water bodies with short residence time. Small-size species predominated.

Extraordinarily high were: biomass (standing crop) and production of zooplankton, and especially that of phytoplankton (Table 2). The ratio of biomass of zooplankton to phytoplankton (0.026) is about 10-fold lower from values typical for polymictic lakes (see: Hilbricht-Ilkowska et al., 1986). Thus zooplankton cannot control the enormous production of phytoplankton, which constitutes the main channel of matter flow, and its transformation in the ecosystem studied.

Annual production of phytoplankton in Druzno Lake equals on average 165 ton per hectare (!), being thus higher than e.g. crops of mangel from intensively fertilized agricultural fields (ca. 70 tons). The annual production of carbon equals 21 700 tons, exceeding the production of macrophytes by one order of magnitude (see below).

Macrophytes

The lake vegetation of Druzno Lake is rich, being represented by about 60 species. Among them is one on the verge of extinction, *Nymphoides peltata* (S. G. Gmel.) Kuntze. The most common species are: *Nuphar lutea* (L.) Sibth., *Nymphaea alba* L., *Phragmites australis* (Cav.) Trin. ex Stend. and *Typha angustifolia* L. Water vegetation covers about 90% of the area of the lake, constituting the habitat for the rich and diverse fauna. By 'covered by macrophytes' we mean those parts of the lake where plants (e.g. leaves of *Nuphar lutea*) covered at least 30% of water surface area. The depth of plant zone was usually 0–1 m. Macrophytes modify the environmental conditions, e. g. by lowering of resuspension. We also suppose that the rich development of macrophytes can limit the development of phytoplankton due to shading, because during the peak of development of macrophytes we noted a decrease of phytoplankton biomass, despite a high concentration of available forms of nitrogen and phosphorus in the water. The growing season was from May to October. Most of the macrophytes disappeared in winter.

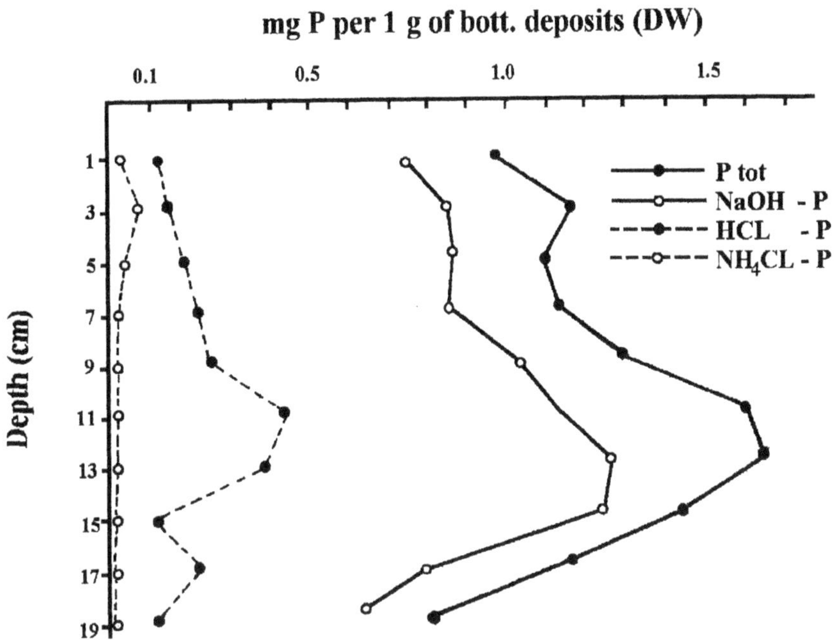

Figure 2. Phosphorus (TP) and its fractions in Druzno Lake bottom deposits.

Table 2. Biomass (standing crop) and production (per year) of plankton.

	Biomass (d. w.)			Production (d. w,) per year		
	$g\ m^{-3}$	$g\ m^{-2}$	t per lake	$g\ m^{-3}$	$g\ m^{-2}$	t per lake
Phytoplankton	1.7	14.6	2.61	13.20	16.50	296.15
Zooplankton	0.3	0.38	6.8	12	15	2.70

We have calculated that the average annual production of above-ground parts of macrophytes in Druzno Lake exceeded 11 thousand tons DW (Table 3).

On the basis of the data of Rejewski (1979) we have assessed that the annual assimilation (cumulation) of the most important elements in the biomass of macrophytes equals (in tons):

N- 176.8, K- 133.1, Ca- 54.7, Na- 38.5, P- 19.5, Mg- 15.3, Fe- 13.3

It is obvious that it is a temporary (for ca. 6 months) accumulation. However, it exerts a significant effect on the dynamics of matter transformation in the lake.

Macrozoobenthos, fish and avifauna

The taxonomic composition of bottom fauna was rather poor. Among Chironomidae, which were studied more in detail, only 22 species and forms were stated. The majority of them constituted eurytopic forms, ubiquitous, typical of shallow, eutrophic lakes and dam reservoirs. Numbers (12.2 thousands ind. m^{-2}) and biomass (25.7 g m^{-2}, fresh weight, without Mollusca) were high. Oligochaeta definitely predominated, similarly as in lowland dam reservoirs (Gizinski et al., 1989).

Dynamics of seasonal changes of macrozoobenthos were typical for pond-type basins; maximum peaks were noted in spring and autumn, and minimum in summer.

The contribution of zoobenthos in the trophic pyramid of Druzno Lake is considerable; annual production (fresh mass) equals 100 g m^{-2}; 1000 kg ha^{-1}, 1800 tons per lake. Secondary production of macrozoobentos constitutes a very good nutrient basis for fishes.

Annual fishing catches (tentative data) are equal to ca. 49 kg ha^{-1}, i.e. they are typical of highly productive lakes, much deeper than Druzno Lake (and thus having a larger capacity of trophogenic layer).

Table 3. Average annual production of above-ground parts of macrophytes in Druzno Lake (in tons).

	Wet weight	Dry weight	C
Nymphaeids	25 571	2 708	367
Helophytes	23 878	8 320	2 515
Together	49 449	11 028	2 882

Table 4. Biomass and production of organic matter in Druzno Lake (in tons).

	Organic matter (biomass, standing crop)	Annual production of organic matter
Macrophytes	9 925	9 925
Phytoplankton	354	39 960
Periphyton	21	2 070
Together Primary producers	10 300	51 955
Zooplankton	9	367
Zoobenthos	42	257
Necton (fishes)	29	19
Together Secondary producers	80	643

* It was assumed, that the biomass of fishes is 3-fold larger, and biological production 2-fold larger than the annual fishing catch (Opuszyński, 1979).

The largest contribution in catches have benthivorous and predatory fishes. Bream (*Abramis brama* L.) definitely predominates, constituting ca. 77% of total catches. Among the predators the most important is pike (*Esox lucius* L.) (6%).

As mentioned above, the lake constitutes the ornithological reserve. Nitecki et al. (1991) stated the occurrence of 207 species of birds (among them 108 nesting species) both on the lake itself, and on marshes surrounding the lake (total ca. 30 km^2). It comprises more than 50% of the avifauna of Poland. The most numerous species is *Larus ridibundus* L. Some species, e.g. *Anas strepera* L. and *Porzana parva* (Scop.) belong to those which are endangered. Their number on the lake constitutes more than 1% of the whole European population. Nitecki et al. (*op. cit*) calculated the consumption of food, and production of faeces by the most numerous populations of birds. The total consumption was equal to 1876 tons, and production of excrements was 1003 tons (WW). It should be mentioned that, according to the cited authors, about 800 tons of food may derive from the land. Therefore it can be concluded that the amount of faeces is slightly lower than the amount of food taken from the lake. The cited authors express an opinion that birds (mainly *Cygnus olor* (Gmel.) and *Anas* spp.) consuming several hundreds tons of aquatic plants per year can inhibit the process of shallowing of Druzno Lake. In our opinion, this consumption constitutes about 15% of annual production of macrophytes, and thus it can rather mechanically diminish the process of excessive overgrowing of the lake.

Tentative assessment of the balance of phosphorus in the lake

The balance of phosphorus (Figure 3) was estimated on the basis of results of analyses of inflowing and outflowing water. The six largest tributaries were studied. The mean concentration of phosphorus (P) was 0.885 mg l^{-1}. It was assumed that the concentration of P in the remaining tributaries and in ground water is the same. The annual input of P was 244 t, output 201 t, and cumulation 43 t (i.e. ca. 18% of the total input).

On the basis of the average rate of sedimentation, Żytkowicz (unpubl. data) assessed that the annual accumulation of phosphorus in Druzno Lake can even exceed 60 tons (~25% of total input). Due to good oxygen conditions in Druzno Lake, processes of P accumulation predominate over the reverse ones. It means that most of the mentioned 60 tons is rather stably deposited in bottom sediments. This confirms the above-presented balance of phosphorus.

Thus, it can be concluded that Druzno Lake acts as a large purification plant, reducing the phosphorus load in throughflowing water by almost 20%. A similar degree of reduction of phosphorus load is also noted in Vistula River, in Włocławek Dam Reservoir (Kentzer & Giziński, 1995).

Conditions of stability of the studied ecosystem

Anxiety was expressed during the 1950's that the lake could become completely overgrown after a few decades, effectively becoming land. However, the lake still exists. Its ecosystem reveals a surprising stability. The results of hydrochemical and hydrobiological studies performed at the beginning of the fifties and at the end of the eighties were quite similar.

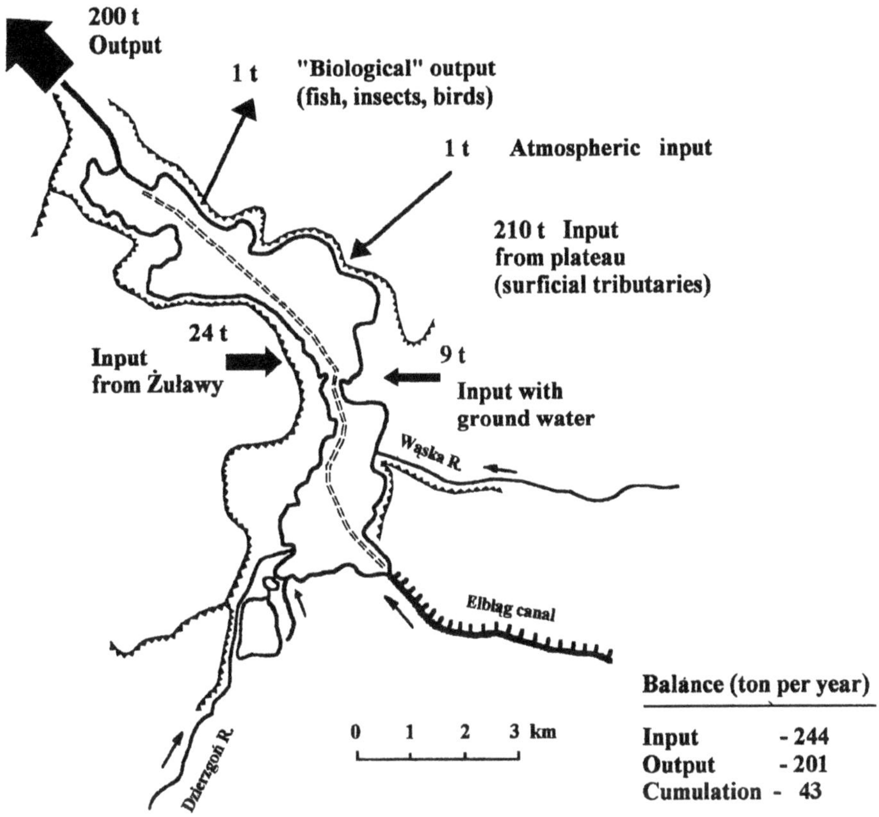

Figure 3. Phosphorus (TP) balance in Druzno Lake.

The contribution of biota to the functioning of the Druzno Lake ecosystem is enormous, which can be confirmed by the high biomass and high production of organic matter (Table 4). There are no dramatic year-by-year changes, fish deaths, or excessive amounts of H_2S, nor are there any other symptoms of disturbance of the ecosystem equilibrium. The lake ecosystem is 'moving into a higher gear', but its functioning is effective.

The main factors of stability and efficiency of the ecosystem studied are as follows:

1. The short water retention time (20 days!) and exposure of the reservoir studied to wind are responsible for the exceptionally high dynamics of water masses: it is an extremely polymictic lake.
2. Due to the reasons mentioned above, and also due to intensive photosynthesis, there are good oxygen conditions in the lake. Water is additionally oxygenated during pumping from polders to the lake. Under good oxygen conditions both allochtonic and autochtonic organic matter undergo quick mineralization, without producing remarkable amounts of H_2S. Due to the good oxygen conditions, the accumulation of phosphorus in the bottom deposits is relatively stable.
3. High fertility, in conjunction with good oxygen conditions, provides suitable conditions for hydrobionts development. The profuse development of macrophytes favours the development of avifauna, whose feeding mechanically retards the overgrowing of the lake by plants.
4. During periods of resuspension in a lake with such an intensive throughflow, a considerable part of resuspended matter is exported. The shallower a lake, the more frequently its sediments undergo resuspension. The more frequent is resuspension, the more seston is exported out of the system, i.e. the process of shallowing of the basin is slowed

down. For lakes having a strong throughflow, resuspension is thus a basic mechanism slowing down their shallowing.

5. Deepening of the navigable channel (repeated every few years) is an additional factor of stability of the Druzno Lake ecosystem. During these manipulations, it is certain that large amounts of suspended matter are being exported beyond the system. Therefore these treatments slow down the processes of aging of the lake and support lake's existence.

References

Atlas Hydrologiczny Polski (Hydrological Atlas of Poland) 1986, Wydawnictwa Geologiczne, Warszawa, 127 pp.

Błędzki, A. L., A. Giziński & A. Kentzer, 1991. Funkcjonowanie ekosystemu Druzno: relacje zlewnia – basen. (Functioning of Druzno Lake Ecosystem: Basin/Lake relationships.) In Drwal, J. (ed.), Basen jeziora Druzno. Regionalny system przyrodniczo-techniczny. (Druzno Lake Basin. The Regional Natural – Technical System.) Univ. of Gdańsk: 69–85.

Cebulak, K., 1991. Ocena i prognoza rozwoju polderowego systemu melioracji basenu jeziora Druzno. (Evaluation and Prognose of Development of the Polder System of Druzno Basin Melioration.) In Drwal, J. (ed.), Basen jeziora Druzno. Regionalny system przyrodniczo-techniczny. (Druzno Lake Basin. The Regional Natural – Technical System.) Univ. of Gdańsk.: 19–38.

Cyberski, J. & Z. Mikulski, 1976. Stosunki Hydrologiczne. (Hydrological Relations.) In Żuławy Wiślane, GTN, Gdańsk: 239–288.

Drwal, J., 1991. Przyrodniczo-techniczne uwarunkowania powstania, trwania i dalszej ewolucji basenu jeziora Druzno. (Natural – Technical Conditions of Origin, Existing and Further Evolution of Druzno Lake Basin.) In Drwal, J. (ed.), Basen jeziora Druzno. Regionalny system przyrodniczo-techniczny. (Druzno Lake Basin. The Regional Natural – Technical System.) Univ. of Gdańsk: 7–17.

Forsberg, C., S. O. Ryding, A. Forsberg & A. Cleasson, 1978. Research and recovery of polluted lakes. I. Improved water quality in lake Boren and Lake Ekoln after nutrient reduction, Verh. int. Ver. Limnol. 20: 825–832.

Giziński, A., L. A. Błędzki, A. Kentzer, M. Rejewski, R. Wiśniewski, J. Żbikowski & R. Żytkowicz, 1991. Hydrobiologiczna charakterystyka jeziora Druzno. (Hydrobiological Characteristics of Druzno Lake.) In Drwal, J. (ed.), Rozpoznanie i ochrona ekosystemow. Artykuły i doniesienia. Część I. (Recognition and protection of ecosystems. Articles & Reports. Part I) Falenty/Elbląg-Olsztyn: 151–163.

Giziński, A., L. A. Błędzki, A. Kentzer, R. Wiśniewski & R. Żytkowicz, 1989. Hydrobiological characterystics of the lowland, rheolimnic Włocłwek Reservoir on the Vistula River. Ekol. pol. 37: 359–403.

Gromadska, M., 1956. Stosunki fizyko-chemiczne wody jeziora Druzno. (Physical-chemical relations of Druzno Lake water.) Ekol. pol. Ser. A, 4: 317–382.

Hieltjes, A. H. M. & L. Lijklema, 1980. Fractionation of inorganic phosphates in calcareous sediments. J. envir. Qual. 9: 405–407.

Hilbricht-Ilkowska, A., J. Ejsmont-Karabin & T. Węgleńska, 1986. Przydatność planktonowych wskaźników stanu trofii jezior do oceny zmian wieloletnich w systemie jezior podgrzanych. (Usefulness of the planktonic indices of trophy in lakes for the evaluation of long-term changes in the system of heated-up lakes.) In Monitoring ekosystemów jeziornych. (Lake ecosystem monitoring.) Ossolineum, Wrocław: 71–81.

Kentzer, A. & A. Giziński, 1995. Bilans i dynamika nutrientów w Zbiomiku Włocławskim (Nutrients balance and dynamics in Włocławek Reservoir.) In Zalewski, M. (ed.), Procesy biologiczne w ochronie i rekultywacji nizinnych zbiorników zaporowych. (Biological processes in the conservation and restoration of lowland reservoirs.) PIOŚ, Łódź: 85–90.

Mikulski, J., 1955. Jezioro Druzno – próba charakterystyki limnologicznej (Druzno Lake – a trial of the limnological characterization.) Ekol. pol., Ser. A, 3: 1–31.

Nitecki, Cz., S. Strawiński & J. Ulatowska, 1991. Stan i perspektywy rozwoju fauny kręgowców jeziora Druzno. (State and perspectives of vertebrate fauna at Druzno Lake.) In Drwal, J. (ed.), Rozpoznanie i ochrona ekosystemów. Część I. (Recognition and protection of ecosystems. Articles & Reports. Part I.) Falenty/Elbląg-Olsztyn: 177–185.

Nurnberg, G. K., 1988. Prediction of phosphorus release rates from total and redoctant soluble phosphorus in anoxic lake sediments. Can. J. Fish. aquat. sci. 145: 453–462.

Opuszyński, K., 1979. Podstawy biologii ryb. (Principles of fish biology.) PWRiL, Warszawa: 480 pp.

Rejewski, M., 1979. The structure and function of macrophytic communities in two eutrophic lakes in the Iława Lake District. Acta Univ. N. Copernici, Limnological Papers 11: 19–34.

Wiśniewski, R., 1995. The regulatory role of sediment resuspension in seston, oxygen and phosphate dynamics in shallow lake Druzno. In Proc. 6th Int. Conf. on the Conserv. and Manage of Lakes. Kasumigaura 1995, Japan: 917–920.

Hydrobiologia **342/343**: 305–310, 1997.
L. Kufel, A. Prejs & J. I. Rybak (eds), Shallow Lakes '95.

Food-web manipulation in shallow, eutrophic lakes: bridging the gap between the whole-lake approach and behavioural and demographic studies

Joanna Pijanowska & Andrzej Prejs
Department of Hydrobiology, University of Warsaw, Banacha 2, 02-907 Warsaw, Poland

Key words: biomanipulation, grazing, herbivores, density, size structure, vertical distribution, behaviour

Abstract

Even in the most thorough biomanipulation studies, which cover the response of all trophic levels to manipulation at the top of the food chain, the mechanisms underlying the success or failure of an experiment often remain obscure. The case of a small eutrophic Lake Wirbel in Poland is an example of how useful it could be to apply demographic and behavioural analysis of dominant planktonic herbivores to explain the sequence of events following biomanipulation measures.

At a certain stage of a seven-year (1988–1994) biomanipulation experiment, rotenone was applied to remove all fish. Following fish removal, summer algal biomass in the 'edible' fraction of phytoplankton ($<30\ \mu$m) was on average 2.5 times lower. Besides the significant increase in the density of herbivorous zooplankton (above all the largest cladoceran *Daphnia cucullata*), we identified three other phenomena underlying the observed reduction in algal biomass which may contribute to the overall zooplankton grazing efficiency: (1) the significant increase in the mean body size of the dominant cladoceran *Daphnia cucullata*, calculated for the whole water column, (2) the clear shift in the vertical distribution of *Daphnia* towards the surface and (3) the increase in mean *Daphnia* size in the surface waters.

We postulate that this type of analysis should be performed along with the measures conventionally used in biomanipulation studies, first, as a tool to fully understand the sequence of events generated by manipulations at the top of the food chain and, second, as a perfect opportunity to study population and individual responses (e.g. morphological, life-history and behavioural) to varying predatory pressure.

Introduction

Most biomanipulation studies aimed at the improvement of water quality are based on top-down manipulations within a food chain. Though principal food web links are conventionally studied in many such manipulation projects, the cascade of events along the entire food chain following a change in the fish stock is seldom used to interpret observed changes in water quality. Such an approach meets the criteria of a type 4 biomanipulation study, from the four types of the whole-lake study proposed by Benndorf (1990). The 'ideal' type 4 study should cover: (1) planktivorous fish density and biomass, the piscivorous/planktivorous fish ratio and the relative share of different species, (2) zooplankton biomass and density, the proportion of herbivorous species and the size structure within the zooplankton assemblage, (3) algal biomass and taxonomic composition, and (4) basic measures of water quality, including nutrient content and water transparency.

In successful experiments which lead to the increase in water transparency associated with reduced algal biomass, the final effects are most often attributed to increased grazing impact from herbivorous zooplankton freed from fish predation pressure (Carpenter & Kitchell, 1993). If it occurs, the decline in algal biomass, is usually attributed to an overall increase in the grazing efficiency of herbivorous zooplankton related to density–biomass–composition changes. In the response to the decreased fish pressure, demographic and behavioural changes within herbivorous zooplankton are most often neglected (see overviews of the most recent whole-lake experiments, Gulati et al., 1990 and Mortensen et al., 1994). However, such studies on herbivorous zooplankton offer a chance to demonstrate whether, or to what extent, the reduction in phytoplankton biomass is a result of the sequence of events cascading over the pelagic food web or of other factors unrelated to zooplankton grazing (see de Melo et al., 1992; Carpenter & Kitchell, 1992). A whole-lake study, carried out for several years in Lake Wirbel – a small eutrophic lake in Poland is an example of the possible advantages of taking the step from community studies to the demographic and behavioural approach.

Biomanipulation rationale

A whole lake experiment was carried out for the seven years 1988–1994 in shallow, eutrophic Lake Wirbel, Poland (11 ha, 1.8 m mean depth, 4.4 m max. depth). In the first phase (1989–1991), repeated introductions of young-of-the-year pike (*Esox lucius*) were used to control the density of juvenile stages of the dominant planktivores (roach *Rutilus rutilus*, white bream *Blicca bjoerkna* and *Leucaspius delineatus*. Successive introductions of juvenile pike were accompanied by selective removal of large pike, roach, bream *Abramis brama* and white bream. In October 1991, rotenone was applied to remove all fish (Prejs et al., 1994).

Density, size structure, fecundity, size at maturity and the vertical distribution of a dominant cladoceran were analyzed along with basic measures of water quality (for details see Prejs et al. (1997) and Pijanowska & Prejs, unpubl. data) at 2–3 week intervals: May–October 1991 (before fish removal) and May–October 1992 (after fish removal).

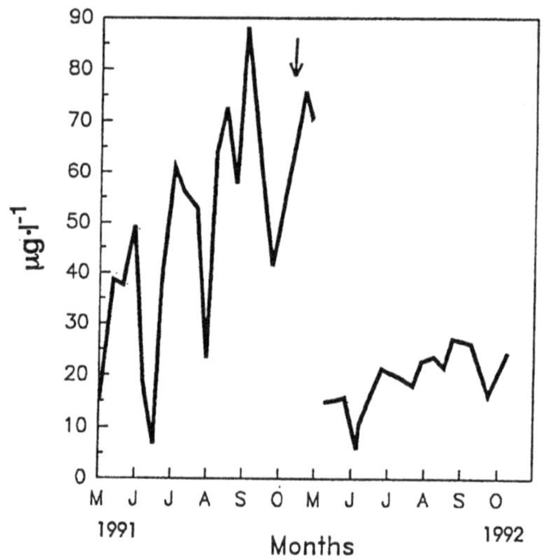

Figure 1. 'Edible' algal biomass expressed as chlorophyll *a* concentration in the fraction <30 µm. Arrow indicates rotenone treatment.

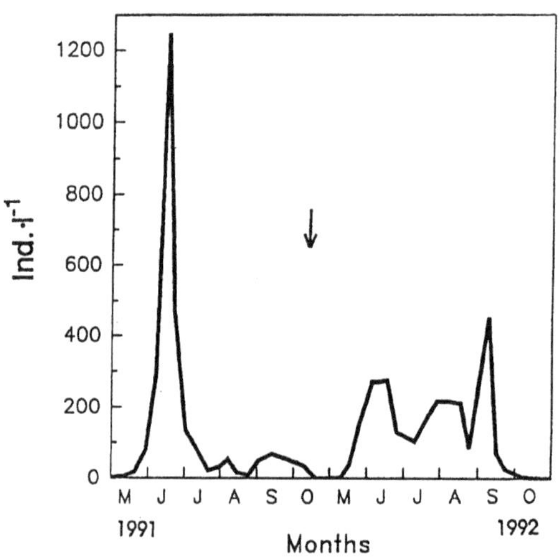

Figure 2. Total density of herbivorous cladocerans.

Algal biomass

After rotenone treatment, there was – in the summer months (June–August) of 1992 – a considerable (on average 2.5-fold) decrease in algal biomass in the 'edible' fraction of the seston particles (smaller than 30 µm), as compared to summer 1991 (Figure 1).

Zooplankton density

The year following the rotenone treatment witnessed no significant ($P<0.05$, ANOVA) increase in the total number of herbivorous plankton (Figure 2). However, the former dominant *Bosmina longirostris* was replaced by *Daphnia cucullata*, which, on average, constituted less than 10% of the total number before, but up to 80% after fish removal. The difference in the *Daphnia* density before and after rotenone treatment (Figure 3) was highly significant ($P=0.001$, $F_{1,34}=13.84$, ANOVA).

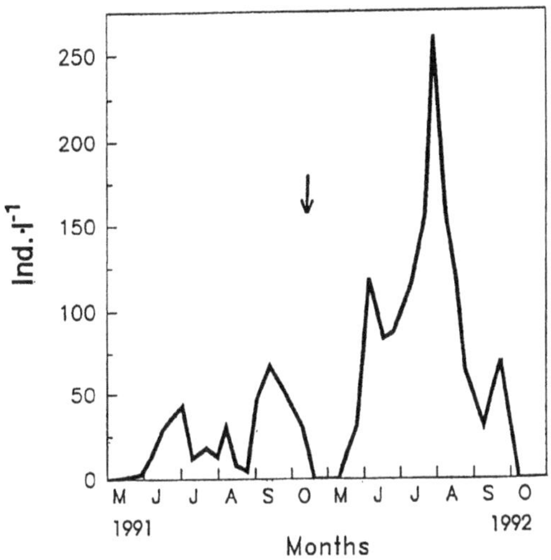

Figure 3. Density of *Daphnia cucullata.*

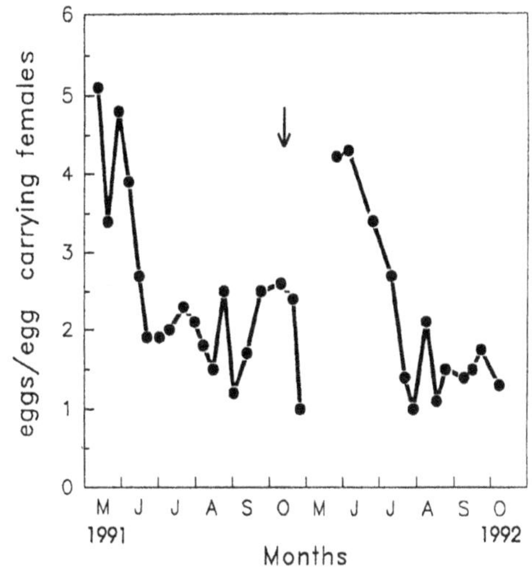

Figure 4. Mean number of eggs carried by individual *Daphnia.*

Does the effect of fish removal cascade over the entire food web?

The increase in *Daphnia* density and the change in the proportions of the dominant species can either be attributed to reduced mortality following fish removal or to enhanced recruitment due to a change in the level of resources (Gliwicz et al., 1981). Under conditions of depleted food supply (Figure 1), *Daphnia cucullata* – as the largest cladoceran – could easily monopolize limiting resources, outcompeting the previously-dominant smaller species. The increase in *Daphnia* density was not related to enhanced reproduction, since a significant decrease in fecundity was observed after fish removal ($P = 0.0089$, $F_{1,34} = 8.01$, ANOVA with body size as a covariate; Figure 4). Simultaneously, a significant ($P = 0.0001$, $F_{1,34} = 31.37$, ANOVA) increase in *Daphnia* body size followed fish removal (Figure 5). There was an apparent contradiction between the increase in mean body size and the simultaneous decline in fecundity (for the relationship between body size and fecundity in planktonic cladocerans see Green 1956). This provided an impulse for consideration of size/fecundity structure, which revealed: (1) an increase in the proportion of individuals larger than 1000 μm following fish removal (Figure 6), (2) an increased proportion of egg carrying females in larger size classess, from 0 to 100% in the two largest size classes (Figure 7); the proportion of egg carrying females in the three smallest size classes was greater when fish were still present in the lake, than after fish removal (<1, 24, 74 before and 0, 2 and 46 afterwards), (3) an increased contribution of females from the four largest size classes (above 1350 μm) to the total number of egg carrying females after fish removal (altogether 42% as compared to 9, Figure 8); females from the first three size classes contributed to the total number of egg carrying females to a greater extent before than after fish removal (<1, 27 and 63% as compared to 0, 2 and 55), (4) an increase in the mean number of eggs per female in the three largest size classes (Figure 9). A significant ($P = 0.0033$, $F_{1,34} = 10.99$, ANOVA) increase in the size at maturity (Figure 10) was observed after fish removal; with no or low mortality risk for large individuals, reproducing at larger size might bring fitness advantages (Stearns 1992).

After fish removal, a change in *Daphnia* vertical distribution was also observed. This entailed a significant ($P = 0.0028$, $F_{1,11} = 23.58$, ANOVA) shift of the mean population depth towards the surface (Figure 11): from 1.67 ± 0.17 m in 1991 to 1.46 ± 0.13 m in 1992 (mean $\pm$ SD for June–August); a shift in the % distribution was observed: from 24 (0–1 m), 36 (1.1–2 m) and 40% (2.1–3.8 m) in summer 1991 to 38, 40 and 22% respectively in summer 1992. There was also a significant ($P = 0.0169$, $F_{1,53} = 6.16$, ANOVA) increase in the mean size of the individuala in the surface water layers (12 m; Table 1).

Taken together, the increase in mean body size (Figure 5), the respectively increased and decreased roles of the largest and the smallest individuals in reproduc-

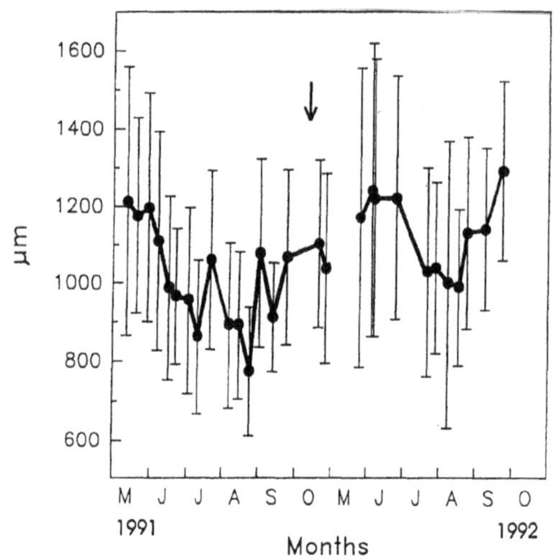

Figure 5. *Daphnia* mean body size (means ± 1SD).

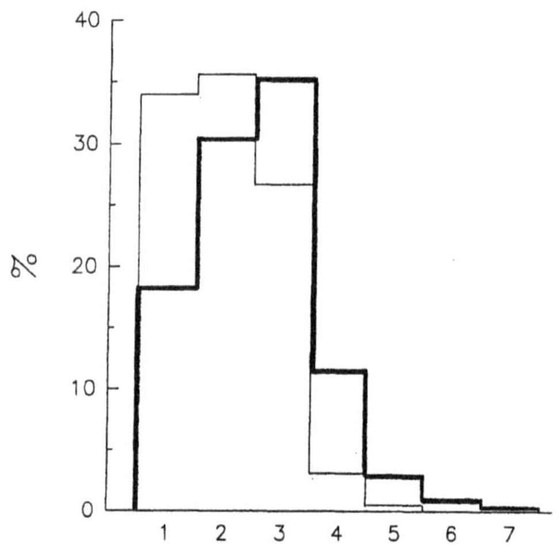

Figure 6. Summer (June–August) size structure (% of total number of individuals before fish removal (solid line) and after it (heavy solid line); means for June–August. Size classes: 1: <815 μm; 2: 816–1078 μm; 3: 1079–1341 μm; 4: 1342–1604 μm; 5: 1605–1867 μm; 6: 1868–2130 μm; 7: >2130 μm.

Table 1. Mean (mm ± SD) body size of *Daphnia cucullata* (at least 50 individuals measured) at different depths, calculated separately for the entire study period and for summer months, in the reference year (1991) and in the year following the rotenone treatment (1992).

Depth (m)	Study periods VI-IX 1991	VI-IX 1992	IV-X 1991	IV-X 1992
0.0–1.0	0.89 ± 0.14	0.99 ± 0.09	0.92 ± 0.15	1.04 ± 0.19
1.1–2.0	0.90 ± 0.08	0.98 ± 0.08	0.95 ± 0.11	1.01 ± 0.15
2.1–3.8	1.10 ± 0.08	1.13 ± 0.13	1.17 ± 0.14	1.15 ± 0.13

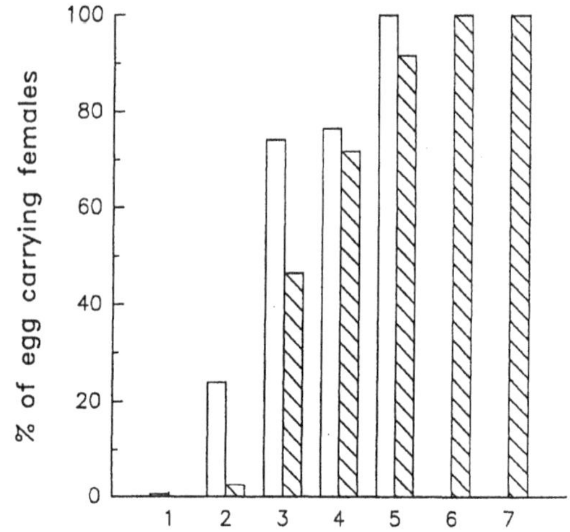

Figure 7. Proportion of egg carrying *Daphnia* in total number of individuals in different size classes before fish removal (open bars) and after it (dashed bars); means for June–August. Size classes as on Figure 6.

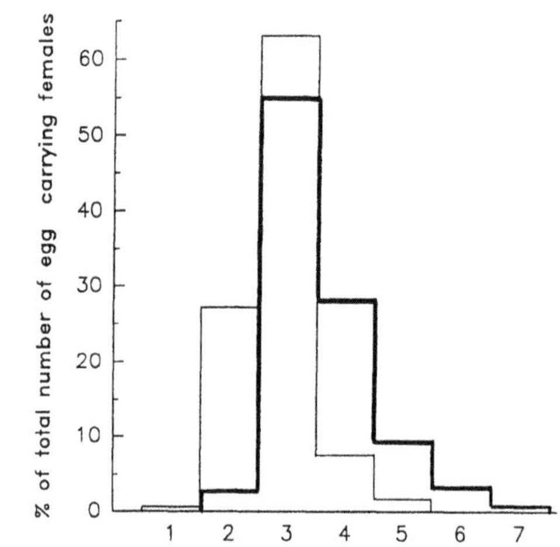

Figure 8. Proportion of egg carrying *Daphnia* in different size classes among total number of egg carrying females, before fish removal (solid line) and after it (heavy solid line); means for June–August. Size classes as on Figure 6.

tion (Figures 6, 7, 8 and 9), the significant increase in the size at maturity after fish removal (Figure 10), the shift in the mean population depth towards the surface (Figure 11) and the increase in the mean size of individuals in the surface layers (Table 1) would seem to make it clear that increased *Daphnia* density (Figure 3) resulted from reduced mortality following fish removal.

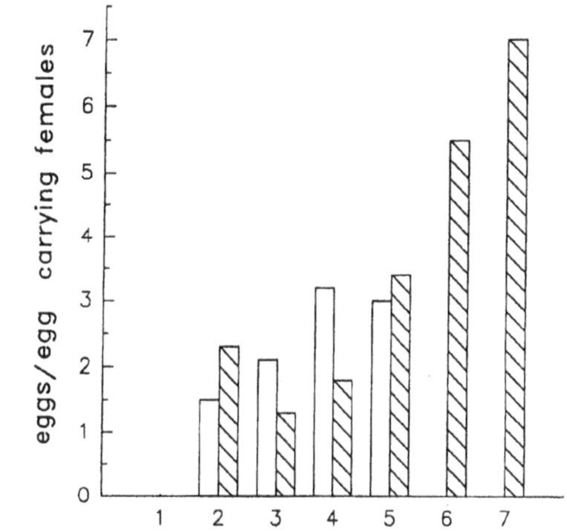

Figure 9. The mean number of eggs carried by individual *Daphnia* in different size classes before fish removal (open bars) and after it (dashed bars); means for June–August. Size classes as on Figure 6.

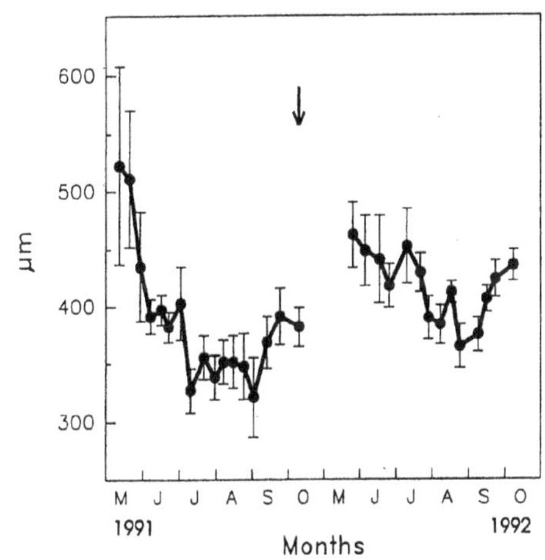

Figure 10. *Daphnia* size at maturity (mean size of 10 smallest adult females, identified according to criteria from Edmondson and Litt (1982) and Stibor & Lampert (1993).

To what extent can *Daphnia* contribute to the decline in algal biomass?

The increases in *Daphnia* density (Figure 3), mean body size (Figure 5) and mean body size in the surface layers (Table 1), combined with the shift in the population mean depth and % distribution towards the surface (Figure 11) could contribute to the decline in phytoplankton biomass (Figure 1). The increase in *Daphnia* density was accompanied by the increase in average mean body size from on average 1016 to 1134 μm after fish removal. This could sufficiently account for

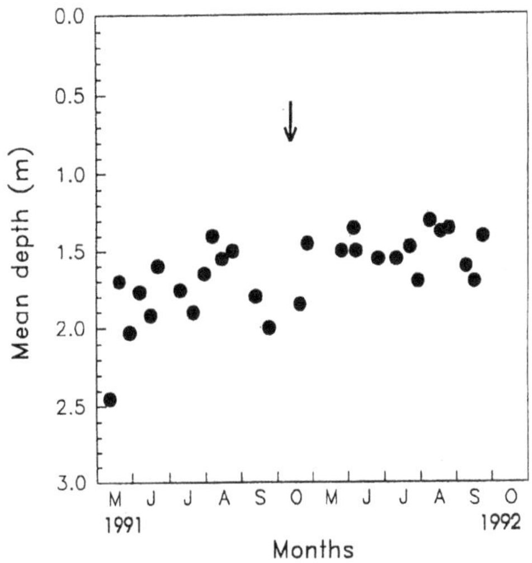

Figure 11. *Daphnia* vertical distribution.

the 1.3-fold (in summer 1.5-fold) decline in algal biomass (the relationship: body size-filtration rate from Burns 1969):

$$F = 0208 \cdot L^{280}$$

where F is filtering rate (ml animal^{-1} h^{-1}) and L is body length in mm. A shift in the distribution of *Daphnia* towards the surface (Figure 11), combined with increased mean size of individuals in the surface waters (Table 1) and increased density at all depths contributed to increased *Daphnia* grazing impact, especially in the surface waters (from 0.8 to 9.6 ml h^{-1}, i.e. 12 times at a depth of 0–1 m, from 1.3 to 9.6 ml h^{-1}, i.e. 7 times at 1–2 m and from 2.5 to 8.7 ml h^{-1}, i.e. 3.5 times in the near-bottom layer.

The observed decline in phytoplankton biomass following fish removal can thus be attributed to demographic and behavioural changes in the population of the dominant grazer. However, the additivity of these effects is not fully recognized and needs further exploration. Without detailed demographic and behavioural analysis, the grazing impact of herbivorous zooplankton could be easily over- or underestimated.

Acknowledgments

Many thanks to Jola Wierzgała, Olo Danielewicz and Tomek Janecki for their help in the field and in the laboratory. Anonymous reviewers are kindly acknowledged for their critical comments. The project was supported financially by KBN Grant 60540 9101 to Andrzej Prejs.

References

Benndorf, J. H., 1990. Conditions for effective biomanipulation: conclusions derived from whole-lake experiments in Europe. Hydrobiologia 200/201: 187–203.

Burns, C., 1969. Relation between filtering rate, temperature and body size in four species of *Daphnia*. Limnol. Oceanogr. 14: 693–700.

Carpenter, S. R. & J. F. Kitchell, 1992. Trophic cascade and biomanipulation: Interface of research and management – A reply to the comment by De Melo et al. Limnol. Oceanogr. 37: 208–213.

Carpenter, S. R. & J. F. Kitchell, 1993 (eds), The trophic cascades in lakes. Cambridge University Press, 385 pp.

de Melo, R., R. France & D. J. McQueen, 1992. Biomanipulation: hit or myth? Limnol. Oceanogr. 37: 192–207.

Edmondson, W. T. & A. H. Litt, 1982. *Daphnia* in Lake Washington. Limnol. Oceanogr. 27: 272–293.

Gliwicz, Z. M., A. M. Gilarov & J. Pijanowska, 1981. Food and predation as major factors limiting two natural populations of *Daphnia cucullata* Sars. Hydrobiologia 80: 205–218.

Green, J., 1956. Growth, size and reproduction in *Daphnia*: (Crustacea: Cladocera). Proc. zool. Soc. Lond. 126: 173–204.

Gulati, R. D., E. H. R. R. Lammens, M.-L. Meier & E. van Donk (eds), 1990. Biomanipulation – Tool for Water Management. Developments in Hydrobiology 61. Kluwer Academic Publishers, Dordrecht. Reprinted from Hydrobiologia 200/201.

Mortensen, E., E. Jeppesen, M. Søndergaard & L. K. Nielsen, 1994. Nutrient Dynamics and Biological Structure in Shallow Freshwater and Brackish Lakes. Developments in Hydrobiology 94. Kluwer Academic Publishers, Dordrecht. Reprinted from Hydrobiologia 275/276.

Prejs, A., Martyniak, S. Boroń, P. Hliwa & P. Koperski, 1994. Food web manipulation in a small eutrophic Lake Wirbel, Poland: effect of stocking with juvenile pike on planktivorous fish. Hydrobiologia 275/276: 65–70.

Prejs, A., J. Pijanowska, P. Koperski, A. Martyniak, S. Boroń & P. Hliwa, 1997. Food web manipulation in a small eutrophic Lake Wirbel (Poland): long-term changes in fish biomass and basic measures of water quality. A case study. Hydrobiologia 342/343: 385–388.

Stearns, S. C., 1992. The evolution of life histories. Oxford University Press, 249 pp.

Stibor, H. & W. Lampert, 1993. Estimating the size at maturity in field populations of *Daphnia*. Freshwat. Biol. 30: 433–438.

Hydrobiologia **342/343:** 311–318, 1997.
L. Kufel, A. Prejs & J. I. Rybak (eds), Shallow Lakes '95.

Pike (*Esox lucius* L.) stocking as a biomanipulation tool 1. Effects on the fish population in Lake Lyng, Denmark

Søren Berg[1], Erik Jeppesen[2] & Martin Søndergaard[2]
[1]*Danish Institute for Fisheries Research, Dept. of Inland Fisheries, Vejlsøvej 39, DK-8600 Silkeborg, Denmark*
[2]*National Environmental Research Institute, Dept. of Freshwater Ecology, Vejløvej 25, P.O. Box 314, DK-8600 Silkeborg, Denmark*

Key words: biomanipulation, pike stocking, predation, 0+ planktivores, pike mortality, pike/perch interaction

Abstract

From 1990–1993 juvenile pike (*Esox lucius*) were stocked each spring in the eutrophic Lake Lyng (9.9 ha, max. depth 7.6 m, mean depth 2.4 m) in densities between 515 and 3616 pike ha^{-1}. In 1989–90 the fish population consisted mainly of roach (*Rutilus rutilus*), rudd (*Scardinius erythrophthalmus*), perch (*Perca fluviatilis*) and ruffe (*Gymnocephalus cernuus*), and total fish biomass was estimated at 477 kg ha^{-1}. Prior to stocking pike was not present in the lake. Following the first year of stocking, the density of roach, rudd and ruffe fry expressed as catch per unit effort decreased significantly by 64 to 97%. In 1991 ruffe disappeared completely. The pike stocking did not affect the density of perch significantly. The growth of pike was high and also the growth of perch increased significantly from 1990 to 1991 ($p<0.001$) and from 1991 to 1994 ($p<0.001$). We found a linear negative relationship between stocking density of pike in May or June and the abundance of juvenile planktivorous fish ($r^2=0.85, p<0.05$) in the littoral zone in August. No relationship was found in the pelagic zone ($r^2=0.21, p>0.4$). Pike survival was low in late August/early September. We suggest that growth of the piscivorous perch increased due to increased Secchi depth and a continuous high density of 0+ planktivores in the pelagic zone of the lake during the years of pike stocking, possibly caused by behaviourial changes and the forcing of the 0+ planktivores into the pelagic zone.

Introduction

Reduction of the planktivorous fish biomass is a well-known method of improving water quality in eutrofi-cated lakes (Shapiro, 1975; Benndorf, 1990). Reduced predation on zooplankton, especially on large clado-cerans, leads to an increased grazing pressure on phyto-plankton. Subsequently, turbidity is reduced and light conditions improved, facilitating the growth of sub-merged macrophytes (Lauridsen et al., 1994).

The bulk of the zooplankton is often consumed by the first two or three year classes of planktivorous fish (e.g. roach (*Rutilus rutilus* L.) and other cyprinids) (Prejs et al., 1994), and reduction of the number of planktivorous fish is consequently essential.

One method of reducing the number of small plank-tivorous fish is to increase the number of predatory fish, a method that has several advantages compared to removal of planktivores by means of traditional fishing gear (trawls, seine nets, pound nets, or electrofish-ing). Firstly, by stocking the right species and the right size of piscivores, as recommended by Urho (1994), the increase can be aimed directly at the small plank-tivores that are the most difficult ones to catch using traditional fishing gear. Secondly, the method improves the planktivores:piscivores ratio, which is an important factor for securing long-term stability in biomanipulat-ed lakes (Jeppesen et al., 1990). In addition, the method is not time-consuming, and the biomass of cyprinids is transformed into more valuable fish. One obvious disadvantage of the method is the possible presence of many older and bigger planktivores that may continue spawning for several years, making repeated introduction of new piscivores during a corresponding number of years necessary (Prejs et al., 1994).

Among predatory freshwater fish, pike (*Esox lucius*, L.) is often used. It is known as a predator (Hunt & Carbine, 1951; Raat, 1988), being able to eat larger prey than most other piscivores of similar size (Benndorf, 1990). Pike is also a cannibal (Hunt & Carbine, 1951; Grimm, 1983; Giles et al., 1986), implying that overpopulation does not result in reduced growth. The natural circumpolar distribution (Raat, 1988) make pike native to most countries in the northern hemisphere experiencing the problems of eutrophication. These may be some of the reasons why pike are one of the most frequently used predators in biomanipulation experiments in Northern Europe (e g. Van Donk, et al., 1989; Benndorf, 1990; Prejs et al., 1994).

The hunting behaviour of pike can be an advantage as well as a disadvantage regarding biomanipulation. Pike is normally an inactive hunter that wait for the prey. As young pike tend to stay in the littoral zone (Raat, 1988) prey fish can therefore more easily avoid pike than active predators such as perch (*Perca fluviatilis* L.) by simply moving away from the pike and into the pelagic zone.

The aim of this experiment was to evaluate how stocking of pike fry in high densities affects fish, zooplankton, phytoplankton and nutrient concentrations in a small, shallow lake. This paper describes the resulting fish population changes following stocking of 0+ pike in various densities during the years 1990–1993. A second paper presents the results of changes in the lower trophic levels and in nutrients (Søndergaard et al., 1997).

Study area

Lake Lyng (Figure 1) is located in Mid-Jutland, Denmark, and is a satellite lake to the Gudenå river system. It is almost surrounded by the town of Silkeborg. The catchment area is only 108 ha of which the town covers 70%, the rest being covered by forest. The estimated hydraulic retention time is more than a year. During summer the lake is stratified. Until 1956 the lake was polluted with untreated sewage from the town. During periods of heavy rainfall the lake occasionally receives sewage, but in substantially smaller amounts than before 1956. Lake Lyng is, however, still very eutrophic (Søndergaard et al., 1997). No submersed macrophytes were present in the lake, but a reed belt consisting almost solely of *Phragmites australis* surrounds most of the lake. At the beginning of the study the natural fish community was dominated by perch, ruffe (*Gymnocephalus cernuus* L), roach, and rudd (*Scardinius erythrophthalmus* L). Pike was not present in the lake prior to stocking.

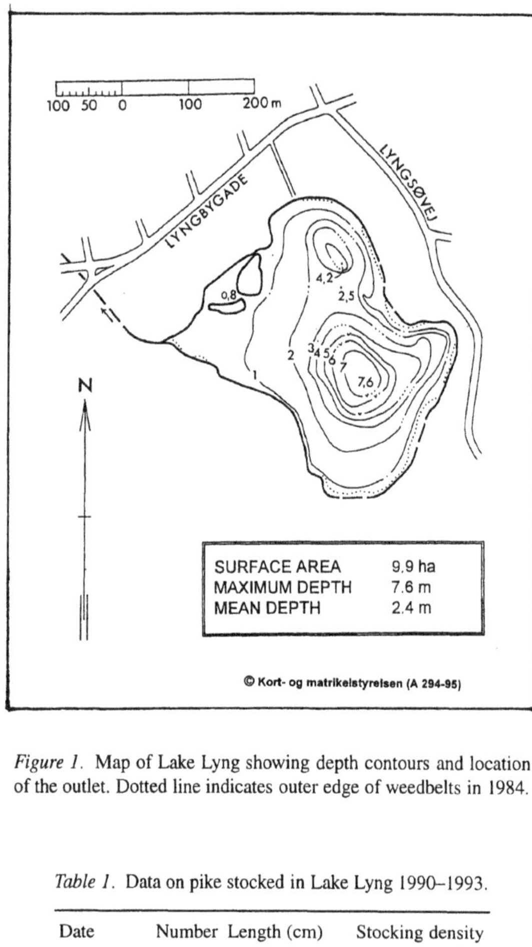

Figure 1. Map of Lake Lyng showing depth contours and location of the outlet. Dotted line indicates outer edge of weedbelts in 1984.

Table 1. Data on pike stocked in Lake Lyng 1990–1993.

Date	Number stocked	Length (cm) mean	Length (cm) range	Stocking density (number ha^{-1})
23/05/1990	16,600	5.9	3.8–9.3	1677
10/06/1991	5,166	3.7	3.1–4.7	
24/06/1991	4,383	5.2	3.5–10.5	
27/06/1991	1,600	2.3	1.9–2.8	
1991 total	11,149	4.1		1121
19/06/1992	5,100	3.3	2.8–4.3	515
26/05/1993	35,771	2.1	1.7–2.6	3616

Methods

During the years 1990 to 1993 a total of 68 600 pike fry were stocked in varying densities each spring (Table 1), the fry being spread as much as possible over the littoral zone just outside the reed belt using a boat. The relative fish population was determined in August/September each year during the period 1989–1994 as catch per unit effort (CPUE) using a Danish standard testfishing programme involving fishing with multiple mesh-sized gill nets (14 mesh sizes, 42 m long, 1.5 m high) and electrofishing (Mortensen et al., 1990). In Lake Lyng the programme included the setting of 15 nets spread over the lake. Fishing was conducted one night at each net, while electrofishing was undertaken along three stretches (300 m each) in the littoral zone.

The population density of fish in the lake was estimated in May 1990, prior to pike stocking. The mark-recapture method in the form of pelvic fin amputation (Youngs & Robson, 1978) was used for perch, ruffe, roach, and rudd. Fish for marking were caught using a handoperated seine net and small fyke nets. Of the total of 884 perch, 66 ruffe, 1895 roach, and 141 rudd that were marked, 11 marked perch, 35 roach, and 5 rudd were recaptured out of a total catch of 264 perch, 0 ruffe, 475 roach and 59 rudd. All species were devided into age groups before population estimate calculations were performed. As no marked ruffe were recaptured the ruffe estimate is based on the number of perch and the ratio between the two species in the catch for marking.

The relation between stocking density of pike and CPUE was tested by linear regression for CPUE in all nets set in the lake and divided into littoral and pelagic nets. The density of 0+ pike were above zero in 1994 due to natural spawning (0+ pike were observed during electrofishing). For this reason the value from 1994 were omitted from regression analysis. No attempt was made to calculate the number of natural spawned pike fry and compensate the stocking density in the years 1991–93.

Scale reading was used for determining the division of perch, roach and rudd into age groups. Pike growth was determined by backcalculating size from scale readings using the linear regression method (Bagenal & Tesch, 1978). The growth figures for each year class of pike were pooled for all years.

Table 2. Population estimate for the fish population in Lake Lyng, May 1990. Figures in brackets indicate percent of population number or biomass. No estimate has been made for the rare species, three-spined stickleback, bream, eel, and crucian carp.

Species	Population estimate (number)	95% C.L.	Biomass ($g\ m^{-2}$)
Perch	93,626 [46.1]	63,856–122,909	15.4 [32.3]
Ruffe	ca. 5,700 [2.8]	4,000–7,500	0.7 [1.5]
Roach	97,844 [48.2]	45,812–151,083	30.1 [63.1]
Rudd	5,859 [2.9]	3,645–8,368	1.5 [3.1]
Total	203,029 [100]	117,313–289,860	47.7 [100]

Results

Population density and biomass

In 1989–90, prior to the pike stocking, perch and roach dominated the fish community in Lake Lyng, while the number and biomass of ruffe and rudd were low (Table 2). During the period 1989 to 1994 one bream (*Abramis brama* L), one crucian carp (*Carassius carassius* L), two eels (*Anguilla anguilla* L) and ten three-spined sticklebacks (*Gasterosteus aculeatus* L) were caught.

The CPUE of the four most abundant fish species in Lake Lyng varied significantly during the study period (Figure 2A–D). For perch $\leq$11 cm (age group 0+) CPUE ranged from 50 to 87 except for a very low CPUE (= 10) in 1992 (Figure 2A). Density of perch >11 cm (age group 1+ and older) decreased from 21 to ca. 10 due to a decreasing number of perch between 11.5 and 18.5 cm (Figure 2G). Division of perch CPUE >11.0 cm into the 1+ group (11.5–18.5 cm) and older perch (>19.0 cm) showed a decreasing density of 1+ perch from 1989 to 1991, whereas the density of older perch remained stable during the period 1989–1994 (Figure 2G).

The CPUE of roach <10 cm (0+) and $\geq$10 cm decreased from 1989 to 1991 by 97 and 88%, respectively (Figure 2B). The CPUE of large roach remained below 30 throughout the rest of the study period, while 0+ roach almost reached the 1989 level in 1992 after which it declined again in 1993. The CPUE of 0+ rudd (<10 cm) and of all age groups of ruffe decreased throughout the study period (Figure 2C, D). In 1991, 1992 and 1994 no 0+ rudd were caught at all. Ruffe disappeared completely in 1991.

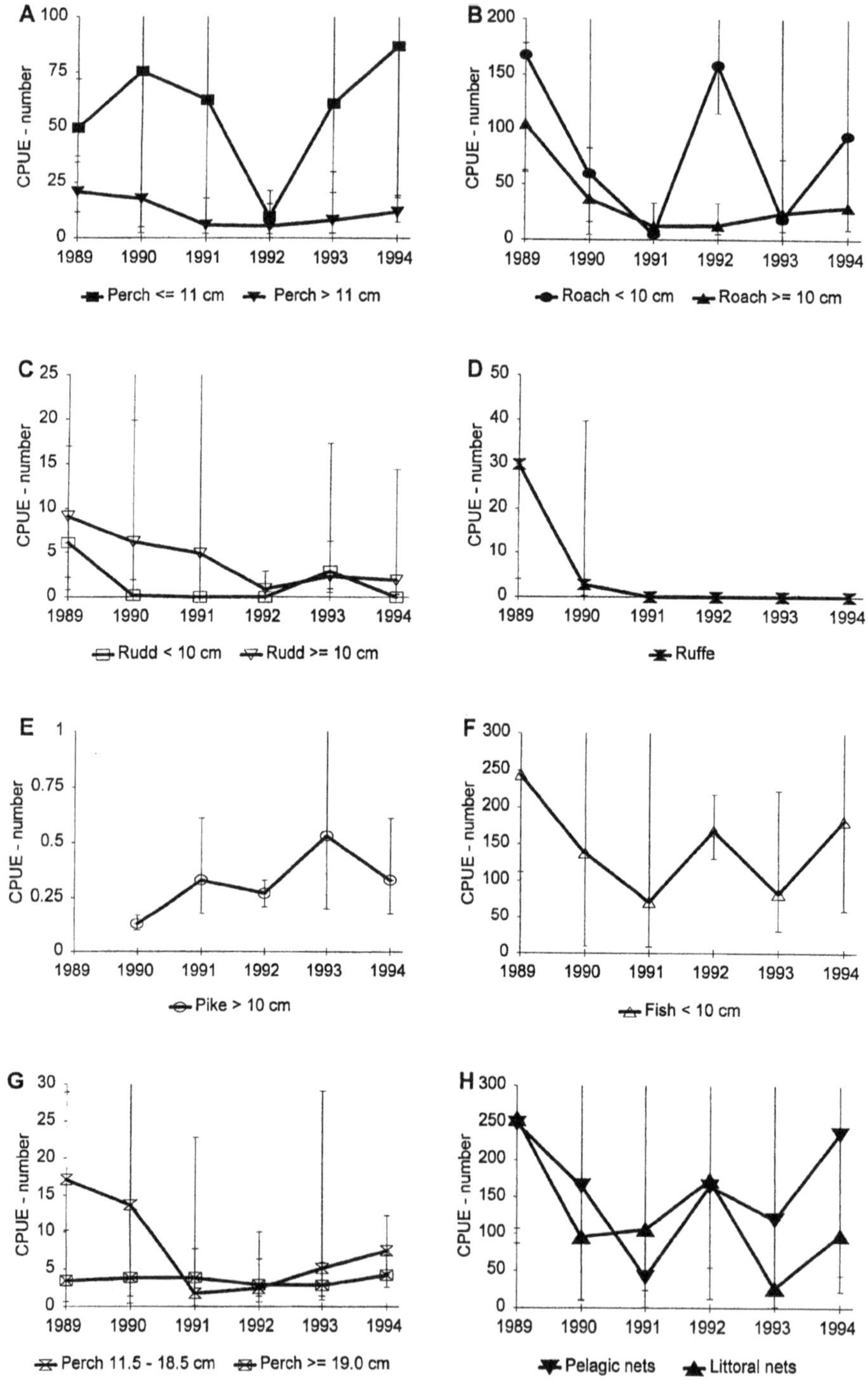

Figure 2. Catch per unit effort (CPUE) for perch (A), roach (B), rudd (C), ruffe (D), pike (E), perch (G), and the combined CPUE for perch, roach, rudd, and ruffe (F and H) caught in multiple mesh-sized gillnets during the years 1989-1994. Perch, roach, and rudd are divided into 0+ (<10 or ≤ 11 cm) and older (≥ 10 or >11 cm) fish. Vertical lines indicate 95% confidence limits.

Table 3. Growth of pike stocked in Lake Lyng.

Age	n	Length	±95% C.L.
1	20	26.1	6.2
2	14	42.8	6.7
3	4	52.6	6.7
4	3	55.5	5.7

The combined CPUE of all 0+ planktivorous fish (Figure 2F) showed much less interannual variation than the individual CPUE variations. Division of the CPUE of the four species into the CPUE for fish caught in gillnets in the pelagic and littoral zones, respectively, showed a tendency towards a larger CPUE reduction in the littoral than in the pelagic zone (Figure 2H). There were only slight variations in pike CPUE from the first stocking in 1990 to 1994 (Figure 2E).

Comparisons made between the CPUE values for the 0+ planktivores and the stocking density (st.d.) of pike in the corresponding years showed an inverse relation between st.d. and CPUE (Figure 3A–C). When the 1994 value were excluded from the linear regression analysis, there was a weak correlation between st.d. and CPUE for the catch of 0+ planktivores comprising all gillnets (Figure 3A, $r^2 = 0.52$, $p<0.2$), a strong correlation between st.d. and CPUE in nets from the littoral zone alone (Figure 3B, $r^2 = 0.85$, $p<0.05$), and no correlation between st.d. and CPUE in nets from the pelagic zone (Figure 3C, $r^2 = 0.21$, $p>0.4$).

Growth of perch and pike

During the first 3 years growth of pike was very high (Table 3), but in the fourth year the growth seemed to decrease. Length-at-age of large (>25 cm) perch changed significantly from 1990 to 1991 ($p<0.001$) and from 1991 to 1994 ($p<0.001$) (Figure 4).

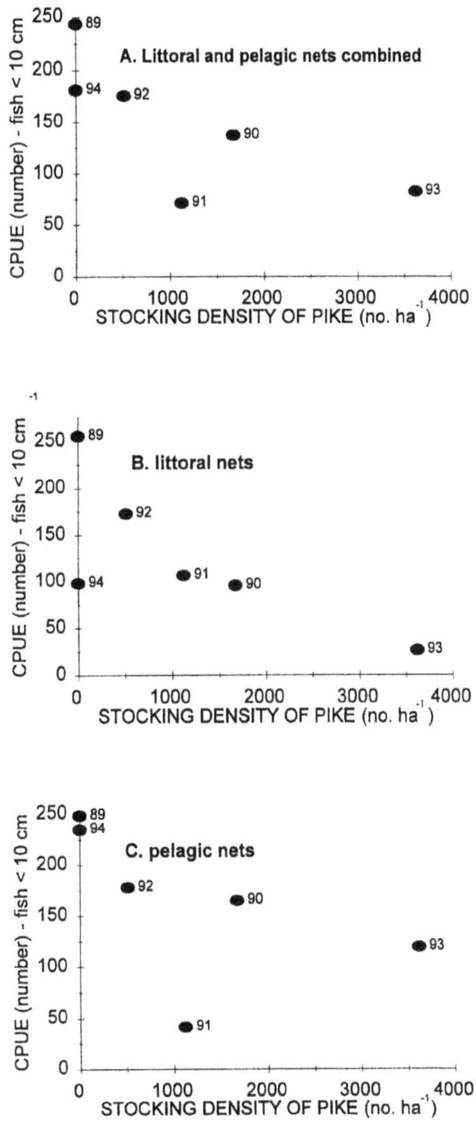

Figure 3. Stocking density of pike vs CPUE (number) for perch, roach, rudd, and ruffe, all <10 cm (perch ≤ 11 cm) for all gillnet set in the lake (A), 6 nets set in the littoral zone (B) and 9 nets set in the pelagic zone (C). Figures beside each point indicate the year.

Discussion

Population density and biomass

The structure of the fish community in Lake Lyng prior to pike stocking was typical for eutrophic lakes: a high total biomass of fish (Table 2), a high proportion of planktivores compared to piscivores (Table 2; Figure 2) (Persson et al., 1988; Jeppesen et al., 1990), and slow growing perch (Figures 2G & 4).

Excepting small planktivorous perch the density of all fish species in the lake changed during the years of pike stocking in accordance with the increased predation pressure resulting from the stocking. The densities of roach, rudd and ruffe decreased during 1989–1991 (Figures 2B, C, D) and apart from roach in 1992 the density of the 0+ age group of these planktivo-

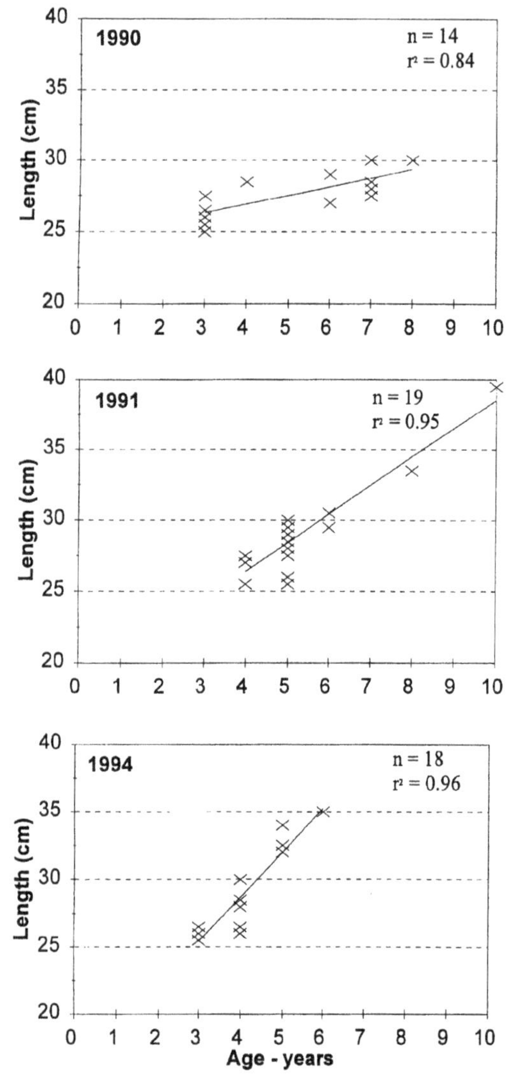

Figure 4. Length-at-age of large (>25 cm) perch in Lake Lyng in 1990, 1991, and 1994. Thin lines indicates the regression line for each year. Sample size (n) and correlation coefficient of the regression line is given in each graph.

rous species stayed low until 1994. Compared with the 1989 density, the density of all three species decreased in the first year following the stocking (Table 1), by ca 64, 97, and 83%, respectively. Rudd was affected more than roach, which can be explained by the difference in habitat choice of the two species. Rudd is connected with the littoral zone more than roach (Bean & Winfield, 1995), and accordingly pike stocked in the littoral zone had easier access to rudd as a prey. Also ruffe were clearly affected by the pike stocking. As ruffe is normally not connected with the littoral zone one would not expect this, but in 1989 twice as many ruffe were caught in littoral nets compared to pelagic ones indicating ruffe in lake Lyng to be living close to the littoral zone (unpubl. results).

Perch <11.0 cm (0+ age group) did not respond like the other three species to the introduction of pike (Figure 2A). Excepting 1992 the density of 0+ perch remained high and even increased slightly during the experiment. Eklöv & Hamrin (1989) found that the preferred prey of pike in open water was rudd rather than perch. The surface-oriented rudd is more conspicuous to predators than perch, and as 0+ roach like rudd appears in the upper layers of the water (Jensen, 1994; Bean & Winfield, 1995), roach will also be more vulnerable to predation than perch. In addition, soft-rayed fish such as rudd and roach are often preferred to spiny-rayed fish such as perch by predators (Beyerle & Williams, 1970; Eklöv & Hamrin, 1989). The reason for the low abundance of 0+ perch in 1992 is unknown. It may, however, be related to unsuccessful hatching or survival of perch larvae due to unfavourable weather conditions in the spring (Kudrinskaya, 1970). This phenomena has been observed in other Danish lakes (Berg et al., 1994; Jensen, 1994). The missing year class of perch corresponded with the year in which the stocking density of pike was lowest (Table 1), indicating that many 0+ roach were ready to respond to the reduced food competition from the missing perch by reducing their own mortality.

Like the 0+ group the density of older (1+ and older) roach, rudd, and ruffe decreased during the experiment, reflecting the reduced density of 0+ fish, i.e. fewer fish are allowed to grow older, and increased predation on 1+ fish by the pike.

As observed in our study (Figure 2E), cannibalism generally seems to be the key factor controlling the density of pike in lakes (Grimm, 1981 & 1983; Prejs et al., 1994). The number of pike by the end of summer will be regulated by the area of suitable habitat, irrespective of the number of pike stocked during the spring. Cannibalism both occurs among the fish of a certain year class (Hunt & Carbine, 1951) as well as among older and larger pike (Grimm, 1983).

Growth and stocking density of pike

As expected from the reduction of the planktivorous fish in Lake Lyng, growth of the pike stocked in the lake was found to be high for the 0+, 1+ and 2+ age groups (Table 3). Also in comparison to the growth generally recorded for European lakes (Raat, 1988), growth figures were high, suggesting a higher consumption of prey fish than usually recorded.

The fact that 80–90% of the consumption by a 0+ pike takes place during June and July (Grimm, 1989) indicates the importance of whether the mortality is highest during June, July, or August. Prejs et al. (1994) assumed that mortality would be highest by the end of July when insufficient food supply would increase cannibalism. Our observation of a linear relationship between the stocking density of pike in May–June and the density of prey fish in late August to early September (Figure 3) indicates that the mortality of 0+ pike was at least partly independent of stocking density within the range used in this experiment. The results of this experiment consequently support the assumption by Prejs et al. (1994). This experiment also showed that pike seemed capable of controlling the density of 0+ planktivores more effectively in the littoral zone (Figure 3B) than in the pelagic zone (Figure 3C).

The results from Lake Lyng suggest that enhancement of predator populations by stocking 0+ pike to reduce the density of 0+ planktivores can be used as a biomanipulation tool in small (<10–15 ha) highly eutrophic lakes with a fish biomass of 4-500 kg ha^{-1}. As the ratio between the littoral and pelagic zone changes with increasing lake size, pike stocking may have less impact on the prey fish population in larger lakes.

Interactions between pike and perch

From 1991 to 1994 a significant increase in the growth of large perch was recorded (Figure 4), indicating considerably improved food conditions for perch. One obvious change was the reduction in turbidity; mean summer Secchi depth increased from 0.72 m in 1989 to 1.4 m or more in the following years (Søndergaard et al., 1997). At the same time the density of O+ planktivores did not decrease as much in the pelagic as in the littoral zone (Figures 2H, 3B & C), which seems indicative of a sufficient density of prey fish for the perch hunting in the pelagic zone. Together with the increased Secchi depth this may explain the improved growth of perch. Another possible explanation of the increase in perch growth is behavioral changes among the 0+ planktivores such as prey fish moving to other habitats in the presence of predators which has been described by others (e.g. Turner & Mittlebach, 1990; Walls et al., 1992; Jacobsen et al., 1997).

Before 1990 pike was not present in Lake Lyng. Consequently the 0+ fish could use the littoral as refuge from predation, but when pike were introduced this refuge disappeared. We therefore hypothize that if piscivorous perch had not been present in the lake, the 0+ planktivores could have moved to the pelagic zone without risk of predation, and the effect of the pike stocking would have been much smaller. The possible change in behaviour by planktivores to the introduction of predators indicates the importance of predators being present in all the different niches of a lake, if stocking of piscivorous fish is to be a successful biomanipulation tool.

References

Bagenal, J. & I. Tesch, 1978. Age and growth. In Bagenal, T., (ed.), Methods for Assessment of Fish Production in Fresh Waters. IBP Handbook No 3, Blackwell Scientific Publications Ltd., Oxford, London, Edinburgh, Melbourne: 101–136.

Bean, C. W. & I. J. Winfield, 1995. Habitat use and activity patterns of roach (*Rutilus rutilus* (L.)), rudd (*Scardinius erythrophthalmus* (L.)), perch (*Perca fluviatilis* L.) and pike (*Esox lucius* L.) in the laboratory: the role of predation threat and structure complexity. Ecol. Freshwat. Fish 4: 37–46.

Benndorf, J., 1990. Conditions for effective biomanipulation; conclusions derived from whole-lake experiments in Europe. Hydrobiologia 200/201: 187–203.

Berg, S., E. Jeppesen, M. Søndergaard & E. Mortensen, 1994. Environmental effects of introducing whitefish, *Coregonus lavaretus* (L.), in Lake Ring. Hydrobiologia 275/276: 71–79.

Beyerle, G. B. & J. E. Williams, 1968. Some observations of food selectivity by northern pike in aquaria. Trans. am. Fish. Soc. 97: 28–31.

Eklöv, P. & S. F. Hamrin, 1989. Predatory efficiency and prey selection: interactions between pike *Esox lucius*, perch *Perca fluviatilis* and rudd *Scardinius erythropthalmus*. Oikos 56: 149–156.

Giles, R., R. M. Wright & M. E. Nord, 1986. Cannibalism in pike fry, *Esox lucius* L.: some experiments with fry densities. J. Fish Biol. 29: 107–113.

Grimm, M. P., 1981. Intraspecific predation as a principal factor controlling the biomass of northern pike (*Esox lucius* L.). Fish. Mgmt 77–80.

Grimm, M. P., 1983. Regulation of biomasses of small (<41 cm) northern pike (*Esox lucius* L.), with special reference to the contribution of individuals stocked as fingerlings (46 cm). Fish. Mgmt 4: 115–134.

Grimm, M. P., 1989. Northern pike (*Esox lucius* L.) and aquatic vegetation, tools in the management of fisheries and water quality in shallow waters. Hydrobiol. Bull. 23: 59–65.

Hunt, B. P. & W. F. Carbine, 1951. Food of young pike, *Esox lucius* L., and associated fishes in Peterson's ditches, Houghton Lake, Michigan. Trans. am. Fish. Soc. 80: 67–83.

Jacobsen, L., M. R. Perrow, F. Landkildehus, M. Hjørne, T. L. Lauridsen & S. Berg, 1997. Interactions between piscivores, zooplanktivores and zooplankton in submerged macrophytes: preliminary observations from enclosure and pond experiments. Hydrobiologia 342/343: 197–205.

Jensen, L. N., 1994. Macrophytes as a structuring factor for juvenile perch (*Perca fluviatilis*) and roach (*Rutilus rutilus*) in Lake Stigsholm. Danmarks Miljøundersøgelser og Biologisk Institut, Odense Universitet. M.Sc.-thesis, 70 pp. (in Danish)

Jeppesen, E., J. P. Jensen, P. Kristensen, M. Søndergaard, E. Mortensen, O. Sortkjær & K. Olrik, 1990. Fish manipulation as a lake restoration tool in shallow eutrophic temperate lakes 2: threshold levels, long-term stability and conclusions. Hydrobiologia 200/201: 219–227.

Kudrinskaya, O. M., 1970. Food and temperature as factors affecting the growth, development and survival of pike-perch and perch larvae. J. Ichthyol. 10: 779–788.

Lauridsen, T. L., E. Jeppesen & M. Søndergaard, 1994. Colonization and succession of submerged macrophytes in shallow Lake Væng during the first five years following fish manipulation. Hydrobiologia 275/276: 233–242.

Mortensen, E., H. J. Jensen, J. P. Müller & M. Timmermann, 1990. Monitoring programme t;sh investigations in lakes. Investigation programme, fishing gear, and methods. Teknisk anvisning fra DMU, nr. 3. National Environmental Research Institute: 60 pp. (in Danish).

Persson, L., G Andersson, S. F. Hamrin & L. Johansson, 1988. Predation regulation and primary production along the productivity gradient of temperate lake exosystems. In Carpenter, S. R. (ed.), Complex interactions in lake communities. Springer Verlag: 45–65.

Prejs, A., A. Martyniak, S. Boron, P. Hliwa & P. Koperski, 1994. Food web manipulation in a small, eutrophic Lake Wirbel, Poland: effect of stocking with pike on planktivorous fish. Hydrobiologia 275/276: 65–70.

Raat, A. P. J., 1988. Synopsis of biological data on northern pike (*Esox lucius* L.). FAO fisheries synopsis (145), 178 pp.

Shapiro, J., V. Lamarra & M. Lynch, 1975. Biomanipulation: an ecosystem approach to lake restoration. In P. L. Brezonik & J. L. Fox (eds), Water quality management through biological control. Report no. ENV-07-75-1, University of Florida, Gainsville, 85 –96.

Søndergaard, M., E. Jeppesen & S. Berg, 1997. Pike (*Esox lucius* L.) stocking as a biomanipulation tool, 2. Effects on the lower trophic levels in Lake Lyng, Denmark. Hydrobiologia 342/343: 319–325.

Turner, A. M. & G. Mittlebach, 1990. Predator avoidance and community structure: interactions among piscivores, planktivores, and plankton. Ecology 71: 2241–2254.

Urho, L., 1994. Removal of fish by predators – theoretical aspects. In Cowx, I. (ed.), Rehabilitation of Freshwater Fisheries. Fishing News Books, Oxford: 93–101.

Walls, M., I. Kortelainen & J. Sarvala, 1990. Prey responses to fish predation in freshwater communities. Ann. Zool. Fennici 27: 183–199.

Van Donk, E., R. D. Gulati & M. P. Grimm, 1989. Food web manipulation in lake Zwemlust: positive and negative effects during the first two years. Hydrobiol. Bull. 23: 19–34.

Youngs, W. D. & D. S. Robson, 1978. Estimation of population number and mortality rates. In Bagenal, T. (ed.), Methods for Assessment of Fish Production in Fresh Waters. IBP Handbook No 3, Blackwell Scientific Publications Ltd., Oxford, London, Edinburgh, Melbourne: 137–164.

Hydrobiologia **342/343**: 319–325, 1997.
L. Kufel, A. Prejs & J. I. Rybak (eds), Shallow Lakes '95.

Pike (*Esox lucius* L.) stocking as a biomanipulation tool 2. Effects on lower trophic levels in Lake Lyng, Denmark

Martin Søndergaard[1], Erik Jeppesen[1] & Søren Berg[2]
[1]*National Environmental Research Institute, Dept. of Lake and Estuarine Ecology, Vejlsøvej 25, P.O. Box 314, 8600 Silkeborg, Denmark*
[2]*Danish Institute for Fisheries Research, Dept. of Inland Fisheries, Vejlsøvej 39, 8600 Silkeborg, Denmark*

Key words: biomanipulation, pike stocking, top-down control, zooplankton, phytoplankton, nutrients, internal loading

Abstract

In order to study how pike stocking affects trophic structure pike fingerlings (0–3600 ha^{-1}) were stocked during six years in eutrophic Lake Lyng (lake area 10 ha), Denmark. Subsequently, marked changes were recorded in the abundance of zooplanktivorous fish, catch per unit effort of roach, which was the dominant fish species, thus varied from 17 to 272. Simultaneously, marked changes were recorded in the abundance and relative composition of zooplankton. *Daphnia* abundance ranged from 7 to 36 ind. l^{-1} and *Bosmina* from 2 to 99 ind. l^{-1}, mean summer abundance of *Daphnia* being highest and *Bosmina* lowest in years with low fish density. Additionally, cyclopoid copepod and rotifer abundance correlated significantly ($p<0.05$, $r^2=0.70$ and 0.83, respectively) and positively with fish abundance. Mean summer chlorophyll *a* correlated significantly ($p=0.04$, $r^2=0.70$) with the abundance of roach, suggesting an impact on phytoplankton level also, disregarding that the phytoplankton biomass was dominated by filamentous cyanobacteria. Total phosphorus concentration was lowest in years with low zooplanktivorous fish abundance, which is hypothesized to be due to lower sedimentation of organic material and lower phosphorus release from the sediment in years with high zooplankton grazing on phytoplankton. It is concluded that pike stocking can be used as a lake restoration tool to increase lake water transparency by creating a trophic cascade. The effect of stocking, however, seems to last only during the season in which it has been undertaken, the impact being most significant at high stocking densities. The method is therefore regarded to be most useful in shallow, turbid lakes in which the nutrient loading has been sufficiently reduced to allow a substantial and permanent macrophyte coverage if clearwater conditions are established.

Introduction

The importance of top-down versus bottom-up control in lakes and the possibilities of creating a trophic cascade by fish manipulation have been amply discussed in recent years (De Melo et al., 1992; Carpenter et al., 1993; Reynolds, 1994). Numerous enclosure and whole-lake experiments have been conducted to investigate the usefulness of fish manipulation as a restoration tool (e.g. review by De Melo et al., 1992; Meijer et al., 1994; Reynolds, 1994). In most cases the starting point has been a direct reduction or removal of the zooplanktivorous fish with a view to increasing the abundance of large-sized and efficient phytoplankton-filtrating species such as *Daphnia*.

More recently, manipulation experiments have been undertaken at the top level of the food chain by introducing carnivorous fish to reduce the number of zooplanktivorous fish. Most frequently, stocking of piscivores has been used as a supplement to fish removal, but in some cases also as the only measure to induce a trophic cascade (Benndorf et al., 1988; He & Wright, 1992; Prejs et al., 1994). Compared to the more traditional biomanipulation methods mentioned above, stocking of piscivorous fish has several advantages (Berg et al. 1997), among these being less time-consuming and less costly. However, the effect of piscivorous fish stocking in terms of stocking density on lower trophic levels is poorly documented.

In the present paper we evaluate how stocking of pike fingerlings in high densities affects the abundance of zooplankton and phytoplankton and nutrient concentrations. More specifically, we wanted to evaluate the impact of varying stocking densities and observe the long-term effects. The investigation was based on a six-year study of a 10 ha large lake into which pike fingerlings were stocked every spring during the four-year period. In another paper we discuss the direct influence of pike on the fish community (Berg et al., 1997).

Methods

Study area

The investigation was conducted in Lake Lyng situated near Silkeborg, Denmark. Lake Lyng (10 ha) has a mean depth of 2.4 m and a maximum depth of 7.6 m and is stratified during summer. Depth is below approximately 5 m (10% of the lake area), and the lake is anoxic from July to September. The lake has no surface inflows and the hydraulic retention time is estimated to be more than one year. Sewage from the town of Silkeborg was diverted in 1956 and except for periods with heavy rainfall there have been no external sources of nutrients to the lake since then. Before the pike stocking mean summer Secchi depth in 1989 was 0.72 m and total phosphorus 0.79 mg P l^{-1} in the epilimnion. See Berg et al. (1997) for further details.

Pike stocking and fish abundance

Stocking of pike juveniles (mean length: 2–6 cm) was conducted from 1990 to 1993 in densities ranging from 500 to 3600 ha^{-1}. Stocking was undertaken in either May or June. Fish abundance was calculated as catch per unit effort (CPUE) once every August or September using multiple mesh size gill nets. For further details see Berg et al. (1997).

Sampling and analysis

Sampling of zooplankton, phytoplankton, nutrients and chlorophyll *a* was undertaken once or twice monthly during summer (1 May to 1 October).

Zooplankton was sampled at a mid-lake station using a 3.31 Patalas sampler. Samples taken at 0.5 m, 2.5 m, 4.5 m and 6.5 m were pooled, filtered through a 20 μm net, and the contents fixed in Lugol (1 ml 100 ml^{-1} tap water). Zooplankton retained on a 140 μm net were counted (at least 150 ind. of the most dominant species) using a Wild microscope, whereas those <140 μm were filtered on a 20 μm net and counted in strips using an inverted microscope (at least 100 ind. of the most dominant species). As the vertical distribution probably is restricted to the epilimnion during most of the summer, the actual zooplankton density in the photic zone above the thermocline is underestimated with up to 100%.

Phytoplankton were sampled at the mid-lake station at a depth of 0.5 m (1994) or as a composite sample from 1/2 Secchi depth, Secchi depth and 2 Secchi depth (other years). If 2 Secchi depth was below the epilimnion this sample was omitted. Phytoplankton biomass was estimated by inverted microscope counting (at least 100 ind. of the dominant species) and converted to biovolume by means of appropriate geometric shapes.

Chemical analyses were conducted on a composite 0.5 m and 2.5 m sample (1994) or from the same depth as phytoplankton (other years). Nutrients and chlorophyll *a* were analysed according to Søndergaard et al. (1990), chlorophyll *a* using ethanol extraction, phosphate PO_4-P by the ascorbic acid molybdene blue method, total phosphorus (tot-P) after persulphate digestion, ammonium (NH_4-N) by the phenol hypochlorite method and total nitrogen (tot-N) after potassium persulphate digestion. Particulate phosphorus (part-P) was calculated as the difference between tot-P and PO_4-P.

The number of zooplankton and phytoplankton samplings varied from 5 in 1993 to 20 in 1989. Samplings for Secchi depth, chlorophyll *a* and nutrients varied from 6–8 in 1990 to 10— 25 in 1989.

Statistical analyses

Mean summer (1 May to 1 October) values of zooplankton, Secchi depth, chlorophyll *a* and nutrients for each of the six years were calculated and correlated to catch per unit effort of roach<10 cm (roach<10), total roach number (roach-total) or, in a multiple regression, to roach-total and perch<10 cm. The regression analyses were performed according to SAS (1989) on untransformed or on log-log transformed data.

Table 1. Catch per unit effort of roach < 10 cm (roach<10), total number of roach (roach-total) and perch <10 cm (perch<10). (See Berg et al. (1997) for further details).

	1989	1990	1991	1992	1993	1994
Roach<10	167	60	5	158	19	94
Roach-total	272	97	17	171	43	123
Perch<10	50	76	67	10	61	87

Table 2. Linear regression between mean summer zooplankton abundance, Secchi depth and nutrient concentrations and catch per unit effort of roach<10 and roach-total. In roach-total + perch<10, stepwise multiple regression was performed including catch per unit effort of perch <10 cm. N.s. means no significant improval (significance level = 0.15) of the relationship.

	Roach<10		Roach-total		Roach-total + perch<10	
	p	r^2	p	r^2	p	r^2
Daphnia	0.42	0.16	0.53	0.11	N.s.	
Bosmina	0.06	0.64	0.21	0.37	N.s.	
Daph./cla. ratio	0.13	0.47	0.22	0.34	N.s.	
Cyc. cop.	0.06	0.63	0.04	0.70	N.s.	
Rotifers	0.01	0.86	0.01	0.83	0.01	0.98
Secchi depth	0.12	0.49	0.05	0.67	N.s.	
Chl a	0.04	0.70	0.04	0.70	N.s.	
Tot-P	0.10	0.53	0.09	0.54	N.s.	
Part-P	0.01	0.87	<0.01	0.93	N.s.	

Results

Fish

The CPUE of zooplanktivorous fish varied greatly during the six years. Thus, roach<10 varied from 5 in 1991 to 167 in 1989 and perch<10 from 10 in 1992 to 87 in 1994 (Table 1). Roach $\geq$ 10 was 105 in 1989 and varied between 12 and 37 in 1990 to 1994. For further details see Berg et al. (1997).

Zooplankton

During the whole sampling period large cladocerans were dominated by *Daphnia cucullata* Sars and *D. galeata* Sars, and small cladocerans mainly by *Bosmina longirostris* O. F. Müller. Cyclopoid copepods were dominated by *Mesocyclops leuckarti* Claus, while calanoid copepods almost exclusively consisted of *Eudiaptomus gracilis* Sars. *Keratella cochlearis* Gosse was the dominant rotifer species during all years.

Mean summer density of *Daphnia* decreased with the density of roach<10, from less than 7 ind. l^{-1} in 1994 to 36 ind. l^{-1} in 1993 (Figure 1). Contrarily, the density of *Bosmina* was highest in years with high roach abundance (Figure 1), the mean being below 20 ind. l^{-1} in years with low roach abundance, but up to 100 ind. l^{-1} in years with high roach abundance. Accordingly, the *Daphnia* to total cladoceran ratio was lowest in years with high roach abundance (Figure 1). Rotifer abundance increased with increasing roach abundance (Figure 1) from less than 2000 ind. l^{-1} in 1991, when the abundance of roach was low, to almost 10 000 in 1989 and 1992 when roach abundance was high. The correlation was significant ($p = 0.01$) to both roach<10 ($r^2 = 0.86$) and roach-total ($r^2 = 0.83$, Table 2). Finally, cyclopoid copepod abundance (adults + copepodites) correlated significantly ($p = 0.04$, $r^2 = 0.70$) and positively with roach-total. Regression analyses performed on log-transformed data did not improve the statistical significance of any of the parameters analysed. Only for rotifers an improved relationship to fish abundance could be established by including the number of perch<10 cm in a multiple regression, r^2, increasing to 0.98 (Table 2).

Phytoplankton and Secchi depth

Phytoplankton biomass was dominated by cyanobacteria from 1989 to 1992, while the community was more diverse in 1993 and 1994 (Table 3). The dominant cyanobacteria were *Anabaena spiroides* Klebahn and *A. flos-aquae* Komarek. *Pediastrum duplex* Meyen and *Dictyosphaerium* sp. were among the dominant green algae in 1989, constituting 24% of the total mean summer biomass. Among the chrysophytes, which were of particular importance for the total phytoplankton biomass in 1991 and 1993, *Syncrypta* sp. (1991) and *Uroglena* sp. (1993) dominated, while picoplankton accounted for a large part of the unspecified phytoplankton biomass in 1994.

A significant ($p = 0.04$, $r^2 = 0.70$) and positive relationship was recorded between chlorophyll *a* and roach<10 and roach-total (Figure 2, Tables 2 and 4). Thus, mean summer chlorophyll *a* varied from 20 μg l^{-1} in 1991 to about 90 μg l^{-1} in 1989 and 1992. Correspondingly, a tendency towards a negative relationship was recorded between Secchi depth and roach<10 ($p = 0.12$, $r^2 = 0.49$) and roach-total ($p = 0.05$, $r^2 = 0.67$, Figure 2, Table 2).

Table 3. Total phytoplankton biomass ($mm^3 l^{-1}$) and major phytoplankton classes as % of total biomass.

Year	Total ($mm^3 l^{-1}$)	Cyanophytes (%)	Cryptophytes (%)	Chrysophytes (%)	Diatoms (%)	Chlorophytes (%)	Others (%)
1989	18.9	61	10	<1	<1	24	5
1990	24.4	89	3	2	<1	1	5
1991	5.1	43	6	27	<1	16	8
1992	20.3	79	16	1	<1	1	3
1993	3.7	4	11	51	<1	18	16
1994	3.9	14	21	2	5	19	39

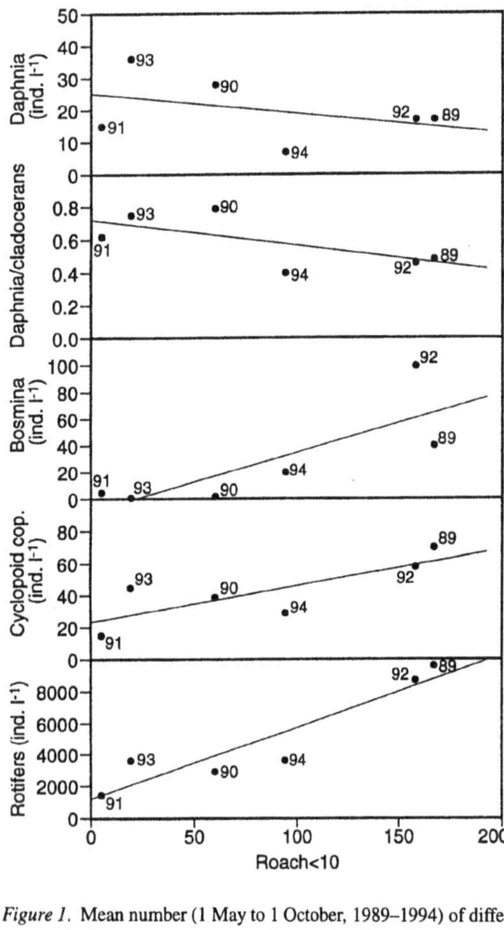

Figure 1. Mean number (1 May to 1 October, 1989–1994) of different zooplankton groups versus catch per unit effort of roach<10.

Table 4. Linear regression between number of cyclopoid copepods (Cyc. Cop., adults+copepodites l^{-1}) rotifers (ind. l^{-1}), chlorophyll a (Chl. *a*, $\mu g\, l^{-1}$) and particulate P (part-P, $\mu g\, P\, l^{-1}$) and roach<10 and roach-total.

Cyc. Cop.	= 21 + 0.18*roach-total,	p=0.038, r^2=0.70
Rotifers	= 1200+ 45*roach<10,	p=0.007, r^2=0.86
Rotifers	= 1040 + 33*roach-total,	p=0.012, r^2=0.83
Chl. *a*	= 24.6 + 0.39*roach<10,	p=0.039, r^2=0.70
Chl. *a*	= 22.3 + 0.29*roach-total,	p=0.037, r^2=0.70
Part-P	= 90 + 0.86*roach<10,	p=0.006, r^2=0.87
Part-P	= 82 + 0.66*roach-total,	p=0.003, r^2=0.94

Nutrients

Mean summer PO_4-P varied from 0.25 to 0.55 mg P l^{-1} from 1989 to 1994, while minimum summer PO_4-P ranged between 0.071 (1989) and 0.340 (1992) mg P l^{-1}. Summer tot-P and PO_4-P were 2–4 times higher than winter concentrations, suggesting a heavy influence of sediment phosphorus release on lake water concentrations.

Mean summer nitrate concentrations never exceeded 0.01 mg N l^{-1} while NH_4-N varied between 0.040 and 0.096 mg N l^{-1}. Tot-N ranged from 0.90 mg N l^{-1} in 1991 to 1.97 mg N l^{-1} in 1992.

A highly significant ($p = 0.01$, $r^2 = 0.87$) and positive relationship was recorded between particulate phosphorus and roach<10, and between particulate phosphorus and roach-total ($p<0.01$, $r^2 = 0.93$, Figure 2, Tables 2 & 4). Furthermore, a tendency towards a positive relationship ($p = 0.10$, $r^2 = 0.53$) was recorded between tot-P and roach<10, and between tot-P and roach-total ($p = 0.09$, $r^2 = 0.54$). Tot-N did not correlate to roach <10 ($p = 0.4$). Changes in mean summer PO_4-P and NH_4-N did not correlate to changes in roach<10 or roach-total ($p>0.3$).

Discussion

Zooplankton composition and abundance in Lake Lyng along with the changes in the planktivorous fish abundance mediated by the introduction of pike fingerlings

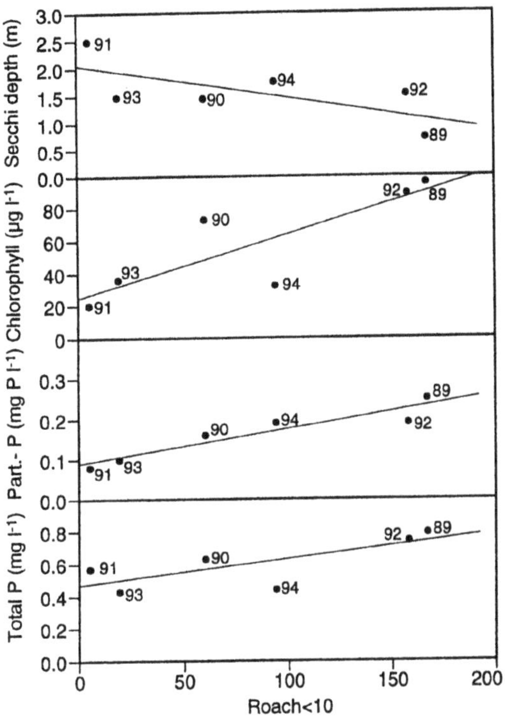

Figure 2. Mean (1 May to 1 October, 1989–1994) Secchi depth, chlorophyll *a* concentration (chla), total phosphorus (total-P) and particulate phosphorus concentration (part-P) versus catch per unit effort of roach<10.

(Berg et al., 1997) showed a number of characteristic features that indicate a marked top-down control in accordance with the classical size-efficiency theory of Brooks & Dodson (1965). Firstly, low planktivorous fish abundance was accompanied by relatively high abundance of *Daphnia*, which generally is considered to be a sign of low grazing pressure by fish (Post & McQueen, 1987; Reinertsen et al., 1990). Secondly, *Bosmina* abundance was high at the highest fish densities. This is typically observed in lakes with a relatively high zooplanktivorous fish stock (Stensson et al., 1978; Benndorf et al., 1988) and is considered to reflect the improved growth conditions for *Bosmina* at low *Daphnia* densities. Consequently, the *Daphnia* to total cladoceran ratio was highest in years with low fish densities. Thirdly, high cyclopoid copepod abundance as in Lake Lyng is also mainly observed in lakes with high fish densities (Benndorf et al., 1988; Hansen & Jeppesen, 1992). Finally, high rotifer abundance, as observed during years with low *Daphnia* abundance in Lake Lyng, is usually associated with a high chlorophyll *a* level (Canfield & Watkins, 1984), and is mainly observed when *Daphnia* density is low as a result of changes in the competition for food and interference (Gilbert, 1985; May & Jones, 1989; Telesh, 1993). Often marked changes in rotifer abundance may be related to changes in fish abundance, the highest abundance of rotifers occurring at the highest fish densities (Reinertsen et al., 1990; Christoffersen et al., 1993; Ronneberger et al., 1993).

Generally, no or only minor improvements were achieved in the correlations by using the total number of roach instead of roach <10 cm. This supports earlier observations (He & Wright, 1992; Jeppesen et al., 1997) that it is the YOY planktivorous fish in particular that create a high predation pressure on the large cladocerans. Furthermore, the lack of impact when including perch<10 cm in the correlation shows that the effect on zooplankton can mainly be ascribed to YOY roach.

The suitability of cyanobacteria and other large-sized algae as a zooplankton food source has been amply investigated and discussed (Hawkins & Lampert, 1989; Bernardi & Giussani, 1990; McQueen, 1990). However, the significant positive relationship recorded in Lake Lyng between roach number and chlorophyll *a* strongly indicates that the effect of changed fish abundance is transferred down to the phytoplankton level. Zooplankton were thus able to suppress phytoplankton abundance in years with low density of planktivorous fish despite dominance of filamentous cyanobacteria during summer, cyanobacteria biomass being, for instance, reduced from 22 $mm^3 l^{-1}$ in 1990 to 2 $mm^3 l^{-1}$ in 1991.

Changes in the planktivorous fish abundance were also accompanied by changes in the nutrient regime, which most likely reflects changes in the biological structure, as changes in external loading were not recorded. As expected, particulate P was highly significantly related to roach abundance, reflecting the correspondingly good correlation to the phytoplankton biomass. However, the substantial year to year variability also in total phosphorus indicates that increasing abundance of zooplanktivores seems able to provoke changes in the internal recycling of phosphorus, causing increased phosphorus concentrations. This suggests a strong cascading effect of changed fish structure all the way down to the nutrient level. Changes in nutrient concentrations were recorded in other biomanipulation experiments. There are various explanations of the phenomenon such as increased sedimentation due to precipitation of calcium carbonate (Andersson et al., 1978), excretion of P below the thermocline by migrating large-bodied zooplankton (Shapiro & Wright, 1984), increased sedimentation of detrital phosphorus (Boers et al., 1991), increased sedimentation of zooplankton and reduced phosphorus release from the sediment because of the lowered pH (Reinertsen et al., 1990), and lower number of cyprinids foraging on the sediment, thereby decreasing release of P from the sediment (Persson & Hamrin, 1994). As the Lake Lyng tot-P concentration in the epilimnion during summer is considerably higher than during winter a substantial internal loading takes place. A decrease in the release of phosphorus from the sediment due to increased sedimentation of organic matter and reduced oxygen demand seems therefore to be the most likely explanation of the reduced tot-P at decreased abundance of zooplanktivorous fish in Lake Lyng. This assumption is supported by the fact that about 50% of the phosphorus in the surface sediment of Lake Lyng is bound to iron and that the phosphorus release rate experimentally has been shown to be very sensible to changes in oxygen and nitrate concentrations (unpubl. data) as has been found in other shallow iron-rich lakes (Jensen et al., 1992). The strong cascading effect down to the nutrient level in this eutrophic lake contradicts the hypothesis of McQueen et al. (1986) that the top-down effect is weak in eutrophic lakes, but it is in accordance with Sarnelle (1992).

Based on the correlation between tot-P and roach it could be argued that the changes in phytoplankton abundance simply reflect changes in the nutrient regime as total phosphorous also increased with density of roach. Orthophosphate, however, was high during the whole experimental period and is therefore not likely to have been a limiting factor for phytoplankton growth. Inorganic nitrogen, on the other hand, may periodically constitute a limiting factor and although mean summer NH_4-N never went below 0.04 mg N l^{-1}, it cannot be excluded that changes in the internal flux of nitrogen, as in phosphorus, may have occurred, which may have influenced the abundance of phytoplankton.

In conclusion, the investigations in Lake Lyng strongly indicate that it is possible to create a top-down control on phytoplankton biomass by introducing pike fingerlings. However, the effects of pike stocking on lower trophic levels seem to last only during the season in which stocking is undertaken. This is in agreement with the finding that the carrying capacity of pikes in lakes is largely determined by the area covered by vegetation (Grimm & Backx, 1990) and in agreement with the finding by Berg et al. (1997) that the autumn pike biomass was not increased by the stocking. To achieve a long-term effect the stocking therefore has to be repeated every year or until more permanent changes in the biological structure have been established. We therefore hypothesize that this management method is most useful in lakes in which changes in other stabilizing elements can be achieved following an increase in water transparency, i.e. particularly in relatively shallow lakes and lakes in which the nutrient loading has been reduced sufficiently to enable the permanent establishment of submerged macrophyte vegetation.

Acknowledgments

The assistance of the technical staff at the National Environmental Research Institute, Silkeborg, is gratefully acknowledged. Field and laboratory assistance was provided by K. Jensen, L. Nørgaard, J. Glargaard, B. Laustsen, J. Stougaard-Pedersen and L. Hansen. Layout and manuscript assistance was provided by K. Møgelvang, A. M. Poulsen and A-D. Matharu.

References

Andersson, G., H. Berggren, G. Cronberg & C. Gelin, 1978. Effects of planktivorous and benthivorous fish on organisms and water chemistry in eutrophic lakes. Hydrobiologia 59: 9–15.

Benndorf, J., H. Schulz, A. Benndorf, R. Unger. E. Penz, H. Kneschke, K. Kossatz, R. Dumke, U. Hornig, R. Kruspe & S. Reichel, 1988. Food-web manipulation by enhancement of piscivorous fish stocks: long-term effects in the hypertrophic Bautzen reservoir. Limnologica: 97–110.

Berg, S., E. Jeppesen & M. Søndergaard, 1997. Pike (*Esox lucius* L.) stocking as a biomanipulation tool. 1. Effects on the fish population in Lake Lyng, Denmark. Hydrobiologia 342/343: 311–318.

Bernardi, R. de & G. Guissani, 1990. Are blue-green algae a suitable food for zooplankton? An overview. Hydrobiologia 200/201: 29–41.

Boers, P., L. Van Ballegooijen & J. Uunk, 1991. Changes in phosphorus cycling in a shallow lake due to food web manipulations. Freshwat. Biol. 25: 9–20.

Brooks, J. L. & S. I. Dodson, 1965. Predation, body size, and composition of plankton. Science 150: 28–35.

Canfield., D. & C. E. Watkins, 1984. Relationships between zooplankton abundance and chlorophyll *a* concentrations in Florida lakes. J. Fresh. Ecol. 2: 335–343.

Carpenter, S. R., X. He., J. F. Kitchell & P. A. Sorrano, 1993. Whole-lake experiments: is grazer functional response stable? Verh. int. Ver. Limnol. 25: 398–400.

Christoffersen, K., B. Riemann, A. Klysner & M. Søndergaard, 1993. Potential role of fish predation and natural populations of zooplankton in structuring a plankton community in eutrophic lake water. Limnol. Oceanogr. 38: 561–573.

De Melo, R., R. France & D. J. McQueen, 1992. Biomanipulation: Hit or myth? Limnol. Oceanogr. 37: 2–207.

Gilbert, J.J., 1985. Competition between rotifers and *Daphnia*. Ecology 66: 43–50.

Grimm, M. P. & J. J. G. M. Backx, 1990. The restoration of shallow eutrophic lakes, and the role of northern pike, aquatic vegetation and nutrient concentration. Hydrobiologia 200/201: 557–566.

Hansen, A-M. & E. Jeppesen, 1992. Changes in the abundance and composition of cyclopoid copepods following fish manipulation in eutrophic Lake Væng, Denmark. Freshwat. Biol. 28: 183–93.

Hawkins, P. & W. Lampert, 1989. The effect of *Daphnia* body size on filtering rate inhibition in the presence of a filamentous bacterium. Limnol. Oceanogr. 34: 1084–1089.

He, X. & R. Wright, 1992. An experimental study of piscivore-planktivore interactions: population and community responses to predation. Can. J. Fish aquat. Sci. 49: 1176– 1183.

Jensen, H. S., P. Kristensen, E. Jeppesen & A. Skytthe, 1992. Iron: phosphorus ratio in surface sediment as an indicator of phosphate release from aerobic sediments in shallow lakes. Hydrobiologia 235-236: 731–743.

Jeppesen, E., J. P. Jensen, M. Søndergaard, T. Lauridsen, L. J. Pedersen & L. Jensen, 1997. Top-down control in freshwater lakes: the role of nutrient state, submerged macrophytes and water depth. Hydrobiologia 342/343: 151–164.

May, L. & D. H. Jones, 1989. Does interference competition from *Daphnia* affect populations of *Keratella cochlearis* in Loch Leven, Scotland. J. Plankton Res. 11: 455–461.

McQueen, D. J., 1990. Manipulation lake community structure: where do we go from here? Freshwat. Biol. 23: 613–620.

McQueen, D. J., R. France & C. Kraft, 1992. Confounded impacts of planktivorous fish on freshwater biomanipulation. Arch. Hydrobiol. 125: 1–24.

McQueen, D. J., J. R. Post & E. L. Mills, 1986. Trophic relations in freshwater pelagic ecosystems. Can. J. Fish. aquat. Sci. 43: 1571–1581.

Meijer, M-L., E. Jeppesen, E. van Donk, B. Moss, M. Scheffer, E. Lammens, E. van Nes, J. A. van Berkum, G. J. de Jong, B. A. Faafeng & J. P. Jensen, 1994. Long-term response to fish-stock reduction in small shallow lakes: interpretations of five-year results of four biomanipulation cases in The Netherlands and Denmark. Hydrobiologia 200/201: 457–466.

Meijer, M-L., E. Jeppesen, E. van Donk, B. Moss, M. Scheffer, E. Lammens, E. van Nes, J. A. Persson & S. F. Hamrin, 1994. Effects of cyprinids on the release of phosphorus from lake sediment. Verh. int. Ver. Limnol. 25: 2124–2127.

Persson, A. & S. F. Hamrin, 1994. Effects of cyprinids on the release of phosphorus from lake sediment. Verh. int. Ver. Limnol. 25: 2124–2127.

Post, J. R. & D. J. McQueen, 1987. The impact of planktivorous fish on the structure of a plankton community. Freshwat. Biol. 17: 79–89.

Prejs, A., A. Martyniak, S. Boron, P. Hliwa & P. Koperski, 1994. Food web manipulation in a small eutrophic Lake Wirbel, Poland: effect of stocking with juvenile pike on planktivorous fish. Hydrobiologia 275/276: 65–70.

Reinertsen, H., A. Jensen, J. I. Koksvik, A. Langeland & Y. Olsen, 1990. Effects of fish removal on the limnetic ecosystem of a eutrophic lake. Can. J. Fish aquat. Sci. 47: 166–173.

Reynolds, C., 1994. The ecological basis for the successful biomanipulation of aquatic communities. Arch. Hydrobiol. 130: 1–33.

Ronneberger, D., P. Kasprzak & L. Krienitz., 1993. Long-term changes in the rotifer fauna after biomanipulation in Haussee (Feldberg, Germany, Mecklenburg-Vorpommern) and its relationship to the crustacean and phytoplankton communities. Hydrobiologia 255/256: 297–304.

Sarnelle, O., 1992. Nutrient enrichment and grazer effects on phytoplankton in lakes. Ecology 73: 551–560.

Sas Institute Inc., 1989. SAS/STAT User's Guide, Version 6, 4th edn., Vols 1 & 2. SAS Institute Inc., Cary, NC, 943 & 846 pp.

Shapiro, J. & D. I. Wright, 1984. Lake restoration by biomanipulation: Round Lake, Minnesota, the first two years. Freshwat. Biol. 14: 371–383.

Søndergaard, M., E. Jeppesen, E. Mortensen, E. Dall, P. Kristensen & O. Sortkjær, 1990. Phytoplankton biomass reduction after planktivorous fish reduction in a shallow, eutrophic lake: a combined effect of reduced internal P-loading and increased zooplankton grazing. Hydrobiologia 200/201: 229–240.

Stensson, J. A. E., T. Bohlin, L. Henrikson, B. I. Nilsson, H. G. Nyman, H. G. Oscarson & P. Larsson, 1978. Effects of fish removal from a small lake. Verh. int. Ver. Limnol. 20: 794–801.

Telesh, I., 1993. The effect of fish on planktivorous fish. Hydrobiologia 255/256: 289–296.

Hydrobiologia **342/343:** 327–333, 1997.
L. Kufel, A. Prejs & J. I. Rybak (eds), Shallow Lakes '95.

Food-web manipulation in the Maltański Reservoir

Ryszard Gołdyn, Anna Kozak & Wanda Romanowicz
Department of Water Protection, A. Mickiewicz University, Drzymały 24, 60-613 Poznań, Poland

Key words: biomanipulation, whole-lake study, water quality, plankton, physico-chemical parameters

Abstract

The Maltański Reservoir (64 ha, mean depth 3.1 m) was emptied in September 1992, and all fish were harvested. Following refilling in the spring of 1993, fry of eel (*Anguilla anguilla*), pike (*Esox lucius*) and wels (*Silurus glanis*) were introduced, in total 236 fry ha^{-1}. In 1994 the reservoir was additionally stocked with adult pike in spring (10 kg ha^{-1}), with pike fry in summer (156 specimens ha^{-1}), and pike and pikeperch (*Stizostedion lucioperca*) fry in the autumn (413 specimens ha^{-1}). High water transparency and low concentration of chlorophyll-*a* were observed throughout 1993 except for a short period of water bloom caused by *Aphanizomenon flos-aquae* in summer. In the spring of 1994 only a short water bloom caused by diatoms was observed. Not only zooplankton grazing but also low phosphorus content limited the growth of phytoplankton biomass in this period. At the beginning of summer cladoceran biomass decreased, leading to a strong water bloom dominated by *Planktothrix agardhii*. During this event water transparency decreased to 0.3 m. This was a result of insufficient introduction of predatory fish, which could not ensure effective control of the rapidly increasing plankivorous fry numbers.

Introduction

The Maltański Reservoir, which is located within the city of Poznań (mid-west Poland) covers an area of 64 ha, with an average depth of 3.1 m. It is not only important for inhabitants of the city (bathing, swimming, water sports), but is also a place where international canoeing and rowing competitions are held. In the 1980's the reservoir was completely restored as its bottom sediments were removed. To improve water quality, the sewage load was diverted and a cascade of four preliminary reservoirs on the tributary, the Cybina River, were constructed (Gołdyn, 1994). After the reservoir was refilled with water in 1990, no fish were introduced. Thanks to this, water transparency was high during the first two years. Unfortunately, in the third year zooplankton biomass and filtration rate decreased gradually, allowing for intensive water blooms (Gołdyn et al., in press). This was caused by increasing numbers of fish which entered this water body from the Cybina River and spawned inside the reservoir (Mastyński & Klimaszyk, 1994). After reservoir drainage in September 1992 all fish were harvested and examined. It was found that the stock had been dominated by common bream (*Abramis brama*), roach (*Rutilus rutilus*) and small-sized perch (*Perca fluviatilis*), while the total fish biomass amounted to 399 kg ha^{-1} (Mastyński & Klimaszyk, 1994).

In April 1993 the reservoir was refilled and stocked with predatory fish in order to enhance the control of planktivorous fish populations and thereby the zooplankton grazing pressure on phytoplankton, according to the classic theory of biomanipulation (Shapiro et al., 1975). In the spring of 1993 fry of eel (*Anguilla anguilla*), pike (*Esox lucius*) and wels (*Silurus glanis*) were introduced (186, 19 and 31 specimens ha^{-1} respectively). During experimental fishing in the autumn of 1993, apart from the introduced predatory fish, large numbers of roach at the age of 0+ were caught in the haul, reaching 246 specimens m^{-2} of fry trawl net (10 mm mesh openings) on average (Mastyński & Klimaszyk, 1994). In 1994, stocking of the Maltański Reservoir with predatory fish was intensified. Adult pike (10 kg ha^{-1}) as well as summer pike fry, autumn pike and pikeperch (*Stizostedion lucioperca*) fry were introduced (156, 21 and 392 specimens ha^{-1} respectively). Nevertheless, results of intensive net fishing in November 1994 revealed that small-sized perch and roach still dominated the fish fauna, since they constituted 64.2% and 30.3% of the catch respectively.

This paper presents changes in selected physico-chemical and biological parameters of the Maltański Reservoir during the first two years after refilling and stocking of predatory fish (1993–1994).

Materials and methods

Water samples were taken monthly from the central part of the reservoir at the surface and at the depths of 1, 2 and 3 m. Between April and August 1994 samples were taken more often, i.e. every fortnight. Water transparency was measured by means of Secchi disc, chlorophyll-*a* concentration was assessed with Lorenzen method using acetone extraction, seston was weighted after filtration through Whatman glass fiber filters and dried at 105 °C, while zooplankton was identified and enumerated under microscope in Sedgwick-Rafter cells, using 10 l samples filtered through a 40-μm mesh net. Phosphorus (both total and dissolved) was determined with vanadomolybdophosphoric acid, nitrate with sulfosalicylic acid, ammonia with the phenate method, organic nitrogen with the Kjeldahl method after mineralization with selenium, while oxygen with iodometric method. All biological analyses were carried out according to Wetzel & Likens (1991), while physico-chemical – according to Polish Standards (Hermanowicz et al., 1976). In this paper mainly mean values from vertical profile are presented.

Results

In the spring of 1993, when the Maltański Reservoir was being refilled with water, a relatively low abundance and biomass of zooplankton was observed (Figure 1a) except for a large number of rotifers in April. Among cladocerans, large-sized species (*Daphnia magna* Straus, *D. similis* Claus, *D. longispina* O. F. Müller and *D. cucullata* Sars) were most numerous and had the highest biomass. Phytoplankton biomass in terms of chlorophyll-*a* concentration was very low at that time (Figure 1b) and Secchi depth reached down to the very bottom (Figure 1c). Disappearance of *Daphnia magna* and *D. similis* in July was followed by a rapid increase in chlorophyll-*a* (water bloom caused by *Aphanizomenon flos-aquae* (L.) Ralfs). In August cladoceran biomass rose rapidly. *Diaphanosoma brachyurum* (Lievin), *Daphnia cucullata* and *D. longispina* were dominant species in this period. Water bloom disappeared suddenly between 3rd and 10th of August (field observation), causing an increase in water transparency (Figure 1c). Till the end of May 1994 chlorophyll-*a* concentration was relatively low, not exceeding 20 mg m^{-3}, except for March, when it rose to 22 mg m^{-3} as a result of abundant appearance of chrysophytes (*Chrysococcus* sp., *Ochromonas* sp.). The appearance of small-sized planktonic algae was accompanied by a significant increase in rotifer biomass, which reached its maximum in the first half of May (Figure 1a). At the beginning they were dominated by *Synchaeta pectinata* Ehrenberg, later replaced by *Polyarthra vulgaris* Carlin and *Keratella quadrata* (Müller). Also biomass of *K. cochlearis* (Gosse) and *Brachionus angularis* Gosse increased at that time. At the beginning of May *Asplanchna priodonta* Gosse became the dominating species, followed by an increase in cladoceran and copepod biomass, mainly *Bosmina longirostris* O. F. Müller, *B. coregoni* Baird, *Daphnia cucullata* and *Mesocyclops leucarti* (Claus).

Between March and May 1994, the content of soluble reactive phosphorus (SRP) in water was completely depleted (Figure 1d). In this period the concentration of dissolved inorganic nitrogen (DIN) was very high (Figure 2a) as a result of flow of nitrates from diffuse sources into the Cybina River and, consequently, into the reservoir. Starting from the end of May 1994, marked differences in water quality between the near bottom layer and the surface layer of water were observed (Figures 2b and 2c). The concentration of dissolved phosphorus (Figure 1d) and chlorophyll-*a* (Figure 1b) increased then, leading to a short diatom bloom dominated by *Fragilaria crotonensis* Kitton.

At the end of June and at the beginning of July, a low value of cladoceran biomass was recorded (Figure 1a). During this period chlorophyll-*a* rose considerably (Figure 1b) as a result of mass appearance of cyanobacteria, particularly *Planktothrix agardhii* (Gom.) Anagnostidis & Komarek. In addition, a decrease in the value of TN:TP ratio (Figure 2d) and a successive increase in SRP concentration in water (Figure 1d) were recorded, particularly near the bottom of the reservoir, while the DIN level was relatively high (Figure 2a). The maximum difference in SRP values (amounting to 0.866 mg l^{-1}) between surface and near-bottom layers of water was recorded at the beginning of August, when the water near the bottom of the reservoir was completely

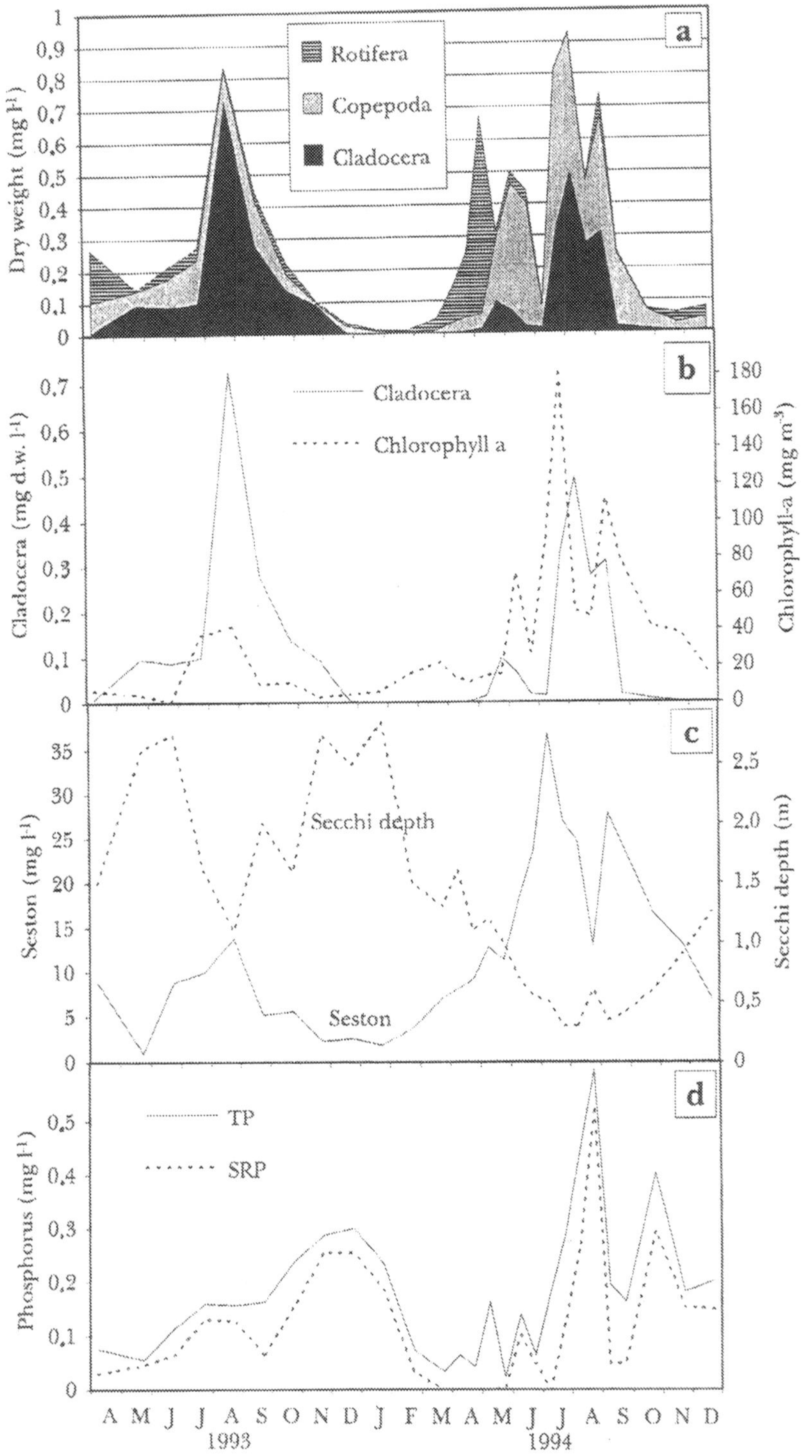

Figure 1. Zooplankton biomass (a), Cladocera dry weight and chlorophyll-*a* (b), Secchi depth and seston (c), total and soluble reactive phosphorus (d), in the water of the Maltański Reservoir in the years 1993–1994.

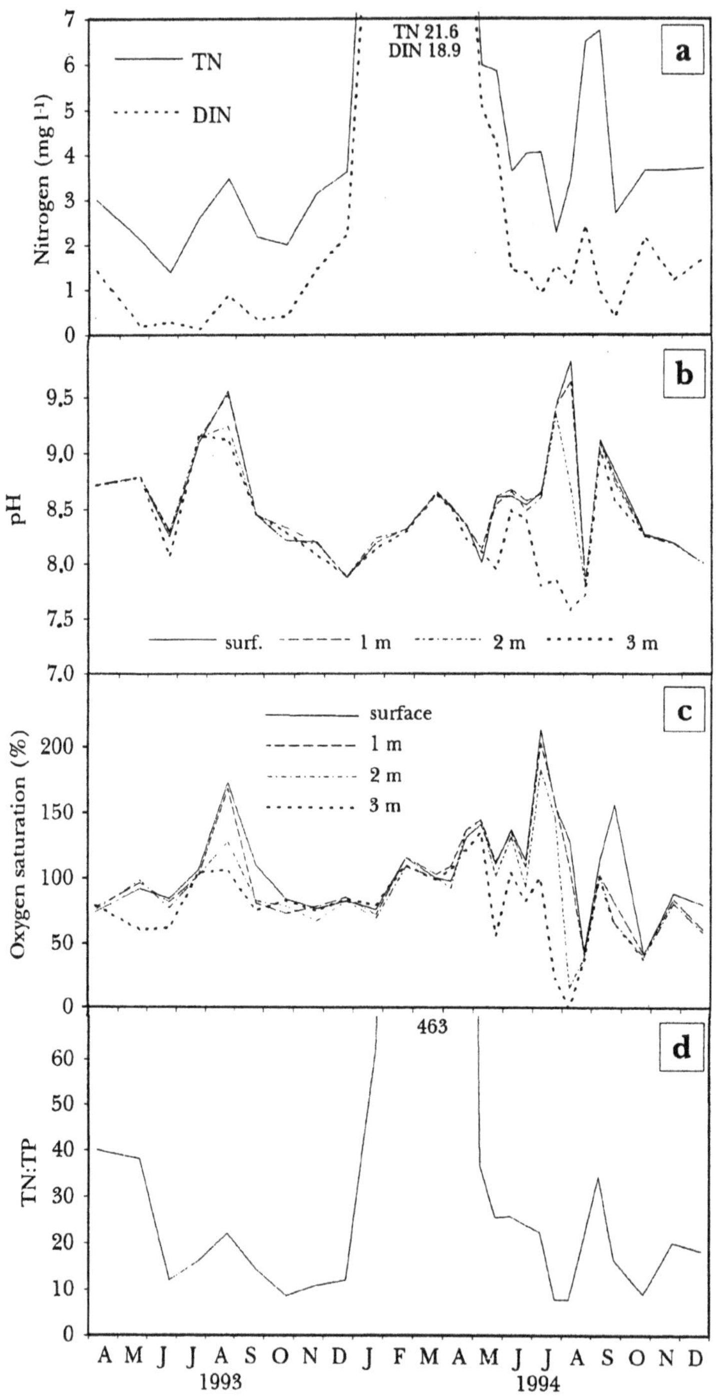

Figure 2. Values of total and dissolved inorganic nitrogen (a), pH (b), oxygen saturation (c), total nitrogen to phosphorus ratio (d), in the water of the Maltański Reservoir in the years 1993–1994.

deoxygenated. Both in July and in August total-P concentration in the tributary was lower than near the bottom of the reservoir. In July it was 0.366 and 0.408 mg l^{-1} P respectively, while at the beginning of August it amounted to 0.719 and 1.046 mg l^{-1} P. These changes show that growing amounts of phosphorus were being released from bottom sediments.

Between April and August Secchi depth was decreasing continuously until it reached 0.3 m (Figure 1c), a value characteristic of hypereutrophic water bodies (Zdanowski, 1982). During the whole period of studies, a strong reverse relationship between water transparency and seston content (Figure 1c) was observed. It may be expressed by the following exponential regression equation: $y = \exp\ (5.528\text{-}0.068x)$ ($\alpha<0.001$, $r=0.9$).

At the end of July and in August zooplankton biomass rose again, reaching its highest value at the beginning of August (Figure 1a). Rotifers were then dominated by *Keratella cochlearis* f. *tecta* (Gosse), copepods by *Mesocyclops leucarti*, while cladocerans by *Diaphanosoma brachyurum*. Zooplankton biomass decreased markedly in September (as water temperature fell to ca. 16 °C), and stayed at low levels till the end of the year. Chlorophyll-*a* concentration was falling much less rapidly (Figure 1b), resulting in rising water transparency (Figure 1c). The decline of *Planktothrix agardhii*, which caused the summer water bloom, contributed to a considerable reduction of oxygen content in water (Figure 2c) and an increase in the concentration of nutrients, particularly of phosphorus (Figure 1d).

Discussion

Biomanipulation experiment in the Maltański Reservoir was only partly successful, because high water transparency, which is characteristic of top-down control, was retained only in 1993. Even then, however, a short but strong water bloom was caused by *Aphanizomenon flos-aquae*. Large-bodied daphnids, especially *Daphnia magna*, which are crucial for biomanipulation experiments (Knoechel & Holtby, 1986; Lyche, 1989; Gliwicz, 1990), were observed only in the spring of 1993. A changed composition of cladoceran zooplankton in June and July resulted in uncontrolled reproduction of the colonial cyanobacterium. Similar situations are often observed in biomanipulation experiments and are regarded as a feed-back effect (Lyche, 1989; Benndorf, 1990; Gliwicz, 1990; Søndergaard et al., 1990; Kasprzak et al., 1993; Ronneberger et al., 1993). The above-mentioned changes in zooplankton composition were probably caused by roach juveniles which were very abundant in the reservoir. It means that the amount of predatory fish introduced in 1993 was too low to control the influence of planktivorous fry on large daphnids. The increased abundance of cladocerans in August could be explained by a gradual reduction of fish fry density during the summer (e.g. due to predation by the predators introduced). Among cladocerans, however, *D. magna* was missing. This may be linked with predation by young roach, which selectively predate large daphnids (Hessen, 1985). Consequently, zooplankton seemed unable to control phytoplankton as effectively as in the spring, in spite of a higher biomass. As a result, despite the disappearance of *A. flos-aquae* bloom, water transparency improvement was not as distinct as in the spring.

Early in the spring of 1994 phytoplankton was under a pressure of abundant rotifers. This was possible because the phytoplankton were dominated by nanoplanktonic forms, which are easily digested by rotifers, both macrofilter-feeders (*Synchaeta* sp., *Polyarthra* sp.) and microfilter-feeders (*Keratella quadrata*) (Karabin, 1985; Telesh, 1993). Nevertheless, dissolved phosphorus may have been the major factor limiting phytoplankton biomass since between March and the middle of May SRP was not detected in the reservoir (Figure 1d). This is supported by the fact, that concentrations of chlorophyll-*a* observed in the Maltański Reservoir in the same period of 1991 (up to 56.6 mg m^{-3}) and of 1992 (up to 80.3 mg m^{-3}) were higher, even though the composition and biomass of rotifers were similar, while SRP concentration was high (Gołdyn et al., 1994, in press).

In 1994 a clear top-down control took place as late as at the end of May, when cladoceran biomass and dissolved phosphorus concentration increased. However, since large-bodied daphnids were lacking, colonial diatoms escaped control and formed a short water bloom. This was a typical feed-back effect of biomanipulation, similar to that observed in the summer of 1993.

The decrease in cladoceran biomass in June 1994 was not caused by food limitation for no clear-water phase typical of this season was detected (Secchi depth was about 0.5 m). Most probably it was an effect of grazing on zooplankton by the abundant juvenile fish (mainly cyprinids and percids), as it is frequently observed in eutrophic lakes (Mamcarz, 1988). As a result, a strong water bloom was caused by *Plank-*

tothrix agardhii, which replaced the *Aphanizomenon flos-aquae* bloom of the previous year. This replacement may have been due to lack of the limiting impact of nitrogen, although there are many reports on N_2 fixation by cyanobacteria lacking heterocysts, including the genus *Oscillatoria* in the broad sense (Paerl, 1988). Although the TN:TP ratio was low (7.9) at the end of July, the concentration of DIN did not fall below 20 mg m^{-3}, which is regarded as the threshold of N-limitation (Ryding & Rast, 1989). At that time the rapidly increasing phosphorus concentration in the reservoir was much more important, because as it was concluded by Trimbee & Prepas (1987), Paerl (1988), Faafeng & Hessen (1993), high phosphorus inputs tended to favour cyanobacterial dominance and bloom formation. Its significance for the biomass of cyanobacteria in the Maltański Reservoir was important only when cladoceran biomass was not very high. This is supported by the fact that abundant reappearance of cladocerans at the beginning of August caused a strong decrease in chlorophyll-*a*. As a result of a rapid deterioration of weather conditions (decrease of water temperature from 27.7 °C to 18.1 °C and thorough water mixing), cladoceran biomass slightly decreased, allowing cyanobacteria to escape grazing pressure. Therefore chlorophyll-*a* content at the end of August increased once more in spite of a relatively high cladoceran biomass.

In conclusion, the expected top-down control took place only in the first year of the experiment and shortly in May and August of the second year. Even in the first year, though, a short summer bloom of *Aphanizomenon flos-aquae* was observed. In the spring of the second year not only zooplankton grazing but also phosphorus had an influence on phytoplankton biomass. A decreased cladoceran biomass in summer led to a strong water bloom. This was a result of insufficient introduction of predatory fish which could not ensure effective control of the rapidly increasing plankivorous fry numbers. According to Sondergaard et al. (1996) and Prejs et al. (1994) stocking of c. 1000 piscivorous fry ha^{-1} is needed to achieve a strong effect on lower trophic levels, while introduction in the Maltański Reservoir was only 236 ha^{-1} in 1993 and 156 ha^{-1} in the spring of 1994. Further stocking with pike and pikeperch fry (413 ha^{-1}) in the autumn of 1994, had little effect on planktivorous fry in this year.

Acknowledgments

The research was supported by the Polish Committee for Scientific Research (Grant No 6 P205 063 06 to Ryszard Gołdyn).

References

Benndorf, J., 1990. Conditions for effective biomanipulation; conclusions derived from whole-lake experiments in Europe. Hydrobiologia 200/201: 187–203.

Faafeng, B. A. & D. O. Hessen, 1993. Nitrogen and phosphorus concentrations and N:P ratios in Norwegian Lakes: Perspectives on nutrient limitation. Verh. int. Ver. Limnol. 25: 465–469.

Gliwicz, Z. M., 1990. Why do cladocerans fail to control algal blooms? Hydrobiologia 200/201: 83–97.

Gołdyn, R., 1994. The role of preliminary reservoirs in the protection of the Maltański Reservoir. Arch. Hydrobiol. Beih. Ergebn. Limnol. 40: 253–262.

Gołdyn, R., A. Kozak & K. Kostka, 1994. Causes of changes in the water quality of the restored Maltański Reservoir in Poznań. Poznańskie Towarzystwo Przyjaciół Nauk, Prace Komisji Biologicznej 74: 33–58.

Gołdyn, R., A. Kozak, W. Romanowicz, K. Grześkowiak & B. Starczewska-Porada, in press. Phytoplankton-zooplankton interaction in the newly-restored Maltański Reservoir. Verh. int. Ver. Limnol. 26:

Hermanowicz, W., W. Dożańska, J. Dojlido & B. Koziorowski, 1976. Fizyczno-chemiczne badanie wody i ścieków [Physico-chemical analyses of water and sewage]. Arkady, Warsaw, 847 pp.

Hessen, D. O., 1985. Selective zooplankton predation by pre-adult roach (*Rutilus rutilus*): The size-selective hypothesis versus the visibility-selective hypothesis. Hydrobiologia 124: 73–79.

Karabin, A., 1985. Pelagic zooplankton (Rotatoria + Crustacea) variation in the process of lake eutrophication. II. Modifying effect of biotic agents. Ekol. pol. 33: 617–644.

Kasprzak, P., L. Krienitz & R. Koschel, 1993. Changes in the plankton of Lake Feldberger Haussee (Germany, Mecklenburg-Vorpommern) in response to biomanipulation. Arch. Hydrobiol. 128: 149–168.

Knoechel, R. & L. B. Holtby, 1986. Cladoceran filtering rate: body length relationships for bacterial and large algal particles. Limnol. Oceanogr. 31: 195–200.

Lyche, A., 1989. Plankton community response to reduction of planktivorous fish populations. A review of 11 case studies. Aquat. Fennica 19: 59–66.

Mamcarz, A., 1988. Czy sadzowy podchów ryb głąbielowatych zmienia strukturę zooplanktonu w jeziorze? [Does cage rearing of Coregonids change the zooplankton structure of a lake?]. Acta Acad. Agricult. Techn. Olst., Protectio Aquarum et Piscatoria 16: 3–14.

Mastyński, J. & P. Klimaszyk, 1994. Importance of fish fauna for the water quality of the Maltański Reservoir. Poznańskie Towarzystwo Przyjaciół Nauk, Prace Komisji Biologicznej 74: 75–82.

Paerl, H. W., 1988. Growth and reproductive strategies of freshwater blue-green algae (cyanobacteria). In: Sandgren, C. D. (ed.), Growth and Reproductive Strategies of Freshwater Phytoplankton. Cambridge University Press: 261–315.

Prejs, A., A. Martyniak, S. Boroń, P. Hliwa & P. Koperski, 1994. Food web manipulation in a small, eutrophic Lake Wirbel,

Poland: effect of stocking with juvenile pike on planktivorous fish. Hydrobiologia 275/276: 65–70.

Ronneberger, D., P. Kasprzak & L. Krienitz, 1993. Long-term changes in the rotifer fauna after biomanipulation in Haussee (Feldberg, Germany, Mecklenburg-Vorpommern) and its relationship to the crustacean and phytoplankton communities. Hydrobiologia 255/256: 297–304.

Ryding, S.-O. & W. Rast (eds), 1989. The control of eutrophication of lakes and reservoirs. Man and the Biosphere Series 1. UNESCO, Paris, 314 pp.

Shapiro, J., V. Lamarra & M. Lynch, 1975. Biomanipulation, an ecosystem approach to lake restoration. In Brezonik, P. L. & J. L. Fox (eds), Proc. Symp. 'Water Quality Management Through Biological Control', Univ. Florida Gainsville: 85–96.

Søndergaard, M., E. Jeppesen & S. Berg, 1997. Pike (*Esox lucius* L.) stocking as a biomanipulation tool. 2. Effects on lower trophic levels in Lake Lyng (Denmark). Hydrobiologia 342/343: 319–325.

Søndergaard, M., E. Jeppesen, E. Mortensen, E. Dall, P. Kristensen & O. Sortkjær, 1990. Phytoplankton biomass reduction after planktivorous fish reduction in a shallow, eutrophic lake: a combined effect of reduced internal P-loading and increased zooplankton grazing. Hydrobiologia 200/201: 229–240.

Telesh, I. V., 1993. The effect of fish on planktonic rotifers. Hydrobiologia 255/256: 289–296.

Trimbee, A. M. & E. E. Prepas, 1987 Evaluation of total phosphorus as a predictor of the relative biomass of blue-green algae with emphasis on Alberta lakes. Can. J. Fish. aquat. Sci 44: 1337–1342.

Wetzel, R. G. & G. E. Likens, 1991. Limnological analyses. Springer-Verlag, New York, 391 pp.

Zdanowski, B., 1982. Variability of nitrogen and fosforus contents and lake eutrophication. Pol. Arch. Hydrobiol. 29: 541–597.

Hydrobiologia **342/343:** 335–349, 1997.
L. Kufel, A. Prejs & J. I. Rybak (eds), Shallow Lakes '95.

Effects of biomanipulation in the large and shallow Lake Wolderwijd, The Netherlands

Marie-Louise Meijer & Harry Hosper
Institute for Inland Water Management and Waste Water Treatment, P.O. Box 17, 8200 AA Lelystad, The Netherlands

Key words: eutrophication, lake restoration, biomanipulation, fish, zooplankton, phytoplankton, macrophytes, transparency

Abstract

Since the early seventies, Lake Wolderwijd (2650 ha, mean depth 1.5 m) suffered from cyanobacterial blooms, turbid water and a poor submerged vegetation as a result of eutrophication. From 1981 onwards the lake was flushed with water low in phosphorus and high in calcium bicarbonate. Total-P and chlorophyll *a* in the lake more than halved, but Secchi depth in summer increased from 0.20 to 0.30 m only. In the hope of triggering a shift from the algae-dominated turbid water state to a macrophyte-dominated clear water state, the lake was biomanipulated during winter 1990/1991. The fish stock, mainly bream (*Abramis brama*) and roach (*Rutilus rutilus*), was reduced from 205 to 45 kg ha^{-1}. In May 1991, 575 000 (217 ind ha^{-1}) pike fingerlings (*Esox lucius*) were introduced. In spring 1991 the lake cleared as a result of grazing by *Daphnia galeata*. The clear water phase lasted for only six weeks. Macrophytes did not respond as expected and most of the young pike died. However, from 1991 to 1993 the submerged vegetation is gradually changing. Characeae began to spread over the lake (from 28 ha in 1991 to 438 ha in 1993). The water over the *Chara* meadows was clear, probably as a result of increased net sedimentation in these areas. It is hypothesized that expansion of the *Chara* meadows might ultimately result in a shift of the whole lake to a lasting clear water state. In order to promote the *Chara*, the fish stock reductions aimed at a spring clear water phase should be continued.

Introduction

Most shallow lakes in the Netherlands are characterized by turbid water and a lack of macrophytes, primarily as a result of eutrophication. The lake restoration strategy has been focused on the reduction of the external phosphorus loading. However, so far this strategy has not resulted in the desired clear water state (Van Liere & Gulati, 1992). Recovery is hampered by the internal phosphorus loading from the sediments (Sas, 1989; Van der Molen & Boers, 1994) and by the abundance of benthivorous and planktivorous fish (Hosper, 1989; Hosper & Jagtman, 1990). The turbid water state appears to be extremely stable and additional measures are necessary to get the recovery process started. Biomanipulation, i.e. removing or stocking of fish, has been suggested as a promising tool (Shapiro et al., 1975; Benndorf et al., 1984; Carpenter et al., 1987; Gulati et al., 1990). Biomanipulation as a tool for restoration of shallow lakes is mainly based upon the concept of 'alternative stable states'. Within certain limits of nutrient loading, there are two alternative stable states: a phytoplankton-dominated turbid water state and a macrophyte-dominated clear water state (Scheffer, 1989, 1990; Moss, 1990; Scheffer et al., 1993). A drastic reduction of the planktivorous and benthivorous fish stock may induce a shift from the turbid water state to the clear water state (Hosper & Jagtman, 1990). This approach proved to be successful in the restoration of small lakes (<50 ha) in the Netherlands (Meijer et al., 1990a, 1990b; Van Donk et al., 1990a; Meijer et al., 1994a; Van Berkum et al., 1995; Van Donk & Gulati, 1995), Denmark (Jeppesen et al., 1990a) and the UK (Philips & Moss, 1994). Substantial fish stock reduction (>70%) resulted in increased zooplankton grazing, decreased sediment resuspension, clear water and as a consequence the return of submerged macrophytes. Submerged macrophytes are believed to play a key role in sustaining the clear water state by several mechanisms, such as competing with algae for nutrients, reducing the resuspension of the sediment or providing refuge for zooplankton (Moss, 1990; Jeppesen et al., 1990b; Scheffer et al., 1993; Meijer et al., 1994a).

Stimulated by the successful case studies in small lakes, biomanipulation was suggested to restore the large Lake Wolderwijd (2650 ha, $z = 1.5$ m). It is clear that the results from small lakes cannot be applied to large lakes, without due consideration. In Lake Breukeleveen (180 ha) fish reduction failed to result in clear water, possibly due to the insufficient removal of fish, the abundance of filamentous cyanobacteria and the strong impact of wind-induced resuspension of the peaty lake sediments (Van Donk et al., 1990b, 1994). Filamentous cyanobacteria, such as *Oscillatoria agardhii* Gomont, are a less favourable food source for zooplankton (Gliwicz, 1990). In comparison to Lake Breukeleveen, Wolderwijd offers better chances for biomanipulation, due to the lower abundance of filamentous cyanobacteria and the predominantly sandy lake sediments. Furthermore, enclosure experiments (2 × 1 ha) in Wolderwijd, showed a rapid colonization by *Chara* spp. after removal of the fish stock (unpubl. results).

Our objective was to test biomanipulation as a tool for restoring a large and shallow lake. In early 1991, the fish stock was substantially reduced and pike (*Esox lucius* L.) fingerlings were stocked to support the control of young-of-the-year (YOY) cyprinids. We expected an increase in transparency, due to high zooplankton grazing and reduced sediment resuspension by fish, and consequently an increase in the abundance of submerged macrophytes. The new clear water state would be sustained by a developing pike stock and an extensive submerged vegetation. Responses on transparency, nutrients and the biotic components of the lake ecosystem were studied. Three years after the main intervention (1991–1993) were compared to two preceding years (1989–1990).

Study site

Wolderwijd is a large and shallow lake in the central part of the Netherlands with an area of 2650 ha (Figure 1). Water depth varies from 0.5 m to 2.5 m, mean depth is 1.5 m and maximum depth is 5.0 m. The lake was created in 1968 along with the construction of the polder of Flevoland in IJsselmeer (Meijer et al., 1994b). Several small streams from a predominantly agricultural area flow into the lake. Sluices in the south of the lake provide the outflow into Nijkerkernauw. A sluice in the north connects Wolderwijd with Veluwemeer. The first year of its existence the water of the lake was clear. Since 1970 the lake is characterized by blooms of *Oscillatoria agardhii* (Berger & bij de Vaate, 1983) and a poorly developed submerged vegetation of *Potamogeton pectinatus* L. and *P. perfoliatus* L. The external P-loading is 0.5–1.0 g P m^2 y^{-1} and varies with the inflow from streams. From 1981 onwards the lake has been periodically flushed with excess polder water, that is low in phosphorus (<100 μg l^{-1}) and high in calcium bicarbonate (Figure 2). In the adjacent Veluwemeer the flushing with polder water started already in 1979 and was very successful (Hosper, 1984; Hosper & Meijer, 1986; Jagtman et al., 1992). Internal P-loading strongly decreased due to the pH buffering effect of the calcium bicarbonate. Particularly during cold winters with long periods of ice-cover, the number of cyanobacteria in Veluwemeer significantly reduced by the flushing (Jagtman et al., 1992). Although the flushing of Wolderwijd was less intense, total-P and chlorophyll *a* more than halved (Figure 4) and the phytoplankton became more diverse. Summer Secchi depth increased from 0.20 to 0.30 m only (Figures 3 and 4).

Materials and methods

Biomanipulation

The objective of the fish stock removal was to reduce the planktivorous fish stock to < 20 kg ha^{-1} and the benthivorous fish stock to <25 kg ha^{-1}. These objectives were based upon the following empirical data: (1) results from enclosure experiments (McQueen & Post, 1988) indicated that above 20 kg ha^{-1} of planktivorous fish, the abundance of zooplankton is strongly depressed and (2) preliminary results from pond experiments (Meijer et al., 1990a) showed that significant effects on turbidity can be expected above 50 kg ha^{-1} of benthivorous fish. This level was reduced to 25 kg ha^{-1} to account for the reinforcing effect of wind-induced waves on turbidity in large lakes. The fishery was carried out between November 1990 and July

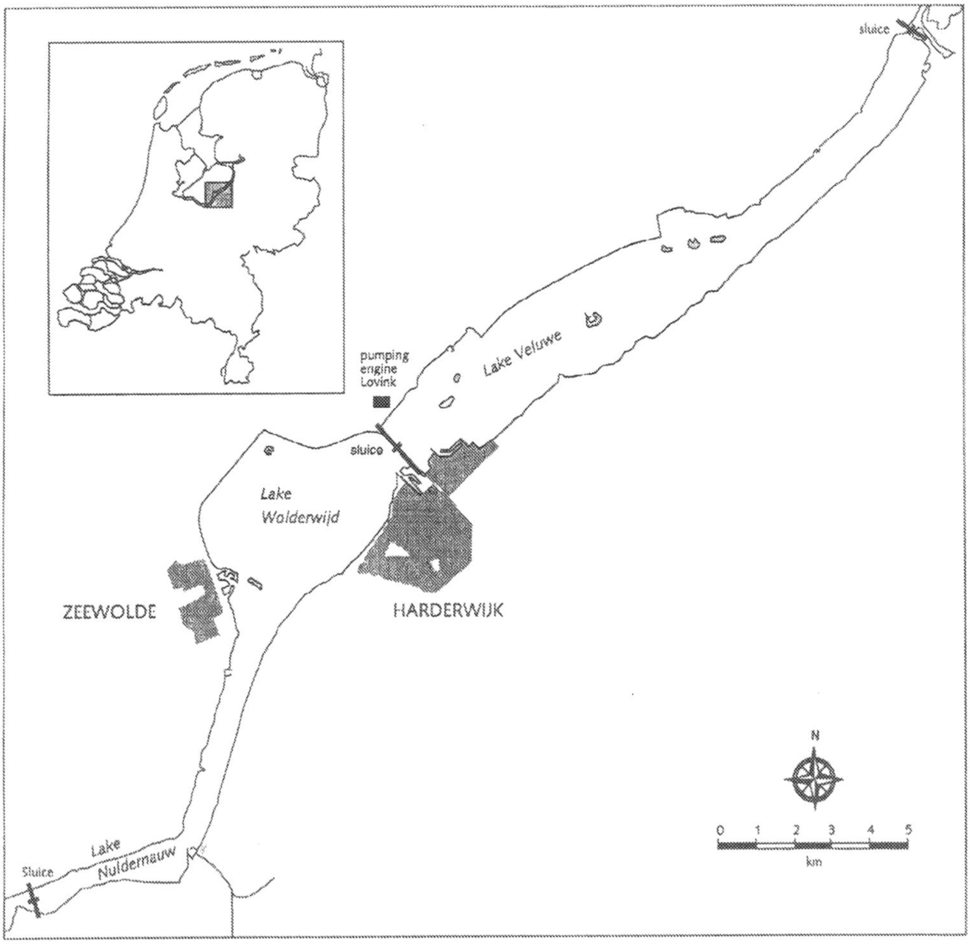

Figure 1. Lake Wolderwijd in the Netherlands.

1991, using large seine nets (180 m, 770 m and 1200 m in length), trawls and fykes. The initial fish stock of about 205 kg ha^{-1} (190–220 kg ha^{-1}) was reduced to about 45 kg ha^{-1} (33–63 kg ha^{-1}). Approx. 425 000 kg of fish were removed, of which about 95 000 kg of bream (*Abramis brama* L.) and roach (*Rutilus rutilus* L.) were restocked elsewhere. The rest was transported to a fish processing plant. The original objectives were realized. For detailed information on the planning and implementation of the fish removal, see Grimm & Backx (1994) and Backx & Grimm (1994). On 8 May and 28 May 1991 in total about 575 000 pike fingerlings (3–4 cm) were introduced. The next year, from January to May 1992, additional removal of small fish (<15 cm) from the harbours and the deeper parts of the lake was undertaken. In May 1992, 42 kg ha^{-1} of large bream (>25 cm) were removed from the spawning areas by fyke nets. From February to August 1993, small fish (<15 cm) and larger bream were again targeted and about 73 kg ha^{-1} was removed.

Lake flushing

To reduce the number of filamentous cyanobacteria and thus enhancing the chances for biomanipulation, flushing of the lake with polder water was intensified from 1990 onwards. Flushing increased from monthly 1–2.10^6 m^3 in 1988–1989, to 4–7.106 m^3 in 1991–1992 (Figure 2). As a result of the flushing, the hydraulic residence time of Wolderwijd, on the basis of inflow,

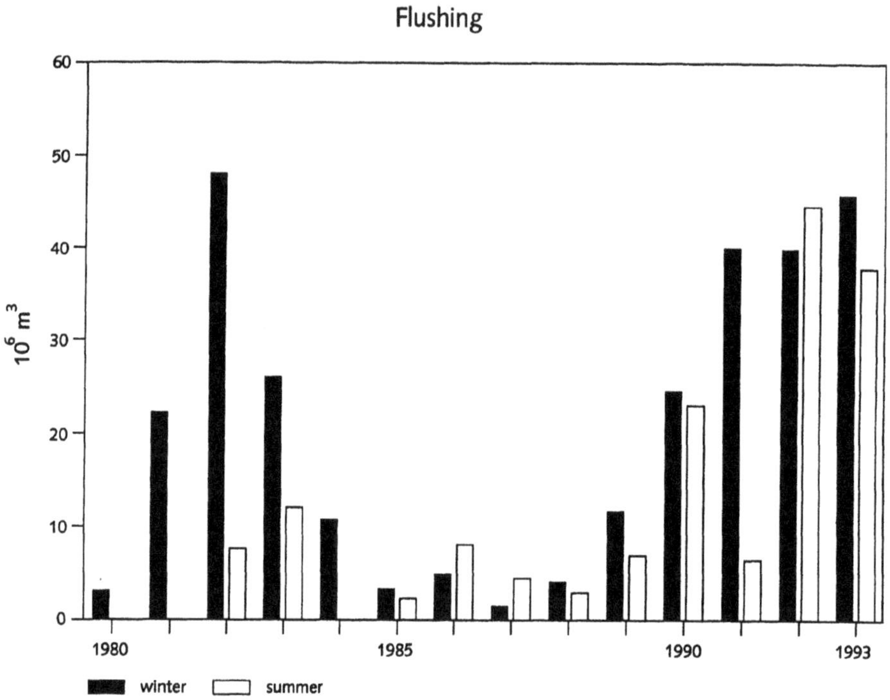

Figure 2. Flushing of Wolderwijd. Total flushing in winter (October–March) and summer (April–September) from 1975 to 1993.

Table 1. Water quality inside and outside the *Chara* vegetation in June–July 1993. Standard deviation within brackets (n = 4 or 5).

Vegetation	Inside vegetation	Outside
Secchi depth * in m	0.6 (0.13)***	0.3 (0)
Chlorophyll a* in μg l^{-1}	12 (5)	38 (5)
Total-P in mg l^{-1}	0.06 (0.02)	0.10 (0.03)
Total-N in mg l^{-1} in mg l^{-1}	1.37 (0.25)	1.8 (0.05)
SRP* in mg l^{-1}	0.003 (0)	<0.001 (0)
Dissolved-N** in mgN l^{-1}	0.045 (0.015)	0.06 (0.013)
Inorg. susp. solids in mg l^{-1}	2.75 (1.3)	7 (1.2)

* Significant difference between inside and outside the vegetation (p<0.05).
** Dissolved N = $NO_3 + NO_2 + NH_4$.
*** Secchi depth to the lake bottom.

reduced from about 1 year to 0.5 year. To prevent fish from entering the lake, the water was flushed through an iron grid (mesh size 3 × 0.5 cm) in the sluice. During summer the water was pumped into the lake after sieving through stones and a Nicolon net (mesh size <1 mm).

Sampling and analysis

Every fortnight, three parts of the lake were sampled (deep, shallow and centre of the lake). In each part 10–15 samples were taken over a length of 500 m with a transparent 1.5 m perspex tube (ø 5 cm). Analysis showed no significant difference between the parts (Van Nes et al., 1992). Only above the Characeae vegetation (after 1991) differences were found, which are separately presented in Table 1. In the rest of this paper data are presented from the centre of the lake. The 10–15 samples were mixed to a composite sample of 25-30 l for analysis of nutrients, chlorophyll a and phytoplankton. Nitrogen, phosphorus, suspended solids and chlorophyll a were measured according to the Netherlands standard methods (NEN), which comply with International Standards (ISO). For analysis of zooplankton a composite sample of 25 l was filtered over a net with mesh size 55 μm and fixed with 96% ethanol. Density, length and fecundity (number of eggs adult ind^{-1}) of *Daphnia* spp. were determined. The composition of phytoplankton was determined by taking a 1 l sample that was then fixed with lugol solu-

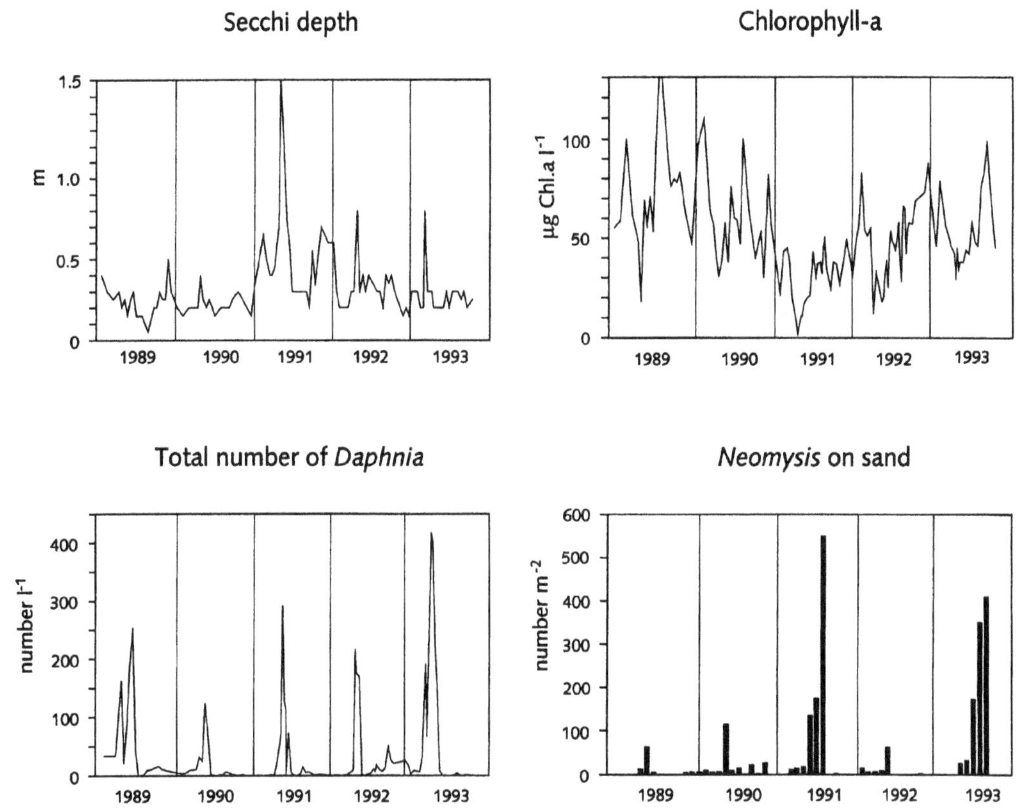

Figure 3. Secchi depth, Chlorophyll *a*, *Daphnia* and *Neomysis* in 1989–1993, before and after the drastic fish stock reduction (November 1990 to June 1991).

tion. At least 200 individuals were determined. The vegetation was estimated in three density classes (0–15%, 16–50% and 51–100% coverage of the lake bottom) by extrapolating from gridpoints in the lake (grid 50 × 100 m). The density of the mysid shrimp, *Neomysis integer* Leach, was monitored monthly by sampling 10 parts of the lake (different sediment type and depth) with five replicates each. The samples were taken by pulling a net with an opening of 0.50 × 0.25 m and a mesh size of 0.5–2.0 mm over 40 m^2. The fish stock was estimated by trawling (35–70 trawls covering 45–70 ha), using the catch per surface area and estimates for the efficiency of the nets (Grimm & Backx, 1994). The fish biomass has an estimated inaccuracy of 10–25% (Backx & Grimm, 1994; Backx, unpubl. data).

Results

Transparency, nutrients and phytoplankton

In 1989 and 1990, before the fish removal, the summer Secchi depth was about 0.30 m. After the fish removal in May 1991, an extraordinary high Secchi depth of 1.7 m was observed throughout the lake (Figure 3). This clear water phase lasted for six weeks. The high transparency coincided with extremely low concentrations of chlorophyll *a* (3 $\mu g\, l^{-1}$, Figure 3) and inorganic suspended solids (4 mg l^{-1}, not in figure). The next year, in May 1992, the transparency reached a peak value of 0.9 m, but this clear water phase lasted for two weeks only. However, in the part of the lake covered with Characeae (see below), the water remained clear until August. In 1993 only during one week a Secchi depth of 0.9 m was observed in the open water, but again above the Characeae, the water remained clear for about six weeks. A sharp distinction was observed

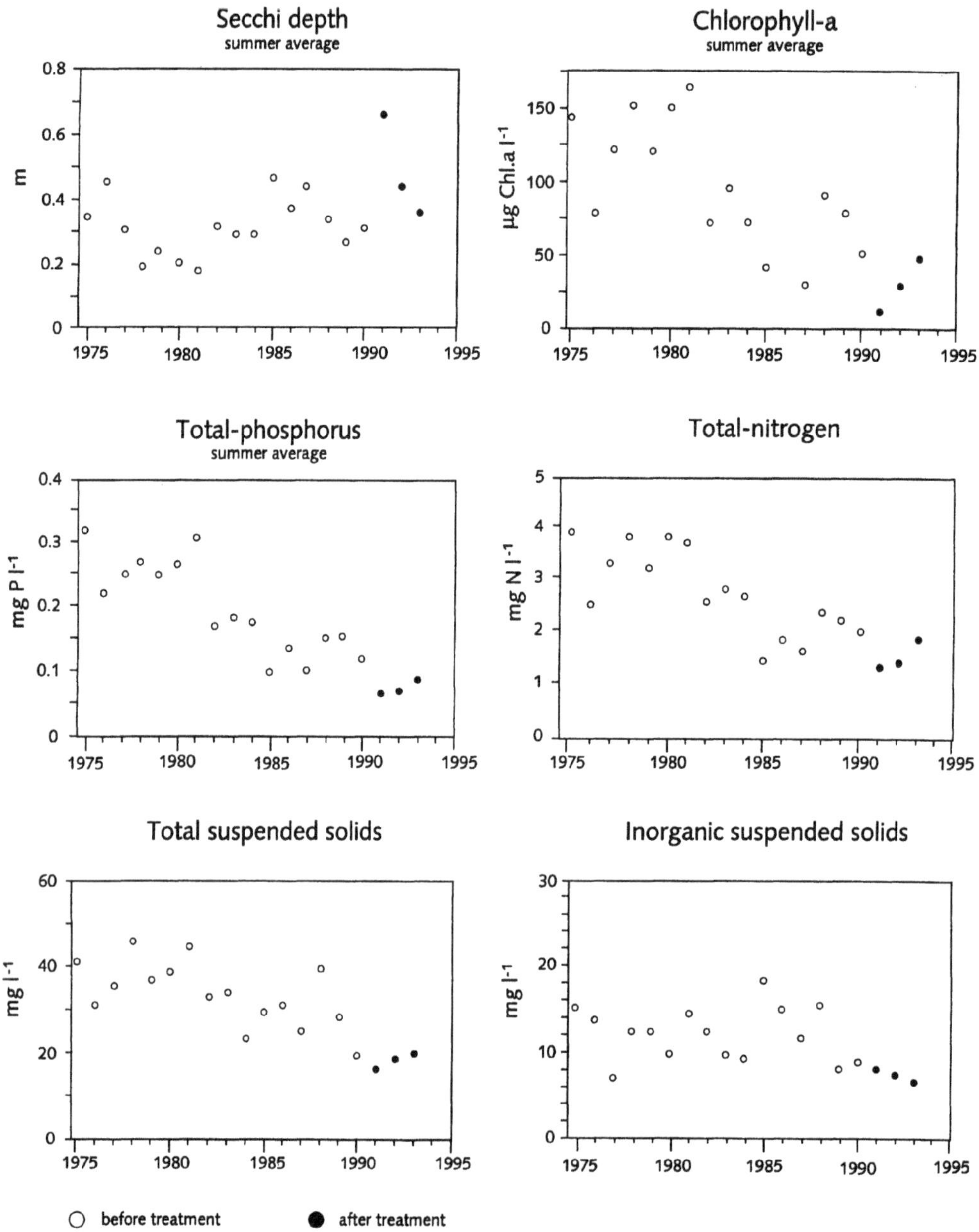

Figure 4. Secchi depth, Chlorophyll *a*, total-P, total-N, total suspended solids and inorganic suspended solids in the open water. Mean values (April–September) for 1975–1993, before and after the drastic fish stock reduction (November 1990 to June 1991).

between the turbid water and the clear water and aerial photographs showed that the transition zone may be less than 10 m (Scheffer et al., 1993). In summer 1991 the mean Secchi depth in the open water was significantly higher ($p<0.05$) than in all other years (Figure 4). In May–June, at the start of the growing season for vegetation, Secchi depth was about twice as high as in all other years (Figure 5). In 1991 the concentration of total phosphorus and total nitrogen was

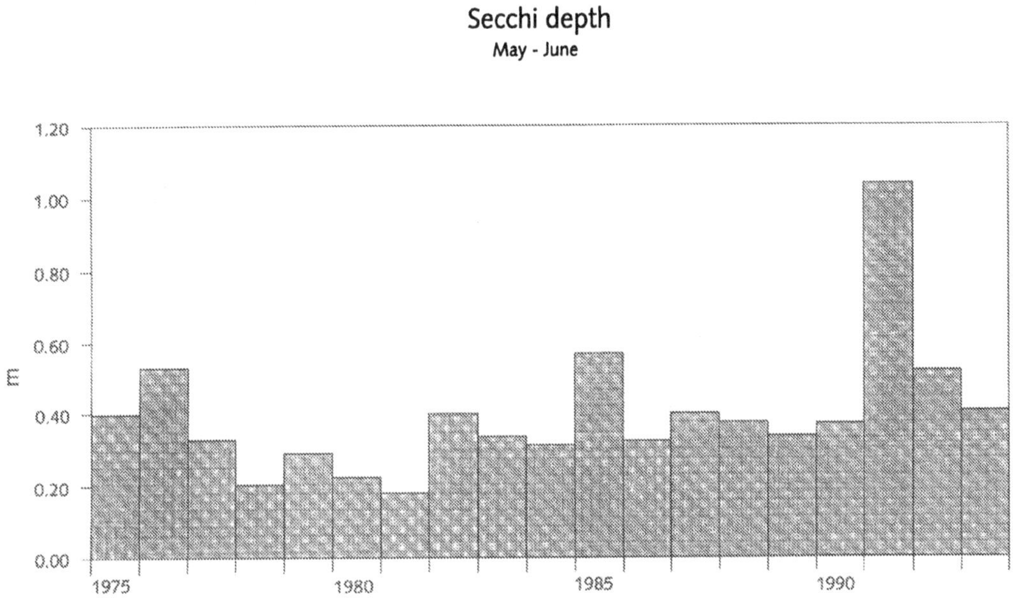

Figure 5. Secchi depth in May–June from 1975 to 1993, before and after the drastic fish stock reduction (November 1990 to June 1991).

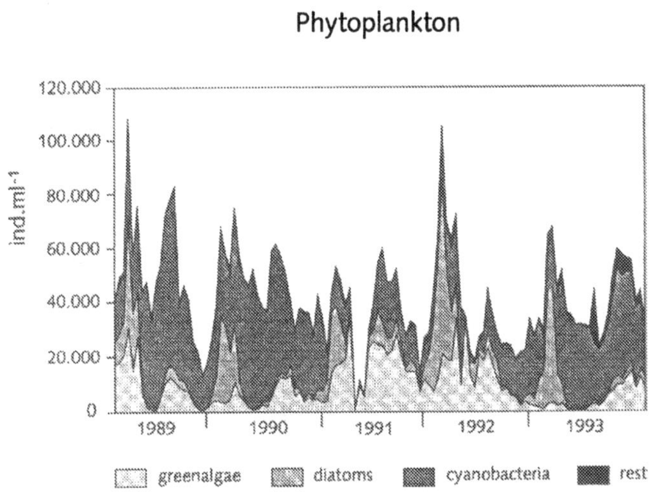

Figure 6. Phytoplankton in 1989–1993, before and after the drastic fish stock reduction (November 1990 to June 1991).

relatively low (resp. 0.07 mg P l^{-1} and 1.4 mg N l^{-1}, Figure 4).

In 1989 and 1990 the phytoplankton was dominated by cyanobacteria, predominantly *Oscillatoria agardhii*, with densities of 10000–70000 ind ml^{-1}. The dominance of cyanobacteria was broken after the fisheries in 1991, but returned in the summer of 1993 (Figure 6).

Daphnia and Neomysis

Each year, *Daphnia* developed a high density in spring (Figure 3). Before and after the fish reduction the *Daphnia* population was mainly composed of *Daphnia galeata* (Sars). In 1991 and 1992 the highest density coincided with the clear water phase. Only in 1992 *Daphnia* slightly recovered in summer, while in all

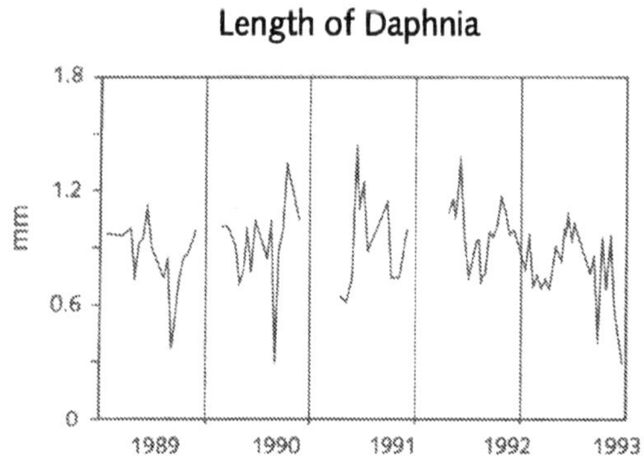

Figure 7. Length of *Daphnia* in 1989–1993, before and after the drastic fish stock reduction (November 1990 to June 1991).

the other years *Daphnia* was almost absent from July onwards. Each summer the average length of *Daphnia* decreased (Figure 7), probably as a result of selective fish predation by YOY fish (Lammens et al., 1985).

During the clear water phase in 1991, the density nor the length of Daphnia was higher than in the years before the fish reduction. In July 1991, however, the average length was relatively high. The mysid shrimp *Neomysis integer* reached high densities in July 1991 and 1993 (over 400 ind m^{-2} on sand) (Figure 3). In other years *Neomysis* was present, but the densities were low (<100 ind m^{-2}).

Macrophytes

Over the years 1989–1993, the total vegetated area of the lake remained the same, about 1000 ha (Figure 8), but a shift in species and density was observed. Before the fish removal (1989–1990) the main species were *Potamogeton pectinatus* and *P. perfoliatus*. In 1992–1993 the dense parts of the *Potamogeton* vegetation almost disappeared and became replaced by Characeae (Figures 8 and 9). The *Chara* vegetation consisted mainly of *Chara* contraria *A. Braun* ex Kütz and *C. vulgaris* L. In 1991 *Chara* occupied 28 ha, in 1992 141 ha and in 1993 the area had expanded to 438 ha of the lake. In 1993, the water remained clear in an area of 200–300 ha above the Characeae (see above). The concentration of chlorophyll *a* and inorganic suspended solids was lower within the *Chara* meadows than outside the vegetation (Table 1). No significant difference was found in total-N and total-P, but the concentration of SRP was higher inside the vegetation area ($p<0.05$, Wilcoxon signed rank test).

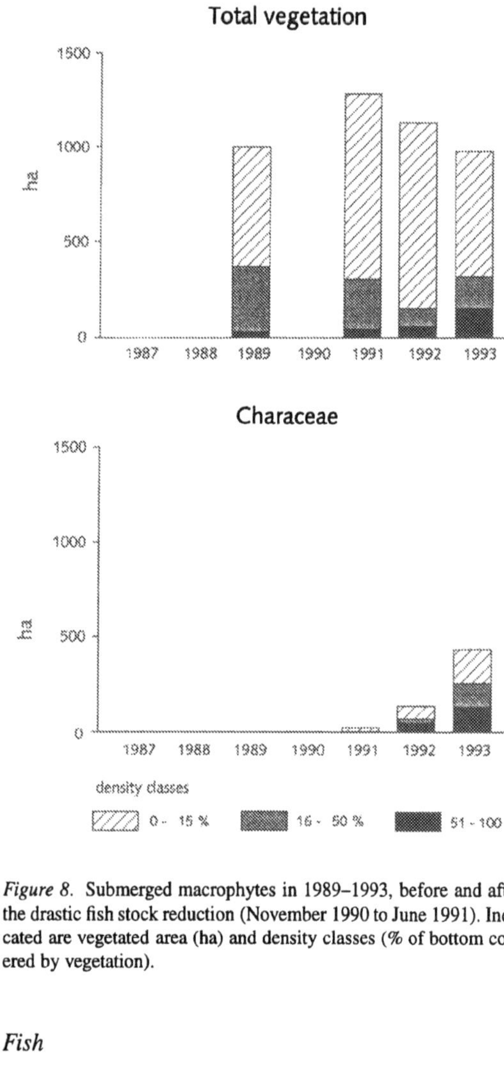

Figure 8. Submerged macrophytes in 1989–1993, before and after the drastic fish stock reduction (November 1990 to June 1991). Indicated are vegetated area (ha) and density classes (% of bottom covered by vegetation).

Fish

The initial fish stock of 205 kg ha^{-1} was reduced to 45 kg ha^{-1} (Figure 10). During the summer of 1991, the fish stock increased again to 100 kg ha^{-1} by September, mainly as a result of the production of young fish (YOY), of which 30 kg ha^{-1} consisted of ruffe (*Gymnocephalus cernua* L.). The dominant species of the fish stock in September were ruffe (30%) and bream >25 cm (23%). In June 1992, after the additional fish removal, the remaining population was estimated at 70 kg ha^{-1}, which at the end of the growing season, in September 1992, had increased to 105 kg ha^{-1}. During the summer 45 kg ha^{-1} of YOY recruited, dominated by perch (*Perca fluviatilis* L.) (58%). In January 1993 the total fish stock had increased to 165 kg ha^{-1}, partly due to immigration of bream >25 cm. In winter 1993 mainly large bream was removed. The remaining population in June 1993 was about 85 kg ha^{-1}, while at the end of the growing season the total biomass had hardly changed. In September 1991, electro-fishing indicated a pike biomass of less than 3 kg ha^{-1}.

Discussion

The objective of this study was to test the concept of biomanipulation in a large and shallow lake. Biomanipulation has been successfully applied in small lakes (<50 ha) (Van Donk et al., 1990a; Jeppesen et al., 1990a; Meijer et al., 1994a; Van Berkum et al., 1995; Philips & Moss, 1994). Apart from the technical feasibility of the fish removal, in larger lakes factors such as wind-induced resuspension of sediments and the more difficult recolonization by macrophytes, may reduce the chances for successful biomanipulation (Hosper & Meijer, 1993; Reynolds, 1994).

The results of 1991 showed that it was technically feasible to remove the bulk of the fish stock and to realize the objectives for planktivorous and benthivorous fish. In 1992 and 1993 the objective of the fish reduction was not achieved because the fish reduction started too late in the season. For optimal fish reduction the fishery should start in November at the latest (Perrow et al., this volume) in order to spread the fishery over a longer period with more intervals without fisheries.

Wolderwijd cleared in spring 1991, demonstrating that attainment of clear water is possible in a large and shallow lake. This was also shown in the large Lake Christina (1650 ha; Hanson & Butler, 1994), Finjasjön (950 ha; Annadotter, 1993) and Sätofta Bay of Ringsjön (400 ha; Hamrin, 1993). Apparently, in Wolderwijd the impact of wind-induced resuspension on turbidity is not as dominant as in other large lakes in the Netherlands, such as Lake Breukeleveen, where biomanipulation was not successful (Van Donk et al., 1990b, 1994). Based on results of biomanipulation in smaller lakes, it was expected that the reduction of benthivorous fish would lead to a reduction of the resuspension of sediment (Meijer et al., 1990b; Breukelaar et al., 1994; Van Berkum et al., 1995). However, the original biomass of benthivorous fish in Wolderwijd (90 kg ha^{-1}) was low as compared to the biomass in the small lakes (300–450 kg ha^{-1}) (Meijer et al., 1994a). This explains why no significant decrease in inorganic suspended solids was found after the fish reduction in Wolderwijd. In this discussion attention will be focused at the dynamics of *Daphnia*, the role of predation by piscivorous fish, the development of Characeae and the high transparency of the water above the Characeae.

Daphnia grazing

In most biomanipulation case studies the reduction of the fish stock led to an increase in large-bodied *Daphnia* in spring (Van Donk et al., 1990a; Faafeng & Braband, 1990; Hanson & Butler, 1994; Meijer et al., 1994a). However, in Wolderwijd no change in *Daphnia* species, nor a significant difference in length or density was found in the *Daphnia* population in that important period. Nevertheless, the obtained clear water period, coincided with high *Daphnia* densities. Possible explanations for the minor difference in *Daphnia* length and density between 1989 and 1991 are food limitation of *Daphnia* in 1991 and a non-representative sampling of *Daphnia*. Food limitation is indicated by the extremely low chlorophyll *a* levels and the low fecundity of the *Daphnia* (<2 eggs ind^{-1} adult *Daphnia*) (Taylor, 1985), during the clear water phase. In that period also the grazing exceeded two to eight times the primary production (Meijer et al., 1994b). Another option is that large *Daphnia* were underestimated during sampling, because they might have been concentrated near the bottom during the clear water period, as was shown in enclosure experiments in Denmark (Schriver et al., 1995).

Apart from the high *Daphnia* peak during the clear water phase, also other factors contributed to the low algal biomass and high transparency in May–June 1991, such as the long period with ice-cover in February 1991 (28 days), the intensified flushing and the low phosphorus concentration. However, in earlier years with a long ice-cover (1979, 1982, 1985 and 1987) or flushing (1981–1983), but without fish reduction, no large increase of transparency was observed in May–June (Figure 5).

Each year *Daphnia* was absent from July onwards and the algal biomass increased. Considering the summer chlorophyll *a* levels, food limitation is out of the question and therefore predation by YOY fish (perch and bream) and invertebrates, such as *Neomysis*, is the most likely explanation for the absence of *Daphnia* during the summer. Unfortunately, due to the minor differences in zooplankton biomass between the years and the few fish data (only 4 years), we are not able to demonstrate a correlation between the density of planktivorous fish and *Daphnia*.

In summer 1991 and 1993 high densities of *Neomysis* were found. An earlier paper stressed the impor-

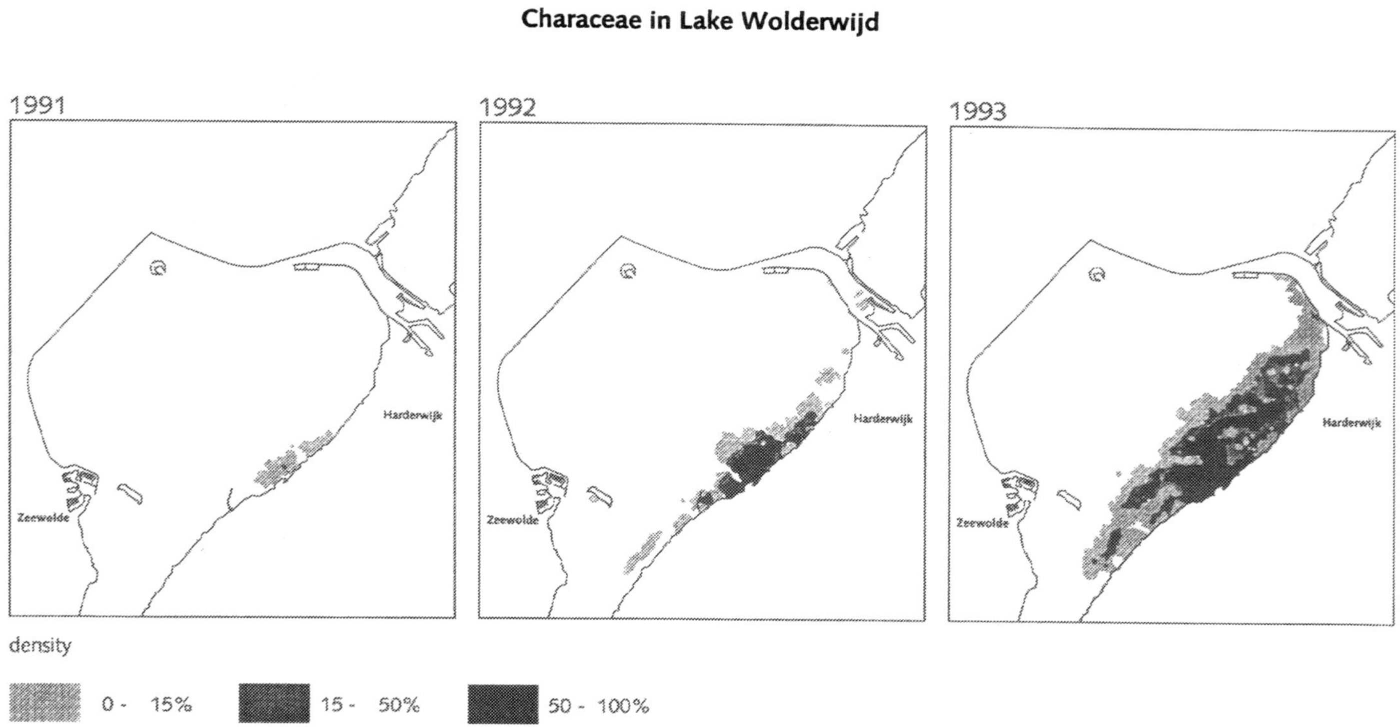

Figure 9. Characeae in 1990–1993. In 1989: no Characeae were present. In 1990: 1 ha.

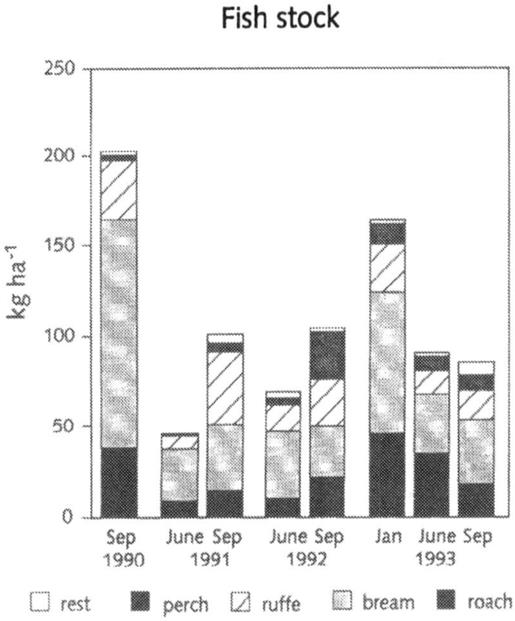

Figure 10. Fish stock in 1990–1993. Drastic fishery was carried out from November 1991 to June 1991. Additional fisheries took place from January to May 1992 and February to August 1993.

tance of *Neomysis* for predation on *Daphnia* in 1991 (Meijer et al., 1994b). Recent results, however, indicate that in periods with very high densities the biomass of Neomysis was formed for 60% by juveniles, which do not consume *Daphnia* (Siegfried & Kopache, 1980). Therefore, in that period Neomysis probably consumed 8% of the biomass of *Daphnia* per day, compared to 20% mentioned earlier (Meijer et al., 1994b). Consequently the total predation pressure decreases from 26% to 14% of the *Daphnia* biomass per day, which is in the range of the daily production of *Daphnia*.

Role of piscivorous fish

The increase of the fish biomass in summer was mainly caused by the production of YOY of perch and ruffe. The biomass of YOY of bream and roach was relatively low. Also the share of piscivorous fish was very low. A survey of 300 Danish lakes indicated that at total-P levels of 0.07–0.10 mg l^{-1}, a stable situation with a high percentage of piscivorous fish (30%) is possible (Jeppesen et al., 1990b). However, in Wolderwijd with total-P of 0.07–0.09 mg l^{-1}, the proportion of piscivorous fish only increased from 2.5% in 1990 to 5–8% in 1991–1993. The biomass of pike was low, probably because of the lack of suitable habitat, such as emergent and submerged vegetation (Grimm & Backx, 1990). In the Danish lakes the piscivorous fish population is mainly composed of perch. In Wolderwijd the biomass of piscivorous perch has slightly increased, but is still quite low (6 kg ha^{-1}). According to Persson (1994), perch can only become important if the phosphorus load is low and macrophytes are abundant. In most of the Danish lakes with a high percentage of piscivorous fish also the coverage of macrophytes was high. Possibly, in Wolderwijd not the nutrient levels, but the limited abundance of submerged and emergent vegetation restrict the development of pike and perch. Therefore, on the short term we do not expect these predators to be able to control the planktivorous fish. Besides, there is hardly any evidence from the field, that demonstrates the role of pike as a regulator of the food web in lakes. In Finjasjön the pike biomass is also quite low (about 3 kg ha^{-1}) and as in Wolderwijd the perch was mainly small and planktivorous (Annadotter, 1993).

Even in biomanipulation experiments in small lakes with higher biomasses of pike, the pike was not able to control the production of planktivorous fish after the fish reduction (Meijer et al., 1994a, 1995). This is partly caused by the lack of emergent vegetation but also by the higher nutrient levels and thus a higher production of planktivorous fish in those small lakes (Klinge et al., 1995). Only in Lake Noorddiep with a more varied vegetation and a pike stock which was already present before the measures, some control of planktivores by piscivores occurred (Meijer et al., 1995). Apart from fish removal also measures should be taken to reduce the spawning success of the cyprinids by removal of eggs after spawning on artificial spawning substrate or the removal of fish from the spawning grounds (Perrow et al., 1997; Grimm & Backx, 1994).

Despite the high biomasses of planktivores in Lake Zwemlust en Bleiswijkse Zoom, the water in these small lakes remained clear during summer, due to the abundant macrophyte growth (Van Donk et al., 1990a; Meijer et al., 1994a).

Development of Characeae

In Wolderwijd no increase in the total area covered by macrophytes was found, but a shift from *Potamogeton* species to Characeae occurred. *Chara* is known to react to increased water transparency and these species are particularly able to persist in large, shallow wind-exposed lakes (Blindow, 1992; Blindow et al., 1993).

In Wolderwijd the area covered by Characeae increased from 28 ha in 1991 to 438 ha in 1993. Probably, the availability of oospores limited the colonization by *Chara* in 1991. Experiments in 1993 showed a marked response after the introduction of oospores or plants (Stam, unpubl. data). The density of the oospores appears to be high inside the vegetated areas, but drops sharply with increasing distance from the *Chara* meadows: at a distance of >100 m the density if 10–100 fold lower than inside the meadows (Stam, unpubl. data). It is likely that the probability of a dense canopy of *Chara* is highest on sites with a high spore density. Given the limited temporal window for establishment (the clear water phase in spring), expansion of the meadows can therefore only occur along the borders of the existing area. The expansion rate will probably increase with a higher availability of light. As soon as a reasonable area of *Chara* is present, *Chara* itself creates a high transparency (Blindow, 1991; Blindow et al., 1993).

Macrophytes and high transparency

In Wolderwijd, high transparency is restricted to the area above the *Chara* vegetation. In smaller lakes the presence of macrophytes over at least 50% of the surface area was sufficient to keep the water clear, also in areas without macrophytes (Meijer et al., 1994a). Obviously, in Wolderwijd the mechanisms causing clear water are limited to the vegetated locations. It is quite difficult to identify the key mechanisms that cause the differences in transparency observed. Many mechanisms may interact and this multiple causality prevents a simple sorting out of hypotheses (Scheffer & Beets, 1994). Possible mechanisms include: macrophytes reduce resuspension of bottom material and increase sedimentation of suspended matter (James & Barko, 1990; Petticrew & Kalff, 1992); macrophytes can act as refuge for zooplankton (Timms & Moss, 1984); macrophytes can reduce algal growth by nutrient competition (Ozimek et al., 1990) or by allelopathy in particular by Characeae (Hootsmans & Blindow, 1993) and macrophytes may stimulate the development of small herbivorous organisms living among plants such as bryozoans and protozoans. In smaller lakes (with less wind resuspension) the water stayed clear in the presence of macrophytes like *Potamogeton pectinatus* (Bleiswijkse Zoom), *Elodea nutalli* or *Ceratophyllum demersum* (Zwemlust), Characeae (Duinigermeer), *Ceratophyllum demersum* and *Nuphar lutea* (Noorddiep). For Lake Zwemlust competition for nitrogen between macrophytes and algae, seems the most likely explanation for the low algal biomass in the presence of macrophytes (Van Donk et al., 1993). Also in the Bleiswijkse Zoom and Noorddiep a decrease in nitrogen is found at the moment that the macrophytes become abundant (Meijer et al., 1994a). However in those small lakes, the processes which keep the water clear are not necessarily the same as in the large and wind-exposed Wolderwijd. In the large Lake Krankesjön (Sweden, 3200 ha) a decrease of the water level resulted in an abundance of Characeae followed by an increase in transparency (Blindow et al., 1993). Here increased sedimentation and reduced availability of phosphorus may have contributed to the high transparency (Blindow et al., 1993). In Wolderwijd there is no evidence for a lower availability of phosphorus within the vegetation. In Lake Christina (USA, 1600 ha), the fish stock reduction also led to an increase in macrophytes. In the first two years only a spring clear water phase was obtained, but in the third year macrophytes kept the water clear. An increased net sedimentation was suggested as one of the reasons for the clear water above the macrophytes (Hanson & Butler, 1990; 1994). As in Wolderwijd the vegetated areas in this Minnesota lake show a higher transparency than the areas without vegetation.

The clear water in Wolderwijd occurred especially in the areas covered by Characeae, while in the parts with *Potamogeton pectinatus* or *P. perfoliatus* the water was turbid. The reason for this difference may stem from the canopy structure of the plants. The canopy of Characeae covers the lake bottom completely, whereas the *Potamogeton* species leave a lot of open spaces. Monitoring in 1994 showed that during high wind speeds the water above the Characeae became turbid. After a storm the water cleared within one day, indicating that increased net sedimentation within the *Chara* meadows may explain the difference in transparency.

Conclusions

This case study showed that in large and shallow lakes drastic fish stock reduction may lead to clear water. However, in Wolderwijd this was limited in time and space. The open water area was clear only during the *Daphnia* spring peak. The limited period of clear water stimulated the expansion of Characeae. In the area with Characeae the water stayed clear as long as plants were present. In the current situation, with the lack of spawning and hiding places the abundance of piscivores will probably remain too low to control the production of planktivores. Therefore, a prolonging of the presence of *Daphnia* throughout the summer cannot be expected. The most promising way to achieve a shift from the turbid water state into the clear water state will be the expansion of the area with a dense *Chara* vegetation. Above the *Chara* plants the water remains clear, probably due to increased net sedimentation. To stimulate *Chara*, spring clear water periods are necessary. For that reason, the lake manager is recommended to continue the fish stock reductions. The fish removal should be substantial in order to get a spring clear water phase. Additionally, removal of eggs of cyprinids after the spawning, may lower the production of planktivores. Moreover, the nutrient load should be reduced further in order to get a lower production of planktivorus fish and less problems with cyanobacteria. On the short term the best obtainable situation is a period with clear water in spring and locally large areas with clear water above the Characeae.

Acknowledgments

We thank the commercial fishermen, the angling clubs and Map Grimm and Joost Backx of Witteveen + Bos, Consulting Engineers for the fruitful co-operation. Without them this project would have been impossible. Rijkswaterstaat – Directorate IJsselmeergebied, the lake manager, did the time-consuming organization of the fisheries and a great deal of the sampling and analysis. Valuable contributions to the project came from Kees Bruning of AquaSense, Ramesh Gulati of the NIE-Centre for Limnology and our colleagues from the RIZA Lake Restoration Research Programme. Thanks are due to Eddy Lammens, Carolien Breukers, Thea Helmerhorst and Hugo Coops for comments on and contributions to the manuscript. Wim Stam is thanked for the data on the oospores and Marcel van de Berg and Roel Doef for the data on Characeae.

References

Annadotter, H., 1993. The restoration of Lake Finjasjon. Report of a meeting with the expert panel, 22 November 1993. The municipality of Hässleholm, Sweden, 1–33.

Backx, J. J. G. M. & M. P. Grimm, 1994. Mass-removal of fish from Lake Wolderwijd (2700 ha), The Netherlands. Part II: Results of the fishing. In Cowx, I. G. (ed.), Rehabilitation of Freshwater Fisheries. Fishing News Books. Hull, UK: 401–414.

Benndorf, J., H. Kneschke, K. Kossatz & E. Penz, 1984. Manipulation of the pelagic food web by stocking with predacious fishes. Int. Revue ges. Hydrobiol. 69: 407–428.

Berger, C. & A. bij de Vaate, 1983. Limnological studies on the eutrophication of Lake Wolderwijd, a shallow hypertrophic *Oscillatoria*-dominated lake in The Netherlands. Schweiz. Z. Hydrol. 45: 458–479.

Blindow, I., 1991. Interaction between submerged macrophytes and microalgae in shallow lakes. Ph. D. Thesis of University of Lund, 108 pp.

Blindow, I., 1992. Long and short term dynamics of submerged macrophytes in two shallow lakes. Freshwat. Biol. 28: 15–27.

Blindow, I., G. Andersson, A. Hargeby & S. Hansson, 1993. Long-term pattern of alternative stable states in two shallow eutrophic lakes. Freshwat. Biol. 30: 159–167.

Breukelaar, A. W., E. H. R. R. Lammens, J. G. P. Klein Breteler & I. Tatrai, 1994. Direct and indirect effects of benthivorous bream (*Abramis brama*) and carp (*Cyprinus carpio*) on turbidity, chlorophyll *a* and nutrients. Freshwat. Biol. 32: 113–121.

Carpenter, S. R., J. F. Kitchell, J. R. Hodgson, P. A. Cochran, J. J. Elser, M. M. Elser, D. M. Lodge, D. Kretchmer, X. He & C. N. Von Ende, 1987. Regulation of lake primary productivity by food web structure. Ecology 68: 1863–1876.

Faafeng, B. A. & A. Braband, 1990. Biomanipulation of a small, urban lake. Removal of fish exclude bluegreen blooms. Verh. int. Ver. Limnol. 24: 597–602.

Gliwicz, Z. M., 1990. Why do cladocerans fail to control algal blooms? Hydrobiologia 200/201: 83–97.

Grimm, M. P. & J. J. G. M. Backx, 1990. The restoration of shallow eutrophic lakes and the role of northern pike, aquatic vegetation and nutrient concentration. Hydrobiologia 200/201: 557–566.

Grimm, M. P. & J. J. G. M. Backx, 1994. Mass-removal of fish from lake Wolderwijd (2700 ha), The Netherlands. Part I: Planning and strategy of a large scale biomanipulation project. In Cowx, I. G. (ed.), Rehabilitation of Freshwater Fisheries. Fishing News Books. Hull, UK: 390–400.

Gulati, R. D., E. H. R. R. Lammens, M.-L. Meijer & E. van Donk, 1990. Biomanipulation, Tool for Water Management. Hydrobiologia 200/201: 1–628.

Hamrin, S. F., 1993. Lake restoration by cyprinid control in Sätofta Bay (Lake Ringsjön). Verh. int. Ver. Limnol. 25: 487–493.

Hanson, M. A. & M. G. Butler, 1990. Early responses of plankton and turbidity to biomanipulation in a shallow prairie lake. Hydrobiologia 200/201: 317–328.

Hanson, M. A. & M. G. Butler, 1994. Responses to food web manipulation in a shallow waterfowl lake. Hydrobiologia 279/280: 457–466.

Hootsmans, M. J. M. & I. Blindow, 1993. Allelopathic limitation of algal growth by macrophytes In van Vierssen, W., M. Hootsmans & J. Vermaat (eds), Lake Veluwe, a Macrophyte Dominated System under Eutrophication Stress. Kluwer Academic Publishers, Dordrecht/Boston/London: 175–192.

Hosper, S. H., 1984. Restoration of Lake Veluwe, The Netherlands, by reduction of phosphorus loading and flushing. Wat. Sci. Tech. 17: 757–768.

Hosper, S. H., 1989. Biomanipulation, new perspective for restoring shallow, eutrophic lakes in The Netherlands. Hydrobiol. Bull. 23: 5–11.

Hosper, S. H. & E. Jagtman, 1990. Biomanipulation additional to nutrient control for restoration of shallow lakes in The Netherlands. Hydrobiologia 200/201: 523–534.

Hosper, S. H. & M.-L. Meijer, 1986. Control of phosphorus loading and flushing as restoration methods for Lake Veluwe, The Netherlands. Hydrobiol. Bull. 20: 183–194.

Hosper, S. H. & M.-L. Meijer, 1993. Biomanipulation, will it work for your lake? A simple test for the assessment of chances for clear water, following drastic fish stock reduction in shallow, eutrophic lakes. Ecol. Engineering 2: 63–72.

Jagtman, E., D. T. van der Molen & S. Vermij, 1992. The influence of flushing on nutrient dynamics, composition and densities of algae and transparency in Veluwemeer, The Netherlands. Hydrobiologia 233: 187–196.

James, W. F. & J. W. Barko, 1990. Macrophyte influences on the zonation of sediment accretion and composition in a north-temperate reservoir. Arch. Hydrobiol. 120: 129–142.

Jeppesen, E., J. P. Jensen, P. Kristensen, M. Søndergaard, E. Mortensen, O. Sjortkjær & K. Olrik, 1990b. Fish manipulation as a lake restoration tool in shallow, eutrophic, temperate lakes 2: threshold levels, long-term stability and conclusions. Hydrobiologia 200/201: 219–227.

Jeppesen, E., M. Søndergaard, E. Mortensen, P. Kristensen, B. Riemann, H. J. Jensen, J. P. Müller, O. Sortkjær, J. P. Jensen, K. Christoffersen, S. Bosselmann & E. Dall, 1990a. Fish manipulation as a lake restoration tool in shallow, eutrophic, temperate lakes 1: cross-analysis of three Danish case-studies. Hydrobiologia 200/201: 205–218.

Klinge, M., M. P. Grimm & S. H. Hosper, 1995. Eutrophication and ecological rehabilitation of Dutch lakes: Presentation of a new conceptual framework. Wat. Sci. Tech.: 207–218.

Lammens, E. H. R. R., H. W. de Nie, J. Vijverberg & W. L. T. van Densen, 1985. Resource partitioning and niche shifts of bream (*Abramis brama*) and eel (Anguilla anguilla) mediated by predation of smelt (*Osmerus eperlanus*) on *Daphnia hyalina*. Can. J. Fish. aquat. Sci. 42: 1342–1352.

McQueen, D. J. & J. R. Post, 1988. Limnological studies of cascading trophic interactions. Verh. int. Ver. Limnol. 23: 739–747.

Meijer, M-L., M. W. de Haan, A. W. Breukelaar & H. Buiteveld, 1990b. Is reduction of the benthivorous fish an important cause of high transparency following biomanipulation in shallow lakes? Hydrobiologia 200/201: 303–315.

Meijer, M-L., E. Jeppesen, E. van Donk, B. Moss, M. Scheffer, E. H. R. R. Lammens, E. H. van Nes, B. Faafeng & J. P. Jensen, 1994a. Long term responses to fish stock reduction in small shallow lakes. Interpretation of five year results of four biomanipulation cases in The Netherlands and Denmark. Hydrobiologia 275/276: 457–466.

Meijer, M-L., E. H. R. R. Lammens, A. J. P. Raat, M. P. Grimm & S. H. Hosper, 1990a. Impact of cyprinids on zooplankton and algae in ten drainable ponds: preliminary results. Hydrobiologia 191: 275–284.

Meijer, M-L., E. H. R. R. Lammens, A. J. P. Raat, J. G. P. Klein Breteler & M. P. Grimm, 1995. Development of fish communities in lakes after biomanipulation. Neth. J. aquat. Ecol. 29: 91–101.

Meijer, M-L., E. H. Van Nes, E. H. R. R. Lammens, R. D. Gulati, M. P. Grimm, J. J. G. M. Backx, P. Hollebeek, E. M. Blaauw & A. W. Breukelaar, 1994b. The consequences of a drastic fish stock reduction in the large and shallow lake Wolderwijd, The Netherlands. Can we understand what happened? Hydrobiologia 275/276: 31–42.

Moss, B., 1990. Engineering and biological approaches to the restoration from eutrophication of shallow lakes in which aquatic plant communities are important components. Hydrobiologia 200/201: 367–377.

Ozimek, T., R. D. Gulati & E. van Donk, 1990. Can macrophytes be useful in biomanipulation of lakes, the Lake Zwemlust example. Hydrobiologia 200/201: 399–408.

Perrow, M. R., M.-L. Meijer, P. Dawidowicz & H. Coops, 1997. Biomanipulation in shallow lakes: state of the art. Hydrobiologia 342/343: 355–365.

Persson, L., 1994. Natural shifts in the structure of fish communities: Mechanisms and constraints on perturbation sustenance. In Cowx, I. G. (ed.): Rehabilitation of Freshwater Fisheries. Fishing News Books, Blackwell Scientific Publ. Oxford, Hull, UK: 421–434.

Petticrew, E. L. & J. Kalff, 1992. Water flow and clay retention in submerged macrophyte beds. Can. J. Fish. aquat. Sci. 49: 2483–2489.

Philips, G. & B. Moss, 1994. Is biomanipulation a useful technique in lake management? National Rivers Authority R and D note 276.

Reynolds, C., 1994. The Ecological basis for the successful biomanipulation of aquatic communities. Arch. Hydrobiol. 130: 1–33.

Sas, H., 1989. Lake restoration by reduction of nutrient loading: Expectations, Experiences, Extrapolations. St. Augustin 1: Akademia Verl., Richarz. ISBN 3-88345379-X.

Scheffer, M., 1989. Alternative stable states in eutrophic shallow freshwater systems: a minimal model. Hydrobiol. Bull. 23: 73–85.

Scheffer, M., 1990. Multiplicity of stable states in freshwater systems. Hydrobiologia 200/201: 475–487.

Scheffer, M. & J. Beets, 1994. Ecological models and the pitfalls of causality. Hydrobiologia 275/276: 115–124.

Scheffer, M., S. H. Hosper, M-L. Meijer, B. Moss & E. Jeppesen, 1993. Alternative equilibria in shallow lakes. Trends Ecol. Evol., 8: 276–279.

Schriver, P., J. Bogestrand, E. Jeppesen & M. Søndergaard, 1995. Impact of submerged macrophytes on the interactions between fish, zooplankton and phytoplankton: large scale enclosure experiments in a shallow lake. Freshwat. Biol. 33: 255–270.

Shapiro, J., V. Lamarra & M. Lynch, 1975. Biomanipulation: an ecosystem approach to lake restoration. In Brezonik, P. L. & J. L. Fox (eds), Water Quality Management through Biological Control. Rep. no. ENV-07-75-1, Univ. of Florida, Gainesville: 85–96.

Siegfried, C. A. & M. E. Kopache, 1980. Feeding of *Neomysis mercedis* Holmes. Biol. Bull. 159: 193–205.

Taylor, B., 1985. Effect of food limitation on growth and reproduction of *Daphnia*. Arch. Hydrobiol. Beih. Ergebn. Limnol. 21: 285–296.

Timms, R. M. & B. Moss, 1984. Prevention of growth of potentially dense phytoplankton populations by zooplankton grazing in presence of zooplanktivorous fish in a shallow wetland ecosystem. Limnol Oceanogr. 29: 472–486.

Van Berkum, J. A., M. Klinge & M. P. Grimm, 1995. Biomanipulation in the Duinigermeer; first results. Neth. J. aquat. Ecol. 29: 81–90.

Van der Molen, D. T & P. C. M. Boers, 1994. Influence of internal loading on phosphorus concentration in shallow lakes before and after reduction of the external loading. Hydrobiologia 275/276: 379–389.

Van Donk, E., M. P. Grimm, R. D. Gulati, P. G. M. Heuts, W. A. de Kloet & E. Van Liere, 1990b. First attempt to apply whole-lake food-web manipulation on a large scale in The Netherlands. Hydrobiologia 200/201: 291–302.

Van Donk, E., M. P. Grimm, R. D. Gulati & J. P. G. Klein Breteler, 1990a. Whole-lake food-web manipulation as a means to study community interactions in a small ecosystem. Hydrobiologia 200/201: 275–289.

Van Donk, E., M. P. Grimm, P. G. M. Heuts, G. Blom, K. Everards & O. F. R. van Tongeren, 1994. Use of mesocosms in a shallow eutrophic lake to study the effects of different restoration measures. Arch. Hydrobiol. Beih. Ergebn. Limnol. 40: 283–294.

Van Donk, E. & R. D. Gulati, 1995. Transition of a lake to turbid state six years after biomanipulation: Mechanisms and pathways. Wat. Sci. Tech. 32: 197–206.

Van Donk, E., R. D. Gulati, A. Iedema & J. T. Meulemans, 1993. Macrophyte-related shifts in nitrogen and phosphorus contents of the different trophic levels in a biomanipulated shallow lake. Hydrobiologia 251: 19–26.

Van Liere, E. & R. D. Gulati, 1992. Restoration and recovery of shallow eutrophic lake ecosystems in The Netherlands: epilogue. Hydrobiologia 233: 283–287.

Van Nes, E. H., M-L. Meijer, A. W. Breukelaar, P. Hollebeek, R. W. Doef, E. H. R. R. Lammens, H. Coops, R. Noordhuis & E. C. L. Marteijn, 1992. Wolderwijd-Nuldernauw 1991. Gevolgen van de uitdunning van de visstand. RIZA nota 92.063, Lelystad, 134 pp.

Hydrobiologia **342/343:** 351–354, 1997.
L. Kufel, A. Prejs & J. I. Rybak (eds), Shallow Lakes '95.

The process of biomass formation as the key point in the restoration of tropical eutrophic lakes

Eduardo von Sperling
Department of Sanitary and Environmental Engineering, Federal University of Minas Gerais, Av. Contorno, 842/706-30110-060, Belo Horizonte, Brazil

Key words: eutrophication, tropical lakes, recovery, macrophytes

Abstract

The paper presents initially some specific features of tropical lakes (high recycling rates, elevated primary production during the whole year, high nutrient assimilation, high settling velocity of nutrients, intense organic matter decomposition, high grazing rates), which leads to the establishment of a very dynamic ecosystem. Due to these characteristics the warm water lakes show completely different reactions to the impacts derived from the phenomenon of eutrophication in comparison with lentic systems situated in temperated regions. Moreover in shallow lakes occurs an almost permanent mixing process, which in its turn allows the resuspension of nutrients and a consequent increase in the primary productivity. The key point at the study of the eutrophication of tropical lakes is the comprehension of the complex relationship between nutrient concentration and biomass formation. The paper presents the result of some studies carried out by the author in Brazilian lakes and reservoirs, leading to the conjecture that the nutrient mass in a water body is often not directly linked with the phytoplankton growth. The strategies for the recovery of degraded tropical lakes should hence be directed towards a control of the formed biomass and not to a limitation in the nutrient input, as is usually the case in cold water lakes.

Introduction

One of the major issues in the water pollution control refers to eutrophication problems in lentic ecosystems (lakes and reservoirs). In tropical regions there is often a lack of infrastructure regarding sewage collection and treatment. Therefore the tropical lakes and reservoirs, mainly those situated in urban areas, are generally subjected to severe eutrophication problems.

The water bodies located in the tropical belt of our planet present some typical features that are summarized below:

(a) The intense solar radiation and the high water temperature values accelerate the nutrient uptake by the algae.

(b) The phytoplanktonic population peaks are less frequent in comparison with temperate aquatic environments.

(c) The high nutrient assimilation, associated with high recycling rates, lead to the prevalence of an intense degree of productivity.

(d) The nutrient concentrations are generally low; as a consequence many water bodies are erroneously classified as oligotrophic in spite of their high productivity.

(e) There is the frequent occurrence of low phytoplanktonic densities, which are however associated with high growth rates.

(f) The high mineralization rates lead to an accelerated oxygen depletion and to the formation of sediments poor in organic matter; consequently there is no direct connection between the content of organic matter in the sediment and the productivity of the water body.

(g) Due to the high temperature the water viscosity is low, leading to a higher settling velocity of organisms and nutrients (von Sperling, 1992).

Eutrophication dynamics in tropical lakes

For the evaluation of the trophic state of tropical lakes a very clear comprehension of the structure and function of these aquatic systems is necessary. As a matter of fact there is a limited amount of information about tropical water bodies, resulting in difficulty in understanding the dynamics of the eutrophication process in these environments (Payne, 1986).

The characteristics of tropical aquatic systems indicate the existence of very complex relationships between the presence of nutrients and the plant biomass formation in the water body. It should be pointed out that the eutrophication problems actually originate from the formation of biomass and not from the mere presence of nutrients. It is the decomposition of the biomass that leads to oxygen depletion in the water body and to the resulting severe constraints in the utilization of the water resource. On the other hand it is also known that the quantity of biomass formed by unity of nutrient depends on the growth rate of the phytoplankton (Sommer, 1993). In tropical lakes and reservoirs high growth rates are needed in order to compensate for the population losses due to sedimentation and grazing, which means less biomass formed by unity of nutrient.

The problematic nature of the determination of the trophic state of tropical lakes and reservoirs induces to reflections about the most convenient way of expressing the trophic degree of these environments. The adoption of physico-chemical and biological indicators presents strong limitations due to the fact that they are not capable of exhibiting faithfully the real processes that conduct to the formation of the several trophic levels. Necessary here is the adoption of criteria that are related to the ecological aspect of the water body. Consequently, the preference is the use of criteria which reflect the association and interaction of organisms and the performance of the several organic functions. The utilization of these criteria provides more reliable information about the behavior of the aquatic environment by reflecting in a systemic and integrated way the complex dynamic of the water body. In a first analysis it is not appropriate to establish a trophic classification presenting a rigid division in groups (oligotrophic, mesotrophic, eutrophic). It is known that a given water body can, in a period of 24 hours, show variable conditions, shifting from an oligotrophic to a eutrophic level. This is very common in shallow lakes situated in tropical climates, which are subjected to frequent alternations of periods of stratification and circulation. In this sense it is completely inadequate to consider for instance as mesotrophic an aquatic environment which alternates characteristics of oligotrophy and eutrophy.

Case study

Lake Pampulha (Figure 1), located within the urban area of Belo Horizonte, Brazil, is considered to be one of the most important tourist attractions of the region. This man-made lake was constructed originally with the objective of water supply, but the degradation of the water quality prevented the further use of this source as such. Presently Lake Pampulha is suitable only for some restricted recreational activities.

The lake has an area of approximately 2.6 km^2 and a volume of about 14 million m^3. Its maximum depth is 16 m and the mean depth is 5 m. The drainage basin occupies the significant area of 97 km^2. The relationship between these figures indicates that the lake receives a considerable influence from the activities developed in the watershed. The main sources of pollution are the discharge of untreated sewage into the lake and the strong erosion of many areas in the drainage basin, leading to a high volume of settlement material in the waterbody.

Several studies have been conducted by the local authorities in order to recover the limnological quality of Lake Pampulha, but their implementation was frequently obstructed due to political or economic influences. In the recent years, the area occupied by the algae and macrophytes has been expanding and has caused serious concern considering the known difficulties involved in controlling eutrophication and silting in urban lakes. In 1989 the local administration finally began the implementation of a recovery program, which consists of the following basic activities:

- management of the watershed (erosion control, sewage diversion);
- construction of sedimentation basins in the tributaries;
- macrophyte (mainly water hyacinth) removal;
- sediment removal.

One of the most important aspects of the restoration program refers to the biomass removal. Considering the strong difficulties coupled with the algae removal (suction, centrifugation, microstraining) emphasis was given to the macrophyte removal.

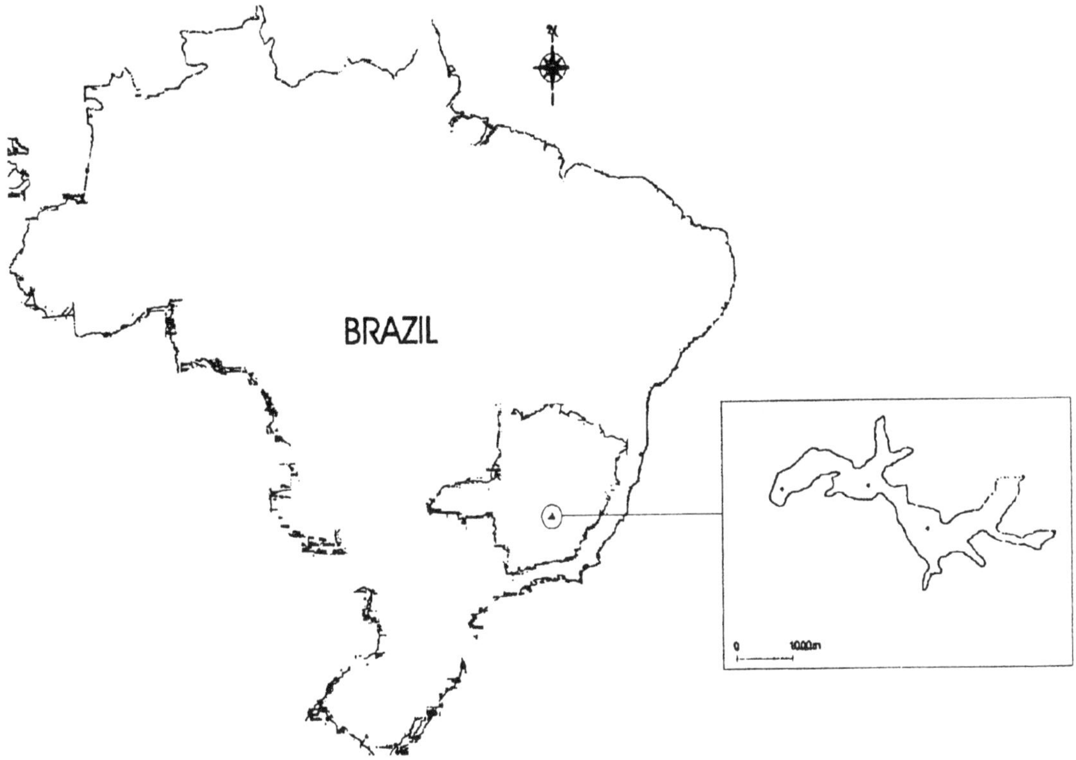

Figure 1. Location of Lake Pampulha

Blooms of aquatic plants are frequently observed in the waters of Lake Pampulha. In many cases the green carpet, formed mainly by water hyacinths (*Eichhornia crassipes*), covers over one third of the surface of the reservoir. Several attempts were made in order to remove efficiently this considerable mass of plants. The problem lies in the very high growth rate of water hyacinths, which, in most of the cases, is far greater than the removal capacity of the equipments. A definitive solution for this inconvenience will only take place after the complete diversion of the untreated sewage that is discharged to the reservoir.

Recently another removal action was undertaken, using appropriate boats and conveyors that transport the water hyacinths to trucks. The final destination of the removed material are the islands created within the reservoir as a consequence of a dredging operation. This last removal program was carried out from August 1993 till February 1994, which is the period of the most intensive vegetation growth in the reservoir. As consequence of this corrective measure it was observed a strong reduction in total inorganic nitrogen concentrations. However, the most outstanding effect was an abrupt change in the zooplankton community and a remarkable growth of some blue-green algae, specially *Microcystis aeruginosa*. The macrophyte and algae populations are supposed to stay permanently in competition for nutrient uptake. The removal of one of them would consequently favor the remaining community. One remediation measure, that is here currently adopted, is to leave always a certain amount (c. 10%) of macrophytes to act as competitor and thus guarantee a desirable ecological equilibrium. With this procedure the algal blooms were also significantly reduced.

Conclusion

The key point in the control of the eutrophication of tropical lakes is the process of biomass formation and its relationship with the nutrient concentration in the water body. Consequently emphasis should be given to the restoration techniques involving biomass (algae and macrophyte) removal. For the success of this corrective procedure it is necessary a broad comprehension of the functioning of the aquatic system together with the development of appropriate equipments to perform efficiently the harvesting function.

References

Payne, A., 1986. The Ecology of Tropical Lakes and Rivers, John Wiley & Sons, New York, 301 pp.

Sommer, U., 1993. The scientific basis of eutrophication management: reconciling basis physiology and empirical biomass models, Proceedings of the 5th Int. Conf. on the Conservation and Management of Lakes, Istituto Italiano di Idrobiologia, Stresa, Italy, 6–8.

von Sperling, E., 1992. Determination of the settling velocity of phosphorus in an aerated lake. Water Sci. Technol. 26: 1973–1979.

Hydrobiologia **342/343:** 355–365, 1997.
L. Kufel, A. Prejs & J. I. Rybak (eds), Shallow Lakes '95.

Biomanipulation in shallow lakes: state of the art

Martin R. Perrow[1*+], Marie-Louise Meijer[2*], Piotr Dawidowicz[3] & Hugo Coops[2]
[1]*Department of Freshwater Ecology, NERI, P.O. Box 314, Vejlsøvej 25, DK-8600 Silkeborg, Denmark*
[2]*RIZA, Institute for Inland Water Management and Waste Water Treatment, P.O. Box 17, 8200 Lelystad, The Netherlands*
[3]*Department of Hydrobiology, Universiy of Warsaw, Banacha 2, PL 02-097, Warszawa, Poland*
[+]*Present address: ECON, School of Biological Sciences, University of East Anglia, Norwich, Norfolk NR4 7TJ, UK*
*(*authors for correspondence)*

Abstract

The current state of biomanipulation was the subject of much discussion at Shallow Lakes '95. This led to a workshop focusing on the factors influencing the establishment of macrophytes and the mechanisms responsible for their stability following biomanipulation. The purpose of the current paper is to distil current knowledge on biomanipulation in shallow lakes gleaned from discussions at the conference and recent literature.

Biomanipulation should be used in the theoretical context of two altemative stable equilibria, as the extreme perturbation required to move from a phytoplankton dominated state to one dominated by macrophytes. Understanding the nature of the factors and mechanisms responsible for turbid water, is critical if biomanipulation is to be appropriate. We suggest that with sufficient information, particular components of the fish community may be targeted and precise figures for removal, designed to exceed critical threshold values, may be set. Without this knowledge, a 'play-safe' strategy should be adopted and at least 75% of the fish removed. Stocking with piscivores may be a useful additional measure to fish removal.

The principal objective of biomanipulation in shallow lakes is to generate a period of clear water of sufficient length to allow macrophytes to establish. To this aim, as well as for technical reasons, biomanipulation is best undertaken in winter and early spring to generate clear water as early as possible in the season. In the cases where grazing is important, this coincides with the spring peak of *Daphnia* spp. Biomanipulation may have to be repeated if macrophytes do not colonise effectively within the first season. The factors responsible for the lag in response of macrophytes in some cases and the potential mechanisms contributing to the maintenance of clear water in macrophyte beds are discussed.

From empirical data sets from many lakes, both a relative increase in the piscivorous fish stock and a reduction in nutrient levels are thought to be important in stabilising the system in the long-term. Whether biomanipulation may lead to alternative equilibria (i.e. high diversity macrophyte communities with piscivorous fish at high P) is unknown. Further study of exceptional cases, theoretical modelling and development and analysis of more long-term (>10 years) case histories is recommended.

Introduction

Biomanipulation may be defined as the re-structuring of the biological community to achieve a favourable response, usually reduction in algal biomass, attainment of clear water and promotion of a diverse biological community. The term originally encompassed a range of techniques (Shapiro, 1990), but now is typically applied to top-down manipulation of fish communities, i.e. enhancement of piscivores or reduction of zooplanktivores and/or benthivores (Lammens et al., 1990).

The concept of biomanipulation is rooted in ecological theory and based on the principal of two alternative stable trophic states in shallow lakes (Irvine et al., 1989; Scheffer, 1990; Scheffer et al., 1993). At high nutrient concentrations, phytoplankton is abundant, water is turbid and the fish community is dominated by populations of zooplanktivores and/or benthivores. In Europe, this is usually cyprinids, such as roach *Rutilus rutilus* and bream *Abramis brama*. At low nutrient concentrations, with high light availability in the clear water, submerged macrophytes dominate and the fish community has a high piscivore/planktivore-benthivore ratio, with increased representation of species such as perch *Perca fluviatilis* and pike *Esox lucius* (Persson et al., 1991; Jeppesen et al., 1997). At intermediate nutrient concentrations (suggested as 0.1–0.15 mg l^{-1} P or higher if N is limiting-Hosper & Jagtman, 1990; Jeppesen et al., 1990a), either altemative may exist. The shift from one state to the other is resisted by intemal biological stabilising mechanisms. For example, in the turbid state, zooplanktivorous fish may indirectly reduce grazing pressure on phytoplankton through predation on grazing zooplankton (Brooks & Dodson, 1965) and benthivorous fish may resuspend sediment and pump P into the water column directly enhancing turbidity as well as indirectly fuelling algal growth (Tatrai & Istvanovics, 1986; Breukelaar et al., 1994). Thus, in the face of nutrient reduction i.e. through external and/or intemal control, to a range where macrophytes may be expected to be abundant, phytoplankton may continue to dominate (Horppila & Kairesalo, 1990; Van Liere et al., 1990). Biomanipulation is then used as an extreme perturbation to push one stable equilibrium to another. Most commonly, the manipulation of fish populations (removal of zooplanktivores/benthivores and/or stocking of piscivores) cascades down the trophic levels.

The restoration of a stable macrophyte-dominated state is of clear value to conservation managers; biological diversity increasing with the presence of macrophytes (de Nie, 1987; Giles, 1992). However, for managers responsible for water supply or for providing an exploited fish stock, the reduction of algal biomass and/or enhanced water transparency may be the goal.

In this paper, we focus on the attainment of the stable macrophyte-dominated equilibrium as the objective of biomanipulation within a restoration programme. To this aim, we recognise three distinct phases:

- establishing clear water
- maintaining clear water
- stabilising the system

We outline a planning scheme that may be adopted in any potential programme of biomanipulation and discuss the need for drastic fish stock reduction (>75more targeted approach (i.e. to exceed a particular critical threshold) may be suitable. In maintaining clear water, macrophytes are suggested to be ultimately essential (Moss, 1990; Jeppesen et al., 1990a; Hanson & Butler, 1994; Meijer et al., 1994a). The factors influencing the establishment of macrophytes and the mechanisms promoting their stability are outlined and their relative importance discussed. Finally, on the question of longterm stability, the importance of nutrient concentrations is discussed.

Establishing clear water

Biomanipulation is likely to have dramatic ecosystem effects, not all of which may be predictable (Benndorf, 1992; Meijer et al., 1994a; Moss et al., 1996) as well as being expensive of resources (time, labour and money). Moreover, there has been much discussion of whether biomanipulation is effective or not (Reynolds, 1994). We consider that there is little doubt that biomanipulation works, it is more a case of whether it is applicable or not in a given situation. Further, there is a great difference between getting a short period of clear water and not achieving clear water at all. Even a short period of clear water may lead to the desired biological changes (e.g. the establishment of macrophytes) and be regarded a 'success' although the latter will always be considered a 'failure'. We therefore suggest that many 'failures' reported in the literature are generally a result of poor planning, of inappropriate or inaccurate seffing of targets, technical and not ecological constraints and little understanding of the processes involved.

Assessing the limiting factors

First, it is important to establish which factors are responsible for turbid water i.e. whether it is algae or suspended sediment. These may be assessed with targeted experiments (e.g. use of enclosures or mesocosms with or without particular fish-Van Donk et al., 1994a) or perhaps through results of extended monitoring.

If resuspension is important, is this caused by benthivorous fish or wind and waves? The latter may be important in large lakes particularly those with loose sediment. In such cases, removal of fish will obviously have little impact. For example, in the Netherlands, both Lake Breukeleveen (Van Donk et al., 1990a;

1994a) and Lake Klein Vogelezang (Van der Vlugt et al., 1992) failed to respond favourably, and with the benefit of hindsight this may have been because wind resuspension was a significant factor. However, even if fish are not the direct cause, it is possible that after fish removal, the sediment may be more compacted (Jeppesen, pers. comm.), thus reducing the effect of wind and waves, but the importance of this is unknown. If benthivorous fish are the ultimate cause of turbid water, these will be the focus of biomanipulation (Meijer et al., 1990).

On the other hand, if turbidity is due to algae, and the density of large *Daphnia* is low then zooplanktivorous fish will be the target of biomanipulation. However, it may be that a particular component of the zooplanktivorous fish stock has the critical impact. For example, in the Norfolk Broads in the UK, *Daphnia* are typically present in spring, often promoting a clear water phase, and it is only after recruitment of YOY fish that these decline rapidly leading to a high standing crop of algae (Cryer et al., 1986; Perrow & Irvine, 1992; Stansfield et al., 1997). The recruitment of YOY fish must therefore be either prevented (i.e. by removal of the adult stock prior to/during spawning and/or interference with spawning and removal of eggs) or be reduced through predation (i.e. through stocking with 0+ piscivores particularly pike). High stocks of older fish (>1+), even accounting for a reduction in foraging efficiency (Perrow & Irvine, 1992) may be sufficient to prevent the development of significant large-bodied *Daphnia* populations in spring. This seems to be the case in some Dutch lakes (Lammens et al., 1992), and it is therefore necessary to target older fish as well as YOY.

We accept that targeting particular functional groups, species or age classes may be difficult in practice, not least as many species, particularly cyprinids, show considerable dietary plasticity and may switch from benthic to plankton-feeding in periods of abundant *Daphnia* (Lammens et al., 1985). Moreover, many species show ontogenetic shifts in diet, and non-target species such as large bream and piscivorous perch produce zooplanktivorous offspring. Also, if YOY are the cause of the problem, the most efficient way to tackle them may be to remove the adult brood stock. However, we suggest that if sufficient information is available, leaving behind non-target species (e.g. tench in the Norfolk Broads-Stansfield & Perrow, 1994) may help control undesirable indirect effects such as a lack of benthic stirring inhibiting germination of macrophytes (Tatrai et al., 1994) as well as reducing conflict with recreational or commercial interests. Moreover, piscivorous fish may be retained (e.g. in the Norfolk Broads, all pike are routinely marked and released), although in the Netherlands most large pike are also removed to avoid cannibalism of small pike (Grimm, pers comm.).

With good information on the fish stock (abundance and biomass), age structure, behaviour and ecology (e.g. patterns of distribution and migration), targets for removal may be set (e.g. Grimm & Backx, 1994). If functional relationships are known, these may aim to simply reduce the fish stock to below a critical fish density/biomass. The amount of fish required may be surprisingly low. For example, only 16% fish removal with no recruitment of YOY, resulted in clear water in Lake Sobygard (Jeppesen et al., 1990b). However, in most cases fish stock reduction usually aims at >75% (Hosper & Meijer, 1993). This is a 'play-safe' strategy as many of the 'failures' are apparently due to insufficient fish removal (Van Donk et al., 1994a; Van der Vlugt et al., 1992). Moreover, the chance of failure is intuitively likely to increase with the size of the lake. Technical problems of catching fish increase in big systems with an increased likelihood of other environmental factors such as wind disturbance, playing an important role. Even so there are a number of notable successes where fish removal apparently triggered clear water in large (>1000 ha) lakes including Lake Christina (1620 ha-Hanson & Butler, 1994), Lake Finjasjon (Strand, pers comm.), and Lake Wolderwijd (Meijer et al., 1994b; Meijer & Hosper, 1997).

The timing of biomanipulation

The question of when to biomanipulate, depends partly on the nature of target processes. For example, the increase phase of zooplankton is generally in spring. Therefore, where zooplankton are the target, biomanipulation should be conducted before this time, particularly if a spring clear water phase is not typical for the lake in question. High *Daphnia* spp. densities in spring may help prevent the build-up of blue-green algae and the subsequent development of blooms in the summer months, where these are a problem. This may occur even where the edibility of blue-green is low (de Bernardi & Giussani, 1990). However, if blue-green algae are already present at high density in spring, the ability of *Daphnia* spp. to control their biomass may be in question (Dawidowicz, 1990; Gliwicz, 1990).

Fishing is best undertaken in winter to prepare for the potential spring clear water phase. Fish are less active at this time and may be concentrated in selected areas (e.g. in deeper areas; e.g. Backx & Grimm 1994, or in littoral margins and dykes-Perrow, 1994). Large catches in a short period of time makes the fishing more cost-efficient and improves the chances of overall success. Fish also exhibit lower mortality as a result of capture and handling in cold water, which may be important where fish are to kept alive and transported to other waters. Winter removal is often supplemented by fishing in spring particularly on the spawning grounds, to remove adult fish as they attempt to spawn (Backx & Grimm, 1994) or by putting in artificial substrate which may be removed once covered with eggs (Perrow, 1994).

There is some discussion whether continuous fishing might be effective. For example, an increase in transparency was noted through this method in Lake Ringsjon (Hamrin, 1993) and Finjasjon (Strand, pers comm.). However, although it may be necessary to extend the period of fishing during the summer to target particular groups e.g. 0+ or large adults, we suggest that continuous fishing is not generally applicable. This is because biomanipulation should be seen as an extreme perturbation to the system to generate a desired response, preferably within one season. Gradual removal may extend the response time and the critical threshold of fish removed may not be achieved until late in the season when there is little opportunity for *Daphnia* or macrophytes to respond.

Stocking with piscivores

In general, stocking with piscivores is necessary to enhance natural populations, unless means are found to raise the natural productivity (e.g. enhancing spawning grounds-Grimm, 1991). The target prey fish should be clearly defined and the species, size, and stock levels of predators tailored accordingly. Focussing on the recruitment of 0+ fish where these are the cause of the problem, with stocking of 0+ pike, seems the preferred strategy. However, the response is variable. For example, stocking with pike at low densities (200–400 ha^{-1}) had no apparent effect in Lake Wolderwijd (Meijer et al., 1995). Recent experiments with high densities (to 3600 ha^{-1}) in Lake Wirbel in Poland (Prejs et al., 1994) and Lake Lyng in Denmark (Berg et al., 1997; Søndergaard et al., 1997) may be more effective.

Stocking with species such as pike has problems as they are cannibalistic and stocking must be repeated annually. In addition, small pike typically favour emergent vegetation (Grimm, 1994) and may be more or less restricted to the littoral margins (Cook & Bergersen, 1988). The edge to surface area ratio may therefore influence the effectiveness of pike and it may be that the chances of success may be higher in small rather than large lakes. We suggest that at this stage, stocking with piscivores is best considered as a supplementary technique in a programme of biomanipulation, especially massive stocking by 0+ pike to control the production of 0+ fish to prolong a clear water phase.

The influence of nutrient levels

Once biomanipulation is complete, achieving clear water seems to be independent of nutrient levels. For example, Lake Zwemlust (TP = 1.5 mg l^{-1} – Van Donk et al., 1990b), Alderfen Broad (TP to 2.4 mg l^{-1} – Perrow et al., 1994) and Little Mere (TP to 5.3 mg l^{-1} – Carvalho, 1994) all showed elevated levels of phosphorus and all became clear as a result of biomanipulation or fish-kill.

Maintaining clear water

In some lakes, the clear water phase created by fish reduction was ended by migration of fish into the lake (e.g. Lake Helgetjern, Faafeng & Braband, 1990). Prevention of immigration is obviously very difficult if navigation is preserved. In addition, the desire to gain entry may be enhanced during spawning or wintering migrations. In the Norfolk Broads, a riverine system where flooding may be extensive at certain times of year, the coincidence of high tides with spawning or wintering migrations may lead to breaching of fish barriers or access through flooded fen or woodland (Perrow, 1994).

Even without migration of fish into the lake, it may prove difficult to control the production of zooplanktivores in the summer. This is partly due to the enhanced fecundity of adult fish and perhaps enhanced spawning conditions (Meijer et al., 1995). In this case, renewed fishing effort may be required. However, there are examples in the UK (e.g. Perrow, 1994; Moss et al., 1996) to show that control of 0+ production is possible. Further, the chances of controlling 0+ production may be improved in the rare case of reasonable stocks of piscivores already being present, thus leading to a high piscivore/zooplanktivore ratio after biomanipulation (e.g. Lake Væng-Meijer et al., 1994a). Alternatively, piscivores may have to be stocked (see above) as inevitably the chances of a natural improvement in piscivore stock is directly related to some change in water transparency and the presence of macrophytes.

The presence of macrophytes appears to be the key to maintaining water clarity, as even where zooplanktivorous fish successfully recruit to a high density the water remains clear in the presence of macrophytes (Meijer et al., 1994a). The objective of biomanipulation in most circumstances is thus to get a clear water period of sufficient length to allow macrophytes to establish a sufficient cover or biomass. For example, Schriver et al. (1995) suggest even 30% PVI (per cent volume infested) has a significant beneficial impact on zooplankton populations (see below). Biomanipulation may therefore have to be repeated on an annual basis to provide a suitable light climate for macrophytes to allow this colonisation which may take several years (e.g. over 4 successive years in Cockshoot Broad – Perrow, 1994; Moss et al., 1996). The fact that biomanipulation is repeatable illustrates scope for generating clear water with low algal standing crops for water supply or recreational purposes, even where nutrient levels remain high, regardless of whether macrophytes are produced or not. This may prove more cost-effective than treating undesirable algae.

Establishing macrophytes

After a suitable light climate is achieved, macrophyte populations may establish quickly. For example in the Dutch lakes, Bleiswijkse Zoom (Meijer et al., 1994a), Duinigermeer (Van Berkum et al., 1995) and Ijzeren Man (Driessen et al., 1994) within two months of fish removal, more than 50% of the lake was covered with macrophytes, where none were present prior to biomanipulation. Some restored macrophyte populations have proved to be unstable in the longer-term (Perrow et al., 1994; Van Donk & Gulati, in press). Alternatively, the recovery of viable macrophyte populations may take several years (Lauridsen et al., 1994; Moss et al., 1996) relying on colonisation from external sources or from spread from an existing population centre (e.g. Lake Wolderwijd – Meijer & Hosper, 1997 and Lake Finjasjon – Strand, unpubl. data). At present, studies are being undertaken to explain these differences. These are focussing on the following factors, which may be considered in a sequential order.

The presence of a viable propagule bank or the production or inflow of propagules is intuitively conditional for rapid development of dense vegetation. The propagule bank may be reduced through removal or burial where remedial action such as suction-dredging has been undertaken. The seeds of several species are known to be long-lived (Ridley, 1930), but may be at too low a density to stock the lake in the first generation. Colonisation may be enhanced through transportation of propagules by waterbirds, fish or water currents (e.g. Agami & Waisel, 1986), but as yet, the relative importance of these mechanisms in promoting colonisation is unknown.

The nature of the sediment chemistry may determine germination and the fate of young seedlings. For example, high hydrogen sulphide concentrations may be toxic in reduced anaerobic sediments (Koch et al., 1990; Smolders, 1995). Once seedlings have germinated, growth may be impaired depending on the substrate. For example, Jupp & Spence (1977); Lauridsen et al. (1993), both show sediment structure and organic content may have a marked effect. Loose lake sediments, particularly those containing a large organic component can be disturbed regularly by turbulence caused by wind or fish (Ten Winkel & Meulemans, 1984) and may thus lead to poor seedling survival.

The role of filamentous algae, growing as benthic algal mats and as epiphytes, in reducing the fitness of macrophyte seedlings is thought to be important (Phillips et al., 1978; Perrow et al., 1997). Molluscs, as grazers on macrophyte surfaces, have been shown to depress periphyton loads and promote macrophyte growth (Underwood et al., 1992). The abundance of molluscs in turn, depends on the relative predation pressure by benthivorous fish (e.g. tench – Bronmark, 1994). The mollusc-epiphyte-macrophyte interaction seems to be important in promoting the success of macrophytes and deserves further attention.

The impact of grazing birds, principally coot (*Fulica atra*) and swans (especially Mute swan *Cygnus olor*) has been widely discussed as the limiting factor in the colonisation, development and subsequent stability of macrophytes (Jupp & Spence, 1977; Lauridsen et al., 1993, 1994; Perrow et al., 1994; Moss et al., 1996; Søndergaard et al., in press). However, information relating to the impact of grazing by birds frequently relies on experiments using macrophytes transplanted into the site (Lauridsen et al., 1993; Moss et al., 1996; Søndergaard et al. in press), rather than the impact on naturally occurring populations (Anderson & Low, 1976; Esler, 1989). Reviewing the available information with data from experiments and monitoring of macrophyte and bird communities, Perrow et al. (1997) argue that the bulk of grazing pressure is outside the macrophyte growing season and although possibly affecting the success of some species (particularly ones that overwinter as adult plants or buds) in subsequent growing seasons (Van Donk et al., 1994b), the colonisation of macrophytes is primarily limited by other factors.

Mechanisms responsible for clear water amongst macrophytes

Detailed knowledge of the mechanisms responsible for maintaining clear water and low algal populations in lakes dominated by macrophytes is critical if restoration of shallow lakes, is to be achieved in a predictable and cost-effective manner (Shapiro, 1990). Several stabilising mechanisms perpetuating the macrophyte-dominated state have been suggested. Whilst all may exist, we were interested in pooling knowledge on which mechanism(s) were generally thought to be of most importance and therefore should be the focus of future research.

Turbidity may be created by phytoplankton or suspended solids. In the case of the latter macrophytes may increase sedimentation through calming of the water column (James & Barko, 1990). However, this is likely to have most effect in large very shallow lakes (e.g. many Dutch lakes), where wind and waves and/or a high biomass of benthivorous fish create high resuspension (Coops & Doef, 1996). Where benthivorous fish are the problem, increased macrophyte coverage may simply limit access to the sediments. Increased sinking of algae may be an important mechanism on a local scale. For example, in Lakes Wolderwijd and Veluwe where clear water occurs over dense *Chara* beds, but is turbid with open-water phytoplankton in the majority of the lake (Scheffer et al., 1994). In macrophyte-dominated lakes, sinking of algae is not thought to be an important mechanism as a shift to slow-sinking taxa which are better adapted to the quiescent waters in macrophyte beds is to be expected (Jeppesen, pers comm.).

Macrophytes may compete with phytoplankton for available resources particularly light and nutrients. Competition for N, with macrophytes maintaining a competitive advantage over algae through: luxury uptake (Balls et al., 1989); acting as a sink for available nutrients (in both water column and sediments) (Ozimek et al., 1990; Van Donk et al., 1993); creating anaerobic conditions or acting as a substrate for denitrification (Denny, 1987); have been the focus of attention in recent years.

Allelopathy, the production of compounds limiting algal growth by the macrophytes themselves, has been proven in laboratory studies (Wium-Anderson et al., 1982), but requires testing in the field situation. It is possible that the production of inhibiting compounds may have evolved in response to epiphytic rather than planktonic algae and is thus of limited value for water transparency (a similar conclusion may be reached for the macrophyte-mollusc interaction above, which also focuses on control of epiphyte loads). Further, allelopathy may be limited to a few taxa e.g. charophytes, and may thus only be of importance where charophytes are an integral part of the macrophyte community. Combined with the possibility that the fine structure and dense growth form of charophytes enhances algal sedimentation (see above), and with the observation of few cladoceran grazers associating with *Chara* beds (Meijer & Hosper, 1997), it is possible that the driving mechanisms responsible for clear water in charophyte beds are different from those in other types of macrophyte.

The provision of refuges for grazing zooplankton amongst macrophytes has recently received much attention (Timms & Moss, 1984; Phillips et al., 1996; Stansfield et al., 1997). Part of the appeal lies with the ability of *Daphnia* spp. to undertake diel migrations from refuges to feed in open water (Timms & Moss, 1984; Lauridsen et al., in press), illustrating potential to promote water clarity in the open water surrounding macrophyte stands. There has been some debate as to the suitability of macrophytes as refuges for *Daphnia* spp. (Pennak, 1973; Dorgelo & Heykoop, 1985). Recent experiments indicate that *Daphnia* spp. overcome the repellency for macrophytes in the presence of fish (Lauridsen & Lodge, 1996). The response of *Daphnia* is presumably through olfactory cues, although whether fish odour alone or the scent of crushed *Daphnia* consumed by the fish, indicative of actual predation risk, is used under natural conditions is untested.

In the field, a variety of grazing cladocerans including *Daphnia* spp. have been shown to use refuges (Schriver et al., 1995; Stansfield et al., 1997). However, the ability of *Daphnia* spp. to use refuges over time is in question, as they seem to have to migrate out of refuges at night to forage in open water (Stansfield et al., 1997) where they suffer at least some predation. With increasing macrophyte cover, *Daphnia* spp. are replaced by other species such as *Ceriodaphnia* spp., *Simocephalus* spp. and *Sida* spp. (Stansfield et al., 1997). The ability of these species in grazing out algal populations is relatively unknown but seems promising (Turner & Mittelbach, 1992).

Overall, macrophytes clearly act as refuges for zooplankton, and this seems to knock-on to reduced algal populations, although more field studies are required to further elucidate the interactions between different zooplankton and fish species with differential foraging abilities, within structurally variable habitats. It is also likely that many of the important fish-zooplankton interactions may occur at night, when *Daphnia* spp. in particular emerge from refuges to feed, and this should also be a target for future research.

An increase in macrophyte cover is known to increase the abundance of piscivores (de Nie, 1987). This is due to improved spawning and foraging (clear water with ambush sites) conditions. Alternatively, an increase in the proportion of piscivores may also be due to a reduction of the relative proportion of zooplanktivores as a result of increased structural complexity changing foraging success (Winfield, 1986).

Direct predation by piscivores may reduce the biomass of prey fish (Spencer & King, 1984; Grimm & Backx, 1990) or the age structure in favour of large individuals (Bronmark et al., 1995). The latter illustrates the potential gape-limitation of predators. The mere presence of piscivores may also impose a perceived risk of predation, which may significantly change the distribution of prey populations in time and space (Gliwicz & Jachner, 1992; Werner et al., 1983; Persson & Eklov, 1995), which may knock on to enhanced populations of zooplankton grazers (Hambright et al., 1986; Turner & Mittelbach, 1992) or simply allow zooplankton to utilise particular habitats (Jacobsen et al., 1997).

However, whether control of the production of zooplanktivorous and/or benthivorous fish by piscivores is a general mechanism applicable in a variety of lake types and across a range of different fish communities is still under discussion. This is because natural populations may be limited in time and space and preferences for particular prey species (Mauck & Coble, 1973; Eklov & Hamrin, 1989) or sizes (Hart & Hamrin, 1988; Van Densen, 1994) may be exhibited.

Stabilising the system

After establishing clear water and macrophytes, the biological, chemical and physical structure of the lake will change. The fish community is likely to shift with increased representation of species such as tench (*Tinca tinca*) and rudd (*Scardinius erythrophthalmus*) as well as an increase in the absolute and relative abundance of piscivores such as perch and pike (Perrow, 1991; Meijer et al., 1995; Jeppesen et al., 1997). The dependence of young fish on zooplankton may decline with an increased availability of macroinvertebrate prey (Diehl, 1993). The macrophyte community itself is likely to display a succession of dominants within the growing season (Coops & Doefs, 1996) and the number of herbivorous birds is likely to increase with an increase in potential food supply (Mitchell et al., 1988; Giles, 1992; Perrow et al., 1997). Theoretically, such a high diversity community should become stable, unless a severe perturbation occurs (e.g. fish kill – Bronmark & Weisner, 1992). However, in many restoration projects, only one or two species of macrophyte dominate (e.g. *Ceratophyllum* and *Elodea* – Van Donk et al., 1994b; Lauridsen et al., 1994; Perrow et al., 1994; Moss et al., 1996). Dependence on a limited suite of species increases the chance of stochastic events or selective factors (e.g. bird grazing – Van Donk et al., 1994b; periphyton loading – Van Donk & Gulati, in press) leading to the disappearance of the macrophytes. As these start to disappear, the clear water state may begin to deteriorate during the summer months (Meijer et al., 1994a; Van Donk & Gulati, in press), although it is not always clear what is the cause and effect of increasing turbidity and decreasing macrophyte growth. It thus seems likely that increased stability will be achieved with a higher diversity of macrophytes. In particular, with high coverage, higher species richness and seasonal succession of different species.

It is often thought that stability over the long term, is only possible at lowered nutrient levels (Hosper & Jagtman, 1990; Benndorf & Miersch, 1991; Jeppesen et al., 1997). For example, comparison of several Danish case studies showed that at decreasing nutrient levels, the % cover of macrophytes was higher and the share of piscivorous fish increased (Jeppesen et al., 1990a). Further, the production of cyprinids is thought to be reduced at lower nutrient levels (Grimm & Backx, 1990). Several lakes are known to alternate between turbid and clear water states (e.g. Lakes Tåkern and Krankesjon – Blindlow et al., 1993) and it may be that biomanipulated lakes that go through turbid periods (e.g. as a result of the successful recruitment of 0+ fish or macrophyte collapse for a variety of reasons), are more likely to recover if nutrient levels are lower and the factors discussed above are in place (e.g. Lake Væng – Meijer et al., 1994a).

We rely on empirical studies on natural lakes outlining what is possible at different nutrient concentrations, and thus domination by macrophytes apparently depends on lowered nutrient levels. However, there is little evidence of any mechanism that prevents establishment of macrophytes at high nutrient levels once clear water has been attained, apart from perhaps conditions at the sediment surface in heavily enriched lakes being unsuitable for germination and growth of many macrophytes (see above). Further, discussion at the workshop revealed that some workers held the belief that the piscivorous/planktivorous fish ratio may be the key to stability. Again, although often correlated with nutrient concentrations (Jeppesen et al., 1997), there is no obvious direct mechanism as to why piscivores cannot be abundant at high nutrient concentrations (e.g. as in Lake Ring – Berg et al., 1994).

Through biomanipulation, we may create new 'equilibria' of which there are few or no natural examples, particularly if regular management (e.g. fish removal) is conducted to maintain such a condition. It may therefore be possible to establish desirable biological components (e.g. macrophytes and piscivorous fish or simply high zooplankton biomass and lowered phytoplankton biomass) at high nutrient loading. Reduction of nutrient load, which is highly expensive and technically difficult, may therefore prove to be of low priority, in some circumstances. Studying exceptional cases (e.g. where diverse macrophyte communities are present at high nutrient levels) and further modelling studies incorporating particular management regimes (e.g. removal/addition of certain components of the fish community) as well as increasing the number of long-term (>10 years) studies may help to give us a better understanding of what is possible.

Acknowledgments

We thank the participants of Shallow Lakes '95, particularly those in the workshop, for providing frank and stimulating discussion. We are also grateful to the organisers of the conference for providing the necessary climate for the exchange of ideas and for encouraging us to undertake a workshop.

References

Agami, M. & Y. Waisel, 1986. The role of mallard ducks (*Anas platyrhynchos*) in distribution and germination of seeds of the submerged hydrophyte *Najas marina* L. Oecologia 68: 473–475.

Anderson, M. G. & J. B. Low, 1976. Use of Sago pondweed on the Delta Marsh, Manitoba. J. Wildl. Mgmt 40: 233–242.

Backx, J. J. G. M. & M. P. Grimm, 1994. Mass removal of fish from Lake Wolderwijd, The Netherlands. Part II: Implementation phase. In Cowx, I. G. (ed.), Rehabilitation of Freshwater Fisheries. Fishing News Books, Blackwell Scientific Publications Ltd., Oxford, England.

Balls, H., B. Moss & K. Irvine, 1989. The loss of submerged plants with eutrophication I. Experimental design, water chemistry, aquatic plant and phytoplankton biomass in experiments carried out in ponds in the Norfolk Broadland. Freshwat. Biol. 22: 71–87.

Benndorf, J., 1992. The control of the indirect effects of biomanipulation. In Sutcliffe, D. W. & J. G. Jones (eds), Eutrophication: Research and Application to Water Supply. Freshwater Biological Association, Far Sawrey, Ambleside, Cumbria, England.

Benndorf, J. & U. Miersch, 1991. Phosphorus loading and efficiency of biomanipulation. Verh. int. Ver. Limnol. 24: 2482–2488.

Berg, S., E. Jeppesen & M. Søndergaard, 1997. Pike (*Esox lucius* L.) stocking as a biomanipulation tool 1. Effects on the fish population in Lake Lyng, Denmark. Hydrobiologia 342/343: 311–318.

Berg, S., E. Jeppesen, M. Sondergaard & E. Mortensen, 1994. Environmental effects of introducing whitefish, *Coregonus lavaretus* (L.) in Lake Ring. Hydrobiologia 275/276: 71–79.

Blindlow, I., G. Andersson, A. Hargeby & S. Johansson, 1993. Long-term pattern of alternative stable states in two shallow eutrophic lakes. Freshwat. Biol. 30: 159–167.

Breukelaar, A. W., E. H. R. R. Lammens, J. G. P. Klein Breteler & I. Tatrai, 1994. Effects of benthivorous beam (*Abramis brama* L.) and carp (*Cyprinus carpio* L.) on sediment resuspension and concentrations of nutrients and chlorophyll *a*. Freshwat. Biol. 32: 113–121.

Bronmark, C., 1994. Effects of tench and perch on interactions in a freshwater, benthic food chain. Ecology 75: 1818–1828.

Bronmark, C., C. A. Paszkowski, W. M. Tonn & A. Hargeby, 1995. Predation as a determinant of size structure in populations of crucian carp (*Carassius carassius*) and tench (*Tinca tinca*). Ecol. Freshwat. Fish 4: 85–92.

Bronmark, C. & S. E. B. Weisner, 1992. Indirect effects of fish community structure on submerged vegetation in shallow, eutrophic lakes: an alternative mechanism. Hydrobiologia 243/244: 293–301.

Brooks, J. L. & S. I. Dodson, 1965. Predation, body size, and composition of plankton. Science 150: 28–35.

Carvalho, L., 1994. Top-down control of phytoplankton in a shallow hypertrophic lake: Little Mere (England). Hydrobiologia 275/276: 53–63.

Cook, M. F. & E. P. Bergersen, 1988. Movements, Habitat Selection, and Activity Periods of Northern Pike in Eleven Mile Reservoir, Colorado. Trans. am. Fish. Soc. 117: 495–502.

Coops, H. & R. W. Doef, 1996. Submerged vegetation development in two shallow, eutrophic lakes. Hydrobiologia 340: 115–120.

Cryer, M., G. Peirson & C. R. Townsend, 1986. Reciprocal interactions between roach *Rutilus rutilus*, and zooplankton in a small lake: Prey dynamics and fish growth and recruitment. Limnol. Oceanogr. 31: 1022–1038.

Dawidowicz, P., 1990. Effectiveness of phytoplankton control by large-bodied and small-bodied zooplankton. Hydrobiologia 200/201: 43–47.

de Bernardi, R. & G. Giussani, 1990. Are blue-green algae a suitable food for zooplankton? An overview. Hydrobiologia 200/201: 29–41.

de Nie, A. W., 1987. The decrease in aquatic vegetation in Europe and its consequences for fish populations. EIFAC Occasional Paper 19. FAO, Rome, 88 pp.

Denny, P., 1987. Mineral cycling by wetland plants – a review. Arch. Hydrobiol. Beih. Ergebn. Limnol. 27: 1–25.

Diehl, S., 1993. Effects of habitat structure on resource availability, diet and growth of benthivorous perch, *Perca fluviatilis*. Oikos 67: 403–414.

Dorgelo, J. & M. Heykoop, 1985. Avoidance of macrophytes by *Daphnia longispina*. Verh. int. Ver. Limnol. 22: 3369–3372.

Driessen, O., B. Pex & H. Tolkamp, 1993. Restoration of a lake: First results and problems. Verh. int. Ver. Limnol. 25: 617–621.

Eklov, P. & S. F. Hamrin, 1989. Predatory efficiency and prey selection: interactions between pike *Esox lucius*, perch *Perca fluviatilis* and rudd *Scardinius erythrophthalmus*. Oikos 56: 149–156.

Esler, D., 1989. An assessment of American coot herbivory of *Hydrilla*. J. Wildl. Mgmt 1147–1149.

Faafeng, B. A. & A. Braband, 1990. Biomanipulation of a small urban lake-removal of fish exclude bluegreen blooms. Verh. int. Ver. Limnol. 24: 597–602.

Giles, N., 1992. Wildlife after gravel: twenty years of practical research by the Game Conservancy and ARC. Game Conservancy, Fordingbridge, UK.

Gliwicz, Z. M., 1990. Why do cladocerans fail to control algal blooms? Hydrobiologia 200/201: 83–97.

Gliwicz, Z. M. & A. Jachner, 1992. Diel migrations of juvenile fish: a ghost of predation past or present? Arch. Hydrobiol. 124: 385–410.

Grimm, M. P., 1991. Water quantity management of Dutch polders, adversely affecting the recruitment of northern pike (*Esox lucius* L.). Verh. int. Ver. Limnol. 24: 2443–2445.

Grimm, M. P., 1994. The influence of aquatic vegetation and population biomass on recruitment of 0+ and 1+ northern pike (*Esox lucius* L.). In Cowx, I. G. (ed.), Rehabilitation of Freshwater Fisheries. Fishing News Books, Blackwell Scientific Publications Ltd., Oxford, England.

Grimm, M. P. & J. J. G. M. Backx, 1990. The restoration of shallow eutrophic lakes, and the role of northern pike, aquatic vegetation and nutrient concentration. Hydrobiologia 200/201: 557–566.

Grimm, M. P. & J. J. G. M. Backx, 1994. Mass removal of fish from Lake Wolderwijd, The Netherlands. Part I: Planning and strategy of a large-scale biomanipulation project. In Cowx, I. G. (ed.), Rehabilitation of Freshwater Fisheries. Fishing News Books, Blackwell Scientific Publications Ltd., Oxford, England.

Hambright, K. D., R. J. Trebatoski & R. W. Drenner, 1986. Experimental study of the impacts of Bluegill (*Lepomis macrochirus*) and Largemouth Bass (*Micropterus salmoides*) on pond community structure. Can. J. Fish. aquat. Sci. 43: 1171–1176.

Hamrin, S. F., 1993. Lake restoration by cyprinid control in Satofta Bay (Lake Ringsjon). Verh. int. Ver. Limnol. 25: 487–493.

Hanson, M. A. & M. G. Butler, 1994. Responses to food web manipulation in a shallow waterfowl lake. Hydrobiologia 279/280: 457–466.

Hart, P. & S. F. Hamrin, 1988. Pike as a selective predator. Effects of prey size, availability, cover and pike jaw dimensions. Oikos 51: 220–226.

Horppila, J. & T. Kairesalo, 1990. A fading recovery: the role of roach (*Rutilus rutilus* L.) in maintaining high phytoplankton productivity and biomass in Lake Vesijarvi, southern Finland. Hydrobiologia 200/201: 153–165.

Hosper, S. H. & E. Jagtman, 1990. Biomanipulation additional to nutrient control for restoration of shallow lakes in The Netherlands. Hydrobiologia 200/201: 523–534.

Hosper, S. H. & M.-L. Meijer, 1993. Biomanipulation, will it work for your lake. A simple test for the assessment of chances for clear water, following drastic fish-stock reduction in shallow eutrophic lakes. Ecol. Engin. 2: 63–72.

Irvine, K., B. Moss & H. Balls, 1989. The loss of submerged plants with eutrophication II. Relationships between fish and zooplankton in a set of experimental ponds, and conclusions. Freshwat. Biol. 22: 89–107.

Jacobsen, L., M. R. Perrow, F. Landkildehus, M. Hjørne, T. L. Lauridsen & S. Berg, 1997. Interactions between piscivores, zooplanktivores and zooplankton in submerged macrophytes: preliminary observations from enclosure and pond experiments. Hydrobiologia 342/343: 197–205.

James, W. F. & J. W. Barko, 1990. Macrophyte influences on the zonation of sediment accretion and composition in a north-temperate reservoir. Arch. Hydrobiol. 2: 129–142.

Jeppesen, E., J. P. Jensen, P. Kristensen, M. Søndergaard, E. Mortensen, O. Sortkjær & K. Olrik, 1990a. Fish manipulation as a lake restoration tool in shallow, eutrophic, temperate lakes 2: threshold levels, long-term stability and conclusions. Hydrobiologia 200/201: 219–227.

Jeppesen, E., J. P. Jensen, M. Søndegaard, T. Lauridsen, L. J. Pedersen & L. Jensen, 1997. Top-down control in freshwater lakes: the role of nutrient state, submerged macrophytes and water depth. Hydrobiologia 342/343: 151–164.

Jeppesen, E., M. Søndergaard, E. Mortensen, P. Kristensen, B. Riemann, H. J. Jensen, J. P. Müller, O. Sortkjær, J. P. Jensen, K. Christoffersen, S. Bosselmann & E. Dall, 1990b. Fish manipulation as a lake restoration tool in shallow, eutrophic temperate lakes 1: cross-analysis of three Danish case-studies. Hydrobiologia 200/201: 205–218.

Jupp, B. P. & D. H. N. Spence, 1977. Limitations of macrophytes in a eutrophic lake, Loch Leven. II. Wave action, sediments and waterfowl grazing. J. Ecol. 65: 431–446.

Koch, M. S., I. A. Mendelssohn & K. L. McKee, 1990. Mechanism for the hydrogensulphide-induced growth limitation in wetland macrophytes. Limnol. Oceanogr. 35: 399–408.

Lammens, E. H. R. R., P. J. Boesewinkel-de Bruyn, H. Hoogveld & E. van Donk, 1992. P-load, phytoplankton, zooplankton and fish stock in Loosdrecht Lake and Tjeukemeer: confounding effect of predation and food availability. Hydrobiologia 233: 87–95.

Lammens, E. H. R. R., H. W. de Nie, J. Vijverberg & W. L. T. van Densen, 1985. Resources Partitioning and Niche Shifts of Bream (*Abramis brama*) and Eel (*Anguilla anguilla*) Mediated by Predation of Smelt (*Osmerus eperlanus*) on *Daphnia hyalina*. Can. J. Fish. aquat. Sci. 42: 1342–1351.

Lammens, E. H. R. R., R. D. Gulati, M.-L. Meijer, E. van Donk, 1990. The first biomanipulation conference: a synthesis. Hydrobiologia 200/201: 619–628.

Lauridsen, T. L., E. Jeppesen & F. Ostergaard Andersen, 1993. Colonization of submerged macrophytes in shallow fish manipulated Lake Vaeng: impact of sediment composition and waterfowl grazing. Aquat. Bot. 46: 1–15.

Lauridsen, T. L., E. Jeppesen & M. Søndergaard, 1994. Colonization and succession of submerged macrophytes in shallow Lake Vaeng during the first five years following fish manipulation. Hydrobiologia 275/276: 233–242.

Lauridsen, T. L. & D. M. Lodge, 1996. Avoidance by *Daphnia magna* of fish and macrophytes: chemical cues and predator-mediated use of macrophyte habitat. Limnol. Oceanogr. 22: 805–810.

Lauridsen, T. L., L. J. Pedersen, E. Jeppesen & M. Søndergaard (in press). The importance of macrophyte bed size for composition and horizontal migration of cladocerans in a shallow lake. J. Plankton Res.

Mauck, W. L. & D. W. Coble, 1973. Vulnerability of Some Fishes to Northern Pike (*Esox lucius*) Predation. J. Fish. Res. Bd Can. 28: 957–969.

Meijer, M.-L., A. W. Breukelaar & S. H. Hosper, 1994a. Mass removal of fish from Lake Wolderwijd, The Netherlands. Part III: Effects on water quality. In I. G. Cowx (ed.), Rehabilitation of Freshwater Fisheries. Fishing News Books, Blackwell Scientific Publications Ltd., Oxford, England.

Meijer, M.-L., M. W. de Haan, A. W. Breukelaar & H. Buiteveld, 1990. Is reduction of the benthivorous fish an important cause of high transparency following biomanipulation in shallow lakes? Hydrobiologia 200/201: 303–315.

Meijer, M.-L. & H. Hosper, 1997. Effects of biomanipulation in the large and shallow Lake Wolderwijd, The Netherlands. Hydrobiologia 342/343: 335–349.

Meijer, M.-L., E. Jeppesen, E. van Donk, B. Moss, M. Scheffer, E. Lammens, E. van Nes, J. A. van Berkum, G. L. de Jong, B. A. Faafeng & J. P. Jensen, 1994b. Long-term responses to fish-stock reduction in small shallow lakes: interpretation of five-year results of four biomanipulation cases in The Netherlands and Denmark. Hydrobiologia 275/276: 457–466.

Meijer, M.-L., E. H. R. R. Lammens, A. J. P. Raat, J. G. P. Klein Breteler & M. P. Grimm, 1995. Development of fish communities in lakes after biomanipulation. Neth. J. aquat. Ecol. 29: 91–101.

Mitchell, S. F., D. P. Hamilton, W. S. MacGibbon, P. K. B. Nayar & R. N. Reynolds, 1988. Interactions between phytoplankton, submerged macrophytes, black swans and zooplankton in a shallow lake. Int. Revue ges. Hydrobiol. 73: 145–170.

Moss, B., 1990. Engineering and biological approaches to the restoration from eutrophication of shallow lakes in which aquatic plant communities are important components. Hydrobiologia 200/201: 367–377.

Moss, B., J. H. Stansfield, K. Irvine, M. R. Perrow & G. L. Phillips, 1996. Progressive restoration of a shallow lake – a twelve-year experiment in isolation, sediment removal and biomanipulation. J. appl. Ecol. 33: 71–86.

Ozimek, T., R. D. Gulati & E. van Donk, 1990. Can macrophytes be useful in biomanipulation of lakes? The Lake Zwemlust example. Hydrobiologia 200/201: 399–407.

Pennak, R. W., 1973. Some evidence for aquatic macrophytes as repellents for a limnetic species of *Daphnia*. Int. Revue ges. Hydrobiol. 58: 569–576.

Perrow, M. R., 1991. Reversing the effects of eutrophication upon fish communities: lessons from Broadland. In Lucas, M. C., I. Diack & L. Laird (eds), Interactions Between Fisheries and the Environment. Proceedings of the Institute of Fisheries Management, Nottingham, England: 111–125.

Perrow, M. R., 1994. Practical aspects of the biomanipulation of fish populations. In Pitt, J.-A. & G. L. Phillips (eds), The Development of Biomanipulation Techniques & Control of Phosphorus Release from Sediments, EC LIFE project 92-3/UK/031, NRA Report No. 475/2/A. National Rivers Authority/Broads Authority, Bristol, UK.

Perrow, M. R. & K. Irvine, 1992. The relationship between cladoceran body size and the growth of underyearling roach (*Rutilus rutilus* (L.)) in two shallow lowland lakes: a mechanism for density-dependent reductions in growth. Hydrobiologia 241: 155–161.

Perrow, M. R., B. Moss & J. H. Stansfield, 1994. Trophic interactions in a shallow lake following a reduction in nutrient loading: a long term study. Hydrobiologia 275/276: 43–52.

Perrow, M. R., J. Schutten, J. R. Howes, T. Holzer, F. J. Madgwick & A. J. D. Jowitt, 1997. Interactions between coot (*Fulica atra*) and submerged macrophytes: the role of birds in the restoration process. Hydrobiologia 342/343: 241–255.

Persson, L., S. Diehl, L. Johansson, G. Andersson & S. F. Hamrin, 1991. Shifts in fish communities along the productivity gradient of temperate lakes-patterns and the importance of size-structured populations. J. Fish. Biol. 38: 281–293.

Persson, L. & P. Eklov, 1995. Prey refuges affecting interactions between piscivorous perch and juvenile perch and roach. Ecology 76: 70–81.

Phillips, G. L., D. F. Eminson & B. Moss, 1978. A mechanism to account for macrophyte decline in progressively eutrophicated freshwaters. Aquat. Bot. 4: 103–126.

Phillips, G. L., M. R. Perrow & J. H. Stansfield, 1996. Manipulating the fish-zooplankton interaction in shallow lakes: a tool for restoration. In Greenstreet, S. P. R. & M. L. Tasker (eds), Aquatic Predators and Their Prey. Blackwell Scientific Publications Ltd., Oxford. England 174–183.

Prejs, A., A. Martyniak, S. Boron, P. Hliwa & P. Koperski, 1994. Food web manipulation in a small, eutrophic Lake Wirbel, Poland: effect of stocking with juvenile pike on planktivorous fish. Hydrobiologia 275/276: 65–70.

Reynolds, C. S., 1994. The ecological basis for the successful biomanipulation of aquatic communities. Arch. Hydrobiol. 130: 1–33.

Ridley, H. N., 1930. The dispersal of plants throughout the world. L. Reeve & Co. Ashford, Kent, England. Scheffer, M., 1990. Multiplicity of stable states in freshwater systems. Hydrobiologia 200/201: 475–486.

Scheffer, M., S. H. Hosper, M.-L. Meijer, B. Moss & E. Jeppesen, 1993. Alternative Equilibria in Shallow Lakes. TREE 8: 275–279.

Scheffer, M., M. van der Berg, A. Breukelaar, C. Breukers, H. Coops, R. Doef & M.-L. Meijer, 1994. Vegetated areas with clear water in turbid shallow lakes. Aquat. Bot. 49: 193–196.

Schriver, P., J. Bøgestrand, E. Jeppesen & M. Søndergaard, 1995. Impact of submerged macrophytes on fish-zooplankton-phytoplankton interactions: large-scale enclosure experiments in a shallow lake. Freshwat. Biol. 33: 255–270.

Shapiro, J., 1990. Biomanipulation: the next phase-making it stable. Hydrobiologia 200/201: 13–27.

Søndergaard, M., L. Bruun, T. L. Lauridsen, E. Jeppesen & T. Vindbaek Madsen, (in press). The impact of grazing waterfowl on submerged macrophytes: *in situ* experiments in a shallow eutrophic lake. Aquat. Bot.

Søndergaard, M., E. Jeppesen & S. Berg, 1997. Pike (*Esox lucius* L.) stocking as a biomanipulation tool 2. Effects on the lower trophic levels in Lake Lyng, Denmark. Hydrobiologia 342/343: 319–325.

Spencer, C. N. & D. L. King, 1984. Role of fish in Regulation of Plant and Animal Communities in Eutrophic Ponds. Can J. Fish. aquat. Sci. 41: 1851–1855.

Stansfield, J. H. & M. R. Perrow, 1994. The influence of benthivorous fish upon benthic communities. In Pitt, J.-A. & G. L. Phillips (eds), The Development of Biomanipulation Techniques & Control of Phosphorus Release from Sediments, EC LIFE project 92-3/UK/031, NRA Report No. 475/2/A. National Rivers Authority/Broads Authority, Bristol, UK.

Stansfield, J. H., M. R. Perrow, L. D. Tench, A. J. D. Jowitt & A. A. L. Taylor, 1997. Submerged macrophytes as refuges for grazing Cladocera against fish predation: observations on seasonal changes in relation to macrophyte cover and predation pressure.

Tatrai, I. & V. Istvanovics, 1986. The role of fish in the regulation of nutrient cycling in lake Balaton, Hungary. Freshwat. Biol. 16: 417–424.

Tatrai, I., E. H. R. R. Lammens, A. W. Breukelaar & J. G. P. Klein Breteler, 1994. The impact of mature cyprinid fish on the composition and biomass of benthic macroinvertebrates. Arch. Hydrobiol. 131: 309–320.

Ten Winkel, E. H. & J. T. Meulemans, 1984. Effects of fish upon submerged vegetation. Hydrobiol. Bull. 18: 157–158.

Timms, R. M. & B. Moss, 1984. Prevention of growth of potentially dense phytoplankton populations by zooplankton grazing, in the presence of zooplanktivorous fish, in a shallow wetland system. Limnol. Oceanogr. 29: 472–486.

Turner, A. M. & G. G. Mittelbach, 1992. Effects of grazer community composition and fish on algal dynamics. Can J. Fish. aquat. Sci. 49: 1908–1915.

Underwood, G. J. C., J. D. Thomas & J. H. Baker, 1992. An experimental investigation of interactions in snail-macrophyte-epiphyte systems. Oecologia 91: 587–595.

Van Berkum, J. A., M. Klinge & M. P. Grimm, 1995. Biomanipulation on the Duinigermeer: first results. Neth. J. aquat. Ecol. 29: 472–486.

Van Densen, W. L. T., 1994. Predator enhancement in freshwater fish communities. In Cowx, I. G. (ed.), Rehabilitation of Freshwater Fisheries. Fishing News Books, Blackwell Scientific Publications Ltd., Oxford, England.

Van der Vlugt, J. C., P. A. Walker, J. van der Does & A. J. P. Raat, 1992. Fisheries management as an additional lake restoration measure: biomanipulation scaling-up problems. Hydrobiologia 233: 213–225.

Van Donk, E., E. de Deckere, J. G. P. Klein Breteler & J. T. Meulemans, 1994b. Herbivory by waterfowl and fish on macrophytes in a biomanipulated lake: effects on long-term recovery. Verh. int. Ver. Limnol. 25: 2139–2143.

Van Donk, E., M. P. Grimm, R. D. Gulati, P. G. M. Heuts, W. A. de Kloet & L. van Liere, 1990a. First attempt to apply whole-lake food-web manipulation on a large scale in the Netherlands. Hydrobiologia 200/201: 291–301.

Van Donk, E., M. P. Grimm, R. D. Gulati & J. P. G. Klein Breteler, 1990b. Whole-lake food-web manipulation as a means to study community interactions in a small ecosystem. Hydrobiologia 200/201: 275–289.

Van Donk, E., M. P. Grimm, P. G. M. Heuts, G. Blom, K. Everards & O. F. R. van Tongeren, 1994a. Use of mesocosms in a shallow eutrophic lake to study the effects of different restoration measures. Arch. Hydrobiol. Beih. Ergebn. Limnol. 40: 283–294.

Van Donk, E. & R. D. Gulati, in press. Transition of a lake to turbid state six years after biomanipulation: mechanisms and pathways. Wat. Sci. Tech.

Van Donk, E., R. D. Gulati, A. Iedema & J. T. Meulemans, 1993. Macrophyte-related shifts in the nitrogen and phosphorus contents of the different trophic levels in a biomanipulated shallow lake. Hydrobiologia 251: 19–26.

Van Liere, L., R. D. Gulati, F. G. Wortelboer & E. H. R. R. Lammens, 1990. Phosphorus dynamics following restoration measures in the Loosdrecht Lakes (The Netherlands). Hydrobiologia 191: 87–95.

Werner, E. E., J. F. Gilliam, D. J. Hall & G. G. Mittelbach, 1983. An experimental test of the effects of predation risk on habitat use in fish. Ecology 64: 1540–1548.

Winfield, I. J., 1986. The influence of simulated aquatic macrophytes on the zooplankton consumption rate of juvenile roach, *Rutilus rutilus*, rudd, *Scardinius erythrophthalmus*, and perch, *Perca fluviatilis*. J. Fish. Biol. 29: 37–48.

Wium-Anderson, S., U. Anthoni, C. Christophersen & G. Houen, 1982. Allelopathic effects on phytoplankton by substances isolated from aquatic macrophytes (Charales). Oikos 39: 187–190.

Hydrobiologia **342/343:** 367–376, 1997.
L. Kufel, A. Prejs & J. I. Rybak (eds), Shallow Lakes '95.

Lake Volkerak-Zoom: a lake shifting from the clear to the turbid state

Carolien P. M. Breukers[1], Edith M. van Dam[2] & Sjoerdtje A. de Jong[3]
[1] *RIZA, Institute for Inland Water Management and Waste Water Treatment, P.O. Box 17, 8200 AA Lelystad, The Netherlands*
[2] *Koeman en Bijkerk bv., P.O. Box 14, 9750 AA Haren, The Netherlands*
[3] *Rijkswaterstaat Directorate Zeeland, P.O. Box 5014, 4330 KA Middelburg, The Netherlands*

Key words: phytoplankton, zooplankton, eutrophication, fish

Abstract

Lake Volkerak-Zoom (area 6000 ha, mean depth 5 m) was created in 1987 by isolating a part of the Eastern Scheldt estuary. The new freshwater lake was clear despite a high phosphorus loading a few years after enclosure. However, from 1990 onwards the transparency dropped.

This paper describes the transition from the clear to the turbid state in Lake Volkerak-Zoom. The processes responsible for the final lowering of the transparency in the period 1988–1994 are discussed and also the measures necessary to meet the target situation: clear water with a transparency of 2 m. From 1990 on transparency decreased due to an increasing algal biomass, which was probably caused by reduced grazing of the zooplankton. It is possible that deterioration of the food quality (by increased cyanobacterial blooms) reduced the zooplankton grazing. However, there are indications that increased mortality, caused by fish predation, played the most important role in the reduction of the zooplankton grazing. In 1992 the recruitment of fish, especially of roach (*Rutilus rutilus*), was high (c. 40 kg ha^{-1}). This caused a high predation pressure on the zooplankton and is shown by a decrease of the mean length and vertical migration of *Daphnia* in 1992. In the same year the largest zooplankton species, *Daphnia pulex*, disappeared from the lake. Smaller zooplankton species with a lower grazing capacity remained. Because of the decreased grazing, the algal biomass increased and the transparency dropped. We expect that in the future the lake will become more turbid. In addition to P-reduction, removal of fish is suggested to attain the target situation: a lasting clear water state.

Introduction

Since turbid lakes, characterized by high algal biomass, often have a reduced species diversity-index and water quality problems, much research has been carried out to find management tools to restore these lakes. Restoration of turbid lakes by means of nutrient reduction (Cullen & Forsberg, 1988; Boers & van der Molen, 1993) and also biomanipulation as an additional measure (Shapiro & Wright, 1984; Jeppesen et al., 1990; Giussiani et al., 1990; Van Donk et al., 1989; Meijer et al., 1994) have been studied extensively. Most studies concentrate on turbid lakes, which are characterized by high algal biomass, shifting into a clear water state, characterized by aquatic macrophyte vegetation. However, little is known about the processes taking place in a relatively clear lake shifting towards the turbid state.

This paper describes the transition from the clear to the turbid state of a newly created freshwater lake, Lake Volkerak-Zoom in The Netherlands. Questions to be answered are: what processes have caused the final lowering of the transparency in the period 1988–1994? Will the lake become more turbid in the future? What measures are needed to meet the target situation: clear water with a transparency of 2 m?

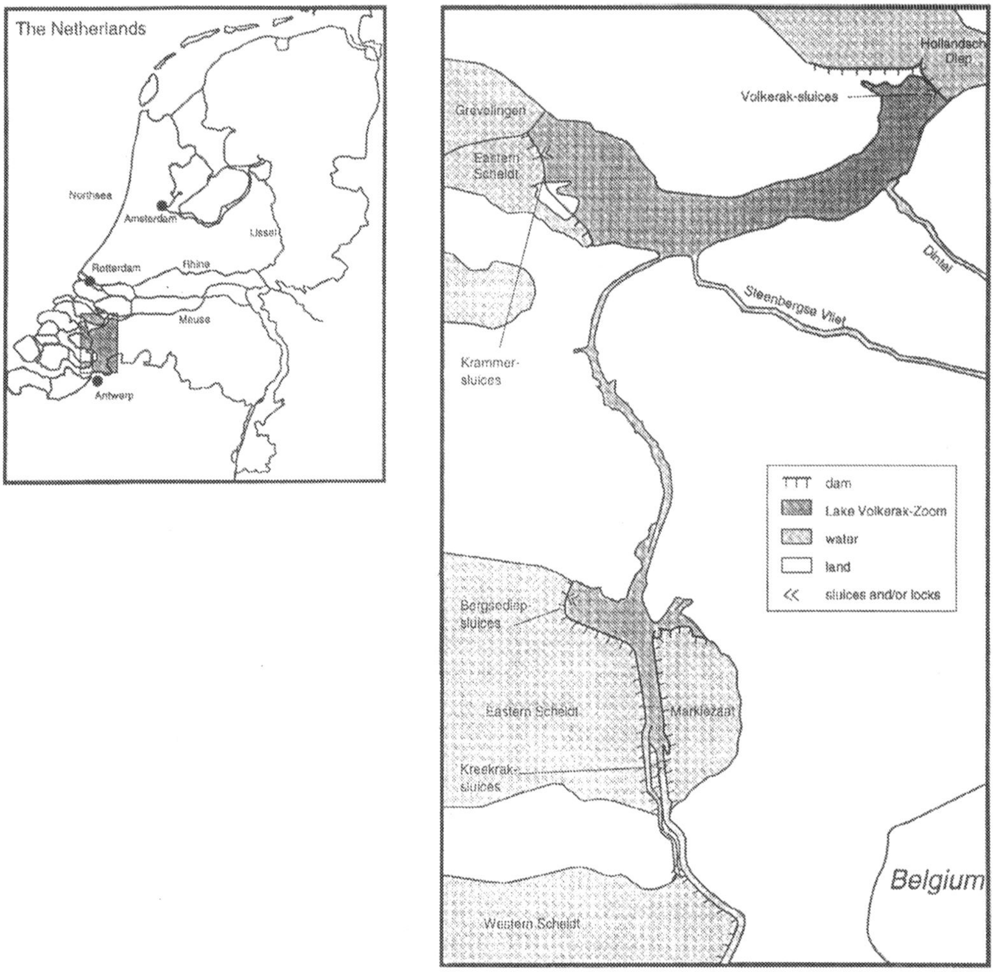

Figure 1. Map of Lake Volkerak-Zoom.

Study site

Lake Volkerak-Zoom (area 6000 ha, mean depth 5 m) is a stagnant freshwater system in the southwest of The Netherlands (Figure 1). It was created in 1987 by isolation of a part of the Eastern Scheldt estuary. The most important water and nutrient inputs are Hollandsch Diep, via the Volkerak sluices, and the river Dintel (Figure 1). The Dintel discharge cannot be influenced, but inflow via the Volkerak sluices is controlled. The mean retention time in 1988–1994 of the northern part of Lake Volkerak-Zoom (Lake Volkerak) is 0.30 yr^{-1} and of the southern part (Lake Zoommeer) 0.16 yr^{-1}.

Flushing with fresh and nutrient-rich water altered the system in about nine months. From a tidal brackish water system it became a freshwater lake. The water level was kept at an almost constant level. Freshwater species very quickly replaced the marine species (Schutten et al., 1994; Frantzen et al., 1994; Van Nes & Smit, 1993).

Methods

Chemical, physical and biological parameters were monitored in Lake Volkerak-Zoom from 1988 until 1994.

Water quality and plankton samples were taken weekly during the summer and biweekly during the rest of the year at three different locations in the lake.

Secchi disk transparency was measured with a 25 cm diameter disk. Water samples were taken from approximately 1 m below the water surface. Total phosphorus and total nitrogen were measured according to Dutch Standard Methods (NEN) which comply with International Standards (ISO).

Phytoplankton samples were collected with a 1.5 m long tube sampler from 1991 until 1994. Five samples were taken and lumped. In 1991 4 samples of 1 litre each were collected at 0.5, 1.0, 1.5 and 2 m depth and lumped. The chlorophyll *a* content in 2 litres of the water sample was extracted by ethanol and measured according to Dutch Standard Methods (NEN). One litre of the sample was fixed with Iodine formaldehyde (in 1991 and 1992) or a Lugol solution (in 1993 and 1994) to determine the phytoplankton species composition. At least 100 individuals were identified using an inverted microscope .

Zooplankton samples were collected with a water sampler in 1990–1994. Five samples of 5 litres each were taken at a depth 1, 2,.. and 5 m, lumped and sieved through a 55 μm meshed net. The residue was preserved with 96% ethanol (in 1991 and 1992) or with a Lugol solution (in 1993 and 1994). Zooplankton was identified and counted using an inverted microscope. The length of 50 individuals of the mesa zooplankton was measured from the upper side of the eye to the mid of the tail spine.

The fish biomass and composition were determined by trawling in November in 1989–1994. Thirty-minute hauls were made at night with a speed of 3.5 km h^{-1}, covering c. 55 ha. Fork length of the fish was measured with an accuracy of 0.5 cm.

Macrophyte distribution was mapped by divers in 1989–1994. Mapping took place on approximately 350 plots, representative for the depth zone from 0 to 5 m. The vertical projected coverage of the bottom by all macrophyte species was estimated in a 10 by 10 m area. The mean coverage per depth zone was multiplied by the surface area of that depth zone, thus giving an estimate of the macrophyte cover.

Results

During the first years of existence (1988 until 1990) the transparency of Lake Volkerak-Zoom increased (Figure 2). In 1990, the mean summer Secchi depth was

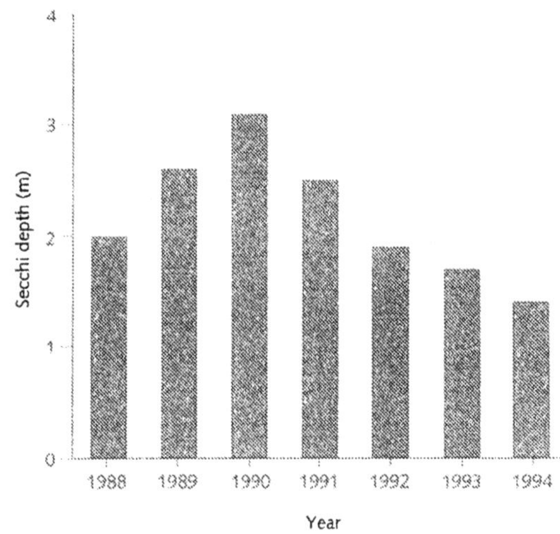

Figure 2. Transparency (m, summer average) of Lake Volkerak-Zoom 1988–1994.

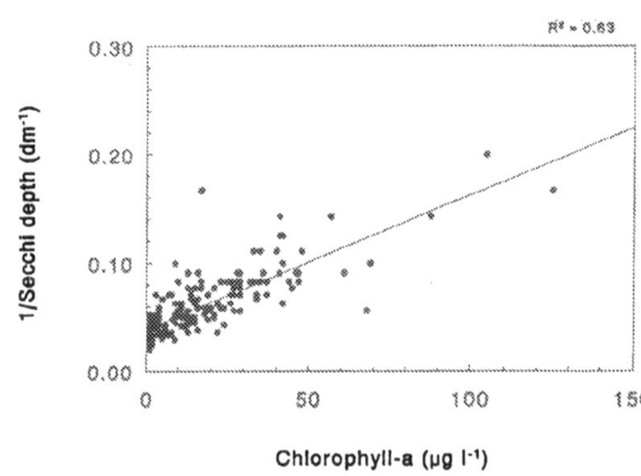

Figure 3. Chlorophyll *a* concentration versus transparency in Lake Volkerak-Zoom 1991–1994.

more than 3 m. In the period 1990–1994 the transparency decreased to a summer average of 1.4 m in 1994. Transparency was inversely correlated with the chlorophyll *a* concentration in the period 1991–1994 (Figure 3).

The P-load decreased from 7.8 g P m^{-2} yr^{-1} in 1988 to 2.4 g P m^{-2} yr^{-1} in 1991 and then increased to 4.2 g P m^{-2} yr^{-1} in 1994 (Figure 4). The most important sources of P in Lake Volkerak-Zoom are the Dintel (including the river Vliet) and Hollandsch Diep. In 1994 total-P concentrations of these sources were 0.26 mg P l^{-1} (Dintel) and 0.20 mg P^{-1} (Hollandsch Diep). In the period 1988–1990 the total-P concentration (summer average) in Lake Volkerak-Zoom dropped from 0.16 to 0.08 mg P l^{-1} (Figure 5). From

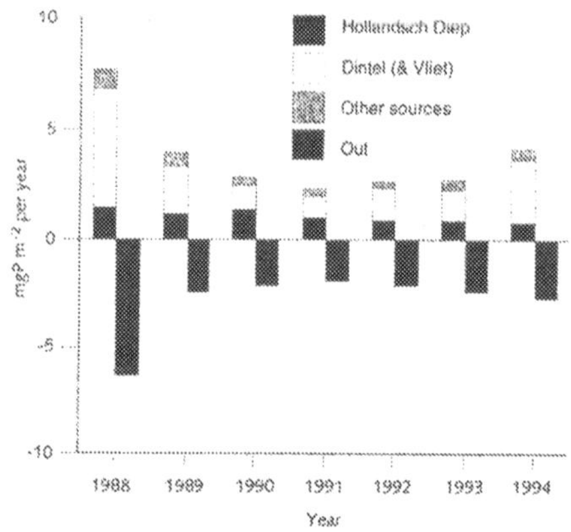

Figure 4. External P-load in Lake Volkerak-Zoom, 1988–1994.

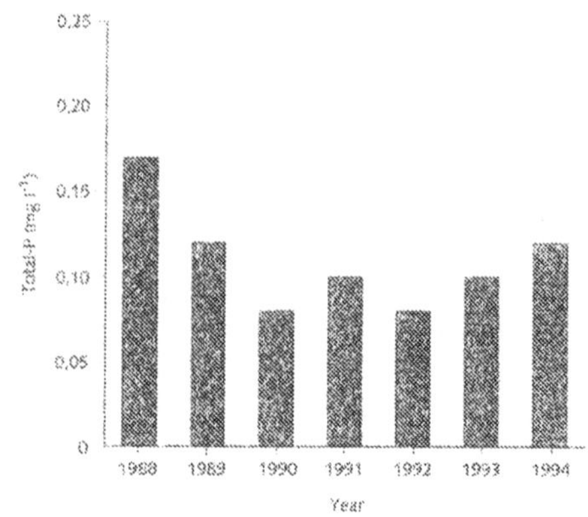

Figure 5. Total-P concentration (summer average) in Lake Volkerak-Zoom 1988–1994.

Table 1. Mean length of *Daphnia* (mm) in Lake Volkerak-Zoom in 1991–1994.

	1991	1992	1993	1994
Mean length of *Daphnia* (mm)	1.44	1.29	1.15	1.11

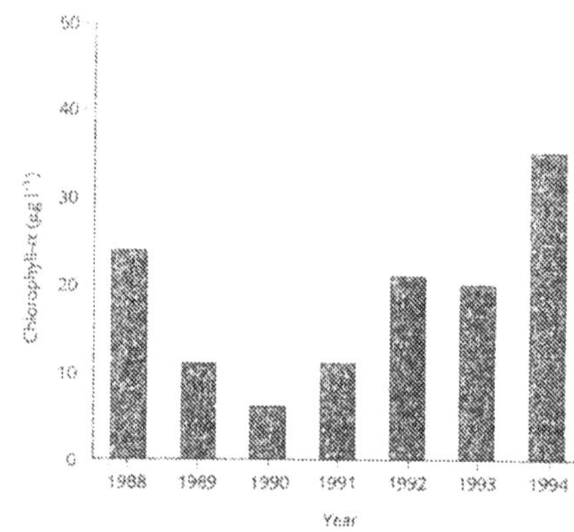

Figure 6. Chlorophyll *a* concentration (summer average) in Lake Volkerak-Zoom 1988–1994.

1990 to 1993 the total-P concentration (summer average) remained around 0.10 mg P l^{-1}. The total-P concentration is slightly higher (0.12 mg P l^{-1}) in 1994.

The total-N load increased sharply from 1990 (91 g N m^{-2} yr^{-1}) to 1992 (192 g N m^{-2} yr^{-1}), due to an increase in the total-N concentration of the Dintel (including the river Vliet), the main N-source of Lake Volkerak-Zoom. The reason for the sudden increase is unknown (it was possibly due to agricultural influences). Total-N load in 1993 was 180 g N m^{-2} yr^{-1} and the total-N concentration 5 mg N l^{-1}.

The chlorophyll *a* concentration decreased from 24 μg l^{-1} in 1988 to 6 μg l^{-1} (summer average) in 1990 and then increased to 35 μg l^{-1} in 1994 (Figure 6). Algal species composition changed in the period 1987–1994. After flushing in 1987 the abundance of freshwater species, like *Scenedesmus* sp., quickly increased. Diatoms (*Stephanodiscus* and *Skeletonema*) have dominated in spring since 1988. Cyanobacteria (like *Aphanizomenon* sp., Figure 7) became increasingly abundant after colonisation in 1988. *Aphanizomenon* sp. and *Microcystis* sp. have dominated the summer algal composition since 1991 (Bijkerk, 1996).

Daphnia pulex was the most abundant Daphnia species until 1992. It disappeared in July of that year. Since *Daphnia pulex* was the largest zooplankton species present at that time, the disappearance caused a sharp drop of the mean length from 1991 to 1993 (Table 1). *Daphnia galeata* and *Daphnia cucullata* (and hybrids) then became the dominant species.

The total fish biomass increased from 42 kg ha^{-1} in 1989 to 137 kg ha^{-1} in 1994 (Figure 8). Main predators in Lake Volkerak-Zoom were perch (*Perca fluviatilis*) and pike perch (*Stizostedion lucioperca*) (Figure 9). The biomass of roach (*Rutilus rutilus*) was high and consisted mainly of 0+ roach (18 kg ha^{-1}) in 1992. They formed c. 45% of the total recruitment of fish in 1992. This total recruitment in 1992 was high (40 kg ha^{-1}, Table 2). The recruitment by bream (*Abramis*

Table 2. Biomass of O+ fish in Lake Volkerak-Zoom, 1989–1994.

	1989	1990	1991	1992	1993	1994
O+ Fish (kg ha^{-1})	27	4	9	40	4	21

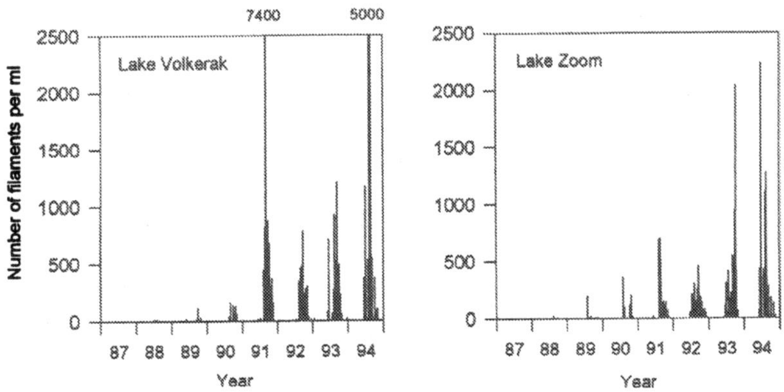

Figure 7. Development of *Aphnanizomenon flos-aquae* var. *klebahnii* at the northern part of Lake Volkerak-Zoom (Lake Volkerak) and the southern part (Lake Zoommeer), 1987–1994.

brama) over the years 1989–1994 was low (less than 2 kg ha^{-1}). The biomass of bream in 1994 consisted mainly of adult bream (60 kg ha^{-1}) and formed about 40% of the total fish biomass in that year (Figure 9). Since there is no relation between size and age structure of large (>25 cm) bream between 1992 and 1993 and between 1993 and 1994, it is thought that they have moved into the lake from elsewhere (Lammens, pers. comm.).

The area covered by submerged plants increased from 1989 till 1992 (Figure 10). From 1992 onwards the area decreased, although the maximum depth for submerged plants remained approximately 3.5 m. Until 1991 *Potamogeton pusillus* was the dominant species in Lake Volkerak-Zoom. After 1991 the coverage of *P. pusillus* decreased and other species became more abundant. *Zannichellia palustris*, *Potamogeton pectinatus*, *P. pusillus* and Characea were the most common species in 1994.

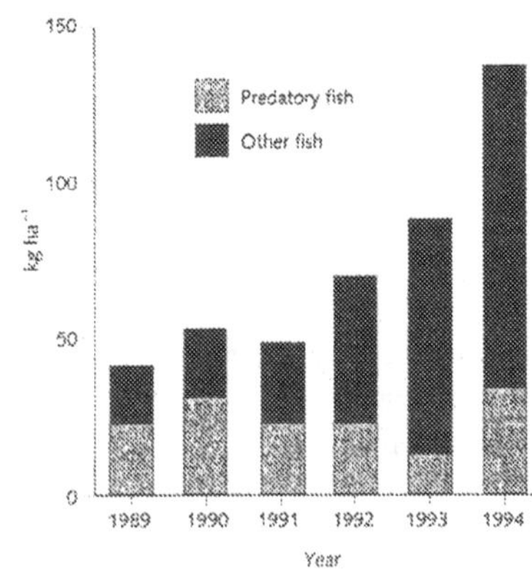

Figure 8. Total fish biomass and biomass predatory fish in Lake Volkerak-Zoom, 1989–1994.

Discussion

Shift from clear to turbid

The first years after existence the transparency increased in Lake Volkerak-Zoom, but then the transparency decreased from 3 m (summer average) in 1990 to 1.4 m in 1994. This was caused by an increase in algal biomass (Figure 3). The main cause of this increase in algal biomass is a reduced grazing by zooplankton and not an increase in nutrients or light conditions, as will be shown below.

The lowering of the P-concentration in 1988–1990 (Figure 5) was partly caused by international, national and regional emission reduction measures carried out to reduce the nutrient loading. Besides these, climatological conditions also influenced the P-concentration in the lake. In wet years, such as 1988, 1993 and 1994, the nutrient load on the lake was higher because of more surface runoff and drainage of the agricultural land. An opposite effect on the P-concentration can be expected from the retention time: in wet years the retention time is lower. However, in Lake Volkerak-Zoom this effect is limited (Van den Hark et al., 1992). CUWVO (Coördinatiecommissie Uitvoering Wet Verontreiniging Oppervlaktewateren) (CUWVO, 1987) has determined some theoretical relationships between total-P and chlorophyll *a* and between total-N and chlorophyll *a* concentrations (Figures 11 and 12) using data of

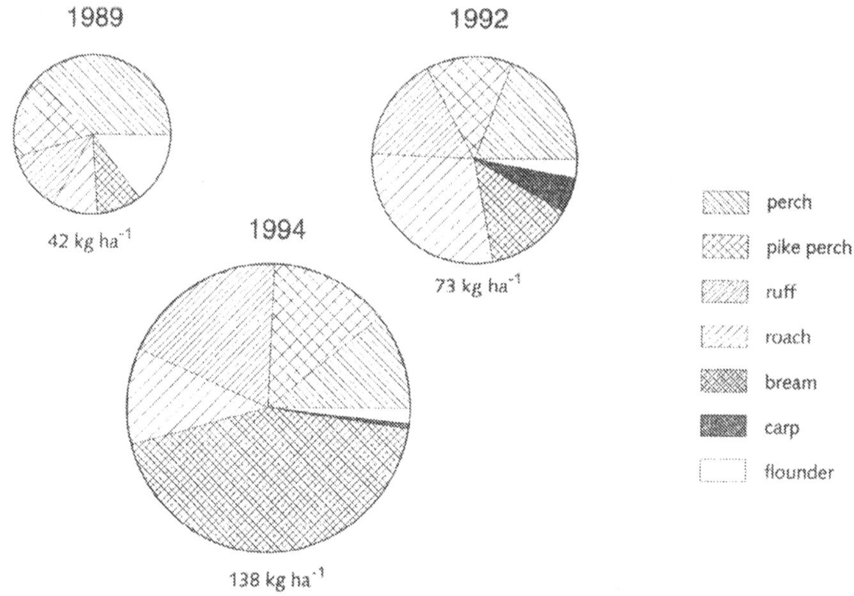

Figure 9. Fish composition and biomass in Lake Volkerak-Zoom 1989, 1992 and 1994.

25 Dutch lakes. The solid lines ('CUWVO-line') show the upper limits for the chlorophyll *a* concentration at the corresponding nutrient concentration (summer average) for eutrophic lakes in The Netherlands. The chlorophyll *a* concentration in Lake Volkerak-Zoom never reached the maximum amount corresponding to the total-P and total-N concentrations in 1988–1994. This means that nutrients are not limiting algal growth.

When light limits the algal biomass the maximum chlorophyll *a* concentration (summer average) can be estimated using the equation (CUWVO, 1987):

$$A = 1250/H - 100/S_0;$$

A = maximum chlorophyll *a* concentration (μg l^{-1});

H = mixing depth (m);

S_0 = Secchi depth (m), when chlorophyll *a* concentration = 0 μg l^{-1}.

The mixing depth of Lake Volkerak-Zoom is 10 m (Van den Hark, 1993) and $S_0 = 1/0.035$ dm (Figure 3). This means that a maximum chlorophyll *a* concentration of 90 μg l^{-1} can be expected in Lake Volkerak-Zoom. This is more than the chlorophyll *a* concentration of 35 μg l^{-1} in 1994, so light is not limiting algal growth either.

As neither nutrients nor light are limiting, zooplankton grazing is probably the major factor controlling the algal biomass. In that case zooplankton grazing must have decreased since 1990, because since then, the algal biomass has increased.

An indication for this might be the inverse relationship between Cladocerans and algal biomass (Figure 13). Reduced grazing may have been caused by food limitation or an increased mortality of the zooplankton because of fish predation, or a combination of those.

(a) Food limitation is possible since the number of cyanobacteria has been increasing in the period 1991–1993 (Figure 7). A reason for this might be that the number of resting cyanobacterial cells in winter has been increasing over the years. These algae are a less suitable food for zooplankton. Two main factors are important: the size or shape of the algae and their biochemical properties (De Bernardi & Giussani, 1990).

(b) There are indications that the presence of planktivorous fish affects the zooplankton in Lake Volkerak-Zoom. Changes in the planktivorous fish abundance will induce vertical migration by zooplankton (Stirling et al., 1990; Ringelberg, 1991). Field studies in Lake Volkerak-Zoom (Schutten & Van der Velden, 1994) showed that in August 1992, when young fish were abundant, large *Daphnia* fled to greater depths during daytime. In May, when young fish were absent, large and small *Daphnia* were distributed evenly over the

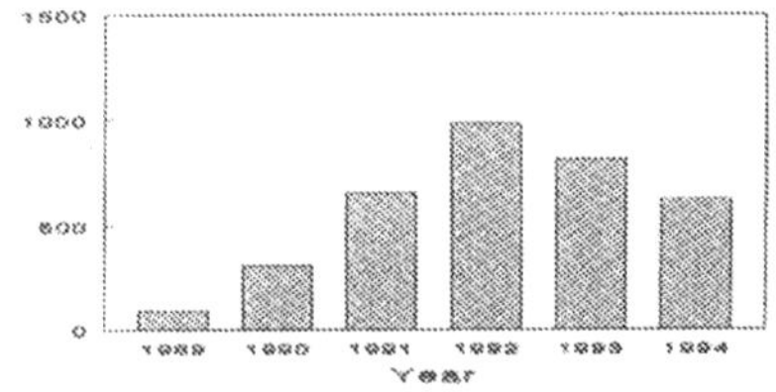

Figure 10. Submerged aquatic plants in Lake Volkerak-Zoom (ha 100% covered), 1989–1994. Only species covering >10 ha are represented.

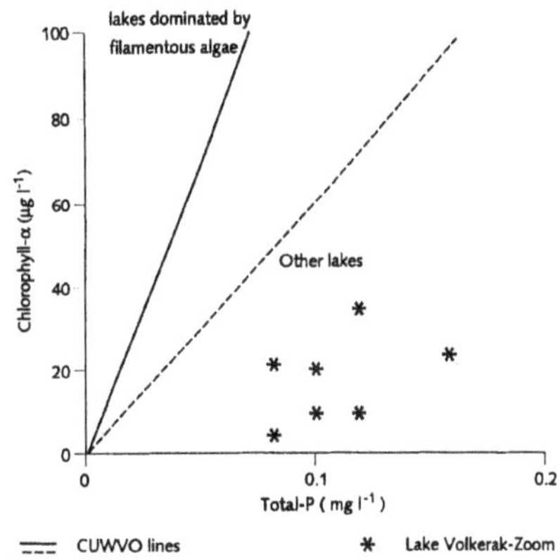

Figure 11. Total-P versus chlorophyll *a*, both summer average, in Lake Volkerak-Zoom 1988–1994. The line expresses the maximum chlorophyll *a* concentration at a certain total-P concentration (CUWVO, 1987).

water column. Vertical migration reduces the feeding time and will cost *Daphnia* more energy. This means that the condition of *Daphnia* will decrease although their chance of survival will increase because of lower fish predation.

(c) The influence of predation by fish is shown by the decreasing mean length of *Daphnia* from 1991 to 1994 (Table 1). When young fish are abundant, the mean length of the zooplankton decreases (Mills et al., 1987) since the fish feeds selectively on the larger zooplankton individuals (Jacobs, 1978; McQueen & Post, 1988; Vijverberg et al., 1990). In July 1992 the *Daphnia* mean length decreased sharply (Figure 14). In that year the recruitment of fish was high (40 kg ha^{-1}) of which 0+ roach formed almost 50%. In good recruitment years for roach the density and mean length of the Cladocerans can drop dramatically (Cryer et al., 1986) since roach prefers Cladoceran prey (Cryer et al., 1986; Winfield & Townsend, 1988). So, the high density of 0+ fish in 1992 will have affected the zooplankton population. This probably caused the disappearance of *Daphnia pulex*, thus the lowering of the mean length of *Daphnia* (Table 1), and so reducing a great deal of the grazing pressure on algae. Grazing of algae by zooplankton became less, algal biomass increased and transparency decreased. There are indications that cyanobacteria can be controlled by zooplankton at the beginning of the algal growing season when the algal biomass is low (De Bernardi & Giussani, 1990; Dawidowicz et al., 1988). Therefore, the increasing number of cyanobacteria (Figure 7) may also be an indication of decreased grazing by *Daphnia*, possibly because of fish predation.

It is possible that both food limitation and fish predation have played a role in the lowering of the grazing pressure on algae by zooplankton. The decrease in *Daphnia* size started almost simultaneously with the increase of the cyanobacteria (Figure 14) and coincided with a strong recruitment of fish. It is therefore difficult to discriminate between these processes.

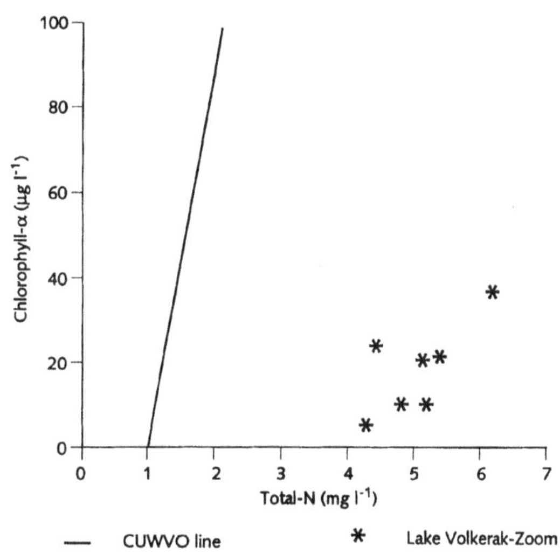

Figure 12. Total-N concentration versus chlorophyll *a*, both summer average, in Lake Volkerak-Zoom, 1988–1994. The line expresses the maximum chlorophyll *a* concentration at a certain total-N concentration (CUWVO, 1987).

Future

The expectation is that without measures the lake will become more turbid in the future.

The algal biomass may increase further, since nutrients are not the limiting factors for the algal growth in the short term, although P-reducing measures (region-

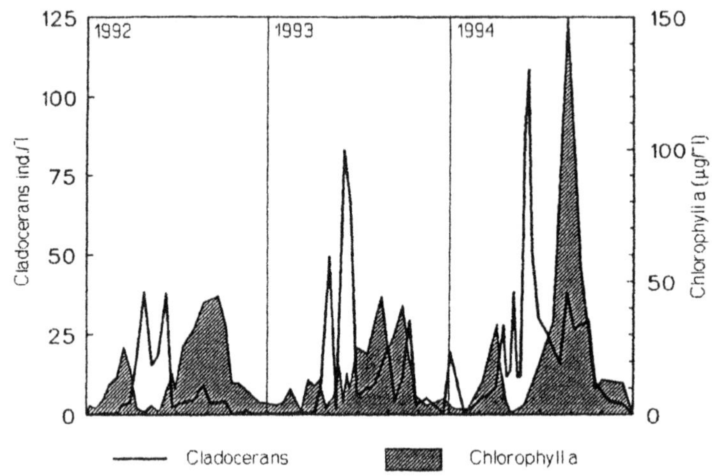

Figure 13. Density of Cladocerans versus chlorophyll *a* concentration in Lake Volkerak-Zoom, 1992–1994.

al, national and international) are being carried out in the area to reduce the P-load to the lake. Since the target situation for Lake Volkerak-Zoom is a transparency of 2 m, it is expected that it will take a long time – or it may even be impossible to reduce the P-load to such an extent that the total-P concentration will be the limiting factor (c. 0.03 mg P l^{-1}) for algal growth (Van den Hark et al., 1992).

The lowering of the water transparency may have made the water system ideal for the settlement of bream: in turbid open water bream (next to pike perch) is a dominant fish species (Lammens, 1989). This is a possible reason for large bream (>25 cm) moving into the lake. Bream is a very efficient predator of the most common food sources (zooplankton, chironomids) (Lammens, 1989). When eating benthivorous food it can contribute to the turbidity of the water by resuspension of sediments (Breukelaar et al., 1994). So, if the number of large bream continues to increase, the transparency may drop further. Adult bream is hardly edible because of its size and shape (Lammens, 1989).

Simulation studies reveal that the predation pressure by pike perch and perch with a density and length as observed in the period 1988–1994 in Lake Volkerak-Zoom, can keep the planktivorous fish biomass below 10 kg ha^{-1} if the recruitment of planktivorous fish is less than 1–2 ind m^{-2}. The recruitment by roach was higher in 1992 and control of the planktivorous fish by predatory fish therefore impossible (Lammens, pers.comm.). This might happen again causing a further increase of the predation pressure on the zooplankton and thus an increase in algal biomass. So, without management measures, the future of Lake Volkerak-Zoom appears to be the turbid state.

Management measures

The aim of the lake manager, Rijkswaterstaat Directorate Zeeland, is to create a sustainable freshwater system characterised by a high biodiversity and a stable clear water state (Iedema et al., 1992). To achieve the target situation Secchi depth should be 2 m or more (summer average). However, in 1994 the transparency was lower (Figure 2). Macrophytes can play an important role keeping the water clear: they compete for nutrients and light with algae (Ozimek et al., 1990), reduce resuspension of sediments and form a refuge for zooplankton (Shapiro, 1990; Moss, 1990). It is therefore alarming that the covered area of macrophytes has been decreasing since 1992. Since it is expected that the transparency will drop further, measures are needed to meet the goals set for clear water in Lake Volkerak-Zoom and to prevent the water system to become more turbid.

The target for total-P in Lake Volkerak-Zoom is a total-P concentration lower than 0.10 mg P l^{-1} (summer average). The chances for a stable clear water state increase below this concentration (Jeppesen et al., 1990; Hosper & Jagtman, 1990). However, to achieve this goal in Lake Volkerak-Zoom, more P-reducing measures are needed since the P-load on the lake is still too high. As the grazing by zooplankton is decreasing, it seems necessary to take additional measures

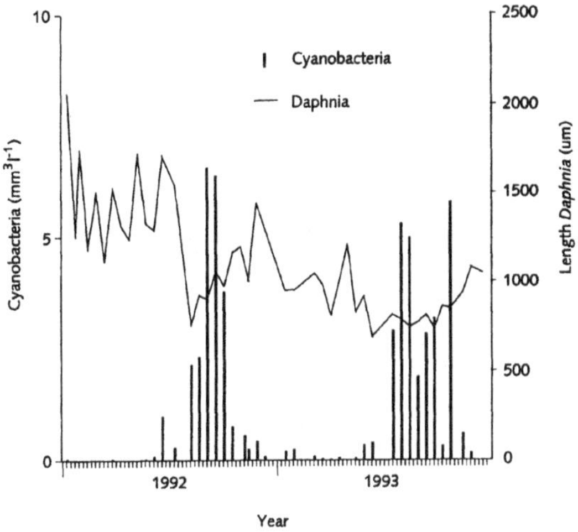

Figure 14. Length of *Daphnia* versus biovolume of cyanobacteria at station VZ2 in Lake Volkerak-Zoom, 1992–1993.

such as biomanipulation. Several options are possible: including the biomass of the fish species changing by improving the amount of piscivorous fish by stocking or creating more habitat for them or removing the other (planktivorous and/or benthivorous) fish. Creation of pike habitat (shallow areas with vegetation for spawning and hiding) will be put into practise in Lake Volkerak-Zoom at the end of 1995 by means of a more natural water level (high in winter, low in summer) (Iedema et al., 1992; Ligtvoet & de Jong, 1995). It is expected, however, that it will take time to affect the total biomass of piscivorous fish, so other additional measures are needed for the meantime. Reducing the number of planktivorous and benthivorous fish might be the best management option for Lake Volkerak-Zoom. Fishing is necessary to remove large bream from the lake, Reduction of planktivorous and benthivorous fish may lead to a chain of events: more zooplankton, fewer algae, higher transparency, more pike, etc., resulting in a more diverse ecosystem.

Acknowledgments

We would like to thank Eddy Lammens, Marie-Louise Meijer, Harry Hosper, Paul Boers and Ronald Bijkerk for their valuable comments on the manuscript. Andrew Spink corrected the English.

References

Bijkerk, R., 1996. Fytoplankton. In Breukers, C. P. M., A. Storm, E. M. van Dam & M. C. M. Van Oirschot (eds), Watersysteemrapportage Volkerak-Zoommeer 1987–1994. Rijkswaterstaat RIZA notanr. 95.003.

Boers, P. C. M. & D. T. Van der Molen, 1993. Control of eutrophication in lakes: the state of the art in Europe. European Water Pollution Control 3: 19–25.

Breukelaar, A. W., E. H. R. Lammens, J. G. P. Klein Breteler & I. Tatrai, 1994. Effects of benthivorous bream (*Abramis brama*) and carp (*Cyprinus carpio*) on sediment resuspension and concentrations of nutrients and chlorophyll *a*. Freshwat. Biol. 32: 113–121.

Coördinatiecommissie Uitvoering Wet Verontreiniging Oppervlaktewateren (CUWVO), 1987. Vergelijkend onderzoek naar de eutrofiëring in Nederlandse meren en plassen; resultaten van de derde eutrofiëringsenquête. Rijkswaterstaat DBW/RIZA, Lelystad. 55 pp.

Cryer, M., G. Peirson & C. R. Townsend, 1986. Reciprocal interactions between roach, *Rutilus rutilus*, and zooplankton in a small lake: prey dynamics and fish growth and recruitment. Limnol. Oceanogr. 31: 1022–1038.

Cullen, P. & C. Forsberg, 1988. Experiences with reducing point sources of phosphorus to lakes. Hydrobiologia 170: 321–336.

Dawidowicz, P., Z. M. Gliwicz & R. D. Gulati, 1988. Can *Daphnia* prevent a blue green algal bloom in hypertrophic lakes? Limnologica, Berlin 19: 21–26.

De Bernardi, R. & G. Giussani, 1990. Are blue-green algae a suitable food for zooplankton? Hydrobiologia 200/201: 29–41.

Frantzen, N. M. L. H. F., J. de Visser & E. H. Van Nes, 1994. Colonization and succession of macroinvertebrates in recently freshened Lake Volkerak-Zoom (The Netherlands). Hydrobiologia 275/276: 323–334.

Giussiani, G., R. De Bernardi & T. Ruffoni, 1990. Three years of experience in biomanipulating a small eutrophic lake: Lago di Candia (Northern Italy). Hydrobiologia 200/201: 357–366.

Hosper, S. H. & E. Jagtman, 1994. Biomanipulation additional to nutrient control for restoration of shallow lakes in The Netherlands. Hydrobiologia 200/201: 523–534.

Iedema, C. W., J. Schutten & J. A. Van der Velden, 1992. Tough measures for a weak system. In Hooghart, J. C. & C. W. S. Posthumus (eds), How an estuary changed into a fresh water lake. Proceedings and Information, CHO/TNO, 46: 85–98.

Jacobs, J., 1978. Influence of prey size, light intensity, and alternative prey on the selectivity of plankton feeding fish. Verh. int. Ver. Limnol.: 2461–2466.

Jeppesen, E., J. P. Jensen, P. Kristensen, M. Søndergaard, E. Mortensen, O. Sortkjaer & K. Olrik, 1990. Fish manipulation as a lake restauration tool in shallow, eutrophic temperate lakes 2: threshold levels, long-term stability and conclusions. Hydrobiologia 200/201: 219–227.

Lammens, E. H. R. R., 1989. Causes and consequences of the success of bream in Dutch eutrophic lakes. Hydrobiol. Bull. 23: 11–18.

Ligtvoet, W. & S. A. De Jong, 1995. Ecosystem development in lake Volkerak-Zoom: concept and strategy. Wat. Sci. Tech. 31: 239–244.

McQueen, D. J. & J. R. Post, 1988. Limnocorral study of cascading trophic interactions. Verh. int. Ver. Limnol. 23: 739–747.

Meijer, M.-L., E. Jeppesen, E. Van Donk, B. Moss, M. Scheffer, E. Lammens, E. Van Nes, J. A. Van Berkum, G. L. De Jong, B. A. Faafeng & J. P. Jensen, 1994. Long-term responses to fish-stock reduction in small shallow lakes: interpretation of five-year results of four biomanipulation cases in The Netherlands and Denmark. Hydrobiologia 275/276: 457–466.

Mills, E. D., D. M. Green & A. Schavione, 1987. Use of zooplankton size to assess the community structure of fish populations in freshwater lakes. North. am. J. Fish. Mgmt 7: 369–378.

Moss, B., 1990. Engineering and biological approaches to the restoration from eutrophication of shallow lakes in which aquatic plant communities are important components. Hydrobiologia 200/201: 367–377.

Ozimek, T., R. D. Gulati & E. Van Donk, 1990. Can macrophytes be useful in biomanipulation of lakes? The Lake Zwemlust example. Hydrobiologia 200/201: 399–407.

Ringelberg, J., 1991. The relation between ultimate and proximate aspects of diel vertical migration in *Daphnia hyalina*. Verh. int. Ver. Limnol. 24: 2804–2807.

Schutten, J. & J. A. Van der Velden, 1994. Verticale en horizontale verspreiding van zooplankton in het Volkerak-Zoommeer. Rijkswaterstaat RIZA werkdokumentnr. 92.126X.

Schutten, J., J. A. Van der Velden & H. Smit, 1994. Submerged macrophytes in the recently freshened lake Volkerak-Zoom (The Netherlands), 1987–1991. Hydrobiologia 275/276: 207–218.

Shapiro, J., 1990. Biomanipulation: the next phase-making it stable. Hydrobiologia 200/201: 13–27.

Shapiro, J. & D. I. Wright, 1984. Lake restoration by biomanipulation: Round lake Minnesota, the first two years. Freshwat. Biol. 14: 371–383.

Stirling, D. G., D. J. McQueen & M. R. S. Johannes, 1990. Vertical migration in *Daphnia galeata mendotae* (Brooks): demographic responses to changes in planktivore abundance. Can. J. Fish. aquat. Sci.47: 395–400.

Van den Hark, M. H. C., 1993. Eutrofiëringsonderzoek Volkerak-Zoommeer; ontwikkelingen van 1988–1990 en prognoses. Rijkswaterstaat RIZA notanr. 92.027.

Van den Hark, M., J. E. W. De Hoog & C. W. Iedema, 1992. Lake Volkerak/Zoom, eutrophic and clear? In Hooghart, J. C. & C. W. S. Posthumus (eds), How an Estuary Changed into a Fresh Water Lake. Proceedings and Information, CHO/TNO, 46: 43–56.

Van Donk, E., R. D. Gulati & M. P. Grimm, 1989. Food web manipulation in lake Zwemlust: positive and negative effects during the first two years. Hydrobiol. Bull. 23: 19–34

Van Nes, E. H. & H. Smit, 1993. Multivariate analysis of macrozoobenthos in lake Volkerak-Zoommeer (The Netherlands): changes in an estuary before and after enclosure. Arch. Hydrobiol. 127: 185–203.

Vijverberg, J., M. Boersma, W. L. T. Van Densen, W. Hoogenboezem, E. H. R. R. Lammens & W. M. Mooij, 1990. Seasonal variation in the interactions between piscivorous fish, planktivorous fish and zooplankton in a shallow eutrophic lake. Hydrobiologia 207: 279–286.

Winfield, I. J. & C. R. Townsend, 1988. Factors affecting prey selection by young bream *Abramis brama* and roach *Rutilus rutilus*: insights provided by parallel studies in laboratory and field. Envir. Biol. Fishes 21: 279–292.

Hydrobiologia **342/343**: 377–381, 1997.
L. Kufel, A. Prejs & J. I. Rybak (eds), Shallow Lakes '95.

Food-web manipulation in a small, eutrophic Lake Wirbel, Poland: the effect of replacement of key predators on epiphytic fauna

Andrzej Prejs, Paweł Koperski & Krystyna Prejs†
Department of Hydrobiology, University of Warsaw, Banacha 2, 02-097 Warsaw, Poland

Key words: key-predator replacement, fish, invertebrate carnivores, epiphytic fauna, Chironomidae

Abstract

The effect of fish removal on the invertebrate fauna associated with *Stratiotes aloides* was studied in a shallow, eutrophic lake. The biomass of invertebrate predators was approximately 2.5 times higher in the invertebrate-dominated year (1992) than in the fish-dominated year (1991), while the density of non-predatory invertebrates in 1991 was ca half that in the invertebrate-dominated year. The decrease was due to a sharp fall in the density of epiphytic chironomids, with the density of plant-mining chironomids being far less affected. Marked declines in the density of non-predatory invertebrates in the invertebrate-dominated year were most probably caused by invertebrate predators. Once freed from suppression induced by fish, invertebrate predators were able to control the density of epiphytic prey more effectively than fish.

Introduction

One of the characteristic features of the biotic structure in the majority of freshwaters is that the only numerous vertebrate carnivores i.e. fish are gape-limited and, as the majority are secondary predators, required to share their invertebrate prey with size-limited invertebrate predators. In consequence, fish, once absent or removed, can be effectively replaced as key predator by far smaller invertebrate carnivores, whose density, freed from control by fish, may increase very substantially (Macan, 1965a, 1965b; Henriksson & Oscarsson, 1978; Stenson, 1978; Gliwicz & Pijanowska, 1989). Such replacement can either have natural causes, e.g. when fish are killed by prolonged drought or oxygen depletion or can be produced by intentional manipulations in food webs that are often aimed at reducing populations of invertebrate-eating fish.

In open water, the effect of the elimination of planktivorous fish has been recognized since the sixties (Hrbaček, 1962; Brooks & Dodson, 1965): their absence drives zooplankton towards domination by larger species. This shows indirectly that predatory invertebrates differ from fish in their ability to control their prey and, that as summarized by Zaret (1980), predation by them is directed mainly towards small and medium-sized zooplankton. Hence, according to the dominance of the particular predator type, either fish- or invertebrate-shaped zooplankton assemblages can exist in open water.

Studies demonstrating the effect of fish replacement by an invertebrate predator on littoral prey (bottom or epiphytic fauna) are rare (but see Diehl, 1992), and there is no evidence to support expansion of Zaret's conclusion to include epiphytic or benthic assemblages. Multispecies assemblages of littoral predatory invertebrates, being more diverse in size, feeding modes and preferences than open water predators, may cover practically the whole spectrum of prey exploited by fish. Thus, the predation effects of both fish and invertebrates could be similar.

The biomanipulation experiment carried out in Lake Wirbel, where almost all fish were removed (Prejs et al., 1994), gave an opportunity to test this assumption by exploring the response of invertebrate carnivores and herbivores to the sudden disappearance of fish as well as the response of invertebrate herbivores to the possible expansion of predatory invertebrates. Chosen

† Deceased

as an appropriate study habitat were extensive stands of *Stratiotes aloides* L. that harbour a rich, diverse invertebrate fauna and offer a refuge for numerous small cyprinids (Prejs et al., 1994).

Study site and methods

The study was conducted over two years (1991–1992) in a small and shallow eutrophic lake Lake Wirbel (northern Poland) (11 ha, mean depth 1.2 m, maximum depth 4 m). The successive steps in the biomanipulation process were: intensive stocking with pike (Esox lucius L.) (1988–1991) and rotenone treatment in October 1991. These were aimed at decreasing the planktivorous fish stock and, eventually, at the improvement of water quality (for more details see Prejs et al., 1994; Prejs et al., 1997).

The macrovegetation, which covered ca 19% of the lake surface, was composed mainly of emergent plants (*Carex* sp., *Phragmites australis* Trin. and *Typha latifolia* L. – ca 7% of the lake surface), floating-leaved plants (*Nymphaea alba* L., *Nuphar luteum* (L.) – ca 10.4% of the lake surface) and *S. aloides* (ca 1.5% of the lake surface) (Prejs et al., 1994). The largest (ca 1500 m^2, depth 0.3–0.5 m) of several areas overgrown with *S. aloides* was chosen as a study site. In each sampling year, whole plants with their epiphytic fauna were collected from early May to late October, every two or three weeks. A sample consisted of 6–10 plants collected by hand, randomly from the edge to the center of the patch, placed separately in plastic bags and analyzed individually. Within the next 1–3 hours periphyton was carefully removed from the plant surface. Epiphytic animals were separated from sediments and periphyton with a sieve of mesh-size 0.35 mm, preserved in 70% methyl alcohol, and then counted. The body lengths of individuals of the dominant taxa were measured and their biomass estimated using length/mass relationships (Prejs & Colomine, 1981). Plant-mining larvae were carefully removed from the insides of leaves using forceps. The density of free-swimming, air-breathing carnivores - Corixidae, *Notonecta glauca* L. (Heteroptera) and *Argyroneta aquatica* Cl. (Araneae), was estimated several times a year by counting the animals from a boat, at a few sites within the stands of 10 plants within a standard unit of time. Water depth at this study site varied between 0.3 and 0.5 m and the Secchi depth during the sampling period never dropped below 0.6 m (Prejs et al., this volume). Fish biomass was taken from Prejs et al. (1994) and Prejs et al. (1997).

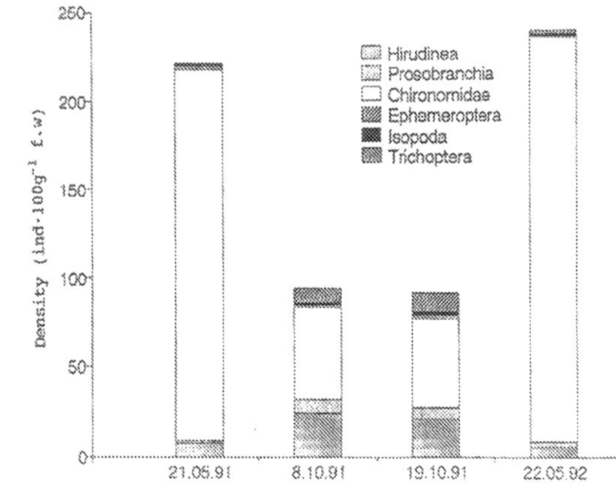

Figure 1. Density of dominant invertebrates associated with *S. aloides* two days before (8.10.1991) and seven days after (19.10.1991) rotenone treatment, and in May of both years.

Results

Rotenone did not reduce the density of the main invertebrate taxa associated with *S. aloides*. Densities of dominant groups did not differ significantly in samples collected just before and a few days after rotenone treatment (Figure 1). The densities and taxonomic composition of epiphytic invertebrates (sampled in May – the starting point for lake vegetation) were also almost identical in the years before and after rotenone treatment (Figure 1).

In summer 1991, prior to fish removal, the total biomass of predators was estimated at 15 g f.w m^{-2} of which fish accounted for 90%. In summer 1992, after rotenone treatment, the total biomass of predators was 10 g f.m. m^{-2}, of which approximately 90% was made up by invertebrates. The increase in the biomass of invertebrate predators in 1992 resulted from sharp rises in the biomass of Odonata (Anisoptera and Zygoptera) larvae, Heteroptera and *Argyroneta aquatica* (Figure 2), which were particularly distinct in late spring (June), early summer (July) (Heteroptera and *A. aquatica*) and summer (July–September) (Odonata). In the case of Odonata biomass and density, the significance of differences between 1991 and 1992 was confirmed by Student's test (biomass: $P=0.01$, density: $P=0.003$). There were no significant differences in the biomass and density of other epi-

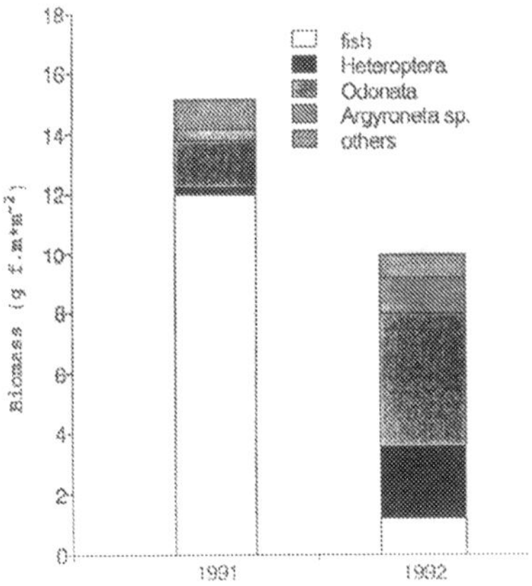

Figure 2. Biomass of predators associated with *S. aloides* in 1991 and 1992 (data for June–September).

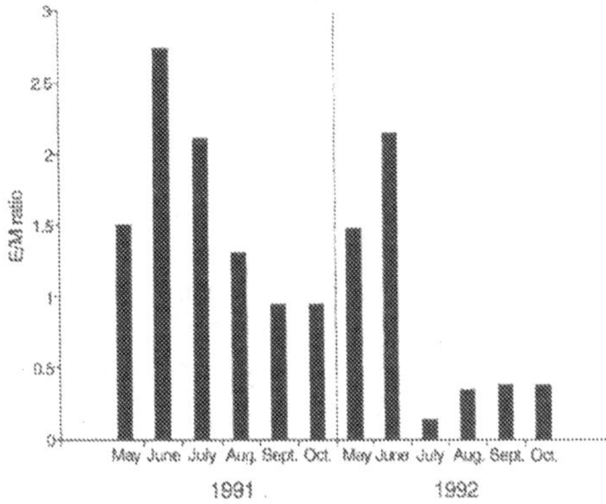

Figure 4. Ratio: density of epiphytic (E) to mining (M) chironomid larvae in 1991 and 1992.

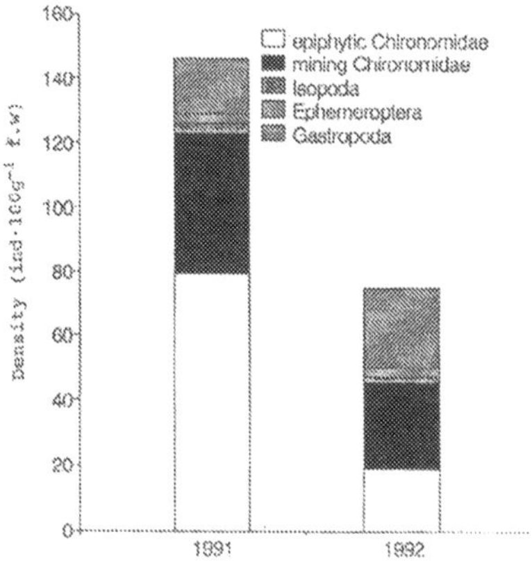

Figure 3. Mean density of non-predatory invertebrates associated with *S. aloides* in 1991 and 1992 (data for June–September).

phytic predators (Trichoptera, mainly *Cyrnus flavidus* McLachlan; Hirudinea, mainly *Erpobdella octoculata* L. and *Helobdella stagnalis* (L.); Turbellaria, mainly *Polycelis* sp.) between 1991 and 1992.

The non-predatory invertebrate assemblage was dominated numerically by chironomid larvae, with 25 epiphytic species (most numerously genera: *Cricotopus, Polypedilum, Endochironomus* and *Glyptotendipes*) and 9 plant-mining species (most commonly: *G. gripekoveni* Kieff. and *E. tendens* Fabr.), as well as Gastropoda (particularly in 1992); Ephemeroptera and the isopod *Asellus aquaticus* Racov. occurred in smaller numbers (Figure 3).

The total density of non-predatory invertebrates in summer 1992 was about half that in 1991 a fish-dominated year. The decrease resulted mainly from a drastic decline in the density of epiphytic chironomids, whose mean abundance in summer 1991 was about 4.4 times greater than in summer 1992 (Student's test, $P = 0.0001$). The differences in the density of plant-mining chironomids and other epiphytic fauna (Figure 3) were not significant. In 1991, the ratio of the density of epiphytic to mining larvae fluctuated between 1.5–2.8 (May–June) and 0.9 (September–October), while in 1992, its value, after reaching 1.4 and 2.1 in May and June, dropped sharply to 0.1 in July and remained below 0.5 throughout the rest of the season (Figure 4). The mean ratio estimated for summer months of 1991 was approximately 4 times greater than that for the 1992.

Discussion

It has been shown repeatedly, that large predatory invertebrates (e.g. Heteroptera, Odonata, Coleoptera, *Sialis* sp.) are highly vulnerable to fish predation (Macan, 1965b; Henriksson & Oscarsson, 1978; Nilsson, 1981; Morin, 1984; Cook & Streams, 1984; Martin et al., 1991; Diehl, 1992). Accordingly, we believe that a reduction in fish biomass triggered a strong response from invertebrate predators with their biomass in 1992, freed from control by fish, increasing very substantially. In turn, in 1992, when the predator guild was distinctly dominated by invertebrates, the density of small invertebrate prey, mainly of epiphytic Chironomidae, remained far lower than in 1991, a fish-dominated year.

There is evidence that large, predatory invertebrates have the capacity to reduce density of their prey (e.g. Lawton, 1971; Benke, 1978; Peckarsky, 1984; Thorp & Cothran, 1984; Johnson et al., 1987; Cloarec, 1990; Diehl, 1992; Prejs & Prejs, 1992). We lack direct evidence that the difference in the density of epiphytic Chironomidae between the summers of 1991 and 1992 was caused by the replacement of fish by invertebrate predators. The difference could, for example, be a result of 'natural' year-to-year fluctuations, which may occur without any significant changes to factors such as predation. In such a case, however, the difference in population density between two consecutive summers is often preceded by a difference of more or less the same order of magnitude between spring densities (Głowacka et al., 1976; Soszka, 1975). In turn, Pieczyński (1973) found that in spring (May–mid June), the differences between the biomass of Chironomidae associated with *S. aloides* in two parts of Lake Warniak, differing distinctly in their fish stocks did not exceed 30%. Some three weeks later, in July – the period of most intensive fish feeding (Prejs, 1973), the biomass of epiphytic Chironomidae in the stocked part of the lake was almost three times lower than in the part with natural fish density.

In Lake Wirbel (Figure 5) too, the difference between the densities of epiphytic Chironomidae was small in spring (May) 1991 and 1992 but substantially greater in July. If linked to predation, a decline in the density of epiphytic Chironomidae in 1992 suggests that invertebrate predators are more effective than fish in regulating the density of epiphytic chironomids. This idea may find support in a sequence of changes in the ratio of the densities of epiphytic to mining chironomids which was noted throughout 1992 – the invertebrate dominated year (Figure 4). The sharp drop in July which resulted from a decrease in the abundance of epiphytic larvae, could be linked to growing predation pressure from the dominant heteropteran – *N. glauca* and from Odonata. In June–July *N. glauca* nymphs reach the last instar stage. Reaching this stage, nymphs include more and more chironomid prey in their diet (Giller, 1986; Prejs, unpubl. data). July in Lake Wirbel was also the period of the most intensive feeding by Odonata on chironomid larvae (Koperski, unpublished).

It is well-known that macrophytes can provide invertebrates with a refuge from large predators (e.g. Werner et al., 1983; Gilinsky, 1984), and that increasing macrophyte density reduces the foraging efficiency of fish (Crowder & Cooper, 1982; Savino & Stein,

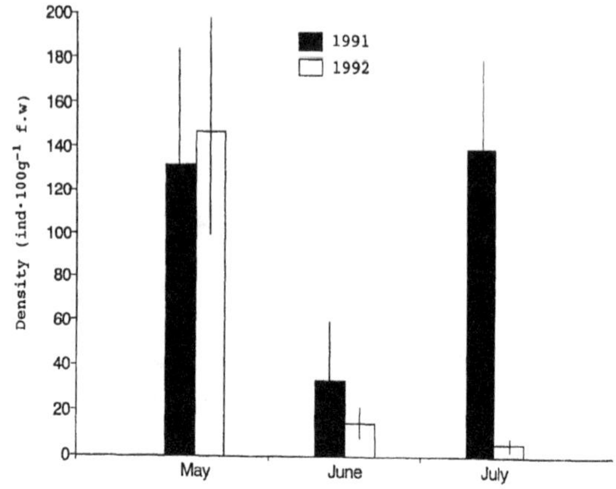

Figure 5. Density (± SD) of epiphytic Chironomidae in spring and early summer 1991 and 1992.

1982; Leber, 1985). As a plant with tough, spiny leaves, partially inaccessible to fish and with a tendency to form very dense stands, *S. aloides* offers shelter against foraging fish, where small herbivores and detritivores are more secure than large, mobile carnivores. In contrast, the foraging efficiency of predatory invertebrates (especially macrophyte-dwelling forms) seems to be less related to the structural complexity of the habitat (Thompson, 1987; Convey, 1988). Furthermore, the fish associated with *Stratiotes* stands were mostly juveniles of only 2–3 cyprinid species (Prejs et al., 1994), which did not differ much in swimming performance, searching abilities, foraging modes and gape limitations. In contrast, the invertebrate predator guilds associated with dense macrophyte stands were far more diverse than the fish in their taxonomy, functional design, foraging modes etc. These two reasons, i.e. the greater skill of large invertebrates in exploiting such habitat, and the broader spectrum of potential prey, may contribute to the higher efficiency of these species in controlling prey density.

Acknowledgments

We thank Joanna Pijanowska for helpful comments. Eliza Penconek is also acknowledged for field and laboratory assistance. This work was supported financially by KBN grant (60540) to Andrzej Prejs.

References

Benke A. C., 1978. Interactions among coexisting predators – a field experiment with dragonfly larvae J. anim. Ecol. 47: 335–350.

Brooks, J. L. & S. I. Dodson, 1965. Predation, body size and the composition of plankton. Science 150: 28–35.

Cloarec, A., 1990. Factors influencing the choice of predatory tactics in a water bug, *Diplonychus indicus* Venk & Rao (Heteroptera, Belostomatidae). Anim. Behav. 40: 262–271.

Convey, P., 1988. Competition for perches between larval damselflies: the influence of perch use on feeding efficiency, growth rate and predator avoidance. Freshwat. Biol. 19: 15–28.

Cook, W. L. & F. A. Streams, 1984. Fish predation on *Notonecta* (Hemiptera): relationship between prey risk and habitat utilization. Oecologia 64: 177–183.

Crowder, L. B. & W. E. Cooper, 1982. Habitat structural complexity and the interaction between bluegills and their prey. Ecology 63: 1802–1813.

Diehl, S., 1992. Fish predation and benthic community structure: the role of omnivory and habitat complexity. Ecology 73: 646–1661.

Gilinsky, E., 1984. The role of fish predation and spatial heterogeneity in determining benthic community structure. Ecology 65: 455–468.

Giller, P. S., 1986. The natural diet of the Notonectidae: field trials using electrophoresis. Ecol. Ent. 11: 163–172.

Gliwicz, Z. M. & J. Pijanowska, 1989. The role of predation in zooplankton succession. In Sommer, U. (ed.), Plankton Ecology: Succession in Plankton Communities. Springer, Heidelberg: 253–269.

Głowacka, I., G. J. Soszka & H. Soszka, 1976. Invertebrates associated with macrophytes. In Pieczyńska, E. (ed.), Selected Problems of Lake Littoral Ecology. Warsaw University Press, Warsaw: 97–122.

Henriksson, L. & H. G. Oscarson, 1978. Fish predation limiting abundance and distribution of *Glaenocorisa p. propinqua*. Oikos 31: 102–105.

Hrbaček J., 1962. Species composition and the amount of zooplankton in relation to the fish stock. Rozpr. csl. Akad Ved. 72: 1–116.

Johnson, D. M., T. H. Pierce, T. H. Martin, C. N. Watson, R. E. Bohanan & P. H. Crowley, 1987. Prey depletion by odonate larvae: combining evidence from multiple field experiments. Ecology 68: 1459–1465.

Lawton, J. H., 1971. Maximum and actual field feeding rates in larvae of the damselfly *Pyrrhosoma nymphula* (Sulzer) (Odonata, Zygoptera). Hydrobiologia 36: 33–52.

Leber, K. M., 1985. The influence of predatory decapods, refuge and microhabitat selection on seagrass communities. Ecology 66: 1951–1964.

Macan, T. T., 1965a. The fauna in the vegetation of a moorland fishpond. Arch. Hydrobiol. 63: 273–310.

Macan, T. T., 1965b. Predation as a factor in the ecology of water bugs. J. anim. Ecol. 34: 691–698.

Martin, T. H., D. M. Johnson & R. D. Moore, 1991. Fish-mediated alternative life history strategies in the dragonfly *Epitheca cynosura*. J. N. am. Benthol. Soc. 10: 271–279.

Morin, P. J., 1984. Odonate guild composition: experiments with colonization history and fish predation. Ecology 65: 1866–1873.

Nilsson, B. I., 1981. Susceptibility of some Odonata larvae to fish predation. Verh. int. Ver. Limnol. 21: 1612–1615.

Peckarsky, B., 1984. Predator-prey interactions among aquatic insects. In Resh, V. H. & D. M. Rosenberg (eds), The Ecology of Aquatic Insects. Praeger, New York, Sydney: 196–251.

Pieczyński E., 1973. Experimentally increased fish stock in the pond type Lake Warniak. XII. Numbers and biomass of the fauna associated with macrophytes. Ekol. pol. 21: 595–610.

Prejs, A., 1973. Experimentally increased fish stock in the pond type Lake Warniak. IV. Feeding of introduced and autochtonous, non-predatory fish. Ekol. pol. 21: 465–505.

Prejs, A. & G. Colomine, 1981. Metodos para el Estudio de los Alimentos y las Relaciones Troficas de los Peces. Instituto de Zoologia Tropical. Caracas, 129 pp.

Prejs, A., A. Martyniak, S. Boroń, P. Hliwa & P. Koperski, 1994. Food web manipulation in a small, eutrophic Lake Wirbel, Poland: effect of stocking with juvenile pike on planktivorous fish. Hydrobiologia 275/276: 65–70.

Prejs, A., J. Pijanowska, P. Koperski, A. Martyniak, S. Boroń & P. Hliwa, 1997. Food web manipulation in a small, eutrophic Lake Wirbel, Poland: long-term changes in fish biomass and basic measures of water quality. A case study. Hydrobiologia 342/343: 383–386.

Prejs, K. & A. Prejs, 1992. Importance of predation in regulating density of meio- and macrofauna in seasonal tropical waters. Hydrobiologia 242: 77–86.

Savino, J. F. & R. A. Stein, 1982. Predator-prey interactions between largemouth bass and bluegills as influenced by stimulated, submersed vegetation. Trans. am. Fish. Soc. 111: 255–266.

Soszka, G. J., 1975. Ecological relations between invertebrates and submerged macrophytes in the lake littoral. Ekol. pol. 23: 393–415.

Stenson, J. A E., 1978. Relations between vertebrate and invertebrate zooplankton predators in some Arctic lakes. Astarte 11: 21–26.

Thompson, D. J., 1987. Regulation of damselfly populations: the effects of weed density on larval mortality due to predation. Freshwater Biol. 17: 367–371.

Thorp, J. H. & M. L. Cothran, 1984. Regulation of freshwater community structure at multiple intensities of dragonfly predation. Ecology 65: 1546–1555.

Werner, E. E., G. G. Mittelbach, D. J. Hall & J. F. Gilliam, 1983. Experimental tests of optimal habitat use in fish: The role of relative habitat profitability. Ecology 64: 1525–1539.

Zaret, T. M., 1980. Predation and freshwater communities. Yale University Press, New Haven, 187 pp.

Hydrobiologia **342/343:** 383–386, 1997.
L. Kufel, A. Prejs & J. I. Rybak (eds), Shallow Lakes '95.

Food-web manipulation in a small, eutrophic Lake Wirbel, Poland: long-term changes in fish biomass and basic measures of water quality. A case study

Andrzej Prejs[1], Joanna Pijanowska[1], Paweł Koperski[1], Andrzej Martyniak[2], Sławomir Boroń[2] & Piotr Hliwa[2]
[1]*Department of Hydrobiology, University of Warsaw, Banacha 2, 02-097 Warsaw, Poland*
[2]*Department of Fisheries, Academy of Agriculture and Technology, 10-957 Olsztyn-Kortowo, Poland*

Key words: biomanipulation, whole-lake study, top predators, trophic cascade, water quality

Abstract

A whole-lake experiment was carried out for seven years (1988–1994) in a shallow, eutrophic lake. In the first phase (1989–1991), repeated introductions of young-of-the-year pike (*Esox lucius*) were used to control the density of juvenile stages of the dominant planktivores (roach *Rutilus rutilus*, white bream *Blicca bjoerkna* and *Leucaspius delineatus*). The successive introductions of juvenile pike were accompanied by selective removal of large pike, roach, bream *Abramis brama* and white bream. No visible improvement in water quality was registered in the first three years. Assessments made after the lake was treated with rotenone, revealed the high efficiency of juvenile pike in controlling prey of vulnerable size. Though the first three age-classes were nearly exterminated, older fish (3+ to 6+) remained abundant enough to keep their planktonic prey in check. Indeed, the non-altered density of dominant cladocerans indicated that they were still heavily preyed upon. Algal biomass remained high.

In October 1991, rotenone was applied to remove all fish. Immigrating fish were gradually recolonizing the lake from spring 1992 onwards. However, the process was kept under control by continued stocking with juvenile pike and intensive fishing. The years following the rotenone treatment witnessed a considerable improvement in water quality, with water transparency 30% higher, a significant 2.2-fold decrease in seston dry weight and 2.8-fold decrease in algal biomass, averaged for the summer months. These events can largely be attributed to substantial changes in the herbivorous zooplankton, above all increased density of Lake Wirbel's largest cladoceran, *Daphnia cucullata*.

Introduction

To improve water quality in a shallow, eutrophic lake a long-term biomanipulation experiment was carried out. Fish stock was intentionally changed to study the response of the entire food web. Basic limnological characteristics were monitored in the course of the experiment, and special attention was paid to the effect of the introduction of juvenile pike on the density and age structure of the resident planktivores. The results concerning fish assemblages are published elsewhere (Prejs et al., 1994), while this paper provides a general overview of events following the applied measures.

Study site

The study was carried out in Lake Wirbel in the Mazurian Lakeland, north-east Poland. This is an 11 ha eutrophic and polymictic lake, with mean depth 1.8 m and max. depth. 4.4 m. It is surrounded by forest and arable land. During the whole study period, its fish assemblage was dominated by cyprinids: roach *Rutilus rutilus*, white bream *Blicca bjoerkna* and *Leucaspius delineatus*. The predatory fish present were pike *Esox lucius* and perch *Perca fluviatilis*. For more details on the study site see Prejs et al. (1994).

The experiment

Biomanipulation measures

The three biomanipulation measures applied were: repeated introductions of 0+ pike, repeated extensive fishing and rotenone treatment. The sequence of such treatments is summarized in Figure 1a. In the first phase of the experiment (1988–1991), introductions of young-of-the-year pike were used to control the density of juvenile stages of the dominant planktivores (roach and white bream) as well as the whole size range of small sized *Leucaspius delineatus*. In parallel, extensive catching was carried out to control older fish. The whole sequence of biomanipulation measures preceding rotenone treatment is described elsewhere (Prejs et al., 1994). No improvement in water quality was observed throughout this phase of the experiment (Figure 1b). In October 1991, rotenone (0.6 mg l^{-1}) was applied to remove all fish (Prejs et al., 1994). From spring 1992 onwards, immigrating fish gradually recolonized the lake, although the process was kept under control by continued stocking with juvenile pike and by intensive fishing (Figure 1a).

Basic measures of water quality

Throughout the study, water transparency, as measured by Secchi depth, was regularly monitored from late April to early November, on average once a week. Starting from a reference year before the rotenone treatment (1991) and for two seasons following it, particularly intensive studies were carried out, including measurements of seston dry weight (in the 'edible' fraction of particles >30μm), and chlorophyll *a* concentration in two size fractions of algae (<30μm, as a measure of the edible fraction, and >30μm, as a measure of interfering algae).

The density and species composition of herbivorous zooplankton was monitored throughout the study.

Results

During the first phase of the study (May 1988–October 1991) no persistent increase in water transparency was seen (Figure 1a), with Secchi readings between 70 and 90 cm, and seasonal changes following the typical patterns of changes in a temperate zone eutrophic lake. In 1992, the year following rotenone treatment, water transparency increased almost 40%, as compared with the reference year 1991. The effect was persistent with readings between 90 and 100 cm up to October 1994. In the four years preceding rotenone treatment, nearly all Secchi readings had been below 90 cm, whereas after the treatment all but one reading (September 1994) was above 90 cm. The mean values for Secchi depth in the summer months (June–September) were on average 80.9 ± 9.9 cm for the four years preceding rotenone treatment and 104.2 ± 28.2 cm for the three years following treatment (a difference statistically significant at $P = 0.0013$; ANOVA). The total phosphorus concentration decreased from 144-166μg l^{-1} in the two summer seasons preceding treatment to 71-93μg l^{-1} in the three following years.

The year-to-year changes in seston concentration followed those in water transparency (Figure 1c). In 1992, the mean seston concentration dropped on average twice as compared to the year preceding the treatment, and this effect persisted until October 1993.

Expressed as chlorophyll *a* content in two fractions of algae (<30 and >30μm), changes in algal biomass followed those in seston concentration (Figure 1d). In 1992, algal biomass decreased almost twice, as compared to the year preceding the treatment. This effect was observed until October 1993. Of total algal biomass an average of 75 ± 8% was 'edible' in 1991, and 85 ± 6% in 1992 and 1993. This difference, however, was not statistically significant at $P<0.05$ (ANOVA). The three-year pattern of changes in total algal biomass followed changes in the 'edible' algal fraction (Figure 1d).

Phytoplankton, dominated prior to the treatment by filamentous blue greens (on average 51 ± 27% of total algal biomass in June-September vs the 9 ± 6% contribution of green algae), shifted in the summer after the treatment towards domination by green algae (on average 53 ± 27% of total algal biomass vs the 9 ± 10% contribution of blue-greens).

Whereas in the years following rotenone treatment (1992-1994), the total number of herbivorous Cladocera (*Daphnia cucullata, Bosmina longirostris, Ceriodaphnia quadrangula, Chydorus sphaericus, Diaphanosoma brachyurum*) did not change substantially, the density of the largest cladoceran *Daphnia cucullata* did 5-fold increase, on average, from summer 1991 to 1992–1993 (Figure 1e).

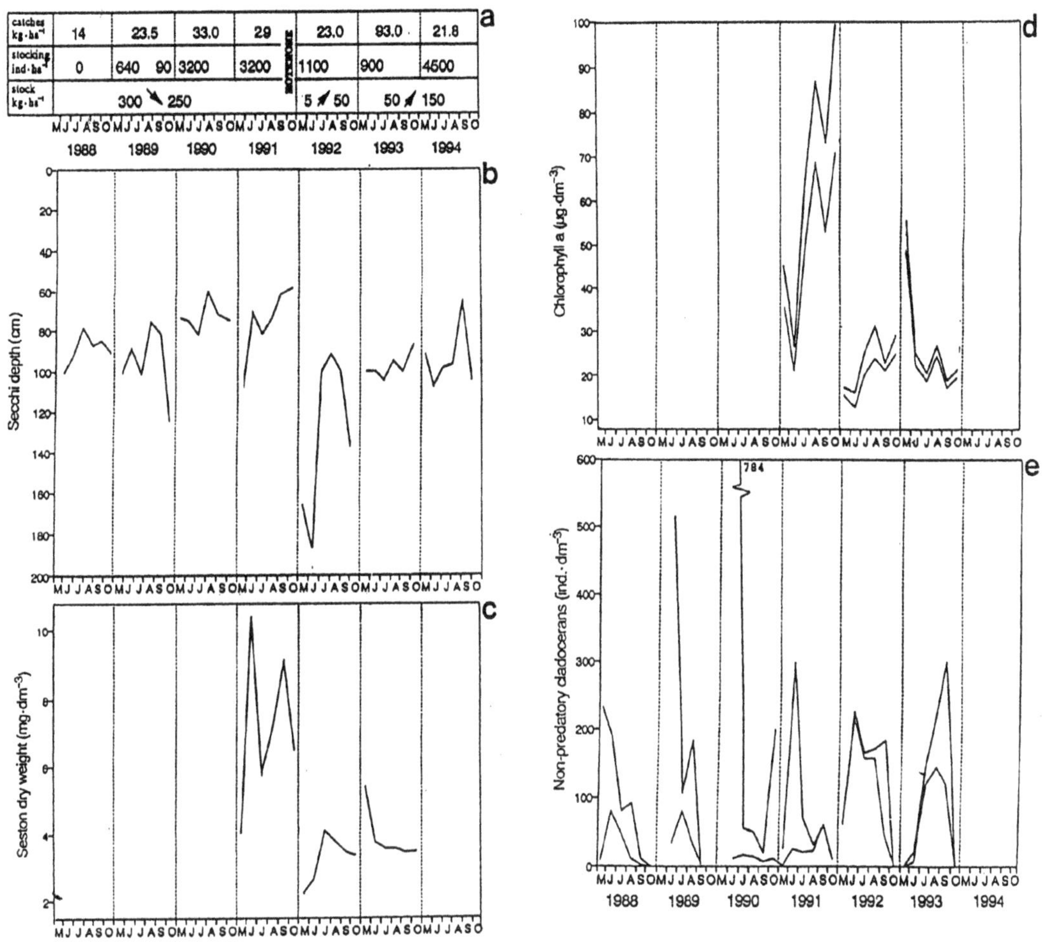

Figure 1. Sequence of biomanipulation events and a summary of basic measures of water quality.
(a) Selective catches of large planktivorous fish, successive introductions of pike fry and fish standing stock; spring introductions: fry of ca 30 mm (except for 1994 when newly hatched larvae were introduced), and autumn introduction: juveniles of ca 170 mm),
(b) Water transparency measured as Secchi depth,
(c) Seston dry weight, in the fraction of seston particles <30μm,
(d) Chlorophyll *a* concentration: total (upper line) and <30μm (lower line),
(e) Density of herbivorous cladocerans: total (upper line) and *Daphnia cucullata* (lower line).

Discussion

The rotenone treatment represented a major breakthrough within the whole sequence of biomanipulation activities. However, the pike introductions and selective extensive fishing preceding rotenone treatment did not bring any substantial improvement in water quality (Figure 1). The observed increase in water transparency, decline in algal biomass and increase in cladoceran density initiated by drastic reduction of fish biomass, follow the well known scenario (for a review see Reynolds, 1994). A deeper insight into the possible mechanisms involved in the decrease in algal biomass is provided by Pijanowska & Prejs (1997).

The successive introductions of predatory fish and the selective catches preceding the rotenone treatment did not produce any visible response at the lower trophic levels, but did cause immense changes within the fish community, with the drastic extermination of small planktivores being the most spectacular (Prejs et al., 1994). However, at ca 200 kg ha^{-1}, with roach as a distinct dominant (Prejs et al., 1994), older planktivorous fish remained abundant enough to keep their planktonic prey in check (Figure 1e), and high phytoplankton biomass persisted (Figure 1d). Thus, the trophic cascade was 'arrested' at the very top of the food chain.

From analysis of the age/size structure of dominant planktivores in the fish sample taken at the time of poisoning (Prejs et al., 1994) it seems clear that, to achieve the desired structure of the fish community, there should be at least one further year of stocking with juvenile pike accompanied by selective catches of older fish. The successive 0+ age-class would then be suppressed, promoting a further shift in the fish size-structure towards the dominance of the oldest, less abundantly-represented classes. This would certainly increase chances of activating a sequence of desired events along the food chain, without the need to take drastic and controversial measures.

Acknowledgements

Our students Jola Wierzgała, Olo Danielewicz and Tomek Janecki offered help at each stage of the project. The Parakiewicz family is kindly acknowledged for their hospitality and technical assistance in the field. Comments by two anonymous reviewers helped to clarify the reasoning. This work was supported financially by KBN grant 60540 9101 to Andrzej Prejs.

References

Pijanowska, J. & A. Prejs, 1997. Food web manipulation in shallow eutrophic lakes: bridging the gap between the whole-lake approach and behavioural and demographic studies. Hydrobiologia 342/343: 305–310.

Prejs, A., A. Martyniak, S. Boroń, P. Hliwa & P. Koperski, 1994. Food web manipulation in a small, eutrophic Lake Wirbel, Poland: effect of stocking with juvenile pike on planktivorous fish. Hydrobiologia 275/276: 65–70.

Reynolds, C. S., 1994. The ecological basis for successful biomanipulation of aquatic communities. Arch. Hydrobiol. 130: 1–33.

Hydrobiologia **342/343:** 387–394, 1997.
L. Kufel, A. Prejs & J. I. Rybak (eds), Shallow Lakes '95.

Eutrophication processes in a shallow, macrophyte-dominated lake – nutrient loading to and flow through Lake Łuknajno (Poland)

Irena Kufel & Lech Kufel
Hydrobiological Station, Institute of Ecology, Polish Academy of Sciences 11-730 Mikołajki, Poland

Key words: nitrogen, phosphorus, drainage waters, sediments, sorption

Abstract

Lake Łuknajno, a shallow (max. depth 3 m) and relatively large (630 ha) water body situated in the Great Masurian Lakes system (Poland) is strongly affected by its agricultural watershed. The total volume of drainage waters pumped into the lake in 1993 amounted 2.84×10^6 m^3 which is equivalent to 65% of the lake volume. In spring and autumn drainage waters were extremely rich in nitrogen (especially nitrates) and the N:P ratio reached 57 in spring. In summer, drainage waters contained relatively more phosphorus so the N:P ratio decreased to nearly 10.

Nutrient concentrations in lake water followed the changes in inflowing waters. High concentrations of nitrates and ammonia were noted in lake water in spring but summer concentrations of these ions were close to zero. Soluble reactive phosphorus varied between 10 and 20 μg P l^{-1} throughout the season. A constant outflow of nitrogen to the lower Lake Śniardwy was assumed based on a permanent concentration gradient between waters of these two lakes.

Lake Łuknajno is a hard water lake. Co-precipitation of phosphorus with calcium carbonate is likely to occur though no significant P accumulation in bottom sediments was found. Sorption of phosphorus on sediments as measured under experimental conditions has minor effects on P cycling. Bottom sediments composed mainly of calcareous gyttja do not accumulate either organic matter or mineral forms of nutrients.

Introduction

Shallow lowland lakes are characterised by a high variability of physical and chemical parameters (Hellström, 1991; Nixdorf et al., 1992; Kozerski et al., 1993), Continuous mixing and aeration of the whole water column are the primary driving forces for the whole lake system. These main physical factors affect nutrient cycling and finally influence primary production in a shallow lake. Aerobic conditions enable mineralisation and prevent considerable accumulation of organic matter. Oxygen in water overlying sediments influences the redox-sensitive iron-bound fraction of phosphorus and can thus control internal P-loading providing there is enough iron in the bottom sediments (H. S. Jensen et al., 1992). Polymixis and resuspension of surface sediment particles may increase nutrient release to water (Jensen & Andersen, 1992; Behrendt et al., 1993) but can also enhance denitrification through introducing nitrates into deeper sediment layers (J. P. Jensen et al., 1992).

Increased nutrient loading shifts the trophic status of a shallow lake from a clear water, macrophyte dominated to a turbid water state where the main producers are planktonic algae. Within a range of nutrient concentrations these two states may occur alternatively (Scheffer et al., 1993). The shift between two alternative states, though possible and described in case studies (e.g. Blindow et al., 1993), is still poorly understood. Therefore, for both theoretical and practical reasons it is necessary to study the fate of nutrients delivered to shallow lakes and the consequences of elevated nutrient loading to these particular water bodies.

This paper, as a part of a larger project, aims at analysing nutrient loading to and nùtrient dynamics in a shallow Lake Łuknajno (Poland) which, being affected by considerable watershed impact, still preserves some features of mesotrophic lakes.

Site

Lake Łuknajno (630 ha, 3 m maximum depth, 0.6 m mean depth and total volume 4.35×10^6 m^3) is situated in the southern part of the Great Masurian Lakes (Poland). Through a short channel it delivers its waters to Lake Śniardwy with a replacement rate of approx. 1.5 y^{-1}. The immediate lake watershed, of area of 48.4 km^2, is composed of arable lands (55%) pastures and meadows (26%) and forests (18%). Sandy, permeable soils dominate in the watershed. Land input of nutrients to the lake is enabled also by a dense (1.5 km km^{-2}) network of amelioration channels draining meadows in the watershed (Jeziora MPK, 1989). Drainage waters from the channels are occasionally pumped out to the lake in order to prevent flooding of surrounding lands.

A considerable part of lake shoreline is of swamps, peat bogs and forests which isolate the lake basin from diffuse seepage. The same function may be exerted by wide belts of emergent macrophytes (mainly common reed and cattail) surrounding Lake Łuknajno.

Because of abundant and diversified waterfowl Lake Łuknajno is protected as a biosphere reserve and as one of the eight Ramsar Convention sites in Poland.

Table 1. Volume of drainage waters pumped to Lake Łuknajno and nutrient concentration in drainage water.

	March 23– April 30	June 11– June 24	August 8– October 18	Annual load
Volume (10^6 m^3)	1.17	0.24	1.43	2.84
SRP (μg P l^{-1})	34	152	36	127.7 (kg)
DP (μg P l^{-1})	44	169	77	202.2 (kg)
TP (μg P l^{-1})	48	268	99	262.1 (kg)
N-NH$_4$ (mg l^{-1})	0.07	0.10	0.20	391.9 (kg)
DKN (mg N l^{-1})	1.03	0.91	1.06	2939.3 (kg)
TKN (mg N l^{-1})	1.29	1.16	1.16	3446.5 (kg)
N-NO$_3$ (mg l^{-1})	1.46	0.29	0.64	2693.0 (kg)
TN (mg N l^{-1})	2.75	1.45	1.80	6139.5 (kg)
Seston (mg l^{-1})	3.5	8.1	2.4	9472 (kg)

Material and methods

Water samples for chemical analyses were taken once a week from April to October from three sites in Lake Łuknajno. Samples of water pumped from amelioration channels to the lake were taken in April, June and September. Temperature and oxygen concentration were measured with an oxygen probe in the field. In the lab, water samples were filtered through Milipore 0.45 μm membrane filters for analyses of soluble reactive phosphorus – SRP (molybdenum blue method with $SnCl_2$ as a reducing agent), ammonia-nitrogen – N-NH$_4$ (indophenol blue method after Solórzano, 1969) and nitrates – N-NO$_3$ (phenyldisulphonic acid method). Calcium was determined complexometrically (Standard Methods, 1960). Dissolved phosphorus (DP) and total phosphorus (TP) were analysed in filtered and unfiltered water, respectively, after mineralisation with conc. perchloric acid (1 ml/50 ml of sample). Dissolved (DKN) and total Kjeldahl nitrogen (TKN) were analysed from filtered and unfiltered water samples, respectively after standard Kjeldahl wet digestion. The sum of total Kjeldahl nitrogen and nitrate-nitrogen gave total nitrogen (TN) in water samples.

Samples of lake water were filtered through Whatman GF/C filters for seston measurements (by weight after drying at a temperature of 105 °C) and for chlorophyll determinations (acetone extraction and spectrophotometric measurements after Golterman, 1969). Concentrations of particulate forms of nitrogen and phosphorus were calculated as a difference between total and dissolved forms of the respective elements.

Once in a summer bottom sediments were taken from 10 stands in Lake Łuknajno. After drying at 105 °C, samples of sediments were mineralised in a mixture of nitric and perchloric acid (1:4 v/v) and in the resulting solution, phosphorus, nitrogen and calcium were analysed as described above.

Phosphorus exchange between water and bottom sediments was determined by experiment. Undisturbed samples of bottom sediments were taken in plexiglass tubes, 8 cm in diameter and 40 cm high. Water from above the sediment was carefully siphoned off, filtered through a Whatman GF/C filter, enriched in phosphates (in the form of KH_2PO_4) and poured over the sediment. Experimental tubes were kept in the dark at a temperature of 20 ± 2 °C. After 10–24 h the change in concentration of soluble reactive phosphorus was determined in the overlying water. From other sediment samples, interstitial water was separated by centrifugation and phosphorus and calcium analysed as described above for lake waters.

All results given in this paper represent mean values from three independent samples.

Results

Water pumped to Lake Łuknajno seems to be the main source of external nutrient loading. Nitrates constituted more than a half of the nitrogen pool discharged to Lake Łuknajno from amelioration channels in spring (Table 1). Concentration of ammonium in drainage waters at the same period was negligible. In June and September, the concentration of nitrates in drainage waters was considerably lower. Other forms of nitrogen, including ammonium, were relatively stable throughout the season. Phosphorus in drainage waters was present mainly in the form of SRP. Maximum concentrations of phosphorus were found in June (Table 1).

Phosphorus load delivered through the drainage system was equivalent to 38.5 mg P m^{-2} of lake area. Atmospheric loading of phosphorus (both wet and dry deposition) may vary in the area between 14.5 and 53.2 mg P m^{-2} (Kufel, in press). Summing up both sources, total P loading to Lake Łuknajno was in the range 53–91.7 mg P m^{-2}. These values are 2–3 times higher than permitted by Vollenweider's (1976) criteria for vulnerability to creation of a eutrophic state.

Other nutrient sources to Lake Łuknajno seem to be negligible – water discharge from natural streams was small (reduced in summer to zero) and nutrient concentrations there did not differ significantly from those in lake water.

The loading pattern of nutrients was reflected in the seasonal dynamics of nitrates in lake water (Figure 1), however, there was no such relationship between ammonia in drainage and in lake water. The former is obviously related to agricultural use of part of the watershed and spring surface runoff of mineral fertilisers. Summer depletion of mineral nitrogen in lake water was followed by increase of both nitrates and ammonia during autumn (Figure 1). Accumulation of ammonia in lake water in autumn could not be explained entirely by external loading. Total autumn input of ammonia would have increased N-NH_4 in lake water by 0.07 mg N l^{-1} while the recorded increase was approx. 0.5 mg N l^{-1} (Figure 1). Therefore, internal nitrogen loading must be assumed, probably from decaying macrophyte biomass. Particulate nitrogen in lake water varied irregularly throughout the season and therefore total N presented no regular pattern of seasonal dynamics. Total N was always higher in Lake Łuknajno than in Lake Sniardwy suggesting constant outflow of nitrogen from the former (Figure 2).

No such relationship was found between phosphorus loading in drainage waters and in-lake P concentration. Total P in lake water showed high variability but summer values were usually higher than those in spring and autumn. SRP varied between 10 and 20 μg P l^{-1} throughout the season (Figure 3). At low water flow in June, drainage waters delivered only 25% of the annual P load in spite of nearly fivefold increase in phosphorus concentration in the drainage waters. The distinct peak in P concentration in drainage waters can probably be explained by the release from sediments in the channels. In June sestonic P constituted 37% of total P pumped out to Lake Łuknajno as compared to 8% in spring. This fact, together with the change of N:P ratio in drainage waters (57 in April and 5 in June) suggests mobilisation of runoff P in algal biomass developing in drainage channels.

Chlorophyll concentrations showed three maxima – at the beginning of April followed by a clear-water phase, the highest in June and a smaller, summer maximum in July (Figure 4). Algae were the main component of seston in Lake Łuknajno, the regression of chlorophyll on suspended solids (chl $a = 0.460 + 2.321 * \text{seston}$) was statistically significant ($n = 26$, $r = 0.78$, $p < 0.05$). Seasonal variability of seston was, however, higher than that of chlorophyll and, except in periods of chlorophyll maxima, did not show any regular pattern (Figure 4).

Variability in the seston could be attributed to mineral particles resuspended from bottom sediments.

Bottom sediments were composed of calcareous gyttja. Their calcium content was 23–28% of sediment dry weight except for one stand overgrown by Nitellopsis (Table 2). The sediments are rather poor in nutrient content and, again except for the *Nitellopsis* stand, their composition does not depend on the type of overgrowing plant community.

High calcium content in bottom sediments suggests a capacity for sorption of mineral P on the sediment surface. Sorption of SRP, as shown in the experiment (Figure 5) was related to the initial phosphate concentration. The process was, however, slow and the equilibrium concentration was attained after 90–100 hours.

Discussion

High concentrations of mineral nitrogen in lake water (because of the composition of drainage waters) resulted in a high N:P ratio, which reached 57 in spring and

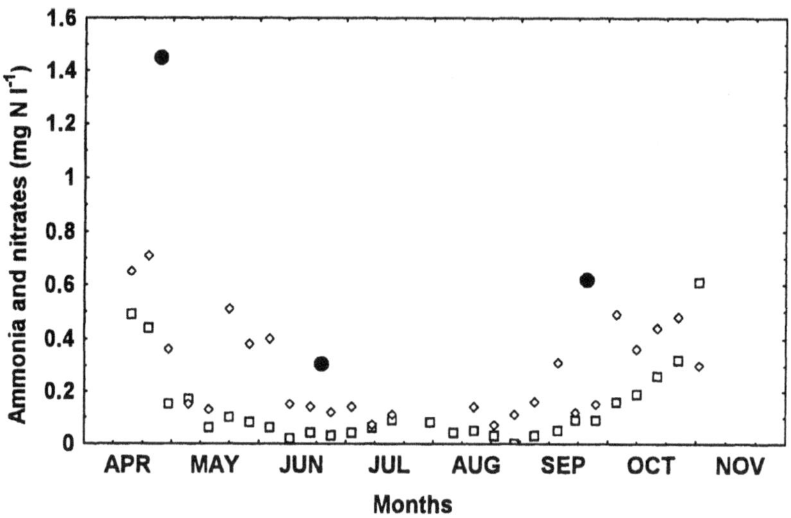

Figure 1. Seasonal changes of ammonia (squares) and nitrates (rhombs) in Lake Łuknajno and nitrates in drainage waters (filled circles)

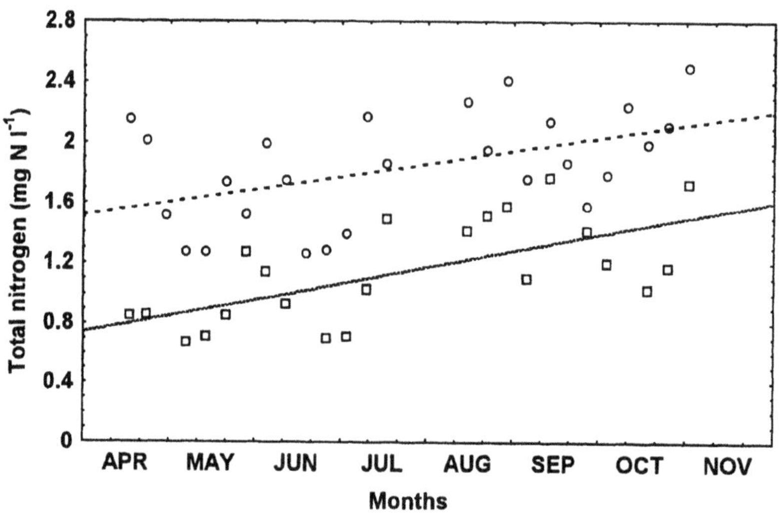

Figure 2. Total nitrogen concentration in Lake Łuknajno (circles, dashed line) and in Lake Śniardwy (squares, solid line).

Table 2. Calcium, phosphorus and nitrogen (in mg g^{-1}) in samples of bottom sediments from stands overgrown by different macrophyte species.

Plants	No. of stands	Ca	P	N
No vegetation	3	248 ± 16	0.79 ± 0.56	10.97 ± 2.11
Chara spp.	5	248 ± 19	0.77 ± 0.24	12.23 ± 2.65
Stratiotes aloides	1	244	0.55	10.86
Nitellopsis obtusa	1	152	0.30	1.11

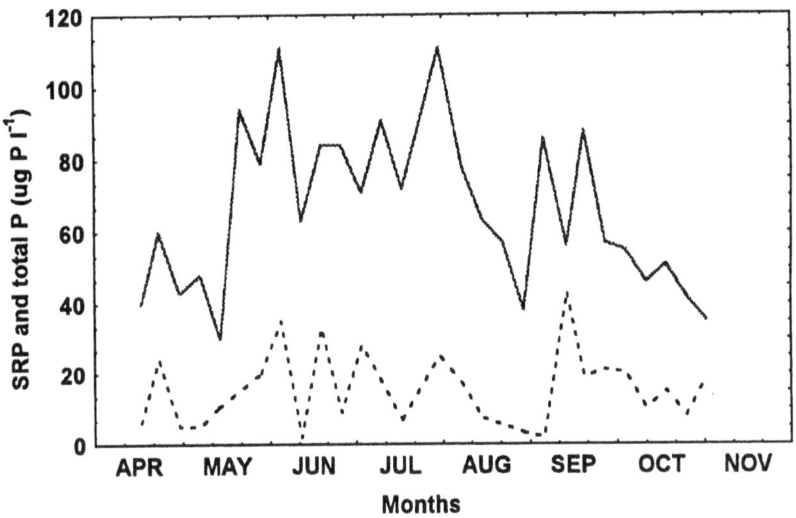

Figure 3. Seasonal dynamics of soluble reactive phosphorus (dashed line) and total phosphorus (solid line) in Lake Łuknajno.

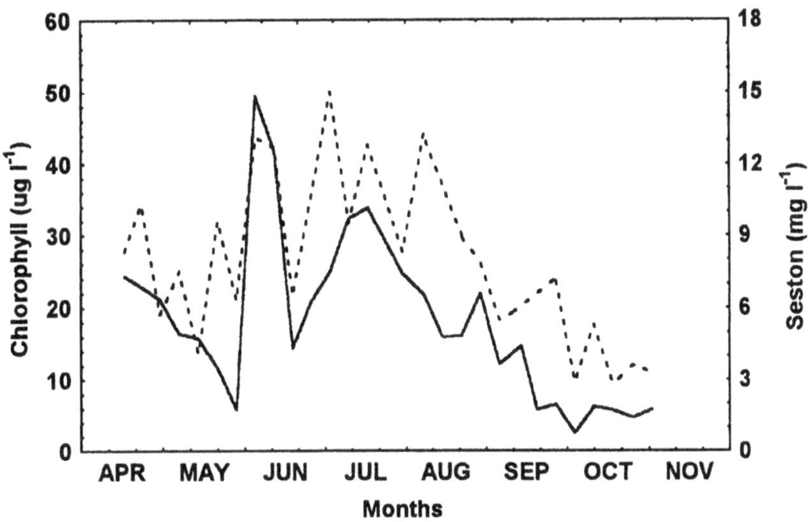

Figure 4. Seasonal course of chlorophyll (solid line) and seston (dashed line) in water of Lake Łuknajno.

over 70 in autumn (Figure 6). Summer values, due to lower N concentrations, were much lower. Sestonic N:P ratio varied between 10 and 20 in spring and summer (Figure 6) and together with the existence of a surplus of mineral nitrogen (high N:P in water) suggests moderate phosphorus limitation throughout the growth season (Healey & Hendzel, 1979; 1980, Tezuka 1985). At the same time, however, SRP in water was relatively high (up to 20 μg P l^{-1}) and alkaline phosphatase activity – low (see Kalinowska (1997)). Thus, it is possible that algae were limited by light attenuated by resuspended mineral particles and/or by self-shading. Light limitation was suggested by Smith (1982), Havens (1994) and Seip (1994) while considering TN and TP effect on algal production. That is probably why phosphorus inputs to Lake Łuknajno in summer have not exerted significant effects through algal blooming (the summer chlorophyll peak was lower than that in June and was noted nearly 1 month after

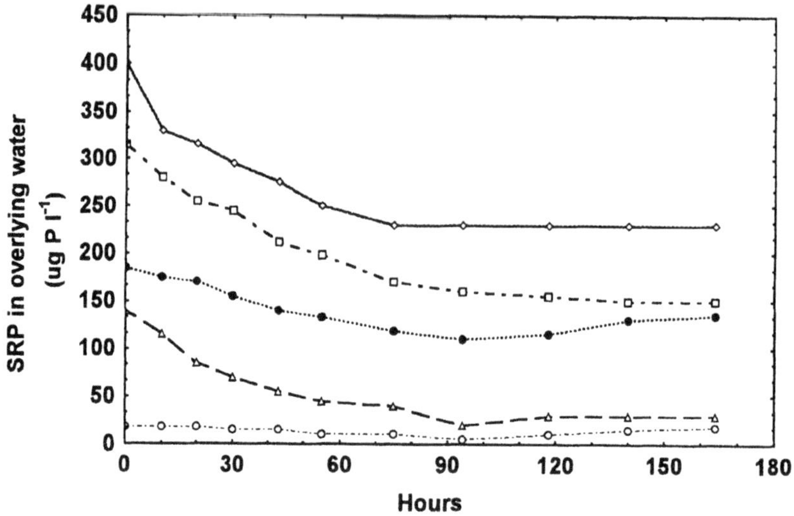

Figure 5. Sorption of phosphates by undisturbed sediments from Lake Łuknajno as a decrease in SRP concentration in overlying water. Different lines represent different initial SRP concentrations.

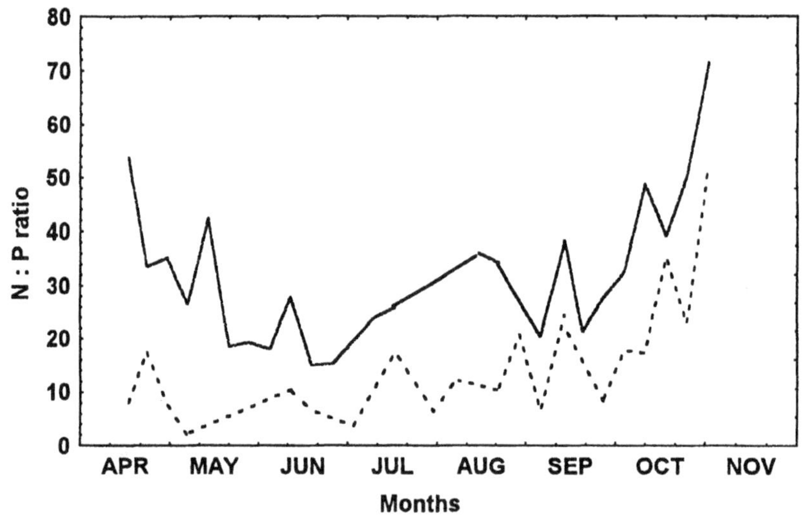

Figure 6. Seasonal changes of water (solid line) and sestonic (dashed line) N:P ratio in Lake Łuknajno.

the increased P input from drainage waters – see Figure 5).

In October, sestonic N:P increased to 50. An autumn increase in sestonic N:P was caused by the enrichment in nitrogen since sestonic P concentration was relatively stable (4–6 mg P g^{-1}). The reason for the increase of sestonic N:P in autumn is not clear. Decomposition processes would have released both nitrogen and phosphorus at similar rates and resuspension of bottom sediments (with N:P ratio = 16) would not increase sestonic N:P up to 50. Bacterial colonisation of detritus and hence elevated nitrogen content of particulate matter could possibly explain observed phenomenon, analysis of bacterial abundance was, however, out of the scope of this paper.

Bottom sediments, as compared with settling seston, were poor in nitrogen and phosphorus (Table 2). It seems that intensive mineralisation took place during

Table 3. Calculated (from the solubility product of hydroxyapatite) vs. recorded SRP concentrations in lake water and in interstitial water.

	Ca (mg l^{-1})	pH	SRP_{calc} (μg P l^{-1})	SRP_{rec} (μg P l^{-1})
Lake water	45.8	8.3	43	0
	53.7	8.0	101	12
Interstitial water	101.9	8.3	12	168

settling (and resuspension) and particulate matter was rather a source of nutrients to overlying water.

Sorption of phosphorus on the surface of bottom sediments may be of some importance for the nutrient budget in Lake Łuknajno. The Freundlich adsorption isotherm ($S = 0.15\ c^{0.90}$, where S is adsorption in mg P m^{-2} of sediment and c is initial SRP concentration in μg P l^{-1}) well described the binding capacity of sediments. At ambient SRP concentration = 20 μg P l^{-1} adsorption equals 2.22 mg P m^{-2} or 19% of actual P content in a 0.6 m average water column. Sorption, under experimental conditions, was slow and probably limited by diffusion (see Figure 5) but in a lake diffusion can be enhanced by frequent mixing.

In a hard water lake, phosphorus cycling can be regulated by calcium (Golterman, 1982). Ca concentration in Lake Łuknajno varied around 50 mg Ca l^{-1}. Calculations of SRP concentrations predicted from the solubility product of hydroxyapatite differed from those measured in lake water and also in interstitial waters taken from sediments (Table 3). Thus, it is coprecipitation of phosphorus with calcium carbonate rather than hydroxyapatite formation that determines phosphorus storage in bottom sediments.

Conclusions

Waters of Lake Łuknajno are strongly affected by the surrounding lands and concentration of e.g. nitrates in lake water is a simple reflection of the seasonal loading dynamics. Except for macrophyte uptake (see Krolikowska, this volume) there is no considerable accumulation of nutrients in the lake. Sediments are highly mineralised and poor in nitrogen and phosphorus. In spite of relatively long retention time (up to 1.5 y depending on weather conditions) Lake Łuknajno seems to be a flow-through basin for nutrients as could be seen from the comparison of nitrogen in Lake Łuknajno and in lower Lake Śniardwy. Nutrient outflow takes place most intensively in spring and autumn during high drainage water discharge. In that way nitrogen (mostly in mineral form) and phosphorus (both in soluble and particulate form) is transported out of Lake Łuknajno.

Acknowledgments

This work was financed through the Committee for Scientific Research within the grant PB 0499/P2/93/4. Authors are indebted to Mrs Grażyna Pośpiech and Mr Tadeusz Krache for help in field work and laboratory analyses. Thanks are also due to two anonymous rewievers, whose remarks improved the final version of this paper. Data on the volume of drainage waters were kindly supplied by Mr Franciszek Mroczek.

References

Behrendt, H., B. Nixdorf & W.-G. Pagenkopf, 1993. Phenomenological description of polymixis and influence on oxygen budget and phosphorus release in Lake Müggelsee. Int. Revue ges. Hydrobiol. 78: 411–421.

Blindow, I., G. Andersson, A. Hargeby & S. Johansson, 1993. Long-term pattern of alternative stable states in two shallow eutrophic lakes. Freshwat. Biol. 30: 159–167.

Golterman, H. L., 1969. Methods for Chemical Analysis of Fresh Waters. Blackwell Scientific Publications, Oxford and Edinburgh, 172 pp.

Golterman, H. L., 1982. Loading concentration models for phosphate in shallow lakes. Hydrobiologia 91/92: 169–174.

Havens, K. E., 1994. Seasonal and spatial variation in nutrient limitation in a shallow sub-tropical lake (Lake Okeechobee, Florida) as evidenced by trophic state index deviations. Arch. Hydrobiol. 131: 39–53.

Healey, F. P. & L. L. Hendzel, 1979. Indicators of phosphorus and nitrogen deficiency in five algae in culture. J. Fish. Res. Bd Can. 36: 1364–1369.

Healey, F. P. & L. L. Hendzel, 1980. Physiological indicators of nutrient deficiency in lake phytoplankton. Can. J. Fish. aquat. Sci. 37: 442–453.

Hellström, T., 1991. The effect of resuspension on algal production in a shallow lake. Hydrobiologia 213: 183–190.

Jensen, H. S. & F. Ø. Andersen, 1992. Importance of temperature, nitrate and pH for phosphate release from aerobic sediments of four shallow, eutrophic lakes. Limnol. Oceanogr. 37: 577–589.

Jensen, H. S., P. Kristensen, E. Jeppesen & A. Skytthe, 1992. Iron:phosphorus ratio in surface sediments as an indicator of phosphate release from aerobic sediments in shallow lakes. Hydrobiologia 235/236: 731–743.

Jensen, J. P., E. Jeppesen, P. Kristensen, P. B. Christensen & M. Søndergaard, 1992. Nitrogen loss and denitrification as studied in relation to reductions in nitrogen loading in a shallow, hypertrophic lake (Lake Søbygaard, Denmark). Int. Revue ges. Hydrobiol. 77: 29–42.

Jeziora Mazurskiego Parku Krajobrazowego. Stan eutrofizacji, kierunki ochrony. 1989. A. Hillbricht-Ilkowska (ed.) [Lakes of the Masurian Landscape Park. Trophic status and protection measures.] Ossolineum, Wroctaw, Warszawa, 166 pp.

Kalinowska, K., 1997. Eutrophication processes in a shallow, macrophyte dominated lake – alkaline phosphatase activity in Lake Łuknajno (Poland). Hydrobiologia 342/343: 395–399.

Kozerski, H. P., J. Gelbrecht & R. Stellmacher, 1993. Seasonal and long-term variability of nutrients in Lake Müggelsee. Int. Revue ges. Hydrobiol. 78: 423–437.

Królikowska, J., 1997. Eutrophication processes in a shallow, macrophyte-dominated lake – species differentiation, biomass and the distribution of submerged macrophytes in Lake Łuknajno (Poland). Hydrobiologia 342/343: 411–416.

Kufel, I. Opad atmosferyczny jako źródło zasilania jezior w azot, fosfor i siarkę. [Atmospheric precipitation as a source of nitrogen, phosphorus and sulphur for lakes.] In Hillbricht-Ilkowska, A. & R. Wigniewski (eds), Funkcjonowanie systemów rzeczno – jeziornych w krajobrazie rzeka Krutynia na Poj. Mazurskim. [Functioning of river-lake systems in a postglacial landscape.] Ossolineum, Wroctaw, Warszawa. (in press)

Nixdorf, B., W.-G. Pagenkopf & H. Behrendt, 1992. Diurnal patterns of mixing depth and its influence on primary production in a shallow lake. Int. Revue ges. Hydrobiol. 77: 349–360.

Scheffer, M., S. H. Hosper, M.-L. Meijer, B. Moss & E. Jeppesen, 1993. Alternative equilibria in shallow lakes. Trends in Ecology and Evolution 8: 275–279.

Seip, K. L., 1994. Phosphorus and nitrogen limitation of algal biomass across trophic gradients. Aquat. Sci. 56: 16–28.

Smith, V. H., 1982. The nitrogen and phosphorus dependence of algal biomass in lakes: An empirical and theoretical analysis. Limnol. Oceanogr. 27: 1101–1112.

Solórzano, L., 1969. Determination of ammonia in natural waters by the phenylhypochlorite method. Limnol. Oceanogr. 14: 799–800.

Standard methods for the examination of water and waste-water. 1960. Am. Publ. Health Assoc. Inc., New York, 626 pp.

Tezuka, Y., 1985. C:N:P ratios of seston in Lake Biwa as indicators of nutrient deficiency in phytoplankton and decomposition process of hypolimnetic particulate matter. Jap. J. Limnol. 46: 239–246.

Vollenweider, R. A., 1976. Advances in defining critical loading level for phosphorus in lake eutrophication. Mem. Ist. ital. Idrobiol. 33: 53–83.

Hydrobiologia **342/343:** 395–399, 1997.
L. Kufel, A. Prejs & J. I. Rybak (eds), Shallow Lakes '95.

Eutrophication processes in a shallow, macrophyte dominated lake – alkaline phosphatase activity in Lake Łuknajno (Poland)

Krystyna Kalinowska
Institute of Ecology PAS, Hydrobiological Station, ul. Leśna 13, 11-730 Mikołajki, Poland

Key words: alkaline phosphatase, chlorophyll *a*, phosphorus, algae, phosphorus regeneration

Abstract

Alkaline phosphatase activity (APA) of unfiltered and fractionated (algae, bacteria, dissolved) surface water samples was measured using a spectrophotometric method in Lake Łuknajno from May to September 1993. The total enzymatic activity varied between 0.44–2.35 μmol l^{-1} h^{-1}. Algae were major producers of alkaline phosphatase activity in surface water of the lake from May to August, and their activity constituted on the average 67% of the total APA activity in the water. APA activity in bacterial size fraction was low and constituted 9.6% (mean) of the total activity. The activities of free (dissolved in water) enzymes were relative high during the time of study. This study show that hydrolysis of organic phosphomonoesters by alkaline phosphatase was negligible in the recycling of phosphorus in Łuknajno Lake.

Introduction

Phosphorus is the nutrient limiting phytoplankton growth in many freshwaters (Wetzel, 1975), and its concentration can be used to predict the total biomass of phytoplankton that will develop in a lake (Dillon & Rigler, 1974). Phosphorus availability also influences the dynamics of most individual algal species (Peterson et al., 1974). The major form in which microorganisms acquire phosphorus is dissolved inorganic phosphate (DIP), the ions $H_2PO_4^-$ + HPO_4^{2-} (Chróst & Overbeck, 1987). When soluble inorganic P concentrations within the water column are low, plankton may produce phosphatase, enzymes which hydrolyse organic P compounds with the release of inorganic P (Kuenzler & Perras, 1965; Reichardt, 1971). This enzyme could play an important role in the phosphorus supply during phosphate depletion. High phosphatase activity is generally thought to indicate condition of phosphate deficiency (Fitzgerald & Nelson, 1966; Perry, 1972; Healey, 1973) and the level of alkaline phosphatase activity indicates the degree of eutrophication of lakes (Jones, 1972). Phosphatase activity has been used before as an indicator of P-deficiency in both algal cultures (Healey & Hendzel, 1979) and natural phytoplankton populations (Healey & Hendzel, 1980).

In the present study I analysed the seasonal course of alkaline phosphatase activity in various size fractions of seston (phytoplankton and bacteria) and of free APA dissolved in water in the context of seasonal variations of chlorophyll *a* concentration, phosphorus concentration and abundance of phytoplankton. The purpose of this study was also to estimate the significance of hydrolysis of organic-P by alkaline phosphatase in the regeneration of phosphate in Lake Łuknajno.

Material and methods

The study was carried out on the shallow (maximum depth 3 m, average depth 0.6 m) relatively large (surface area 630 ha) Lake Łuknajno dominated by macrophytes. Lake Łuknajno has been nature reserve since 1937 and a biosphere reserve since 1977 for its very rich ornithofauna.

Samples of lake water were taken once a month from the surface layer during the period May to September 1993. Enzyme activity was divided into three fractions. Total enzymatic activity was measured in unfiltered lake water. Filtrate from Whatman GF/C filters was assumed to contain bacterial + free enzymes (Gajewski et al., 1993) and filtrates from Milipore 0.45 μm membrane filters – free dissolved enzyme. Enzyme activities associated with algae and bacteria were calculated as differences between the activities measured in respective filtrates. Alkaline phosphatase activity (APA) was determined spectrophotometrically as the release of *p*-nitrophenol from *p*-nitrophenyl phosphate (pNPP). Four ml pNPP (0.5 g l^{-1}) and 4 ml 0.1 M TRIS buffer (adjusted to pH 8.5 with acetic acid) were added to 50 ml of water samples. The final concentration of substrate in assay was 67 μmol. Samples were incubated at 37 °C for 15–18 hours, depending on the enzyme activity. The amount of pNP produced was measured as the increase in absorbance at 395 nm. Enzyme activity was expressed as μmol of *p*-nitrophenol released per liter per hour. Different concentrations of pNPP substrate were added to the samples in order to establish enzyme-substrate saturation and to enable the calculation of the kinetics parameters (V_{max} and K_m). Computer programme 'Enzfitter' was used for the calculation of data. The rate of hydrolysis of organic P compounds was estimated from the Michaelis-Menten equation. Organic-P concentration was measured as the difference between dissolved phosphorus (DP) and soluble reactive phosphorus (SRP). APA hydrolysable phosphorus fraction was not analysed in lake water, therefore organic P was taken as an actual substrate for enzyme activity. This assumption could result in overestimation of the real rate of hydrolysis. The velocity calculated from Michaelis-Menten equation was taken as the rate of release of P-PO_4 by alkaline phosphatase and was corrected for lake ambient temperature.

Results

The total alkaline phosphatase activity during the studies ranged between 0.44 and 2.35 μmol l^{-1} h^{-1} The highest value of the total APA was noted in July. At this time 82% of the total activity was derived from algal size fraction. Algae were the major producer of alkaline phosphatase in Lake Łuknajno from May to August and their activity varied from 0.99 to 1.92 μmol l^{-1} h^{-1} (Figure 1) and during this period constituted from 63-82% (mean 67%) of the total activity in lake water. From July activities in algal size fraction gradually decreased to reach 36% of the total APA at the end of September. This study showed that algal APA presented similar dynamics to that of chlorophyll-a concentrations (Figure 2). Concentration of chlorophyll *a* in water of Lake Łuknajno increased from 11.7 μg l^{-1} in May to 32 μg l^{-1} in July and subsequently gradually decreased to 2.5 μ l^{-1} in September. The high activity in algal size fraction preceded intensive increase of cyanophyta numbers in July/August (Figure 3). Blue-green algae contributed in 80% to the total number of phytoplankton and in 60% to the total biomass in the middle of August (Kornatowska, pers. com.). From August the algal biomass decreases; the reason of the drastic decrease was a heavy pressure of zooplankton preventing long-term blooms of algae (Ejsmont-Karabin et al., unpubl. data).

Alkaline phosphatase activity associated with bacterial size fraction fluctuated considerably during the studies and ranged from 0.50 μmol l^{-1} h^{-1} in June to 0.006 μmol l^{-1} h^{-1} in July (Figure 1). The highest bacterial activity was noted in June when phosphate content in water was low and amounted 9 μ l^{-1}. Free, dissolved APA were a significant fraction of the total activity and accounted for 16–58%. Their activity varied from 0.25 to 0.45 μmol l^{-1} h^{-1} (Figure 1).

The highest V_{max} values of the total APA (3.4 μmol l^{-1} h^{-1}) were noted in June and July. During autumn total APA activity was relatively low and V_{max} varied between 0.4 and 1.9 (mean 1.1) μmol $^{-1}$ h^{-1}. The Michaelis constant values (K_m) of the total APA did not vary significantly during the study period and the K_m values were in the range 0.98–3.42 (mean 2.49) μmol l^{-1} and increased with an increase of DIP content in water. The highest amount of K_m refer to the low concentration of phosphate (9 μg l^{-1}) The real activity of the total APA in surface waters of Lake Łuknajno as calculated from the Michaelis-Menten equation was low due to low concentration of organic hydrolyzable P compounds which varied from 4–47 μ P l^{-1}. The rate by which phosphate was released to water ranged between 1.6 and 17.0 (mean 10.7) nmol l^{-1} h^{-1}. The release of PO_4^{3-} by APA hydrolysis increased in July with decrease (down to 9 μg P l^{-1}) in the ambient concentration of phosphate in water. Similarly, minimum rate was observed at the end of study, when phosphate content was highest (21 μg P l^{-1}).

Discussion

Alkaline phosphatase activity (APA) is one of the major mechanisms to release mineral phosphorus from organic P compounds. When all external and internal phos-

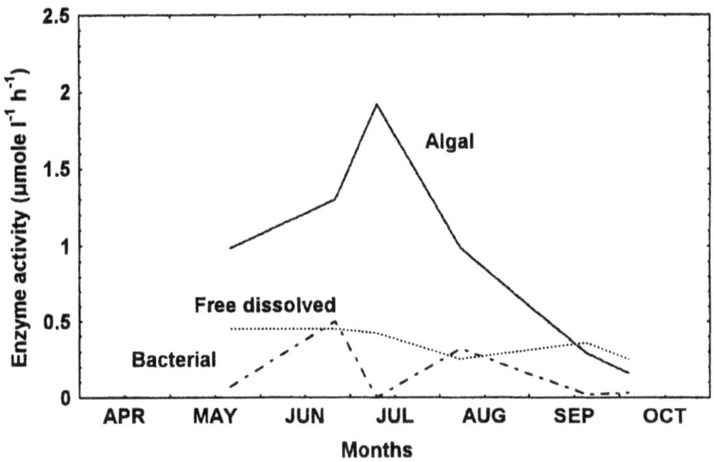

Figure 1. Alkaline phosphatase activity in various size fractions in Lake Łuknajno.

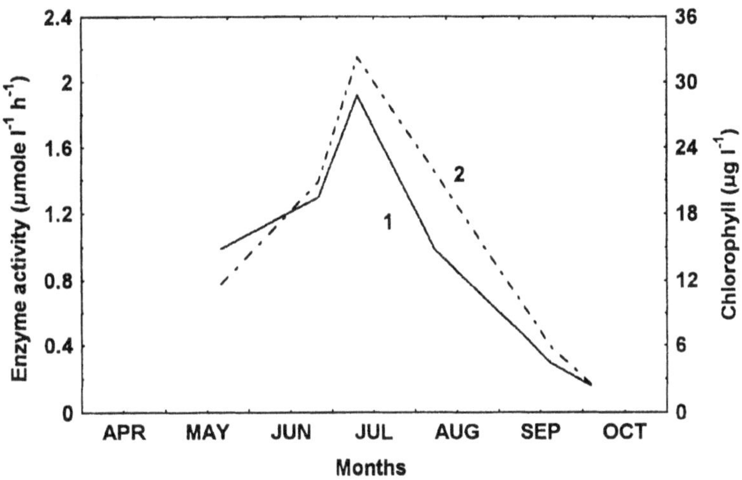

Figure 2. Seasonal variation of alkaline phosphatase activity in algal size fraction (1) and chlorophyll *a* concentration (2) in Lake Łuknajno.

phorus reservoirs are depleted, microplankton must produce phosphatases and hydrolyse phosphoesters as an alternative P source to compensate for the lack of DIP (Chróst & Overbeck, 1987). Studies suggest the strong competition for P between phytoplankton and bacteria (Currie & Kalff, 1984a). Rhee (1972), who examined the algal and bacterial growth in monospecific and mixed cultures found that the presence of bacteria reduced algal biomass, therefore bacteria are superior competitors for phosphorus.

My study demonstrated that the majority of enzymatic activity during the phytoplankton blooms was associated with algal size fraction (>1 μm) of seston (63–82%). Enzymatic activity of APA in this fraction probably also originated from bacteria, because some portion of bacteria was retained on glass fiber filters. On the other hand, bacteria play a minor part, owing to their very small biomass (as compared with algae). The effects of their activity are masked by the high physiological activity of algae (Tiwari & Mishra, 1982). Thus more of the phosphatase activity in size fraction >1 μm are of algal origin. However some authors define bacteria as size fraction 0.2 to 0.8 μm (Palmer, 1994) or 0.1 to 1.0 μm (Kirchman et al., 1992; Lee et al.,1993). After intensive growth of phytoplankton, the APA activity in this fraction decreased (36–43%).

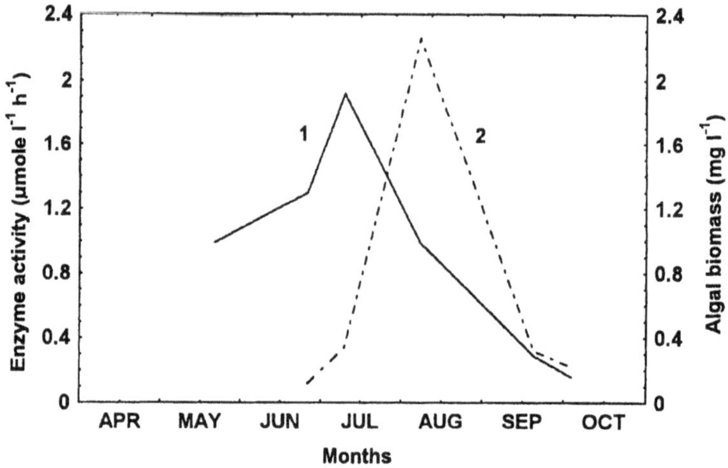

Figure 3. The comparison of alkaline phosphatase activity in algal size fraction (1) and biomass of blue-green algae (2) in water of Lake Łuknajno.

During the bloom of phytoplankton (mainly blue-green algae) the surplus phosphorus content in lake water was rather high and oscillated irregularly around 20 μg P l^{-1} (Kufel & Kufel, 1997). Blue-green algae often occur in lakes when ortho-phosphate is at very low concentrations (Hutchinson, 1967), indicating that ortho-phosphate is being made available to the plankton community by external loading, by recycling, or by internal loading (Heath & Cooke, 1975). Some phytoplankton species, when growing in phosphate limiting conditions in the laboratory, adaptatively produce alkaline phosphatase which is capable of releasing ortho-phosphate from phosphomonoesters (Fitzgerald & Nelson, 1966). Recycling processes may be an additional and significant source of ortho-phosphate to phytoplankton populations (Heath & Cooke, 1975). Alkaline phosphatase activity in this fraction could also depend on phytoplankton species composition. Olsson (1990) observed significant APA in eutrophic Swedish lakes when diatoms and cyanobacteria dominated the phytoplankton biomass. Similarly, Hino (1988) found the highest APA in some diatoms and cyanobacteria species. Bacterial APA was relatively low and constituted maximal 22% of the total activity. The activity in bacterial size fraction increased when algal activity decreased. Bacterial APA production usually increased rapidly when the concentration of phosphate in the water decreased substantially. In Lake Łuknajno during the study phosphate concentrations varied between 9 and 21 μ P l^{-1} (Kufel & Kufel, 1997) and were sufficient for the bacterial P requirement, thus bacteria were not limited by phosphate, but their activity and growth might be limited by amount of directly utilizable carbon (Chróst et al., 1986; Cole et al., 1982). Therefore bacteria produced phosphatases to hydrolyze organophosphoric compounds but probably utilized the organic moiety as an organic carbon source. Free, dissolved APA showed constant, relatively high activity and probably originated from bacteria that passed through the 0.45 μm filters and from cell lysis (Berman, 1970) or damage during filtration. This fraction of enzyme could take part in destruction of organic matter. The existence of free enzymes is well documented in the literature (Reichard et al., 1967; Chróst & Overbeck, 1987). The goal of this study was to determine the significance of alkaline phosphatase in phosphorus regeneration. The enzymatic release of phosphate by the phosphatase may be one of the process of internal DIP load of the lake (Chróst et al., 1984). Taking concentrations of available organic substratum into account 0.05–0.53 μ P l^{-1} h^{-1} could have been released on that way into the surface waters. The respective turnover times for P recycling were thus 17–420 h, the shortest in June during SRP depletion and algal blooming. This load of P is lower than P calculated on the basis of regeneration by zooplankton. Zooplankton regenerate from 21 μ P l^{-1} 24 h^{-1} to 405 μg P l^{-1} 24 h^{-1} (Ejsmont-Karabin et al., unpubl. data). The entire DOP fraction in the water cannot be hydrolysed by alkaline phosphatase; part of the DOP in natural waters is composed of a high molecular weight fraction resistant to alkaline phosphatase (Feuillade et al., 1990). My results indicate that hydrolysis by alkaline phosphatase was not a significant process in the recycling of phosphorus in Lake Łuknajno during the time of this study. Taking into consideration that total DOP was not hydrolysed, the real rate of APA hydrolysis and release of inorganic-P by this process could be even lower than evaluated in this study.

Acknowledgments

I thank my colleagues for their technical assistance during chemical analysis of water samples, R. Kornatowska for information on phytoplankton composition, Dr L. Kufel for his comments and English correction and two anonymous referees for their comments and careful English correction. The research was supported by a grant from Polish Research Committee (Grant No. PB0499/P2/93/04).

References

Berman, T., 1970. Alkaline phosphatases and phosphorus availability in Lake Kinneret. Limnol. Oceanogr. 15: 663–674.

Chróst, R. J., W. Siuda, D. Albrecht & J. Overbeck, 1986. A method for determining of enzymatically hydrolyzable phosphate (EHP) in natural waters. Limnol. Oceanogr. 31: 662–667.

Chróst, R. J., W. Siuda & G. Z. Hałemejko, 1984. Longterm studies on alkaline phosphatase activity (APA) in a lake with fish – aquaculture in relation to lake eutrophication and phosphorus cycle. Arch. Hydrobiol. Suppl. 70: 1–32.

Chróst, R. J. & J. Overbeck, 1987. Kinetics of alkaline phosphatase activity and phosphorus availability for phytoplankton and bacterioplankton in Lake Plußsee (North German eutrophic lake). Microb. Ecol. 13: 229–248.

Cole, J. J., G. E. Likens & D. L. Strayer, 1982. Photosynthetically produced dissolved organic carbon: An important carbon source for planktonic bacteria. Limnol. Oceanogr. 27: 1080–1090.

Currie, D. J. & J. Kalff, 1984a. A comparison of the abilities of freshwater algae and bacteria to acquire and retain phosphorus. Limnol. Oceanogr. 29: 298–310.

Dillon, P. J. & F. H. Rigler, 1974. The phosphorus – chlorophyll relationship in lakes. Limnol. Oceanogr. 19: 767–773.

Feuillade, J., M. Feuillade & P. Blanc, 1990. Alkaline phosphatase activity fluctuations and associated factors in a eutrophic lake dominated by *Oscillatoria rubescens*. Hydrobiologia 207: 233–240.

Fitzgerald, G. P. & T. C. Nelson, 1966. Extractive and enzymatic analyses for limiting or surplus phosphorus in algae. J. Phycol. 2: 32–37.

Gajewski, A. J., R. J. Chróst & W. Siuda, 1993. Bacterial lipolytic activity in an eutrophic lake. Arch. Hydrobiol. 128: 107–126.

Healey, F. P., 1973. Characteristics of phosphorus deficiency in Anabaena. J. Phycol. 9: 383–394.

Healey, F. P. & L. L. Hendzel, 1979. Indicators of phosphorus and nitrogen deficiency in five algae in culture. J. Fish Res. Bd Can. 36: 1364–1369.

Healey, F. P. & L. L. Hendzel, 1980. Physiological indicators of nutrient deficiency in lake phytoplankton. Can. J. Fish. aquat. Sci. 37: 442–453.

Heath, R. T. & G. D. Cooke, 1975. The significance of alkaline phosphatase in a eutrophic Lake. Int. Ver. Theor. Angew. Limnol. Verh. 19: 959–965.

Hino, S., 1988. Fluctuation of algal alkaline phosphatase activity and the possible mechanisms of hydrolysis of dissolved organic phosphorus in Lake Barato. Hydrobiologia 157: 77–84.

Hutchinson, G. E., 1967. A Treatise on Limnology. 2. Introduction to Lake Biology and the Limnoplankton. Wiley & Sons, New York, 1115 pp.

Jones, J. G., 1972. Studies on freshwater microorganisms: phosphatase activity in lakes of differing degrees of eutrophication. J. Ecol. 60: 777–791.

Kirchman, D. L., J. Moss & R. G. Keil, 1992. Nitrate uptake by heterotrophic bacteria. Does it change the f-ratio? In Bjoernsen, P. K., B. Riemann (eds), Microbial Ecology of Pelagic Environments. FGR Schweiz.-Sche-Verlagsbuchhandlung, Stuttgart, 37: 129–138.

Kuenzler, E. J. & J. P. Perras, 1965. Phosphatases of marine algae. Biol. Bull. 128: 271–284.

Kufel, I. & L. Kufel, 1997. Eutrophication processes in a shallow macrophyte-dominated lake – nutrient loading to and flow through Lake Łuknajno (Poland). Hydrobiologia 342/343: 387–394.

Lee, T., E. Barg, D. Lal & F. Azam, 1993. Bacterial scavenging of super (234) Th in surface ocean waters. Mar. Ecol. Prog. Ser. 96: 109–116.

Olsson. H., 1990. Phosphatase activity in relation to phytoplankton composition and pH in Swedish Lakes. Freshwat. Biol. 23: 353–362.

Palmer, R., E. Drinan & T. Murphy, 1994. A previously unknown disease of farmed Atlantic salmon: Pathology and establishment of bacterial aetiology. Dis. aquat. Org. 19: 7–14.

Perry, M. J., 1972. Alkaline phosphatase activity in subtropical Central North Pacific waters using a sensitive fluorometric method. Mar. Biol. 15: 113–119.

Peterson, B. J., J. P. Barlow & A. E. Savage, 1974. The physiological state with respect to phosphorus of Cayuga Lake phytoplankton. Limnol. Oceanogr. 19: 396–404.

Reichardt, W., 1971. Catalytic mobilization of phosphate in lake water and by Cyanophyta. Hydrobiologia 38: 377–394.

Reichardt, W., J. Overbeck & L. Steubing, 1967. Free dissolved enzymes in lake waters. Nature 216: 1345–1347.

Rhee, G. Y., 1972. Competition between an alga and an aquatic bacterium for phosphate. Limnol. Oceanogr. 17: 505–514.

Tiwari, B. K. & R. R. Mishra, 1982. A study on biological activity measurements and heterotrophic bacteria in small freshwater lake. Hydrobiologia 94: 257–267.

Wetzel, R. G., 1975. Limnology. W. B. Sanders Co., Philadelphia, 743 pp.

Hydrobiologia **342/343:** 401–409, 1997.
L. Kufel, A. Prejs & J. I. Rybak (eds), Shallow Lakes '95.

Eutrophication processes in a shallow, macrophyte-dominated lake – factors influencing zooplankton structure and density in Lake Łuknajno (Poland)

Andrzej Karabin, Jolanta Ejsmont-Karabin & Rozwita Kornatowska
Hydrobiological Station, Institute of Ecology, Polish Academy of Sciences, 11-730 Mikołajki, Poland

Key words: Rotifera, Crustacea, phytoplankton, seasonal dynamics, eutrophication

Abstract

Influence of abiotic (temperature, concentration of nutrients) and biotic (macrophytes, phytoplankton, fish predation) factors on the species composition and seasonal dynamics of the zooplankton community (Rotifera and Crustacea) was analyzed. Strong interrelations between phyto- and zooplankton as well as the long-term effect of predation by fish on zooplankton were observed. An analysis of the qualitative and quantitative structure of zooplankton was used to assess the trophic state of Lake Łuknajno.

Introduction

Lake Łuknajno, surrounded with arable lands, pastures and meadows, receives large inputs of nutrients from the watershed: nitrogen in spring and autumn, phosphorus – in summer (I. Kufel & L. Kufel, 1997). Total annual load of nutrients exceeds that permissible (according to Vollenweider's criteria) so that the lake should become eutrophic. However, observations made for last ten years of study have not revealed any signs of increasing trophic status of the lake and the lake sustains its meso-eutrophic character.

Besides factors supporting low trophic state of lakes, such like adsorption of phosphorus by mineral and organic particles and sedimentation (Lampertus, 1993), also biological factors may play an important role. 'Clearing' of lake water by large cladocerans is among the best known mechanisms of the kind (Lampert, 1978; Lampert et al., 1986). *Daphnia* species seem to play especially important role in the process (Kerfoot et al., 1988; Jürgens, 1994). However, the fact that Lake Łuknajno due to dense communities of Characeae and relatively low trophic status is a good habitat for planktivorous fish seems to contradict a possibility of a development of communities of large Cladocera. It can be presumed that alternative mechanisms of zooplankton control of phytoplankton have developed in the study lake.

An analysis of a character and functioning of zooplankton communities in a shallow, macrophyte-dominated lake and an impact of the communities on phytoplankton is a trial of finding these alternative control mechanisms. Besides the main aim of the complex studies, i.e. analysis of the rate of the eutrophication process, the purpose of this paper was also to find and analyse biological mechanisms other than feeding by large-sized Cladocera on algae that could prevent phytoplankton blooms and thus help to maintain a relatively low of trophic status of Lake Łuknajno.

Study area and methods

Studies were carried out on Lake Łuknajno, a shallow lake (mean depth – 0.6 m; maximal – 3.0 m) with surface area of 623 ha. Due to rich ornithofauna the lake has been a nature reserve since 1937 and a biosphere reserve since 1977.

The studies on plankton were a part of complex studies conducted in 1993 and covering the physical and chemical characteristics of lake water (I. Kufel & L. Kufel, 1997), phytoplankton, zooplankton (rotifers and crustaceans), macrophyte (Krolikowska, 1997) and waterbird communities.

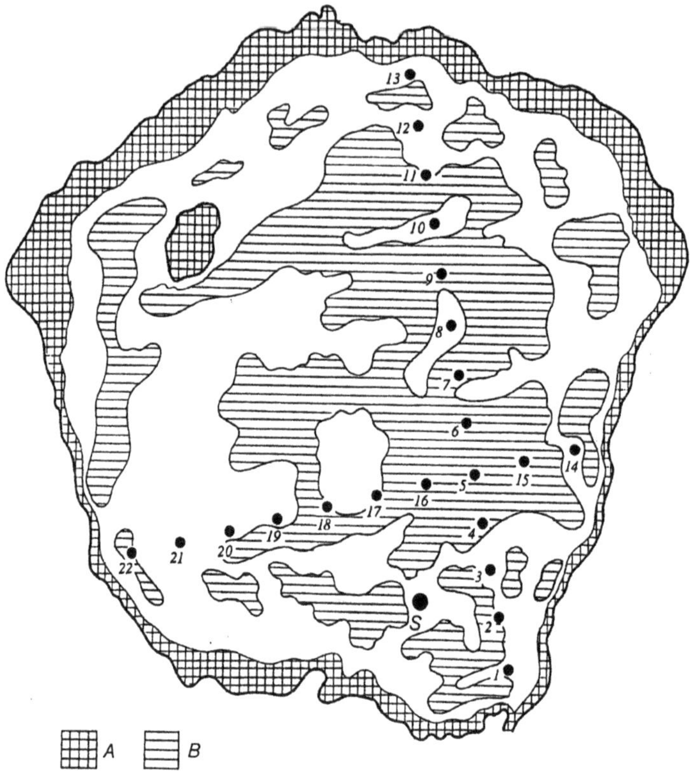

Figure 1. Lake Łuknajno: A – emerged macrophytes; B – submerged macrophytes; Sampling stations of plankton; 1–22 – transect stations.

Zooplankton and phytoplankton samples were collected weekly, from 16 April to 28 October 1993, at the deepest site of the lake (Figure 1).

The species composition, numbers and biomass of phytoplankton (based on measurements of cells) were determined from a sample representative for the whole water column, composed of subsamples of 100 ml taken from each consecutive 0.5 m. Samples were preserved with Lugol's solution and formalin and analyzed under the miscroscope after sedimentation.

Zooplankton samples were collected with a 5-litre sampler at every 0.5 m depth, from the surface to the bottom, and then pooled and concentrated using a plankton net of a mesh size 30 μm. All the phytoplankton and zooplankton samples were taken from a water layer above submerged macrophytes and analyzed for seasonal succession.

Additionally, the lake was sampled on 3 August 1993, at 22 stations along two transects in order to determine the horizontal distribution of zooplankton in the lake (Figure 1). At stations with the bottom covered with macrophyte stands samples were taken from the lake surface to the vegetation level (i.e. 0.5 or 1.0 m depth) and at the stations devoid of macrophytes – from the surface to the bottom of the lake.

Some results are also used of the project concerning monitoring of the Great Masurian Lakes. Lake Łuknajno being one of these lakes was monitored in the years 1985, 1986, 1988, 1990 and 1991, once a year, in summer.

The length:wet weight relationship (Petzhen, 1965; Klekowski & Shushkina, 1966; Karabin, 1974) was used to derive the mean body weight of individuals of particular crustacean species. Modified formulas by Ruttner-Kolisko (1977) were used to calculate the mean individual weight in rotifers.

Results

Seasonal succession of phyto- and zooplankton

59 rotifer and 27 crustacean species were identified in Lake Łuknajno during this study. *Polyarthra* species dominated in the rotifer community and were persistent. They constituted over 50% of biomass of non-predatory rotifers. *Keratella* species were also an important component of the community (Figure 2). These two genera together comprised a very major portion of the numbers and biomass of nonpredatory rotifers.

The crustacean community was strongly dominated by few species as well. The most abundant species, *Mesocyclops leuckarti* (Claus), occurred in high numbers throughout the study period, forming about 50% of crustacean biomass in spring and summer (Figure 2). The remaining important species were periodically common. The population of *Bosmina longirostris* (O. F. Müller) dramatically increased in spring. In summer *Chydorus sphaericus* (O. F. Müller) made up the majority of the crustacean community and in autumn *Ceriodaphnia quadrangula* (O. F. Müller) was the most abundant species (Figure 2). Low occurrence of Calanoida and *Daphnia* species was a striking feature of zooplankton in the study lake. *D. cucullata* (Sars), which is a common and abundant species in Masurian water bodies of different morphology and trophic status, in Lake Łuknajno was found occasionally and in numbers never exceeding 5 ind. l^{-1}.

During the whole year zooplankton was dominated by euplanktonic organisms. Though littoral-periphytic species of the rotifer genera *Colurella, Notholca, Lecane, Lepadella, Cephalodella* and crustaceans *Ectocyclops, Acanthocyclops, Alona, Alonella, Scapholeberis, Rhynchotalona* were frequent, they occurred in lake pelagial in low numbers.

Horizontal distribution of zooplankton did not reveal a significant impact of submerged macrophytes on density of the community in an open water zone.Mean biomass of zooplankton at stations with the bottom covered with macrophytes was 2.96 mg WW l^{-1} (SD = 0.73), and at stations devoid of macrophytes it was 2.73 mg WW l^{-1} (SD = 1.02). Species composition of zooplankton and contribution of littoral and periphytic species to the community were also similar at both these types of stations. On the other hand, littoral and periphytic forms were absent at two stations with the highest biomass of macrophytes.

Trophic conditions and temperature are the most important factors influencing the rate and character of changes in zooplankton communities (Herzig, 1987). At the very beginning of the study temperature of water was low, about 7 °C. Rotifer community was then dominated by cold water, adapted to high oxygen concentration species, *Polyarthra dolichoptera* Idelson, *Filinia terminalis* (Plate), *Synchaeta pectinata* Ehrenberg. At the same time crustacean community was dominated by wintering in a stadium of copepodite Cyclopoida, whereas Cladocera were scarce.

At the end of April temperature rapidly increased from 7 °C to 18 °C within 7 days. In May temperature was still rising (Figure 3A), resulting a rapid increase in numbers and biomass of *Bosmina longirostris*, i.e. up to 1902 ind. l^{-1} and 24.98 mg WW l^{-1} (Figure 3C). At the same time densities of cold water species rapidly declined. It is evident that ambient temperatures influenced this change in zooplankton species structure, as Cladocera start their development at temperatures above 10 °C.

The comparison of parallel changes in biomass of non-predatory zooplankton (dominated by *B. longirostris*) with chlorophyll concentrations in April and May revealed a heavy grazing pressure on phytoplankton (Figure 3A, C) for chlorophyll concentrations fell sharply and reached minimum a week after peak abundance of zooplankton. Such low values were not found until autumn.

Relationships between phyto- and zooplankton at the time of the rapid development of the latter could not be investigated due to a lack of data on species structure of phytoplankton. However, at the beginning of the study phytoplankton was gradually growing both in numbers and biomass, whereas during a maximum of zooplankton density and just after it, phytoplankton biomass was very low (Figure 3B). No notable changes of phytoplankton species structure were observed at that time. The biomass of the assemblage was dominated by diatoms and species of the genera *Scenedesmus, Rhodomonas, Cryptomonas, Cyclotella* were the most common small (<30 μm) algae both before and after the maximum of zooplankton biomass. The rapid decrease in phytoplankton abundance seems to have led to a deterioration of feeding conditions of zooplankton. This could be a reason of the equally high reduction in density of *B. longirostris*.

The strong relationship between phyto- and zooplankton is evidenced by a comparison of further changes in density of both communities – a rapid growth of chlorophyll concentration and biomass of

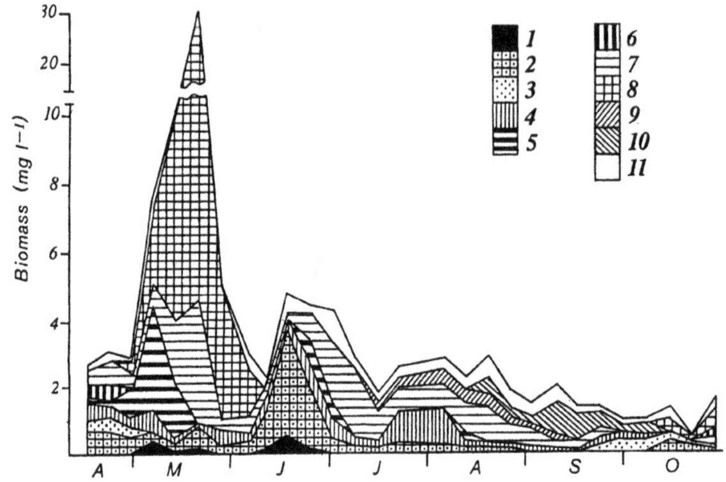

Figure 2. Seasonal changes in the biomass of dominant zooplankton taxons: 1 – *Keratella cochlearis*; 2 – *Polyarthra* spp.; 3 – *Synchaeta* spp.; 4 – other non-predatory rotifers; 5 – *Asplanchna priodonta*; 6 – *Cyclops strenuus*; 7 - *Mesocyclops* spp.; 8 – *Bosmina longirostris*; 9 – *Chydorus sphaericus*; 10 – *Ceriodaphnia quadrangula*; 11 – other Crustacea.

phytoplankton at the end of May and beginning of June, when zooplankton biomass was relatively low and a decrease again in the middle of June, when the increasing zooplankton biomass was dominated by *Polyarthra* spp. (Figure 3C), feeding exclusively on algae with maximum size of 20–30 μm (Pourriot, 1977). Small-sized phytoplankton was then dominated by the genera *Rhodomonas, Cryptomonas, Cyclotella* as well as *Chrysodalis* and *Chromulina.*

This kind of mutual relations between phyto- and zooplankton was observed up to mid July, when a cardinal change in a qualitative structure of phytoplankton took place. This previously highly diversified community, became dominated by cyanobacteria forming 45–60% of phytoplankton biomass till the end of September. From this time, in spite of a relatively high biomass of phytoplankton (Figure 3B), biomass (and numbers) of non-predatory zooplankton was slowly decreasing (Figure 3C). This tendency was very well seen in the rotifer biomass, which dropped to the minimum value. At this time the crustacean community was dominated by *Chydorus sphaericus*, and then by *Ceriodaphnia quadrangula.*

Rapid decrease observed in spring and early summer in the biomass of both 'edible' and 'net' algae deserves mentioning. This decrease of the latter might have been caused by self-shading. Large diatoms which dominated the spring biomass of large algae in Lake Łuknajno most likely decreased by sedimentation, thus explaining the algal decline (Sommer et al., 1986). This hypothesis is supported by changes in trophic structure of non-predatoryzooplankton, i.e. a distinct increase of the importance of detritovores in period of decreasing biomass of 'net' algae (Figure 4).

Zooplankton density decreased until the end of September, when water temperature dropped below 10 °C (Figure 3A) and zooplankton structure became similar to that usually observed in spring. 'Spring species' like *Synchaeta* spp. (*S. pectinata* among others), *Bosmina longirostris* and *Cyclops* spp. appeared again and even dominated. Phytoplankton was then dominated by small algae, which comprised over 90% of the total phytoplankton biomass (Figure 3B).

Evaluation of eutrophication of Lake Łuknajno based on a structure of rotifer and crustacean communities

Lake Łuknajno is one of the Great Masurian Lakes. 28 lakes of this lake complex have been monitored since 1985. Zooplankton was sampled always in summer, i.e. between 15 July and 15 August. Research done in 1976–1978 of the zooplankton communities of a large group of lakes of different trophy in north-eastern Poland has revealed parameters of structure of both rotifer and crustacean communities, that can be used as indices of lake trophy. These were: rotifer numbers, percentage of the tecta form in *Keratella cochlearis* biomass, ratio of biomass to numbers in rotifers, percentage of Cyclopidae in Crustacea biomass and ratio of Cyclopidae to Cladocera biomass.

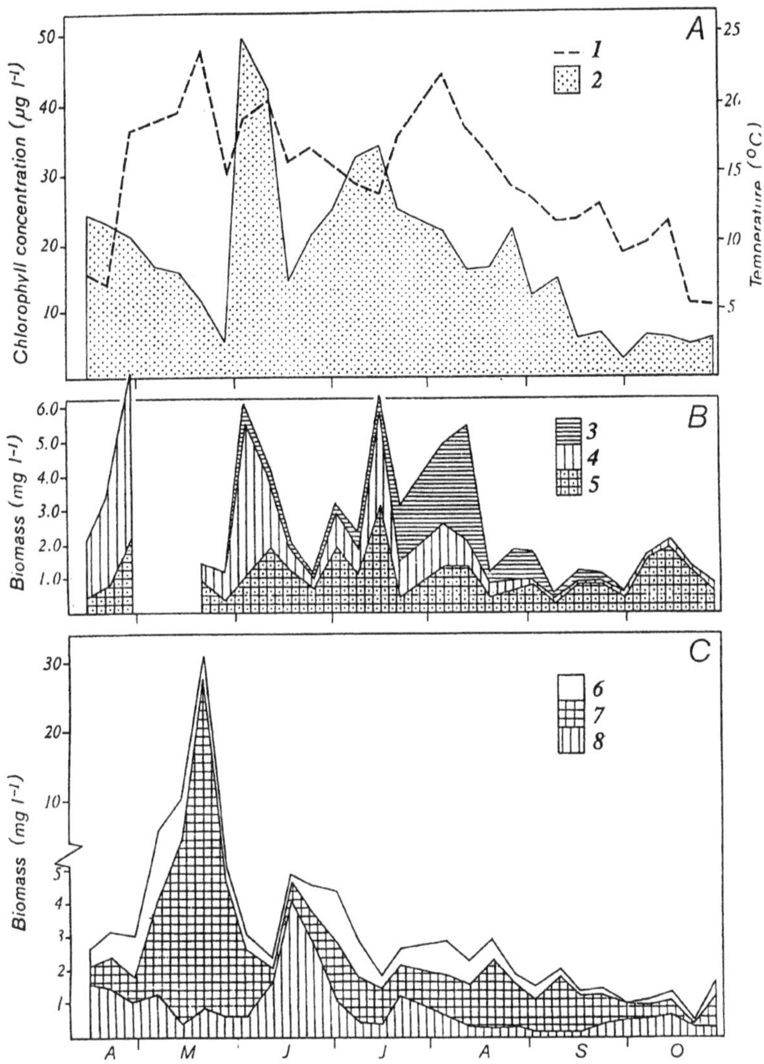

Figure 3. Seasonal changes in water temperature and concentration of chlorophyll (A), phytoplankton biomass (B) and zooplankton biomass (C) in Lake Łuknajno. 1 – water temperature, 2 – chlorophyll concentration, 3 – blue-greens, 4 – remaining net-algae, 5 – small algae (<35 μm), 6 – predatory zooplankton, 7 – non-predatory crustaceans, 8 – non-predatory rotifers.

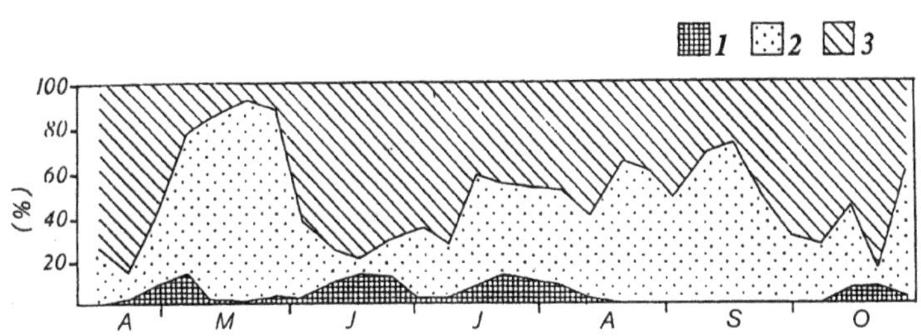

Figure 4. Trophic structure of non-predatory zooplankton biomass in Lake Łuknajno: 1 – detrivores; 2 – microfiltrators; 3 – macrofiltrators–raptators.

Additionally, ecological groups of species typical of low and high lake trophy have been distinguished. Relations between biomass of these two groups are of an indicatory character (Karabin, 1985a). Using the above mentioned relationships and trophic state index (TSI) introduced by Carlson (1977) formulas enabling calculation of TSI from quantitative data on zooplankton have been established.

Mean for this period values of particular zooplankton indices of lake trophy are shown in Table 1. They are compared with results of many-years' monitoring of the lake. Trophic state index (TSI) values of 35 to 45 are characteristic of mesotrophic lakes (i.e. high mesotrophy), 45–55 – meso-eutrophic, 55–65 eutrophic, and above 65 – hypertrophic ones.

Noteworthy is a difference in the trophic state index based on parameters of rotifer and crustacean communities. In the years 1985–1991, the rotifer communities were characteristic of mesotrophy or meso-eutrophy, whrereas the crustacean communities were typical of eutrophy, or even hypertrophy. Predation by planktivorous fish may explain the trophic differences. Being visual predators, the fish eliminate first of all big zooplankters (Brooks & Dodson, 1965; Węgleńska, 1979), mostly species indicatory of low trophy. Thus, eutrophication processes and fish predation influence crustacean communities in similar way (i.e. by exclusion of big individuals), what means that changes in the community structure indicating an increase in lake trophy, may actually reflect an impact of fish on the community. No influence of the kind was observed in rotifers, as they are rather accidental food of fish. Therefore, 'rotifer indices' were used for evaluation of eutrophication of the lake.

Values of the indices increased during 9 years' study. This gradual increase is especially marked in ratio of biomass of ecological groups and rotifer numbers. In 1993 values of all 'rotifer indices' and even the most conservative one – the percentage of *tecta* form in *Keratella cochlearis* population – were the highest and close to eutrophy rather than mesotrophy (Table 1). Thus, the results indicate advanced eutrophication of Lake Łuknajno. The process seems to go rather slowly, as the lake is still of a transitory (meso-eutrophic) character (Table 1).

Discussion

Temperature of water played an important role mainly in spring, when as a result of its rapid increase a simultaneous and mass hatching of ephippia caused an exponential development of the population of *Bosmina longirostris*. The same factor most likely influenced changes in zooplankton structure in spring, i.e. disappearance from the community of *Cyclops strenuus, Synchaeta pectinata, Filinia terminalis* and *Polyarthra dolichoptera*, which earlier dominated.

The most important factor deciding on a character of Lake Łuknajno are macropytes, especially submerged ones covering about 50% of the bottom area (Królikowska, 1997). As the lake is very shallow and susceptible to mixing (for its flat and uncovered shore) and a water layer above macrophyte stands is 20 to 50 cm thick, the macrophytes were expected to have a significant impact on density and/or species structure of zooplankton, but it is not supported by the results obtained. Though during the study year rotifers and crustaceans typical of littoral, periphyton and benthos were present in pelagial, they did not have any impact on zooplankton density or species structure. Studies on the horizontal distribution of zooplankton confirmed the lack of any impact of macrophytes on zooplankton. There were not found statistically significant relationships between abundance and species structure of zooplankton on the one hand and occurrence, composition and abundance of submerged macrophytes on the other hand.

A pressure of planktivorous fish seems to be the another factor having a significant impact on zooplankton, and especially its crustacean component. The pressure cannot be quantitatively evaluated due to a lack of data on fish stock. However, the above conclusion may be drawn from a discrepancy in evaluation of trophic state of the lake based on structure of rotifer and crustacean communities. Structure of the community of Crustacea is characteristic of highly eutrophic lakes, whereas the community of Rotifera is typical of mesotrophic ones. Size selective feeding by fish (i.e. preference of big size) as well as a reduction of the fecundity of Cladocera (due to a preference for egg-bearing females) (Vijverberg & Richter, 1982) lead to changes in species structure of the crustacean community, leading to an increase in density and domination of small species. Thus effects of fish pressure and eutrophication on zooplankton are similar. Some authors are convinced that fish cannot have an impact on changes in crustacean communities in Masurian Lakes (Gliwicz & Prejs, 1977; Hillbricht-Ilkowska et al., 1979; Karabin, 1985b) but their studies were restricted to large and deep lakes with more or less controlled fish stock due to intense fishery. In Lake

Table 1. Based on zooplankton indices analysis of trophic state of lake Łuknajno. 1 - contribution of species typical of eutrophy to the biomass of indicator group of Rotifera (in %); 2- percentage of form *tecta* in the population of *Keratella cochlearis*; 3- rotifer numbers (ind. l^{-1}); 4- biomass to numbers ratio (mg WW); 5- the same as 1- for Crustacea; 6- percentage of Cyclopoida in the biomass of Crustacea; 7- Cyclopoida to Cladocera biomass ratio; TSI - trophic state index.

Years	Rotifera								Crustacea					
	1		2		3		4		5		6		7	
	value	TSI	value	TSI	value	TSI	value	TSI	value	TSI	value	TSI	value	TSI
1985	0.7	36	1.4	46	318	49	0.00030	46	87.9	63	58.6	65	1.85	71
1986	0.3	35	1.0	46	264	48	0.00041	43	81.2	60	21.1	49	0.34	48
1988	16.4	45	2.5	46	584	52	0.00023	49	96.0	70	65.9	67	1.93	72
1990	16.6	45	4.4	46	879	54	0.00023	49	100.0	70	60.0	65	1.81	71
1991	37.2	44	6.2	46	3971	57	0.00022	50	92.2	67	44.7	60	0.82	60
1993	88.6	51	22.4	50	7476	64	0.00016	52	98.2	70	73.2	69	4.46	85

Łuknajno catches of fish are low and fish also enters the lake from the neighbouring Lake Śniardwy. It is then possible that the density of planktivorous fish in the lake is higher than in large, stratified Masurian lakes (M. Kragiel, pers. com.). Low occurrence of large-sized species like *Daphnia* spp. and *Diaptomus* spp. might have been an effect of the above mentioned selective feeding by fish. As a result, in Lake Łuknajno the only group dominating during the whole season were small-sized species – *Bosmina, Chydorus* and *Ceriodaphnia*.

Thus, a typical of many different lakes (Sommer et al., 1986) seasonal succession from large *Daphnia* in spring to small Cladocera in summer is not observed in the lake. This feature of zooplankton of Lake Łuknajno is permanent as during 10 years of monitoring of the lake *Daphnia* and *Calanoida* were scarce or absent. It may be hypothesized that the pressure of planktivorous fish in the lake is 'long-term' in that due to a direct elimination of individuals as well as reduction of population fecundity it causes low density of large-sized species in the subsequent year.

Trophic conditions are among the main factors influencing quantitative changes in the zooplankton community in a course of season. Phytoplankton density is low in Lake Łuknajno, and typical blooms are not observed. Except during a 'clear water phase', in highly eutrophicated lakes phytoplankton seldom limit zooplankton abundance. In the studied lake the situation is similar to that observed in lakes of low trophy – food seems to be a limiting factor as too strong pressure of consumers leads to 'hunger conditions' (Gulati, 1982) and then to equally strong reduction of consumers. An analysis of parallel changes in biomass of zooplankton, concentration of chlorophyll and phytoplankton numbers shows that strong relations exist between these two communities – in spring and early summer each decrease in the numbers of algae is preceded by an increase in zooplankton density. Nutrients do not seem to determine the changes in phytoplankton as phosphorus concentrations were near 20 μg P l^{-1} during the whole season and their fluctuations were not correlared with the quantitative changes in phytoplankton. Concentrations of nitrogen were not correlated with the phytoplankton density as well (Kufel & Kufel, 1997).

Allelopathic inhibition of algal production by dominating in the study lake Charales (Wiüm-Anderson et al., 1982), may be one of the reasons of low density of phytoplankton in Lake Łuknajno as well. However, this impact cannot explain rapid changes in phytoplankton density and structure throughout the season.

Calculated by Ejsmont-Karabin et al. (in press) consumption rate of 'edible' algae by herbivorous zooplankton indicates a strong pressure of the community on phytoplankton in Lake Łuknajno. Daily consumption was several times higher than the phytoplankton biomass, and in May, i.e. period of the exponential growth of *Bosmina longirostris*, consumption time of all the edible fraction of phytoplankton was only 14 minutes (Ejsmont-Karabin et al., in press). At this time chlorophyll and phytoplankton declined to a level typical of late autumn, thus causing 'clear water phase' and, in consequence – reduction of the population of herbivores. A lack of data on phytoplankton in May makes impossible an exact determination of duration of the phase, but according to the existing data it may be estimated as no more than 2 weeks, which is in agreement with observations reported by Lampert et al. (1986) and Sommer et al. (1986).

The population of cladocerans is then replaced by herbivorous rotifers. In mid June biomass of species of the genus *Polyarthra* amounts to as much as 4 mg WW l^{-1}. Consumption time is then as short as during the peak of *Bosmina longirostris* (Ejsmont-Karabin et al., in press). The increase of the biomass of herbivores coincides with a consecutive decrease in chlorophyll and phytoplankton,

Thus, even zooplankton community devoid (due to fish pressure) of large-sized herbivorous zooplankton is able to control phytoplankton in Lake Łuknajno as large cladocerans are replaced by smaller zooplankton (cladocerans and rotifers).

This mutual relations between phyto- and zooplankton were observed till the mid July only. Nevertheless, trophic factors played a significant role till the end of the season, thought since July the relation was rather single-acting, i.e. population of consumers was always limited by food availability. During the period of a rapid increase of density of blue-green algae (dominated by filamentous and colonial species – *Oscillatoria* sp., *Gomphospheria* sp., *Aphanotece* sp., *Microcystis* sp.) the biomass of zooplankton gradually decreased. At this time species diversity of cladocerans was very low and only *Chydorus sphaericus* dominated. It supports the hypothesis that large filter-feeders are more sensitive to disturbances produced by large-sized algae (Gliwicz & Siedlar, 1980; Richman & Dodson, 1983). This disturbing effect does not concern rotifers which are not susceptible to atoxic influence of blue-green algae (Kirk & Gilbert, 1992). Thus,the strong decrease in rotifer density observed during the second half of summer was probably caused by other factors (e.g. low concentration of their food).

Poor quality of food source of herbivorous zooplankton is also evidenced by trophic structure of the community. In eutrophic lakes detrivores are very abundant and in summer their share in the biomass of zooplankton is high (Karabin, 1985b). In Lake Łuknajno both density of the community and its role in food consumption were low (Ejsmont-Karabin et al., in press). Limited food resources (low biomass of phytoplankton and a lack of typical blooms) and consequently, strong relations beetwen phyto- and zooplankton often leading to hunger conditions as well as relatively low importance of detrivores in zooplankton seem to indicate a transitory, meso-eutrophic character of Lake Łuknajno. However, many years' studies of zooplankton community structure indicate that the process of eutrophication of the lake is slowly but inevitably proceeding.

References

Brooks, J. L. & S. Dodson, 1965. Predation, body size and composition of the plankton. Science, N. Y. 150: 28–35.

Carlson, R. E., 1977. A trophic state index for lake. Limnol. Oceanogr. 22: 361–369.

Ejsmont-Karabin, J., A. Karabin, R. Kornatowska, J. Królikowska & T. Ozimek. The effect of zooplankton on phosphorus cycling in a shallow lake coveredwith macrophytes. Ekol. pol. (in press).

Gliwicz, Z. M. & A. Prejs, 1977. Can planktivorous fish keep in check planktonic crustacean population? A test of size efficiency hypothesis in typical Polish lakes. Ekol. pol. 25: 567–591.

Gliwicz, Z. M. & E. Siedlar,1980. Food size limitation and algae interfering with food collection in *Daphnia*. Arch. Hydrobiol. 88: 155–177.

Gulati, R. D., 1982. Zooplankton and its grazing as indicators of trophic status in Dutch lakes. Envir. Monit. Ass. 3: 343–354.

Herzig, A., 1987. The analysis of planktonic rotifer populations: a plea for longterm investigations. Hydrobiologia 147: 163–180.

Hillbricht-Ilkowska, A., I. Spodniewska & T. Węgleńska, 1979. Changes in the phytoplankton – zooplankton relationship connected with the eutrophication of lakes. Symp. Biol. Hung. 19: 59–75.

Júrgens, K., 1994. Impact of *Daphnia* on planktonic microbial food weebs – A review. Mar. Microbial Food Webs 8: 295–324.

Karabin, A., 1974. Studies on the predatory role of the cladoceran *Leptodora kindtii* (Focke) in secondary production of two lakes with different trophy. Ekol. pol. 22: 295–310.

Karabin, A., 1985a. Pelagic zooplankton (Rotatoria + Crustacea) variation in the process of lake eutrophication. I. Structural and quantitative features. Ekol. pol. 33: 567–616.

Karabin, A., 1985b. Pelagic zooplankton (Rotatoria + Crustacea) variation in the process of lake eutrophication. II. Modifying effect of biotic agents. Ekol. pol. 33: 617–644.

Kerfoot, C. W., C. Levitan & W. R. DeMott, 1988. *Daphnia*-phytoplankton interactions: density-dependent shifts in resource quality. Ecology 69: 1806–1825.

Kirk, K. L. & J. J. Gilbert, 1992. Variation in herbivore response to chemical defenses: zooplankton foraging on toxic caynobacteria. Ecology 73: 2208–2217.

Klekowski, R & E. A. Shushkina, 1966. Energetitzheskij balans *Mesocyclops albidus* (Jur.) v period ego razvitia. (The energetic balance of *Mesocyclops albidus* (Jur.) during the period of its development). In Bieljaev, G. M. (ed.), Ekologia vodnych organizmov. Izd. 'Nauka', Moskva: 125–136.

Królikowska, J., 1997. Eutrophication processes in a shallow, macrophyte-dominated lake – species differentiation, biomass and the distribution of submerged macrophytes in Lake Łuknajno (Poland). Hydrobiologia 342/343: 411–416.

Kufel, I. & L. Kufel, 1997. Eutrophication processes in a shallow, macrophyte-dominated lake – nutrient loading to and flow through Lake Łuknajno (Poland). Hydrobiologia 342/343: 387–394.

Lampert, W., 1978. Climatic conditions and planktonic interactions as factors controlling the regular succession of spring algal bloom and extremely claer water in Lake Constance. Verh. int. Ver. Limnol. 20: 969–974.

Lampert, W., W. Fleckner. H. Rai & B. E. Taylor, 1986. Phytoplankton control by grazing zooplankton: a study on the spring clear-water phase. Limnol. Oceanogr. 31: 478–490.

Lampertus, L., 1993. Considerations in modeling the sediment-water exchange of phosphorus. Hydrobiologia 253: 219–231.

Petzhen, G. A., 1965. Produkcia vetvistousykh rakoobraznykh ozernogo zooplanktona. (Production of cladocerans of lake zooplankton). Gidrobiol. Zh. 1: 19–26.

Pourriot, R., 1977. Food and feeding habits of Rotifera. Arch. Hydrobiol. Beih. (Ergebn. Limnol.) 8: 243–260.

Richman, S. & I. Dodson, 1983. The effect of food quality on feeding and respiration by *Daphnia* and *Diaptomus*. Limnol. Oceanogr. 28: 948–956.

Ruttner-Kolisko, A., 1977. Suggestions for biomass calculation of plankton rotifers. Arch. Hydrobiol. Beih. (Ergebn. Limnol.) 8: 71–76.

Sommer, U., Z. M. Gliwicz, W. Lampert & A. Duncan, 1986. The PEG-model of seasonal succession of planktonic events in fresh waters. Arch. Hydrobiol. 106: 433–471.

Vijverberg, J. & A. F. Richter, 1982. Population dynamics and production of *Daphnia hyalina* Leydig and *Daphnia cucullata* Sars in Tjeukemeer. Hydrobiologia 95: 235–259.

Węgleńska, T., K. Dusoge, J. Ejsmont-Karabin, I. Spodniewska & J. Zachwieja, 1979. Effect of winter-kill and changing fish stock on the biocenose of the pond-type lake Warniak. Ekol. pol. 27: 39–70.

Wiüm-Anderson, S., V. Anthoni, C. Christopherson & G. Jouen, 1982. Allelopathic effects on phytoplankton by substances isolated from aquatic macrophytes (Charales). Oikos 39: 187–190.

Hydrobiologia **342/343:** 411–416, 1997.
L. Kufel, A. Prejs & J. I. Rybak (eds), Shallow Lakes '95.

Eutrophication processes in a shallow, macrophyte-dominated lake – species differentiation, biomass and the distribution of submerged macrophytes in Lake Łuknajno (Poland)

Joanna Królikowska
Hydrobiological Station, Institute of Ecology, Polish Academy of Sciences, Leśna 13, 11-730 Mikołajki, Poland

Key words: macrophytes, *Chara*, plant biomass, nutrients

Abstract

In Lake Łuknajno submerged macrophytes occupied 50% of lake area and their mean biomass was 350 g dry weight m^{-2} in the 1993 year. Maximum biomass values of charophytes reached high levels above 1000 g dry weight m^{-2} (in a few samples). During the last 20 years changes in submerged macrophytes were mainly quantitative. In that period the area occupied by plants decreased twice, presumable by eutrophication. In 1993 Lake Łuknajno was still a *Chara*-type. Seven species of Characeae were recorded, *Chara aculeolata* and *C. tomentosa* being the most abundant. The contribution of charophytes to total dry weight of all submerged macrophytes was 90%. At stations covered with *Chara*, phosphorus, nitrogen and chlorophyll concentrations were markedly lower than at stations devoid of plants.

Introduction

Macrophyte structure and abundance in lakes several depend on different factors – trophic state, depth of light penetration, and water movements being the most important. For macrophytes the total pool of nutrients is less important than an amount of its exchangeable fraction (Rejewski, 1981). The amount of the fraction depends on ecosystem efficiency and – to some extent – on macrophytes occupying the water body as they both depend on and create a nutrient budget of their environment.

The role of macrophytes in functioning of lake ecosystems is often emphasized in literature (e.g. Boyd, 1971; Hutchinson, 1975; Pieczyźska & Ozimek, 1976; Engel, 1988). Declining of submerged vegetation is one of the symptoms of an advanced eutrophication of lakes (Ozimek, 1992; Phillips et al., 1970).

The aim of the present investigations was to estimate changes in species composition of macrophytes of Lake Łuknajno that have taken place for the last 20 years. An attempt has also been made to evaluate the role of dominant species of macrophytes in the nutrient budget of the lake.

Area and methods

Studies were carried out in a shallow (mean depth – 0.6 m; maximum – 3.0 m) and small (surface area – 623 ha) Lake Łuknajno situated in the southern part of the Great Masurian Lakes (53° 49′N, 21° 38′E). Due to its rich ornithofauna the lake has been a nature reserve since 1937 and a biosphere reserve since 1977. Main components of lake water (June) were: Ca, 45.8 mg l^{-1}; hardness, 3.38 mval l^{-1}; pH, 8.32; TP, 57 μg l^{-1}. Mean summer (June–September) concentration in lake water of P total was 52 μg l^{-1}, soluble reactive phosphorus 8 μg l^{-1}, chlorophyll 10.7 μg l^{-1} (Kufel & Ozimek, 1994).

In the year 1993 a detailed study was conducted of macrophytes, mainly macrophytic submerged vegetation. For this purpose the lake surface was divided into squares of 100 × 100 m side. By means of the squares 1179 qualitative samples of submerged plants were collected (every 25 m), 704 among them containing macrophytes. Species composition and vegetation cover at the bottom were determined in the field and then maps were drawn up of spatial distribution and species structure of plant communities. Biomass of the

Table 1. Frequency and mean dry weight of particular species of submerged macrophytes in Lake Łuknajno (August, 1993).

Species	Frequency (%)	Dry weight ($g\ m^{-2}$)
Potamogeton pectinatus	41.96	8.2
Chara aculeolata	37.50	231.3
Chara tomentosa	35.27	383.5
Chara aspera	27.68	220.6
Chara contraria	16.96	172.3
Nitellopsis obtusa	9.82	118.5
Ceratophyllum demersum	6.70	43.4
Batrachium circinatum	5.36	4.0
Stratiotes aloides	4.91	235.5
Nitella flexilis	1.34	2.1
Chara fragilis	1.34	10.3
Myriophyllum spicatum	0.45	6.1
Elodea canadensis	0.45	3.0
Potamogeton perfoliatus	0.45	3.1
Potamogeton mucronatus[1]	–	–

[1] Only one individual.

plants was determined in samples collected every 50 m from an area of 0.16 m^2 with a Bernatowicz's sampler. 607 samples were collected in that way, as 381 stations were devoid of macrophytes. In July the subsamples were taken of eight plant species to determine phosphorus and calcium content. Chemical composition of plants were made after Jackson (1958), total P (colorimetric), Ca (flame photometer). Concentrations of total phosphorus, nitrogen and calcium were also analysed in lake water and bottom sediments at station with and without macrophytes and the same time: N – Kjeldahl's, P – colorimetrically, Ca – complexometrically (Standard methods, 1960). For detailes dealing with methods, see Kufel & Kufel (this volume).

The area covered with helophytes was determined planimetrically, as a difference between the shoreline of the lake and its open water zone – both shown on a map of spatial distribution of submerged macrophytes.

Results

Submerged vegetation occupying about 50% of the bottom area consisted of 8 species of vascular plants and 7 species of Charophyta. The most frequent among the vascular submerged macrophytes was *Potamogeton pectinatus* L. (Table 1). The associations of submerged plants dominated by *Chara tomentosa* L., *C. aculeolata* Kutz. and *C. aspera* (Deth.) Willd occupied 86–90% of the overgrown bottom area of Lake Łuknajno (Figures 1 & 2). From among the remaining charophytic species *Nitellopsis obtusa* (Desvaux) J. Groves was the only one forming dense, one-species patches covering a total area of ca 16 ha. *Chara contraria* Kutz, *C. fragilis* Desvaux (= *C. globularis*) and *Nitella flexilis* (L.) Agardh. occured sporadically.

From among the observed species of the Characeae four to be mentioned remain green all the year. The species *Chara tomentosa* and *C. aculeolata* reached the maximum (among submerged macrophytes) mean biomass and frequency (383 g DW m^{-2} and 35.27% and 281 g DW m^{-2} and 37.50%, respectively, Table 1). A maximum values (July 1993) of dry weight of these species in very dense stands were 1200 g DW m^{-2} and 1030 g DW m^{-2}, for covered bottom area of the lake. High biomass per unit area was produced by *Stratiotes aloides* L., though its frequency was below 5% (Table 1). This species formed three dense patches, and single young individuals were found at nearly half of the lake area. *Elodea canadensis* Rich. and *Potamogeton mucronatus* Schrad. were another species found as single individuals at a few sites in Lake Łuknajno. The peak biomass of *Chara aspera* (a species remaining green under ice cover) was observed in June, whereas it was nearly 2.5 times lower in early spring and 5 times lower in late autumn. In the peak of vegetation period (July/ August) mean biomass of submerged macrophytes reached 350 g DW m^{-2} of covered area and 130 g DW m^{-2} of lake area. Characeae formed the dominant component of submerged plants constituting 90% of total biomass (Figure 3).

Mean phosphorus content in dry weight of six from among eight plant species was almost equal, whereas two remaining species markedly varied: the content was twice lower in *Nitellopsis obtusa* and nearly four times higher in *Ceratophyllum demersum* L. Calcium content in *Chara* species was markedly higher than in vascular plants (Table 2).

Concentrations of phosphate – P, nitrate – N and ammonia – N were found to be lower in water from stations with *Chara* than those without plants. In the neighbourhood of stands of *Nitellopsis obtusa* concentrations of phosphate – P and ammonia – N were higher. At the same stations the lowest concentrations of phosphorus and nitrogen in bottom sediments were observed (Kufel & Kufel, 1997). Phosphorus, nitrogen and calcium content in submerged vegetation of Lake Łuknajno totalled respectively – 1.2, 2.8 and 17.9 tons (Table 3).

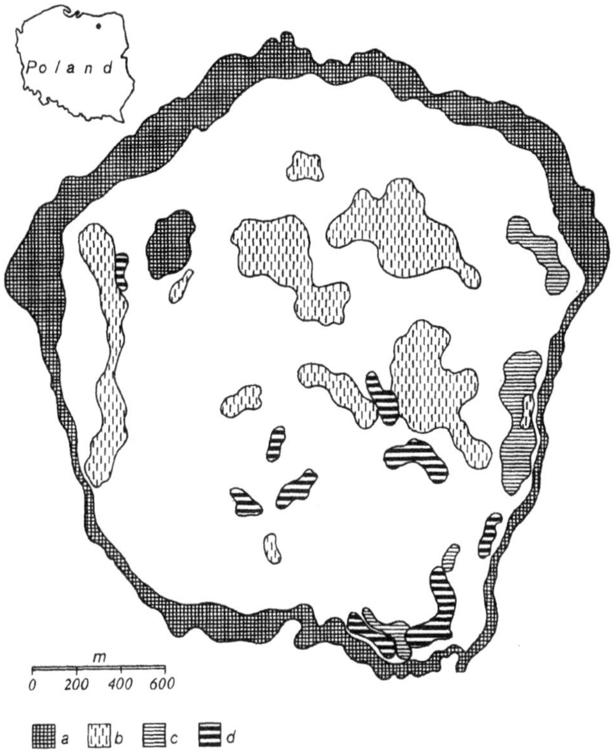

Figure 1. Distribution of emergent and submerged plants in Lake Łuknajno. a – emergent macrophytes, b – *Chara tomentosa*, c – *Nitellopsis obtusa*, d – *Potamogeton pectinatus*.

Table 2. Mean concentration of phosphorus and calcium in dry weight of particular species of submerged macrophytes (Lake Łuknajno, August, 1993).

Species	P mg g^{-1} ± SD	P g m^{-2}	Ca mg g^{-1} ± SD	Ca g m^{-2}
Chara aspera	1.52 ± 0.20	0.32	243 ± 15	52.05
Chara aculeolata	1.44 ± 0.10	0.38	239 ± 17	62.02
Chara tomentosa	1.18 ± 0.35	0.43	234 ± 15	85.59
Chara contraria	1.16 ± 0.21	0.19	246 ± 8	39.63
Nitellopsis obtusa	0.88 ± 0.13	0.10	216 ± 10	24.28
Ceratophyllum demersum	4.98 ± 0.52	0.10	19 ± 5	0.37
Potamogeton pectinatus	1.73 ± 0.31	0.01	50 ± 14	0.27
Stratiotes aloides	1.63 ± 0.24	0.38	32 ± 12	7.40

Table 3. Biomass and nitrogen and calcium content of submerged macrophytes (Lake Łuknajno, August, 1993).

Lake area occupied (ha)	Dry weight (t) occupied per 1 ha	total	Nutrient content N kg ha^{-1}	P kg ha^{-1}	Ca kg ha^{-1}	N total (t)	P total (t)	Ca total (t)
262.66	3.4	893	10.6	4.7	68.3	2.8	1.2	17.9

Figure 2. Distribution of macrophytes in Lake Łuknajno. a – emergent macrophytes, b – *Chara aculeolata*, c – *Chara aspera*, d – *Ceratophyllum demersum*, e – *Stratiotes aloides*.

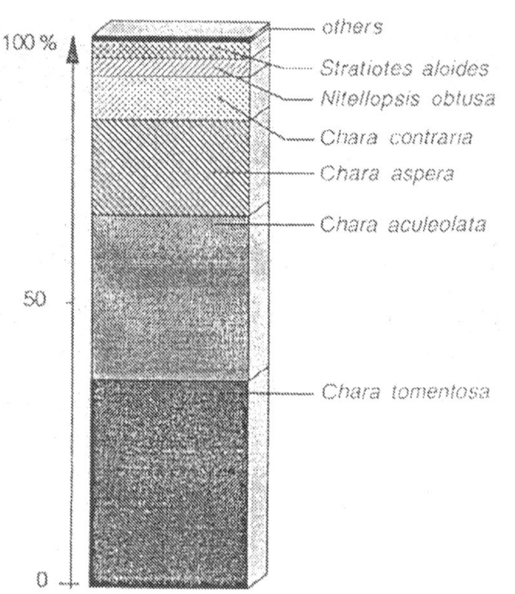

Figure 3. Percentage contribution of selected species to the total biomass of submerged plants in Lake Łuknajno (July–August, 1993).

Along the shoreline of Lake Łuknajno occurs a broad and dense belt of rushes. Communities of helophytes cover about 15% of lake area. The primary dominant species of these shoreline rushes is *Phragmites australis* (Cav.) Trin. ex Steud. (90%), which forms large monospecific associations. However, in some of these patches *Typha angustifolia* L. and *Schoenoplectus lacustris* (L.) Palla play quite an important role. Small patches of *Acorus calamus* L. accompanied in 20% by *Phragmites australis* are sometimes observed at the land edge of the reedbelt.

Discussion

Area of the bottom of Lake Łuknajno covered with submerged vegetation has decreased twice since observations made by Polakowski et al. (1973) 20 years earlier. Area formerly covered with *Chara aculeolata* and *C. contraria* markedly decreased, while that occupied by *C. tomentosa* increased. Lake Łuknajno can be still classified as a floral type 'CHARA' (according to typology by Samuelsson, 1925; Bernatowicz, 1960).

The belt of rushes did not change significantly. Associations of emergent macrophytes are still slightly poorer in species than most of meso- and eutrophic lakes of Masurian Lakeland (Kocół, 1953; Zarzycki, 1958). According to Polakowski et al. (1973) this phenomenon may be explained as a result of ecological homogeneity which is characteristic to Scirpo-Phragmitetum association in the reserve. Rushes are less numerous in species at the open water side of the belt and more species are usually found at the land side. Phragmites australis is a distinct dominant among emergent macrophytes of Lake Łuknajno, such situation is in most of the lakes in Poland.

Species composition of submerged macrophytes has not changed significantly for over 20 years. Nevertheless, *Chara rudis* L., *Potamogeton lucens* L., *Elodea canadensis* and *Potamogeton mucronatus* Schrad. seems are just withdrawing from the lake. These are species more sensitive to poor light conditions (Radziej, 1992). Withdrawing of these species and – at the same time – appearance of highly expansive and productive *Stratiotes aloides* is observed. The most frequencies of *Potamogeton pectinatus* may indicate slow eutrophication of the lake.

The highest biomass of dominating in Lake Łuknajno *Chara tomentosa* and *C. aculeolata* were observed at stations often visited by large numbers of swans. It shows that fertilization of water at these sites with faeces of the birds may stimulate an increment of plant biomass (Halba R., pers. com.) even if the same birds grazing on macrophytes decrease their biomass.

Submerged macrophytes are able to exploit both ambient water and bottom sediments as a source of nutrients (Barko et al., 1991; Rattmay et al., 1991). Phosphorus content in the biomass of Characeae in Lake Łuknajno fall within lower and nitrogen – upper limits as given by Riemer & Toth (1968). *Ceratophyllum demersum*, when coexisting with *Chara*, is able to take up more phosphorus than the latter, resulting in a markedly higher phosphorus content in this plant than in the remaining submerged macrophytes analysed in this paper. Accumulation of phosphorus in biomass of particular species is more differentiated if expressed per area unit than per weight unit.

In the peak of the vegetation season mineral phosphorus and nitrogen content in lake water at stations with *Chara* cover were lower than at neighbouring stations devoid of vegetation. Release of nutrients accumulated in macrophytes depends on their conditions and phenological cycles. Submerged macrophytes which remain green all the year (*Chara aspera*, for instans) supply small organisms with both refuges and habitats (Pereyra-Ramos, 1981) and retain most nutrients for a long time. Charophytes seem to have a slow decomposition rate as shown by Bastardo (1978). After Hugh & Putt (1988) a high calcium and low phosphorus concentrations in water are indispensable conditions for the optimum growth of charophytes. Blindow (1991) found that charophytes can be superior competitors and can successfully colonise shallow lakes previously dominated by phytoplankton due to higher biomass and accumulation of larger amounts of nutrients than angiosperms. Results presented by Kufel & Ozimek (1994) dealing with charophytes in the Lake Łuknajno demonstrated that such population of *Chara* may itself stabilize moderately eutrophic conditions by rapid uptake and large accumulation of phosphorus. However, most of submerged macrophytes found in Lake Łuknajno return nutrients to lake water and sediments at the end of the vegetation season. Only small amounts are retained in turions (Ozimek & Rybak, 1993). Data in literature about the role of charophytes concerning e.g. the response of these plants to elevated concentrations of phosphorus in lake water is still a controversy (Forsberg, 1964; Hough & Putt, 1988; Blindow, 1988; Kufel & Ozimek, 1994). Results obtained by above cited authors and these obtained from investigation in lake Łuknajno shows that in the structure and function of the lakes littoral make charophytes an important factor in controlling a nutrient budged and turnover in lakes. Present situation of charophytes in Lake Łuknajno and contrary data in literature cited above exacting a more detailed investigation about a role of charophytes in Lake Łuknajno (together with an experimental investigations).

Acknowledgments

This research was supported by a grant from the Polish Scientific Committee (Grant No PB 0499/P2/93/04). The author is grateful to Dr T. Ozimek for analytical work dealing with a nutrient budget, Dr J. Ejsmont-Karabin for advice and assistance, and Dr A. Karabin for help with drawings, and Mr T. Krache for assisting with field works.

References

Barko, J. W., D. Gunnison & S. R. Carpenter, 1991. Sediment interactions with submersed macrophyte growth and community dynamics. Aquat. Bot. 41: 41–65.

Bastardo, H., 1978. Laboratory studies on decomposition of littoral plants. Ph. D. Thesis, University of Warsaw, Warsaw, 45 pp.

Bernatowicz, S., 1960. Charakterystyka jezior na podstawie roślin naczyniowych. [The characteristics of lakes on the base of vascular plants] Roczn. Nauk Roln. 77-B-1: 79–101.

Blindow, I., 1988. Phosphorus toxicity in *Chara*. Aquat. Bot. 32: 393–395.

Blindow, I., 1991. Interactions between submerged macrophytes and microalgae in shallow lakes. Doctoral dissertation, Lund University, Lund, 112 pp.

Boyd, C. E., 1971. The limnological role of aquatic macrophytes and their relationships to reservoir management. Reserv. Fish. Limnol. 8: 153–166.

Engel, S., 1988. The role and interactions of submersed macrophytes in a shallow Wisconis Lake. J. Freshwat. Ecol. 4: 329–341.

Forsberg, C., 1964. Phosphorus, a maximum factor in the growth of Characeae. Nature 201: 517–518.

Hough, R. A. & D. A. Putt, 1988. Factors influencing photosynthetic productivity of *Chara vulgaris* L. in a moderately productive hard-water lake. J. Freshwat. Ecol. 4: 411–418.

Hutchinson, G. E., 1975. A treatise on limnology. Limnological Botany. Wiley & Sons, New York, 660 pp.

Jackson, M. L., 1958. Soil chemical analysis. Prentice-Hall Publishing Inc., Englewood Clifts, New Jersey, 498 pp.

Kocót, L., 1953. Rozmieszczenie roślinności w litoralu jeziora Tajty. [Distribution of vegetation in the littoral of Lake Tajty] Roczn. Nauk Roln., D, 67: 67–84.

Kufel, I. & L. Kufel, 1997. Eutrophication processes in a shallow, macrophyte-dominated lake – nutrient loading to and flow through Lake Łuknajno, Poland. Hydrobiologia 342/343: 387–394.

Kufel, L. & T. Ozimek, 1994. Can *Chara* control phosphorus cycling in Lake Łuknajno (Poland)? Hydrobiologia 275/276: 277–283.

Ozimek, T., 1992. Makrofity zanurzone i ich relacje z glonami w jeziorach o wysokiej trofii. [Submerged macrophytes and their relationship with algae in eutrophic and hypertrophic lakes] Wiad. ekol. 38: 13–34.

Ozimek, T. & J. I. Rybak, 1993. Macrophytes of small water bodies of the Suwałki Landscape Park (North-Eastern Poland). Ekol. pol. 41: 157–171.

Pereyra-Ramos, E., 1981. The ecological role of Characeae in the lake littoral. Ekol. pol. 29: 167–209.

Philips, G. L., D. Eminson & B. Moss, 1970. A mechanism to account for macrophyte decline in progressively eutrophicated freshwater. Aquat. Bot. 4: 103–126.

Pieczyńska, E. & T. Ozimek, 1976. Ecological significance of lake macrophytes. Int. J. Ecol. envir. Sci. 2: 115–128.

Polakowski, B., J. Dziedzic & E. Polakowska, 1973. Roślinność rezerwatu 'Jezioro Łukniany' na Pojezierzu Mazurskim. [The vegetation of the 'Łukniany Lake' nature reserve in the Masurian Lake District] Ochr. Przyr. 38: 85–114.

Radziej, J., 1992. Roślinność naczyniowa w kilkunastu jeziorach Wigierskiego Parku Narodowego. Zdanowski, W. B. (ed.), Jeziora Wigierskiego Parku Narodowego. Ossolineum-Wrocław, Zesz. Nauk. 'Człowiek i Środowisko' 3: 73–77.

Rattmay, M. R., C. Howard-Williams & J. M. A. Brown, 1991. Sediment and water as sources of nitrogen and phosphorus for submerged rooted aquatic macrophytes. Aquat. Bot. 40: 225–237.

Rejewski, M., 1981. Roślinność jezior rejonu Laski w Borach Tucholskich. [Lake vegetation of the Laska region in the Tuchola Forest (Bory Tucholskie)] Ph. D. thesis, Uniwersytet M. Kopernika, Toruń, 178 pp.

Riemer, D. N. & S. J. Toth, 1968. A survey of aquatic plants in New Jersey, N. J. Agr. Stat. Coll. Agric. envir. Sci. Rudger Univ. Bull. 820: 14 pp.

Samuelsson, G., 1925. Untersuchungen über die hohere Wasserpflanzen von Dalarne. Svenska Vaxtsoc. Sallskap. Handl. 9: 1–31.

Standard methods for the examination of water and waste water. 1960. Am. Publ. Health Assoc. Inc., New York, 626 pp.

Zarzycki, K., 1958. Die wichtingsten Grunlandgesellschaften des oberen Weichseltales und die Grundwasser-Ganglinien. Acta Soc. Bot. Pol. 27: 383–420.

www.ingramcontent.com/pod-product-compliance
Lightning Source LLC
LaVergne TN
LVHW062109200926
848484LV00007B/1626
9780792344216